W9-AGN-508

Proceedings

of the

Fifth International Symposium

on the

Reproductive Physiology of Fish

The University of Texas at Austin, Austin, Texas, U.S.A.
2-8 July 1995

Edited by: F.W. Goetz and P. Thomas

Published by Fish Symposium 95, Austin

Proceedings of the Fifth International Symposium on the Reproductive Physiology of Fish, The University of Texas at Austin, 2-8 July 1995

F. Goetz and P. Thomas (Editors)

Order from FishSymp 95, The University of Texas at Austin, Marine Science Institute, 750 Channelview Drive, Port Aransas, TX 78373 USA

Printed in the USA by The University of Texas at Austin Printing Department

Book cover artwork by Dinah Bowman, Dinah Bowman Studio Gallery, 312 5th Street, Portland, TX 78374.

Book layout by Patty Webb, The University of Texas at Austin, Marine Science Institute.

FOREWARD

The papers contained in this book were presented at the Fifth International Symposium on Reproductive Physiology of Fish, held July 2-8, 1995, at The University of Texas, Austin, Texas. The purpose of the Fifth Symposium was similar to that of the previous meetings held at Paimpont (1977), Wageningen (1982), St. Johns (1987) and Norwich (1991); to serve as a forum for the communication of recent advances on all aspects of the reproductive physiology of fish. The meeting had eight oral sessions covering the pituitary and gonadotropins, the hypothalamus and brain, aquaculture, environmental influences on reproduction, reproductive life histories, reproductive behavior, gonadal physiology and gametogenesis. Each session began with a keynote talk followed by four shorter oral presentations. The five-page papers in this book represent keynote talks while shorter presentations are contained in three-page papers. Besides oral presentations, there were two large poster sessions containing over 180 poster presentations. These are represented in the book by one-page papers. In keeping with past symposia in this series, the papers were all reviewed at the time of the meeting by symposium participants. We would like to thank the numerous reviewers for their time and effort. Their comments helped authors in revising approximately 100 of the symposium manuscripts.

We are especially grateful to the following people on the **Local Organizing** and **Scientific Committees** for their invaluable help in organizing the meeting:

Gloria Callard
Duncan MacKenzie
Reynaldo Patiño
Paul Rosenblum
Carl B. Schreck
Peter Sorensen
Craig Sullivan
Penny Swanson
Yonathan Zohar

We gratefully acknowledge the financial support of the following agencies and universities which enabled us to subsidize the registration fees of nearly half of the participants and the costs of publishing this volume.

Sponsors

Aquatic Eco-Systems, Inc.
Fisheries Society of the British Isles
National Science Foundation, Integrative Animal Biology Program
North Carolina State University
Northwest Fisheries Science Center, NOAA Fisheries
Texas A&M University
- **Department of Biology**
- **Institute of Biosciences and Technology Center for Animal Biotechnology**

Texas A&M University Sea Grant
Texas Cooperative Fish & Wildlife Research Unit
University of Minnesota Sea Grant
University of Notre Dame
The University of Texas at Austin
- **Marine Science Institute**
- **College of Natural Sciences**

USDA, Enhancing Reproductive Efficiency Program

Besides the Local Organizing and Scientific Program Committees, many other individuals contributed their time and effort in organizing and running the meeting. We would like to thank these individuals for their generous help:

Session Chairs

Rickard Bjerselius
Jim Cardwell
Martin Fitzpatrick
Henk Goos
Hamid Habibi
Bernard Jalabert
Lyndal Johnson
David Kime
Charles Laidley
John Leatherland
Ned Pankhurst
Richard Peter
L.H. Ran
Carel Richter
Sandy Scott
Craig Sullivan
Graham Young

Technique Workshop Organizers

D. Doering
R. Klassy
K. Maxey

Symposium Secretaries

Patty Webb
Diane Breckenridge-Miller

Symposium Helpers

David Campbell
Shampa Ghosh
Beth Hawkins
Margaret Gryzinski
Andrea Johnson
Jennifer Johnson
Izhar Khan
Kevin Leiner
Angie Lott
Rudy Ortiz
Jonathon Pinter
Karen Rogowski
Todd Sperry
Jacqueline Thomas
Janet Ungerer
Cinnamon Van Putte
Yong Zhu

Finally, we would especially like to thank the participants. The attendance at this meeting by over 325 individuals from 35 countries clearly shows that this symposium series is a pivotal gathering for fish reproductive biologists.

Frederick Goetz
Peter Thomas
(Co-organizers)

CONTENTS

Pituitary/Gonadotropins

Hypothalamus/Brain

Aquaculture

Environmental Influences on Reproduction

Reproductive Life History

Behavior

Gonadal Physiology

Gametogenesis

Pituitary/Gonadotropins

PHYSIOLOGICAL, MORPHOLOGICAL, AND MOLECULAR ASPECTS OF GONADOTROPINS IN FISH WITH SPECIAL REFERENCE TO THE AFRICAN CATFISH, *CLARIAS GARIEPINUS*.

R.W. Schulz, J. Bogerd, P.T. Bosma, J. Peute, F.E.M. Rebers, M.A. Zandbergen, H.J.Th. Goos

Faculty Biology, Research Group Comparative Endocrinology, University of Utrecht, Padualaan 8
NL-3584 CH Utrecht, The Netherlands

Summary

After introducing the duality of gonadotropins (GTH) in teleost fish, this review will deal with GTH receptors (GTH-Rs) and with the biological activities of teleost GTHs. Attention will then focus on the regulation of GTH gene expression, before cell biological aspects of the gonadotropes are presented (post-translational processing, intracellular storage, and eventually secretion or intracellular breakdown of GTH). Finally, we suggest that some teleosts may show a single GTH only.

In representatives of four teleost orders, two pituitary GTHs have been characterised, one, GTH I, which resembles the tetrapod FSH and the other, GTH II, which resembles LH. The shared biological activities of GTH I and II in the salmon can be attributed to the low GTH specificity of one of the two types of GTH-R. In the African catfish, a species possibly expressing a GTH II-like hormone only, a cDNA encoding a putative testicular GTH-R predicts traits of both mammalian FSH and LH receptors. Much information is available regarding the steroidogenic activities of fish GTHs whereas their role in the regulation of germ cell development is largely unknown. In catfish and trout, a limited expression of the GTH II β-subunit and a selective elimination of the α-subunit may reflect intracellular mechanisms affecting the amount of biologically active GTH II. GTH secretion is mainly regulated by gonadotropin-releasing hormone (GnRH) and dopamine while gonadal steroids are involved in the regulation of GTH gene expression. While there is little information on the regulation of GTH I release, a large assembly of compounds modulates GTH II release directly or indirectly. In the catfish, chicken GnRH-II and catfish GnRH is found in the pituitary, the latter being a 150-fold less potent GTH II secretagogue but present in 700-fold higher concentrations. Both GnRHs interact with a single type of GnRH receptor that apparently is restricted to the gonadotropes in the catfish. Ca^{2+} plays a prominent role in the secretory process and GnRHs induce rapid, transient elevations of $[Ca^{2+}]_i$ in fish gonadotropes.

The Duality of Teleost Gonadotropins

Since the pioneering work on salmonids (Suzuki et al. 1988; Swanson et al. 1991), additional species representing four teleost orders have been shown to produce two types of GTH. These heterodimeric glycoproteins consist of a common α-subunit and a hormone specific β-subunit which allows the classification of the GTHs into a FSH-like GTH I and a LH-like GTH II. Certain heterogeneity exists within the GTHs in a given species. Thus, for instance, different molecular varieties of the α-subunit are found (Kitahara et al. 1988; Huang et al. 1991); salmon GTH I occurs in forms that do or do not dissociate easily into subunits (Suzuki et al. 1988; Swanson et al. 1991); and GTH II may have several isoforms (Huang et al. 1981; Swanson et al. 1987; Banerjee et al. 1993). Although the duality of GTHs is well established for certain orders, this duality may not apply for all teleost (see below and Quérat; this volume).

Receptor Interaction and Steroidogenic Activity

Binding studies on coho salmon gonadal membrane preparations demonstrate the presence of two types of GTH-Rs (Yan et al. 1991; 1992). GTH-R type I show affinity to both GTH I and II, with a slightly higher affinity to GTH I, while GTH-R type II specifically bind GTH II. During vitellogenesis, GTH I, the predominant type of GTH secreted during this period (Swanson 1991), interacts with GTH-R-I in the theca and granulosa layers (Miwa et al. 1994). Special theca cells produce testosterone (T) which is aromatised to estradiol-17ß (E2) by granulosa cells (Kagawa et al. 1982). Yet, GTH-R-I do not appear to be involved in an acute stimulation of ovarian aromatase activity (Kanamori & Nagahama 1988). GTH-R-II are not detectable during vitellogenesis. In postvitellogenic/preovulatory ovaries, when GTH I plasma levels are decreasing and those of GTH II rise to peak values (Swanson 1991), GTH II interacts with GTH-R-I in the theca layer to stimulate the production of 17α-hydroxyprogesterone. The latter is converted to 17α-hydroxy,20β-dihydroprogesterone (17,20βP; Young et al. 1986) by granulosa cells that express GTH-R-II. This appears to involve the GTH II-stimulated *de novo* synthesis of 20β-hydroxysteroid dehydrogenase (reviewed by Nagahama 1994). GTH-R-I were also found in the interstitium of postvitellogenic/preovulatory ovaries.

In male salmon, GTH-R-I are associated with Sertoli cells throughout all stages of development (Miwa et al. 1994); however, their functional significance awaits elucidation. Although GTH-R-I cannot be demonstrated by autoradiography on Leydig cells of spermatogenetic testis, GTH-R-I nevertheless appear to be associated with Leydig cells as GTH I and II are equipotent in stimulating androgen (T and 11-ketotestosterone - 11KT) and 17,20βP secretion (Planas & Swanson 1995). Towards the end of spermatogenesis the GTHs' steroidogenic activities are discernible as GTH II becomes progressively more potent than GTH I in stimulating testicular steroid secretion. This is probably related to the appearance of GTH-R-II on Leydig cells of spawning fish. Part of the 17,20βP production in advanced testis can be attributed to spermatozoa (Ueda et al. 1984). Taken together, the period of gonadal growth in the salmon is associated with elevated levels of GTH I and the presence of GTH-R-I whereas the spawning period is characterised by the additional presence of GTH-R-II and elevated levels of GTH II. These conditions enable the two salmon GTHs to develop distinct domains of biological activity.

In perciform fish (Okada et al. 1994, Copeland & Thomas 1993; Tanaka et al. 1993), data on the GTHs' steroidogenic activities are too limited to be compared with the situation in salmonids. In the common carp (van der Kraak et al. 1992), GTH I and II share the same spectrum of biological activities: stimulation of steroid secretion in vitellogenic as well as preovulatory ovaries and induction of final oocyte maturation. In the African catfish, a cDNA clone encoding a putative GTH-R has been characterised. From the deduced sequence of amino acids and potential sites of glycosylation, the catfish testicular GTH-R bears similarity to both LH- and FSH-Rs in mammals. We have no information yet regarding the cellular localisation of GTH-R in catfish. However, testicular androgen secretion *in vitro* is responsive to gonadotropic stimulation in prepubertal, pubertal, and adult males (Schulz et al. 1994a,b), suggesting that Leydig cells express GTH-R during these stages of development.

GTH Effects on Germ Cell Development

In both sexes, information is scarce regarding GTH effects on germ cell development that are not related to the GTHs' steroidogenic activity. However, GTH I stimulates vitellogenin sequestration by trout oocytes, GTH II being hardly effective (Tyler et al. 1991). In goldfish, both carp GTH I and II stimulate vitellogenin sequestration *in vitro* (Nunez-Rodriguez et al. 1992), again suggesting differences between cyprinid and salmonid fish. The mechanism by which GTH stimulates vitellogenin sequestration remains to be clarified. Attention should also be devoted, for instance, to the possible role of GTHs in the recruitment of previtellogenic follicles into vitellogenesis.

The distribution of GTH-Rs in salmon testis may lead to the speculation that, like in mammals (e.g. Ackland et al. 1992; Pescovitz et al. 1994), the effects of androgens and FSH on germ cells are mainly mediated by paracrine, juxtacrine, and autocrine factors, such as insulin-like growth factor (IGF), transforming growth factor, activin/inhibin, serotonin, or endorphin. Indeed, in the trout, IGFs stimulate the proliferation of spermatogonia *in vitro* (Loir 1994; Loir & LeGac 1994). The proliferation is counteracted by IGF binding protein, and by Sertoli cells or Sertoli cell-conditioned medium from maturing and mature testis but not from testis about to resume spermatogenesis. Thus, Sertoli cells apparently undergo changes in their secretory activity, express GTH-R-I and may produce IGF and IGF binding protein, as has been shown in mammals (Dombrowicz et al. 1992). Trout testes also express IGFs (Chen et al. 1994) but the cellular source and possible regulation of IGF secretion remains to be elucidated in detail. However, GTH II does not stimulate spermatogonial proliferation in co-culture with Sertoli cells (Loir 1994). A highly active steroidogenic system in immature trout and catfish (Schulz & Blüm 1990; Schulz et al. 1994a) suggests that steroids too are involved in the initiation of spermatogenesis. In the Japanese eel hCG-stimulated 11-KT production by Leydig cells triggers Sertoli cells to stimulate spermatogenesis, which may be mediated by Sertoli cell-derived activin B (Miura et al. 1991; Nagahama 1994). Electron-microscopical studies again suggest that para- or autocrine interactions between somatic and germ cells are important considering that an increase in the number of Leydig cells is particularly obvious at sites of spermatogonial multiplication (Kanamori et al. 1985; Nakamura & Nagahama 1989).

Evidently the specific effects of gonadotropins on spermatogenesis in fish require more research. This area receives much attention throughout the vertebrates and the cystic organisation of spermatogenesis in fish and the possibility to study germ cell development *in vitro* (Miura et al. 1991; Loir 1994) offer important advantages over mammalian experimental models.

GTH Gene Expression: Steroid Hormones and GnRH

T and E2 stimulate the low or undetectable GTH II β gene expression in cultured gonadotropes from immature trout (Xiong et al. 1994a,b); the stimulatory effect is attributable to the interaction of ligand-activated oestrogen receptors with a proximal oestrogen responsive element (ERE), lifting the inhibitory effect of a silencer sequence. The glycoprotein α- and the GTH I β-subunit expression is not affected by steroids and the respective mRNAs are detectable in gonadotropes of immature trout without steroid treatment. In the coho salmon, however, the steady-state level of the α-subunit mRNA increases following steroid treatment although the pituitary GTH I content remains unchanged (Dickey & Swanson; this volume). In immature male African catfish, the GTH II β-subunit mRNA was detectable only after PCR amplification while the α-subunit mRNA was detected on routine Northern blots. RIA analysis of pituitary extracts demonstrated a large excess of the α-subunit over complete GTH II and its β-subunit. This excess is reduced after the beginning of spermatocyte formation by the augmented (10-fold) increase in the content of the GTH II β-subunit. Thus, the GTH II β-subunit appears to be a limiting factor for the amount of complete GTH II in immature catfish.

In mature trout, T or E2 are not sufficient to stimulate GTH II β gene expression *in vitro* (Xiong et al. 1994a), but the effect and mechanism of action of steroids that are prominent at more advanced stages of development (for example 11-KT and 17,20βP) are yet unknown. In addition, the intracellular signalling pathways used by steroids and GnRHs might interact in mature salmon, possibly through a cluster of distal EREs, one of which may be sensitive to phorbol ester-induced transcription as well (Xiong et al. 1994a). This feature is of interest in context with the stronger effects of GnRHs on the GTH II subunit mRNA levels in mature as compared to regressed goldfish (Khakoo et al. 1994), considering that GnRHs make use of the protein kinase C pathway (Levavi-Sivan & Yaron 1989; Jobin & Chang 1992). Further data on the effects of steroids on the GTH mRNA levels are presented by Dickey & Swanson, Gur et al., and Huggard & Habibi (all this volume). Moreover, seasonal changes of GTH I and II subunit expression are described by Elizur et al. and Weil et al. (this volume).

Aspects of Gonadotrope Cell Biology and Regulation of GTH Secretion

GTH I and II are produced by two distinct types of gonadotropes in the trout (Nozaki et al. 1990; Naito et al. 1991; 1993), and changes in staining intensities, reflecting immunoreactivity or mRNA, show a pattern that correlates with the reproductive cycle. In catfish, on the contrary, all pituitary cells showing the ultrastructural characteristics of gonadotropes contain GTH II β-subunit immunoreactive material, irrespective of the ontogenetic stage (Schulz et al. 1995). A colocalisation of the two subunits is confined to the secretory granules (Zandbergen et al. 1993), whereas the GTH II β- but not the α-subunit is detectable in globules and irregular, membrane-bound masses (IMs); globules and IMs probably arise from granules during a crinophagic pathway and do not appear to be involved in GTH secretion (Sharp-Baker et al. 1995). The intracellular distribution of GTH II subunits in trout gonadotropes is similar to that in the catfish (Naito et al. 1995). In GTH I-producing cells, secretory granules too show a colocalization of both subunits whereas the globules are devoid of the GTH I β-subunit (Naito et al. 1993). Thus, intracellular mechanisms appear to selectively modify or eliminate one of the GTH subunits in fish gonadotropes. An antiserum against the recombinant, non-glycosylated GTH II β-subunit of the catfish revealed immunogold-labeling in the rough endoplasmic reticulum, in the Golgi cisternae, and - surprisingly - also in the secretory granules while globules and IMs remained unlabeled. Interestingly, E2 augments LH glycosylation in rat (Liu & Jackson 1990), suggesting that the post-translational processing of GTH subunits is yet another regulatory site of relevance.

In salmonids, GTH I secretion is responsive to GnRH stimulation at early stages of gonadal development (Kawauchi et al. 1989; Swanson 1992), and treatment with T or E2 reduces circulating GTH I levels (Dickey & Swanson; this volume). Other publications dealing with the regulation of GTH secretion in fish are restricted to GTH II secretion, where GnRHs, dopamine, and sex steroids are key players (reviewed by Peter et al. 1991; Kah et al. 1993; Yaron 1995). In addition, a large assembly of compounds modulates directly or indirectly the release of GTH II, thus being subject to an intricate regulatory network. For example; NPY has a dual stimulatory role on GnRH and GTH II release (Peng et al. 1993); activin and inhibin stimulate GTH II release (Ge & Peter 1994); or bombesin/gastrin-releasing peptide shows GTH II release potency (Himick & Peter 1995). The situation is further complicated by other factors such as dietary effects on GnRH synthesis and release (e.g. Kah et al. 1994), or the effect of social stimuli on the activity of GnRH neurones (eg. Francis et al. 1993).

In several teleost species, two or more GnRH peptides have been identified which are synthesised by distinct groups of neurones (Sherwood et al. 1994). One group is scattered in the ventral forebrain, gives rise to a prominent hypophysiotropic fibre tract, and appears to be of primary relevance for the regulation of GTH secretion (Kah et al. 1993). These neurones produce salmon GnRH, catfish GnRH, sea bream GnRH, or mammalian GnRH. A second group of GnRH neurones is situated in the midbrain and produces chicken GnRH-II (cGnRH-II). In a number of species including the African catfish, cGnRH-II is also present in the pituitary. In the catfish, however, cGnRH-II fibres connecting the midbrain to the pituitary could not be detected (Zandbergen et al. 1995), so that cGnRH-II might reach the pituitary *through* the circulation, for example. The GTH II release potency of cfGnRH is a 150-fold lower than that of cGnRH-II while the pituitary content of cfGnRH is 700-fold higher than that of cGnRH-II. Thus, despite its low potency, cfGnRH appears to be the principal regulator of GTH II secretion in the catfish. This may allow a fine-tuning of GTH II release in a species showing mostly low GTH II levels and a testicular steroidogenic system that is sensitive to minute changes of plasma GTH II levels (Schulz et al. 1993). An alike situation may exist in the gilthead sea bream pituitary where sea bream GnRH is less potent but more abundant than salmon GnRH (Zohar et al. 1995). Studies on goldfish in which the olfactory tract was severed indicate that the GnRH neurones of the terminal nerve system are unlikely to be involved in reproductive processes (Kobayashi et al. 1994).

The post-receptor action of GnRH on gonadotropes is mediated through intracellular events spanning the influx of Ca^{2+} and its mobilisation from intracellular sources, phosphoinositide turnover and the formation of IP3 and diacylglycerol, activation of PKC, the synthesis of arachidonic acid or its metabolites, and the synthesis of cAMP and activation of PKA (reviewed by Chang et al. 1993; Levavi-Sivan & Yaron 1993). Different GnRHs may operate through somewhat different signal transduction cascades in the goldfish, although the GnRHs bind to the same receptor type (Chang et al. 1993). Regarding the cytosolic free calcium concentration $[Ca^{2+}]_i$, catfish gonadotropes do not show, in contrast to their mammalian counterparts, spontaneous or GnRH-induced oscillations of $[Ca^{2+}]_i$. However, rapid and transient increases in $[Ca^{2+}]_i$ are observed upon GnRH stimulation and the dose-dependent response of GTH II release to cfGnRH or cGnRH-II is parallel to the amplitude of $[Ca^{2+}]_i$ increase (Rebers et al; this volume). This increase is not shared by catfish somatotropes which appear to be devoid of GnRH receptors (Bosma et al.; this volume). Also goldfish (Kah et al. 1992) and trout (Flores; this volume) gonadotropes show GnRH-induced, rapid and transient $[Ca^{2+}]_i$ increases.

Is there a Single GTH in the African Catfish?

GTH I-like hormones could not be isolated from the European eel (Quérat; personal communication), from several cypriniform species (Lo et al. 1991), or from the siluriform African catfish (Koide et al. 1992). Regarding the catfish, the GTH II content in pituitary extracts accounts for all the extracts' steroidogenic activity (Koide et al. 1992; Schulz et al. 1994a), and all gonadotropes contain GTH II β-subunit immunoreactivity (Schulz et al. 1995). Moreover, no evidence was found for a GTH I β-subunit-like compound when screening a pituitary cDNA library. Thus, a putative GTH I-like hormone in catfish would either fully cross-react with antibodies against the GTH II β-subunit, shares its steroidogenic activities, and would be synthesised in the same type of gonadotrope as GTH II, or it would be devoid of steroidogenic activity and produced in a cell type not showing the ultrastructural characteristics of gonadotropes. At present, we can only suggest that a GTH I-like hormone is not expressed in the African catfish pituitary.

Conclusion

The duality of teleost GTHs is well established for several taxa but may not apply for teleosts in general. It is anticipated that the implementation of molecular biology tools in comparative endocrinology will keep accelerating the research on many aspects of the gonadotropins' physiology. For the nearby future, significant progress is expected regarding the gonadotropes' physiology (e.g. cloning of and research on receptors mediating the modulation of GTH secretion; GTH gene expression; post-translational processing of GTH subunits), and regarding the actions of GTH (e.g. GTH-R gene expression; role of GTHs for germ cell development).

References

Ackland JF, Schwartz NB, Mayo KE & Dodson RE 1992. Physiol Rev 72:731-787

Banerjee PP, Banerjee S, Shen S-T, Kao Y-H & Yu JY-L 1993. Comp Biochem Physiol B 104:241-253

Chang JP, Jobin RM, & Wong AOL 1993. Fish Physiol Biochem 11:25-33

Chen TT, Marsh A, Shamblott M, Chan K-M, Tang Y-L, Cheng CM & Yang B-Y 1994. In: Fish Physiology, Vol XIII, Farrell AP & Randall DJ (series eds), Sherwood NM & Hew CL (eds), Acad Press, San Diego, pp 179-212

Copeland PA & Thomas P 1993. Gen Comp Endocrinol 91:115-125

Dombrowicz D, Hooghe-Peters EL, Gothot A, Sente B, Vanhaelst L, Closset J & Hennen G 1992. Arch Int Physiol Biochim Biophys 100:303-308

Francis RC, Soma K & Fernald RD 1993. Proc Natl Acad Sci USA 90:7794-7798

Ge W & Peter RE 1994. Zool Sci 11:717-724

Himick BA & Peter RE 1995. Neuroendocrinology 61:365-376

Huang FL, Huang CJ, Lin SH, Lo TB & Papkoff H 1981. Int J Peptide Protein Res 18:69-78

Huang CJ, Huang FL, Chang GD, Chang YS, Lo CF, Fraser MJ & Lo TB 1991. Proc Natl Acad Sci USA 88:7486-7490

Jobin RM & Chang JP 1992. Cell Calcium 13:531-540

Kagawa H, Young G, Adachi S & Nagahama Y 1982. Gen Comp Endocrinol 47:440-448

Kah O, Audy M-C, Mollard P 1992. In: Abstracts, 2nd Int Symp Fish Endocrinol, St. Malo (France, June 1-4, 1992), P29

Kah O, Anglade I, Leprêtre E, Dubourg P & de Montbrison D 1993. Fish Physiol Biochem 11:85-98

Kah O, Zanuy S, Pradelles P, Cerda JL & Carrillo M 1994. Gen Comp Endocrinol 95:464-474

Kanamori A, Nagahama Y & Egami N 1985. Zool Sci 2:707-712
Kanamori A & Nagahama Y 1988. Gen Comp Endocrinol 72:39-53
Kawauchi H, Suzuki K, Itoh H, Swanson P, Naito N, Nagahama Y, Nozaki M, Nakai Y & Itoh S 1989. Fish Physiol Biochem 7:29-38
Khakoo Z, Bhatia A, Gedamu L & Habibi HR 1994. Endocrinology 134:838-847
Kitahara N, Nishizawa T, Gatanaga T, Okazaki H, Andoh T & Soma GI 1988. Comp Biochem Physiol 91B:551-556
Kobayashi M, Amano M, Kim MH, Furukawa K, Hasegawa Y & Aida K 1994. Gen Comp Endocrinol 95:192-200
Koide Y, Noso T, Schouten G, Bogerd J, Peute J, Zandbergen MA, Schulz RW, Kawauchi H, Goos HJTh 1992. Gen Comp Endocrinol 87:327-341
Kraak G van der, Suzuki K, Peter RE, Itoh H & Kawauchi H 1992. Gen Comp Endocrinol 85:217-229
Levavi-Sivan B & Yaron Z 1989. Gen Comp Endocrinol 75:187-194
Levavi-Sivan B & Yaron Z 1993. Fish Physiol Biochem 11:51-59
Liu TC & Jackson GL 1990. Neuroendocrinology 51:642-648
Lo TB, Huang FL, Liu CS, Chang YS, Huang CJ 1991. Bull Inst Zool Acad Sin Monogr 16:19-38
Loir M 1994. Mol Cell Endocrinol 102:141-150
Loir M & LeGac F 1994. Biol Reprod 51:1154-1163
Miura T, Yamauchi K, Takahashi H & Nagahama Y 1991. Proc Natl Acad Sci USA 88:5774-5778
Miwa S, Yan L, Swanson P 1994. Biol Reprod 50:629-642
Nagahama Y 1994. Int J Dev Biol 38:217-229
Naito N, Hyodo S, Okumoto N, Urano A & Nakai Y 1991. Cell Tissue Res 266:457-467
Naito N, Suzuki K, Nozaki M, Swanson P, Kawauchi H & Nakai Y 1993. Fish Physiol Biochem 11:241-246
Naito N, Koide Y, Amano M, Ikuta K, Kawauchi H, Aida K, Kitamura S & Nakai Y 1995. Cell Tissue Res 279:93-99
Nakamura M & Nagahama Y 1989. Fish Physiol Biochem 7:211-219
Nozaki M, Naito N, Swanson P, Dickhoff WW, Nakai Y, Suzuki K & Kawauchi H 1990. Gen Comp Endocrinol 77:358-367
Nunez-Rodriguez J, Suzuki K, Peter RE, Kawauchi H 1992. In: Abstracts, 2nd Int Symp Fish Endocrinol, St. Malo (France, June 1-4, 1992), L51
Okada T, Kawazoe I, Kimura S, Sasamoto Y, Aida K & Kawauchi H 1994. Int J Pept Prot Res 43:69-80
Peng C, Chang JP, Yu KL, Wong AOL, Goor F van, Peter RE & Rivier JE 1993. Endocrinology 132:1820-1829
Pescovitz OH, Srivastava CH, Breyer PR & Monts BA 1994. Trends Endocrinol Metab 5:126-131
Peter RE, Trudeau VL, Sloley BD, Peng C & Nahorniak CS 1991. In: Scott AP, Sumpter JP, Kime DE, Rolfe MS (eds), Reproductive Physiology of Fish. Sheffield: FishSymp, pp 30-34
Planas JV & Swanson P 1995. Biol Reprod 52:697-704
Schulz RW, Blüm V 1990. Gen Comp Endocrinol 80:189-198
Schulz RW, Paczoska-Eliasiewicz H, Satijn DGPE & Goos HJTh 1993. Fish Physiol Biochem 11:107-115
Schulz RW, van der Corput L, Janssen-Domerholt J, Goos HJTh 1994a. J Comp Physiol B 164:195-205
Schulz RW, Sanden MCA van der, Bosma PT & Goos HJTh 1994b. J Endocrinol 140:265-273
Schulz RW, Renes IB, Zandbergen MA, Dijk W van, Peute J & Goos HJTh 1995. Biol Reprod, in press
Sharp-Baker HE, Peute J, Diederen JHB, Brokken L 1995. Cell Tissue Res 280:113-122
Sherwood NM, Parker DB, McRory JE & Lescheid DW 1994. In: Fish Physiology, Vol XIII, Farrell AP & Randall DJ (series eds), Sherwood NM & Hew CL (eds), Acad Press, San Diego, pp 3-67
Suzuki K, Kawauchi H & Nagahama Y 1988. Gen Comp Endocrinol 71:292-301
Swanson P, Dickhoff WW, Gorbman A 1987. Gen Comp Endocrinol 65:267-287
Swanson P 1991. In: Scott AP, Sumpter JP, Kime DE, Rolfe MS (eds), Reproductive Physiology of Fish. Sheffield: FishSymp, pp 2-7
Swanson P, Suzuki K, Kawauchi H & Dickhoff WW 1991. Biol Reprod 44:29-38
Swanson P 1992. Am Zool 32:20A
Tanaka H, Kagawa H, Okuzawa K & Hirose K 1993. Fish Physiol Biochem 10:409-418
Tyler CR, Sumpter JP, Kawauchi H & Swanson P 1991. Gen Comp Endocrinol 84:291-299
Ueda H, Kambegawa A, Nagahama Y 1984. J Exp Zool 231:435-439
Xiong F, Liu D, Ledrean Y, Elsholtz HP & Hew CL 1994a. Mol Endocrinol 8:782-793
Xiong F, Suzuki K, Hew CL 1994b. In: Fish Physiology, Vol XIII, Farrell AP & Randall DJ (series eds), Sherwood NM & Hew CL (eds), Acad Press, San Diego, pp 135-158
Yan LG, Swanson P & Dickhoff WW 1991. J Exp Zool 258:221-230
Yan LG, Swanson P & Dickhoff WW 1992. Biol Reprod 47:418-427
Yaron Z 1995. Aquaculture 129:49-73
Young G, Adachi S, Nagahama Y 1986. Dev Biol 118:1-8
Zandbergen MA, Branden van den CAV, Schulz RW, Janssen-Dommerholt J, Ruijter JM, Goos HJTh & Peute J 1993. Fish Physiol Biochem 11:255-263
Zandbergen MA, Kah O, Bogerd J, Peute J & Goos HJTh 1995. Neuroendocrinology, submitted
Zohar Y, Elizur A, Sherwood NM, Powell JFF, Rivier JE & Zmora N 1995. Gen Comp Endocrinol 97:289-299

STRUCTURAL RELATIONSHIPS BETWEEN "FISH" AND TETRAPOD GONADOTROPINS

B. Quérat.

Laboratoire de Physiologie Générale et Comparée, MNHN, URA 90 Evolution des Régulations Endocriniennes, CNRS, 7, rue Cuvier, 75231 Paris Cedex 05, France.

Introduction

The pituitary gonadotropins (GTH1 and GTH2 in "Fish", luteinizing hormone [LH] and follicle stimulating hormone [FSH] in Tetrapods) together with thyrotropin (TSH) are members of the glycoprotein hormone family. In Primates, Equids and probably in some other Mammals, this family also includes a chorionic gonadotropin (CG) (Pierce and Parsons, 1981). LH and FSH have been found in all groups of Tetrapods except in Squamates where LH has not (yet) been characterized. The GTH2-type gonadotropin has been fully characterized in one Chondrostean and in many Teleostean species (*cf.* Quérat, 1994). The presence of the GTH1-type has so far only been proven (by protein sequences) in relatively modern Teleosts (Salmonids and Percomorphs). Numerous attempts to purify a second gonadotropin in primitive Teleosts such as eel or catfish have been unsucessfull, but a pituitary gonadotropin with biochemical properties close to those of a GTH1-type hormone has been characterized in carp (Van der Kraak *et al.*, 1992), suggesting that this hormone may have appeared at least at the level of Euteleostei. What are the parental relationships existing between these "Fish" (Actinopterygian) gonadotropins and those from Tetrapods (Sarcopterygies), and how are they linked to the thyrotropins? The reconstruction of the history of these glycoprotein hormones would not only provide clues to understanding thyrotropin and gonadotropin functions, but would also allow more accurate revision of the present nomenclature of the gonadotropins in Fish. To address these questions, a tentative phylogenetic tree is proposed, based on the topologies obtained by means of two different algorithms with all the sequences currently available.

Results and discussion

All the glycoprotein hormones are composed of two glycosylated subunits, α and ß, the latter bearing the hormonal specificity. In any given species, the α subunit is unique but two different subunits have been found in certain Fish species such as carp (Chang *et al.*, 1988), salmon (Suzuki et al., 1988; Gen *et* al., 1993) and bonito (Koide *et al.*, 1993). Both subunits are strongly cross-linked by dissulfide bonds: 5 for α and 6 for ß. The crystal structure of the human CG has recently been obtained (*cf* Patel, 1994), revealing the assignment of the disulfide bonds *i.e.* α 1-4, 2-7, 3-8, 5-9, 6-10 and ß 1-6, 2-7, 3-12, 4-8, 5-9, 10-11 (numbers referring to the position of the half-cystines from the N-terminal end). Three of them are involved in the formation of a cystine-knot motif, very similar in both subunits. The ß 3-12 bonding is completed after the association between the α and the ß subunits has occured. This particular structure should probably be adopted by all members of the glycoprotein hormone family with the possible exception of the GTH1-type hormones in which the cysteine arrangement is modified (the position of the 3rd cysteine is not conserved). This might explain why the dissociation characteristics of the GTH1 are quite different from those of other members of the family (Suzuki *et al.*, 1988, Koide *et al.*, 1993, Van der Kraak *et al.*, 1992).

Alignment of α subunit sequences shows a high degree of homology for this subunit from primitive Teleosts to recent Teleosts on the one hand, and to Mammals and Birds on the other. In addition to the half-cystines and to the sugar-bearing asparagines (with a few exceptions), numerous amino acids are conserved, especially in the central part and in the carboxy-terminal part of the molecule.

A phylogenetic tree was obtained by using the maximum parsimony method (Paup), indicating that the duplications leading to the duality of the α subunits in carp, salmons and bonito have occurred recently and independently in these three genera.

All ß subunits may easily be aligned, provided that gaps are introduced, particularly in the gonadotropin ß-subunits, so as to take into account the two additional amino acids present between the 5th and 6th cysteines of the TSH ß-subunits. The fact that Teleost ß-TSH share these two insertions with the Tetrapod ß-TSH indicates that the separation of this lineage from all gonadotropin ß-subunit lineages took place before the separation

I am grateful to Drs. H. Philippe and G. Lecointre for help in the construction and analysis of the dendrograms. This work was supported by the GDR 1005 Systématique moléculaire.

between Actinopterygies and Sarcopterygies. Additional important characteristics revealed by the alignment is the modified arrangement of the cysteines in the GTH1 ß-subunits, and the number of deletions that have to be introduced for best alignment of these subunits showing that they have evolved rapidely. As an example, the percentage of homology between a Salmonid (*Oncorhynchus keta)* and a Scombridae (*Katsuwonus pelamis*) ß-subunit drops from 67 when comparing the GTH2 ß-subunits to 42 for the GTH1 ß-subunits. This raises the question of the functional significance of such a rapid evolution of this gonadotropin as compared to the others.

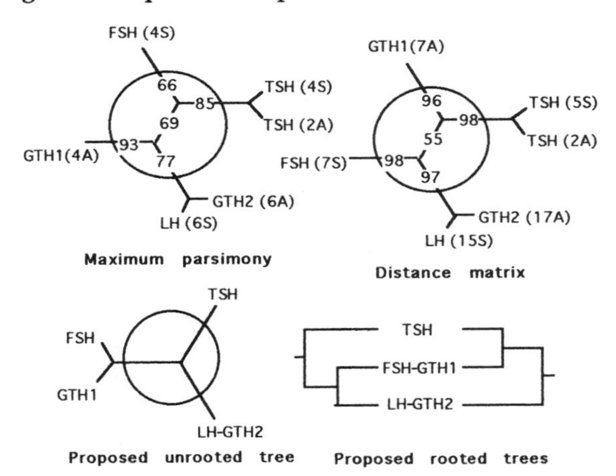

Fig. : Dendrograms of the phyletic relationships between glycoprotein hormone ß-subunits.
Top : trees were constructed from the protein sequences taken between the two external cysteines, by mean of a maximum parsimony method (left) and a distance matrix method (right). As all the represented groups of subunits are monophyletic, only the initial nodes of each lineage are presented. The number of sequences used for each lineage is indicated in brackets. Numbers on the branches represent the Bootstrap or the NJBoot value of the nodes (in %). The separation between Actinopterygies (A) and Sarcopterygies (S) took place outside the circles. Bottom : Unrooted and rooted trees are proposed that take into account the different data (see text).

Based on the protein sequences, dendrograms were constructed using both maximum parsimony (only representative sequences of each groups were used) and distance matrix (Neibhor-joining) methods, together with a statistical analysis (Bootstrap or NJBoot) of the robustness of the nodes. Since there is no reliable information about the order of appearance of the different lineages of ß-subunits, the trees were unrooted. Whatever the method used, GTH2 sequences appear to form a monophyletic group with LH, *i.e.* they share a common direct ancestor. Each of GTH1, FSH and TSH also form monophyletic groups, and only the initial embranchment of each group (the node at the base of each lineage) is represented in the dendrograms (figure, top). However, trees obtained by using the two different methods exhibit one major difference in topology. Both topologies are inconsistent with the data. With the maximum parsimony, the GTH1 lineage branches out with the LH-GTH2 lineage, indicating that the separation between these two lineages took place before the separation of Actinopterygies from Sarcopterygies. With the distance matrix, GTH1 and TSH appear as neighbors and, as indicated above, the TSH lineage diverged before the separation of Actinopterygies and Sarcopterygies. If these topologies are true, a GTH1-type hormone would be present (in addition to FSH) in all the Sarcopterygies. Conversely, a FSH type hormone would be present (in addition to GTH1) in all Actinopterygies. The problem in positioning the GTH1 lineage probably results from a rapid evolution of this lineage, and from the lack of data about primitive Teleosts and Chondrosteans. To solve this problem, the most parsimonious solution would be to combine the GTH1 lineage with the FSH lineage, giving them a common direct ancestor. Consistent with such a parental relationship is the fact that the shifted cysteine of the GTH1 ß-subunits located in the N-terminal part is aligned with a cysteine present in the signal peptide of the Tetrapod FSH ß-subunits (but not in the ß-TSH). According to this tentative topology, there would be three lineages at the time of the separation between Actinopterygies and Sarcopterygies, one TSH lineage, and two gonadotropic (FSH-GTH1 and LH-GTH2) lineages. The GTH1 sub-lineage, with its special cysteine arrangement would have appeared by mutations on the FSH-GTH1 precursor sometime after the emergence of the Actinopterygies, even though this gene might able to find one gene belonging to the FSH-GTH1 lineage in primitive (but present) Actinopterygies, eventhough this gene might have lost its ability to be expressed.

The respective roles of GTH1 and GTH2 in the control of reproduction of Fish is not clearly understood, though increasing data arising from studies mainly done on the salmonid model support the idea that GTH1 would act, like FSH, to promote spermatogenesis and follicular growth, whereas GTH2, like LH, would ellicit spermiation and ovulation (*cf.* Shulz, this volume). So both functional and structural data are in agreement with the existence of a direct parental relationship between GTH1 and FSH on the one hand, and GTH2 and LH on the other. These data may provide a reasonable basis for the revision of the nomenclature of Fish gonadotropins.

One problem that still remains unsolved and is critical for the understanding of the evolution of the glycoprotein hormone family, is the order of appearance of the three lineages. Obviously there are three possibilities but two of them seem more likely (figure, bottom). In the first hypothesis, the initial precursor has duplicated to give the TSH and a gonadotropic hormone, the latter then duplicating to give the two gonadotropic (LH-GTH2 and FSH-GTH1) lineages. In the second hypothesis, the first duplication has led to one gonadotropin (LH-GTH2) and a second hormone still bearing both gonadotropic and thyrotropic potentialities. A second duplication has given rise to the TSH and FSH-GTH1 lineages. The choice of the LH-GTH2 lineage rather than the FSH-GTH1 lineage as the first gonadotropic lineage that would have arisen relies on receptor studies in salmon : GTH2 is able to bind to a highly specific receptor whereas the GTH1 binding is displaced by GTH2 (Yan *et al.*, 1992). This result argues for the idea that GTH2 (the LH-GTH2 lineage) is more ancient than GTH1 (the FSH-GTH1 lineage).

When did these events occur? The separation of Osteichtyes in Actinopterygies and Sarcopterygies took place 400 million years ago when there were already three glycoprotein hormone lineages. The last duplication that occured in this family took place about 60 million years ago, at the time of the emergence of the Primates, with the ß-LH gene to give rise to a ß-CG gene, that is at least 350 million years after the preceding duplication. The first Vertebrates appeared some 500 million years ago. So it would not be very surprising if at least the first duplication of the ß-gene arose before the emergence of the Vertebrates. This duplication would have been preferentially selected for the physiological advantage of having control of the thyrotropic function (growth and development) clearly separated from control of reproduction.

References

Chang, Y.-S., Huang, C.-J., Huang, F.-L., and Lo, T.-B., 1988. Primary structures of carp gonadotropin subunits deduced from cDNA nucleotide sequences. Int. J. Peptide Protein Res. 32: 556-564.

Gen, K., Maruyama, O., Kato, T., Tomizawa, K., Wakabayashi, K. and Kato, Y. 1993. Molecular cloning of cDNAs encoding two types of gonadotrophin α subunit from the masu salmon, *Onchorhynchus masou*: construction of specific oligonucleotides for the $\alpha 1$ and $\alpha 2$ subunits. J. Mol. Endocrinol. 11: 265-273.

Koide Y., Itoh, H., and Kawauchi, 1993. Isolation and characterization of two distinct gonadotropins, GTHI and GTHII, from bonito (*Katsuwonus plelamis*) pituitary glands. Int. J. Peptide Res. 41: 52-65.

Patel, D., 1994. Glycoprotein hormones: a clapsed embrace. Nature 369: 438-439.

Pierce, J. G., and Parsons, T. F., 1981. Glycoprotein hormones: structure and function. Ann. Rev. Biochem. 50: 465-495.

Quérat, B. 1994. Molecular evolution of the glycoprotein hormones in Vertebrates. *In* Perspectives in Comparative Endocrinology, K.G. Davey, R.E. Peter and S.S. Tobe eds., coll. NRCC, Ottawa, 27-35.

Suzuki, K., Kawauchi, H. and Nagahama, Y., 1988b. Isolation and characterization of subunits from two distinct salmon gonadotropins. Gen. Comp. Endocrinol. 71: 302-306.

Van Der Kraak, G., Suzuki, K., Peter, R. E., Itoh, H., and Kawauchi, H., 1992. Properties of common carp gonadotropin I and gonadotropin II. Gen. Comp. Endocrinol. 85: 217-229.

Yan, I., Swanson, P. and Dickhoff, W.W., 1992. A two-receptor model for salmon gonadotropins (GTH I and GTH II). Biol. Reprod. 47: 418-427.

STEROIDOGENIC ACTIVITIES OF TWO DISTINCT GONADOTROPINS IN RED SEABREAM, *PAGRUS MAJOR*

H.Tanaka[1], H.Kagawa[1], and K.Hirose[2].

[1]National Research Institute of Aquaculture, Nansei, Mie, 516-01, Japan
[2]National Research Institute of Fisheries Science, Fukuura, Kanazawa, Yokohama, 236, Japan

Summary

Steroidogenic activities of PmGTH I and II in the ovarian follicles at various developmental stages and the testicular fragments in red seabream were investigated *in vitro*. Both PmGTH I and II stimulated estradiol-17β production in the vitellogenic follicles, though PmGTH II caused about a 5-fold increase in estradiol-17β over that of PmGTH I. Estradiol-17β production by PmGTH II decreased when oocytes reached about 500 μm in diameter. Estradiol-17β production in the follicles of oocytes at the migratory nucleus stage and the mature stage was not stimulated by PmGTH I or II. These results indicate that both PmGTH I and II stimulate *in vitro* estradiol-17β production in vitellogenic follicles, and PmGTH II is more potent than PmGTH I. In contrast, both PmGTH I and II are equipotent in stimulating the production of 11-ketotestosterone (11-KT) in the testicular fragments *in vitro*.

Introduction

We have previously shown the presence of two distinct gonadotropins, PmGTH I and II, in red seabream pituitary (Tanaka *et al.* 1993). However, biological activities of the GTHs still have not been shown in detail.

In vitro steroidogenic activities of GTH I and II were investigated in amago salmon ovarian follicles (Suzuki *et al.* 1988) and in the ovaries and testes of juvenile coho salmon (Swanson *et al.* 1989). Recently, Planas *et al.* (1993) revealed the biological activities of GTH I and II in male coho salmon *in vitro*. However, detailed studies on the steroidogenic activities of GTH I and II have been restricted to salmonid fish.

We have previously shown that PmGTH I and II stimulated *in vitro* estradiol-17β production in ovarian fragments of the red seabream (Tanaka *et al.* 1993). However, detailed information on estradiol-17β production by PmGTH I and II in various developmental stages of oocytes have not been elucidated yet. The present study aims to clarify the biological activities of PmGTH I and II in estradiol-17β production of the ovarian follicles at various developmental stages in red seabream. Stimulation of *in vitro* 11-ketotestosterone (11-KT) production in testicular fragments by PmGTH I and II were also examined.

Materials and Methods

PmGTH I and II were highly purified by ion-exchange chromatography and gel filtration followed by rpHPLC on Asahipak C4P-50 at pH 6.8. Female red seabream, which were spawning daily, were sacrificed at different times of the day. Ovaries were removed and cut into small pieces. Oocytes with their surrounding follicular layers intact were dispersed from the ovarian pieces by pipetting and oocytes at various developmental stages were collected by sieving. About 20-30 mg (wet weight) of oocytes were incubated in 24-well culture plates containing 1 ml L-15 medium, pH 7.4, supplemented with 2.3 g HEPES, 100 mg streptomycin and 100,000 IU penicillin per litter, with or without various doses of PmGTH I and II. After incubation for 18 h at 20°C, the concentrations of estradiol-17β in the media were measured by radioimmunoassay. For testicular incubation, mature males were killed and their testes removed and chopped in a petri dish on ice with a razor blade, and washed with incubation medium three

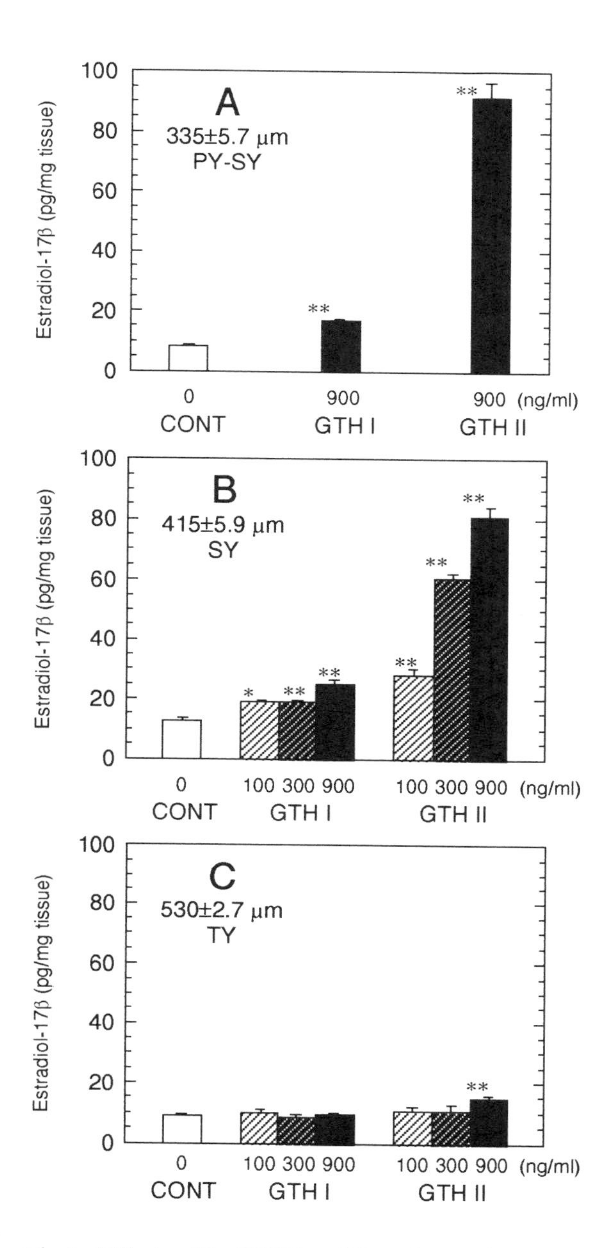

Fig.1. Effects of PmGTH I and II on the production of estradiol-17β in red seabream ovarian follicles incubated for 18 h at 20°C. Follicles at the various stages of development (A, 335±5.7 µm in diameter: primary-secondary yolk stage; B, 415±5.9µm: secondary yolk stage; C, 530±2.7µm: tertiary yolk stage) were incubated in medium alone (Control) or medium with various doses of PmGTH I or II (100-900 ng/ml). Each bar represents the mean ± SEM of 3 replicate incubations. Asterisks denote that the mean values significantly differed from the controls (*P<0.05, **P<0.01).

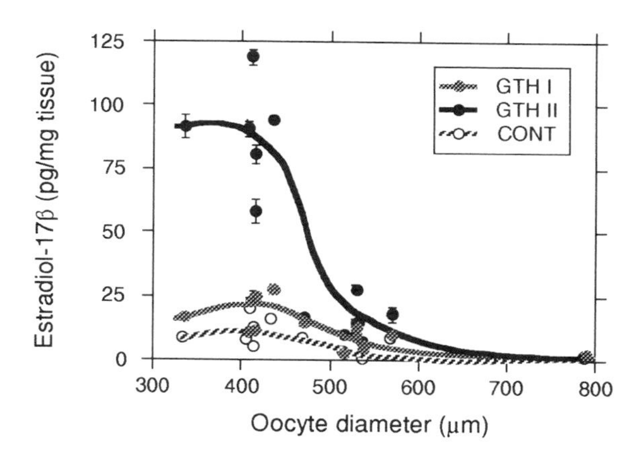

Fig.2. Summary of changes in estradiol-17β production stimulated by PmGTH I and II during oocytes development. Each point represents the mean ± SEM of 3 replicate incubations of oocytes at various size in medium alone (control) or medium with 900 ng/ml of hormone (GTH I and II).

times. About 20-40 mg (wet weight) of testicular tissues were incubated as described for oocytes cultured under gentle shaking. After incubation for 18 h at 20°C, media were collected and the concentrations of 11-KT inmedia were measured by radioimmunoassay. Remaining oocytes and testicular fragments were lyophilized and weighed. Measured values of steroid concentrations were divided by the dried tissue weights. All data are presented as means ± SEM (pg/mg tissue) of 3 or 6 replications. Differences between control and treated incubations were analyzed using ANOVA followed by Dunnett multiple comparisons test after log transformation.

Results and Discussion

Both PmGTH I and II significantly stimulated estradiol-17β production in early vitellogenic stage ovarian follicles (oocytes 335±5.7 µm in diameter corresponding to the primary to secondary yolk stage) (Fig.1A). They also stimulated estradiol-17β production in mid-vitellogenic stage follicles (415±5.9 µm, secondary yolk stage) (Fig.1B). However, no substantial stimulations were observed after oocytes completed vitellogenesis (530±2.7µm,

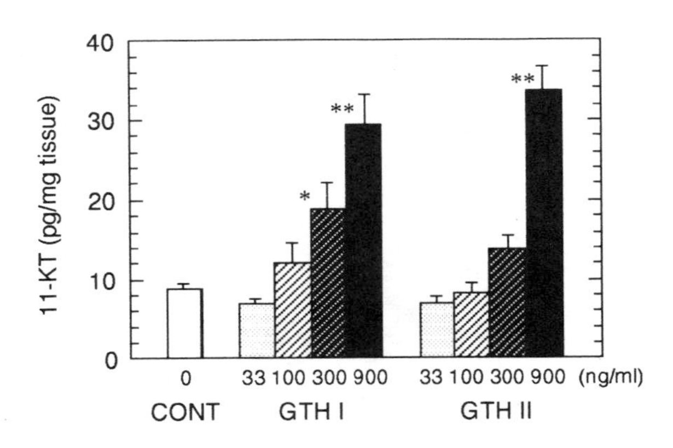

Fig.3. Effects of PmGTH I and II on the production of 11-ketotestosterone in red seabream testicular tissues incubated for 18 h at 20°C. Testicular tissues were incubated in medium alone (Control) or medium with various doses of PmGTH I or II (33-900 ng/ml). Each bar represents the mean ± SEM of 6 replicate incubations. Asterisks denote that the mean values which significantly differed from the control (*$P<0.05$, **$P<0.01$).

tertiary yolk stage) (Fig.1C).

Changes in estradiol-17β production stimulated by PmGTH I and II during oocyte development are presented in Fig.2. PmGTH I (900 ng/ml) caused approximately a 2-fold increase in estradiol-17β over the controls, whereas PmGTH II (900 ng/ml) caused 6 to 10-fold increase in estradiol-17β over the controls in vitellogenic follicles of 300-500 μm in diameter. Stimulation of estradiol-17β production by PmGTH II decreased when oocytes reached about 500 μm in diameter. Estradiol-17β production in follicles of oocytes at the migratory nucleus stage and the mature stage (>550 μm in diameter) was not stimulated by PmGTH I or II. These results indicate that both PmGTH I and II stimulate *in vitro* estradiol-17β production in vitellogenic follicles, and that PmGTH II is more potent than PmGTH I.

In testicular tissue, both PmGTH I and II stimulated 11-KT production in a dose dependent manner (Fig.3). The two GTHs (900 ng/ml) both caused about a 3-fold increase in 11-KT over the control.

The present study shows that both PmGTH I and II have stimulatory activities in estradiol-17β production in vitellogenic follicles and 11-KT production in testicular fragments *in vitro*.

This study was supported in part by a grant-in-aid (Bio Media Program) from the Ministry of Agriculture, Forestry, and Fisheries (BMP 95-II-2-4).

References

Planas,J.V., P.Swanson and W.E.Dickhoff, 1993. Regulation of testicular steroid production *in vitro* by gonadotropins (GTH I and GTH II) and cyclic AMP in coho salmon (*Oncorhynchus kisutch*). Gen. Comp. Endocrinol. 91:8-24.

Suzuki,K., Y.Nagahama and H.Kawauchi, 1988. Steroidogenic activities of two distinct salmon gonadotropins. Gen. Comp. Endocrinol.71: 452-458.

Swanson,P., M.Bernard, M.Nozaki, K.Suzuki, H.Kawauchi and W.W.Dickhoff, 1989. Gonadotropins I and II in juvenile coho salmon. Fish Physiol. Biochem. 7: 169-176.

Tanaka,H., H.Kagawa, K.Okuzawa and K.Hirose, 1993. Purification of gonadotropins (PmGTH I and II) from red seabream (*Pagrus major*) and development of a homologous radioimmunoassay for PmGTH II. Fish Physiol. Biochem. 10: 409-418.

SEABREAM GONADOTROPINS: SEXUAL DIMORPHISM IN GENE EXPRESSION

A. Elizur, I. Meiri, H. Rosenfeld, N. Zmora, W.R. Knibb, and Y. Zohar[1].

National Center for Mariculture, IOLR, P.O.Box 1212, Eilat Israel and Center of Marine Biotechnology, University of Maryland, Baltimore, MD 21228, USA[1]

Abstract

The temporal profile of the gilthead seabream ßGtH-I and ßGtH-II gene expression was measured over a 28 month period. Both ßGtH-I and ßGtH-II were expressed throughout the year, however absolute levels of transcripts were higher for ßGtH-II. Transcript levels varied also between sexes with βGtH-I levels generally higher in males and βGtH-II levels generally higher in females. The effect of administration of native seabream forms of GnRH on GtH subunits and GH gene expression was examined. None of the native seabream GnRH forms caused a statistically significant increase in transcripts levels at 8 and 24 hrs following administration however large differences between ßGtH-I and ßGtH-II was observed in males and females.

Introduction

The gilthead seabream, *Sparus aurata*, is a protandrous hermaphrodite. All individuals are males by the end of the first year and subsequently most change into females which have asynchronous development of oocytes and spawn over a three month period. Three forms of gonadotropin releasing hormones have been identified recently in the brain of the gilthead seabream (Powers *et. al.*, 1994), namely salmon GnRH, (sGnRH), chicken II GnRH (cGnRH-II) and a novel form, seabream GnRH (sbGnRH). We have cloned the α, βI and βII gonadotropin (GtH) subunit cDNAs and used them as probes to study the temporal profile of GtH gene expression over a two year period and in response to the administration of the native seabream GnRH forms.

Methods

Temporal profile: Seabream were maintained in two cohorts. At the start of the study, cohort 1 fish were 7 month old and all males while cohort 2 fish were over 18 month old with both sexes present. Each month, 10 fish (5 males and 5 females when relevant) were sacrificed from each cohort. Pituitary RNA was extracted from each individual separately (Chomczynski and Sacchi 1987) and analyzed by dot blot hybridizations using seabream βGtH-I and βGtH-II probes. The autoradiograms were scanned with a Molecular Dynamics computing densitometer (model 300A). The absolute levels of transcripts were determined by loading predetermined amounts of sense strand βGtH-I and II RNA standards.

Effect of GnRH: Seabream spermiating males and preovulatory females were injected with 5 and 25 μ/kg of native GnRH forms and the D-Ala^6 mGnRH (Bachem), as described in Zohar *et. al.*, (1995). The fish were sacrificed 8 and 24 hours after injections (3 males and 3 females / treatment) and their pituitary RNA extracted and used for hybridizations as above, except that in addition hybridizations with seabream αGtH and growth hormone (GH) and ß-actin cDNA probes were carried out and the loadings corrected by ß-actin readings.

Statistical analysis were carried out using analysis of variance (ANOVA) models.

Results and Discussion

Temporal profile of βGtH-I and βGtH-II gene expression.

The levels of βGtH-I and II transcripts were measured in the samples collected from the two cohorts over a period of 28 months. Both genes are expressed throughout the year however there is a clear rise in gene expression towards the reproductive season. In the fish from cohort 1 (data not shown), levels of βGtH-I transcript / pituitary ranged between 1pg to 60 pg and values peaked during the spawning season (considering all 13 time points, one-way ANOVA, $F_{[12,110]} = 7.36$, $P < 0.001$). Corresponding βGtH-II levels were 1 pg and 160 pg, and values also peaked during the spawning season (one-way ANOVA, $F_{[12,113]} = 8.33$, $P < 0.001$). In cohort 2, the absolute levels of βGtH-I again peaked during the spawning time (one-way ANOVA, $F_{[13,122]} = 5.78$, $P < 0.001$, Figure 1), and were correlated with their gonadal somatic index (not shown). In this regard a similar pattern was observed for βGtH-II (one-way ANOVA, $F_{[13,130]} = 5.41$, $P < 0.001$, Figure 1). However, one difference between βGtH-I and βGtH-II was that βGtH-I expression peaked towards December, when spawning season began while βGtH-II levels peaked towards the end of the spawning season. This latter profile is in close agreement with the yearly profile of the seabream GtH secretion levels (Zohar, unpublished).

Examination of the expression data in relation to sex reveals a clear and distinct sexual dimorphism (Figure 2). Usually βGtH-I levels were higher in males than in females (considering only known males and females and adjusting for the effect of time, two-way ANOVA, $F_{[1,99]} = 9.57$, $P < 0.01$) while the reverse was apparent for βGtH-II (two-way ANOVA, $F_{[1,107]} = 5.98$, $P < 0.05$). These trends were most prominent for fish in their third reproductive season which is also the first season that females spawn reliably in captivity.

The effect of native seabream GnRH forms on GtH gene expression.

Previously it was shown that three native seabream GnRH forms, sbGnRH, sGnRH and cGnRH-II as well as the mammalian D-Ala[6] GnRH analog stimulate GtH secretion in female seabream (Zohar *et. al.*, 1995) and to a lesser extent in males (unpublished). However, no information was available on the effect of these forms on gene expression. Here we examined the effects of these GnRH forms on mRNA levels of αGtH, GH, βGtH-I and βGtH-II. Levels of αGtH and GH were not significantly different among treatments, nor did they vary with sex when considering each time point (8hr and 24hr after administration) separately or combined (data not shown). Levels of βGtH-I and βGtH-II transcripts also did not significantly differ among treatments when considering each time point separately or combined (Figure 3 shows combined data). Khakoo *et al.*, (1994) reported an increase of βGtHII mRNA levels in goldfish in response to GnRH forms at 6 and 24 hrs after administration with cGnRH-II having a stronger effect then the salmon GnRH (in sexually mature fish). Hassin *et. al.*, (1995) observed an increase in both αGtH and βGtHII in response to mGnRHa 9 hrs after administration in the striped bass, (*Morone saxatilis*). It remains possible that these apparent discrepancies result from the few degrees of freedom available in our statistical analysis and indeed we note that the cGnRH-II form results in the highest seabream βGtH-II levels (Figure 3) and highest GtH releasing activity (Zohar *et. al.*, 1995).

For our present data on the seabream, levels of both βGtH-I and βGtH-II were statistically significantly different between sexes when considering each time point separately or combined (considering combined time points and adjusting for the effect of analogs, two-way ANOVA, $F_{[1,44]} = 61.12$, $P < 0.001$; $F_{[1,44]} = 40.85$, $P < 0.001$, respectively, Figure 3). In that βGtH-I levels were higher in males while βGtH-II levels were higher in females. Our results were consistent with the those obtained previously for the temporal profiles.

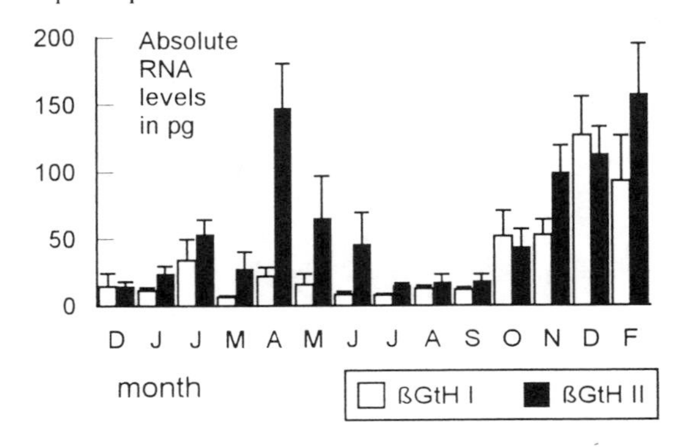

Fig.1 The temporal expression of βGtH-I and βGtH-II in cohort 2.

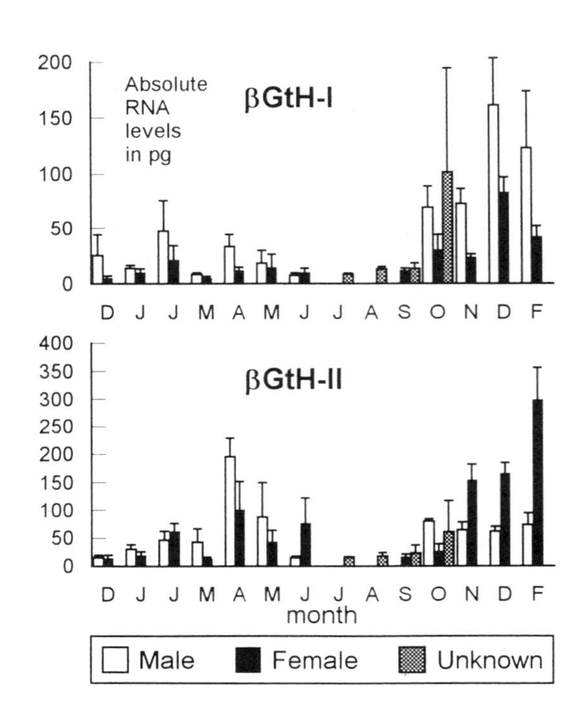

Fig. 2 . Sexual dimorphism in βGtH-I and βGtH-II gene expression (cohort 2)

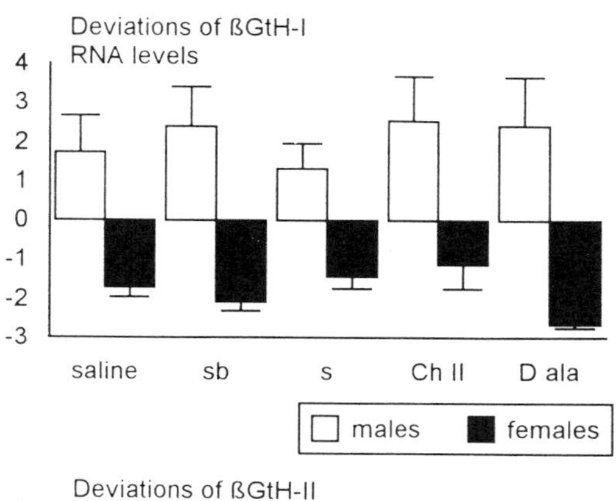

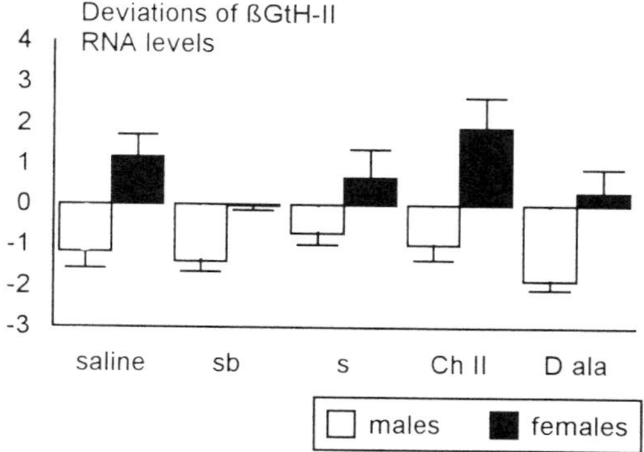

Fig. 3. The effect of GnRH on GtH subunits and GH gene expression. sb, seabream GnRH, s, salmon GnRH, Ch II, cGnRH-II and D ala, mammalian D ala^6 GnRH analog. To combine the 8 hr and 24 hr time points, values were expressed as deviations from the mean of the value of the saline controls for each experiment separately.

References

Chomczynski, P. and Sacchi, N. (1987). Single-step method of RNA isolation by acid guanidinium thiocyanate phenol chloroform extraction. *Anal. Biochem.* **162**, 156-159.

Hassin, S., Elizur, A. and Zohar, Y. (1995). Molecular cloning and sequence analysis of striped bass (*Morone saxatilis*) gonadotropins I and II subunits. *Journal of Mol. Endocrinol. In press.*

Khakoo, Z., Bhatia, A., Gedamu, L., and Habibi H.R. (1994). Functional specificity for salmon gonadotropin releasing hormone (GnRH) and chicken GnRH-II coupled to the gonadotropin release and subunit messenger ribonucleic acid level in the goldfish pituitary. *Endocrinology* **134**, 838-847.

Powell, J.F.F., Zohar, Y., Elizur, A., Park, M., Fischer, W.H., Craig, A.G., Rivier, J.E., Lovejoy, D.A. and Sherwood, N.M. (1994). Three forms of gonadotropin-releasing hormone characterized from brains of one species. *Proc. Natl. Acad. Sci.* USA. **91**, 12081-12085.

Zohar, Y., Elizur, A., Sherwood, N.M., Powell, J.F.F. Rivier, J.E. and Zmora, N. (1995). Gonadotropin-releasing activities of the three native forms of gonadotropin-releasing hormone present in the brain of gilthead seabream, *Sparus aurata*. *Gen. Comp. Endocrinol.* **97**, 289-299.

PRELIMINARY OBSERVATIONS ON GtH 1 AND GtH 2 mRNA LEVELS DURING GONADAL DEVELOPMENT IN RAINBOW TROUT, Oncorhynchus mykiss.

C. Weil (1), M. Bougoussa-Houadec (1), C. Gallais (2), S. Itoh (3), S. Sekine (4) and Y. Valotaire (2).

(1)INRA, Laboratoire de Physiologie des Poissons, Campus de Beaulieu, 35042 Rennes Cedex France,(2) Laboratoire de Biologie Moléculaire, Université de Rennes I, Campus de Beaulieu, 35042 Rennes Cedex, France, (3) Tokyo Research Laboratories, Kyowa Hakko Kogyo Co., Ltd., 3-6-6, Asahimachi, Machidashi, Tokyo 194, Japan, (4) Pharmaceutical Research Laboratories, Kyowa Hakko Kogyo Co., Ltd., 1188 Shimotogari, Nagaizumi-Cho, Sunto-Gun, Shizuoka-Ken 411, Japan.

Summary

In this preliminary study, the variations of α and ß GtH 1 and 2 gene expression were measured during gonadal development in male and female rainbow trout. Slot blots of total RNA, prepared from pools of pituitaries, were hybridized with chum salmon α and ß GtH 1 and 2 probes. In males and females, ßGtH 1 predominates in early stages of gonad development (spermatogonia A and previtellogenesis), ßGtH 2 being weakly expressed. Both ßGtH 1 and 2 are expressed during prespermiation, spermiation and the periovulatory period with a predominance of ßGtH 2. In both sexes αGtH variations follow ßGtH 2 variations.

Introduction

In salmonids, the existence of two gonadotropins, GtH 1 and GtH 2, controlling gonadal development, is now well established. They have distinct chemical characteristics (Suzuki *et al.* 1988 b,c) and the complete amino acid sequences of the ß- (Itoh *et al.*,1988) and α-subunits (GtH 1α1 and α2, GtH 2α - Itoh *et al.*, 1990) and the nucleotide sequences of the complementary deoxyribonucleic acids (cDNAs) encoding the α and ß-subunits (Sekine *et al.*, 1989 for chum salmon) have been determined.

At the pituitary level, these hormones are characterized by a distinct cellular distribution (Nozaki *et al*, 1990a) and a different ontogeny : GtH 1 appears early in larval development (Saga *et al.*, 1993), whereas GtH 2 appears alongside GtH 1 at the onset of spermatogenesis and vitellogenesis (Nozaki *et al.*, 1990b). Concerning plasma GtH 1 and 2 levels in various salmonid species, in immature fish they are low or undetectable, respectively (Swanson *et al.*, 1989). GtH 1 increases first, when gametogenesis is beginning and GtH 2 increases later, around ovulation and spermiation, remaining high in recently ovulated animals while GtH 1 decreases (Suzuki *et al.*, 1988a ; Swanson, 1991).

The present study examines the variations of α and ß GtH 1 and 2 subunit gene expression, during gonadal development in male and female rainbow trout. This was accomplished by measuring the mRNA levels present in pituitaries, using slot blot hybridization of a constant amount (2μg) of total RNA prepared from pooled pituitaries, probed with chum salmon α and ß GtH 1 and 2 cDNAs (0.7 to 3 10^9 dpm/μg). The gonadal stages studied included immature stages to ovulation and spermiation. So far, only one study has reported the changes in gene expression of the GtH subunits during ovarian development), using *in situ* hybridization combined with immunochemistry (Naito *et al.*, 1991).

Results

Variation in GtH gene expression with ovarian development (figure 1 A)

The female stages studied and the number of pituitaries pooled for the corresponding RNA preparations were as follows:

* F1: previtellogenic oocytes, n=16
* F2 : beginning of vitellogenesis, n=4
* F3 : end of vitellogenesis, n=5
* F4 : oocyte with peripheral germinal vesicle, n=2
* F5 : 0-8 days post-ovulation, n=6

In previtellogenic fish, βGtH 1 mRNA levels are the highest and predominate when compared with ß GtH 2 levels. When vitellogenesis starts, ßGtH 1 mRNA levels decrease to remain at the same level at the end of vitellogenesis and during the periovulatory period. On the other hand, during these latter stages, a dramatic increase occurs in the ß GtH2 mRNA levels indicating that GtH 2 is predominant, as compared to ßGtH1 mRNA levels. Variations in αGtH are observed, with higher levels around ovulation (F3 to F5) when ß GtH2 is predominant.

Variation in GtH gene expression with testicular development (figure 1 B).

The male stages studied and the number of pituitaries used for each RNA preparation were as follows :

*M1 : immature (A spermatogonia), n=8

*M2 : beginning of spermatogenesis (A and B spermatogonia), n=2

* M3 : prespermiation, n=4

*M4 : beginning of spermiation, n=5

*M5 : end of spermiation, n=6

In immature males, ßGtH1 mRNA predominates over ßGtH2 mRNA. ßGtH2 mRNA increases at prespermation and spermiation , the levels being the highest at the beginning of spermiation compared to prespermiation and the end of spermiation. αGtH mRNA levels do not show dramatic variations, except in fish at the beginning of spermiation in which the highest levels are observed.

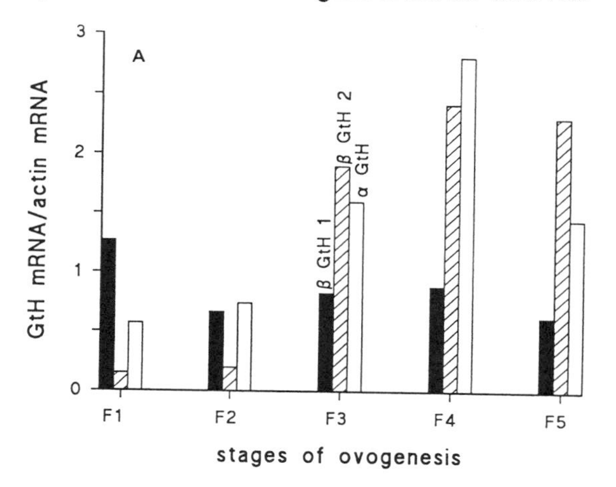

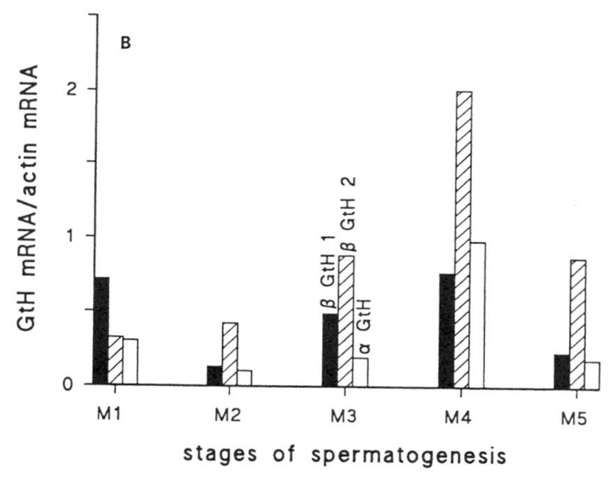

Figure 1 : Variation with ovarian **(A)** and testicular **(B)** development of pituitary mRNA levels of the β subunit of GtH 1 and 2 and the common α subunit of GtH 1 and 2 in rainbow trout.

Subunits were measured by slot blot hybridization analysis of 2µg of total RNA prepared from pools of pituitaries and scanning densitometry. Results are expressed as arbitrary units of GtH subunit/actin mRNA ratios. The gonadal stages studied and the number of pooled pituitaries, are indicated in the text (Results).

Discussion

The aim of the present work was to give preliminary results on the changes in gene expression of GtH 1 and 2 subunits during ovarian and testicular development. The method used employed slot blot hybridization of a constant quantity of total RNA (2 µg) probed with chum salmon cDNA (Sekine *et al.*, 1989). The α-subunit cDNA probe used was identical to the cDNA encoding the 1α2-subunit, the α-subunit common to both GtH1 and GtH2. As a result, we cannot differentiate the proportion due to GtH 1 or 2 and it is not known if the 1α1-subunit mRNA can be hybridized with the probe used. The values given do not correspond to absolute values of each mRNA corresponding to the different GtH subunits. However, as the specific activity of the different probes used is of the same range of magnitude, the values obtained for each mRNA can be compared and give an estimate of whether one gene is more or less expressed.

In the fish used, the sexual differentiation had already occurred and the most immature stage studied were females with previtellogenic oocytes and males with testis containing only A spermatogonia. At these two stages, the expression of GtH 1 is predominant compared to GtH 2 although a weak signal is observed in both sexes. On the other hand, in previtellogenic and prespermatogenic (presence of spermatogonia only) rainbow trout, Nozaki *et al.*, (1990b) identified only GtH1 cells immunocytochemically, using antibodies raised against coho salmon GtH 1 and GtH 2 ß subunits. Furthermore, Naito *et al.* (1991), using *in situ* hybridization found no significant signal above background for ßGtH 2 mRNA in cells considered to be putative GtH2 cells. These authors observed a significant signal only at a more advanced stage, during early vitellogenesis. However, in the present study we revealed a slight increase in ßGtH 2 mRNA levels at this time. This increase in ßGtH2 mRNA levels might be induced by the increase in oestradiol production. Indeed, in immature eel, *in vivo* treatment with testosterone or oestradiol increases ßGtH 2 mRNA levels (Quérat *et al.*, 1991) as does testosterone in immature trout (Trinh *et al.*, 1986). By the end of vitellogenesis, when plasma oestradiol levels are maximal (Billard *et al.*, 1978) we observed, as did Naito *et al.* (1991) using *in situ* hybridization, a dramatic increase in ßGtH 2 mRNA levels. αGtH mRNA levels present the same pattern of change. In fish, oestradiol increases slightly αGtH 2 subunit mRNA levels (Quérat *et al.*, 1991) but data are not available for the αGtH 1 subunit. Plasma oestradiol levels decrease from the end of vitellogenesis to ovulation (Fostier and Jalabert, 1986) but ßGtH 2 and αGTH mRNA levels remained high in our study. These high levels could be due to the presence at these stages of elevated synthesis and release of endogenous GnRH (Breton *et al.*, 1986), since exogenous GnRH has recently been shown to increase ß and αGtH 2 mRNA levels (Khakoo *et al.*, 1994) in sexually mature goldfish of mixed sex. This could also explain the high increase observed in spermiating males for ßGtH 2 and αGtH, since Amano *et al.*, 1993 observed variations in brain and pituitary GnRH at the time of testis maturation. The lowest levels, observed at prespermiation and at the end of spermiation, could be related to lower GnRH synthesis and release.

References

Amano, M., Aida, K.,Okumoto, N. and Y. Hasegawa, 1993. Changes in levels of GnRH in the brain and pituitary in male masu salmon, Oncorhynchus masou, from hatching to maturation. Fish Physiol. Biochem. 11, 233-240.

Billard, R., Breton, B., Fostier, A., Jalabert, B. and C. Weil, 1978. Endocrine control of the teleost reproductive cycle and its relation to external factors : Salmonid and Cyprinid models. *In* Gaillard P.J. and Boer H.H. eds. Comparative Endocrinology : 37-48. Elsevier, Amsterdam.

Breton, B., Motin, A., Billard R., Kah, O., Geoffre, S. and G. Precigoux. 1986. Immunoreactive gonadotropin-releasing hormone -like material in the brain and the pituitary gland during the preovulatory periods in the brown trout (Salmo trutta L.) : relationships with the plasma and pituitary gonadotropin. Gen. Comp. Endocrinol. 61, 109-119.

Fostier, A. and B.Jalabert, 1986. Steroidogenesis in rainbow trout (Salmo gairdneri) at various preovulatory stages : changes in plasma hormone levels and in vivo and in vitro responses of the ovary to salmon gonadotropin. Fish Physiol. Biochem. 2, 87-89.

Itoh, H., Suzuki, K. and H. Kawauchi, 1988. The complete amino acid sequences of ß-subunits of two distinct chum salmon GtHs. Gen. Comp. Endocrinol. 71, 438-451.

Itoh, H., Suzuki, K., and H. Kawauchi, 1990. The complete amino acid sequences of α subunits of chum salmon gonadotropins. Gen. Comp. Endocrinol. 78, 56-65.

Khakoo, Z., Bhatia, A., Gedamu, L. and H. R. Habibi, 1994. Functional specificity for salmon gonadotropin releasing hormone (GnRH) and chicken GnRH-II coupled to the gonadotropin release and subunit messenger ribonucleic acid level in the goldfish pituitary. Endocrinology 134, 838-847.

Naito, N., Hyodo, S., Okumoto, N., Urano A. and Y.Nakai, 1991. Differential production and regulation of gonadotropins (GtH I and GtH II) in the pituitary gland of rainbow trout, Oncorhynchus mykiss during ovarian development. Cell Tissue Res. 266, 457-467.

Nozaki, M., Naito, N., Swanson, P., Miyata, K., Nakai, Y., Suzuki, K. and H. Kawauchi, 1990a. Salmonid pituitary gonadotrophs. I. Distinct cellular distributions of two gonadotropins, GtH I and GtH II. Gen. Comp. Endocrinol. 77, 348-357

Nozaki, M., Naito, N., Swanson, P., Dickhoff, W. W., Nakai, Y., Suzuki, K. and H. Kawauchi, 1990b. Salmonid pituitary gonadotrophs II Ontogeny of GtH I and GtH II cells in the rainbow trout (Salmo gairdneri irideus). Gen. Comp. Endocrinol. 77, 358-367.

Quérat, B., Hardy, A. and Y.A. Fontaine, 1991. Regulation of the type-II gonadotrophin α and ß subunit mRNAs by oestradiol and testosterone in the European eel. J. Mol. Endocrinol. 7, 81- B. 86.

Saga, T., Oota, Y., Nozaki, M. and P. Swanson, 1993. Salmonid pituitary gonadotrophs. III. Chronological appearance of GtH I and other adenohypophysial hormones in the pituitary of the developing rainbow trout (Oncorhynchus mykiss irideus). Gen. Comp. Endocrinol. 92, 233-241.

Sekine, S., Saito, A., Itoh, H., Kawauchi, H. and S. Itoh,.1989. Molecular cloning and sequence analysis of chum salmon gonadotropin cDNas. Proc. Natl. Acad. Sci. USA. 86, 8645-8649.

Suzuki, K., Kanamori, A., Nagahama, Y. and H. Kawauchi, 1988a. Development of salmon GtH I and II radioimmunoassays. Gen. Comp. Endocrinol. 71, 459-467.

Suzuki,K., Kawauchi, H. and Y. Nagahama, 1988b. Isolation and characterization of two distinct gonadotropins from chum salmon pituitary glands. Gen. Comp. Endocrinol. 71, 292-301.

Suzuki, K., Kawauchi, H., and Y. Nagahama, 1988c. Isolation and characterization of subunits from two distinct salmon gonadotropins. Gen. Comp. Endocrinol. 71, 302-316.

Swanson, P., Bernard, M., Nozaki, M., Kawauchi, H. and W. W. Dickhoff, 1989. Gonadotropins I and II in juvenile coho salmon. Fish Physiol. Biochem. 7, 169-176.

Swanson, P. 1991. Salmon gonadotropins : reconciling old and new ideas. *In* : Scott, A.P. ; Sumpter, J.P.; Kime , D. E. & Rolfe, M.S. eds. Proceedings of the Fourth International Symposium on the Reproductive Physiology of fish. Norwich 1991 : 2-7. The University of East Anglia, Sheffield, U.K

Trinh, K. Y., Wang, N.C., Hew, C.L. and L. W. Crim,.1986. Molecular cloning and sequencing of salmon gonadotropin ß subunit. Eur. J. Biochem. 159, 619-624.

EFFECT OF TESTOSTERONE ON GROWTH HORMONE AND MATURATIONAL GONADOTROPIN SUBUNIT GENE EXPRESSION IN THE CULTURED GOLDFISH PITUITARY, *IN VITRO*

D. Huggard and H.R. Habibi

Department of Biological Sciences, University of Calgary, Calgary, Alberta, Canada T2N 1N4.

Introduction

Teleosts are seasonal spawners which undergo annual reproductive cycles in response to environmental cues. While the growth of fish is controlled primarily by pituitary growth hormone (GH), the reproductive cycle is regulated by the pituitary gonadotropin hormones (GtHs) (Habibi and Peter, 1991). The control of growth and reproduction, specifically of GtH and GH production, is complex and involves multiple factors including the gonadal steroids. Sexual maturation (gonadal recrudescence) in goldfish involves increases in the circulating levels of the GtHs (Kobayashi et al., 1986), GH (Marchant and Peter, 1986), and of the gonadal steroids (Kobayashi et al., 1986). Because these hormones change with the season in a similar manner, the potential for interactions among these hormones is great. It has been demonstrated in many species that the steroids act via a classical negative feedback loop on GtH release, as gonadectomy leads to a rapid increase in the plasma gonadotropin levels. This rapid increase can then be returned to near normal levels following the administration of gonadal steroids (Gay and Bogdanove, 1969). In goldfish, both positive and negative effects of the gonadal steroids have been observed (Kobayashi et al., 1989; 1990). Studies in fish demonstrated negative feedback of the gonadal steroids primarily in *mature* animals (Billard et al., 1977; Billard, 1978; Peter, 1982; Kobayashi et al., 1990), whereas positive feedback effects were observed in *regressed* or *immature* fish (Crim et al., 1981; Crim and Evans, 1983; Dufour et al., 1983; Counis et al., 1987; Trudeau et al., 1991; Kobayashi et al., 1989).

While there is considerable data documenting the effects of the gonadal steroids on GtH release, information on the effects of the gonadal steroids on GtH and GH synthesis is limited. As well, the mechanisms by which the gonadal steroids act to regulate growth and reproduction are not fully understood. The purpose of this study was to investigate the effect of testosterone on GtH and GH synthesis in the sexually immature goldfish pituitary.

Materials and Methods

Hormones and other chemicals

Testosterone (Sigma) was solubilized in a 1:1 solution of propylene glycol:ethanol at the concentration of 10 µg/20 µl and stored at -20 C. Appropriate concentrations of testosterone were prepared by dilution immediately prior to use. The trout α-tubulin cDNA fragment (1.5 kilobases (kb) in length) was provided by Dr. G. Dixon (Department of Medical Biochemistry, University of Calgary, Calgary, Alberta, Canada) and was used as an internal standard to control for loading. The carp GtH-II-α cDNA fragment (0.8-0.9 kb), the carp GtH-II-β cDNA fragment (0.7-0.8 kb), and the carp Growth Hormone fragment (1.1 kb) were provided by Dr. F.L. Huang (Institute of Biological Chemistry, Academia Sinica, Taipei, Taiwan).

Perifusion of pituitary fragments

Experiments to determine the direct effect of testosterone on GtH-II subunit and GH mRNA levels were carried out *in vitro* using a modified superfusion system as described previously (Khakoo et al., 1994). Goldfish, *Carassius auratus,* of mixed sex at the early recrudescence stage where purchased from Aquatic Imports (Calgary, AB). Fish were anaesthetized and sacrificed according to the animal care regulations of the University of Calgary. The pituitary glands were removed and fragmented. Eight pituitary equivalents were placed in each column and treated continuously for 15 hours at a flow rate of 5 ml/hour. Upon completion the pituitary fragments were removed and total RNA was extracted as described below.

Determination of GtH-II subunit and GH mRNA levels

Total RNA was extracted using the acid guanidinium thiocyanate-phenol-chloroform extraction method (Chomczyski and Sacchi, 1987). Three micrograms of total RNA were loaded into separate wells and resolved on a 1.2% agarose/formaldehyde gel and transferred onto a

Supported by a Natural Sciences and Engineering Research Council of Canada Grant U-0037946 to H.R.H.

Hybond N^+ membrane (Amersham) in the presence of 20X SSPE using capillary transfer. Purified cDNA fragments were labelled with [α-^{32}P] deoxycytidine 5'-triphosphate (dCTP) using the T7 quick prime random primer method (Pharmacia). Membranes were prehybridized with rapid hybridization buffer (Amersham) and subsequently hybridized with the specific probes. The membranes were then washed in a series of increasing stringency washes up to 0.1X SSC in the presence of 0.1% SDS. The autoradiograms were scanned using a computer densitometry program provided by the NIH (Bethesda, MD). The quantified mRNA levels for the GtH-II subunit and GH mRNA levels were corrected for loading errors with respect to corresponding α-tubulin levels. It should be noted that the α-tubulin levels are not affected by treatment with steroids. The relative mRNA levels shown in Figures 1-2 represent changes relative to control levels in each treatment group and are expressed as means +/- SE. These values were statistically analyzed using Student's t test and were determined to be statistically significant.

Results

Experiments were carried out *in vitro* to investigate the direct action of testosterone on GtH-II subunit and GH mRNA levels in the goldfish pituitary. Previous validation experiments demonstrated that pituitary fragments cultured in this way remain viable for 24 hours (results not shown). In our experiments where the pituitary fragments were incubated for 15 hours there was no evidence of degradation.

Treatment with 20 ng/ml of testosterone significantly ($P<0.05$) increased GtH-II α and β mRNA levels in the goldfish pituitary. The results after quantification with respect to the α-tubulin standard indicated a 7-9 fold stimulation compared to control (Fig. 1). A similar stimulation in GH mRNA levels following treatment with 20 ng/ml of testosterone was observed (Fig. 2). Previous experiments demonstrated a biphasic effect of testosterone *in vivo* resulting in an inhibition of GtH-II subunit and GH mRNA levels when injected at concentrations of 20 µg/fish or greater (unpublished results). For this reason we tested testosterone at 5 µg/ml *in vitro* and the results clearly show a stimulatory effect at concentrations which cause inhibition *in vivo* (Fig. 1 and 2).

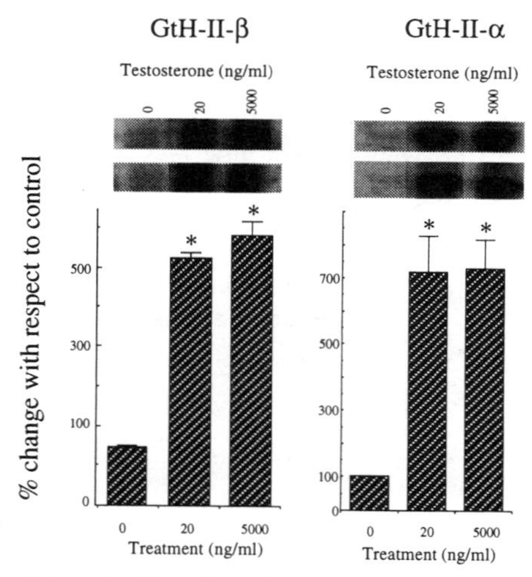

Figure 1. The effect of tesosterone on GtH-II subunit mRNA levels in sexually immature goldfish *in vitro*. Data represensts means +/- SE of two observations, each obtained from 8 pooled pituitaries. The quantified values are corrected with respect to alpha tubulin standard.
* Significantly different ($P<0.005$) from the control.

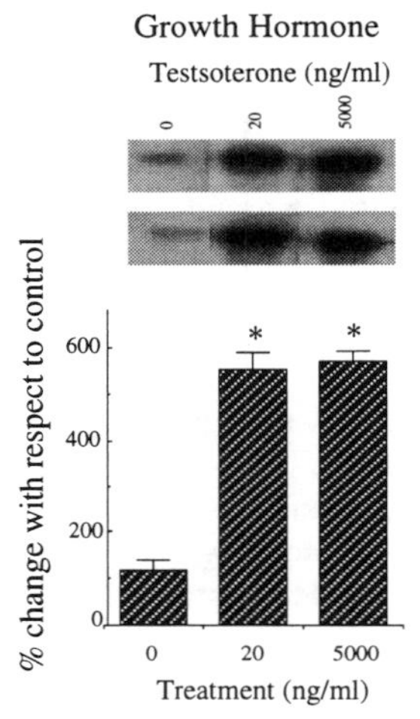

Figure 2. The effect of tesosterone on Growth Hormone mRNA level in sexually immature goldfish *in vitro*. Data represensts means +/- SE of two observations, each obtained from 8 pooled pituitaries. The quantified values are corrected with respect to alpha tubulin standard.
* Significantly different ($P<0.05$) from the control.

Discussion

While there is considerable data available on the effects of the gonadal steroids on GtH and GH release, information concerning the effects of the gonadal steroids on GtH and GH synthesis is limited. In addition, the mechanims by which the gonadal steroids act to regulate GtH-II subunit and GH mRNA levels are not fully understood. In this study the direct effects of testosterone on GtH-II subunit and GH mRNA levels at the level of the pituitary gland were investigated in sexually immature goldfish which contain very low levels of circulating sex steroid levels. The results provide evidence for a direct action of testosterone at the level of the pituitary gland in goldfish since addition of the steroid to the incubation medium significantly stimulated GtH subunit and GH mRNA levels.

Recent studies in our laboratory have demonstrated a biphasic effect of testosterone on both GtH-II subunit and GH mRNA levels in sexually immature goldfish *in vivo* (unpublished results). The present results clearly demonstrate that testosterone acts in a stimulatory manner directly at the level of the pituitary gland to increase both GtH-II subunit and GH mRNA levels. These findings suggest that the inhibitory effects of testosterone previously observed *in vivo* are not the result of the steroid acting directly at the level of the pituitary gland, but are likely to be through indirect action on other neuroendocrine factors, or receptors.

In summary, these results demonstrate for the first time that testosterone acts in a direct manner at the level of the pituitary gland to stimulate GtH-II subunit and GH mRNA levels in the sexually immature goldfish.

References

Billard R. 1978. Testicular feedback on the hypothalamo-pituitary axis in rainbow trout (Salmo gairdneri R.). *Ann Biol Anim Biochem Biophys.* 18:813-818.

Billard R, Richard M, and Braton B. 1977. Stimulation of gonadotropin secretion after castration in rainbow trout. *Gen Comp Endocrinol.* 33:163-165

Chomczyski P, and Sacchi N. 1987. Single-step method of RNA isolation by acid guanidinium thiocyanate-phenol-chloroform extraction. *Anal Biochem.* 162:156-159.

Counis R, Dufour S, Ribot G, Querat B, Fontaine YA, and Jutisz M. 1987. Estradiol has inverse effects on pituitary glycoprotein hormone alpha-subunit messenger ribonucleic acid in the immature European eel and the gonadectomized rat. *Endocrinology.* 121:1178-1184.

Crim LW, and Evans DM. 1983. Influence of testosterone and/or luteinizing hormone releasing hormone analogue on precocious sexual development in the juvenile rainbow trout. *Biol Reprod.* 29:137-142.

Crim LW, Peter RE, and Billard R. 1981. Onset of gonadotropic hormone accumulation in the immature trout pituitary gland in response to estrogen or aromatizable androgen steroid hormones. *Gen Comp Endocrinol.* 44:374-381.

Dufour S, Delerue-LeBelle N, and Fontaine Y-A. 1983. Effects of steroid hormones on pituitary immunoreactive gonadotropin in European freshwater eel, *Anguilla anguilla*, L. *Gen Comp Endocrinol.* 52:190-197.

Gay VL, and Bogdanove EM. 1969. Plasma and pituitary LH and FSH in the castrated rat following short-term steroid treatment. *Endocrinology.* 84:1132-1142.

Habibi HR, and Peter RE. 1991. Gonadotropin-releasing hormone (GnRH) receptors in teleosts. *Proceedings of the Fourth International Symposium on the Reproductive Physiology of Fish.* FishSymposium. pp. 109-113.

Khakoo Z, Bhatia A, Gedamu L, and Habibi HR. 1994. Functional specificity for salmon gonadotropin-releasing hormone (GnRH) and chicken GnRH-II coupled to the gonadotropin release and subunit messenger ribonucleic acid level in the goldfish pituitary. *Endocrinol.* 134:838-847.

Kobayashi M, Aida K, Hanyu I. 1986. Annual changes in plasma levels of gonadotropin and steroid hormones in goldfish. *Bull Jpn Soc Sci Fish.* 5:1153-1158.

Kobayashi M, Aida K, Hanyu I. 1989. Induction of gonadotropin surge by steroid hormone implantation in ovariectomized and sexually regressed female goldfish. *Gen Comp Endocrinol.* 73:469-476.

Kobayashi M, and Stacey NE. 1990. Effects of ovariectomy and steroid hormone implantation on serum gonadotropin levels in female goldfish. *Zool Science.* 7:715-722.

Marchant TA, and Peter RE. 1986. Seasonal variations in body growth rates and circulating levels of growth hormone in the goldfish, *Carassius auratus*. *J Exp Zool.* 237:231-239.

Peter RE. 1982. Neuroendocrine control of reproduction in teleosts. *Can J Fish Aquat Sci.* 39:48-55.

Trudeau VL, Peter RE, and Sloley BD. 1991. Testosterone and estradiol potentiate the serum gonadotropin response to gonadotropin-releasing hormone in goldfish. *Biol Reprod.* 44:951-960.

BLOCKS ALONG THE HYPOTHALAMO-HYPOPHYSEAL-GONADAL AXIS IN IMMATURE BLACK CARP, *MYLOPHARYNGODON PICEUS*

Z. Yaron[1], G. Gur[1], P. Melamed[1], B. Levavi-Sivan[1], A. Gissis[2], D. Bayer[1,3], A. Elizur[3], C. Holland[4], Y. Zohar[4] and M.P. Schreibman[5]

[1]Department of Zoology, Tel-Aviv University, Tel Aviv 69978, Israel, [2]Fish Hatchery, Kibbutz HaMa'apil, Israel, [3]National Center for Mariculture, Eilat, Israel, [4]Center of Marine Biotechnology, University of Maryland, Baltimore, USA, [5]Department of Biology, Brooklyn College, New York, USA

Summary

The objective of the present work was to localize sites along the hypothalamo-hypophyseal-gonadal axis which are not functional in immature, 2 y old black carp. Circulating cGTH in fish injected with sGnRHa+metoclopramide (10 µg/kg + 20 mg/kg), and estradiol in fish injected with carp pituitary extract (CPE; containing 350 µg/kg cGTH) did not differ from those in saline-injected controls. cGTH release from culture of dispersed pituitary cells did not change after exposure to sGnRH (1pM-1µM), however, a 300% increase occurred in response to TPA (12.5 nM) indicating a block proximal to PKC activation, probably at the GnRH receptor level. Exposure of pituitary cells to testosterone (T; 0.1-100 µM) for 2 days in culture resulted in an increase in both basal and GnRH-stimulated cGTH release. This would indicate that T stimulates directly the pituitary cells to increase the synthesis of GTH and the formation of GnRH receptors. Estradiol (E_2) secretion from the rudimentary gonad in incubation was stimulated by dbcAMP (0.3-3 mM) but not by CPE containing 0.1-4 µg/ml. This would indicate the presence of steroidogenic enzymes in the immature gonad and a block at the level of the GTH receptors. The retroperitoneal fat-pad underlying the rudimentary gonad is able to secrete E_2 *in vitro*. The secretion from the fat-pad, but not from visceral adipose tissue, could be stimulated by dbcAMP. Due to its large mass, the fat-pad should be considered to be a major source of sex steroids during the process of puberty.

Introduction

The black carp, *Mylopharyngodon piceus* is a molluscivorous fish; its potential to reduce excessive populations of snails has been demonstrated in the main water reservoirs in Israel (Leventer, 1979). Production of black carp fry for stocking is hampered by the difficulty in maintaining broodstock of fish which reaches sexual maturation only after 6 years.

A project has been undertaken to find what steps along the hypothalamo-hypophyseal-gonadal axis are functioning or blocked in the immature fish, and whether such blocks can be lifted by endocrine manipulations. This paper reports on the *in vivo* and the *in vitro* experiments carried out in order to examine the functional state of the pituitary and the gonads in 2 year old black carp.

Materials and Methods

In vivo challenge experiments: The experiments were carried out in June during the natural spawning season of adult fish. Fish of both sexes (0.7-2 kg bw; 12 in each group) were injected with sGnRHa + metoclopramide (GnRH+MET; 10 µg/kg+20 mg/kg), carp pituitary extract (CPE; containing 350 µg/kg cGTH) or saline. Blood was sampled before injection, and 4, 8, 12 and 24 h thereafter. Two pituitaries and samples of all gonads were taken for histology.

In vitro challenge experiments: The saline-injected fish of the previous experiment were used in these experiments. Cells from 5 pituitaries were dispersed, pooled and cultured for 2 days at 28°C and then challenged for 2 h with sGnRH, cGnRH II (1pM-1µM), TPA (12.5 nM) or forskolin (0.1 µM). The medium was collected for cGTH determination by RIA using as standard the common carp GTH (GTH II, B. Breton, Rennes or E. Burzawa-Gerard, Paris). The use of cGTH RIA is justified by the parallel dilution curves obtained from black carp pituitary extract and plasma and that of the cGTH standard. In addition, the amino acid sequence, as deduced from the black carp GTH-IIß cDNA, was found to have 97% homology with that of the common carp (Bayer and Elizur, unpublished). Other details of the culture were as in Levavi-Sivan and Yaron (1992).

In another experiment, pituitary cells were exposed for 2 days in culture to testosterone (T; 1-100 nM) and then challenged for 2 h either with sGnRH or cGnRH II, both at 100 nM.

Supported by Grant No. IS-2149-92 from BARD, The US-Israel Binational Agricultural R & D Fund.

Segments of the rudimentary gonads or adipose tissues were preincubated for 3 h and then exposed for 3 h to either IBMX alone (0.1 mM), or in combination with CPE (containing 0.1-4 μg/ml cGTH), dbcAMP (0.3-3 mM) or hCG (1 and 10 IU/ml). Other details were as in Bogomolnaya and Yaron (1984).

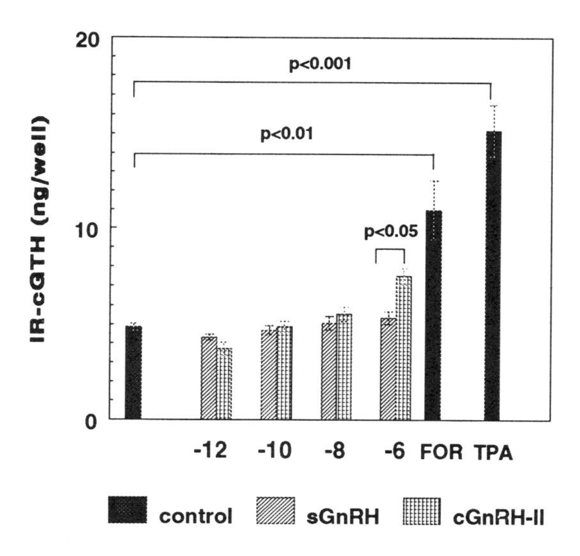

Fig. 1. cGTH release from culture of pituitary cells of 2 y old black carp in response to sGnRH, cGnRH II (concentrations in log M), Forskolin (10 μM) or TPA (12.5 (nM). Mean±SEM, n=6, Student's t-test.

Results

Histological examination of the rudimentary gonads have shown the presence of well-differentiated oocytes up to 73 μm in diameter or primary spermatogonia arranged in cysts.

A small amount (*ca.* 30 μg/gland) of cGTH was present in the pituitary of control fish. Circulating levels of cGTH in fish injected with GnRH+MET did not differ from those in the controls throughout the 24 h experiment. E_2 levels in the CPE-treated fish were similar to that of the controls (not shown).

cGTH release from static culture of dispersed pituitary cells did not change at all after 2 h exposure to sGnRH (1 pM-1 μM). A slight but significant increase (43%) in cGTH release was evident in response to the highest dose of cGnRH II. A two-fold increase in the release occurred in response to forskolin (10 μM), and a three-fold increase was noted in response to TPA (12.5 nM; Fig. 1)

Pituitary cells exposed for 2 days in culture to testosterone (1-100 nM) released amounts of cGTH higher than cells not exposed to the androgen, and the release was further augmented in response to either sGnRH or cGnRH II (100 nM; Fig. 2).

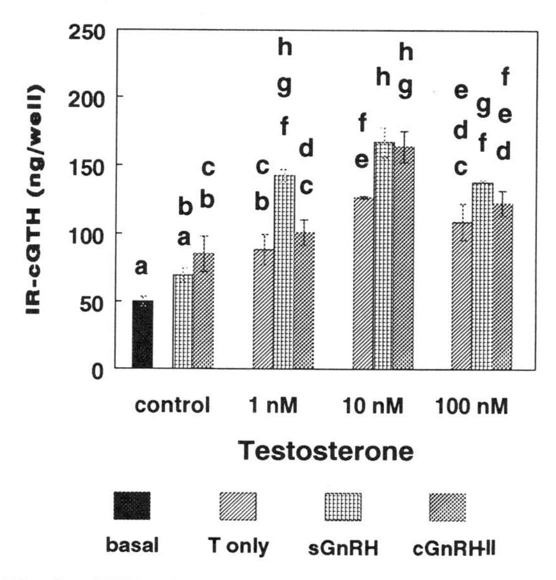

Fig. 2. cGTH release from pituitary cells exposed for 2 days to T and then challenged for 2 h with sGnRH or cGnRH II (100 nM). Mean±SEM, means marked by the same letter do not differ significantly .

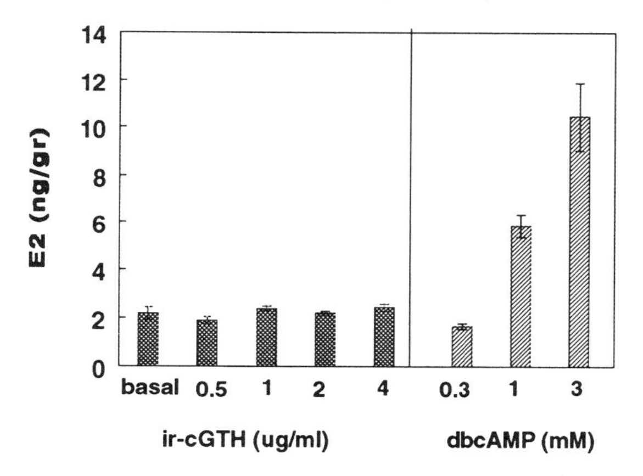

Fig. 3. Estradiol secretion from gonadal segments of 2 y old black carp in response to CPE calibrated to contain cGTH at the concentrations marked (left panel) or to dbcAMP (right panel) all in the presence of 0.1 mM IBMX. Mean ±SEM, n=4.

E_2 secretion from gonadal segments was stimulated by IBMX alone (0.1 mM), and in a dose-dependent manner by dbcAMP (0.3 - 3 mM) in the presence of IBMX. However, the secretion of the estrogen remained at the basal level when exposed to IBMX together with CPE containing cGTH (Fig. 3).

E_2 secretion from the retroperitoneal fat-pad underlying the rudimentary gonads was stimulated by either IBMX alone (0.1 mM), or in combination with dbcAMP (0.3 and 1 mM) or hCG (1 and 10 iu/ml). Although the secretion in response to these agonists was always higher than the basal secretion, the response to the lower concentration of dbcAMP, or to hCG at both concentrations did not exceed that in response to IBMX alone (Fig. 4).

In order to ascertain that the steroidogenic capacity is unique to this fat-pad, visceral adipose tissue from around the gut was examined in parallel. None of the agonists used were able to stimulate the secretion of E_2 from this tissue (Fig. 4).

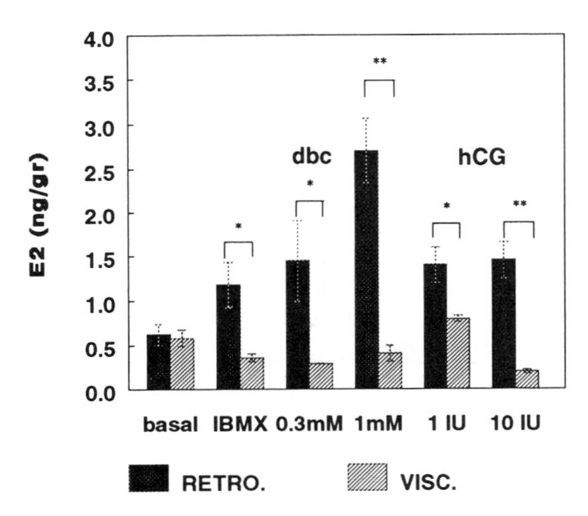

Fig. 4. Estradiol secretion from the retroperitoneal fat pad (RETRO.) and from visceral adipose tissue (VISC.) in response to IBMX alone (0.1 mM) or in combination with dbcAMP (dbc) or hCG. Mean ±SEM, n=3. *$p<0.05$; **$p<0.01$, Student's t-test.

Discussion

Injection of sGnRHa in combination with metoclopramide did not elicit any increase in the circulating level of cGTH in the immature 2 y old black carp. However, activation of protein kinase C (PKC) or protein kinase A (PKA) by TPA and dbcAMP, respectively, in dispersed pituitary cells of fish at the same age was effective in releasing the stored gonadotropin. These results, together with the finding that the pituitary does contain a small amount of cGTH, would indicate that the effect of GnRH in the 2 y old black carp is blocked at a site proximal to the formation of PKC and PKA, probably at the GnRH receptor level. It also indicates that the post-receptor components of the GnRH signal transduction cascades are already present in the gonadotrophs of black carp 4 years before the normal age of puberty, and that they can become functional when activated externally.

The two-day presence of T in the culture medium rendered the pituitary cells responsive to both GnRHs (Fig. 2). This provides evidence that T has a direct effect on the pituitary cells probably by increasing the number of GnRH receptors. Furthermore, as the basal release of cGTH from cells exposed *in vitro* to T was higher than in the controls, it is assumed that the androgen also stimulates the synthesis of GTH. This assumption is corroborated by the results attained *in vivo* by Gur et al. (this volume). Their results show that the administration of T-containing microspheres into 2 y old black carp, maintaining a high level of the androgen for 4 weeks, resulted in an increased cGTH cell content, elevated basal secretion rate and a higher steady-state level of GTH IIß mRNA.

The fact that the rudimentary gonad did not respond to gonadotropic stimulation but did respond to dbcAMP or IBMX would point to a lack of GTH receptors in the gonads at this stage (Fig. 3). The retroperitoneal fat-pad showed a similar response to these agonists, whereas the capacity of the visceral fat to produce estradiol was very limited and could not be stimulated by cAMP agonists. It is assumed that the retroperitoneal fat-pad derives from the mesoderm of the genital ridge as in amphibians, and thus relates in its origin to the endocrine cells of the gonads. Due to its large mass, in relation to the rudimentary gonad, this fat-pad should be considered to constitute a major source of sex steroids during the process of puberty.

References

Bogomolnaya, A. and Yaron, Z. (1984). Stimulation *in vitro* of estradiol secretion by the ovary of the cichlid fish, *Sarotherodon aureus*. Gen. Comp. Endocrinol. 53, 187-196.

Levavi-Sivan B. and Yaron, Z. (1992). Involvement of cyclic adenosine monophosphate in the stimulation of gonadotropin secretion from the pituitary of the teleost fish, tilapia. Mol. Cell. Endocrinol. 85, 175-182.

Leventer, H. (1979). "Biological Control of Reservoirs by Fish". Mekoroth Water Co. Nazareth. 71 pp.

ISOLATION AND MORPHOLOGICAL IDENTIFICATION OF GONADOTROPIC CELLS FROM THE MEDITERRANEAN YELLOWTAIL (*Seriola dumerilii* Risso 1810) PITUITARY

B. Agulleiro, M.P. García Hernández, A. López Ruiz and A. García Ayala

Department of Cell Biology. Faculty of Biology. University of Murcia. 30100 Murcia, Spain

Summary

Pituitary cells from Mediterranean yellowtail, *Seriola dumerilii*, were dispersed and separated in a Percoll discontinuous density gradient into five fractions. 9, 37, and 54% of the cells observed in bands 2, 3, and 4, respectively, were gonadotrops. Gonadotropic cells showed ultrastructural features similar to those of intact tissue.

Introduction

To facilitate the study of neuroendocrine regulation and of the intracellular mechanisms mediating pituitary hormone release by the different pituitary cell types, several cell enrichment methods have been developed. The aim of this study was to optimize dispersion and fractionation methods in order to obtain enriched fractions of functional gonadotropic cells from M. yellowtail.

Results

Pituitary cells from M. yellowtail were dispersed using a slightly modified trypsin/DNAase treatment procedure described by Chang et al. (1990). A cell suspension with a viability of approximately 95% was obtained and fractionated by application of a discontinuous density gradient of 40, 50, 60, and 70% Percoll. After centrifugation, five cell bands were visible, one at each of the interfaces between the Percoll densities and one at the bottom (band 5). The bands consisted of 3.25×10^6 (1), 9.28×10^6 (2), 13.5×10^6 (3), and 10^6 (4) cells, the fifth band being discarded due to the scarcity of cells. The percentage of gonadotropic cells in each band was tested by the immunocytochemical peroxidase-antiperoxidase (PAP) technique, using an antiserum raised against chum salmon ß-GTH I (provided by Prof. Kawauchi, Iwate, Japan), in cell samples spread on microscopic slides. Band 1 contained no gonadotropic cells, while 9, 37, and 54% of the cells found in the second, third and fourth band, respectively, were gonadotrops.

Gonadotropic cells had secretory granules of variable size and electron density and a very developed and dilated rough endoplasmic reticulum. In some cells, enormous vacuoles resulting from dilatation of the reticulum cisternae were found. Some cells showed very scarce, small secretory granules. Gonadotropic cells had, after dispersion and fractionation, the same ultrastructural features as those of intact tissue. Only some points of the cell membrane were slightly damaged.

Discussion

A discontinuous gradient of 40, 50, 60, and 70% Percoll gave good results for the fractionation of pituitary cells from the M. yellowtail. Slightly varied gradients were used for other teleosts (de Leeuw et al. 1984; Van Goor et al. 1994). The highest percentage of gonadotropic cells from M. yellowtail was 54%, which is similar to that found in *Carassius auratus* (van Goor et al. 1994) and lower than that obtained for *Clarias lazera* (de Leeuw et al. 1984). In M. yellowtail intact pituitaries, the antiserum used by the latter authors (the anti-catfish α,ß-GTH) recognized, in addition to TSH cells, cells that were not revealed by the anti-chum salmon ß-GTH I and which probably corresponded to the gonadotropic cells with very scarce and small secretory granules. In addition anti-chum salmon ß-GTH I did not react with TSH cells (unpublished results).

References

Chang, J.P., Cook, H., Freedman, G.L., Wiggs, A.J., Somoza, G.M., de Leeuw, R., and Peter, R.E., 1990. Use of a pituitary cell dispersion method and primary culture system for the studies of gonadotropin-releasing hormone action in the goldfish, *Carassius auratus*. Gen. Comp. Endocrinol. 77:256-273.

Van Goor, F., Goldber, J.I., Wong, A.O.L., Jobin, R.M., and Chang J.P., 1994. Morphological identification of live gonadotropin, growth-hormone, and prolactin cells in goldfish (*Carassius auratus*) pituitary-cell cultures. Cell Tissue Res. 276:253-261.

De Leeuw, R., Goos, H.J.Th, Peute, J., van Pelt, A.M.M., Burzawa-Gérard, E., and van Oordt, P.G.W.J., 1984. Isolation of gonadotrops from the pituitary of the African catfish, *Clarias lazera*. Morphological and physiological characterization of the purified cells. Cell Tissue Res. 236:669-675.

MODIFICATION OF 3'-RACE (RANDOM 3'-RACE): PARTIAL CLONING OF THE ESTROGEN RECEPTOR OF THE AFRICAN CATFISH, *Clarias gariepinus*

J. Bogerd and J.J.L. Jacobs

Department of Experimental Zoology, Utrecht University, Padualaan 8, 3584 CH Utrecht, The Netherlands

Summary

The 3'-RACE method (Frohman, 1990) has been modified to partially clone the cDNA encoding the estrogen receptor of the African catfish (cfER).

Introduction

Degenerate primers were used to amplify the cDNA region encoding the DNA-binding domain (DBD) of several nuclear receptors of the catfish pituitary and brain (Fig. 1).

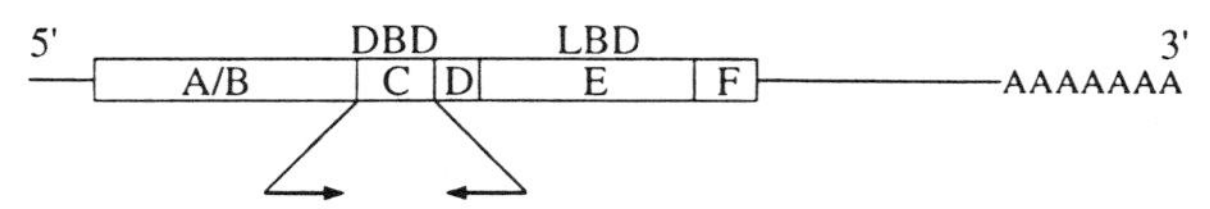

Fig. 1. PCR strategy for cloning cDNAs encoding the DBDs of several nuclear receptors.

DNA sequence analysis of cloned DBD cDNA sequences showed that at least 13 of them were derived from cDNA sequences encoding different DBDs, one of them most closely resembling known ER-DBDs. Since 'traditional' 3'-RACE (rapid amplification of cDNA 3'-ends) failed to amplify the estimated 5 kb cDNA region between the cfER-DBD and its poly(A)tail, we modified 3'-RACE (using oligo dT-adaptor primed cDNA) to **random 3'-RACE** (using FoFEiR-adaptor primed cDNA; see below).

Results and Discussion

Random 3'-RACE (see Fig. 2) takes advantage of the limited presence of specific nucleotide sequences in an RNA template. The FoFEiR-adaptor primer, used for initiation of cDNA synthesis in random 3'-RACE, consists of (from 5' to 3') a known adaptor sequence, eight random nucleotides (to facilitate annealing) and four fixed nucleotides (here: 5'-CTAG-3'). These four fixed nucleotides are crucial since they determine the incidence of initiation of cDNA synthesis, allowing this primer to anneal within 100 to 1,500 nucleotides with a probability of 0.324 to 0.997, respectively (using the formula: $1-\{(1-0.25^n)^y\}$; n=the number of fixed nucleotides (=4) and y=the number of bases within which this primer should not anneal). Nested PCRs were performed on FoFEiR-adaptor-primed pituitary as well as brain cDNA, using cfER-DBD-specific primers I and II (based on the cloned cfER-DBD cDNA sequence; see above) and adaptor primers I and II, respectively. DNA sequence analysis of cloned PCR products (1.2, 1.0, 0.85 and 0.7 kb in length) generated 1.1 kb of new sequence information, spanning the area from the cfER-DBD via its ligand-binding domain (LBD) to the stopcodon. A phylogenetic tree of ERs (Fig. 3), clearly showed that the cfER is more homologous to the rainbow trout ER than to other ERs identified so far (Pakdel *et al.*, 1989).

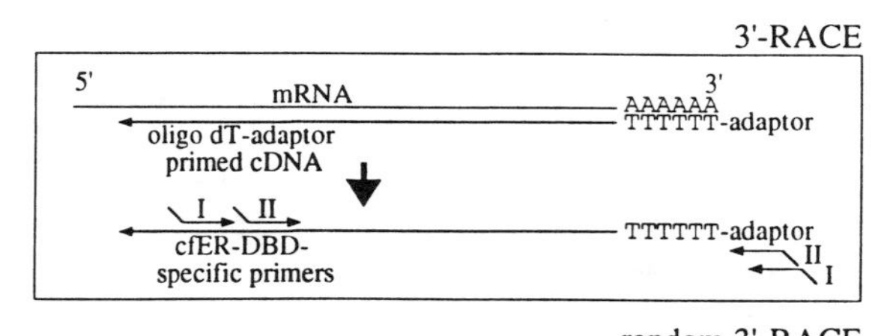

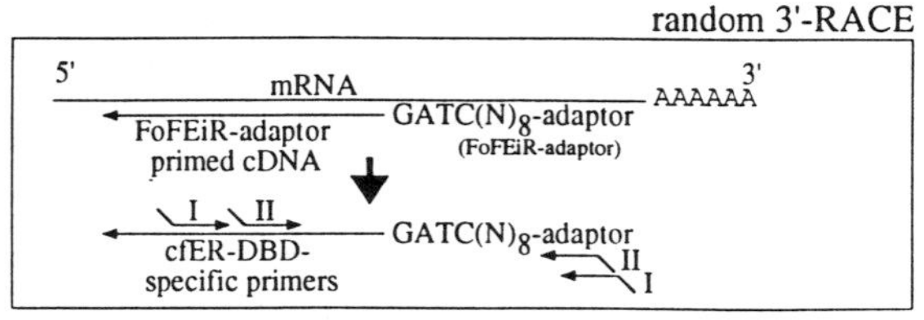

Fig. 2. The site of initiation of cDNA synthesis determines the difference between 3'-RACE (using oligo dT-adaptor primed cDNA) and random 3'-RACE (using FoFEiR-adaptor-primed cDNA).

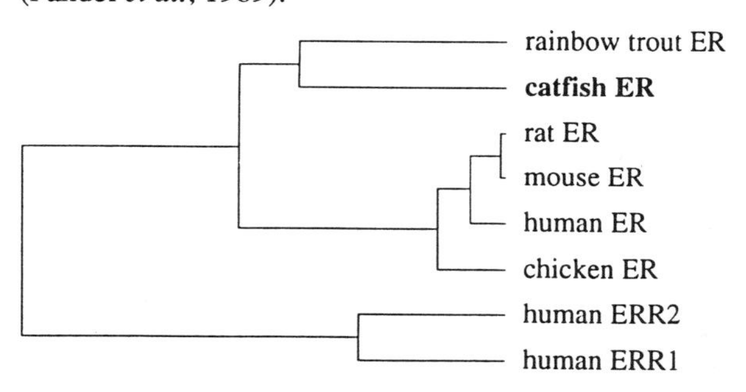

Fig. 3. UPGMA-tree (unweighted pair group method with arithmetic mean-tree) of ERs, based on the amino acid sequences of the ER-DBDs to their stopcodon, was constructed by the method of Nei (1987).

References

Frohman, M.A., 1990. In: *PCR protocols: a guide to methods and applications* (Innis, D.H., Gelfand, J.J. and White, T.J., eds.) pp.28-38, Academic Press, San Diego, USA.

Pakdel, F., Le Guellec, C., Vaillant, C., Le Roux, M.G. and Valotaire, Y., 1989. Identification of estrogen induction of two estrogen receptors (ER) messenger ribonucleic acids in the rainbow trout: sequence homology with other ERs. Mol. Endocrinol. 3:44-51.

Nei, M., 1987. Molecular Genetics, pp.293-298, Columbia University Press, New York, USA.

CLONING OF TWO PROLACTIN (PRL) COMPLEMENTARY DNAs (cDNAs) FROM GOLDFISH, *CARASSIUS AURATUS*.

Y.H.Chan[1], K.W.Cheng[2], K.L.Yu[2], and K.M.Chan[1].
[1]Department of Biochemistry, The Chinese University of Hong Kong, Shatin, N.T., Hong Kong; [2]Department of Zoology, The University of Hong Kong, Hong Kong.

Summary

We have identified two different cDNA clones each encoding goldfish PRL (gfPRL) from a pituitary cDNA library. The coding regions of these clones contain four silent mutations giving rise to identical amino acid (aa) sequences. The 3'-untranslated regions (UTRs) of these clones show only 72% nucleotide (nt) sequence identity. The two genes each encoding goldfish PRL might have derived from a recent gene duplication, before the divergence of goldfish from other Cypriniforms.

Introduction

PRL, somatolactin (SL) and growth hormone (GH) are three pituitary hormones whose genes are considered to have evolved from a common ancestral gene [1]. As a first step towards the development of a sensitive assay for studying the differential regulation of these hormones in goldfish, we cloned cDNAs encoding these hormones. We report here the identification of two cDNA sequences, encoding the same gfPRL aa sequence.

Results

From a goldfish pituitary cDNA library, we have isolated two different gfPRL cDNA sequences. Eleven clones were isolated using a carp PRL-specific probe. This probe (cloned in pUC18) carries the coding region of the carp PRL and is a PCR product amplified from a goldfish pituitary first strand cDNA pool using primers designed from a bighead carp PRL sequence [2]. Plaque-purified clones were first classified by restriction enzyme mapping into two types (Fig. 1). Two clones, P1A & P8A, representing these two types of cDNA clones were analyzed by sequencing.

The nt sequences of these two clones were found to code for the same precursor comprising 23 aa of signal peptide and 188 aa of mature hormone, with four silent mutations (nt) found in a transversional manner. The deduced gfPRL aa sequences share sequence identities of 95-97% with other carp PRLs, 73-74% with salmon PRLs, 70% with eel PRL, only 65% and 57% with tilapia PRL-I ($tPRL_{188}$) and PRL-II ($tPRL_{177}$), respectively.

Discussion

The major function of PRL in fish is osmoregulation for adaptation in fresh water [3], however, the molecular mechanism of PRL actions in teleosts is not very well studied partly because of the lack of purified PRL preparations. Genomic and cDNA clones for fish PRLs have been characterized from a variety of teleost species including carp [2, 4], salmon [5, 6], eel [7] and tilapia [8]. In tilapia, the two forms of tPRLs ($tPRL_{177}$ & $tPRL_{188}$) are expressed equally well in *Oreochromis mossambicus*, but they were found to have differential expression in tissues of *O. niloticus* [3]. Our study of goldfish pituitary hormones have led to the discovery of two distinct GH precursors (Law *et al.*, this proceeding), whereas one gfPRL precusor is encoded by two cDNA sequences. The two gfPRL cDNAs share nt sequence identities of 99% and 72% in their coding regions and 3'-UTRs respectively. The duplicate loci for gfPRL might have emerged from recent gene duplication prior to the divergence of goldfish and carps from their common ancestor.

Fig.1 Restriction map of gfPRL cDNA clones, P1A & P8A.

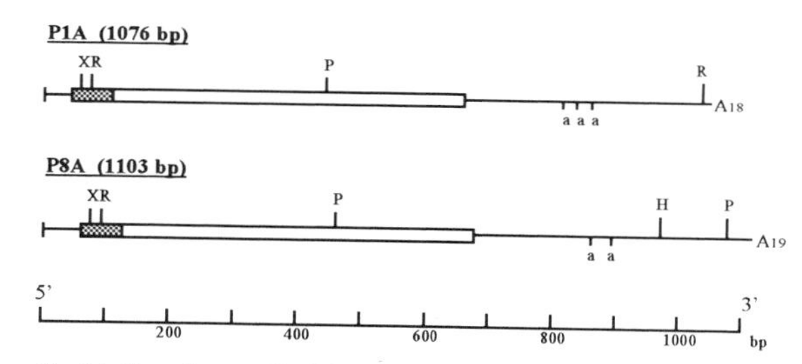

Restriction sites -- X : *Xba* I ; R : *Rsa* I ; P : *Pst* I ; H : *Hind* III
"a" -- polyadenylation signal (AATAAA)
signal peptide
mature hormone

References

[1]. Chen, T.T. *et al.*, 1994. *Fish Physiology*. XIII, 179-209.
[2]. Chang, Y.S. *et al.*, 1992. *Gen. Comp. Endocrinol.*87, 260-265.
[3]. Ayson, F.G. *et al.*, 1993. *Gen. Comp. Endocrinol.* 89, 138-148.
[4]. Chen, H.-T. *et al.*, 1991. *Biochim. Biophys. Acta.* 1088, 315-318.
[5]. Song, S. *et al.*, 1988. *Eur. J. Biochem.* 172, 279-285.
[6]. Xiong, F. *et al.*, 1992. *Mol. Mar. Biol. Biotech.* 1,155-164.
[7]. Querat, B. *et al.*, 1994. *Mol. Cell. Endocrinol.* 102, 151-160.
[8]. Rentier-Delrue, F. *et al.*, 1989. *DNA* 8, 261-170.

[Supported by Hong Kong Research Grant Council Earmarked Grants to KLY (HKU391/94M) and KMC (CUHK 17/93M)]

DEVELOPMENT OF RNASE PROTECTION ASSAYS FOR QUANTIFICATION OF GONADOTROPIN (GTH I AND GTH II) SUBUNIT TRANSCRIPT LEVELS IN COHO SALMON (*ONCORHYNCHUS KISUTCH*)

J. T. Dickey[1] and P. Swanson[2]

[1]School of Fisheries, University of Washington, Seattle, WA 98195
[2]National Marine Fisheries Service, 2725 Montlake Blvd. E., Seattle, WA, USA.

Summary

RNase protection assays were developed to quantify subunit transcript levels for coho salmon (*Oncorhynchus kisutch*) gonadotropins (GTH I and GTH II). Using these assays, the feedback effects of estradiol-17β (E) on mRNA levels of the GTH subunits were examined. In sexually immature coho salmon, *in vivo* E treatment increased GTH II β transcript levels, but did not alter those of the GTH I β or α subunits.

Introduction

In salmonids, the mechanisms involved in controlling the differential synthesis and secretion of GTH I and GTH II are poorly understood. To investigate factors involved in regulating GTH subunit synthesis at the transcriptional level, RNase protection assays (RPAs) were developed to quantify pituitary mRNA levels for GTH α (α1 and α2) , GTH I β and GTH II β subunits.

Methods and Results

Coho salmon GTH α-2, GTH I β and GTH II β subunit cDNAs were obtained by RT-PCR of pituitary total RNA. Transcript sizes for the subunits were approximately 1 kilobase as determined by Northern blot analysis of total RNA from pituitaries of vitellogenic and spermiating fish. Sense RNA standards and antisense RNA probes were prepared from cDNA clones spanning the signal peptide and a portion of the mature protein. The RPAs were performed according to the method of Duan et al. (1993) and were found to be specific by cross-hybridization to sense RNA standards for the other GTH subunits. The standard curves were linear in the range of 3 to 200 pg RNA.

Using these RPAs, the effects of *in vivo* E treatment on pituitary GTH transcript levels were examined in sexually immature coho salmon given intraperitoneal injections of E (0.1 mg/100 g body wt.) in coconut oil. Plasma samples were taken at the initiation of the experiment and 7 days post-treatment when fish were sacrificed and pituitaries were collected. The E treatment caused a significant increase in steady state mRNA levels for GTH II β, but did not alter either GTH α or GTH I β subunit mRNA levels (Fig. 1).

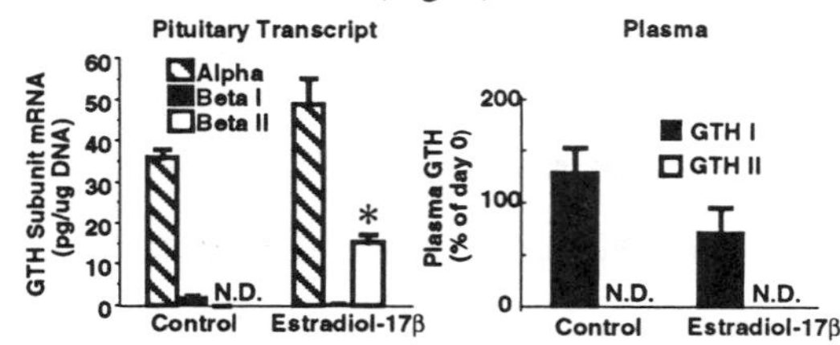

Fig. 1. Effect of *in vivo* E treatment on pituitary GTH subunit transcript levels and plasma levels of GTH I and GTH II in sexually immature salmon.

GTH II was undetectable in plasma before and after the E treatment; plasma levels of GTH I ranged from 3-12 ng/ml and were not significantly altered by E treatment.

Discussion

The coho salmon GTH subunit cDNAs are 97-98% identical to sequences reported for masu salmon GTHs (Kato et al., 1993; Gen et al., 1993). The stimulatory effect of E on GTH II β subunit transcript levels and the lack of effect of E on either GTH α or GTH I β subunit transcript levels is similar to that found previously in rainbow trout (Xiong et al., 1994). This result is not surprising due to the presence of the steroid response element in the promoter for the GTH II β subunit gene (Xiong et al., 1994).

In previous studies with maturing coho salmon, negative feedback effects of E on plasma GTH I levels were found. However, in the present study, no significant alterations in either plasma GTH I or GTH I β transcript levels were detected. The inability to detect a negative feedback of E on GTH I may be due to the state of maturity of the fish. Further studies will be required to determine the feedback effects of gonadal factors on GTH I and GTH II synthesis, and whether they change during gametogenesis.

References

Duan, C., Duguay, S.J., Plisetskaya, E.M. 1993a Insulin-like growth factor I (IGF-I) mRNA expression in coho salmon, *Oncorhynchus kisutch* : Tissue distribution and effects of growth hormone/prolactin family hormones. Fish Physiol. Biochem. 11:371-379.

Gen, K., Maruyama, O., Kato, T., Tomizawa, K., Wakabayashi, K., Kato, Y. 1993. Molecular cloning of cDNAs encoding two types of gonadotrophin α subunit from the masu salmon, *Oncorhynchus masou*: construction of specific oligonucleotides for the α-1 and α-2 subunits. J. Molec. Endocrinol. 11:265-273.

Kato, Y., Gen, K., Maruyama, O., Tomizawa, K., Kato, T. 1993. Molecular cloning of cDNAs encoding two gonadotrophin β subunits (GTH-I β and GTH-II β) from the masu salmon, *Oncorhynchus masou:* rapid divergence of the GTH-I β gene. J. Molec. Endocrinol. 11:275-282.

Xiong, F., Dong, L., Le Drean, Y., Elsholtz, H., Hew, C. 1994. Differential recruitment of steroid hormone response elements may dictate the expression of the pituitary gonadotropin IIβ subunit gene during salmon maturation. Mol. Endocrinol. 8:782-793.

The financial support of the US Dept. of Agriculture (# 93-37203-9409) is gratefully acknowledged.

ISOFORMS OF GTHs FROM TUNA (*Thunnus thynnus*) PITUITARIES ISOLATED BY CHROMATOFOCUSING AND ION EXCHANGE CHROMATOGRAPHY.

A. García-García[1], M.C., Sarasquete[1], J.A., Muñoz-Cueto[2], M.L., Gonzalez de Canales[2], and R.B., Rodríguez[1].

[1] Instituto de Ciencias Marinas de Andalucía (C.S.I.C.) and [2] Dpto. Biología Animal, Vegetal y Ecología, Fac. Ciencias del Mar, Universidad de Cádiz, Pol. Río San Pedro, 11510, Puerto Real, Cádiz, Spain.

Summary

Several isoforms of tuna (*Thunnus thynnus*) GTHs have been obtained by chromatofocusing and ion exchange chromatography, using the fraction adsorbed in Concanvaline-A-Sepharose (ConA-II).

Introduction

Glycoprotein hormones, GTHs and TSHs, isolated from vertebrate pituitaries are heterogeneous. Families of isoforms with different isoelectric point (pI), molecular size, immunoreactivity and bioactivity have been described (1,2). This microheterogeneity is considered to be caused mainly by variations in the sugar moiety of the molecule and/or by slight post-translational differences in the ß-chain amino acid residues. These variations play an important role in the biological and biochemical properties of the glycoprotein hormones (2).

Results and Discussion

Unadsorbed (ConA I) and adsorbed (ConA II) fractions from tuna pituitaries were obtained by affinity chromatography on Concanavaline A-Sepharose. Both fractions were immunoreactive against anti-carp αßGTH (1:500) and stimulated the *in vitro* 17ß-estradiol (E_2) production by ovarian vitellogenic follicles of *Fundulus heteroclitus* (0.02-20 µg protein/ml medium); however, ConA II was much more immunoreactive and bioactive than ConA I. ConA-II was chromatofocused and four zones (I-IV) were eluted at pH from 7.5 to 3, each zone having several elution peaks corresponding to different pIs. These results were similar to those obtained in coho salmon (3). Fractions IIB (pI 6.4), IIC (pI 5.97), IIIA (pI 5.5), IIIB1 (pI 5.4), IIIB2 (pI 4.85), IIIC1 (pI 4.36), IIIC2 (pI 4.35) and IV (eluted after NaCl, pI$<$4) were immunoreactive with a different intensity in Dot-Blot (anti-carp-αßGTH, 1:500)(Table I). For *in vivo* bioassay, 0.1 µg of each fraction/g body weigth were injected to *F. heteroclitus* females and after 24 h, plasma were extracted and E_2 was measured by RIA. IIB, IIIB1, IIIB2, IIIC2 and IV fractions were significantly bioactive (Fig 1). The results of SDS-PAGE and Western-Blot are summarized in Table I.When compared with tuna GTHs and its subunits purified in our laboratory, the bands obtained (35, 30, 19, 16, 12 kMr) were found to corresponded to GTH I, GTH II, GTH I- and GTHII-ß subunits and α subunit, respectively. The fractions from chromatofocusing were ion exchange chromatographied. Characterization of the peaks obtained showed the existence of proteins with the same molecular size as described above and a varying degree of bioactivity and immunoreactivity (data not shown).Although isoforms of GTHs without biological heterogeneity have been described in several species, it has been reported that molecular GTH heterogeneity could result in differences in biological activity (1,2).These variations could also be reflected in modifications of pI, immunoreactivity and bioactivity-immunoreactivity ratio (2), as we have found for tuna pituitary fractions. This polymorphism is considered to be caused by variations in sugar moieties of GTHs. The presence of GTH isoforms seems to be regulated by gonadal and pituitary factors and is dependent on age and physiological status (2,4).

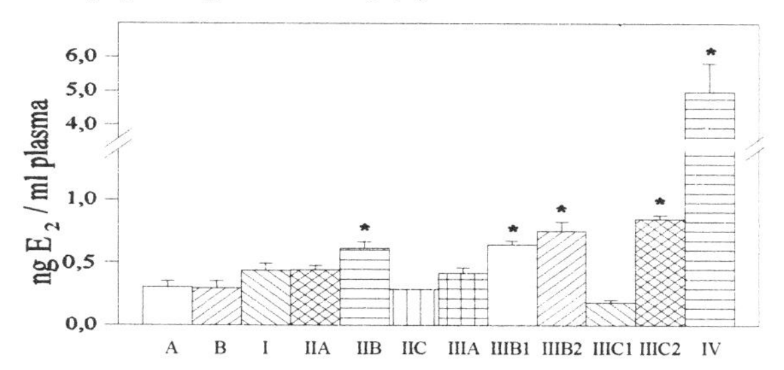

Figure 1: *In vivo* heterologous bioassay of fractions obtained in chromatofocusing. A, non-injected control; B, saline-injected control.

Table I: Results of Dot-Blot, SDS-PAGE and Western Blot (anti-carp αßGTH,1:500) of the fractions obtained from chromatofocusing; intensity of immunoreactivity in Dot-Blot is indicated as --, +, ++,+++.

Fraction	Dot-Blot	SDS-PAGE (bands kMr)	Immunoreactive bands
I	--	13	--
IIA	--	13	--
IIB	+++	12,16,30,35	12,35
IIC	++	12,16,30,35	12,30,35
IIIA	+++	12,16,30,35	12,30,35
IIIB1	+++	12,19,30,35	12,30,35
IIIB2	+++	12,19,30,35	12,19,30,35
IIIC1	+/++	12,19 30,35	12,19,35
IIIC2	+	12,19,30,35	12,19,35
IV	+	12,19,30,35	12,19,35

References

1.-Ando, H., & Ishii, S., 1988. Gen. Comp. Endocrinol., 70: 269-287.

2.-Simoni, M., Jockenhövel, F., & Nieschlag, E., 1994. J. Endocrinol., 141: 359-367.

3.-Swanson, P., Dickhoff, W.W., & Gorbman, A., 1987. Gen. Comp. Endocrinol., 70: 181-192.

4.-Stanton, P.G., Pozvek, G., Burgon, P.G., Robertson, D.M., & Hearn, M.T.W., 1992. J. Mol. Endocrinol., 138: 529-543.

Acknowledgements: We thank Dr. Burzawa-Gérard for the generous gift of antibodies.

ISOLATION AND CHARACTERIZATION OF TWO DISTINCT GONADOTROPINS FROM MEDITERRANEAN YELLOWTAIL (*Seriola dumerilii*, Risso 1810) PITUITARY GLANDS

M.P. García Hernández[1], Y. Koide[2], A. García Gómez[3] and H. Kawauchi[2]

[1]Department of Cell Biology, Faculty of Biology, University of Murcia, 30100 Murcia, Spain. [2]Laboratory of Molecular Endocrinology, School of Fisheries Sciences, Kitasato University, Sanriku, Iwate 022-01, Japan. [3]Laboratorio de Cultivos Marinos, Centro Oceanográfico de Murcia, Instituto Español de Oceanografía, Mazarrón, Murcia, Spain

Summary

Two gonadotropins, GTH I and GTH II, have been isolated from the pituitary of Mediterranean yellowtail on the basis of their different properties, providing further evidence for the duality of gonadotropins in teleosts.

Introduction

The occurrence of two GTHs was demonstrated for the first time in salmonids (Suzuki et al., 1988). However, chemical characterization of two teleost GTHs has been confined to only four teleost species. This study intends to isolate and characterize two GTHs from Mediterranean yellowtail.

Results

M. yellowtatil pituitary glands were extracted with 35% ethanol-10% ammonium acetate and fractionated by ion-exchange chromatography on a DE-52 column by stepwise increase of ammonium bicarbonate concentration. The fraction eluted at 0.2 M ammonium bicarbonate was subjected to Asahipak C4P-50 column using a linear gradient of acetonitrile in 0.05 M ammonium acetate (Fig. 1).

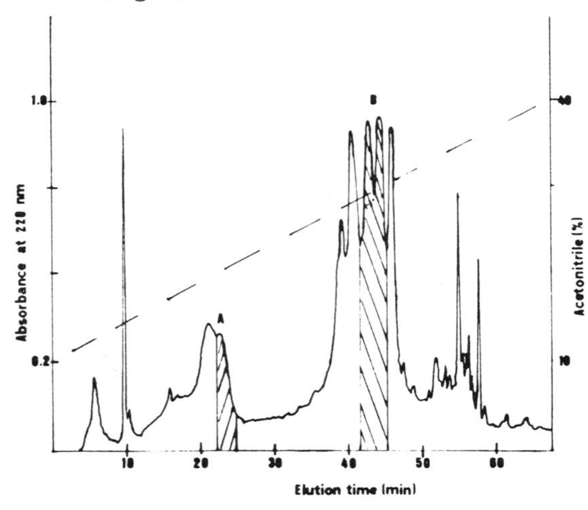

Fig. 1. rpHPLC on a Asahipak C4P-50 column of the fraction eluted at 0.2 M ammonium bicarbonate from the DE-52 column.

GTH I and GTH II were purified by gel filtration of fractions A and B, respectively, on a Superdex 75 column in 0.15 M ammonium bicarbonate.

Molecular weights were estimated to be 47 kDa for GTH I and 29 kDa for GTH II by SDS-PAGE, whereas they were 49 kDa (GTH I) and 42 kDa (GTH II) by gel filtration. GTH II was completely dissociated, while GTH I was partially dissociated by treatment with 0.1% trifluoroacetic acid.

Subunits of GTH I and II were purified by rpHPLC on an ODS-120T column using a linear gradient of acetonitrile in 0.1% TFA. S-pyridil-ethylated-GTH α and GTH II β, and S-carboxymethylated GTH Iβ were digested with lysyl endopeptidase, trypsin and chymotrypsin. The complete amino acid sequences of the subunits were determined by analyzing the peptide fragments. The α, Iβ and Iiβ subunits consisted of 91, 105 and 115 amino acid residues, respectively.

Discussion

Two chemically distinct GTHs have been isolated from the pituitary of M. yellowtail. Physicochemeical properities such as acid-stability and molecular weight of yellowtail GTHs were similar to those of other teleost GTHs. These GTHs showed the highest sequence identities with those of bonito (Koide et al., 1993) and tuna (Okada et al., 1994); 83% (α-subunit), 69% (Iβ-subunit) and 91% (II β-subunit).

References

Koide, Y., Itoh, H., and Kawauchi, H., 1993.Isolation and characterization of two distinct gonadotropins, GTH I and GTH II, from bonito (Katsuwonus plelamis) pituitary glands. Int.J. Peptide Protein Res. 41:52-65.

Okada, T., Kawazoe, I., Kimura, S., Sasamoto, Y., Aida, K., and Kawauchi, H., 1994. Purification and characterization of gonadotropin I and II from pituitary glands of tuna (Thunnus obersus). Int. J. Peptide Protein Res. 43:69-80.

Suzuki, K., Kawauchi, J. and Nagahama, Y., 1988. Isolation and characterization of two distinct gonadotropins from chum salmon pituitary glands. Gen. Com. Endocrinol. 71:292-301.

MOLECULAR CLONING OF THE cDNAs ENCODING Prl, GH, SL AND POMC PRECURSORS OF THE AFRICAN CATFISH, *Clarias gariepinus*

H.J.Th. Goos, M. Ter Bekke, N. Van der Spoel, J.C.M. Granneman and J. Bogerd

Department of Experimental Zoology, Utrecht University, Padualaan 8, 3584 CH Utrecht, The Netherlands

Summary

The cDNAs encoding the Prl, GH, SL and POMC precursors of the African catfish were isolated by grouping via cross-hybridization of 500 randomly picked clones of a pituitary cDNA library.

Introduction

Fish pituitaries consist of six major cell types: prolactin (Prl), proopiomelanocortin (POMC), growth hormone (GH), gonadotropic hormone (GTH), thyroid-stimulating hormone (TSH) and somatolactin (SL) cells. Since these cell types are hormone-producing cells, we argued that the mRNAs encoding these hormones (subunits) are abundantly expressed. Therefore, cDNAs of these transcripts should also be abundantly present in a cDNA library of this tissue. We tested if isolation of these cDNAs could be achieved by grouping via cross-hybridization.

Results

After mass *in vivo* excision of an oligo dT-primed cDNA library of the African catfish pituitary, 500 clones were selected *ad random* for further analysis. The cDNA inserts of all 500 clones were PCR-amplified using primers flanking the *Eco* RI cloning site, and dotted and covalently linked to Hybond-N membranes. In order to select homologous cDNAs in the remaining 499 clones, one of the 500 PCR-amplified cDNA inserts (minus the sequences flanking the *Eco* RI cloning site) was radioactively labeled and used as a probe for hybridization with the membranes. Next, the cDNA insert of the next cDNA clone was used for identical purposes, etc. etc. In this way, four groups of cDNAs were identified.

DNA sequence analysis of the longest cDNA of each group showed that we isolated the cDNAs encoding the precursors of Prl (20 clones), GH (35), POMC (20) and SL (2). Northern blot analysis revealed that the length of the Prl, GH and POMC transcripts was ~1300, ~1300 and ~1700 bases, respectively. No signal could be obtained for the SL transcript. Phylogenetic trees of Prl, GH, SL and POMC precursors are shown in Fig. 1.

Conclusion

Isolation of abundantly expressed mRNAs can be easily achieved by grouping via cross-hybridization of randomly picked cDNA clones.

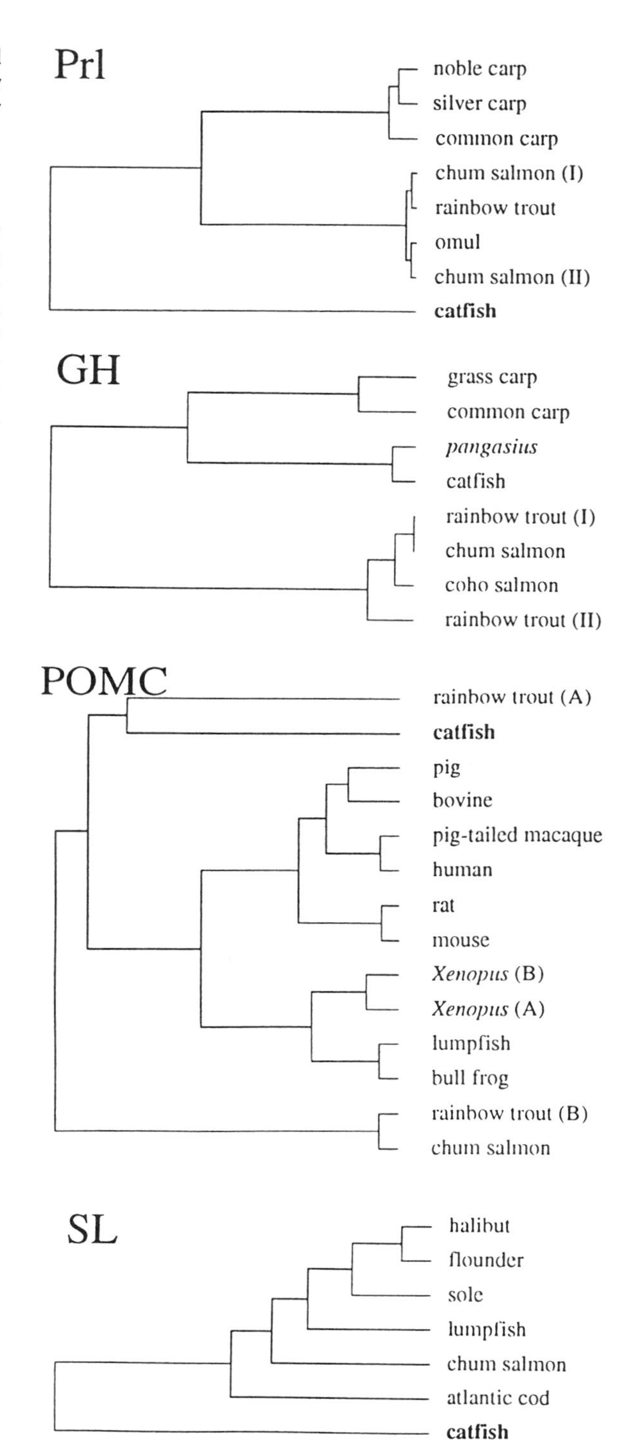

Fig. 1. Phylogenetic trees (UPGMA-trees; unweighted pair group method with arithmetic mean) of selected Prl, GH, POMC and SL precursor amino acid sequences.

LONG-TERM TESTOSTERONE TREATMENT STIMULATES GTH II SYNTHESIS AND RELEASE IN THE PITUITARY OF THE BLACK CARP, *MYLOPHARYNGODON PICEUS*.

G. Gur[1], P. Melamed[1], B. Levavi-Sivan[1], C. Holland[2], A. Gissis[3], D. Bayer[4], A. Elizur[4], Y. Zohar[2] and Z. Yaron[1].

[1]Department of Zoology, Tel-Aviv University, Tel-Aviv. [2]Center of Biotechnology, Baltimore, USA. [3]HaMaapil Hatchery, Israel. [4]National Center for Mariculture, IOLR, Eilat, Israel.

Summary

Microspheres containing testosterone (T; 0.8-3.2 mg/kg) were injected into immature 2 y old fish, which elevated circulating T levels for 28 days. Plasma cGTH levels in T-treated fish did not differ from the controls. Nevertheless, cGTH content and GTH-IIß mRNA levels were higher in T-treated fish. Only pituitary cells of T-treated fish responded to GnRH.

Introduction

Prior *in vitro* exposure of pituitary cells of immature fish to testosterone (T) facilitated basal and GnRH-stimulated cGTH release (see Yaron et al., this volume). These findings have prompted further *in vivo* experiments using T for advancing puberty in the fish.

Materials and Methods

Two year-old fish (1.5-3.2 kg, 11-12 in each group) were injected with T-containing microspheres (Holland et al., this volume) at 0.8, 1.6 or 3.2 mg/kg. Blood was sampled before injection, and every week thereafter for 28 days. The pituitary cells were pooled for culture and for cGTH content determined by RIA (see Yaron *et al.,* this volume). Quantification of GTH-IIß mRNA was carried out by slot blot hybridization in whole pituitaries with a black carp GTH IIß cDNA probe (Bayer et al., unpublished) and corrected for loading using ß actin.

Results and Discussion

Plasma T increased dose-dependently and remained above the initial level at day 28. Circulating cGTH in T-treated fish did not differ from that in the controls. cGTH content in dispersed cells was higher in pituitaries of T-injected fish than in the controls (Fig. 1a). The steady-state level of GTH IIß mRNA in the pituitaries of T-treated fish increased dramatically over the controls (Fig. 2). cGTH release from cells of blank-injected controls did not respond at all to either sGnRH or cGnRH II (0.01-1.0 μM).

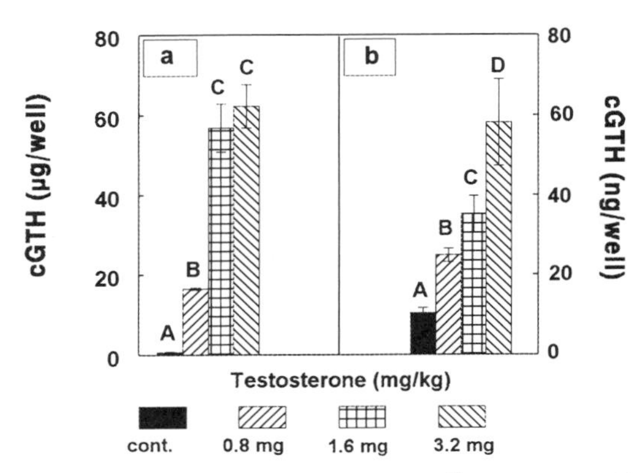

Fig. 1. (a) cGTH content in 0.25x10^{6} pituitary cells of 2 y old fish injected with microspheres containing T as marked. (b) cGTH release in response to 10 nM sGnRH (Mean±SEM; ANOVA & LSD Test).

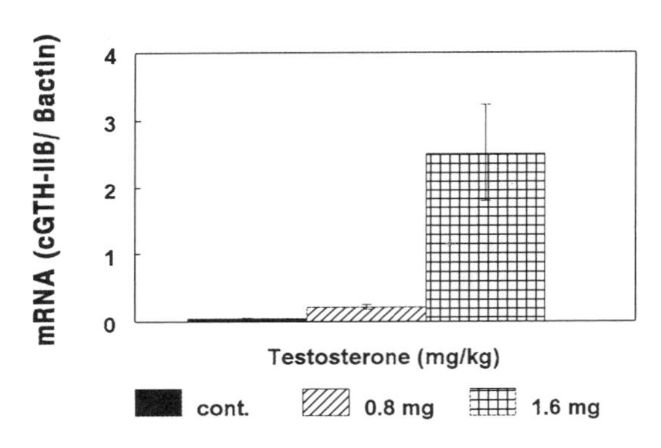

Fig. 2. GTH IIß mRNA level in pituitaries of fish injected as in Fig. 1 (Mean±SEM).

However, pituitary cells of T-injected fish had higher basal release of cGTH, which was further augmented by sGnRH (Fig. 1b).

These results suggest that a long-term T-treatment stimulates the expression of the gene encoding the GTH II ß subunit, or the stabilization of its mRNA. Studies on the other GTH subunits are in progress. Furthermore, the facilitated response to GnRH of pituitary cells from T-injected fish indicates a stimulatory effect of the androgen in increasing the sensitivity of the gonadotrophs to GnRH.

Supported by Grant No. IS-2149-92 from BARD, U.S.-Israel Binational Agricultural R&D Fund

CLONING AND CHARACTERIZATION OF TWO GROWTH HORMONE cDNAs FROM GOLDFISH, *CARASSIUS AURATUS*

M. S. Law[1], K. W. Cheng[1], T. K. Fung[1], Y. H. Chan[2], K. L. Yu[1] and K. M. Chan[2]

[1]Department of Zoology, The University of Hong Kong, Hong Kong; [2]Department of Biochemistry, The Chinese University of Hong Kong, Hong Kong

Summary

We have isolated and characterized two cDNA clones encoding goldfish growth hormone (GH) from a cDNA library prepared from pituitary poly(A)$^+$ RNA. One of the GH cDNA (gfGHI) encodes a polypeptide with five cysteine residues whereas another (gfGHII) encodes a polypeptide with four cysteine residues. Nucleotide sequence comparison shows that gfGHI and gfGHII exhibit remarkably high similarity with their corresponding common carp GH cDNA counterparts. It is suggested that two types of prototypic carp GH gene have emerged from a gene duplication early in bony fish evolution prior to the divergence of goldfish and common carps from their common ancestor.

Introduction

GH is one of the best-studied pituitary hormone in fishes because of its potential applications in aquaculture. Apart from its important role for normal growth and development, GH has also been implicated in puberty, gametogenesis and fertility of fishes [1]. Despite extensive studies on the regulation of GH secretion in goldfish, the primary structure of the goldfish GH has not been determined. In the present study, we have cloned and sequenced two goldfish GH cDNAs to predict the primary structures of the two types of GH.

Results

gfGHI cDNA clone is 1099 nucleotide long and lacks a 5' noncoding region. The 3' noncoding region of gfGHI contains 469 nucleotides that can form a hairpin secondary structure 44 nucleotide away from the stop codon. gfGHII is 1161 nucleotide long with a 5' noncoding region of 40 nucleotides. No hairpin structure is predicted in the 488 bp 3' noncoding region. There are totally 42 nucleotide substitutions in the coding region between gfGHI and gfGHII and 30 of them are located in the wobble position of their respective codons, resulting in a difference of 11 aa residues between the two deduced GH aa sequences. The gfGHI coded polypeptide, similar to the GHs of channel catfish and other carp species [2], contains five cysteine residues which are universal in the reported carp GH sequences. In contrast, the polypeptide encoded by gfGHII, similar to other vertebrates, contains four cysteine residues only and does not possess the fifth cysteine residue found in gfGHI and other carp GH sequences.

	1 gfGHI	2 gfGHII	3 ccGHI	4 ccGHII	5 scGH	6 bcGH	7 gcGH
1.		78.6	88.1	82.6	72.6	73.3	72.6
2.			84.4	92.8	79.1	79.7	79.3
3.				86.5	78.1	78.6	78.4
4.					79.1	78.6	79.0
5.						99.0	99.7
6.							99.0

Table 1. Percentage nucleotide sequence homology of the GH cDNAs of carp species. The GHs for comparison are from goldfish (gf); common carp (cc); silver carp (sc); bighead carp (bc); and grass carp (gc).

Discussion

The isolation of two goldfish GH cDNAs encoding distinct GHs is consistent with the isolation of two distinct GH polypeptides from other fishes eg. Atlantic cod [3], chum salmon [4], Japanese eel [5] and yellowtail [6]. Nucleotide sequence homology among GH cDNAs of goldfish and common carp showed that gfGHI and gfGHII exhibit remarkably high similarity (over 88%) with their corresponding common carp GH cDNA counterparts (Table 1). With reference to our GHs, the common carp GH reported by Koren and coworkers [7] and that by Chao and coworkers [8] can be referred as ccGHI and ccGHII respectively. In addition, hairpin secondary structures are found in both gfGHI and ccGHI. Cluster analysis of 3' noncoding region of GH cDNAs suggested that the prototypic carp GHI and GHII might have been established prior to the divergence of goldfish and common carps from their common ancestor. The biochemical and biological properties of these two goldfish GHs remain to be studied in the future.

References

[1] Le Gac, F. *et al.*, 1993. Fish Physiol. Biochem. 11: 219-232.
[2] Chen, T.T. *et al.*, 1994. Fish Physiology. Vol. VIII, pp. 179-209. Academic Press, New York. Cold Spring Harbor Laboratory Press, New York.
[3] Rand-Weaver, M. *et al.*, 1991. Gen. Comp. Endocrinol. 81: 39-50.
[4] Kawauchi, H.*et al.*, 1986. Arch. Biochem. Biophys. 244: 542-552.
[5] Yamaguchi, K. *et al.*, 1987. Gen. Comp. Endocrinol. 66: 447-453.
[6] Kawazoe, I. *et al.*, 1988. Nippon Suisan Gakkaishi 54: 393-399.
[7] Koren, Y. *et al.*, 1989. Gene. 77: 309-315
[8] Chao, S.C. *et al.*, 1989. Biochim. Biophys. Acta. 1007: 233-236.

FUNDULUS HETEROCLITUS GONADOTROPINS. 4. CLONING AND SEQUENCING OF GONADOTROPIC HORMONES (GTH) α-SUBUNIT

S.W. Limesand, Y-W.P. Lin, D.A. Price, and R.A. Wallace

The Whitney Laboratory, University of Florida, St. Augustine, FL 32086, USA

Summary

Amplification, sequencing, and analysis of GTH α-subunit of *F. heteroclitus* from a cDNA pituitary library shows a 30 amino acid (a.a.) signal peptide and a 95 a.a. mature protein, with 72% identity to the *K. pelamis* and *T. obesus* GTH a.a. sequences.

Introduction

Mammalian GTH (LH and FSH) affect steroidogenesis and gametogenesis to regulate sexual maturation and reproductive function. Early reports suggest fish may possess only one GTH (Burzawa-Gerard, 1982). However, two GTHs have been reported in salmon (Sekine *et. al.*, 1989) and *F. heteroclitus* (Lin *et. al.*, 1992), as well as many other species. These GTH appear to be homologous to the classical mammalian LH and FSH.

Results and Discussion

Degenerate primers based on teleost a.a. sequences were used to generate PCR products from a *F. heteroclitus* pituitary cDNA library (Lin *et. al.* 1992).

The complete nucleotide sequence for the α-subunit of *F. heteroclitus* GTH (ASSEMGEL; PC/GENE) consists of 378 residues coding for 125 a.a. The N-terminal signal sequence was found, and its cleavage site matches the mature protein of *T. obesus* (Okada *et. al.*, 1994) and *K. pelamis* (Kawauchi *et. al.*, 1991). The predicted mature protein of *F. heteroclitus* (95 a.a.) has 72% homology with those of *T. obesus* and *K. pelamis*. Comparison of these a.a. sequences revealed conservation of the cysteine (disulfide bridges) and asparagine (glycosylation sites) residues, that are hallmarks of the GTH α-subunit.

Sequence analysis of 18 cDNAs revealed five point mutations, which resulted in two a.a. differences. These point mutations cause a change from threonine to methionine (a.a. 42) and from alanine to proline (a.a. 65). Such mutations may result from true polymorphism or from PCR amplification.

***Fundulus heteroclitus* GTH α-subunit**

```
ATG GTA TCT GCT GTC ACG ACT ATG GGC TGC ATG AAG GCA GCC GGC  42
 m   v   s   a   v   t   t   m   g   c   m   k   a   a   g  -16
                                             a           b
GTG TCT CTT CTT CTC TTG TAT TTT CTA CTG AAC GC[A] GCT GAT TC[T]  84
 v   s   l   l   l   l   y   f   l   l   n   a   a   d   s  -1

CAT CCC AAC TTT GAT TCC TCA AGC ATG GCC TGT GGG GAA TGC AGC 126
 H   P   N   F   D   S   S   S   M   A   C   G   E   C   S  15

CTG GGA CTG AAC AGG CTT TTT TCT CGG GAT CGT CCG CTC TAC CAG 168
 L   G   L   N   R   L   F   S   R   D   R   P   L   Y   Q  30
        PEL 7 ------------------------------>  c
TGC ATG GGT TGC TGC TTC TCC CGA GCG TAC CCC A[C]G CCT CAA ACA 210
 C   M   G   C   C   F   S   R   A   Y   P   T   P   Q   T  45
                                             <-----------
GCC ATA CAG ACG ATG GCG ATC CCG AAG AAC ATC ACT TCA GAG GCA 252
 A   I   Q   T   M   A   I   P   K   N   I   T   S   E   A  60
------ ROW 37                      ▲
  d              e
AA[G] TGC TGC GTT [G]CG AAG CAC AGC TAC GAG ACA AAA GTA GAT GAC 294
 K   C   C   V   A   K   H   S   Y   E   T   K   V   D   D  75
                                 <-----------------------
ATA ACA GTG AGA AAC CAC ACA GAG TGT CAC TGC AGC ACC TGC TAC 336
 I   T   V   R   N   H   T   E   C   H   C   S   T   C   Y  90
PEL 8            ▲
TAT CAC AAA TTG ATA TGA                                     378
 Y   H   K   L   I   .                                       95
```

Fig.1. Deduced sequence of *F. heteroclitus* GTH α-subunit. Lower case letters of the a.a. sequence indicate signal peptide and upper case letters show mature protein. Oligonucleotide primer positions are indicated by horizontal arrows. Arrowheads below the asparagines (N) indicate possible glycosylation sites. Lower case letters above the nucleic acid sequence signify point mutations (a=A/T; b,c=C/T; d=G/A; e=G/C).

References

Burzawa-Gerard, E. (1982) Can. J. Fish. Aquat. Sci. 39: 80-91.

Kawauchi, H., H. Itoh, Y. Kiode. (1991) Proceeding 4th Int. Symp. on Reprod. Phys. of Fish, Fish Symp 91 Sheffield, UK. 19-21.

Lin, Y.-W. P., B.A. Rupnow, D.A. Price, R.M. Greenberg, and R.A. Wallace. (1992) Mol. Cell. Endocrinol. 85: 127-139.

Okada, T., I. Kawazoe, S. Kimura, Y. Sasamoto, K. Aida, and H. Kawauchi. (1994) Int. J. Peptide Protein Res. 43: 69-80.

Sekine, S., A. Saito, H. Itoh, H. Kawauchi, and S. Itoh. (1989) Proc. Natl. Acad. Sci. USA 86: 8645-8649.

DIRECT POSITIVE EFFECTS OF TESTOSTERONE ON GnRH-STIMULATED GONADOTROPIN RELEASE FROM DISPERSED GOLDFISH (*Carassius auratus*) PITUITARY CELLS.[1]

A. Lo§, J. Emmen¶, H.J.Th. Goos¶, J.P. Chang§.
§Dept. Bio. Sci., U. Alberta, Edmonton, Canada; ¶Dept. Exptl. Zool., U. Utrecht, Utrecht, The Netherlands.

Summary

Prolonged treatments of dispersed goldfish pituitary cells with 0.1μM of testosterone (T) potentiated the gonadotropin (GTH) response to salmon GnRH (sGnRH) and chicken GnRH-II (cGnRH-II) in perifusion studies. However, acute applications of T during GnRH pulses had no effects. Overnight T treatments did not alter cellular GTH content or basal GTH release. The direct positive effects of T on gonadotropes in goldfish pituitary may involve actions at the GnRH receptor or post-receptor level.

Introduction

Gonadal steroids exert both negative and positive feedback actions on GTH secretion in goldfish. Negative feedback actions are demonstrated in ovariectomized and steroid-replaced female goldfish (Kobayashi and Stacey, 1990). However, in gonad-intact goldfish, implantation of T or estradiol potentiates the serum GTH response to GnRH and the GnRH-stimulated GTH release from goldfish pituitary fragments (Trudeau *et al.*, 1991; 1993). In this study, the possible direct actions of T on sGnRH- and cGnRH-II-stimulated GTH release were investigated using dispersed goldfish pituitary cells prepared from sexually regressed male and female goldfish.

Results

Dispersed pituitary cells were prepared according to Chang *et al.* (1990). 15-min pulse applications of 100 nM sGnRH or cGnRH-II at 2.5 h intervals stimulated GTH release in perifusion studies. Overnight (16 hrs) 0.1μM T pretreatment alone, or in combination with continuous T exposure during perifusion, enhanced the GTH response to 100nM sGnRH or cGnRH-II (Fig. 1). Acute 0.1μM T treatment at the time of GnRH application showed no potentiating effects.

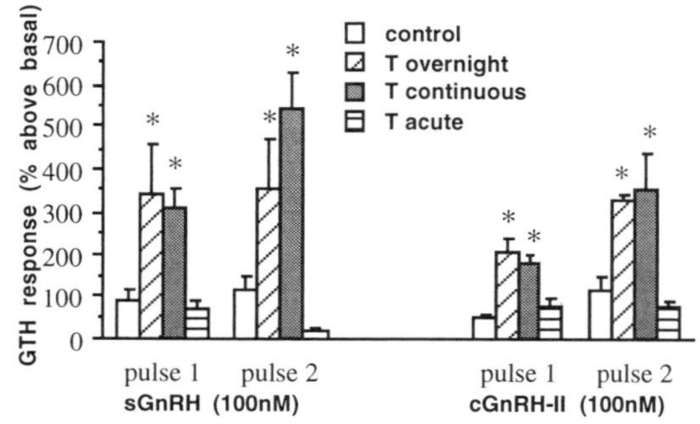

Figure 1. Effects of different T treatments on total net GTH release from perifused goldfish pituitary cells. * indicates significant difference from control cells ($P<0.05$; *t*-test; mean ± SE; n=4-6 columns each).

The effects of overnight 0.1μM T treatment on basal GTH release and cellular GTH content were also investigated using cells in static incubation (Chang *et al.*, 1990). T did not affect either of these parameters. (Basal GTH release from control and T-treated cells were 366 ± 38 and 358 ± 34 ng/0.1 million cells, respectively; GTH cell contents were 398 ± 43 and 510 ± 38 ng/0.1 million cells, respectively; $P>0.05$; n=4).

Discussion

Results from this study indicate, for the first time, that the potentiating effects of prolonged T exposure on GnRH-stimulated GTH release in goldfish can be exerted directly at the pituitary cell level. In contrast, acute exposure to T had no effects on GnRH-stimulated GTH release. This requirement for long term T exposure agrees with earlier results showing the positive effect of T is dependent on protein synthesis (Trudeau *et al.*, 1993). In this study, overnight T treatments did not alter basal GTH release or GTH cell content, suggesting that T did not affect the total GTH pool or tonic GTH release. It is possible that the potentiating effects of T is exerted at the GnRH receptor or post-receptor level. However, Trudeau *et al.* (1993) showed that T enhanced the pituitary sensitivity to GnRH peptides in sexually regressed goldfish without changes in GnRH receptor number or affinity. The mechanisms whereby T directly potentiates the GTH response to GnRH and the seasonality of T effects require further studies. The direct feedback action of T may be an important event in the regulation of changes in serum GTH levels and hormone release responsiveness to GnRH in the goldfish seasonal reproductive cycle.

References

Chang, J.P., Cook, H., Freedman, G.L., Wiggs, A.J., Somoza, G.M., de Leeuw, R. and Peter, R.E. 1990. Use of a pituitary cell dispersion method and primary culture system for the studies of gonadotropin-releasing hormone action in the goldfish, *Carassius auratus*. I. Initial morphological, static, and cell column perifusion studies. *Gen. Comp. Endocrinol.* 77:256-273.

Kobayashi, M. and Stacey, N.E. 1990. Effects of ovariectomy and steroid hormone implantation on serum gonadotropin levels in female goldfish. *Zool. Sci.* 7:715-721.

Trudeau,V.L., Peter, R.E. and Sloley, B.D. 1991. Testosterone and estradiol potentiate serum gonadotropin response to gonadotropin-releasing hormone in goldfish. *Biol. Reprod.* 44:951-960.

Trudeau, V.L., Murthy, C.K., Habibi, H.R., Sloley, B.D. and Peter, R.E. 1993. Effects of sex steroids treatments on gonadotropin-releasing hormone-stimulated gonadotropin secretion from the goldfish pituitary. *Biol. Reprod.* 48:300-307.

[1](Supported by a NSERC grant, a NATO Intl. Collab. Res. Grant and travel grants from U. Alta and U. Utrecht.)

PURIFICATION OF STRIPED BASS (*Morone saxatilis*) GONADOTROPIN II AND DEVELOPMENT OF AN ENZYME IMMUNOASSAY FOR ITS MEASUREMENT

E. Mañanós, P. Swanson[1], J. Stubblefield and Y. Zohar

Center of Marine Biotechnology, University of Maryland, 701 E. Pratt St., Baltimore, MD 21202, USA.
[1]National Marine Fisheries Service, Seattle, WA 98112, USA.

Summary

This work describes the preparation of a highly purified gonadotropin II (stbGtH-II) and its α and ß subunits from pituitaries of hybrid striped bass (*Morone saxatilis x Morone chrysops*). Using specific antibodies against the ß subunit, an enzyme-linked immunosorbent assay (ELISA) was developed. Displacement curves obtained with pituitary and plasma samples from several fish species belonging to the family Moronidae were parallel to the standard curve. This ELISA was used to measure GtH-II levels in striped bass injected with an analog of GnRH.

Introduction

Recently, the existence of two different GtHs (GtH-I and GtH-II, homologous to FSH and LH, respectively) has been demonstrated in several fish species. Nevertheless, the biological function of these GtHs has not been clearly established, although GtH-I seems to be involved in the process of vitellogenesis/spermatogenesis and GtH-II in the processes of oocyte maturation and ovulation/spermiation. Studying the biological function of GtH-I and GtH-II in fish is dependent on the availability of specific measurement techniques. The aim of the present study was to produce a highly purified preparation of striped bass GtH-II and to develop a homologous ELISA to enable its measurement.

Results and discussion

The stbGtH-II was purified from pituitaries of sexually mature fish by gel filtration, ion-exchange chromatography and FPLC, and identified by its *in vitro* estradiol-17β stimulatory activity in a homologous bioassay system, SDS-PAGE, and N-terminal amino acid sequencing. The stbGtH-II α and β subunits were purified by gel filtration and HPLC and their molecular weights (18 KDa and 22 KDa, respectively) determined by SDS-PAGE. A competitive ELISA was developed using antibodies against the β subunit and the intact stbGtH-II for the standard curve. Fig. 1 shows the parallelism between the standard curve (b=-0.98) and displacement curves obtained with serial dilutions of samples from striped bass: pituitary extract (b=-0.97), GnRH-injected male (b=-1.01) and female (b=-0.98) plasma, and mature female plasma (b=-1.03). The sensitivity of the assay was 156 pg/ml (15.6 pg/well) and the intra- and inter-assay variability (at 50% binding) 7.71% and 8.72%, respectively.

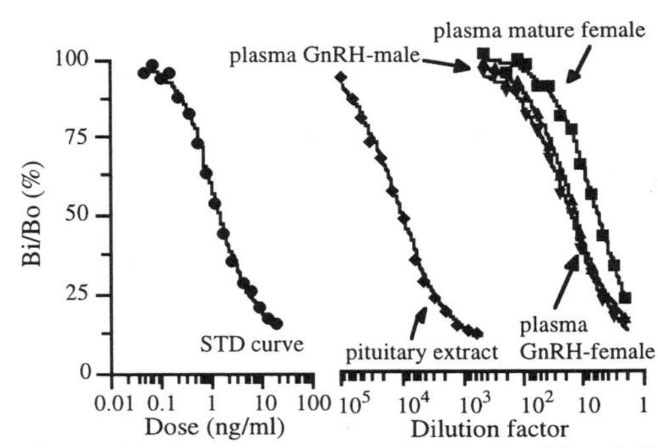

Fig. 1. Displacement curves obtained with stbGtH-II (standard curve) and serial dilutions of pituitary and plasma samples from striped bass.

Using this ELISA, a surge of GtH-II secretion was monitored in striped treated with GnRHa. As shown in fig 2, a significant release of GtH-II was observed 6 hours after GnRHa injection while GtH-II levels in controls remained unchanged.

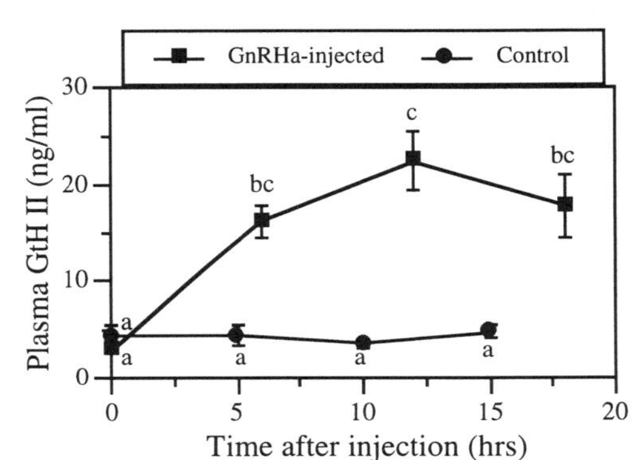

Fig. 2. Plasma GtH II levels (mean ± SEM, n=10) in striped bass at different times after their injection with GnRHa (10 µg/kg BW). Statistical differences are indicated by different letters ($p<0.05$).

Displacement curves obtained with serial dilutions of pituitary and plasma samples from species closely related to striped bass (such as white bass -*Morone chrysops*-, white perch -*Morone americana*- and sea bass -*Dicentrarchus labrax*-) were parallel to the standard curve, allowing the use of this ELISA for measuring GtH-II levels in fish belonging to the family Moronidae. The availability of the stbGtH-II assay will enable the intensification of studies on the reproductive physiology and spawning manipulation of striped bass and its relatives, fish which are important to the aquaculture industry.

PREOVULATORY CHANGES IN GONADOTROPIN GENE EXPRESSION AND SECRETION IN THE GILTHEAD SEABREAM, *SPARUS AURATA*.

I. Meiri[1], Y. Gothilf[2], W.R.Knibb[1], Y. Zohar[2] and A. Elizur[1].

[1]National Center for Mariculture, IOLR, Eilat, Israel and [2]Center of Marine Biotechnology, University of Maryland, Baltimore, USA.

Introduction

The gilthead seabream, *Sparus aurata*, (s.b.), is an economically important marine teleost. Females have a non- synchronous ovarian development. Oocyte maturation, ovulation and spawning occur at 24-hour intervals over a period of three months. In this study, changes in oocyte morphology, gonadotropin gene expression and secretion were monitored at 4-hour intervals, during 24 hours.

Methods

Six groups of seabream were analyzed, each had 8 females and 3 males. Females were bled and sacrificed at 20hr, 16hr, 12hr, 8hr and 4hr preceding spawning and at spawning time. Total RNA was extracted from individual pituitaries (Chomczynski and Sacchi, 1987) and analyzed by dot blot hybridizations using s.b. ßGtHI and II cDNA probes. Loadings of RNA were corrected by s.b. ß actin readings and quantified using ßGtHI and II sense strand RNA standards. The autoradiograms were scanned by a computing densitometer. Levels of s.b. GtHII in plasma samples were measured by radioimmunoassay (Zohar *et al.* 1990). Samples of gonads were fixed for histological studies. Data were statistically analyzed using natural log transformed values and analysis of variance (ANOVA) models.

Results

Levels of ßGtHI mRNA showed a trend of declining towards 12 hr before spawning time (ANOVA $F_{[5,39]} = 2.15$, $P < 0.1$). ßGtHII mRNA levels declined earlier (16 hr before spawning) and increased towards spawning, with the initiation of final oocyte maturation (ANOVA $F_{[5,38]} = 2.57$, $P < 0.05$, Fig. 1). The process of final oocyte maturation was initiated 8 hr before spawning. Ovulation occurred within the 4 hr period preceding spawning. Levels of GtHII in plasma peaked 8 hr before spawning (ANOVA $F_{[5,38]} = 9.07$, $P < 0.001$, Fig. 2) and showed a similar profile to ßGtHII transcript levels..

Discussion

At 20 to 12 hours before spawning, oocytes did not show signs of final maturation. During that period plasma GtH II and pituitary ßGtHII mRNA were low. The process of final oocyte maturation (germinal vesicle migration, coalescence of yolk globules and germinal vesicle brake down) was initiated eight hours before spawning. Ovulation occurs within the 4 hours preceding spawning. βGtH II gene expression and secretion increased with the initiation of final oocyte maturation and remained elevated until ovulation and spawning. Throughout the preovulatory daily cycle, βGtH II mRNA steady state profile closely paralleled GtH II secretion. The presence of βGtHI transcripts during spawning time is consistent with the non-synchronous nature of its ovary and further presents the seabream as an important model in the studies of fish reproduction.

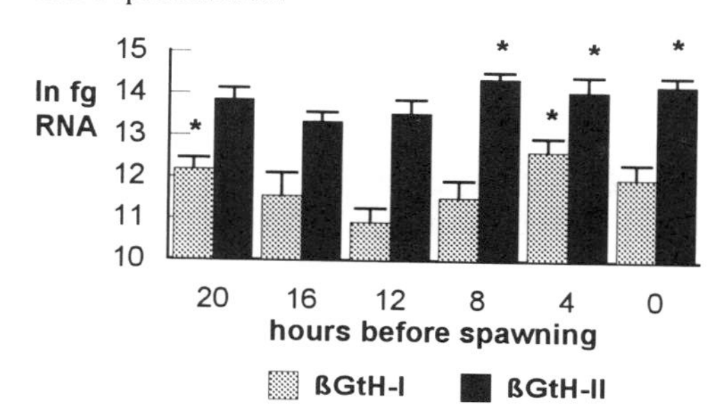

Fig. 1. βGtHI and II transcript levels (ln fg means ± S.E, n=8). Asterics indicate significant differences from lowest value.

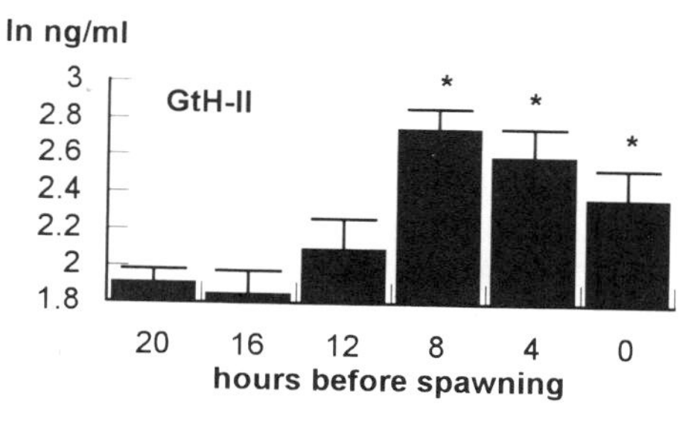

Fig. 2. Plasma levels of GtHII (ln means ± SE, n=8) at different time points around spawning. Asterics indicate significant differences from lowest value.

References

Chomczynski, P. and Sacchi, N. (1987). Anal. Biochem. **162**, 156-159.

Elizur, A., Zmora, N., Rosenfeld, H., Meiri, I., Hassin, S., Gordin, H. and Zohar, Y. (1995). *Submitted.*

Zohar, Y., Breton, B., Sambroni, E., Fostier,E., Tosky, M., Pagelson, G., and Liebovitz, D. (1990). Aquaculture **88**, 189-204.

INTRACELLULAR MEDIATION OF GnRH EFFECT ON TRANSCRIPTION OF THE TILAPIA GtH IIß GENE

P. Melamed[1], G. Gur[1], B. Levavi-Sivan[1], A. Elizur[2] and Z. Yaron[1]

[1] Department of Zoology, Tel-Aviv University, Tel Aviv, 69978, Israel.
[2] National Center for Mariculture, IOLR, Eilat, Israel.

Summary

Activation of either PKA or PKC pathways, which mediate GtH release in tilapia, also increased the IIß mRNA steady-state levels; the effect of the former occurring rapidly. The half-life of the mRNA after exposure to forskolin was nearly double that in control or TPA-exposed cells. Exposure to a specific PKA inhibitor, H89, reduced both basal and sGnRH-stimulated mRNA levels. These results suggest that both PKC and PKA pathways may mediate the GnRH effect on GtH IIß mRNA levels, involving an increase in the rate of transcription and a stabilizing effect.

Introduction

Injection of GnRH increased mRNA levels of goldfish GtH IIα and ß subunits, although sGnRH and cGnRH-II had different potencies (Khakoo et al., 1994). In mammals, GnRH elevates both LH and FSH mRNA; the effect on the LH transcript is mediated by both PKA and PKC pathways (Counis and Jutisz, 1991). An increase in transcript stability has been implicated in the cAMP effect, involving lengthening of the poly (A) tail (Ishizaka et al., 1993). In tilapia, both PKA and PKC pathways stimulate GtH release, therefore elements of these pathways were studied for possible effects on GtH IIß mRNA.

Methods

Experiments were carried out on pituitary cells (1.10^6 cells/well) of tilapia hybrids (*Oreochromis niloticus x O. aureus*) which were exposed to the test substances on the fourth day of culture. Total RNA was then extracted, loaded onto northern or slot blots and hybridized using tilapia GtH IIß cDNA; mRNA levels were standardized with those of ß actin.

Results and Discussion

sGnRH (10 nM) elevated GtH IIß mRNA after 12-24 h (Fig 1). Similarly, exposure of the cells to TPA (12.5 nM) had no effect after 1.5-6 h but increased transcript levels after 10-24 h. Elevation of cAMP levels by forskolin (10 μM) or IBMX (0.2 mM) showed quite different kinetics of response, the mRNA being affected as early as 1.5 h. Mobilization of Ca^{++} by ionomycin (1 μM) had an immediate (after 30 mins) impact on the mRNA levels, which was sustained after 8 h exposure. The involvement of PKA in transcription of GtH IIß was verified when a specific PKA inhibitor, H89 (0.1 μM), reduced both basal and sGnRH-stimulated transcript levels. Effects on the mRNA stability were examined by incubating cells with forskolin or TPA for 24 h and then exposing them to actinomycin D (8 μM) for 4-21 h. The rates of degradation were similar in control and TPA-treated cells ($T_{½}$ = 8.47 h and 8.38 h, respectively), but slower in forskolin-treated cells ($T_{½}$ = 14.1 h). These results suggest that activation of both PKC and PKA pathways may mediate the GnRH stimulatory effect on GtH IIß mRNA steady-state levels, possibly by different mechanisms involving both an increase in transcription rates and a stabilizing effect.

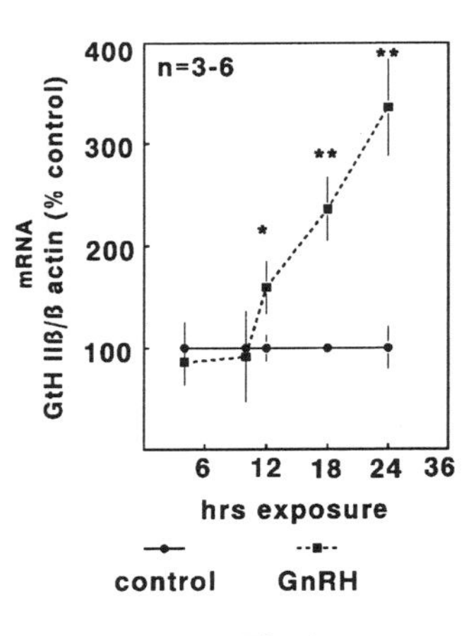

Fig 1

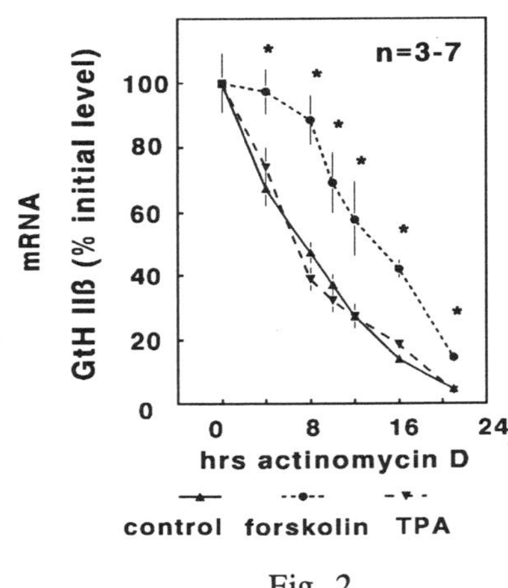

Fig. 2

References

-Counis, R. and M. Jutisz, 1991. Trends Endocrinol. Metab. 2:181-187.
-Ishizaka, K., Tsujii, T. and S.J. Winters, 1993. Endocrinology 133:2040-2048.
-Khakoo, Z., Bhatia, A., Gedamu, L. and H. Habibi, 1994. Endocrinology 134:838-847.

LOCALIZATION OF PITUITARY CELLS BY IMMUNOCYTOCHEMISTRY IN THE "ARGENTINE SILVERSIDE" ODONTESTHES BONARIENSIS.

Vissio,P.G.; Somoza,G.M.*; Maggese,M.C. and Paz,D.A.**.

Laboratorio de Embriología Animal. Departamento de Biología. UBA. Ciudad Universitaria (1428) Buenos Aires. Argentina. (*) INEUCI-CONICET Depto de Biología. UBA. (**) LABINE-CONICET. Argentina

Introduction

Odontesthes bonariensis the "Argentine silverside" (Atheriniformes) is an endemic species from Argentina, with economic importance. We are interested in studying the reproductive biology of this fish and, in this context the localization of different pituitary cell types was analyzed by immunocytocemical techniques. Briefly, the samples were taken from lagoons of Buenos Aires Province at different times of the year. The brains with the pituitaries attached were fixed "in situ" with Bouin's fluid and immunocytochemical procedures according to Miranda *et al.*, 1995 were used.

Results and Discussion

The morphology of the pituitary gland does not differ from the basic scheme of teleost fishes. Three areas can be recognized in the adenohypophysis: rostral pars distalis (RPD), proximal pars distalis (PPD) and pars intermedia (PI). The different cell types were visualized with specific antisera (Table I).

Prolactin cells are restricted to the RPD and were revealed with two different antisera raised against prolactin from two different fish.

ACTH producing cells were identified using an anti human antiserum in the RPD and crossreactivity was seen presumably with MSH cells in the PI. Anti carp GH and anti chum salmon GH identified a group of cells located in the PPD in close association with fibers of the anterior neurohypophysis.

Thyrotropic (TSH) cells reacted with an anti-human ß-TSH in the PPD. In this case a weak immunoreaction was observed in an area corresponding to the presumptive gonadotrophic (GtH) producing cells, probably due to the use of an heterologous antiserum.

The gonadotrops were observed using three different antisera mainly in the PPD and also spreading along the external border of the PI. A similar pattern was seen with these antisera. Taking into consideration that two of them were raised against chum salmon GtH I and GtH II ß-subunits a more detailed analysis has to be done with preabsortion of these antisera with the heterologous ß-subunit and/or using double immunostaining to see differences in localization and relationship with the reproductive status.

Somatolactin (SL) producing cells were identified with anti-chum salmon antiserum and are located in the PI as previously described in other species (Rand-Weaver, *et al.*, 1991). Due to the low **number of samples we obtained a preliminary view of a more extensive SL-immunostained area in sexually** mature when compared to sexually regressed fish. These data would suggest a relationship between SL and the reproductive status in this species. This work will provide a basis for future research on the hypothalamus-pituitary-gonadal axis of the "Argentine silverside".

Table I: Immunohistochemical staining reaction in the pituitary of Odontesthes bonariensis.

Antiserum to	Dilution	P	C	S	T	G	MSH	SL
chumPRL	1:1000	++						
cPRL	1:1000	++						
hACTH	1:1000		++				++	
chumGH	1:500			++				
cGH	1:1000			++				
hßTSH	1:500				++	+		
croakerGTH	1:500				+	++		
chumGTHI	1:500				+	++		
chumGTHII	1:500				+	++		
chumSL	1:1000							++

P, prolactin cells; C, ACTH cells; S, somatotrops; T, thyrotrops; G, gonadotrops; MSH, MSH cells; SL, somatolactin cells. +, weak immunostaining; ++, strong immunostaining

References

Miranda,L. *et al.*, 1995. Immunocytochemical and morphometric study of TSH, PRL, GH and ACTH cells in *Bufo arenarum* larvae inhibited thyroid function. Gen Comp Endocrinol 98:166-176.

Rand-Weaver,M. *et al.*, 1991. Cellular localization of somatolactin in the pars intermedia of some teleost fishes. Cell Tissue Res 263:207-215.

This work was supported by a grant from the University of Buenos Aires (EX-208)

CHANGES IN SERUM AND PITUITARY LEVELS OF PROLACTIN AND GROWTH HORMONE WITH REPRODUCTION AND FASTING IN THE TILAPIA, *OREOCHROMIS MOSSAMBICUS*.

G. M. Weber[1] and E. G. Grau

Hawaii Institute of Marine Biology, University of Hawaii, Kaneohe, HI 96744-1346

Summary

Serum prolactin (PRL) and serum and pituitary growth hormone (GH) levels were elevated in female tilapia late in the brooding phase of the reproductive cycle, compared with females late in vitellogenesis. Serum PRL and serum and pituitary GH levels were also elevated in response to fasting. Comparisons of hormone patterns between fasting and brooding fish suggest GH may have functions intrinsic to reproduction in tilapia.

Introduction

Prolactin and GH have been implicated in regulation of reproduction, metabolism and osmoregulation in teleosts. Female tilapia brood eggs and larvae in their buccal cavity for up to 3 weeks. Feeding is reduced during brooding. Our objective was to determine whether there are changes in serum and pituitary levels of PRL (tPRL177 and tPRL188) or GH during brooding that are associated with regulation of reproduction.

Methods

Serum and pituitary PRL and GH levels were measured during the reproductive cycle of female tilapia (Studies 1-3). Hepatosomatic index (HSI; (liver wt/body wt) X 100) and condition factor (CF; (body wt (g)/ (std L $(cm)^3$)) X 100) were evaluated. Fish were adapted to fresh water (FW; Studies 1 and 2) or seawater (SW; Study 3).

Serum and pituitary PRL and GH levels were measured during fasting in FW-adapted tilapia, (Studies 4 and 5). In Study 4, females with high CF (~3.6) were fasted for 10 and 21 days and males were fasted for 21 days. In study 5, males with low CF (~3.2) were fasted for 10, 21 and 31 days. Hormone levels were measured by homologous radioimmunoassays (Ayson *et al.*, 1993).

Results

Serum tPRL177 levels increased late in brooding in Study 1 and early in Study 2, but never exceeded levels in non-brooding females by more than 60%. Serum tPRL177 was not detectable in SW-adapted females. Serum levels of tPRL188 and pituitary content of both PRLs did not change or did not change consistently among studies. Serum GH was increased ~4-fold and pituitary GH by ~50-70% in females brooding post-yolksac larvae compared with fish late in vitellogenesis. Serum GH was first elevated in fish brooding yolksac larvae, and pituitary GH was first elevated in females brooding post-yolksac larvae.

Compared with fed fish, serum tPRL177 levels were ~3-fold higher in females fasted for 10 and 21 days and males for 21 days (Study 4). Pituitary content of tPRL177 was ~60% higher in females fasted for 21 days than in fed females or females fasted for 10 days. There was no change in males. Serum levels of tPRL188 were 67% higher in males fasted for 21 days than in fed males. There were no differences in serum tPRL188 levels in females, or pituitary levels in either sex. Pituitary but not serum levels of GH were elevated in males and females fasted for 21 days (~40-50%).

In Study 5, serum tPRL177 and tPRL188 were elevated in males fasted for 10, 21, and 31 days. There was no change in pituitary content of tPRL177 but tPRL188 was reduced in males fasted for 31 days (~37%). Pituitary content of GH was elevated in males fasted for 21 and 31 days (63 and 80% respectively). Serum GH was elevated in males fasted for 31 days (~3.5 ng/ml) but not 21 days (~1 ng/ml).

Discussion

Increases in serum and pituitary levels of GH and serum levels of both PRLs after fasting suggest that these hormones act in regulation of metabolism in tilapia. Furthermore, the earlier increase in serum PRL compared with GH suggests the PRLs and GH may have different roles in regulating metabolism.

Decreases in HSI and CF suggest that metabolic state changes with brooding. Altered metabolic state during fasting cannot account for all changes in serum and pituitary GH and PRL levels during brooding. The rise in serum tPRL177 was more pronounced with fasting and occurred earlier than the rise in GH, while the reverse was true with brooding. Serum concentrations of GH increased prior to changes in pituitary content with brooding, whereas the reverse was true during fasting.

Serum PRL was generally highest in females brooding post-yolksac larvae, but changes in serum and pituitary levels during the reproductive cycle were small and inconsistent. Thus, changes in serum concentrations of the PRLs do not appear to be driving parental behavior or ovarian function in the female tilapia. By contrast, the pattern of serum and pituitary GH levels during the brooding phase of the reproductive cycle was consistent among the studies. Thus, GH may have functions intrinsic to tilapia reproduction.

References

Ayson, F.G., Kaneko, T., Tagawa, M., Hasegawa, S., Grau, E.G., Nishioka, R.S., King, D.S., Bern, H.A. and Hirano, T. 1993. Gen. Comp. Endocrinol. 89:138-148.

Supported by NSF Grant DCB 91-04494 and NOAA/Sea Grant No. NA36RG05097/R/AQ-37 to E. G. Grau.

[1]Present address: Dept. of Zoology, North Carolina State University Raleigh, NC 27695-7617

MOLECULAR CLONING OF cDNA ENCODING TWO GONADOTROPIN β SUBUNITS (GTH I β AND II β) FROM THE GOLDFISH

Y. Yoshiura[1], M. Kobayashi[1], Y. Kato[2], and K. Aida[1].

[1]Dept. of Fisheries, Fac. of Agriculture, The University of Tokyo, Bunkyo, Tokyo 113; [2]Biosignal Research Center, Institute for Molecular and Cellular Regulation, Gunma University, Maebashi, Gunma 371, Japan.

Summary

Two types of cDNAs (GTH I β and II β) encoding the β subunit of goldfish gonadotropin were cloned and sequenced. In particular, the amino acid sequence of goldfish GTH I β showed considerable differences (40-49% homology) from those of other teleosts. Genomic Southern blot analysis showed that cyprinid species harbor genes which are homologous to goldfish GTH I β and II β. These results demonstrated the duality of GTH in cyprinid fishes as has been shown in other teleost fishes. Northern blot analysis of GTH I β and II β mRNAs from fish at differing stages of ovarian maturity (juvenile, maturing, mature, and regressed) showed that mRNA levels of both GTH I β and II β are correlated with ovarian maturity. Thus, we demonstrated for the first time the presence and expression of GTH I β and II β genes in cyprinids.

Introduction

In recent reports, cDNAs encoding the β subunits of GTH I and II have been cloned and sequenced in several teleost fishes (Sekine *et al.*, 1989; Lin *et al.*, 1992; Kato *et al.*, 1993), thus providing definitive proof that two distinct GTHs are produced in the pituitary in teleosts. However, the significance of the two GTHs in regulating teleost reproductive processes remains unclear. In order to obtain better understanding of mechanisms regulating teleost reproduction, we cloned two types of cDNAs (GTH I β and II β), and examined the levels of mRNA encoding GTH I β and II β in the pituitary at differing stages of ovarian maturity in goldfish.

Results and Discussion

Two types of GTH β subunits

Two types of cDNAs (GTH I β and II β) encoding the β subunit of GTH were cloned using the polymerase chain reaction and a cDNA library prepared from goldfish pituitary mRNAs. The nucleotide sequences showed that the GTH I β cDNA was 616 bp long, encoding 135 amino acids, and that the GTH II β cDNA was 546 bp long, encoding 140 amino acids. When compared to the amino acid sequences of known teleost GTH β subunits, goldfish GTH II β showed high homology (99%) to carp GTH II β with only one differing amino acid. Goldfish GTH I β showed low homology (ranging from 40-49%) with known teleost GTH I β, but had well-conserved regions specific to teleost GTH I β. These results indicate that goldfish have two distinct forms of GTH. Genomic Southern blot analysis for goldfish and other cyprinid species showed positive bands for goldfish GTH I β and II β, thus demonstrating the duality of GTH in cyprinid fishes as has been shown in other teleost fishes.

The expression of GTH I β and II β mRNA at differing stages of ovarian maturity

In order to examine the pituitary GTH I β and II β mRNA levels in fish of various ovarian stages (juvenile, maturing, mature, and regressed), Northern blot analysis was carried out using cloned I β and II β cDNA probes. The hybridized membranes were scanned by a FUJIX BAS 1000Mac Bio-Imaging Analyzer and quantified using a recommended program.

The mRNA levels of GTH I β and II β were low in juvenile fish, and increased with the progression of maturity. The mRNA levels of I β and II β in mature fish were 11- and 27-fold higher, respectively, compared to those in juvenile fish. Sexually-regressed fish showed levels which had declined. Thus, the mRNA levels of both GTH Iβ and IIβ were correlated with ovarian maturity, but expression levels of GTH Iβ were much lower than those of GTH Iiβ at all stages. It remains to be elucidated how GTH I and GTH II are involved in the regulation of gonadal development in goldfish.

References

Kato, Y., Gen, K., Maruyama, O., Tomizawa, and Kato, T. (1993) *J. Mol. Endocrinol.* 11, 275-282.

Lin, Y. W. P., Rupnow, B. A., Price, D. A., Greenberg, R. M., and Wallance, R. A. (1992) *Mol. Cell. Endocrinol.* 85. 127-139.

Sekine, S., Saito, A., Itoh, H., Kawauchi, H., and Itoh, S. (1989) *Proc. Natl. Acad. Sci. U.S.A.* 86, 8645-8649.

PLASMA SOMATOLACTIN CONCENTRATIONS IN ATLANTIC CROAKER DURING GONADAL RECRUDESCENCE

Y.Zhu and P. Thomas.

The University of Texas at Austin, Marine Science Institute, Port Aransas, Texas 78373.

Introduction

Somatolactin (SL) is a peptide hormone recently isolated from the pituitaries of several teleost species. It has been suggested that SL may be involved in the regulation of reproduction based on the observation that plasma SL levels are elevated in coho salmon during gonadal recrudescence (Rand-Weaver *et al.*, 1992). However, this relationship has not been examined in any other teleost species. A radioimmunoassay has recently been developed for SL measurement in two sciaenid fishes, red drum and Atlantic croaker (Zhu and Thomas, 1995). Therefore, in the present study changes in plasma SL concentrations during gonadal recrudescence in wild caught Atlantic croaker were investigated.

Results

The gonadosomatic index (GSI) increased significantly from mid-August to November in both male (0.13 ± 0.06 to 2.2 ± 0.3) and female (0.38 ± 0.02 to 9.7 ± 0.47) Atlantic croaker. There was an overall decline in plasma SL concentrations during gonadal development. However, plasma SL concentrations were only significantly ($P < 0.01$, Tukey's HSD test) different between females with immature (perinucleolar oocytes) and those in pre-spawning (late-yolk globule) condition (Fig.1). Plasma SL levels did not change significantly at the onset of vitellogenesis. In contrast, both plasma estradiol-17β and testosterone were significantly higher in vitellogenic compared to pre-vitellogenic female fish. In males, plasma SL levels did not change significantly during testicular development.

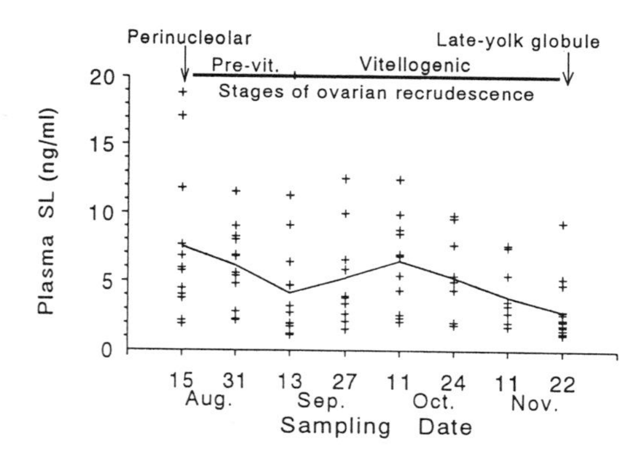

Fig.1. Mean and individual plasma SL concentrations in female Atlantic croaker.

Discussion

In contrast to the observation in coho salmon (Rand-Weaver *et al.*, 1992), plasma SL levels did not increase during gonadal development in Atlantic croaker. Previously, *in vitro* studies have demonstrated a weak steroidogenic activity of SL in coho salmon gonads (Planas *et al.*, 1992). However, in the present study the dramatic increase in circulating levels of gonadal steroids was not associated with an elevation in SL levels in croaker of either sex. These results suggest that SL does not have an important steroidogenic function in croaker during gonadal recrudescence. We have previously observed a minor increase in plasma SL levels in pre-spawning red drum compared to those in gonadally immature fish. In contrast, dramatic changes in plasma SL was observed during background adaptation in red drum (Zhu and Thomas, 1995). Taken together, our results suggest that the primary function of SL may be a non-reproductive one in sciaenid fishes.

References

Planas, J., Swanson, P., Rand-Weaver, M., and Dickhoff, W.W. (1992). Somatolactin stimulates *in vitro* gonadal steroidogenesis in coho salmon, *Oncorhynchus kisutch*. *Gen. Comp. Endocrinol.* 87, 1-5.

Rand-Weaver, M., Swanson, P., Kawauchi, H., and Dickhoff, W.W. (1992). Somatolactin, a novel pituitary protein: purification and plasma levels during reproductive maturation of coho salmon. *J. Endocrinol.* **133**, 393-403.

Zhu, Y. and P. Thomas. (1995). Red drum somatolactin: development of a homologous radioimmunoassay and plasma levels after exposure to stressors or various backgrounds. *Gen. Comp. Endocrinol.* **99**, 275-288.

Hypothalamus/Brain

FUNCTIONAL INTERACTIONS BETWEEN NEUROENDOCRINE SYSTEMS REGULATING GTH-II RELEASE

V.L. Trudeau[1] and R.E. Peter[2]. [1]Dept. of Zoology, University of Aberdeen, Scotland, AB9 2TN; [2]Dept. of Biological Sciences, University of Alberta, Edmonton, Canada, T6G 2E9.

Summary

Gonadotropin-releasing hormone (GnRH) and dopamine (DA) producing neurons are respectively the principal stimulatory and inhibitory systems controlling GTH-II release. Considerable progress has been made in identifying other possible stimulatory neuropeptides and neurohormones; the neurotransmitter γ-aminobutyric acid (GABA) has the most prominent stimulatory actions, and likely modulates both the GnRH and DA systems. For the majority of the factors influencing GTH-II release, seasonal and/or steroidal modulation of the GTH-II response has been noted.

Introduction

The dual stimulatory and inhibitory regulation of GTH-II release in teleosts is well described (Peter et al., 1986; 1991; Kah, 1990). Recent evidence indicates that amino acid neurotransmitters are important for GTH-II release. In goldfish, GABA has clear stimulatory effects on GTH-II release. This results from both increased GnRH release and decreased dopaminergic activity. Thus, the GABAergic cells may be considered as a modulatory neuronal system. Using this example of a functional interaction between neuroendocrine systems, a model is presented for the regulation of seasonal GTH-II release and gonadal development in the adult.

A physiological role for GnRH and DA

The majority of neurohormones have stimulatory effects on GTH-II release (Table 1). Of these, the various natural and synthetic forms of GnRH are the best studied (Peter et al., 1991; Sherwood et al., 1994). The first demonstration of a physiological role for GnRH was by Breton et al. (1971). Recently, the structure-activity relations of GnRH-antagonists in goldfish have been characterized (Murthy et al., 1993; 1994). Sex pheromone (17α,20β,dihydoxy-4-pregnen-3-one; Stacey et al., 1994) and DA antagonist induced GTH-II release can be blocked in vivo by a teleost GnRH antagonist. This establishes for the first time that endogenous GnRH is released by teleost hypophysiotropic neurons in situ to stimulate pituitary GTH-II secretion. Injection of GnRH also stimulates GTH-II (both α- and β- subunits) mRNA accumulation in the pituitary (Khakoo et al, 1994). In vitro GnRH-stimulated GTH-II release is a desensitizing phenomenon associated with a loss of pituitary GnRH receptor numbers (Habibi, 1991). GnRH receptor-signal transduction mechanisms have been reviewed (Chang & Jobin, 1994).

Extensive pharmacological studies indicate that DA is released from nerve terminals in the pituitary where it activates DA_2 receptors on the gonadotroph to restrain GTH-II secretion; DA also inhibits GnRH release (Peter et al., 1986; 1990; Sloley et al., 1991; Yu & Peter, 1990). Sex pheromone-induced GTH-II release is associated with reduced pituitary DA turnover in males, emphasizing the physiological role of DA in GTH-II inhibition (Dulka et al., 1992). With only a few possible exceptions, all other factors studied to date stimulate GTH-II release. Key neurohormones involved in regulating GTH-II release are discussed below.

Amines

Injection of serotonin (5HT) stimulates GTH-II release via a $5HT_2$-like receptor; maximal responses occur in sexually mature animals (Somoza et al., 1988; 1991; Khan & Thomas, 1994). Current evidence suggests that 5HT stimulates GTH-II release by activating the GnRH neuron at either the cell body or terminal levels (Yu et al., 1991; Khan & Thomas, 1993). Periovulatory changes in hypothalamic 5HT turnover in trout (Saligaut et al., 1992) further support a facilatory role for this amine in GTH-II release. 5HT can serve as a precursor for melatonin (MEL), but MEL has no direct effects on pituitary GTH-II release (Somoza et al., 1991). Nevertheless, MEL remains a candidate for involvement in seasonal reproductive cycles, because its secretion is regulated by photoperiod, and MEL receptors are found in neuroendocrine territories of the teleost brain (Kah, 1993; Davies et al., 1994; Martinoli et al., 1991).

Catecholamines (and neuropeptides) in direct retinohypothalamic projection neurons (Holmqvist et al., 1992) could be important for photoregulation of GTH-II release; this hypothesis has not yet been tested. Norepinephrine is found in the preoptic area and ventromedial hypothalamus but not the pituitary (Dulka et al., 1992; Trudeau et al., 1993c). Norepinephrine stimulates GTH-II release via an α-1 receptor mechanism (Chang et al., 1991). Using an elegant in vitro brain explant preparation, Bailhache et al. (1989) have demonstrated that microinjection of NE into the intact pituitary stimulates GTH-II release in rainbow trout. Given that NE innvervation of the pituitary has not been detected (Dulka et al.,1992; Trudeau et al., 1993c), perhaps peripherally released NE acts on GnRH nerve terminals and the gonadotrophs. Centrally NE may activate the GnRH neuron (Peter et al., 1990; 1991, Yu et al., 1991). Estradiol modulates NE turnover in goldfish brain, suggesting that part of the gonadal feedback mechanism may involve NE neurons (Trudeau et al., 1993c).

Amino Acids

Glutamate and taurine are found in substantial amounts in the teleost brain and pituitary, and both have been shown to stimulate GTH-II release in vivo. In goldfish, the glutamate agonist N-methyl-D,L-aspartic acid (NMA) stimulates GTH-II release; this response is not affected by sex steroids (Trudeau et al. 1993b). In sexually immature rainbow trout implanted with testosterone (T), NMA also stimulates GTH-II release; however, NMA does not stimulate GTH-II release in sexually regressed trout. Unfortunately, these studies did not control for T implantation or sexual maturation, and it is therefore difficult to determine the importance of glutamate in this species. Nevertheless, in T-implanted trout, NMA appears to stimulate GTH-II release indirectly through GnRH release (Flett et al.,1994).

Intraperitoneal and brain injection of taurine stimulates GTH-II release in goldfish (Sloley et al., 1992; Trudeau et al., 1993b). Relatively high levels of taurine are needed, which likely reflects the observation that taurine is one of the most abundant amino acids in the vertebrate brain (Huxtable, 1989). A physiological relevance for taurine is suggested since administration of the precursor hypotaurine but not the metabolite isetheonic acid stimulates GTH-II release in vivo (Sloley et al., 1992). The cell types in brain or pituitary producing or accumulating taurine are not defined. Interestingly, plasma levels of taurine increase after feeding in fish (Lyndon et al., 1993), thus raising the interesting possiblility of dietary taurine stimulating GTH-II release.

The neurotransmitter GABA is found in the hypothalamus and GABA neurons directly innervate the anterior pituitary (Kah, 1993; Médina et al., 1994). Intraperitoneal or brain injections of GABA stimulate GTH-II release in goldfish, in contrast to mammals where GABA generally inhibits luteinizing hormone (LH) release (Kah, 1993; Trudeau et al., 1993d). GABA does not affect basal or GnRH-stimulated GTH-II release from dispersed pituitary cells in vitro (Kah et al. 1992). The stimulatory effects of GABA on GTH-II release results from both increased GnRH release and decreased dopaminergic activity (Trudeau et al., 1993d). However, the major stimulatory effect of GABA is via the GnRH system, since DA antagonists and DA synthesis inhibitors do not block, but in fact potentiate, GABA-stimulated GTH-II release. The cellular receptor mediating this stimulatory effect is predominantly a $GABA_A$-like receptor (Trudeau et al., 1993d). Of the two GnRH molecules in goldfish, current evidence suggests that GABA preferentially regulates sGnRH over cGnRH-II (Sloley et al. 1994). Inhibition of the major metabolic enzyme GABA transaminase with γ-vinyl-GABA (GVG), results in GTH-II release in goldfish, indicating a physiological importance of GABA in GTH-II regulation (Trudeau et al., 1993d; Sloley et al., 1994). Moreover, GABA-mediated GTH-II release leads to a functional change in gonadal activity; serum testosterone levels are elevated in both male and female goldfish following treatment with GVG (Sloley et al., 1994).

Neuropeptides

Neuropeptide Y (NPY) stimulates GnRH release both at neurosecretory terminals in the pituitary and centrally, and NPY also has direct stimulatory actions on gonadotrophs (Peng et al., 1991; 1993a,b). Cholecystokinin (CCK) is found in the teleost brain and pituitary, and stimulates GTH-II release in vitro; sulfated CCK (CCK8s) but not non-sulfated CCK (CCK8ns) is the active form (Himick et al., 1993). Pituitary fragments from sexually regressed fish are less responsive than those from recrudescent animals, suggesting that CCK responsiveness may be dependent on sex steroids.

The opiate antagonist naloxone had both stimulatory and inhibitory effects on serum GTH-II levels in goldfish, suggesting that endogenous opioid peptides (EOP) play some undefined modulatory role in regulation of GTH-II secretion (Rosenblum and Peter, 1989).

Table 1. Neurohormones reported to affect GTH-II release in fish

	Seasonal or steroidal modulation	References
STIMULATION		
Amines		
NE	*	Chang et al. '91; Trudeau et al. '93c
5HT	*	Khan & Thomas '94 Somoza et al. '88
Amino acids		
β-Alanine	?	Sloley et al. '92
GABA	*	Kah et al. '92; Trudeau et al. '93b
Glutamate (NMDA)	0/?	Trudeau et al. '93b
Taurine (Hypotaurine)	*	Trudeau et al. '93b
Neuropeptides		
CCK8s/Gastrin17S	*	Himick et al.'93
EOP	?	Rosenblum & Peter '89
GnRH (2+)	*	Habibi et al. '89, Trudeau et al. '91; Trudeau et al.'93a; Weil & Marcuzzi '90
NPY	*	Peng et al. '90; 93b
INHIBITION		
Amines		
DA	*	Sokolowska et al. 1985; Peter et al. '86; Trudeau et al. '93c

Key: (*)- yes; (0)- no; (?)- unknown or equivocal.
Note: Factors tested and found to have no effects: catechol-estrogens, melatonin, cysteic acid, glutamine, isetheonic acid, lysine, CCK8ns, GRF, TRH, SRIF.

Growth Factors

Inhibin and activin are growth factors in the transforming growth factor-β family, acting through serine/threonine kinase receptors. Mammalian inhibin-A and activin-A stimulate both acute and prolonged GTH-II release from goldfish pituitaries in vitro (Ge et al., 1992). Steroid-free gonadal extracts stimulate GTH-II release; extracts from sexually recrudescent goldfish were most active (Ge et al., 1994). Cloning and sequencing of goldfish activin subunit genes from the gonads confirms the presence of inhibin/activin-like peptides (Ge et al., 1993).

Seasonal variations in pituitary GTH-II release and the role of sex steroid feedback

The GTH-II-releasing activity of many neurohormones changes under the influence of sex steroids (Table 1). Each of these modulatory actions of sex steroids contributes to gonadal feedback on the neuroendocrine system. Below we will focus on three systems in which seasonal and/or steroidal modulation has been well studied: (a) GnRH, (b) DA and (c) GABA.

(a) GnRH- Seasonal changes in pituitary responsiveness to GnRH has been noted in several fish species; maximal responses being associated with mature gonads. In goldfish, testosterone (T) through aromatization to estradiol (E_2), enhances GnRH responsiveness both in vivo and in vitro; 5α-dihydrotestosterone (DHT) and 11-ketotestosterone (11KT) had no effects (Trudeau et al. 1991; 1993a). Injection of hCG to sexually mature male goldfish increases steroid production and enhances GnRH responsiveness, directly demonstrating a true positive feedback (+FB) loop. Part of the mechanism of the seasonal increase in GnRH responsiveness in goldfish is the associated elevation of pituitary GnRH receptors capacity (Habibi et al, 1989). There are also seasonal variations in GnRH-stimulated glycoprotein α– and GTH-IIβ–subunit mRNA accumulation in vivo (Khakoo et al., 1994). In sexually regressed goldfish, sGnRH stimulates both mRNAs and cGnRH-II was without effect. In mature animals, both sGnRH and cGnRH-II stimulate α and GTH-IIβ mRNAs; cGnRH-II was approximately 2-fold more effective. The short term +FB effect of T directly at the pituitary is protein synthesis-dependent but does not involve a change in pituitary GTH-II content or GnRH-receptor capacity (Trudeau et al. 1993a).

In female rainbow trout, a direct positive effect of E_2 on GnRH-stimulated GTH-II release has also been demonstrated (Weil and Marcuzzi, 1990). The progestogen 17α-hydroxy, 20β-dihydroprogesterone had a positive effect at previtellogenic and preovulatory stages but a negative effect at the time of ovulation in trout. In African catfish, GnRH responsiveness increases with pubertal maturation but it is not known if responsiveness varies seasonally in the adult (Schulz et al., 1993; 1994). Nevertheless, 11-KT has inhibitory effects on GnRH-stimulated GTH-II release in the adult (Schulz et al., 1993; 1994), presumably reflecting a decrease in GnRH-receptor binding (Habibi et al., 1989).

These results indicate that the sex steroids modulate GnRH responsiveness, predominantly enhancing GTH-II release. Differences in the effects of the various steroids between species may reflect true species differences or differences in the experimental paradigms used.

(b) DA- The inhibitory actions of DA on GTH-II release varies seasonally. Dopaminergic inhibition of GTH-II release through the DA_2 receptor in greatest in sexually mature fish (Sokolowska et al., 1995; Sloley et al.,1991) and sex steroids enhance pituitary DA turnover and DA-inhibition of GTH-II release (Trudeau et al., 1993e). It is likely that the DA neuron is directly responsive to sex steroids since the estrogen receptor is found in tyrosine hydroxylase expressing neurons of trout brain (Kah et al., this symposium). Periovulatory changes in pituitary DA turnover in trout suggest that DA inhibition is maximal when serum GTH-II levels are decreased and E_2 levels are increased (Saligaut et al., 1992). DA contents in telencephalon and hypothalamus of the Venezuelan caribe colorado, Pygocentrus notatus, increases during seasonal gonadal development (Guerrero et al., 1990). The factors responsible for seasonal or steroidal changes in DA inhibitory tone are not totally elucidated. However, evidence in catfish suggests that modulation of synthesis and degradation of the principal catabolic enzyme monoamine oxidase (Sloley et al., 1992) is important (Senthilkumaran and Joy, 1995). Together, these results indicate that the major site for sex steroid negative feedback (-FB) in teleosts is through stimulation of the inhibitory DA neuron.

(c) GABA- The actions of GABA in stimulating GTH-II release are influenced by stage of the seasonal sexual cycle and gonadal steroids in goldfish (Kah et al. 1992; Trudeau et al. 1993b,d). GABA stimulates GTH-II release in the early stages of gonadal recrudescence but not in sexually regressed or mature animals. In sexually regressed goldfish, T enhances GABA action. Estradiol decreases GABA-stimulated GTH-II release in recrudescent animals (Trudeau et al. 1993b). Thus, in the early stages of seasonal gonadal development, at a time when serum T levels are increasing, goldfish gain the ability to respond to GABA, and T potentiates GABA action. As the gonad develops, E_2 levels also increase and appear to exert a -FB effect to reduce GABA responsiveness, possibly indicating an effect on GABA receptor numbers. Sex steroids also affect GABA synthesis. For example, T decreases and E_2 increases pituitary GABA synthesis rates in sexually regressed goldfish (Trudeau et al. 1993d). Thus, the GABA system is very sensitive to changes in steroidal milieu and may act to transduce complex endocrine signals to GnRH and/or DA neurons through variations in both neurotransmitter availability and action.

An interactive model to describe seasonal gonadal development in the female goldfish

In goldfish circulating GTH-II levels increase in the autumn to stimulate the onset of ovarian recrudescence and the production of sex steroids (Fig. 1). Both T and E_2 have +FB effects at the pituitary to enhance GnRH responsiveness and GTH-II release, and in a feed-forward manner stimulate gonadal development. Non-steroidal factors such as activin and inhibin stimulate GTH-II release and contribute to gonadal +FB (Ge et al., 1992; 1994). Notably, sex steroids do not affect basal GTH-II levels in vivo (Trudeau et al., 1991), indicating that there must be activation of a -FB system to regulate GTH-II release. Thus, concurrent activation of the inhibitory DA system keeps basal GTH-II release under tight negative control (Trudeau et al., 1993e). Given this apparent balance between +FB and -FB, how do basal GTH-II levels increase during gonadal development ? We suggest that the GABA system performs a modulatory function to allow GTH-II levels to increase during gonadal development, as follows: **(1)** GABA is the single most robust stimulator of GTH-II release in goldfish identified to date, reflecting its dual action on GnRH and DA; **(2)** GABA-stimulated GTH-II release has an unusual characteristic of a rapid (min) initial increase and a prolonged (hrs-days) secretory response (Trudeau et al., 1993d; Sloley et al. 1994), making the GABA system an ideal neuronal system to transduce environmental or physiological signals into long-term changes in GTH-II release and gonadal function (Fig.1). We also hypothesise that there is a three-way cross-talk between GnRH (S; stimulatory), DA (I; inhibitory) and GABA (M; modulatory) systems to regulate GTH-II release (Fig. 2). Preliminary evidence for an inhibitory effect of DA on GABA-stimulated GTH-II release exists (Trudeau et al. 1993d) and there is anatomical evidence for reciprocal connections between GnRH and DA neurons (Anglade et al. 1991). There may also be interconnections amongst neurons of a given type, as has been suggested for GnRH (Peter et al. 1991, Yu et al. 1990). It is also clear that the gonadal steroids modulate many aspects of the neuroendocrine axis. This model provides a background and framework for future studies on the regulation of seasonal reproduction.

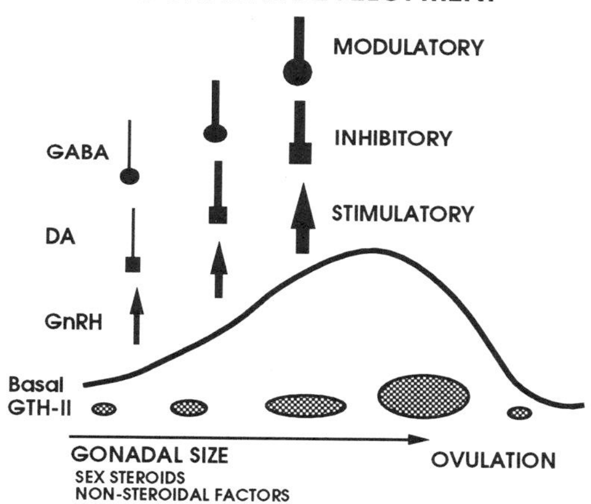

Fig. 1. An interactive model to describe seasonal gonadal development in female goldfish. Larger symbol sizes (arrow, GnRH; square, DA; circle, GABA) indicates increasing activity of the neuroendocrine systems. See text for full explanation.

References

Anglade, I. Tramu, G. & Kah, O. 1991. Proc. 4th Intl. Symp. Reprod. Physiol. Fish. A.P. Scott, J.P. Sumpter, D.E. Kime & Rolfe, M.S., (eds). FishSymp 91. p.60.

Bailhache, T, Salbert, G., Guillet, J.C., Saligaut, C., Breton, B. & Jego, P. 1989. Comp. Biochem. Physiol. 94A: 305-313.

Breton, B., Jalabert, B., Billard, R. & Weil, C. 1971. C.R. Acad. Sci. Paris, Sér. D. 273:2591-2594.

Chang, J.P. & Jobin, RM. 1994. In: Perspectives in Comparative Endocrinology. K.B. Davey, R.E. Peter and S.S. Tobe, (eds). National Research Council of Canada, Ottawa, pp. 41-51.

Chang, J.P., Van Goor, F.& Acharya, S. 1991. Neuroendocrinology 54:202-210.

Davies, B., Hannah, L.T., Randall, C.F., Bromage, N. & Williams, L.M. 1994. Gen. Comp. Endocrinol. 96: 19-26.

Dulka, J.G., Sloley, B.D., Stacey, N.E. & Peter, R.E. 1992. Gen. Comp. Endocrinol. 86: 496-505.

Flett, P.A., Van Der Kraak, G. & Leatherland, J.F. 1994. J. Expt. Zool. 268: 390-399.

Ge, W., Chang, J.P., Vaughan, J., Rivier, J. & Peter, R.E. 1992. Endocrinology. 131: 1922-1929.

Ge, W., Gallin, W.J., Strobeck, C. & Peter, R.E. 1993. Biochem. Biophys. Res. Commun. 193: 711-717.

Ge, W. & Peter, R.E. 1994. Zool. Sci. 11: 717-724.

Guerrero, H.Y., Caceres, G., Paiva, C.L. & Marcano, D. 1990. Gen. Comp. Endocrinol. 80: 257-263.

Habibi, H.R. 1991. Biol. Reprod. 44: 275-283.

Habibi, H.R., De Leeuw, R., Nahorniak, C.S., Goos, H.J.Th., & Peter, R.E. 1989. Fish Physiol. Biochem. 7: 109-118.

Himick, B.A., Golosinski, A.A., Jonsson, A.-C. & Peter, R.E. 1993. Gen. Comp. Endocrinol. 92: 88-103.

Holmqvist, B.I., Ostholm, T, Alm, P. & Ekstrom, P. 1992. In: Rhythms in Fishes, M. Ali, (ed). Plenum Press, New York, pp. 293-318.

Huxtable, R.J. 1989. Prog. Neurobiol. 32: 471-533.

Kah, O. 1993. Fish Physiol. Biochem. 2: 25-34.

Kah, O., Trudeau, V.L., Sloley, B.D., Chang, J.P. , Dubourg, P., Yu, K.L. & Peter, R.E. 1992. Neuroendocrinology 55: 396-404, 1992.

Khakoo, Z., Bhatia, A., Gedamu, L. & Habibi, H. 1994. Endocrinology 14: 838-847.

Khan, I.A. & Thomas, P. 1993. Gen. Comp. Endocrinol. 91: 167-180.

Khan, I.A. & Thomas, P. 1994. J. Expt. Zool. 269: 531-537.

Lyndon, A.R., Davidson. I. & Houlihan, D.F. 1993. Fish Physiol. Biochem. 10: 365-375.

Martinoli, M.G., Williams, L.M., Kah, O., Titchner, L.T. & Pelletier, G. 1991. Mol. Cell. Neurosci. 2: 78-85.

Médina, M., Repérant, J., Dufour, S., Ward, R., Le Belle, N. & Micelli, D. 1994. Anat. Embryol. 189:25-39.

Murthy, C.K., Nahorniak, C.S., Rivier, J.E. & Peter, R.E. 1993. Endocrinology 133: 1633-1644.

Murthy, C.K., Zheng, W., Trudeau, V.L., Nahorniak, C.S., Rivier, J.E. & Peter, R.E. 1994. Gen. Comp. Endocrinol. 96: 427-437.

Peng, C., Chang, J.P., Yu, K.L., Wong, A.O.L., Van Goor, F., Peter, R.E. & Rivier, J.E. 1993a. Endocrinology 132:1820-1829.

Peng, C., Huang,Y.P. & Peter, R.E. 1990. Neuroendocrinology 52:28-34.

Peng, C., Trudeau, V.L. & Peter, R.E. 1993b. J. Neuroendocrinology 5: 273-280.

Peter, R.E., Chang, J.P., Nahorniak, C.S., Omeljaniuk, R.J., Sokolowska, M., Shih, S.H. & Billard, R. 1986. Recent Prog. Horm. Res. 42:

513-548.
Peter, R.E., Trudeau, V.L., Sloley, B.D., Peng, C. & Nahorniak, C.S. 1991. Proc. 4th Intl. Symp. Reprod. Physiol. Fish. A.P. Scott, , J.P. Sumpter, D.E. Kime & M.S. Rolfe, (eds). FishSymp 91. pp. 30-34.
Peter, R.E., K.L. Yu, T.A. Marchant and P.M. Rosenblum. 1990. J. Exp. Zool. Supl. 4: 84-89.
Rosenblum, P.M. & Peter, R.E. 1989. Gen. Comp. Endocrinol. 73: 21-27.
Saligaut, C., Salbert, G., Bailhache, T., Bennani, S. & Jego, P. 1992. Gen. Comp. Endocrinol. 85: 261-268.
Senthilkumaran, B. & Joy, K.P. 1995. Gen. Comp. Endocrinol. 97: 1-12.
Sherwood, N.M., Parker, D.B., McRory, J.E. & Lesheid, D.W. 1994. Fish Physiol. Vol. XIII, 3-66.
Schulz, R.W., Paczoska-Eliasiewicz, Satijn, D.G.P.E. & Goos, H.J.Th. 1993. Fish Physiol. Biochem. 11: 107-115.
Schulz, R., van der Sanden, M.C.A., Bosma, P.T. & Goos, H.J.Th. 1994. J. Endocrinol. 140: 265-273.
Sloley, B.D., Kah, O., Trudeau, V.L., Dulka, J.G. & Peter, R.E. 1992. J. Neurochem . 58: 2254-2262.
Sloley, B.D.,Trudeau, V.L., D'Antoni, M. & Peter, R.E. 1994. Endocrine J. 2: 385-391.
Sloley, B.D.,Trudeau, V.L., Dulka, J.G. & Peter, R.E. 1991. Can. J. Physiol. Pharmacol. 69: 776-781.
Sloley, B.D.,Trudeau, V.L. & Peter, R.E. 1992. J. Exp. Zool. 263: 398-405.
Sokolowska, M., Peter, R.E., Nahorniak, C.S. & Chang, J.P. 1985. Gen. Comp. Endocrinol. 57:472-479.
Somoza, G.M. & Peter, R.E. 1991. Gen. Comp. Endocrinol. 82: 103-110.
Somoza, G.M., Yu, K.L. & Peter, R.E. 1988. Gen. Comp. Endocrinol. 72: 374-382.
Stacey, N.E., Cardwell, J.R., Liley, N.R., Scott, A.P & Sorenson, P.W. 1994. In: Perspectives in Comparative Endocrinology. K.B. Davey, R.E. Peter and S.S. Tobe, (eds). National Research Council of Canada, Ottawa, pp. 438-448.
Trudeau, V.L., Murthy, C.K., Habibi, H.R., B.D. Sloley & R.E. Peter. 1993a. Biol. Reprod. 48: 300-307.
Trudeau, V.L., Peter, R.E. & Sloley, B.D. 1991. Biol. Reprod. 44: 951-960.
Trudeau, V.L., Sloley, B.D. & Peter, R.E. 1993b. J. Neuroendocrinol. 5:129-136.
Trudeau, V.L., Sloley, B.D. & Peter, R.E. 1993c. Brain Res. 624: 29-34.
Trudeau, V.L., Sloley, B.D. & Peter, R.E. 1993d. Am. J. Physiol. 265: R348-R355.
Trudeau, V.L., Sloley, B.D., Wong, A.O.L. & Peter, R.E. 1993e. Gen. Comp. Endocrinol. 89: 39-50.
Weil, C. & Marcuzzi, O. 1990. Gen. Comp. Endocrinol. 79: 483-491.
Yu, K.L. & Peter, R.E. 1990. Neuroendocrinology 52: 276-283.
Yu, K.L., Rosenblum, P.M. & Peter, R.E. 1991. Gen. Comp. Endocrinol. 81: 256-267.

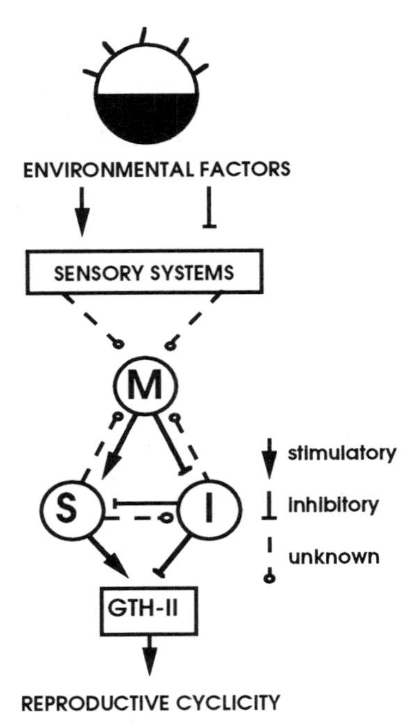

Fig. 2. Model indicating possible reciprocal connections between stimulatory (S), inhibitory (I) and modulatory (M) neurohormones regulating GTH-II release. Demonstrated and hypothetical connections are indicated by solid and dotted lines, respectively.

SOME INSIGHTS INTO SEX STEROID FEEDBACK MECHANISMS IN THE TROUT

B. Linard[1], I. Anglade[1,2], S. Bennani[1], G. Salbert[1], J.M. Navas[2], T. Bailhache[1], F. Pakdel[1], P. Jégo[1], Y. Valotaire[1], C. Saligaut[1], and O. Kah[1,2]

[1]Biologie Cellulaire et Reproduction, URA CNRS 256, Campus de Beaulieu, 35042 Rennes, [2]Neurocytochimie Fonctionnelle, URA CNRS 339, Avenue des Facultés, 33405 Talence, France

Summary

The possible relationships between circulating estradiol levels and the dopaminergic inhibition of GTH2 secretion were investigated in the rainbow trout. It was shown that blocking the synthesis of catecholamines caused increased GTH2 levels, and that this increase is correlated to plasma E2. This effect was most likely due to dopamine (DA) as injections of the DA antagonist pimozide also caused increased GTH2 levels. In addition, the pituitary dopaminergic turnover was reduced at the time of ovulation. Double immunohistochemical studies indicated that a group of preoptic tyrosine hydroxylase-positive neurons, most likely dopaminergic and hypophysiotropic, express the estradiol receptor, whereas GnRH neurons do not. The data are discussed with respect to the hormonal changes over the reproductive cycle in the rainbow trout.

Introduction

It is well documented that gonadal steroids exert positive and negative feedback effects by which the brain-pituitary complex is kept informed on the endocrine status of the animal. These actions are believed to play a crucial role in synchronizing the different steps of the reproductive process and the different organs implicated. However, the precise mechanisms involved, and notably the relationships between brain targets for E2 and the neuronal systems influencing GTH2 secretion, are still largely unknown. In this report, the dopaminergic inhibition of GTH2 secretion was investigated with respect to the estrogenic environment. We also present evidence that estradiol receptors are not co-localized with GnRH in the brain of the rainbow trout (Oncorhynchus mykiss), but are largely expressed in a population of preoptic dopaminergic neurons most likely inhibiting GTH2 release during vitellogenesis.

Results

Dopamine inhibits GTH2 secretion

Injections of a catecholamine synthesis inhibitor (α-methyl-p-tyrosine:240 mg/kg; ip.) in immature, immature E2-implanted (20 mg/kg; 17 days) rainbow trout, and maturing untreated females resulted in a significant stimulation of GTH2 plasma levels in E2-treated immature (600%) and maturing females (300%) with high levels of circulating E2. The same experiment performed in mature females at the end of vitellogenesis only resulted in a two-fold significant increase in GTH2 levels. However, in this latter group, there was a positive correlation between the efficiency of alpha-MPT to increase GTH2 and E2 levels of individual fish (Figure 1).

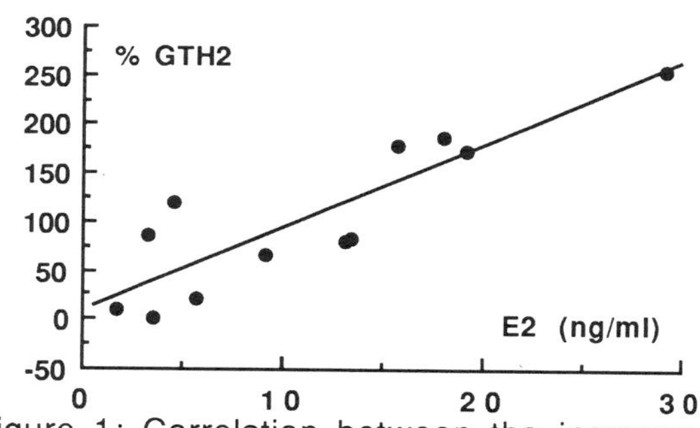

Figure 1: Correlation between the increase in GTH2 levels (expressed in percentage pre-treatment) following α-MPT injection (4 hrs) and E2 levels in individual mature females; R^2= 0.77.

Treatment of vitellogenic females by the DA antagonist pimozide also resulted in increased GTH2 levels (Figure 2).

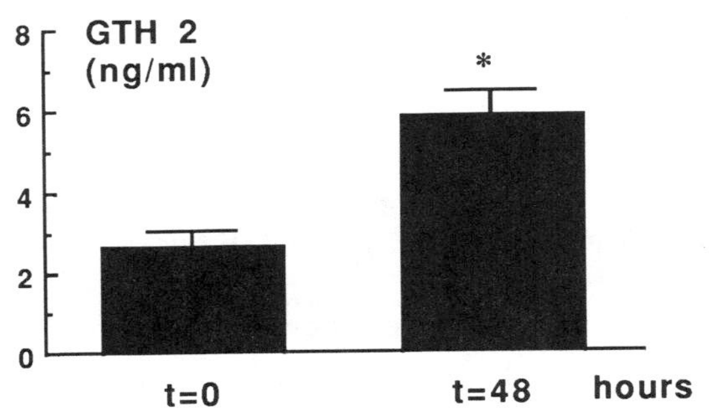

Figure 2: Effects of a single injection of pimozide (10 mg/kg; ip.) on plasma GTH2 levels in vitellogenic females (ANOVA: $p<0.01$; n = 8).

Dopamine turnover during the sexual cycle

The DA turnover was evaluated over the reproductice cycle by high pressure liquid chromatography (Saligaut et al., 1992). The pituitary DA and DA metabolites (3-methoxytyramine: 3MT + dihydroxyphenylacetic acid: DOPAC) contents were monitored and the ratio metabolites/dopamine was calculated to evaluate the DA turnover. The results showed a marked significant decrease of the pituitary DA turnover at the time of the germinal vesicle breakdown corresponding to low E2 levels.

Estrogen receptors and GnRH neurons

Antibodies directed against a fusion protein consisting of glutathione-S-transferase fused in frame with the hormone binding domain of the rainbow trout estrogen receptor were used to localize ER-immunoreactive (ER-ir) cells in the brain of the trout (Anglade et al., 1994; Pakdel et al., 1994). Immunoreactive cell nuclei were detected in the ventral telencephalon, the preoptic region and the mediobasal hypothalamus. Double staining studies carried out on 650 GnRH neurons of vitellogenic females and maturing males failed to indicate that ER are expressed in GnRH neurons in the rainbow trout.

Estrogen receptors are expressed in preoptic catecholaminergic neurons

Double staining studies using monoclonal antibodies to TH and antibodies to the ER indicated that an overwhelming majority of ER-ir cells in the ventral wall of the preoptic recess, referred to as the nucleus preopticus pars anteroventralis (NPOav), were also positive for TH in vitellogenic and mature females (Figure 3).

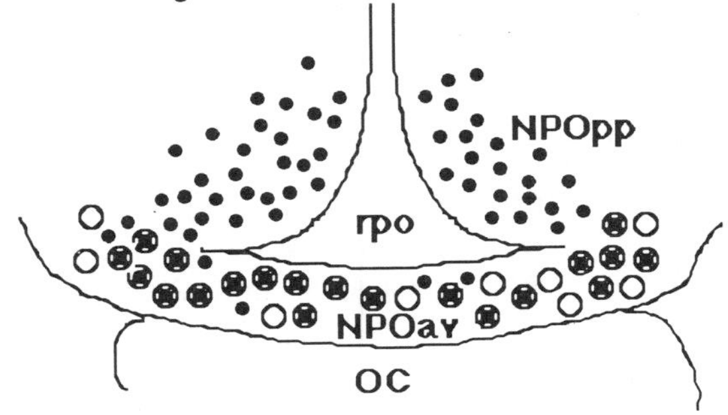

Figure 3: Comparative distribution of TH-ir (open circles), ER-ir (black dots) and TH-ER-ir cells around the preoptic recess (rpo). Colocalization of the two stainings was restricted to cells situated in the ventral wall of the recess.

Presence of dopaminergic hypophysiotropic neurons in the preoptic region

Immunohistochemistry using monoclonal antibodies to dopamine showed that the NPOav contained a population of DA-ir neurons with a distribution overlapping exactly that of the TH-ER-ir cells described above. Morevover, a DA-ir pathway could be traced from the NPOav to the neural lobe of the pituitary. In addition, implantation of a crystal of DiI in the pituitary according to Anglade et al. (1993), consistently resulted in the retrograde labelling of neurons having a size and distribution similar to those of the above-described DA-ir neurons.

Discussion

The present results indicate that, in agreement with the situation in other teleosts, dopamine inhibits GTH2 secretion in the rainbow trout (Breton et al., 1993). However, the fact that a catecholamine synthesis blocker stimulates GTH2 release in immature E2-implanted fish and vitellogenic females, with high levels of circulating E2, and the correlation observed between α-MPT efficiency and E2 levels in mature females would indicate that this dopaminergic inhibition is effective when E2 levels are high, i.e. during vitellogenesis (Fostier et al., 1983; Scott et al., 1983). On the contrary, the periovulatory period is characterized by a drop of E2 levels leading to removal of the dopaminergic inhibition and allowing the preovulatory surge of GTH2 to occur, as strongly suggested by our data showing clear changes in the pituitary dopaminergic turnover at the time of ovulation (Saligaut et al., 1992).

The hypothesis of relationships between E2 levels and the intensity of the dopaminergic inhibition is strongly reinforced by our neuroanatomical studies showing that the anterior ventral preoptic region, known in goldfish for being the source of the dopaminergic inhibition (Peter and Paulencu, 1980; Kah et al., 1987), contains a population of hypophysiotropic dopaminergic neurons most likely direct targets for E2. Indeed, the NPOav contains a group of TH-ir neurons which express estradiol receptors and exhibit a distribution overlapping perfectly that of dopamine-ir and hypophysiotrophic neurons. On the contrary, double staining techniques failed to demonstrate that GnRH neurons in the ventral telencephalon, the preoptic region and the mediobasal hypothalamus expressed ER receptors, indicating that the well documented of estradiol on GnRH synthesis are mediated by other unidentified neurons.

At this stage, the precise molecular targets of E2 in TH-expressing neurons of the preoptic regions are unclear. It is possible that the gene of TH, the rate limiting enzyme of catecholamine synthesis is E2-dependent, however we cannot exclude the possibility that E2 acts on the expression of enzymes of the DA catabolism, or some regulating receptors.

In conclusion, these results may explain in part the hormone changes observed over the periovulatory period in the rainbow trout (Fostier et al., 1983). During vitellogenesis, E2 levels remain elevated to stimulate ER as well as vitellogenin synthesis in the liver, and also to increase GTH2 and GnRH synthesis in the pituitary and the brain respectively. At this stage, the dopaminergic inhibition would prevent the secretion of massive amounts of GTH2. At the end of vitellogenesis, E2 levels drop, triggering removal of the dopaminergic inhibition and allowing the preovulatory surge of GTH2 necessary for induction of final maturation under the influence of 17α,20-β-dihydroxyprogesterone (Fostier et al., 1983). Whether this hypothesis is valid in other teleost species remains to be investigated.

References

Anglade I, Pakdel F, Bailhache T, Petit F, Salbert G, Jégo P, Valotaire Y, Kah O (1994) Distribution of estrogen receptor immunoreac-tive cells in the brain of the rainbow trout (*Oncorhynchus mykiss*). J. Neuroendocrinol. 6: 573-583.

Anglade I, Zandbergen AM, Kah O (1993) Origin of the pituitary innervation in the goldfish. Cell Tissue Res. 273: 345-355.

Breton B, Mikolajczyk T, Popek W (1993) The neuroendocrine control of gonadotropin (GTH2) secretion in teleost fish. In: Aquaculture: fundamental and applied aspects Lalhou B., Vitello P. (eds.) American Geophysical Union, pp. 199-215.

Fostier A, Jalabert B, Billard R, Breton B, Zohar Y (1983) The gonadal steroids. In: "Fish Physiology" vol. IXA; Hoar WS and Randall DJ eds., Academic Press, New York, pp. 277-371.

Kah O, Dulka JG, Dubourg P, Thibault J, Peter RE (1987) Neuroanatomical substrate for the inhibition of gonadotrophin secretion in goldfish: existence of a dopaminergic preoptico-hypophyseal pathway. Neuroendonology 45: 451-458.

Pakdel F, Petit F, Anglade I, Kah O, Delaunay F, Bailhache T, Valotaire Y (1994) Overexpression of rainbow trout estrogen receptor domains in *E. Coli*: characterization and utilization in the production of antibodies for immunoblotting and immunocytochemistry. Mol. Cell Endocrinol. 104: 81-93.

Peter RE, Paulencu CR (1980) Involvement of the preoptic region in gonadotropin inhibition in goldfidsh, *Carassius auratus*. Neuroendocrinology 31: 133-141.

Saligaut C, Salbert G, Bailhache T, Bennani S, Jégo P (1992) Serotonin and dopamine turnover in the female rainbow trout (*Oncorhynchus mykiss*) brain and pituitary: changes during the annual reproductive cycle. Gen. Comp. Endocrinol. 85: 261-268.

THREE FORMS OF GONADOTROPIN-RELEASING HORMONE IN GILTHEAD SEABREAM AND STRIPED BASS: PHYSIOLOGICAL AND MOLECULAR STUDIES.

Y. Gothif, A. Elizur[1], and Y. Zohar

Center of Marine Biotechnology, University of Maryland, 701 East Pratt St., Baltimore, MD 21202.
[1]National Center for Mariculture, IOLR, P.O.Box 1212 Eilat, Israel.

Summary

Seabream and striped bass, two perciform species, possess three GnRHs: salmon GnRH (sGnRH), chicken GnRH-II (cGnRH-II) and seabream GnRH (sbGnRH). When exogenously administered, all three forms stimulate GtH-II secretion in seabream and striped bass. The full length cDNAs encoding the three GnRHs were cloned from seabream brains. Using *in situ* hybridization, only sbGnRH producing cells were found in the preoptic area of the seabream brain, while sGnRH and cGnRH-II were located in the terminal nerve and midbrain tegmentum, respectively. In addition, sbGnRH is the dominant form in the pituitary of sexually mature seabream and striped bass. Considered together, our studies indicate that sbGnRH is the most relevant GnRH form for the regulation of GtH-II release in sexually mature seabream and striped bass.

Introduction

Nine forms of gonadotropin-releasing hormone (GnRH) have been isolated from vertebrate brains and characterized (for review see Sherwood et al. 1994). These highly conserved decapeptides are traditionally named after the species in which they are first discovered. It has been widely accepted that teleost fish possess two forms of GnRH, the ubiquitous cGnRH-II and a species specific form (in most cases sGnRH). Recently, the presence of three forms of GnRH in the brain of the gilthead seabream (*Sparus aurata*) was demonstrated (Powell et al. 1994): cGnRH-II, sGnRH and a novel form, referred to as seabream GnRH (sbGnRH) (Fig. 1). In order to further understand the role of the three GnRHs in the control of reproduction and to enable future studies on their synthesis and its regulation, we have (1) evaluated their *in vivo* bioactivity; (2) isolated and characterized the nucleotide sequences encoding their precursors; and (3) localized the cells producing the three forms in the brain by means of *in situ* detection of messenger RNA.

sb	pGlu	His	Trp	Ser	Tyr	Gly	Leu	Ser	Pro	GlyNH_2
s	pGlu	His	Trp	Ser	Tyr	Gly	Trp	Leu	Pro	GlyNH_2
c-II	pGlu	His	Trp	Ser	His	Gly	Trp	Tyr	Pro	GlyNH_2

Figure 1. Three forms of GnRH in seabream and striped bass brains. sb, sbGnRH; s, sGnRH; c-II, cGnRH-II

Bioactivity in Seabream and Striped Bass.

The hypophysiotropic potency of different GnRH forms has been traditionally examined as an indicator of their possible function. We tested and compared the *in vivo* bioactivities of the three native forms of GnRH in seabream and striped bass females during the spawning season.

Females were injected with various doses of sGnRH, cGnRH-II or sbGnRH, bled before and at variouse intervals after injection, and plasma levels of the maturational gonadotropin (GtH-II) were measured. All three forms of GnRH were found to stimulate GtH-II secretion in reproductively mature seabream (Zohar et al. 1995) and striped bass females (Zohar, Stubblefield, Hassin, Gothilf and Mananos, unpublished). However, differences in the GtH-II release potencies of the three forms were observed: cGnRH-II was the most active and sbGnRH was the least active of the three forms in both seabream (Zohar et al. 1995) and striped bass (data not shown).

Isolation and characterization of GnRH cDNAs

In order to enable further studies on the regulation of GnRH synthesis we isolated and characterized the cDNAs encoding the three forms of GnRH from a seabream brain cDNA library. For this purpose, cDNA fragments were first amplified using PCR, and identified. The PCR products were then used as probes to screen seabream brain cDNA libraries and positive clones were characterized. This is the first report of cloning the cDNAs for three forms of GnRH from one species.

sGnRH cDNA - A fragment of sGnRH cDNA was PCR amplified from striped bass brain mRNA using degenerate primers (kindly donated by Drs. M. Grober and D. Mayers, Cornell University) designed according to conserved regions among published sGnRH cDNA sequences. The PCR product was cloned, identified and used as a probe to isolate the full length cDNA from the seabream brain cDNA library. This cDNA encodes a 90 amino acid primary translation product (Fig. 2) which is composed of the three expected major regions: (1) a 23 amino acid leader sequence, (2) the biologically active sGnRH followed by a processing site (Gly-Lys-Arg), and (3) a 54 amino acid GnRH associated peptide (GAP). This precursor has around 90% amino acid identity with sGnRH precursors of other perciforms and 75% identity with sGnRH precursors of salmonids but only low (20-33%) identity, mainly at the GnRH region, with precursors for other forms of GnRH.

cGnRH-II cDNA - A short fragment, 97bp long, of cGnRH-II cDNA was PCR amplified from seabream brain mRNA using degenerate primers which were designed according to conserved regions among cGnRH-II cDNA sequences of catfish (Bogerd et al. 1994) and African cichlid (White et al. 1994). The PCR product was cloned and identified. Using this PCR product as a probe, we screened a seabream brain cDNA library, isolated, and characterized the full length cDNA. This cDNA encodes a 85 amino acid precursor peptide (Fig. 2) composed of a 23 amino acid leader sequence, the biologically active cGnRH-II followed by a processing site, and a 49 amino acid GAP. This precursor has 66% and 94% identity with cGnRH-II precursors of catfish (Bogerd et al. 1994) and cichlid (White et al. 1994), respectively. With precursors for other forms of GnRH, amino acid identity is only 30-

40% and is mainly at the GnRH and cleavage site regions.

sbGnRH cDNA - The sbGnRH cDNA was isolated and sequenced (Gothilf et al. 1995). This cDNA encodes a 95 amino acid precursor peptide (Fig. 2) which is composed of a 25 amino acid signal peptide, the biologically active sbGnRH followed by a processing site (Gly-Lys-Arg), and a 57 amino acid GAP (Gothilf et al. 1995).

The GnRH precursors characterized in the present study follow the general organization of all other GnRH precursors: the primary translation product contains a signal peptide, the GnRH decapeptide, the conserved cleavage site and the GAP region. The three GnRH precursors isolated in this study share similarity only at the GnRH and cleavage site region while the rest of the precursor is highly diverse. Nevertheless, precursors for the same form of GnRH obtained from diverse species are highly conserved throughout the entire protein. This suggests that the genes encoding for the three GnRHs were formed early in the evolution of the GnRH peptide family.

sGnRH

MEASSRVTVQVLLLALVVQVTLS**QHWSY GWLPGG**KRSVGELEATIRMMGTGGVVSLP EEASAQTQERLRPYNVIKDDSSPFDRKKR FPNK

cGnRH-II

MCVSRLVLLLGLLLCVGAQLSNG**QHWSH GWYPG**GKRELDSFGTSEISEEIKLCEAGE CSYLTPQRRSVLRNILLDALARELQKRK

sbGnRH

MAPQTSNLWILLLLVVVMMMSQGCC**QHW SYGLSP**GGKRDLDSLSDTLGNIIERFPHV DSPCSVLGCVEEPHVPRMYRMKGFIGSER DIGHRMYKK

Figure 2. Deduced amino acids of the GnRH precursors in seabream (One letter code). The GnRH decapeptides are underlined

Distribution.

In general, the function of a neuropeptide is largely related to its anatomical arrangements within the brain. The anatomical distribution of the different forms of GnRH in the brain and their presence in the axon terminals within the pituitary gland may be used as indicators of their relevance in the regulation of gonadotropin release.

In order to localize the cells producing the different GnRHs in the seabream brain, the distinct mRNAs encoding for the three GnRH forms were specifically detected by *in situ* hybridization on brain sections. Single strand DNA probes were used for detection. For each GnRH mRNA, two probes were synthesized: one directed to the GAP coding region and the other to the 3' untranslated region. These regions are not conserved among the three different GnRH forms and, therefore, enabled us to specifically localize each of the forms without the complication of cross reactivity. The probes were 3' end-labeled with a tail of about ten ^{35}S-dATP. The procedures for *in situ* detection were according to Selmanoff et al (1991). Brain locations were considered positive for a certain GnRH only if cells hybridized with both of the two probes which are directed to different regions of a specific GnRH-mRNA.

Cells positive for sGnRH mRNA were detected only in the anterior ventral telencephalon (Fig. 3). The positive cells are large and are found in clusters of 3-10 cells in each section. They appear to be associated with the terminal nerve. These sGnRH-producing cells did not hybridize with the probes directed against the mRNA of other GnRH forms.

Cells containing cGnRH-II mRNA were found only in the midbrain tegmentum, anterior to the valvula cerebellum. (Fig. 3). These cells were large and confined to a defined area but were not as clustered as the sGnRH cells. In each section, 2-7 cGnRH-II cells were detected in the midbrain and no reaction with probes to the other forms of GnRH could be detected in this region. The existence of cGnRH-II cells in this area of the brain is in agreement with studies in many fish species, amphibia and birds.

Messenger RNA for sbGnRH was detected only in cells in the preoptic area (Fig. 3). These cells seem to be smaller in size than the GnRH cells in the anterior telencefalon and midbrain, and appear to be much more scattered. Probes for other forms of GnRH did not hybridize to any cells in this area.

The respective contents of the three GnRHs in the pituitaries of seabream and striped bass was analyzed using combined HPLC/RIA. High levels of the sbGnRH were detected in pituitaries of sexually mature seabream (Powell et al. 1994) and striped bass (Zohar, Gothilf, Powell and Sherwood, unpublished) as compared to sGnRH, which was 500-1000 times lower. cGnRH-II was undetectable.

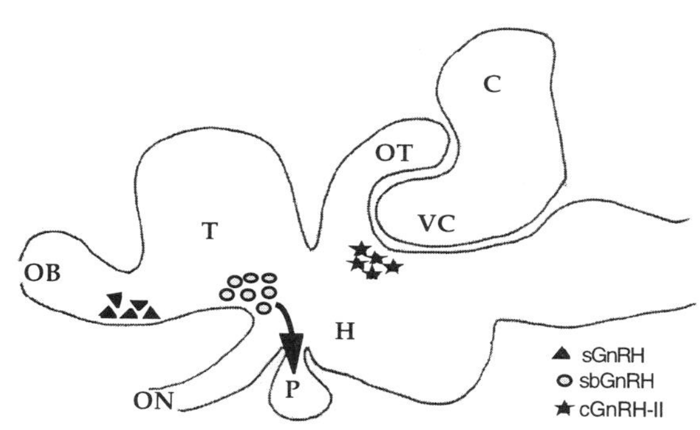

Figure 3. Distribution of GnRH-producing cells in the seabream brain. C, cerebellum; H, hypothalamus; OB, olfactory bulb; ON, optic nerve; OT, optic tectum; P, pituitary; T, telencephalon; VC, valvula cerebellum.

Discussion

This is the first report of cloning and localizing three GnRH cDNAs from one species. We have shown that striped bass and seabream express three distinct forms of GnRH in the brain. When injected, all three forms stimulate gonadotropin secretion in sexually mature females of both species. Nevertheless, the three forms have different gonadotropin releasing potencies, cGnRH-II being the most potent, followed by sGnRH and sbGnRH. The higher bioactivity of cGnRH-II is not unique to striped bass and seabream. In the catfish (Schulz et al. 1993), goldfish (Chang et al. 1991), and chicken (Wilson et al. 1989), cGnRH-II is more potent than the other native forms. The differences in the bioactivities of the three native GnRHs may be due to differences in their affinity to the pituitary GnRH receptors or in degradation and clearance rates (Zohar et al. 1990). However, these differences are rather pharmacological and do not reflect the relevance of the GnRHs in regulating gonadotropin secretion under normal physiological conditions.

Indeed, we believe that sbGnRH is the most relevant form for the control of gonadotropin release. This is based on the distribution of the three peptides in the brain and pituitary. The sbGnRH form is produced in the preoptic area (Fig. 3), a region known to control gonadotropin secretion, and is found at high levels in the pituitary of sexually mature seabream and striped bass. Of the other two forms, cGnRH-II is synthesized only in the midbrain tegmentum and is not found in the pituitary, and sGnRH is synthesized in the terminal nerve in the anterior telencephalon and is found in very low levels in pituitaries of seabream and striped bass, compared with sbGnRH.

The overall organization of the GnRH system in the seabream is similar to that of other fish species, with cells in the telencephalon, preoptic area and midbrain. However, while in all fish examined, one form of GnRH (usually sGnRH) is expressed in both the telencephalon and the preoptic area, in seabream the telencephalic and preoptic cells express two different forms of GnRH (sGnRH and sbGnRH). Whether these two populations of GnRH cells are developmentally related remains to be answered.

It is widely accepted that duplication of the GnRH gene occurred early in the evolution of vertebrates. One of these ancestral forms became the highly conserved cGnRH-II and the other gave rise to mammalina GnRH (mGnRH), sGnRH, Catfish GnRH (cfGnRH) and chicken GnRH-I (cGnRH-I) by base mutations (Sherwood et al. 1994; Millar and King, 1994). Since mGnRH is found in primitive fish, it is thought that mGnRH gave rise to sGnRH, cfGnRH and cGnRH-I by point mutations (Sherwood et al. 1994). The coexistence of three forms of GnRH in a single species is strong evidence that two duplications occurred. Taking into consideration the structure, function and the anatomical distribution of sbGnRH, we propose that the sbGnRH was derived from the mGnRH (found in primitive fish) by point mutations and is evolutionary closer to the mammalian GnRH and cGnRH-I than sGnRH and cGnRH-II are. Further studies on the phylogenetic distribution of sbGnRH are required to support our hypothesis.

Acknowledgments

The financial suport to Y.G. by the Rhoda and Jordan Baruch Fellowship is greatly appreciated. We are grateful to Drs. Magliulo-Cepriano, Schreibman, Selmanoff and Sagrillo for their help and advice. This study was supported by NOAA grant NA90AA-D-SG063 from the Maryland Seagrant College program.

Reference

Bogerd J., Zandbergen T., Andersson E. and H.J.Th. Goos, 1994. Isolation, characterization and expression of cDNA encoding the catfish-type and chicken-II-type gonadotropin-releasing hormone precursors in the African catfish (*Clarias gariepinus*). Europ. J. Biochem. 222:541-549.

Chang, J.P., Wildman, B., and F. Van Goor, 1991. Lack of involvement of arachidonic acid metabolism in chicken gonadotropin-releasing hormone-II (cGnRH-II) stimulation of gonadotropin secretion in dispersed pituitary cells of goldfish, *Carassius auratus*. Identification of a major difference in salmon GnRH and cGnRH-II mechanisms of action. Mol. Cell. Endocrinol. 79:75-83.

Gothilf, Y., Chow. M.M., Elizur, A., Chen, T.T., and Y. Zohar, 1995. Molecular cloning and characterization of a novel gonadotropin-releasing hormone from the gilthead seabream (*Sparus aurata*). Mol. Marine Biol. Biotech. 4(1):27-35.

Millar, R.P. and J.A. King,, 1994. Plasticity and conservation in gonadotropin-releasing hormone structure and function. In: Perspectives in comperative endocrinology, Davey, K.G., Peter, R.E., and Tobes, S.S. (editors). National research council of Canada.

Powell, J.F.F., Zohar, Y., Elizur, A., Park, C., Fischer, W.H., Craig, A.G., Rivier, J.E., Lovejoy, D.A., and N.M. Sherwood, 1994. Three forms ofgonadotropin-releasing hormone characterized from brain of one species. Proc. Natl. Acad. Sci. USA. 91:12081-85.

Schultz R.W., Bosma, P.T, Zandbergen M. A., Van Der Sanden M. C. A., Van Dijk W., Peute J., Bogard J. and H.J.TH. Goos, 1993. Two gonadotropin-releasing hormones in the African catfish, *Clarias gariepinus*: Location, pituitary receptor binding and gonadotropin release activity. Endocrinology 133:1569-77.

Selmanoff M., Shu C., Petersen S.L., Barraclough C.A. and R.T. Zoeller, 1991. Single cell levels of hypothalamic messenger ribonucleic acid encoding luteinizing hormone-releasing hormone in intact, castrated, and hyperprolactinemic male rats. Endocrinology 128: 459-466.

Sherwood, N.M., Parker, D.B., McRory, J.E., and D.W. Lescheid, 1994. Molecular evolution of GHRH and GnRH. In: "Molecular Endocrinology of Fish". Sherwood, N.M. and Hew, C.L. (eds.). Fish Physiology Vol. XIII. Farrel, A.P. and Randall D.J. (series eds.) New York, Academic Press.

White S.A., Bond C.T., Francis R.C., Kasten T.L., Fernald R.D. and J.P. Adelman, 1994. A second gene for gonadotropin-releasing hormone: cDNA and expression pattern in the brain. Proc. Natl. Acad. Sci. USA 90: 1423-27.

Wilson, S.C., Cuningham, F.J., Chairil, R.A., and R.T. Gladwell, 1989. Maturational changes in the LH response of domestic fowl to synthetic chicken LHRH-I and II. J. Endocrinol. 123:311-318.

Zohar, Y., Goren, A., Fridkin, M., Elhanati, E., and Y. Koch, 1990. Degradation of gonadotropin releasing hormones in the gilthead seabream, *Sparus aurata*: II. Cleavage of native salmon GnRH, mammalian LHRH and their analogs in the pituitary, kidney, and liver. Gen. Comp. Endocrinol. 79:306-319.

Zohar, Y., Elizur, A., Sherwood, N.M., Powell, J.F.F., Rivier, J.E., and N. Zmora, 1995. Gonadotropin-releasing activities of three native forms of gonadotropin releasing hormones present in the brain of gilthead seabream, *Sparus aurata*.. Gen. Comp. Endocrinol. 97:289-299.

GnRH RECEPTORS ARE RESTRICTED TO GONADOTROPES IN MALE AFRICAN CATFISH

P.T. Bosma, W. van Dijk, S.F. van Haren, S.M. Kolk, O. Lescroart[1], R.W. Schulz, M. Terlou[2], H.J.Th. Goos.

Research Group for Comparative Endocrinology, Dept. of Experimental Zoology, University of Utrecht, Padualaan 8, NL-3584 CH Utrecht, The Netherlands. [1]Zoological Institute, Catholic University of Leuven, Leuven, Belgium. [2]Dept. for Image Processing and Design, University of Utrecht, Utrecht, The Netherlands.

Summary

The possible involvement of the two forms of gonadotropin-releasing hormone (GnRH) present in the African catfish (*Clarias gariepinus*), chicken GnRH-II ([His^5,Trp^7,Tyr^8]GnRH, cGnRH-II) and catfish GnRH ([His^5,Asn^8]GnRH, cfGnRH), on the release of growth hormone (GH) was studied. Moreover, GnRH receptors were localized on enzymatically dispersed pituitary cells using a combination of autoradiography and immunohistochemistry. GnRH receptors were confined to cells that contained gonadotropin II (GTH II) ß-subunit immunoreactive material. *In vivo* experiments showed that cfGnRH and cGnRH-II, at doses that strongly elevated circulating GTH II levels, did not effect plasma GH levels. However, the dopamine agonist apomorphine led to significant increase in plasma GH level. These results indicate that the two native forms of GnRH in the African catfish are not directly involved in the regulation of the release of GH.

Introduction

In teleost fish, at least two forms of GnRH are present in the brain (Sherwood *et al.* 1994). In brain and pituitary extracts of the African catfish cGnRH-II and cfGnRH have been identified (Bogerd *et al.* 1992; Schulz *et al.* 1993). A well established function of GnRH is the stimulation of pituitary GTH release, of which only a single form, the LH-like GTH II, has been found in the African catfish (Koide *et al.* 1992; Schulz *et al.* this volume). Both endogenous GnRHs stimulate GTH II release, but cGnRH-II is approximately 100-fold more active than cfGnRH. The difference in GTH II release activity correlates with the GnRHs' relative receptor affinities, but may be compensated for by the large excess of cfGnRH over cGnRH-II in the catfish pituitary (Schulz *et al.* 1993).

In goldfish (Marchant *et al.* 1989), rainbow trout (LeGac *et al.* 1993), common carp (Lin *et al.* 1993), and tilapia hybrids (Melamed *et al.* 1995), GnRHs also stimulate the release of GH. Accordingly, GnRH receptors were found on goldfish gonadotropes and somatotropes (Cook *et al.* 1991). The present study was conducted to investigate the possible involvement of cfGnRH and cGnRH-II on the release of GH in the African catfish.

Methods and Results

GnRH receptor localization

Iodinated salmon GnRH analogue ([D-Arg^6,Trp^7,Leu^8,Pro^9-NEt]GnRH, sGnRHa; Habibi *et al.* 1987) was used to detect autoradiographically GnRH receptors on dispersed pituitary cells (De Leeuw *et al.* 1984) in a primary culture. Gonadotropes and somatotropes were identified immunohistochemically, using an antiserum against the ß-subunit of catfish GTH II or against catfish GH, respectively. Cells were stained using the peroxidase-antiperoxidase method, and subsequently covered with a photographic emulsion. Autoradiograms were developped after two weeks of exposure.

Following autoradiography, silver grains were exclusively found to be associated with gonadotropes, indicating the presence of sGnRHa binding sites on gonadotropes (Fig. 1A). No silver grains were associated with somatotropes (Fig. 1B) or with other pituitary cells.

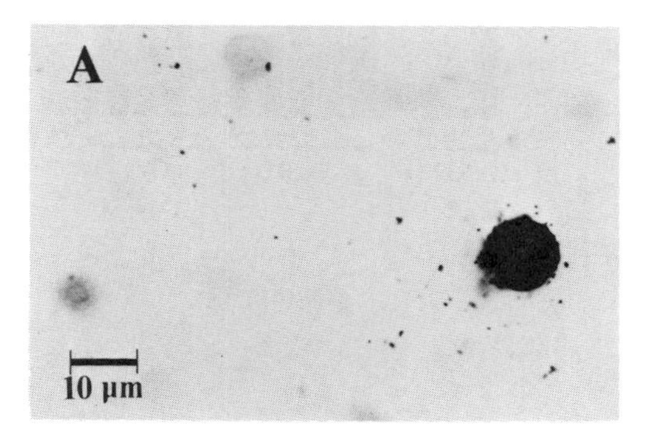

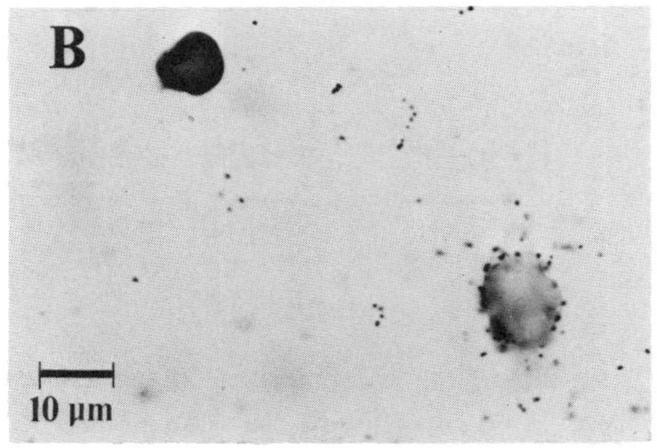

Fig. 1. Autoradiograms of cultured pituitary cells, immunostained for GTH IIß (A), or GH (B).

The silver grain distribution on cultured pituitary cells was quantified using a computerized image analysis system. As silver grains above cells mostly fuse to clusters, the area occupied by these clusters was measured. The labeling of each cell was determined as the ratio between the silver grain area and the cell area. Displacement studies were carried out with radioinert sGnRHa, cfGnRH and cGnRH-II. For each GnRH concentration, 20 immunostained cells were analyzed.

The addition of an excess of radioinert sGnRHa (10 µM) caused an almost complete displacement of ^{125}I-sGnRHa on gonadotropes (Fig. 2). Radioinert cfGnRH and cGnRH-II displaced the label on gonadotropes in a dose-dependent manner. To achieve a half-maximal displacement, 100-fold higher concentrations of cfGnRH then of cGnRH-II were needed (approx. 1 µM cfGnRH *vs.* 10 nM cGnRH-II; Fig. 2). The low binding of ^{125}I-sGnRHa on somatotropes could not be displaced by the unlabeled peptides, and therefore has to be considered as non-specific binding (Fig. 2).

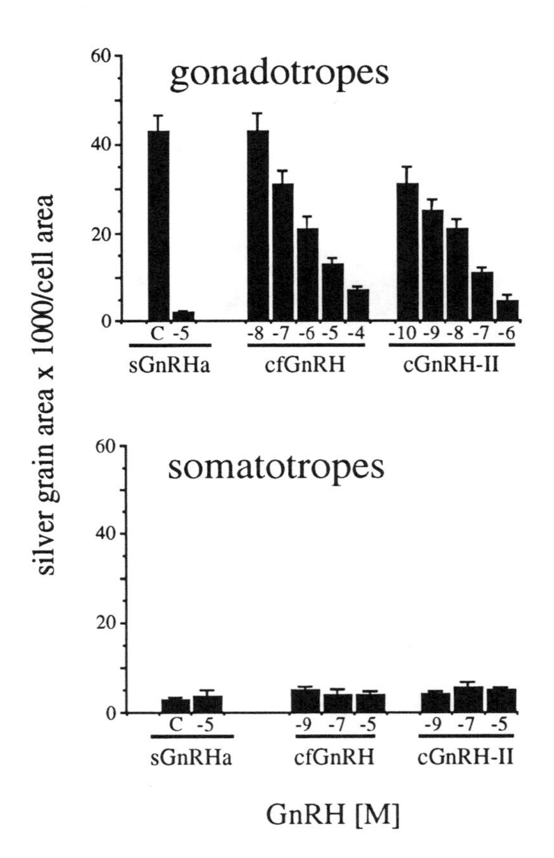

Fig. 2. Results of the computerized image analysis of the silver grain distribution on gonadotropes and somatotropes. C: incubation with ^{125}I-sGnRHa in the absence of radioinert GnRH.

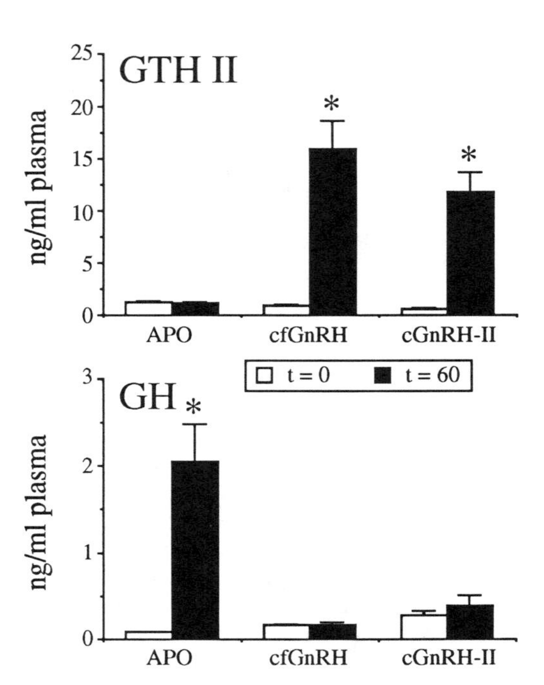

Fig. 3. Plasma levels of GTH II and of GH (ng/ml ± s.e.m.; n = 8-10) before and 1 hour after treatment with apomorphine (APO), cfGnRH, or cGnRH-II. Differences between pre- and post-injection hormone levels were statistically tested by a two-tailed Student's t-test (*$p < 0.0005$).

Plasma levels of GH

Groups of 8-10 mature males were injected ip with apomorphine, cfGnRH or cGnRH-II (10 mg, 250 µg, and 5 µg/kg body weight, respectively). GTH II and GH plasma levels were measured after 1 hour by RIA (Goos *et al.* 1986; Lescroart *et al.* 1994). Apomorphine had no effect on the GTH II plasma levels, whereas both cfGnRH and cGnRH-II significantly elevated GTH II plasma levels (Fig. 3). On the contrary, apomorphine effectively augmented the plasma GH levels, while the two GnRHs were not effective (Fig. 3). Also at other times after injection (0.5-24 h) these GnRHs did not change GH plasma levels (results not shown).

Discussion

After incubations of catfish pituitary cells with ^{125}I-sGnRHa, a significant amount of silver grains was associated with the gonadotropes, while other cell types remained unlabeled. The labeling could be inhibited by an excess of radioinert sGnRHa. The two native GnRHs of the African catfish competed with ^{125}I-sGnRHa for the binding sites on gonadotropes in a dose-dependent fashion. The effective concentrations of cGnRH-II (approx. 10 nM) and of cfGnRH (approx. 1 µM) inducing half-maximal displacement of ^{125}I-sGnRHa are in line with the respective receptor binding affinity, and were found to be in accord with their potency to induce the release of GTH II or cytosolic free calcium ($[Ca^{++}]_i$) increase in the catfish gonadotropes *in vitro* (Schulz *et al.* 1993; Rebers *et al.* this volume). We therefore conclude that the present autoradiographic technique specifically detects GnRH receptors on gonadotropes. These data furthermore indicate that both cGnRH-II and cfGnRH interact with a single type of GnRH receptor on catfish gonadotropes.

Silver grains were never associated with pituitary cells identified as somatotropes, suggesting the absence of sGnRHa binding sites. As we have not studied the binding of radiolabeled cGnRH-II or cfGnRH, it cannot be ruled out that catfish somatotropes possess GnRH binding sites that show affinity to one of the two endogenous catfish GnRHs, but that do not bind sGnRHa. However, cGnRH-II and cfGnRH failed to evoke increases in the plasma levels of GH or $[Ca^{2+}]_i$ (Rebers *et al.*; this volume). We therefore conclude that, in the African catfish, GnRHs are not directly involved in the regulation of GH secretion. These data are not in agreement with results obtained from other fish species (see Introduction).

In mammals, growth hormone releasing factor (GRF) is the primary stimulator of GH release, whereas the GH release in goldfish and carp is stimulated by a group of interacting factors, including GRF, GnRH and dopamine (reviewed by Peter and Marchant 1995). At present, a GRF-like peptide has been characterized in carp only (Vaughan *et al.* 1992), but there is evidence for its presence in other teleosts, including a catfish (McRory *et al.* 1995). Dopamine appears to be involved in the regulation of GH secretion in the African catfish, as the dopamine agonist apomorphine increased GH plasma levels. Thus, the African catfish appears to share the dopaminergic, and possibly also the GRF-mediated, regulation of GH secretion with other teleost species.

In conclusion, GnRH receptors were shown to be confined to gonadotropes in the African catfish. Together with the inability of GnRHs to increase GH plasma levels, this supports the notion that, in contrast to other teleosts, GnRHs are not functioning as GH-releasing factors in the African catfish.

References

Bogerd, J., K.W. Li, C. Janssen-Dommerholt & H. Goos 1992. Two gonadotropin-releasing hormones from African catfish (*Clarias gariepinus*). Biochem. Biophys. Res. Commun. 187:127-134.

Cook, H., J.W. Berkenbosch, M.J. Fernhout, K.L. Yu, R.E. Peter, J.P. Chang & J.E. Rivier 1991. Demonstration of gonadotropin releasing-hormone receptors on gonadotrophs and somatotrophs of the goldfish: an electron microscope study. Reg. Pept. 36:369-378.

De Leeuw, R., H.J.Th. Goos, J. Peute, A.M.M. van Pelt, E. Burzawa-Gérard & P.G.W.J. van Oordt 1984. Isolation of gonadotrops from the pituitary of the African catfish, *Clarias lazera*. Morphological and physiological characterization of the purified cells. Cell Tiss. Res. 236:669-675.

Goos, H.J.Th., R. de Leeuw, E. Burzawa-Gérard, M. Terlou & C.J.J. Richter 1986. Purification of gonadotropic hormone from the pituitary of the African catfish, *Clarias gariepinus* (Burchell), and the development of a homologous radioimmunoassay. Gen. Comp. Endocrinol. 63:162-170.

Habibi, H.R., R.E. Peter, M. Sokolowska, J.E. Rivier & W.W. Vale 1987. Characterization of gonadotropin-releasing hormone (GnRH) binding to pituitary receptors in goldfish (*Carassius auratus*). Biol. Reprod. 36:844-853.

Koide, Y., T. Noso, G. Schouten, J. Peute, M.A. Zandbergen, J. Bogerd, R.W. Schulz, H. Kawauchi & H.J.Th. Goos 1992. Maturational gonadotropin from the African catfish, *Clarias gariepinus*: purification, characterization, localization, and biological activity. Gen. Comp. Endocrinol. 87:327-341.

LeGac, F., O. Blaise, A. Fostier, P.Y. LeBail, M. Loir, B. Mourot & C. Weil 1993. Growth hormone (GH) and reproduction: a review. Fish Physiol. Biochem. 11:219-232.

Lescroart, O., I. Roelants, L.R. Berghman, P. Verhaert, E.R. Kühn, F. Vandesande & F. Ollevier 1994. Purification of a growth hormone from the pituitary gland of the African catfish (*Clarias gariepinus*) and development of a radioimmunoassay. Proc. Int. Workshop on the Biological Bases for Aquaculture of Siluriformes p. 107.

Lin, X.W., H.R. Lin & R.E. Peter 1993. Growth hormone and gonadotropin secretion in the common carp (*Cyprinus carpio* L.): *in vitro* interactions of gonadotropin-releasing hormone, somatostatin, and the dopamine agonist apomorphine. Gen. Comp. Endocrinol. 89:62-71.

Marchant, T.A., J.P. Chang, C.S. Nahorniak & R.E. Peter 1989. Evidence that gonadotropin-releasing hormone also functions as a growth hormone-releasing factor in the goldfish. Endocrinology 124:2509-2518.

McRory, J.E., D.B. Parker, S. Ngamvongchon & N.M. Sherwood 1995. Sequence and expression of cDNA for pituitary adenylate cyclase activating polypeptide (PACAP) and growth hormone-releasing hormone (GHRH)-like peptide in catfish. Mol. Cell. Endocrinol. 108:169-177.

Melamed, P., N. Eliahu, B. Levavi-Sivan, M. Ofir, O. Farchi-Pisanty, F. Rentier-Delrue, J. Smal, Z. Yaron & Z. Naor 1995. Hypothalamic and thyroidal regulation of growth hormone in tilapia. Gen. Comp. Endocrinol. 97:13-30.

Peter, R.E. & T.A. Marchant 1995. The endocrinology of growth in carp and related species. Aquaculture 129:299-321.

Sherwood, N.M., D.B. Parker, J.E. McRory & D.W. Lescheid 1994. Molecular evolution of growth hormone-releasing hormone and gonadotropin-releasing hormone. Fish Physiol. 13:3-66.

Schulz, R.W., P.T. Bosma, M.A. Zandbergen, M.C.A. van der Sanden, W. vàn Dijk, J. Peute, J. Bogerd & H.J.Th. Goos 1993. Two gonadotropin-releasing hormones in the African catfish, *Clarias gariepinus*: localization, pituitary receptor binding, and gonadotropin release activity. Endocrinology 133:1569-1577.

Vaughan, J.M., J. Rivier, J. Spiess, C. Peng, J.P. Chang, R.E. Peter & W. Vale 1992. Isolation and characterization of hypothalamic growth-hormone releasing factor from common carp, *Cyprinus carpio*. Neuroendocrinol. 56:539-549.

DETECTION AND MEASUREMENT OF GONADOTROPIN II (GTH II) SECRETION FROM INDIVIDUAL TROUT, *SALVELINUS FONTINALIS*, PITUITARY CELLS USING A REVERSE HEMOLYTIC PLAQUE ASSAY (RHPA).

Jorge A. Flores.
PO Box 6057, Department of Biology, West Virginia University, Morgantown, West Virginia 26506-6057.

Summary

A specific and sensitive reverse hemolytic plaque assay is described for the measurement of GTH II secretion from single trout gonadotropes. The power of this single cell model allowing measurements of second messengers, gene expression, and exocytosis on the same individual cell is stressed.

Introduction

The relationship between the strength of a stimulus and the integrated secretory response of its target tissue is poorly understood. Upon receptor activation, second messenger generation, gene expression, and exocytosis are all integrated in the secretory response. Lack of an appropriate single cell model in which all these integrated cellular responses can be measured hinders progress in this field. The pituitary gonadotrope is a very valuable model for stimulus-secretion-coupling studies. When stimulated by luteinizing hormone releasing hormone (LHRH), gonadotropes secrete gonadotropin. A RHPA was developed for detecting and measuring GTH II secretion from individual trout, *Salvelinus fontinalis* gonadotropes. Once these plaque forming cells were identified, they were loaded with the calcium indicator fluorescent dye, fura-2AM to study the effects of LHRH on the cytoplasmic concentration of free calcium ions ($[Ca^{2+}]_i$). This single cell model would be ideal for measuring second messengers, gene expression, and exocytosis on the same cell.

Materials and methods

Pituitary glands were collected from spawning and immature male trout at Casta Line Trout Farms (Goshen, Virginia). Fish pituitaries were removed and dispersed in Spinners's Minimum Essential Medium (S-MEM, Gibco, Grand Island, NY) containing trypsin/DNAse (1). A modified protocol published by Smith et al. (2) was used for the RHPA. Briefly, dispersed trout pituitary cells were washed with experimental medium (in mmol: 127 NaCl, 5 KCl, 1.8 $CaCl_2$, 2 $MgCl_2$, 5 $KHPO_4$, 5 $NaHCO_3$, 10 HEPES, 10 glucose, and 0.1 % BSA, pH 7.4) and mixed (1:1) with protein-A-conjugated ovine red blood cells. This cell mixture was plated as a monolayer in Cunningham chambers constructed on poly-L-lysine-coated microscope slides. After a period of 2h the gonadotropes were stimulated with salmon LHRH (sLHRH, Peninsula Laboratories, Inc. Belmont, CA) in the presence of a salmon GTH II specific antibody (3) at a dilution of 1:40. Guinea pig complement (1:10, Gibco Laboratories, Grand Island, NY) was used to develop the plaques. The number of GTH II secreting cells (plaque forming cells expressed as a percentage of total cells) as well as the amount of GTH II secreted (plaque area) were examined in a period of 2h as a function of the LHRH concentration used.

To study the effects of sLHRH on the $[Ca^{2+}]_i$, plaque forming cells (gonadotropes) were loaded with the calcium indicator fluorescent dye, fura-2AM (Calbiochem, San Diego, CA). The cells were loaded with 2 μM fura-2AM in experimental medium for 20 min at room temperature (22-24^oC). Following dye loading, the Cunningham chamber was placed on the stage of a Zeiss Axioplan microscope (Carl Zeiss, Inc., Thornwood, NY) set up for epifluorescence. The excitation light was supplied by a high pressure xenon arc UV lamp. The excitation wavelengths were selected by 360 and 380 nm filters (Corion, Hollinston, MA) mounted in a computer-controlled rotating filter wheel (Electronic Products, Hawthorne, NY). The fluorescence emitted by the fura-2 within the cell was collected by the objective lens and passed through a barrier filter (490-600 nm transmission) to the face of a silicon-intensified target (SIT) camera (Dage-MTI Inc,

Michigan, IN). The resulting video signal was stored on broadcast quality tape (U-matic, Sony Corporation, Tokyo, Japan). The recorded video signal was captured and digitized using a QX-7 image analysis system (Quantex Corporation, Sunnyvale, CA).

Results

The GTH II antibody used at a dilution of 1:40 was effective in a sensitive an specific RHPA to measure GTH II secretion from individual trout gonadotropes (Fig. 1). Omitting the GTH II antibody or using a coho salmon antibody againt GTH I ß (3), an antibody againt the rat LH or a GTH II antibody pre-incubated with coho salmon GTH II ß resulted in a failure to induce plaque formation in the RHPA at all LHRH concentrations tested (data not shown). In mature male trout LHRH

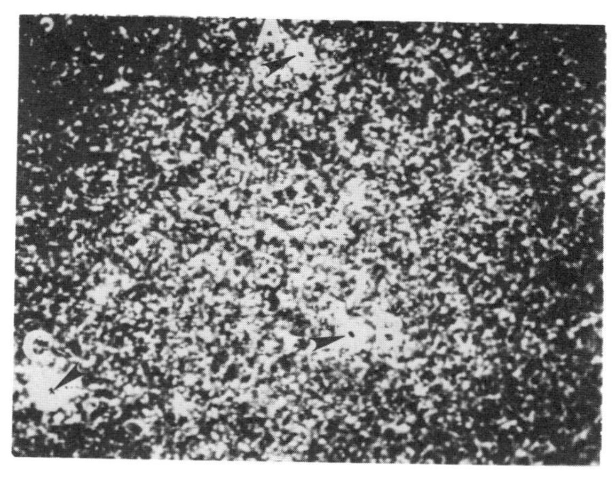

Figure 1. Three trout pituitary cells are shown in the center (arrows) of lysed red blood cells. The stimulus was sLHRH at 10^{-10} M.

increased in a dose dependent manner the amount of GTH II secreted by individual gonadotropes (Fig. 2) as well as the percentage of secreting gonadotropes (Fig. 3). The range of LHRH concentration over which the plaque area was affected was much narrower than that over which the percentage of cells secreting GTH II was affected. In fact, the plaque area remained unchanged above LHRH concentrations of 10^{-10} M (Fig.2). In contrast, the maximal percentage of cells secreting GTH II was not reached until the strength of the stimulus reached 10^{-8} M (Fig.3).

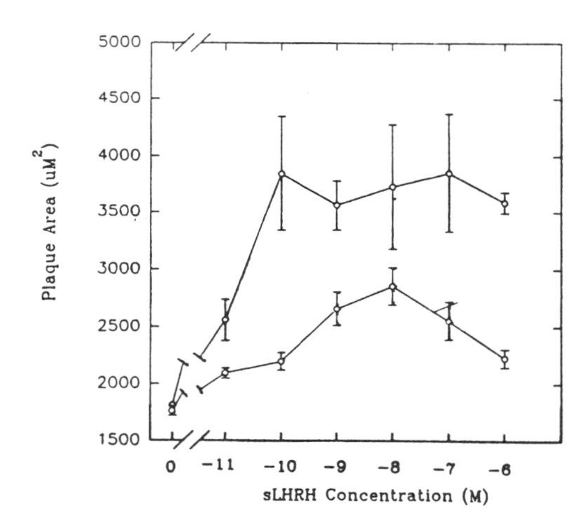

Figure 2. Relationship between the concentration of sLHRH and the plaque area (μm^2) formed by GTH II secreting gonadotropes of mature (upper plot) and immature (lower plot) male trout.

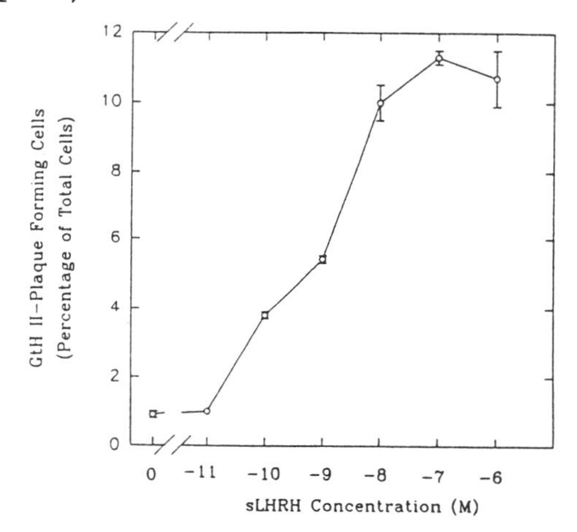

Figure 3. Relationship between the concentration of sLHRH and the percentage of GTH II secreting cells in mature male trout.

In male trout whose testes were histologically characterized by having many spermatogenic cysts but not mature spermatozoa in the lumen (stage III, reference #4), sLHRH was less potent than in mature males in increasing the plaque area (Fig. 2). However, the sLHRH concentration range over which this stimulatory effect was seen was wider than that seen for mature males. Furthermore, at high concentration sLHRH became inhibitory in immature males (Fig. 2).

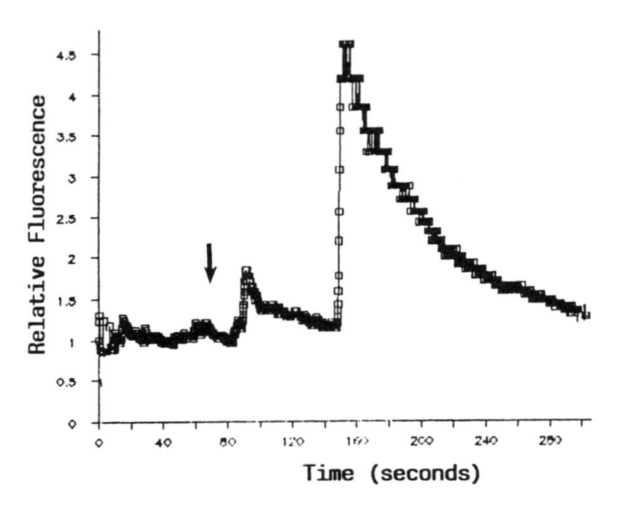

Figure 4. sLHRH stimulated increase in $[Ca^{2+}]_i$ in single trout gonadotrope. sLHRH (10^{-9} M) was added to the cell at the point indicated by the arrow and it was continously present throughout the observation period. Changes in relative fluorescence (380 nm) are shown over the observation time (seconds).

A sLHRH concentration of 10^{-10} M or greater induced rapid and transient rises in $[Ca^{2+}]_i$ of individual gonadotropes previously identified by the RHPA. One example of these responses is shown in Fig 4.

Discussion

The RHPA described here is a sensitive and specific assay to monitor GTH II secretion from individual trout gonadotropes. It was possible to detect basal and sLHRH-stimulated GTH II secretion. The physiological relevance of the sensitivity of this RHPA was demonstrated by the detected differences in the saturation of the effective range of the RHPA of gonadotropes from sexually mature and immature male.

In mature male trout, increasing the strength of the stimulus appeared to have a greater effect on the percentage of cells that secreted GTH II than on the given quantity of GTH secreted per cell. Similar relationship between stimulus strength and secretion has also been found in rat gondotropes (5).

Identified gonatropes could be successfully used to monitor the sLHRH effects on $[Ca^{2+}]_i$. With this single cell approach the fluorescent signal detected is unambiguously specific for an effect of LHRH on gonadotropes. This single cell model is a powerful tool for measuring second messengers, gene expression, and exocytosis on the same cell.

References

1. Chang JP, Cook H, Freeman GL, Wiggs AJ, Somoza GM, de Leeuw R, Peter RE. Use of a cell dispersion method and primary culture system for the studies of gonadotropin-releasing hormone action in the goldfish, Carassius auratus. I. Initial morphological, static and cell column perifusion studies. Gen Comp. Endocrinol 1990;77;256-273.
2. Smith PF, Luke EH, and Neil JD. 1986. Detection and measurement of secretion from individual neuroendocrine cells using a reverse hemolytic plaque assay. Methods Enzymol 124:443-465.
3. Swanson P, et al.1991. Isolation and characterization of two salmon gonadotropins, GTH I and GTH II. Biol Reprod 44:29-38.
4. Schulz R. 1984. Serum levels of 11-oxotestosterone in male and 17ß-estradiolin female raibow trout (Salmo gairdneri) during the first reproductive cycle. Gen Comp Endocrinol 56:111-120.
5. Leong DA, Thorner M. 1990. A potential code of luteinizing hormone-releasing hormone-induced calcium ion responses in the regulation of luteinizing hormone secretion among individual gonadotropes. J.Biol Chem 266:9016-9022.

DOPAMINE ACTIONS ON CALCIUM CURRENT IN IDENTIFIED GOLDFISH (*Carassius auratus*) GONADOTROPIN CELLS.

F. Van Goor, J.I. Goldberg and J.P. Chang

Department of Biological Sciences, University of Alberta, Edmonton, Alberta, Canada, T6G 2E9

Summary

The action of the gonadotropin (GTH) release-inhibitory factor, dopamine (DA), on voltage-sensitive Ca^{2+} channels in identified goldfish GTH cells was investigated. Using Ba^{2+} as the charge carrier through Ca^{2+} channels, isolated Ba^{2+} currents were studied with standard whole-cell, voltage-clamp recording techniques. The general DA receptor agonist apomorphine (1 μM) reduced the Ba^{2+} current amplitude, but did not shift the current-voltage relationship. To determine if DA can inhibit GTH release by acting down-stream of the voltage-sensitive Ca^{2+} channels, the Ca^{2+} ionophore A23187 (10 μM) was used to increase intracellular Ca^{2+} concentrations independent of native Ca^{2+} channels. Apomorphine (1 μM) did not affect A23187-stimulated GTH release. These results suggest that DA inhibits GTH release by acting on voltage-sensitive Ca^{2+} channels rather than on the Ca^{2+}-dependent secretory pathway.

Introduction

The primary neuroendocrine regulators of GTH release from the goldfish anterior pituitary are GTH-releasing hormone (GnRH) and dopamine (Peter *et al.*, 1991). The GTH-releasing action of the two native GnRHs, salmon-GnRH and chicken-GnRH-II, is mediated by a Ca^{2+}/protein kinase C (PKC)-dependent signalling pathway (Chang *et al.*, 1993). Pharmacological studies demonstrate that the activation of voltage-sensitive Ca^{2+} channels and the resulting increase in cytosolic free Ca^{2+} concentration are important events underlying GnRH action (Chang *et al.*, 1993). These studies also suggest that the voltage-sensitive Ca^{2+} channels involved in GnRH-stimulated GTH release are similar to the dihydropyridine-sensitive Ca^{2+} channels characterized in other cell-types (Chang *et al.*, 1993). Recently, whole-cell, voltage-clamp studies have directly demonstrated the presence of dihydropyridine-sensitive Ca^{2+} channels in identified goldfish GTH cells (Van Goor *et al.*, 1994b). These Ca^{2+} channels are similar to the high-voltage activated and inactivation-resistant channels identified in other pituitary cell preparations (Tse and Hille, 1993). Taken together, these results strongly indicate the importance of voltage-sensitive Ca^{2+} channels in the regulation of GTH release by GnRH.

In goldfish, DA inhibits both basal and GnRH-stimulated GTH release via DA D2 receptors. In addition to inhibiting GnRH release and reducing GnRH receptor levels at the pituitary (Peter *et al.*, 1991; Habibi and Peter, 1991), DA may act at different levels of the signalling pathway that mediates GnRH-stimulated GTH release. Preliminary studies suggest that DA reduces GTH release by inhibiting GnRH-stimulated increases in cytosolic Ca^{2+} concentration (Chang *et al.*, 1993). Due to the importance of voltage-sensitive Ca^{2+} channels in GnRH action, it is likely that DA inhibits Ca^{2+} influx through these channels to suppress GTH secretion. In this study, we investigated the possible inhibitory action of DA on voltage-sensitive Ca^{2+} currents in identified goldfish GTH cells using standard whole-cell, voltage-clamp recording techniques. The possibility that DA also acts down-stream of Ca^{2+} influx to inhibit GTH release was also examined.

Methods

Goldfish pituitaries were excised and their cells were dispersed using a controlled trypsin/DNase treatment procedure described in Chang *et al.* (1990). For GTH-release studies, dispersed pituitary cells were cultured overnight in 24-well culture plates under conditions of 5% CO_2, saturated humidity and 28°C. GTH release was measured using an radioimmunoassay specific for maturational GTH. For electrophysiological and immunocytochemical studies, cells were cultured overnight in poly-L-lysine-coated glass-bottom dishes (Van Goor *et al.*, 1994a).

For studies on the possible co-localization of multiple hormones in one cell-type, GTH, growth hormone (GH) and prolactin (PRL) cells were first identified based on their unique morphology when viewed under Nomarsky differential interference contrast microscopy (Van Goor *et al.*, 1994a). Cells were then processed for immunofluores-cence staining using antibodies specific for GTH, GH and PRL according to Van Goor *et al.* (1994a).

For electrophysiological studies, GTH cells were identified according to their unique cellular morphology (Van Goor *et al.*, 1994a). Whole-cell, voltage-clamp recordings (Hamill *et al.*, 1981) of voltage-sensitive Ca^{2+} currents were obtained at room temperature (20°C) using a Dagan 3900 integrating patch-clamp amplifier. Patch electrodes were fabricated from borosilicate glass (World Precision Instruments) using a Flaming Brown horizontal puller (Sutter Instruments) and were subsequently heat polished to a final tip resistance of between 2 and 5 MΩ when filled with (in mM) 120 Cs-glutamate, 20 tetraethylammonia, 2 $MgCl_2$, 11 EGTA, 1 $CaCl_2$, 2.5 Mg-ATP, 0.2 Na-GTP (pH 7.2). Liquid junction potentials were cancelled and series resistance was calculated to be < 27 MΩ. Current records were corrected for linear leakage and capacitance using a P/-4 procedure (Bezanilla and Armstrong, 1977). Pulse generation and data acquisition were carried out using an AT 486 2DX compatible computer equipped with a Digidata 1200 interface in conjunction with pCLAMP software (Axon Instruments). Barium was used as the charge carrier through Ca^{2+} channels to minimize current rundown. The external solution contained (in mM) 120 choline chloride, 1.26 $CaCl_2$, 20 $BaCl_2$, 1.0 $MgCl_2$, 5.5 glucose and 10 HEPES (pH 7.2) and was continuously exchanged using a gravity driven perifusion system which resulted in complete exchange near the cell in < 10 s. Differences between groups were considered to be significant when $p < 0.05$ using an ANOVA followed by Fisher's LSD test.

Supported by grants from NSERC and AHFMR, Canada

Results

Lack of co-localization of GTH, GH and PRL. Since DA had been implicated in the regulation of GTH, GH and PRL release in teleosts (Grau and Helms, 1990; Chang *et al.*, 1993; Chang *et al.*, 1994), it is critical to understand whether DA acts on distinct cell-types or cells containing multiple hormones. We therefore tested for co-localization of homones within identified cells by determining whether cells displaying the morphological characteristics of one cell-type stained for another hormone (Table 1). When GTH and PRL cells were identified by their unique morphologically characteristics as described in Van Goor *et al.* (1994a), no evidence for co-localization was found. In addition, only 4 to 5% of the morphologically-identified GH cells stained for GTH or PRL. Thus, co-localization of PRL or GH did not occur in identified GTH cells. Therefore, the effects of DA on identified GTH cells can only be attributed to the control of GTH, but not PRL or GH release.

Table 1. Lack of co-localization of GH and PRL with identified GTH cells.

	Percentage of identified cells stained for antiserum (as) specific for other hormone. (n)		
Predicted cell-type	GTH-as	GH-as	PRL-as
GTH	—	0.00 (40)	0.00 (50)
GH	4.20 (70)	—	4.62 (65)
PRL	0.00 (62)	0.00 (50)	—

Apomorphine effects on voltage-sensitive Ca^{2+} currents. To investigate the possibility that DA suppresses GTH release by inhibiting voltage-sensitive Ca^{2+} channels, we examined the effect of the general DA receptor agonist apomorphine on Ca^{2+} current. Using Ba^{2+} as the charge carrier for Ca^{2+} through voltage-sensitive Ca^{2+} channels, command potentials more depolarized than -40 mV elicited an inward current which was resistant to inactivation (Figure 1). These results indicate the presence of high-voltage activated Ca^{2+} channels. Application of 1 µM apomorphine significantly reduced the Ba^{2+} current to 52.0 ± 2.7% of control currents (holding potential = -80 mV; test potential = 0 mV). Following washout, the Ba^{2+} current returned to 81.5 ± 2.5% of control levels (mean ± SEM; n = 4; Figure 1 A). Apomorphine did not shift the current-voltage relationship of the voltage-sensitive Ba^{2+} current (Figure 1 B). These results suggest that DA inhibits Ca^{2+} influx through voltage-sensitive Ca^{2+} channels.

Apomorphine effects on Ca^{2+}-induced GTH release. To determine if DA inhibits Ca^{2+}-dependent signaling mechanisms down-stream of the voltage-sensitive Ca^{2+} channels, we investigated the effects of apomorphine on A23187-stimulated GTH release. The Ca^{2+} ionophore, A23187, stimulates GTH release by inducing extracellular Ca^{2+} entry independent of voltage-sensitive Ca^{2+} channels. Apomorphine had no effect on A23187-stimulated GTH release during 2-hour static culture incubations (Figure 2). Together, these data indicate that the inhibitory effects of DA are mediated at the level of the voltage-sensitive Ca^{2+} channels rather than the Ca^{2+}-dependent secretory apparatus.

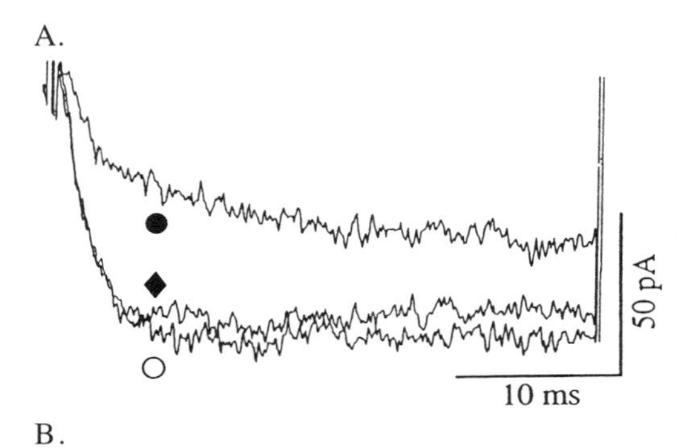

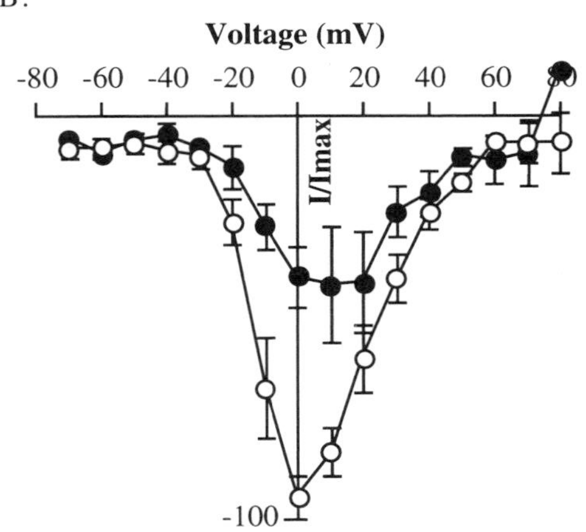

Figure 1. Inhibition of Ba^{2+} current through voltage-sensitive Ca^{2+} channels by apomorphine. A. Whole-cell, voltage-clamp recordings of Ba^{2+} currents elicited during a 40 ms step to 0 mV before (open circles), during (closed circles), and after (diamond) the application of 1 µM apomorphine (holding potential = -80 mV). B. Current-voltage relation of the peak Ba^{2+} current before (open circles) and after (filled circles) the application of 1 µM apomorphine. The current-voltage relation of the Ba^{2+} current was obtained by holding GTH cells at -80 mV and then stepping for 200 ms to command potentials between -70 and +80 mV, in 10 mV increments. Currents were normalized to maximal inward current (mean ± SEM; n = 4).

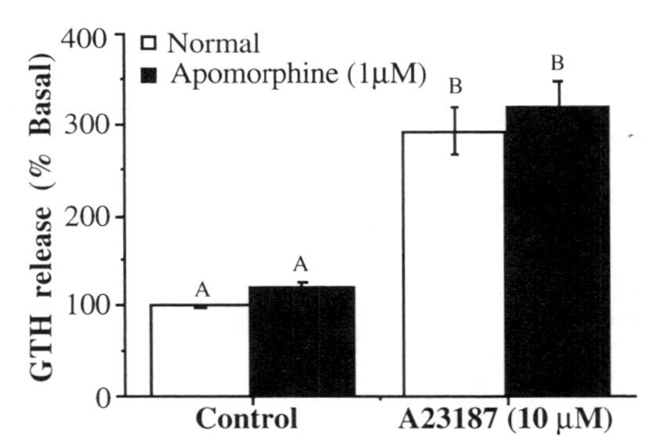

Figure 2. Apomorphine had no effect on A23187-stimulated GTH release. Values were normalized to % control (225.7 ± 30.7 ng/ml; mean ± SEM; n = 16). Groups that are not significantly different from one another are indicated by the same letter.

Discussion

To better understand the mechanisms of action by which DA inhibits GnRH-stimulated GTH release, we examined the action of DA at a single-cell level and at a cell population level. In the present study, the general DA agonist, apomorphine, reduced the peak Ba^{2+} current through voltage-sensitive Ca^{2+} channels, but did not shift the current-voltage relationship of the Ba^{2+} current. Furthermore, the inhibitory action of apomorphine on GTH release did not occur downstream of Ca^{2+} influx, as apomorphine did not affect A23187-stimulated GTH release. These results indicate that the inhibitory action of DA on GTH cells is mediated by the reduction of Ca^{2+} current through voltage-sensitive Ca^{2+} channels. Previously, we demonstrated that a DA D2 agonist inhibits GnRH-induced increases in cytosolic Ca^{2+} from mixed populations of goldfish pituitary cells (Chang *et al.*, 1993). Taken together with the dependence of salmon-GnRH- and chicken-GnRH-II-stimulated GTH release on voltage-sensitive Ca^{2+} channels, these results indicate that the inhibitory action of DA on Ca^{2+} channels is an important event during DA inhibition of GTH release.

Pharmacological studies have implicated the participation of DA D2 receptors in DA-induced inhibition of GnRH-stimulated GTH release from dispersed goldfish pituitary cells (Peter *et al.*, 1991; Chang *et al.*, 1993). Whether the inhibitory action of DA on voltage-sensitive Ca^{2+} currents in identified GTH cells is similarly exerted through DA D2 receptors remains to be confirmed. However, it has been shown in both frog (Vaudry *et al.*, 1994) and rat (Keja *et al.*, 1992) melanotropes, as well as in rat lactotropes (Lledo *et al.*, 1992), that activation of DA D2 receptors inhibits voltage-sensitive Ca^{2+} currents. This inhibition of Ca^{2+} current in rat lactotropes is mediated directly by G-proteins (Lledo *et al.*, 1992). The mechanisms whereby DA inhibits Ca^{2+} currents in goldfish GTH cells remain to be elucidated. DA may suppress Ca^{2+} currents through G-proteins as in rat lactotropes or may act through interactions with PKC. Previously, we have shown that DA also decreases PKC-induced GTH release and PKC likely mediates GnRH-induced activation of voltage-sensitive Ca^{2+} channels in goldfish (Chang *et al.*, 1993).

This study provides the first direct evidence for inhibition of voltage-sensitive Ca^{2+} channels by DA in any GTH cells studied to date. In addition, they further our understanding on how DA interacts with GnRH signalling pathways to inhibit GTH release.

References

Bezanilla, F. and Armstrong, C.M. 1977. Inactivation of sodium channels. I. Sodium current experiments. Journal of General Physiology. 70:549-560.

Chang, J.P., Cook, H., Freedman, G.L., Wiggs, A.J., Somoza, G.M., de Leeuw, R. and Peter, R.E. 1990. Use of a pituitary cell dispersion method and primary culture system for the studies of gonadotropin-releasing hormone action in the goldfish, *Carassius auratus*. I. Initial morphological, static, and cell column perifusion studies. General and Comparative Endocrinology. 77: 256-273.

Chang, J.P., Jobin, R.M. and Wong, A.O.L. 1993. Intracellular mechanisms mediating gonadotropin and growth hormone release in the goldfish, *Carassius auratus*. Fish Physiology and Biochemistry. 11:25-33.

Chang, J.P., Van Goor, F., Wong, A.O.L., Jobin, R.M. and Neumann, C.M. 1994. Signal transduction pathways in GnRH- and dopamine D1-stimulated growth hormone secretion in the goldfish. Chineses Journal of Physiology. 37:111-127.

Grau, E.G. and Helms, L.M.H. 1990. The tilapia prolactin cell-twenty-five years of investigation. In: Progress in Comparative Endocrinology. A. Epple, C.G. Scanes and M.H. Stetson. (Eds.): Wiley-Liss, NY. p. 534-540.

Habibi, H.R. and Peter, R.E. 1991. Gonadotropin-releasing hormone (GnRH) receptors in telesots. In: Reproductive Physiology of Fish. A.P. Scott, J.P. Sumpter, D.E. Kime and M.S. Rolfe. (Eds.): Fish Symposium 91, Sheffield, p. 109-113.

Hamill, O.P., Marty, A., Neher, E., Sakmann, B. and Sigworth, F.J. 1981. Improved patch-clamp techniques for high-resolution current recording from cells and cell-free membrane patches. Pflugers Archiv; European Journal of Physiology. 391:85-100.

Keja, J., Stoof, J.C. and Kits, K.S. 1992. Dopamine D2 receptor stimulation differentially affects voltage-activated calcium channels in rat pituitary melanotropic cells. Journal of Physiology. 450:409-435.

Lledo, P.M., Homburger, V., Bockaert, J. and Vincent, J.D. 1992. Differential G protein-mediated coupling of D2 dopamine receptors to K^+ and Ca^{2+} currents in rat anterior pituitary cells. Neuron. 8:455-462.

Peter, R.E., Trudeau, V.L. and Sloley, B.D. 1991. Brain regulation of reproduction in teleosts. Bulletin of the Institute of Zoology, Academia Sinica. 16:89-118.

Vaudry, H., Lamacz, M., Desrues, L., Louiset, E., Valentijn, J., Mei, Y.A., Chartrel, N., Conlon, J.M., Cazin, L. and Tonon, M.C. 1994. The melanotrope cell of the frog pituitary as a model of neuroendocrine integration. In: Perspectives in Comparative Endocrinology. K.G. Davey, R.E. Peter and S.S. Tobe. (Eds.): National Research Council of Canada, Ottawa, p. 5-11.

Tse, A. and Hille, B. 1993. Role of voltage-gated Na^+ and Ca^{2+} channels in gonadotropin-releasing hormone-induced membrane potential changes in identified rat gonadotropes. Endocrinology. 132:1475-1481.

Van Goor, F., Jobin, R.M., Wong, A.O.L., Goldberg, J.I. and Chang, J.P. 1994a. Morphological identification of live gonadotropin, growth-hormone, and prolactin cells in goldfish (*Carassius auratus*) pituitary-cell cultures. Cell and Tissue Research. 276:253-261.

Van Goor, F., Goldberg, J.I., Chang, J.P. 1994b. Ion channels in identified goldfish gonadotropin cells. In: Programs and Abstracts, 76th Annual Meeting of the Endocrine Society. Abst. 167, p. 242.

MONOAMINE METABOLISM IN THE HYPOTHALAMUS OF THE JUVENILE TELEOST FISH, *Chaetodipterus faber* (PISCES:Ephippidae)

D. Marcano, H. Y. Guerrero, N. Gago, E. Cardillo, M. Requena and *L. Ruiz

Laboratorio de Neuroendocrinología, Departamento de Fisiología, Escuela de Medicina J.M.Vargas and Instituto de Medicina Experimental, Universidad Central de Venezuela. Apartado 47633. Caracas 1041-A. Estación de Investigaciones Marinas de Mochima-FUNDACIENCIA and *Escuela de Ciencias, Universidad de Oriente. Cumaná. Venezuela.

Summary

In *Chaetodipterus faber*, hypothalamic levels of monoamines exhibited significant age variations during development. High concentrations of both serotonin (5-hydroxytryptamine, 5-HT) and noradrenaline (NA) were found, while dopamine (DA) concentrations were about 10-fold lower than NA. A decrease in dopaminergic activity and an increase in noradrenergic activity was observed at puberty. Serotoninergic activity did not show significant changes during gonadal development. It is suggested that a reduction of DA turnover in the hypothalamus appears to serve, at least in part, as a neuroendocrine trigger for puberty-induced gonadotropin release in female *Chaetodipterus faber*. DA inhibition to the pituitary may function in combination with an increased noradrenaline activity and endogenous GnRH systems to promote activation of hypothalamic-pituitary-ovary axis at puberty to acquire reproductive competency.

Introduction

Central monoaminergic pathways are important components of the hypothalamus-pituitary system, both in lower and higher vertebrates. In teleosts, the aminergic cells seem to be highly concentrated in the hypothalamus. Immunocytochemical studies have demonstrated an extensive distribution of serotonin (5-hydroxytryptamine; 5-HT), noradrenaline (NA) and dopamine (DA) in the hypothalamus-pituitary system of some teleosts, and physiological studies have implicated NA, 5-HT and DA in the regulation of several pituitary hormones. In addition, hypothalamic monoamines have been shown to undergo significant seasonal variations in several teleosts (Guerrero et al., 1990; Khan and Joy, 1990; Saligaut et al., 1992; Senthilkumaran and Joy, 1995).

Immunocytochemical studies carried out in our laboratory have shown the presence of immunoreactive DA (ir-DA) neurons in various hypothalamic nuclei of *Chaetodipterus faber*, particularly in the *nucleus lateralis tuberis* (NLT). Moreover, a dense innervation of ir-DA fibers were found in the ventral hypothalamus distributed lateral to the NLT toward the floor of ventricle and the pituitary gland (Gago et al., 1993). In addition, many reports indicate that DA, NA and 5HT are involved in the control of fish gonadotropin (GTH) secretion (see Peter et al., 1986) . In the present study, we present data on DA, NA and 5HT metabolism in the hypothalamus of juvenile *Ch. faber* under farm culture conditions. Amine levels have been correlated with gonadal stages (based on visual inspection). A specific and sensitive HPLC-ED method was used to measure the levels of DA, NA, 5HT, 3,4-dihydroxyphenylacetic acid (DOPAC), homovanilic acid (HVA) methoxyhydroxyphenylglycol (MHPG) and 5-hydroxyindoleacetic acid (5HIAA) and the 5HIAA/5HT, MHPG/NA and (DOPAC+HVA)/DA ratios were calculated. Animals were sacrificed at different ages and gonadal stages (Table 1) were assigned according to criteria derived from Nikolsky (1963) as modified for this species. All data were analyzed with a one-way ANOVA test. Significance was accepted at the 0.05 level.

Table 1. Criteria for macroscopic staging of *Ch. faber*

Stage	Age month	Classification	Macroscopic Appearance
0	6	undifferentiated	undeveloped gonad, undetermined sex
1	9	immature	ovary clear thread, transparent, pink in color
2	12	puberty	ovary small, orange-pale in color without oocyte

Results

In all groups, high NA and 5HT concentrations were observed (Table 2). In contrast, DA concentration was about 10-fold lower than NA concentration (Table 2). Of the two DA metabolites monitored, only DOPAC occurred in measurable concentrations. Levels of HVA were below accurate detection. DA content significantly increased (P<0.05) at stage 1. At puberty, there was a significant decrease of NA (P<0.01), DA (P<0.05) and 5HT (P<0.01) contents as well as a decrease in 5HIAA (P<0.001) and DOPAC (P<0.01) levels when compared to stages 0 and 1 (Table 2).

Table 2. Monoamines and Catabolites in the Hypothalamus of *Ch. faber*.

	Gonadal Stage		
Amines	**0**	**1**	**2**
DA	28.7±16	98.1±20	34.5±17
DOPAC	168.9±39	239.5±52	36.5±16
NA	682.7±140	936.1±143	225.9±134
MHPG	74.7±13	32.0±10	51.1±16
5-HT	456.6±95	557.1±129	103.4±73
5-HIAA	417.0±95	801.3±145	253.3±137

Note. All values are means ±SEM. Concentrations of amines are expressed as pg/mg fresh tissue. N = 9-11 animals.

The monoaminergic activity is shown in Figure 1. A decrease was found in the DOPAC/DA ratio in animals at initiation of puberty (stage 2). However, an increase was observed in the MHPG/NA ratio during the same period. The content of 5HT and 5HIAA showed a decrease at puberty, however, 5HIAA/5HT ratio, an index of brain serotoninergic activity, did not change significantly throughout gonadal development.

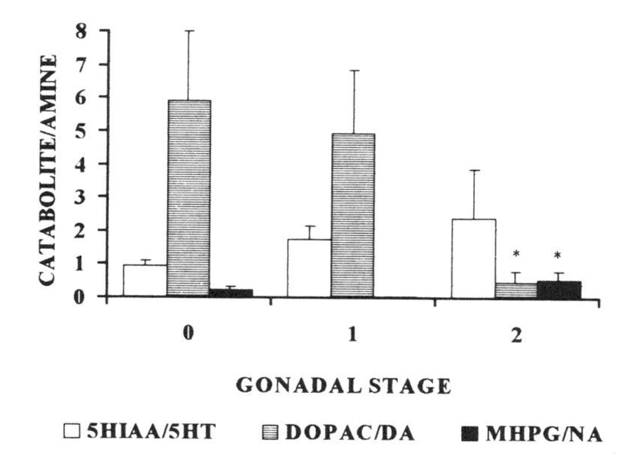

Figure 1. Hypothalamic monoamine activity at various gonadal stages in *Ch. faber*. Values are means ± SEM from 9-11 individuals. Asterisks denote significant differences between stage 1 and stage 2. * P < 0.05 one-way ANOVA test.

Discussion

Puberty is the last phase of the complex process of sexual maturation, a process by which an individual acquires reproductive competency. In juveniles the hypothalamic-pituitary-gonadal axis (HPG axis) is non-functional. It might be that the activation of that axis will initiate reproductive competency.

Our study presents evidence that puberty in *Ch. faber* is associated with a decrease in dopaminergic activity, as reflected by significant decreases in the DOPAC/DA ratio at the level of the hypothalamus. Furthermore, an increase in noradrenergic activity occurs during the same period. A reduction in DOPAC/DA ratio at the hypothalamic level might reflect decreased activity of the DA neurons which project to the gonadotropes (Kah et al., 1987). It may be possible that a reduction in DA inhibition of the pituitary functions may account for activation of the HPG axis at puberty.

Little is known about developmental maturational changes in the neuroendocrine control of the fish reproductive system. In teleosts, environmental conditions and gonadal steroids are known to influence GTH secretion. However, the mechanism through which these factors modulate the GTH release is not clearly understood. One such mechanism that has been pursued in some detail is that these factors may act through the hypothalamic monoaminergic system to modulate GTH secretion (Khan and Joy, 1990). It is probable that changes in the synthesis and release of monoamine neurotransmitters drive increased hypothalamic gonadotropin releasing hormone (GnRH) and/or GTH synthesis and secretion, resulting in enhanced gonadal function at puberty. Besides, GTH release in fish is regulated by the brain through the stimulatory actions of GnRH and the inhibitory actions of DA (Peter et al., 1986).

From the present data it is evident that DOPAC is the predominant metabolite in the hypothalamus of *Ch. faber*. There is considerable controversy in the literature as to what compound is the dominating metabolite of DA catabolism in fish. Nilsson, (1989) using HPLC/ED, reported that HVA was the major DA metabolite in brain of *Carassius carassius*, whereas DOPAC seems to be the predominant DA metabolite in goldfish (Dulka et al., 1992; Sloley et al., 1992). Saligaut et al., (1990) suggested that 3-methoxytyramine (3-MT) was the major DA metabolite in rainbow trout, while Saligaut et al. (1992) showed that the concentration of DOPAC exceeded that of 3-MT in rainbow trout hypothalamus and pituitary. The reason for the apparent discrepancies between studies on DA metabolism in fish is unclear. Apart from the species differences, this apparent disagreement in the DA catabolism may be due to differences in analytical technique used to measured DA and its metabolites. It is also likely that the rates of conjugation and/or clearance of DA metabolites from the brain differ between, or even within species.

Both hypothalamic 5HT and 5HIAA contents decrease during puberty, suggesting decreased 5HT synthesis and release. However, the 5HIAA/5HT ratio did not show significant changes during gonadal development. Further studies are needed to elucidate the importance of 5HT in the neuroendocrine activation of the HPG axis.

References

Dulka, J.G., B.D. Sloley, N.E. Stacey and R.E.Peter, 1992. A reduction in pituitary dopamine turnover is associated with sex pheromone-induced gonadotropin secretion in male goldfish. Gen. Comp. Endocrinol. 86: 496-505.

Gago, N., H.Y. Guerrero, E. Cardillo, M. Requena, L. Ruiz, G. Cáceres-Dittmar, F. J. Tapia, G. Robaina and D. Marcano. 1993. Distribución de la hormona liberadora de las gonadotropinas (GnRH) y dopamina (DA) en el area pre-óptica y el hipotálamo de la paguara, *Chaetodipterus faber*. Acta Cient. Vzlana. 44: Sppl. 1. 251.

Guerrero, H.Y., G. Cáceres-Dittmar, C.L. Paiva, and D. Marcano. 1990. Hypothalamic and telencephalic catecholamine content in the brain of the teleost fish, *Pygocentrus notatus*, during the annual reproductive cycle. Gen. Comp. Endocrinol. 80: 257-263.

Kah, O., J.G. Dulka, P. Dubourg, J. Thibault and R.E. Peter. 1987. Neuroanatomical substrate for the inhibition of gonadotropin secretion in goldfish: Existence of a dopamine preoptic-hypophyseal pathway. Neuroendocrinology 45: 451-458.

Khan, I.A. and K.P. Joy. 1990. Differential effects of photoperiod and temperature on hypothalamic monoaminergic activity in the teleost *Channa punctatus* (Bloch). Fish Physiol. Biochem. 8: 291-297.

Nikolsky, G. 1963. In: The ecology of fishes. Academic Press, London, pp. 145-187.

Nilsson, G.E. 1989. Regional distribution of monoamines and monoamine metabolites in the brain of the crucian carp, *Carassius carassius* L. Comp. Biochem. Physiol. C 94: 223-228.

Peter R.E., J.P. Chang, C.S. Nahorniak, R.J. Omeljaniuk, M. Sokolowska, S.H. Shih and R. Billard. 1986. Interaction of catecholamines and GnRH in regulation of gonadotropin secretion. Recent Prog. Hormone Res. 42: 513-545.

Saligaut, C. , T. Bailhache, G. Salbert, B. Breton and P. Jego. 1990. Dynamic characteristics of serotonin and dopamine metabolism in the rainbow trout brain: a regional study using liquid chromatography with electrochemical detection. Fish Physiol. Biochem. 8: 199-205.

Saligaut, C., G. Salbert, T. Bailhache, S. Bennani and P. Jego. 1992. Serotonin and dopamine turnover in the female rainbow trout (*Oncorhynchus mykiss*) brain and pituitary: Changes during the annual reproductive cycle. Gen. Comp. Endocrinol. 85: 261-268.

Senthilkumaran, B. and K.P. Joy. 1995. Changes in hypothalamic catecholamines, dopamine-ß-hydroxylase, and phenylethanolamine-N-methyltransferase in the catfish *Heteropneustes fossilis* in relation to season, raised photoperiod and temperature, ovariectomy, and estradiol-17ß replacement. Gen. Comp. Endocrinol. 97: 121-134.

Sloley, D.B., V.L. Trudau and R.E. Peter. 1992. Dopamine catabolism in goldfish (*Carassius auratus*) brain and pituitary: lack of influence of catecholestrogens on dopamine catabolism and gonadotropin secretion. J. Exp. Zool. 263: 398-405.

Acknowledgments This work was supported by grants from the Consejo de Desarrollo Científico y Humanístico de la Universidad Central de Venezuela (CDCH 10-12-2770/94) and the CONICIT-2427 to D.M.

A SEROTONERGIC CONTROL OF PITUITARY-GONADAL ACTIVITY IN THE FEMALE CATFISH, *HETEROPNEUSTES FOSSILIS*, UNDER NORMAL AND LONG PHOTOPERIODS

K.P. Joy and B. Senthilkumaran

Department of Zoology, Banaras Hindu University, Varanasi 221 005, India.

Summary

Exposure of *Heteropneustes fossilis* to a long photpoeriod regime (16L : 8D; 17 ± 1°C; 30 days) in gonadal preparatory phase caused significant elevations of plasma gonadotropin (GTH), estradiol - 17β (E_2) and testosterone (T) levels, and gonadosomatic index (GSI). Concurrently, hypothalamic serotonin (5-HT) content and turnover, and monoamine oxidase (MAO) activity were significantly elevated. Administration of para-chlorophenylalanine (p-CPA; a tryptophan hydroxylase inhibitor; 10mg / 100g BW; 10 injections over 30 days) decreased significantly the content and turnover of 5-HT and MAO activity with their daily patterns abolished in both normal and long photoperiod groups. Concomitantly, the treatment also inhibited plasma levels of GTH, E_2 and T, and the GSI; the reductions being greater in the long photoperiod group. Administration of 5-hydroxytryptophan (5-HTP; a 5-HT precursor, 5 mg /100 g BW; 10 injections over 30 days) prevented the effect of p-CPA on all the variables. The results suggest that hypothalamic 5-HT is involved in the photoperiodic stimulation of the pitutary-ovarian axis.

Introduction

Serotonin (5-hydroxytryptamine, 5-HT) is known to stimulate gonadotropin (GTH-II; maturational GTH) secretion in goldfish *Carassius auratus* (Somoza & Peter, 1991) and Atlantic croaker *Micropogonias undulatus* (Khan & Thomas, 1994). Hypothalamic 5-HT is implicated in the mediation of photoperiod effects on the hypothalamo-hypophyseal system of teleosts (de Vlaming & Olcese, 1981; Joy & Khan, 1991). In the catfish, *Heteropneustes fossilis*, we have reported that hypothalamic 5-HT and monoamine oxidase (MAO) exhibit significant annual variations with high values in the recrudescent phase (Senthilkumaran & Joy, 1993). Besides, in the early gonadal preparatory phase when gonads start recrudescing, these variables display significant day-night variations in the hypothalamus and telencephalon. Moreover, the daily patterns were modified by altering the environmental photoperiod (Senthilkumaran & Joy, 1994a). In *Channa punctatus,* the day-night patterns were abolished by a combination of pinealectomy and blinding, and induced in the resting phase by long photoperiods (Khan & Joy, 1990 a,b). Although the above studies do suggest a positive relationship between photoperiod and 5-HT, evidence showing a direct and specific role of the latter in the mediation of photoperiod effects on pituitary - gonadal axis is lacking. This has prompted us to undertake the present study.

Materials and Methods

Acclimatized female *H. fossilis* (30-40g) were divided into 6 groups of 50 each and the following experiments were conducted in the preparatory phase (March). Group 1 and 2 fish were maintained under normal photoperiod and ambient temperature conditions (NP, 11.5L; 12.5D; 17±1°C) and received 0.6% alkaline saline (pH 7.8, vehicle control) and p-chlorophenylalanine (p-CPA, 10 mg/100 g BW), respectively, every third day for a month. Group 3 fish were maintained under similar photoperiodic conditions and received both p-CPA and 5-hydroxytryptophan (5-HTP, 5 mg/100 g BW), every third day for a month. Group 4 and 5 fish were exposed to long photoperiod (LP, 16 L; 8D; 17±1°C) daily with light (40 W fluorescent tube) on from 06 to 22 h by an automatic timer. They received the vehicle medium and p-CPA, respectively, at the same dosage as in group 1 and 2. Group 6 fish were exposed to the long photoperiod regime and received both p-CPA and 5-HTP at the same dosage as in group 3. On completion of the experiment, the fish in each group were divided into 2 subgroups of 25 each for sampling at 12h and 24h. Night sampling was made under dim red light. Each subgroup was further divided into 3 batches of 5, 10 and 10 fish, respectively. The batch 1 fish were used for the estimation of MAO activity. The batch 2 and 3 were given, respectively, 0.6% saline (NaCl) and pargyline (7.5 mg/100 g BW; Sigma), 6h before killing for the estimation of 5-HT content and determination of turnover. Blood was drawn by caudal puncture. Plasma was separated and stored at - 20°C until assayed for GTH, E_2 and T. The procedures and details of assays of MAO, and 5-HT content, and radioimmunoassay (RIA) of the hormones were given previously (Senthilkumaran & Joy, 1994b). The catfish GTH and antiserum were obtained from Prof. H.J.Th. Goos and the E_2 and testosterone antisera were from Prof. G.D Niswender and WHO, Geneva,

Supported by CSIR, New Delhi.

respectively as generous gifts. Statistical analysis involved Student's t test.

Results

Exposure of the catfish to long photoperiod (LP) has elevated gonadal activity with significant increases in GSI and plasma levels of E_2, T and GTH (Figs. 1 & 2). Hypothalamic 5-HT content and MAO activity also increased significantly but their day-night patterns were abolished (Figs. 3 & 4). However, the treatment maintained the 5-HT turnover pattern. Administration of p-CPA in the LP and NP groups produced significant inhibitory effects on all the gonadal parameters as well as hypothalamic 5-HT content and turnover, and MAO activity. The magnitude of reductions of GTH (76%) and GSI (63%) were greater in the LP group than in the NP group (68% and 50%, respectively). The p-CPA treatment abolished the day-night patterns of 5-HT (content and turnover) and MAO in the NP group and that of 5-HT turnover in the LP group. Injection of 5-HTP alongwith p-CPA prevented the latter's inhibitory effects on all the parameters in both NP and LP groups. But it did not restore the LP-induced abolition of day-night pattern of 5-HT content and MAO.

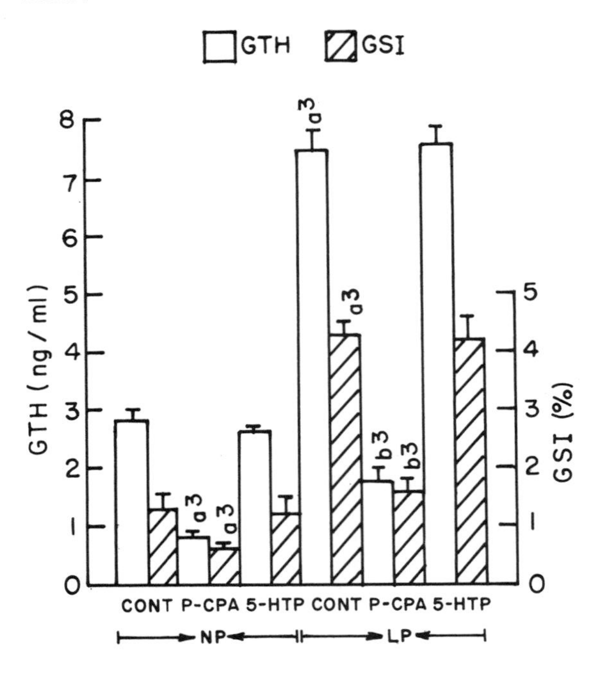

Fig. 1 : Effects of p-CPA alone and in combination with 5-HTP on GSI and plasma GTH level in the catfish (mean ± SEM; n = 5). a- compared with normal photoperiod (NP), b- compared with long photoperiod (LP), cont-control. [3]p<0.001, Student's t test.

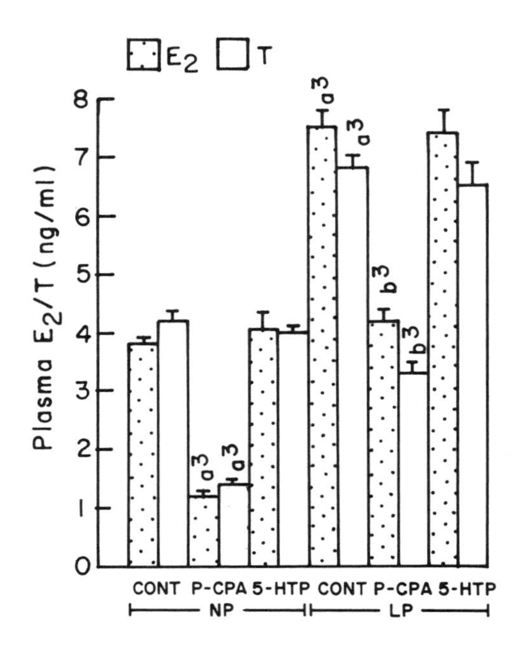

Fig. 2 : Effects of p-CPA alone and in combination with 5-HTP on plasma E_2 and T levels in the catfish (mean ± SEM; n = 5). Other details are as in Fig. 1.

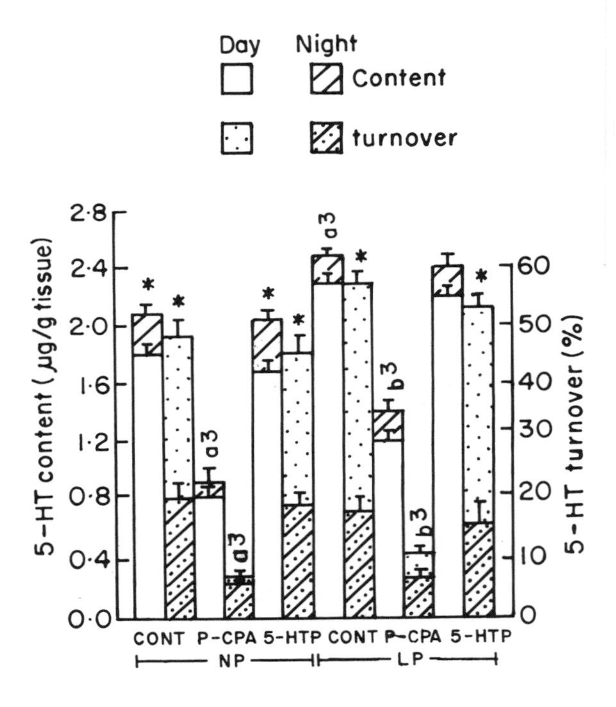

Fig. 3 : Effects of p-CPA alone and in combination with 5-HTP on hypothalamic 5-HT content and turnover in the catfish (mean ± SEM; n = 5).
* - denotes significant day-night patterns. Other details are as in Fig. 1.

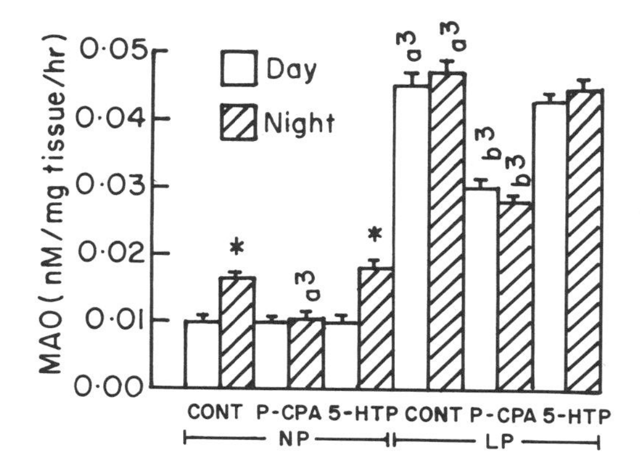

Fig. 4 : Effects of p-CPA alone and in combination with 5-HTP on hypothalamic MAO activity in the catfish (mean ± SEM; n = 5). Other details are as in Figs. 1 & 3.

Discussion

The present study examined whether an impairment of hypothalamic 5-HT in the preparatory phase would inhibit pituitary-gonadal activity and its restoration reinstate this function. The p-CPA treatment had not only inhibited the 5-HT content and turnover, but also abolished their daily patterns. This has led to the inhibition of GTH release and arrest of ovarian recrudescence as compared with that of the saline control group. The fact that the inhibitions were greater in the long photoperiod group shows a strong functional correlation of photoperiod and 5-HT with ovarian recrudescence. The simultaneous administration of 5-HTP reversed the p-CPA effects and restored the pituitary-ovarian function. Thus the present data demonstrate the specific involvement of hypothalamic 5-HT in the photoperiod -mediated effects on the pituitary-gonadal function. The data are consistent with the view that 5-HT stimulates GTH secretion directly or indirectly by stimulating gonadotropin - releasing hormone activity at hypothalamic or pituitary levels (Peter *et al.*, 1991).

The present study also shows that MAO activity was inhibited by the p-CPA treatment. The enzyme inhibition may be a secondary effect and can be attributed either to the decreased titre of plasma E_2 and its resulting feedback effect (Senthilkumaran & Joy, 1994 b) or to decreased 5-HT synthesis, its natural substrate. The fact that 5-HTP treatment failed to restore the day-night patterns in 5-HT content and MAO activity already abolished by the LP treatment, despite restoration of their levels, indicates that the prevailing environmental photoperiod regulates the daily patterns (Senthilkumaran & Joy, 1994 a).

References

de Vlaming, V.L. and J. Olcese, 1981. The pineal and reproduction in fish, amphibians, and reptiles. In: The Pineal Gland, Reproductive Effects, Vol.2. R.J. Reiter (Ed.) : CRC Press, Boca Raton, p. 1-29.

Joy, K.P. and I.A. Khan, 1991. Pineal-gonadal relationship in the teleost *Channa punctatus* (Bloch) : Evidence for possible involvement of hypothalamic serotonergic system. J. Pineal Res. 11: 12-22

Khan, I.A. and K.P.Joy, 1990a. Effects of season, pinealectomy, and blinding, alone and in combination, on hypothalamic monoaminergic activity in the teleost *Channa punctatus* (Bloch). J. Pineal Res. 8: 277-287.

Khan, I.A. and K.P. Joy, 1990b. Differential effects of photoperiod and temperature on hypothalamic monoaminergic activity in the teleost *Channa punctatus* (Bloch). Fish Physiol. Biochem. 8: 291-297.

Khan, I.A. and P. Thomas, 1994. Seasonal and daily variations in the plasma gonadotropin II response to a LHRH analog and serotonin in Atlantic croaker (*Micropogonias undulatus*): Evidence for mediation by 5-HT_2 receptors. J. Expt. Zool. 269 : 531-537.

Peter, R.E., V.L. Trudeau, and B.D. Sloley, 1991. Brain regulation of reproduction in teleosts. Bull. Inst. Zool. Academia Sinica, Monograph 16: 89-118.

Senthilkumaran, B. and K.P. Joy, 1993. Annual variations in hypothalamic serotonin and monoamine oxidase in the catfish *Heteropneustes fossilis* with a note on brain regional differences of day-night variations in gonadal preparatory phase. Gen. Comp. Endocrinol. 90: 372-382.

Senthilkumaran, B. and K.P. Joy, 1994 a. Effects of Photoperiod alterations on hypothalamic serotonin content and turnover, and monoamine oxidase activity in the catfish *Heteropneustes fossilis*. Fish Physiol. Biochem. 13 : 301-307.

Senthilkumaran, B. and K.P. Joy, 1994 b. Effects of ovariectomy and oestradiol replacement on hypothalamic serotonergic and monoamine oxidase activity in the catfish *Heteropneustes fossilis* : A study correlating plasma oestradiol and gonadotrophin levels. J. Endocrinol. 142: 193-203.

Somoza, G.M. and R.E. Peter, 1991. Effects of serotonin on gonadotropin and growth hormone release from in vitro perifused goldfish pituitary fragments. Gen. Comp. Endocrinol. 82: 103-110.

CYTOSOL SEX STEROIDS BINDING IN BRAIN AND ITS LEVELS IN BLOOD OF STURGEON (ACIPENSER GUELDENSTAEDTI BR.) DURING ANADROMOUS MIGRATION.

I.A.Barannikova [1], B.I.Feldkoren [2]

[1] Physiological Institute, St. Petersburg University, St.Petersburg, 194034, Russia
[2] St. Petersburg Research Institute of Physical Culture, St. Petersburg, 197042, Russia

Summary

Cytosol specific binding (SB) of androgens (A) and estrogens (E) in brain and pituitary as well as testosterone (T) and estradiol 17 β (E_2) levels in blood of winter form of sturgeon, migrating into river one year before spawning were studied. A positive correlation between E_2 in blood and SBE in forebrain was recorded. Negative correlation between T in blood and SBA in forebrain and hypothalamus was established; positive correlation of SBA and SBE in forebrain and hypothalamus was revealed.

Introduction

Investigation of sex steroid receptors in brain and comparison of these data with steroid levels in blood are necessary for elucidating the hormonal regulation of reproductive functions. Most data in this direction on lower vertebrates are available for teleosts (Christoforov et al., 1993). Research on sturgeons (fam.Acipenseridae, Chondrostei) is interesting because of their taxonomic position.

The winter form of sturgeon, migrating to the river Volga in spring, one year before spawning, was investigated (Barannikova, 1991). Vitellogenesis is not completed in this period, the germinal vesicle has a central position in oocytes (GSI of female - 13,1±1,2%). Cytosol binding of A and E in brain and in the pituitary using ^{3}H-testosterone and ^{3}H-estradiol was studied. Samples were incubated at 4^{0}C overnight. Unbound steroid was removed by dextran-coated charcoal (For methods see Christoforov & Murza al., 1993). Ligand binding affinity of brain cytosol was determinated by Scatchard analysis. The equilibrium dissociation constant was in the nM range (8,1 nM and 7,6 nM for T and E_2, correspondently). The results of SB determination by single point method are expressed as fmol/mg protein. Radioimmunoassey was used for T and E_2 determination in blood serum.

Results

Specific binding of A and E was detected in different parts of the brain and in the pituitary of sturgeon. The values of SBA and SBE differ in different parts of brain and in the pituitary (Table 1).

Table 1. SBE and SBA in brain and pituitary of sturgeon (fmol/mg protein)

Binding site	SBE	SBA
Forebrain	136,9±29,0	76,3±9,6
Hypothalamus	122,4±10,6	77,3±8,3
Pituitary	50,0±2,9	28,0±2,2

The highest level of SBE was found; it was significantly lower in the pituitary. SBA in hypothalamus and forebrain did not differ significantly from each other however in the pituitary SBA is significantly lower than in the brain of same fish. T concentration in blood serum was 14,8±1,3 ng/ml, E_2 - 1,3±0,17 ng/ml.

Comparison of sex steroids levels in blood with SBE and SBA in brain and pituitary shows significant correlation between these indexes. A positive correlation for E_2 in blood and SBE in the forebrain (Rs=+0,72) as well as between SBE in forebrain and hypothalamus (Rs=+0,73) were found. Negative correlation for T concentration in blood and SBA in forebrain (Rs=-0,86) and hypothalamus (Rs=-0,77) were revealed. (Differences significant, $p<0,05$). Positive correlation for SBA and SBE in hypothalamus and forebrain suggest participation of these structures in regulation of hormonal systems involved in reproductive function. Negative correlations for T in blood and SBA in forebrain and in hypothalamus apparently indicate the active role of this hormone at the beginning of anadromous migration of sturgeon.

References

Barannikova I.A. 1991. Peculiarities of intrapopulational differentiation of sturgeon (Acipenser gueldenstaedti Br.) under present day conditions. In : Acipenser Cemagr.Premier Coll. Intern. sur l'esturgeon, Bordeau, P. Willot ed. ,p.134-142.

Christoforov O.L., Murza I.G., 1993. Deference of tissues specific binding of testosterone and estrodiol from levels of these hormones in blood of atlantic salmon Salmo salar L. In : Biologically active substances and factors in Aquaculture. Moscow, VNIRO, p.71-100.

SEROTONIN METABOLISM IN THE BRAIN OF THE JUVENILE MARINE TELEOST, *Chaetodipterus faber* (PISCES:Ephippidae).

E. Cardillo, N. Gago, H.Y. Guerrero, M. Requena, *L. Ruiz and D. Marcano.

Laboratorio de Neuroendocrinología, Departamento de Fisiología, Escuela de Medicina J.M. Vargas and Instituto de Medicina Experimental, Universidad Central de Venezuela. Apartado 47633. Caracas 1041-A. Estación de Investigaciones Marinas de Mochima-FUNDACIENCIA and *Escuela de Ciencias, Universidad de Oriente. Cumaná. Venezuela.

Introduction

In teleost fishes, immunohistochemical techniques have revealed widespread monoaminergic innervation of different areas of the central nervous system. In the present study, we have investigated the changes of serotonin (5HT) and 5-Hydroxyindoleacetic acid (5HIAA) in the brain of a marine teleost, *Chaetodipterus faber* and its relation to gonadal development. A specific and sensitive HPLC-ED method was used to measure the levels of 5HT and 5HIAA. Animals were sacrificed at the following ages: 6 months, stage 0 (undifferentiated gonads); 9 months, stage 1 (immature gonads); and 12 months, stage 2 (puberty). The brain was removed and dissected; the olfactory bulbs (OB), the cerebellum (Cer), the hypothalamus (H), the telencephalon (Tel) and the optic tectum (OT) were stored at -70°C until HPLC assay. All data were analyzed with one-way ANOVA test and a multiple range test. Differences with $p \leq 0.05$ were considered significant.

Results

Table 1. 5HT and 5HIAA concentration.

BRAIN REGION		GONADAL STAGE: 0	1	2
OB	5HT	UD	UD	UD
	5HIAA	38±15	55±31	77±46
Cer	5HT	UD	UD	UD
	5HIAA	27±13	20±10	7±5
H	5HT	454±95	557±129	$103±73^{ab}$
	5HIAA	417±95	801±145	$253±13^{b}$
Tel	5HT	325±93	341±46	$84±57^{ab}$
	5HIAA	609±82	887±110	$436±16^{b}$
OT	5HT	93±33	132±31	$28±20^{b}$
	5HIAA	393±59	368±42	$111±63^{ab}$

Note. UD= Under detection limit (8 pg/mg wet tissue). Values are expressed as means in pg/mg wet tissue ± SEM of 9-11 animals. a: Significantly different from stage 0, b: Significantly different from stage 1.

In OB and Cer 5HT was not detected however, 5HIAA was present in low concentrations. Levels of 5HT in both H and Tel were significantly higher than those in OT at stages 0 and 1, while there were no differences in stage 2. In H, Tel and OT the content of 5HT and 5HIAA showed a significant decrease at puberty (stage 2); however, as shown in Fig. 1, the 5HIAA/5HT ratio did not change significantly throughout gonadal development in any cerebral region.

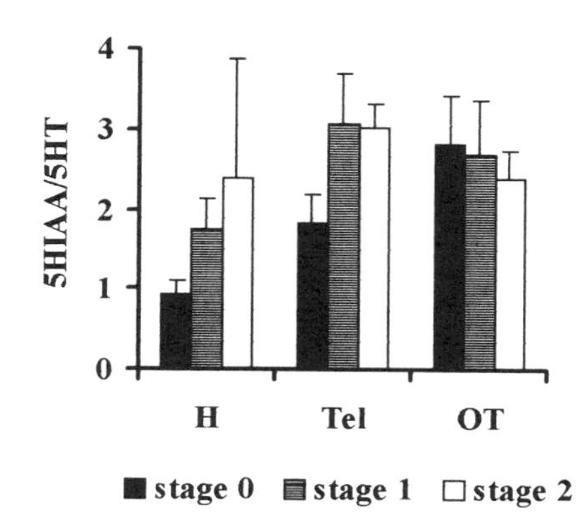

Figure 1. Serotoninergic activity in brain regions of juvenile *Ch. faber* during gonadal development. Values are means ± SEM from 9-11 individuals.

Discussion

Our results show non detectable levels of 5HT and very low concentrations of 5HIAA in OB and Cer, in juvenile individuals suggesting a low metabolic activity. The high 5HT and 5HIAA contents in the Tel and OT suggest a high population of 5HT axonal endings in these areas. The serotonergic innervation of these areas has been described in other teleosts. No significant changes in the 5HIAA/5HT ratio were observed during the three stages of gonadal development in both Tel and OT. These results suggest that 5HT metabolism in these brain areas has little influence on gonad development.

Acknowledgments This work was supported by grants from the Consejo de Desarrollo Científico y Humanístico de la Universidad Central de Venezuela (CDCH 10-12-2770/94) and CONCIT-2427 to D.M.

MOLECULAR CHARACTERIZATION OF THE SEABREAM (SB) GnRH GENE ISOLATED FROM STRIPED BASS, MORONE SAXATILIS.

M. Chow, Y. Gothilf and Y. Zohar

Center of Marine Biotechnology and Agricultural Experimental Station, University of Maryland Biotechnology Institute. 701 East Pratt St., Baltimore, Maryland. 21202

Summary

A novel form of GnRH has previously been identified in gilthead seabream (sbGnRH)[1].We report here the cloning and molecular characteriza-tion of the sbGnRH gene from a striped bass genomic library. The overall genomic organization of the sbGnRH gene is similar to the GnRH genes isolated from salmon (sGnRH), chicken (cGnRH-I), human and mouse (mGnRH). The nucleotide sequence encoding the precursor of the striped bass sbGnRH gene is 80% identical to the cDNA sequence we have reported for seabream. Comparison of the deduced amino acids from the polypeptide precursor of sbGnRH in both sea-bream and striped bass reveals a high degree of conservation. Taken together, our findings demon-strate that the sbGnRH gene is a highly conserved gene in these perciforme species.

Introduction

Gonadotropin-releasing hormone (GnRH) has been implicated as the key molecule for the control of reproduction in all vertebrates. It is a 10 amino acid peptide, synthesized in the hypo-thalamus and, in the case of teleost fish, released directly into the pituitary gland. To date, nine closely related, yet distinct forms have been identified in vertebrates. Seabream (sb) GnRH is the most recent form identified and has been demonstrated to be the most abundant of the three forms present in the pituitary of sexually mature gilthead seabream (*Sparus aurata*) and striped bass (*Morone saxatilis*)[2]. In order to promote investiga-tions of the underlying mechanisms which regulate the synthesis of the reproductively relevant form of GnRH, the sbGnRH gene and its corresponding promoter have been cloned from a striped bass genomic library. This is the first nonsalmonid GnRH gene to be cloned in fish.

Results

Partially digested genomic DNA isolated from a striped bass was used to construct a genomic library in the bacteriophage vector λ Fix II.

Screening of this library with the full-length sbGnRH cDNA isolated from gilthead seabream produced a single positive clone. A 2.0kb fragment containing 82% of the sbGnRH polypeptide precursor and approximately 1 kilobase of the 5' upstream regulatory region was characterized.

Comparison of the striped bass sbGnRH coding region with the sbGnRH cDNA from seabream reveals an 80% identity at the nucleotide level. Our findings demonstrate that the sbGnRH gene is highly conserved in these two species of perciforme fish.

Analysis of the upstream regulatory region reveals a **TATAAAA** at -25 from the putative start site of transcription (based on homology with the seabream sbGnRH cDNA 5' untranslated region). In addition, two putative CAAT boxes are present at -112 (**CAAT**) and -150 (**CAAAT**), respectively. These elements resemble the consensus sequence found in all RNA polymerase II promoters. A slightly modified estrogen response (ERE-like) motif has been identified at position -930. Whether this functions as a true consensus sequence for the binding of an estrogen receptor remains to be proven. Further characteriza-tion of the 5' upstream region may identify other possible regulatory elements.

Conclusions

1) The seabream (sb)GnRH gene has been isolated from striped bass. This is the first non-salmonid GnRH gene clone in fish.
2) The genomic organization of the sbGnRH gene is very similar to the other GnRH genes previously characterized.
3) Comparison of the deduced amino acids encoded by the precursor polypeptide of the sbGnRH sequence isolated from striped bass with the previously isolated seabream cDNA demonstrates that the sbGnRH is a highly conserved gene.
4) A slightly modified "ERE-like" motif has been identified which may be responsible for possible estradiol control of the regulation of the sbGnRH gene in perciforme fishes.

REFERENCES:

[1] Gothilf Y, Chow M, Elizur A,Chen T and Zohar Y(1995) "Molecular cloning and characteriza-tion of a novel gonadotropin-releasing hormone from the gilthead seabream (*Sparus aurata*)". Molecular Marine Biology and Biotechnology 4 (1), 27-35.

[2] Zohar Y, Elizur A, Sherwook N, Powell J, Rivier J and Zmora N(1995) "Gonadotropin-releasing activities of three native forms of gonadotropin releasing hormones present in the brain of gilthead seabream, *Sparus aurata*". General Comparative Endocrinology. 97: 289-299.

NEUROENDOCRINE CONTROL OF GONADOTROPIN II RELEASE IN THE ATLANTIC CROAKER: INVOLVEMENT OF GAMMA-AMINOBUTYRIC ACID

I.A. Khan and P. Thomas.

The University of Texas at Austin, Marine Science Institute, Port Aransas, Texas 78373.

Summary

The effects of gamma-aminobutyric acid (GABA) on GtH II release in Atlantic croaker are dependent on the reproductive stage of the fish; stimulatory in gonadally regressed fish and inhibitory in individuals with fully developed gonads. These effects appear to be mediated primarily by $GABA_A$ receptors.

Introduction

There is considerable evidence for the involvement of monoamine and amino acid neurotransmitters in the control of gonadotropin II (GtH II) secretion in teleosts. It has previously been shown that serotonin stimulates GtH II release (Khan and Thomas, 1992) in the Atlantic croaker (*Micropogonias undulatus*), whereas an inhibitory dopaminergic control mechanism demonstrated in several teleosts is absent in this species (Copeland and Thomas, 1989). Recent studies in goldfish have demonstrated that GABA stimulates GtH II release and this effect is mediated by $GABA_A$ receptors (Trudeau *et al.*, 1993). In addition, GABA has been implicated in the mediation of steroid-feedback control of gonadotropin release in goldfish and rats. Therefore, we have investigated the possible involvement of GABA in the control of GtH II release in Atlantic croaker with regressed or fully developed gonads.

Results

Intraperitoneal administration of GABA (100 μg/g body wt.) elicited a significant elevation in GtH II levels in regressed croaker (Table 1). However, the same dose of GABA significantly inhibited GtH II release in croaker with fully developed gonads. GABA did not significantly alter LHRH analog (LHRHa)-induced GtH II release in regressed or mature fish. Similar results were obtained with muscimol (1 μg/g), a $GABA_A$ agonist, whereas baclofen (1 μg/g), a $GABA_B$ agonist, failed to induce any significant alterations in plasma GtH II levels in either group. Pretreatment with bicuculline (1 μg/g), a $GABA_A$ antagonist, blocked the stimulatory effect of GABA on GtH II release in regressed croaker. None of the GABAergic drugs could influence LHRHa-induced GtH II release in the mature or regressed fish.

Table 1. Effects of GABA on GtH II release.

Treatments	Plasma GtH II levels (ng/ml)	
	Regressed	Mature
Vehicle	0.06±0.03	0.25±0.07
LHRHa*	1.42±0.34[a]	1.82±0.12[a]
GABA	0.69±0.16[a]	0.02±0.02[a]
LHRHa+GABA	1.39±0.35[a]	1.78±0.28[a]

*Doses: 50 ng/g (regressed); 5 ng/g (mature).
[a]Significantly different from vehicle-treated controls.

Discussion

The stimulatory influence of GABA on GtH II release in gonadally regressed croaker is comparable to that observed in goldfish (Trudeau *et al.*, 1993). However, the inhibitory effect in croaker with fully recrudesced gonads differs from the goldfish model in that GABA has no effect on GtH II release in goldfish with fully developed gonads. The stimulatory effect of GABA in regressed croaker is mediated primarily by $GABA_A$ receptors. Our results suggest that the inhibitory effect in fish with fully developed gonads may also be mediated by the same receptor subtype. The lack of an effect of GABA on LHRHa-induced GtH II release is consistent with the hypothesis that the effects of GABA on GtH II release are mediated by altered GnRH release. The differential effects of GABA in Atlantic croaker with regressed and fully developed gonads suggest that gonadal steroids may modulate GABA's actions on GtH II release.

References

Copeland, P.A. and P. Thomas, 1989. Control of gonadotropin release in the Atlantic croaker (*Micropogonias undulatus*): Evidence for a lack of dopaminergic inhibition. Gen. Comp. Endocrinol. 74:474-483.

Khan, I.A. and P. Thomas, 1992. Stimulatory effects of serotonin on maturational gonadotropin release in the Atlantic croaker, *Micropogonias undulatus*. Gen. Comp. Endocrinol. 88:388-396.

Trudeau, V.L., B.D. Sloley and R.E. Peter, 1993. GABA stimulation of gonadotropin-II release in goldfish: Involvement of $GABA_A$ receptors, dopamine, and sex steroids. Amer. J. Physiol. 265:R348-R355.

CHARACTERIZATION OF MOLECULAR VARIANTS OF GnRH IN THE BRAIN OF THE PROTOGYNIC "SWAMP EEL", SYNBRANCHUS MARMORATUS.

F. Lo Nostro, M. Ravaglia, G. Guerrero, M.C. Maggese and G.M. Somoza*.

Laboratorio de Embriología Animal, Depto de Biología, Facultad de Ciencias Exactas y Naturales, Universidad de Buenos Aires. (*) INEUCI-CONICET, Depto de Biología, UBA. Ciudad Universitaria (1428). Buenos Aires. Argentina.

Introduction

Synbranchus marmoratus belongs to the order Synbranchiformes, a small group of teleost fishes, commonly known as "swamp eels". They are widely distributed from Argentina to Mexico throughout almost all Brazil. They are characterized by having aerobic respiration, the lack of paired fins, a reduced caudal fin and the absence of scales. S. m. is a protogynic diandric fish; some of them develop initially as females and then change sex to become functional males (secondary males), others develop initially as functional males (primary males) (Liem, 1968). The physiological basis of sex reversal in S. m. has not been studied yet, but in a related species Monopterus albus it has been shown that GnRH may play an important role in the induction of sex reversal (Tao et al., 1993).

The purpose of the present study was to identify the molecular variants of GnRH in the brain of S. m. using RP-HPLC and RIA with different antisera.

Results and Discussion

By using three different RIA systems with the following antisera: cII678, PBL#45 and PBL#49, we could detect two main ir-GnRH peaks coeluting with cIIGnRH and sGnRH. On the other hand, two minor peaks were detected between cIIGnRH and sGnRH elution position, both of them did not coelute with any of the synthetic GnRH variants assayed. The first eluting minor peak was detected with the three RIA sistems used. The second one was detected by cII678 and PBL#49 but not with the PBL#45 RIA system. Neither of these two peaks showed any crossreactivity in a cIIGnRH homologous assay (cII675 antiserum). Neither did they show crossreactivity in two sGnRH assays (s1668 and Aida-sGnRH antisera).

The ir-GnRH peak coeluting with cIIGnRH was analyzed by serial dilutions with an cII675 cIIGnRH RIA system and parallelism with respect to synthetic cIIGnRH was obtained. A similar approach was followed with the late eluting fractions coeluting with sGnRH using two different homologous RIA systems with s1668 and Aida-sGnRH antisera. In both cases parallelism with respect to synthetic sGnRH was obtained. In summary, HPLC analysis combined with RIA of brain extract of S. m. suggest that this species has at least two different variants of GnRH: cIIGnRH and sGnRH.

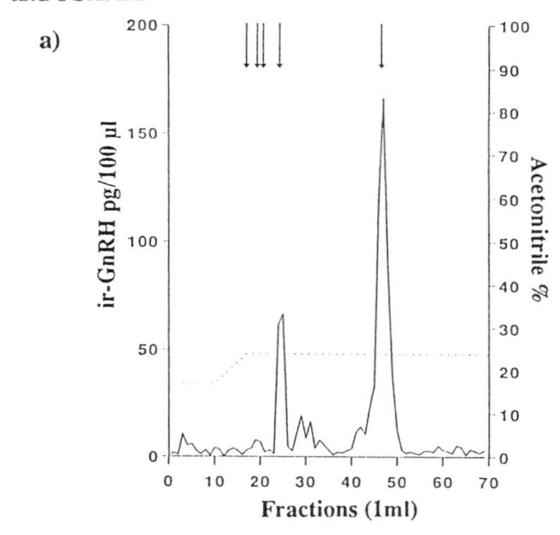

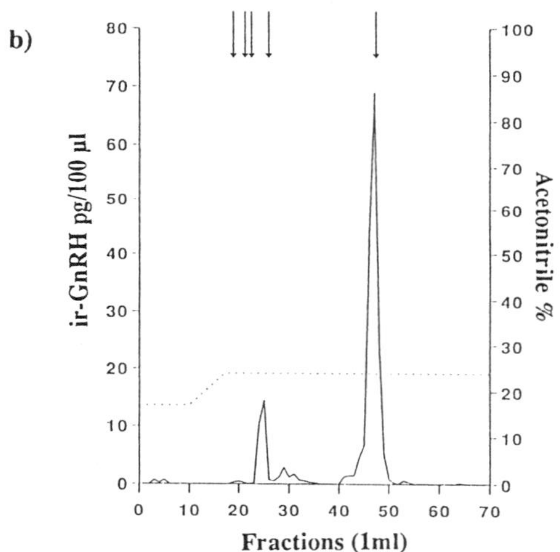

Figure: RP-HPLC of brain extract of S.m. assayed with a) cII678 and b) PBL#49. Arrows show the elution positon from left to rigth: cfGnRH; mGnRH and LIGnRH; cIGnRH; cIIGnRH; sGnRH.

References

Liem,K. (1968). Geographical and taxonomic variation in the pattern of natural sex reversal in the teleost fish order Symbranchiformes. J. Zool, Lond 156:225-238.

Tao,Y.;Lin,H.;Van der Kraak,G. and Peter,R.E. (1993). Hormonal induction of precocious sex reversal in the rice eel, Monopterus albus. Aquaculture 118: 131-140.

This work was supported by a grant from the University of Buenos Aires (EX-095).

The Effect of Pineal Removal and Enucleation on Circulating Melatonin Levels in Atlantic Salmon Parr

Mark Porter, Clive Randall and Niall Bromage

Institute of Aquaculture, University of Stirling, Stirling FK9 4LA, Scotland.

Introduction

As in other vertebrates the pineal gland in salmonids is thought to use photoperiodic information to synchronize both daily and seasonal behavioral and physiological events. It is not yet clear whether this is achieved through neural or endocrine pathways. As melatonin is one of the main endocrine hormones responsible for phototransduction a simple and effective method of removing the two main sites of melatonin secretion was developed for future long-term experiments.

Materials and Methods

In study A, 30 one-year old salmon parr (Salmo salar L.) were maintained under a constant LD (12:12) photoperiod and ambient temperature. Groups of ten fish (mean weight 55.3g) were pinealectomised (pinx), sham-pinealectomised (sham-pinx) or left intact (control). The pinx and sham-pinx fish were anaesthetized and a 5mm horizontal incision was made posterior to the pineal window. A flap of tissue was then lifted anteriorly to reveal the pineal. In the pinx group the pineal stalk was cut at the point of attachment to the diencephalon and the pineal removed with forceps. The pineal was left intact in sham-pinx fish. The overlying tissue was then replaced and an orahesive powder with cicatrin antibiotic applied over the incision. Subsequent experiments on larger fish used sutures to hold the flap of skin in place. Pinealectomy plus enucleation was performed on ten individuals in study B. Enucleation was achieved by incising the cornea and severing the optic nerve to allow removal of the eye; the orbit was then cleaned and tissue glue applied to its surface.

Further work on large numbers of parr revealed a 7% mortality rate for pinealectomised fish and less than 1% mortalities in the control and sham-pinx groups. Twelve weeks after the operation a 0.5 ml blood sample was taken from each fish 3hr after lights out. Samples were taken under a dim red light (wavelength 650-800nm). Two weeks later this procedure was repeated 3hr into the light phase. Samples were stored at -70^{0}C prior to assay for melatonin. The absence of the pineal was confirmed by dissection at autopsy 15 months after the operation.

Results

Diel fluctuations in melatonin levels were found in control and sham-pinx groups. However, after removal of the pineal gland, night-time levels of circulating melatonin were significantly reduced from 598±19.3pg/ml (means ± S.E.M.) and 612±29.7pg/ml in control and sham-pinx groups to 96±6.5pg/ml in pinx fish (Fig.1.). Photophase levels for control, sham-pinx and pinx groups did not differ significantly having means of 63.3±4.5pg/ml, 70.3±3.2pg/ml and 64.2±1.3pg/ml respectively. The enucleation study produced lower overall levels of melatonin in all groups due to seasonal fluctuations (Randall *et. al.* 1995). Day and night-time levels of melatonin in control and pinx groups were significantly different (36.0±5.5 and 38.55±3.5 pg/ml during the photophase with 289.1±31.5 and 36.71±6.1 pg/ml during the scotophase). The enucleation+pinx group revealed significantly lower levels of melatonin during both light and dark phases with 16.92±1.0 and 19.48±1.9 pg/ml respectively

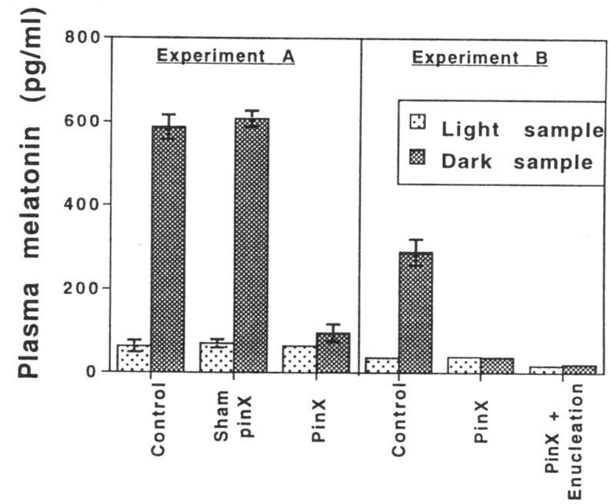

Fig.1. Plasma melatonin levels in pinX, sham-pinX, pinX+enucleated and control groups of Atlantic salmon taken during both phases of the light cycle.

Discussion

These results confirm that the pinealectomy procedure described here is an effective method of reducing nocturnal melatonin levels in young salmon. Further reductions were produced by enucleation, however, the pineal remains the principal source of circulating night-time melatonin. The origin of the remaining melatonin is as yet unknown, however, Huether (1993) identified the gastrointestinal tract as a possible site of melatonin synthesis in higher vertebrates.

Summary

Nocturnal plasma melatonin levels showed a significant decrease in pinealectomised Atlantic salmon parr in comparison with intact control and sham operated fish. Enucleation plus pinealectomy further reduced scotophase melatonin levels.

References

Huether,G. (1993). The contribution of extrapineal sites of melatonin synthesis to circulating melatonin levels in higher vertebrates. *Experentia* **49**, 665-670.

Randall,C.F.,Bromage,N.R., Thorpe,J.E., Miles,M.S. and Muir,J.S. (1995). Melatonin rhythms in Atlantic Salmon (*Salmo salar*) maintained under natural and out-of-phase photoperiods. *Gen. Comp. Endocrinol.* **98**, 73-86.

(Supported by a BBSRC award to M.P. & N.E.R.C. grant to N.B. & C.R)

GnRH-INDUCED CHANGES OF $[Ca^{++}]_i$ IN AFRICAN CATFISH (*CLARIAS GARIEPINUS*) GONADOTROPES.

F.E.M. Rebers*, P.Th. Bosma*, P.W.G.M. Willems‡, C.P. Tensen*, R. Leurs†, R.W. Schulz* and H.J.Th. Goos*.

*Res. Group Comp. Endocrinol., Dept. Exp. Zool., Utrecht University, Padualaan 8, 3584 CH, Utrecht, The Netherlands. ‡ Dept. Biochem., University of Nijmegen, 6500 HB, Nijmegen, The Netherlands. † Dept. Pharmacochem. Vrije Universiteit, De Boelelaan 1083, 1081 HV, Amsterdam, The Netherlands.

Introduction

In brain and pituitary extracts of the African catfish two forms of GnRH have been identified, chicken GnRH-II (cGnRH-II) and catfish GnRH (cfGnRH) Both GnRHs stimulate GTH II release, but cGnRH-II is active at lower concentrations and shows a higher affinity to pituitary GnRH receptors (1). In mammalian gonadotropes GnRH increases the frequency of spontaneous oscillations of the cytosolic free Ca^{++} concentration $[Ca^{++}]_i$ (2). Respective information is missing for fish gonadotropes. We have studied if $[Ca^{++}]_i$ in catfish gonadotropes is modulated by the two endogenous GnRHs, and if cGnRH-II and cfGnRH show differential effects on $[Ca^{++}]_i$ that may explain their different GTH II release potencies.

Material and Methods

Gonadotropes were enriched by Percoll gradient centrifugation from an enzymatically dispersed pituitary cell suspension of mature male catfish. Fraction IV (60-70% gonadotrope, 10-15% somatotrope) was washed and cultured overnight on glass coverslips (L-15, 15 mM HEPES, 26 mM $NaHCO_3$, 1% streptomycin/penicillin, pH 7.4; 25°C, 5% CO_2). Cells were loaded with 10 µM FURA2-AM, 0.02% Pluronic™ for 1 hour, washed with PBS (10 mM, 0.8% NaCl, 1 mM $CaCl_2$, pH 7.4) and perfused with PBS at 1 ml/min. GnRH was added to the cells via the perfusion medium. Dynamic video imaging was carried out using the MagiCal hardware and TARDIS software of Joyce Loebl (UK).

Results

Both GnRHs induce a dose-dependent rise in $[Ca^{++}]_i$ (Fig. 1A). No spontaneous or GnRH-induced $[Ca^{++}]_i$ oscillations were observed. Somato-tropes present in fraction IV did not share this reaction. A prolonged stimulation (10 min) of gonadotropes with 1 nM cGnRH-II, led to a quick rise in $[Ca^{++}]_i$ which decreased to pre-pulse levels in about 2 min. At the end of the GnRH treatment a further decrease of $[Ca^{++}]_i$ was observed.

A similar range of GnRH doses was analysed for the potency in GTH II release experiments (Fig. 1B).

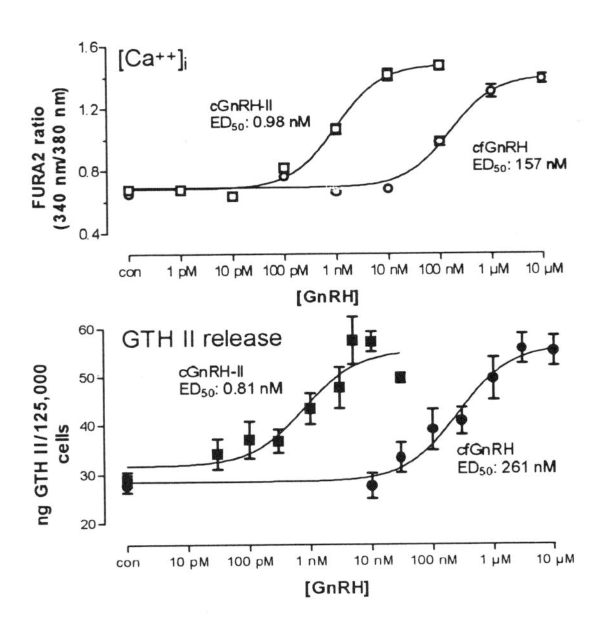

Figure 1: A, Amplitude of $[Ca^{++}]_i$ at the indicated doses of GnRH (mean ± SEM, n=13-68). B, GnRH-stimulated GTH II secretion by dispersed pituitary cells incubated for 30 min with the indicated doses of GnRH (mean ± SEM, n=6).

Discussion and Conclusion

The effects of the two endogenous GnRHs on the GTH II release and $[Ca^{++}]_i$ show a similar dose-response relationship. In both cases, the ED_{50} concentrations for cGnRH-II are lower than those for cfGnRH. In contrast to mammalian gonadotropes (2), catfish gonadotropes do not show spontaneous or GnRH-induced oscillations in $[Ca^{++}]_i$. Thus catfish gonadotropes appear to differ from their mammalian counterparts in the use of $[Ca^{++}]_i$ as a mediator of GnRH signalling.

The present results are in line with previous data (1) on the potency differences between cGnRH-II and cfGnRH as regards their GTH II release potency (both *in vivo* and *in vitro*) and their GnRH receptor binding affinity. Since both GnRHs evoke similar maximal responses on $[Ca^{++}]_i$ and GTH II release, although at different concentrations, we conclude that the difference in receptor binding affinity of the two GnRHs is responsible for their difference in potency.

References
1: Schulz RW et al. (1993) Endocrinol. 133: 1569-77
2: Stojilkovic SS et al. (1994) Endocr. Rev. 15: 462-99

OBSERVATIONS OF THE BRAIN-PITUITARY AXIS IN IMMATURE BLACK CARP

M. P. Schreibman[1], L. Magliulo-Cepriano[2], M. Pennant[1], Z. Yaron[3] and G.Gur[3]

[1]Department of Biology, Brooklyn College, Brooklyn, NY, 11210, Biology Department, [2]State University of NY, Farmingdale, 11735, Zoology Department, [3]Tel-Aviv University, Israel

Introduction

The successful culturing of the black carp, *Mylopharyngodon piceus,* is impeded by their relatively late age of maturation (5-6 years). In order to gain insight into the induction of precocious puberty in these fish we undertook a study of the brain, pituitary gland and olfactory system of the black carp at two ages in their pre-pubertal development. In addition, we have traced the distribution of four molecular forms of gonadotropin releasing hormone (GnRH), two forms of gonadotropin hormone (GTH) and a neuropeptide, FMRF-amide, within the neuroendocrine structures of these animals.

Materials and Methods

Two 4 mos old and two 16 mos old black carp were decapitated after light anesthesia and fixed in Bouin solution. The entire head of the 4 month old specimens were decalcified and processed for analysis. The brain and pituitary of the 16 month old specimens were removed from the cranium before processing them for study. Alternate paraffin-embedded sections (6 micra thick) were stained with routine histological stains (Masson's trichrome and Periodic acid-Schiff) or were treated with the avidin-biotin method (Vectastain, ABC Elite) utilizing polyclonal antibodies generated against variant forms of GnRH, GTH, and FMRF-amide (see Magliulo-Cepriano et al., 1994). Antigenic sites were visualized with 3'3'-diaminobenzedine (DAB). Both absorption and replacement controls resulted in an elimination of immunoreactive staining.

Results

The black carp has a well-developed olfactory system, consisting of highly folded lamellae connected by a short stalk to the olfactory lobe. The pituitary gland is oriented along a dorsal-ventral axis in a depression of the parasphenoid bone.

Immunoreactive (ir)-salmon (s) GnRH was localized in cells and nerve tracts in the olfactory lobe, in the short stalk between the lobe and olfactory epithelium, in the nucleus preopticus periventricularis, in the nucleus lateralis tuberis and in the cells of the caudal pars distalis (CPD) of the pituitary gland..

Ir-chicken II (cII) GnRH was restricted to sparse neural tracts of the olfactory lobe. In younger fish, very pale staining, ir-granules were noted in the pars nervosa.

Ir-lampry (l) GnRH was localized in the cells of the pars intermedia, but was not seen in the brain.

Ir-mammalian (m) GnRH was localized in the same regions as ir-sGnRH and -FMRF-amide but they did not colocalize.

Ir-sGTH I was found in cells throughout the CPD. Ir-sGTH II was not observed in either age group.

Ir-FMRF-amide was distributed throughout the brain with the greatest concentrations in the paraventricular organ (PVO) and the nucleus lkateralis tuberis. Ir-FMRF-amide was not seen in the pituitary glands of the specimens examined.

Conclusions

Ir-GnRH, ir-GTH I and FMRF-amide have been localized in distinct regions in the brain and pituitary gland of immature black carp. These observations indicate that many of the key neurohormonal components involved in the onset of puberty in fish are present long before the event actually occurs. Experimental manipulation of these factors may allow for the induction of early puberty in the black carp. (Supported by BARD #IS-2149-92)

References

Magliulo-Cepriano, L., Schreibman, M.P., and Bluem, V. (1994). Distribution of variat forms of immunoreactive gonadotropin releasing hormone and Beta-gonadotropins I and II in the platyfish, *Xiphophorus maculatus*, from birth to sexual maturity. Gen. Comp. Endocrinol., 94:135-150.

CHARACTERIZATION OF AN ANDROGEN RECEPTOR IN THE BRAIN OF THE ATLANTIC CROAKER, *MICROPOGONIAS UNDULATUS*.

T. Sperry and P. Thomas

University of Texas Marine Science Institute, Port Aransas, TX 78373

Introduction

Androgen receptors have previously been identified in the brain of the goldfish, *Carassius auratus*[1] and the skin of the brown trout, *Salmo trutta*[2]. In this study, a nuclear androgen receptor (AR) specific for testosterone (T) was characterized in the brain of the Atlantic croaker, *Micropogonias undulatus*. This represents the first AR characterization in a perciform fish.

Materials and Methods

The methods described by Pasmanik and Callard[1] were used to assay the AR in the brain of the Atlantic croaker, with only minor modifications.

For Scatchard analysis, 0.25 to 20 nM [^{3}H] T was added to each tube with or without 1 mM unlabeled T. Tubes were incubated at 4°C for at least 5 h. Bound and free steroid were separated by a 5 min incubation with dextran coated charcoal and centrifuged at 5,000 x g for 15 min at 4°C. All one point assays were performed using 5 nM [^{3}H] T with or without 1 mM unlabeled T.

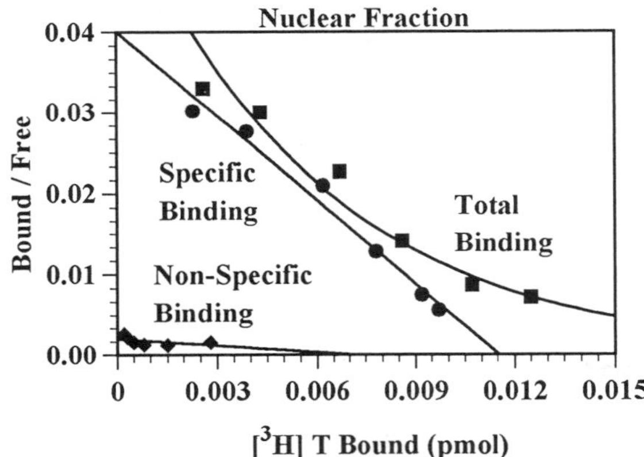

Fig. 1. Representative Scatchard plot from a nuclear fraction of pooled brain tissue, where K_D = 1.21 ± 0.17 nM, B_{max} = 0.58 ± 0.07 pmol/g brain, n = 9.

Results and Discussion

A single class of high affinity, low capacity binding sites for T in both the nuclear (Fig. 1) and cytosolic (K_D = 1.45 ± 0.27 nM, B_{max} = 0.85 ± 0.07 pmol/g brain, n = 6) fractions of Atlantic croaker were found. The AR saturated at 4-5 nM [^{3}H] T in both fractions. The K_D and B_{max} are similar to those found in the goldfish brain[1] and in the brown trout skin[2].

In the nuclear fraction, saturation was reached within 3 h ($T_{1/2}$ = 14.4 min) while complete dissociation occured within 12 h ($T_{1/2}$ = 2.8 h). Similar rates of association and dissociation were found for the cytosolic fraction.

Competition studies indicated that T had the highest affinity for the receptor while dihydrotestosterone, progesterone and methyltestosterone bound with an order of magnitude less affinity. 11-ketotestosterone (11-KT), and the teleost maturation inducing steroids 17α,20β-P and 20β-S had relative binding affinities of less than 1%. The low binding affinity of 11-KT to the AR has also been demonstrated in the goldfish[1], thus raising questions concerning the physiological importance of 11-KT in the brain. No displacement was detected with 11β-hydroxytestosterone, androstenedione, 17β-estradiol and cortisol at concentrations up to 5,000 nM.

T binding moieties were detected in the liver, kidney, swimming and drumming muscles, and testes. Receptor levels were low in these tissues, 10 to 25% of those in the brain, except for the liver cytosol which had comparable levels.

Receptor levels in the nuclear fraction of immature croaker were 3% of those found in mature fish. However, the cytosolic fraction had ten times higher receptor numbers than the nuclear fraction, implying that the receptor may not be activated in the brain during this stage of development.

Gonadally regressed fish had one third the concentration of brain AR of that found in gonadally mature fish. In addition, there was a slight increase in the number of AR's in the brain 1 week after the injection of 1 mg/kg T (ip) to regressed female fish. These results suggest that circulating androgens regulate brain AR levels.

References

1) Pasmanik, M. and Callard, G.V. (1988). A high abundance androgen receptor in goldfish brain: Characteristics and seasonal changes. *Endocrinology*, **123**, 1162-1171.

2) Pottinger, T. (1987). Androgen binding in the skin of mature male brown trout, *Salmo trutta*, L. *Gen. Comp. Endocrinol.* **66**, 224-232.

CHROMATOGRAPHIC AND IMMUNOLOGICAL EVIDENCE FOR A THIRD FORM OF GnRH IN ADDITION TO cIIGnRH AND sGnRH IN THE BRAIN OF ODONTESTHES BONARIENSIS (ATHERINIFORMES).

A.V. Stefano, O. Fridman and G.M. Somoza*.

Fundación CIMAE. Luis Viale 2831 (1416) Buenos Aires. (*) INEUCI-CONICET Depto de Biología. UBA. Ciudad Universitaria (1428). Buenos Aires.Argentina.

Introduction

Molecular forms of GnRH in the brain of Odontesthes bonariensis were studied using a combination of RP-HPLC and RIA with different antisera.

Multiple variants of GnRH molecule have previously been described in the brain of several bony fishes (King & Millar, 1992), but in some other species three different forms have been reported to coexist in the brain of a single species (Somoza et al, 1994). The purpose of the present study was to determine the molecular variants of GnRH in the brain of a South American atheriniforme. Brain extracts were studied using RP-HPLC and RIA with different antisera raised against GnRH variants.

Results and Discussion

Brain extracts were chromatographed as previously described (Somoza et al., 1994). A first screening of GnRH immunoreactivity was done using three different antisera (cII678, PBL#45 and PBL#49) and three GnRH immunoreactive peaks were revealed. The earliest eluting peak eluted in the same position as synthetic cIIGnRH. This peak yielded a serial dilution displacement curve parallel to that of cIIGnRH using a cIIGnRH homologous assay (with cII675 antiserum). The second peak did not coelute with any of the synthetic GnRH variants assayed, and showed crossreactivity in a cIIGnRH homologous assay but not in a sGnRH homologous assay (s1668 antiserum). The third peak eluted in the same position as synthetic sGnRH. This peak yielded a curve parallel to that of sGnRH in two sGnRH homologus RIA systems (with s1668 and sGnRH-Aida antiserum).

A cochromatographic study was performed adding synthetic cIIGnRH and sGnRH to a brain extract and a screening of GnRH immunoreactivity in the HPLC fractions was done with the antiserum PBL#49. The fractions with ir-GnRH immunoreactivity coeluting with cIIGnRH and sGnRH were pooled independently (either in the chromatography of brain extract and brain extract plus standards) and assayed with RIAs homologous for cIIGnRH and sGnRH for comparison. The earliest eluting peak, coeluting with cIIGnRH, was assayed with the antiserum cII675. The equivalent to 1.08 ng was detected in the case of the brain extract, and 3.86 ng were measured in the cochromatographic fractions (difference = 2.78 ng). The difference was due to the synthetic cIIGnRH added to the extract. A similar analysis was done with the latest eluting peak analyzed with two RIAs for sGnRH. In the sGnRH RIA system with s1668 antiserum, 1.16 ng were meassured in the brain extract and 4.14 ng in the cochromatographic fractions (difference = 2.98ng). With the sGnRH RIA system with sGnRH-Aida antiserum the measurements were: 0,8 ng in the brain extract and 3.83 ng in the cochromatographic fractions (difference= 3,03 ng). In both chromatographies the second eluting peak showed no difference (see Figure 1).

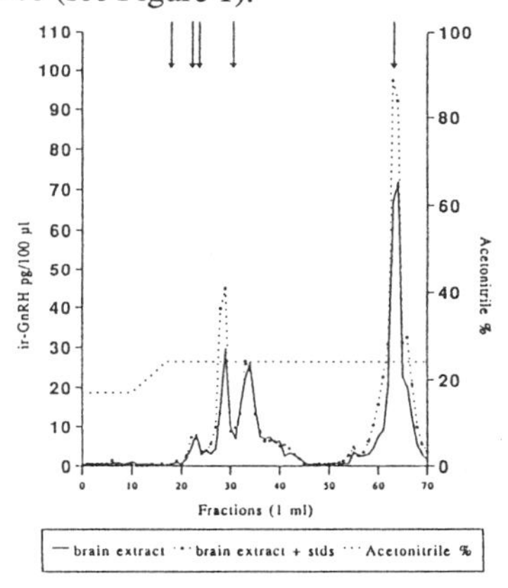

Figure 1: Cochromatographic study of brain extract.

Taking into consideration these results the second eluting ir-GnRH peak could represent a third form of GnRH coexisting with cIIGnRH and sGnRH in the brain of Odontesthes bonariensis.

References

King,J.A. and Millar,R.P. 1992. Evolution of gonadotropin-releasing hormones.Trends Endocrinol Metabol 3:339-346.

Somoza,G.; Stefano,A. et al. 1994. Immunoreactive GnRH suggesting a third form of GnRH in addition to cIIGnRH and sGnRH in the brain and pituitary gland of Prochilodus lineatus (Characiformes). Gen Comp Endocrinol 94:44-52.

DISTRIBUTION OF NEUROPEPTIDE Y-LIKE IMMUNOREACTIVITY IN THE FOREBRAIN AND RETINA OF THE KILLIFISH, *FUNDULUS HETEROCLITUS*

Nishikant Subhedar[1,2], Joan Cerdá[1] and Robin A. Wallace[1]

[1]Whitney Laboratory, University of Florida, St. Augustine FL 32086, USA
[2]Department of Pharmaceutical Sciences, Nagpur University, Nagpur, India

Summary

Neuropeptide Y (NPY)-immunoreactive cells occurred widely along the basal neuraxis, while the fibers were extensively distributed throughout the forebrain. Regional specializations in terms of CSF-contacting sites, innervation of the pineal, retina and the pituitary gland underline the importance of the peptide in a range of central processes, viz., integrative mechanisms, neuroendocrine regulation, photic entrainment, circadian rhythms, etc.

Introduction

NPY has emerged as a chemical messenger of considerable importance in mammals. In teleosts it is known to regulate the secretion of gonadotropins and growth hormone and to mediate the response to sex hormones. However, information with reference to its disposition in the teleostean brain is limited. Therefore, the distribution of the peptide was studied in reproductively active *Fundulus heteroclitus*, a biweekly spawning fish that has its breeding cycle synchronous with lunar tides and that serves as a distinctive teleostean model.

Results

In the forebrain of female killlifish, NPY immunoreactivity was encountered in the cells of the nucleus olfactoretinalis, in the scattered populations of neurons in the lateral part of the ventral telencephalon and in the nucleus entopeduncularis. Isolated neurons were noticed in the hypophysiotropic regions, viz., nucleus preopticus periventricularis (NPP), nucleus preopticus (NPO), nucleus lateralis tuberis (NLT) and also in the nucleus dorsomedialis thalami. While dense network of immunoreactive fibers was seen in the olfactory bulb, the telencephalon showed regional differences, with certain areas having dense terminal fields. In addition, the immunoreactive fibers were seen in the preoptic area, organum vasculosum laminae terminalis, nucleus suprachiasmaticus, tuberal hypothalamus and in the paraventricular thalamic nuclei. Discrete CSF-contacting sites were observed at the level of the NPO, preoptic recess, lateral recess of the third ventricle and in the thalamic region. In the retrochiasmatic area, certain NPY containg fibers and seen to spread on the pial surface. The pituitary gland was richly innervatated by NPY-containing fibers in the neurohypophysis. We also observed distinct fascicles of NPY-containing fibers ascending via the pineal stalk into the pineal gland where they form terminal fields, particularly in the periphery of the gland. In the retina, immunoreactivity was noticed in some amacrine cells; NPY immunoreactivity was also organized into three distinct layers within the inner plexiform layer.

Discussion

In comparison to other teleosts, the NPY system in killifish is relatively well developed and is unique in some respects. For the first time, we report the occurrence of NPY innervation of the pineal. The killifish retina showed NPY reactivity in three distinct layers within the innerplexiform layer, which is quite different from that in goldfish[1]. The presence of NPY in the nucleus olfactoretinalis and its fiber system, suprachiasmatic nucleus, optic nerve, retina and the pineal suggests a role for the peptide in photosensory system and circadian rhythms. There is an emerging possibilty that NPY may colocalize with GnRH and FMRFamide in the NOR and its fiber connectivities to the pineal and retina. The occurrence of the peptide in hypophysiotropic areas like the NPP, NPO and NLT and the rich innervation of the pituitary with NPY fibers underscores its neuroendocrine significance. The nucleus entopeduncularis, which is conspicuous by the presence of large number of NPY containing cells, may also be innervating the pituitary gland. The presence of the peptide in CSF-contacting sites suggests a secretory/sensory role for the peptide.

Reference

1. Osborne, N.N., S. Patel, G. Terenghi. J.M. Allen, J.M. Pollack, and S.R. Bloom (1985) Cell Tiss. Res. 241: 651-656.

Research supported by N.S.F. Grant No. IBN-9306123 awarded to R.A.W.

DETECTION OF OVARIAN AROMATASE mRNA IN GOLDFISH BY RT-PCR ANALYSIS USING BRAIN-DERIVED AROMATASE cDNA PRIMERS

A.Tchoudakova and G.Callard
Department of Biology, Boston University, Boston, MA

Summary

We isolated goldfish ovarian aromatase cDNA fragment by RT-PCR. Partial sequencing of ovarian transcript was practically identical with brain aromatase cDNA.

Introduction

The brain of teleost fish, including the goldfish (*Carassius auratus*), demonstrates exceptionally high aromatase activity in the central nervous system (CNS), with levels from 100- to 1000- times greater than those measured in mammalian CNS (1). By contrast, goldfish ovarian aromatase activity comprises <10% of that in brain, regardless of time of year. Using the aromatase cDNA isolated from goldfish brain, Northern analysis of total RNA (20 μg) isolated from goldfish brain gave a strong hybridization signal with our brain-specific aromatase cDNA; however, when the same probe was used with total (20 μg) or poly(A)-enriched mRNA (5μg) from the ovaries, no signal was detectable. This study was undertaken to determine: (a) if a more sensitive method of analysis (RT-PCR) than RNA blotting would enable us to detect aromatase mRNA in goldfish ovary; and (b) if the sequence of the ovarian transcript differed from that of brain.

Results

RT-PCR of ovarian and brain total RNA using brain-derived aromatase primers MW1 and MW4 resulted in 1000 bp fragments. These were subsequently analyzed by restriction and sequence analyses. Brain clone 1 and ovarian clone 2 had restricton maps coresponding to known goldfish brain aromatase cDNA. Sequencing of brain clone 1 verified that the insert showed to be authentic brain aromatase cDNA. Ovarian cDNA 2 was shown to have a sequence essentially identical to that of brain aromatase cDNA (2) Fig. 1. An ovarian cDNA clone 1 had high sequence similarity to vertebrate lamins. Sequences of the other four ovarian PCR products showed no matches in the BLAST database.

Fig 1. Comparison of nucleotide sequences of ovarian (ov2) and brain (br) aromatase cDNAs.

ov2 GNTCATGAGCTACAGCAGGTTCCTATG
br GATCATGAGCTACAGCAGGTTCCTATG

ov2 GATGGGGATCGGCTCCGCATGCAACTA
br GATGGGGATCGGCTCCGCATGCAACTA

ov2 CTACAATCAAAAATATGGCAGCATTGC
br CTACAATGAAAAATATGGCAGCATTGC

ov2 TCGGGTCTGGATCAGCGGAGAAGAGAC
br TCGGGTCTGGATCAGCGGAGAAGAGAC

ov2 CTTTATACTTAGCAAGTCCTCTGTGGT
br CTTTATACTTAGCAAGTCCTCTGCGGT

ov2 GTATCATGTTCTGAAGAGCAATAATT
br GTATCATGTTCTGAAGAGCAATAATT

Discussion

Goldfish ovarian aromatase transcripts were detected for the first time by RT-PCR, indicating that negative results with Northern analysis were due to insufficient sensitivity. Restriction analysis and partial sequencing of this transcript, which included a region with the presumptive functional domains (Ozols peptide and part of helix regions) revealed no differences between ovarian and brain aromatase mRNA. These data do not rule out the existence of additional non-neural forms of the aromatase message. Also, sequencing of the remaining part of the ovarian clone 2 is required to determine whether differences may exist in other parts of the molecule.

References

1. Callard G, Drygas M and Gelinas D, 1993. Molecular and cellular physiology of aromatase in the brain and retina. J. Steroid Biochem. Molec. Biol. Vol. 44(4-6):541-547.
2. Gelinas D, Pitoc G and Callard G. (manuscript in preparation).

Supported by NSF-DCB89-16809.

THE NEUROENDOCRINE REGULATION OF GONADOTROPIN SECRETION AND OVULATION IN THE BAGRID CATFISH, *MYSTUS MACROPTERUS*

D.S. Wang, H.R. Lin and H.J.Th. Goos*

Department of Biology, Zhongshan University, Guangzhou, P.R. China, *Department of Experimental Zoology, Research Group for Comparative Endocrinology, University of Utrecht, Utrecht, The Netherlands

Summary

LHRH-A (0.1 ug/g body wt) alone stimulated an increase in serum GtH levels significantly at 6 hr after injection, but was ineffective for the induction of ovulation. The dopamine antagonist, domperidone (10 ug/g body wt) alone was ineffective in increase serum GtH level within 24 hr after injection. However, domperidone caused a marked potentiation of the GtH release and ovulation response to LHRH-A,indicating that dopamine functions as a gonadotropin release inhibitory factor. The use of domperidone in combination with LHRH-A provides a reliable practical method for the induction of ovulation in the bagrid catfish.

Introduction

Bagrid catfish (*Mystus macropterus*) are rapidly gaining popularity as cultured fish in China recently. The effects of : (1) a luteinizing hormone-releasing hormone analog, (D-Ala^6,Pro^9-NEt)-LHRH; (2) the dopamine antagonist, domperidone (DOM); and (3) a combination of these substances, on gonadotropin (GtH) secretion and ovulation in the bagrid catfish were investigated.

Sexually mature bagrid catfish (145-195 g of body weight) were collected from the Jialingjiang River, a branch of the Yangtze River. LHRH-A (0.1 ug/g body wt) and domperidone (10 ug/g body wt) were administered by intraperitoneal injection. Blood samples were taken at 6, 24 and 36 hr after the injection. GtH concentration in the serum samples was measured by radioimmunoassay (RIA) using African Catfish GtH as standard and antiserum to African catfish GtH as described by Goos et al.(1986) with minor modification. Duncan's multiple range test was used to determine the differences ($P<0.05$) in the mean GtH levels.

Results

1. Injection of LHRH-A alone at 0.1 ug/g body wt was effective in stimulating an significant increase in serum GtH levels at 6 hr, but not at 24 and 36 hr postinjection (Fig,).The ovulatory response to injection of LHRH-A alone did not differ from the saline injection.

2. Injection of DOM alone at 10 ug/g body wt failed to stimulate an increase in serum GtH levels at 6 hr or 24 hr after injection when compared to the control fish.

3. Combined injections of DOM+LHRH-A resulted in significant higher serum GtH levels than injection of LHRH-A alone at 6 hr postinjection. At 24 hr, serum GtH level in DOM+LHRH-A injected fish were similar to LHRH-A injected fish. The combination of DOM+LHRH-A also resulted in a significant ovulatory response.

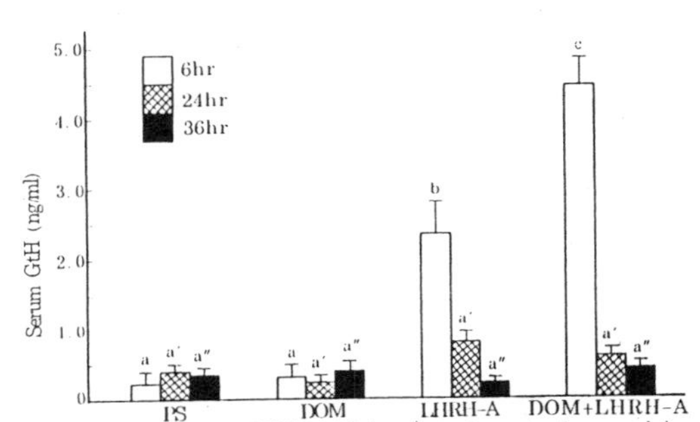

Figure: Effects of LHRH-A(0.1 μg/g body wt) alone and in combination with DOM (10 μg/g body wt) on serum GtH levels in bagrid catfish. GtH values are reported as mean ±SE. At each sampling time, groups with similar serum GtH levels, as determined by Duncan's multiple range test ($p<0.05$) are identified by the same superscript.

Discussion

The present results demonstrated that LHRH-A stimulates GtH release, but was ineffective in inducing ovulation in sexually mature bagrid catfish. DOM alone did not stimulate GtH release, however, DOM potentiated the GtH release response to LHRH-A; the combined treatment was effective in inducing ovulation. This suggests that dopamine does not affect the GtH release directly, but indirectly by blocking the effect of gonadotropin-releasing hormone on GtH secretion, which is consistent with the concept that dopamine functions as a gonadotropin release inhibitory factor in the African catfish (de Leeuw et al., 1985) and other teleosts, such as Chinese loach (Lin et al., 1986).

References

Goos,H.J.Th.,De Leeuw,R.,Burzawa-Gerard,E., Terlou,M. and Richter,C.J.J. 1986. Purification of gonadotropic hormone from the pituitary of the African Catfish, *Clarias gariepinus*(Burehell), and the development of a homologous radioimmunoassay. Gen.Comp.Endocrinol., 63:162-170.

CONCENTRATION OF GONADOTROPIN-RELEASING HORMONES IN BRAIN OF LARVAL AND METAMORPHOSING LAMPREYS OF TWO SPECIES WITH DIFFERENT ADULT LIFE HISTORIES

J.H. Youson,[1] M. Docker,[2] and S.A. Sower.[3]

[1]Division of Life Sciences, Scarborough Campus, University of Toronto, Scarborough, Ontario M1C 1A4; [2] Pacific Biological Station, Department of Fisheries and Oceans, Nanaimo, B. C. V9R 5K6; [3]Department of Biochemistry and Molecular Biology, University of New Hampshire, Durham, NH 03824.

Summary

GnRH-I and -III were found at all periods of the life cycle of L. richardsoni) and P.marinus but the values are much lower in the nonparasitic species. There is a higher GnRH-III: GnRH-I ratio during larval life and throughout early metamorphosis in both species but this situation is reversed during postmetamorphic intervals. An earlier increase in GnRH-I in L. richardsoni during metamorphosis might reflect the time of a major stimulus to the final development of the gonad and be a key event in dictating its nonparasitic adult life history.

Introduction

There are two adult life history types among lampreys, parasitic and nonparasitic, following metamorphosis. Whereas juveniles of parasitic species commence an interval of feeding and somatic growth with little gonadal maturation for 1-2 years, postmetamorphic individuals in nonparasitic species immediately begin sexual maturation without feeding. We wish to test the hypothesis that metamorphosis is the time when a variable stimulus to the reproductive system occurs which ultimately directs the adult life history type. We measured the concentrations of two forms of lamprey gonadotropin-releasing hormones (GnRH-I, Sherwood et al., 1986; GnRH-III, Sower et al., 1993) during larval life and metamorphosis of a nonparasitic (Lampetra richardsoni) and a parasitic (Petromyzon marinus) species.

Materials and Methods

Entire brains were extirpated, rapidly frozen on dry ice, extracted, and fractions from HPLC were assayed for the two GnRHs according to the method described in Fahien and Sower ('90) with modifications as in Youson and Sower (1991). The sensitivity of the RIA was 9.8 pg/tube and the range of binding was 36-43%. Data for hormone concentrations are presented as mean pg or ng/brain ± 1SE.

Results

Lamprey GnRH-I and -III were detected in all brain samples of both species (Figs. 1 and 2) but levels of these hormones were lower in the nonparasitic species. Particularly noteworthy were the detection of the GnRHs in all year classes of larval P. marinus. GnRH-III is the predominate GnRH in larvae and early metamorphosing individuals with GnRH-I concentration increasing during metamorphosis. The apparent equivalence in levels of GnRH-I and -III appears earlier (stage 4) in L. richardsoni than P. marinus (stage 6) but by the end of metamorphosis GnRH-I dominates in both species.

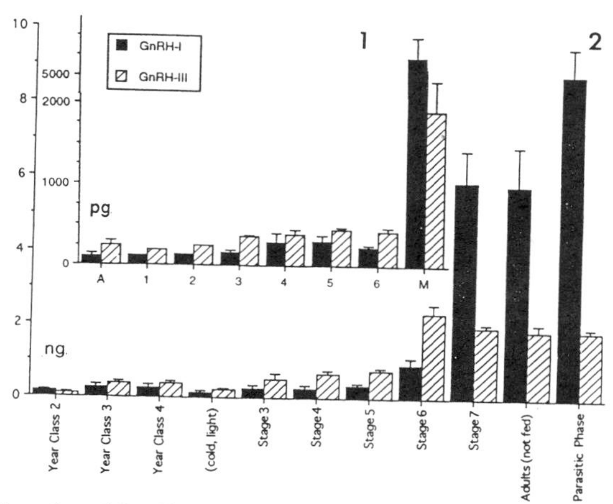

Figs. 1 and 2. Concentrations (mean pg or ng/brain±SE) of two GnRHs (I and III) in the brain at various stages in the life cycle of L. richardsoni (1) and P. marinus (2).

Discussion

In a previous study on anadromous P. marinus we showed that there were low levels of GnRH-I and an additional form of GnRH (now -III) in brains of larvae and early metamorphosing individuals (Youson and Sower, 1991). The relative values we provided indicate that GnRH-III is the dominant form of GnRH throughout larval life and at least in early metamorphosis. GnRH-I is the dominant adult form. The present investigation is the first to report GnRH concentrations in the brains of nonparasitic species but values are low when compared to similar-sized premetamorphic and nonfeeding adult P. marinus. Despite this difference, interspecific comparisons can be made on the timing of the onset of the adult-type profile of GnRH. Equivalent values of the GNRHs are reached and essentially maintained much earlier in L. richardsoni and this may be an indication of some earlier reproductive stimulus in the latter species.

References

Sherwood, N.M., S.A. Sower, D.A. Marshak, B.A. Fraser, and M. J. Brownstein. 1986. Primary structure of gonadotropin-releasing hormone from lamprey brain. J. Biol. Chem. 261: 4812-4818.

Sower, S.A, Y.C. Chiang, S. Lovas, J.M. Conlon. 1993. Primary structure and biological activity of a third gonadotropin-releasing hormone from lamprey brain. Endocrinology 132: 1125-1131.

Youson, J.H., and S.A. Sower. 1991. Concentration of gonadotropin-releasing hormone in the brain during metamorphosis in the lamprey, Petromyzon marinus. J. Exp. Zool. 259: 399-404.

Supported by NSERC and GLFC grants to JHY and a NSF grant to SAS.

Aquaculture

CRYOPRESERVATION OF AQUATIC GAMETES AND EMBRYOS: RECENT ADVANCES AND APPLICATIONS

Krishen Rana

Institute of Aquaculture, University of Stirling, Scotland, FK9 4LA

Summary

Although successful cryopreservation of fish eggs and embryos is still elusive, spermbanks which are currently feasible can play a crucial role in aquaculture and conservation management (McAndrew *et al.*, 1993). Heterogenous results and cryopreservation of small volumes of milt, however, are likely to prevent uptake of this technology. The possible consequences of previously neglected variables such as milt collection techniques and precooling storage, handling and cooling conditions for consistent cryosuccess are therefore considered here together with possible causes of cryoinjury. Recent developments in field and large scale cryopreservation techniques and their applications in aquaculture and conservation are considered.

To increase the likelihood of successful cryopreservation of fish eggs and embryos, concerted research effort is required to identify and address fundamental cryobiological constraints.

Introduction

Following cryopreservation of Atlantic herring (*Clupea harengus*) testes (Blaxter, 1953), the feasibility of successfully cryopreserving spermatozoa has been demonstrated in over 200 fish species (Stein cited by Billard *et al.*, 1995a) notably for the salmonids, carps, tilapias. Although a number of different protocols, even for the same species are advocated in the literature (see reviews by Leung and Jamieson, 1991; McAndrew *et al.*, 1993) the components of cryopreservation are the same. Freshly collected semen is diluted with a balanced salt solution containing either 7-10% dimethyl sulphoxide (Me_2SO), glycerol or methanol, cooled as pellets on dry ice or more commonly in straws over liquid nitrogen (LN), stored in LN and rapidly warmed in a 20-40°C water bath prior to inseminating eggs. The interpretation of published results, however, is fraught with difficulties. Often components of cryopreservation such as, prefreezing milt quality and storage conditions, packaging, cooling, warming, insemination and protocol evaluation vary considerably between and within studies (Billard *et al.*, 1995a; Rana, 1995a&b). This together with the empirical and trial and error approach has led to variable results (Fig. 1).

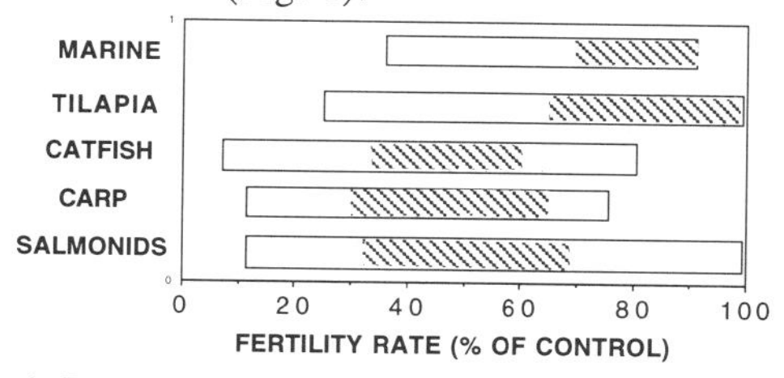

Fig.1 Range of viability of post-thawed spermatozoa. Shaded area = majority data.

Successful cryopreservation of aquatic eggs and embryos is limited to a few groups of invertebrates, notably the Pacific oyster (*Crassostrea gigas*). Although low permeability and surface:volume ratio and high internal water are suggested the fundamental reasons for cryofailure of fish eggs and embryos remains unclear. In this presentation recent information on precooling cryoprotectant toxicity, permeability and osmotically active water will be discussed.

Components of spermatozoa cryopreservation

Collection technique and gamete quality

Sperm fitness and in particular the inorganic and organic seminal plasma composition shows high intra and interspecies variation (Leung and Jamieson, 1991; Billard *et al.*, 1995a; Rana, 1995a&b).

To date semen quality analysis is based on milt samples expressed by abdominal pressure but unavoidable urine contamination can dilute milt by as much as 80% (Rana, 1995a) and may result in false intra-individual variation.

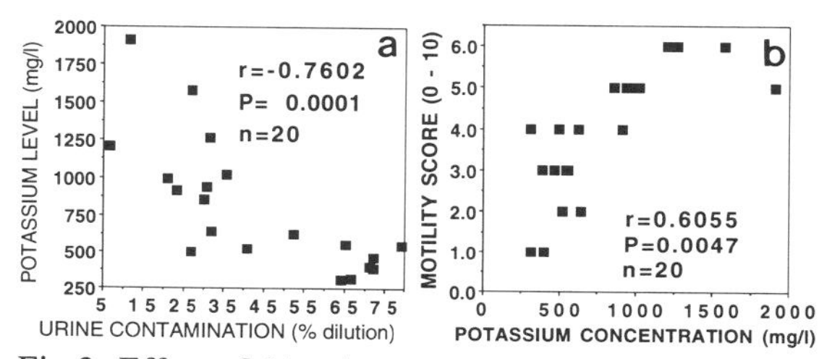

Fig 2. Effect of (a) urine contamination on potassium levels and (b) milt potassium levels on motility.

In Atlantic salmon, *Salmo salar*, the wide range (220-300mOsm/kg) in milt osmolality (Munkittrick and Moccia, 1987) was probably due to urine contamination. The osmolality of catheterised salmon milt is highly homogenous (300-330 mOsm/kg vs 180-290 mOsm/Kg; Rana, 1995a). The heterogeneity in urine contamination (Fig 2a) and K^+ of contaminated milt may also reduce the proportion of active spermatozoa (Fig 2b).

Such variability is likely to be further confounded by intra-male variation in protein and osmolality of urine (Rana, 1995a). The overall, resultant variability in milt may reduce sperm quality and alter cooling properties of the milt in an unpredictable and random manner. In simulated trials the post-thaw motility score (0-10) of Atlantic salmon milt containing 0-25% urine was reduced from 6 to 4 (Rana, unpublished data).

Duration of precooling storage

Although it is widely recommended that spermatozoa be cryopreserved immediately after collection this may not always be practical under hatchery situations. Detrimental effects of precooling storage, however, are unclear and results contradictory. In rainbow trout (*Oncorhynchus mykiss*) a 6 delay in cooling chilled milt reduced post-thaw fertility rates from 74 to 52% (Schmidt-Baulain and Holtz, 1989). Simarily, in Atlantic halibut (*Hippoglossus hippoglossus*), motility was reduced from 80 to 5% within 7h of collection. In contrast, Baynes and Scott (1987) report higher post-thaw fertility following 26h storage at 0°C. The reasons for the variability are unclear. In carp (*Cyprinus carpio*), 8-10h storage at 4°C resulted in a decline in intracellular ATP and morphological changes (Billard *et al.*, 1995b). Recent evidence supports the view that spermatozoa activated prior to cooling retain their potential to be successfully cryopreserved. In *Oreochromis niloticus*, storage of milt at 4°C for up to 6 days had no significant effect on post-thaw fertilization rate of eggs (Rana *et al.*, 1990). In Atlantic salmon, up to 40-50% of spermatozoa diluted in modified Cortlands (Truscott, *et al.*, 1964) containing 5% methanol and held at -4°C for up to 24 days prior to cryopreservation were motile but the extent of recovery was dependent on dilution ratio (Fig. 3).

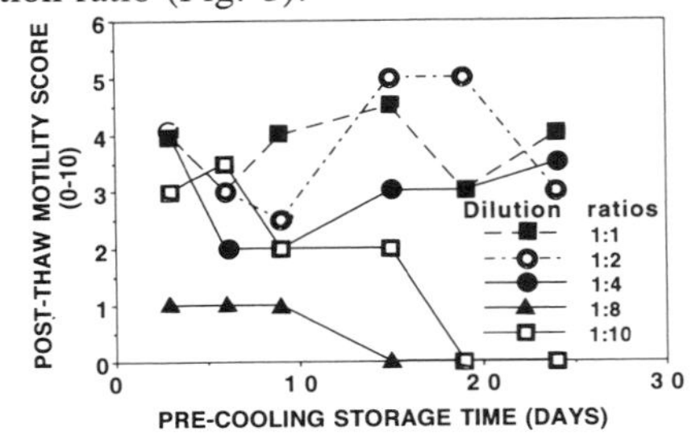

Fig.3 Precooling storage time on post-thaw motility.

Cryodiluents and dilution ratio

Undiluted gametes cannot survive the traumas of cooling and warming. Complex diluents (see reviews by Scott and Baynes, 1980; Leung and Jamieson, 1991; McAndrew *et al.*, 1993) used to dilute semen offer no advantage over simple osmotically balanced solutions such as those containing 0.3-0.6M sucrose and 10% Me_2SO (Holtz *et al.*, 1991) or glycerol (Piironen, 1993). Whilst simple extenders are equally successful it must be emphasised that the results do not suggest any improvement in the efficacy of protocols rather they confirm our limited knowledge of the cryobiological processes associated with spermatozoa cryopreservation. Studies on rainbow trout by Holtz *et al.*(1991) using a 0.6M sucrose cryodiluent reported maximum fertility rate of 87% using pellet technique. These levels of post-thaw fertility rates, however, are achieved using up to 100 times more milt than normal (Billard, *et al.*, 1995a) with up to 200 eggs. Whilst this level of efficacy may be regarded as acceptable for the salmonids which produce copious volumes of milt (Holtz *et al.*, 1991) it is wholly inadequate in, for example, *Haplochromid* spp. which produce maximum milt volumes of 5-10μl.

Milt is usually diluted up to ten fold prior to cooling. Sperm density, however, exhibits high seasonal and intra-individual differences (Leung and Jamieson, 1991). Consequently, sperm density can vary by as much as 300% for any given dilution ratio. High cell densities may also result in compression damage. During cooling water freezes and cells can be compressed in residual water channels. In red blood cells the density of cells within these channels influences post-thaw viability of cells (Mazur and Cole, 1985). Therefore, in fish sperm cryopreservation standardised cell density should dictate the dilution ratio.

Cooling method and rates

Although cooling rate is the most critical phase in cryopreservation (Grout and Morris, 1987) it is the least standardised variable in fish sperm cryopreservation studies.

The cooling rate and its reproducibility varies with the cooling method. In insulated boxes, dewar necks and pellets, the pre- and post freezing phases of the cooling rates are driven by the difference between ambient and straw temperature and therefore the cooling rates between these phases can vary by as much as 500%. Consequently, cooling rates in straws within and between runs can be highly variable (Table 1). Such variation can be minimised by cooling milt in controlled rate coolers or specially designed heat sinks (see below).

Table 1 Mean cooling rates within straws

	Insulation Box[1]		Dewar[2]	
	PrF[3]	PoF[3]	PrF[3]	PoF[3]
°C/min	34	8.1	16.1	17.8
Range	25-41	5-13	6-36	6 - 27
CV(%)	14	35	44	34
No. straws	12	12	28	28

[1]Straws positioned 8cm above LN surface, [2]Goblet placed 2cm below the neck of dewar, [3] Prf and PoF = pre- and post freezing cooling rates, respectively.

Evaluation of protocols

To date there is no convincing correlation of sperm fitness and fertility rate. Recent studies on rainbow trout (Gallant and McNiven, 1991) show that neither motility ratings, LDH leakage nor proportion of permeablised cells have any significant bearing on the fertility rate of eggs (Fig. 4). One possible reason may be that like motility rating samples for evaluation are not taken from the site of fertilization and consequently bear little relation to fertilization.

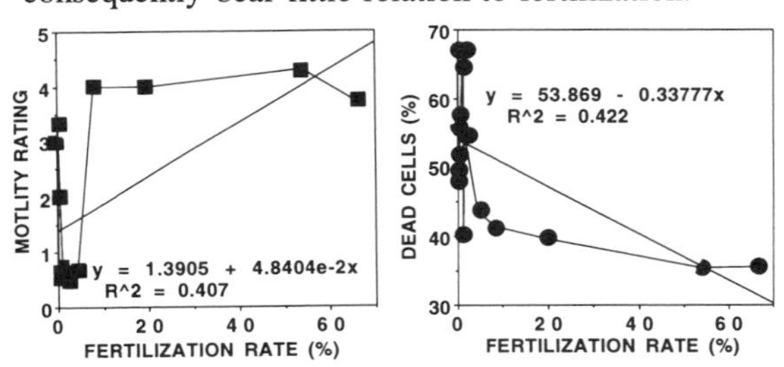

Fig. 4. Relationship between a) permeabilised cells and b) motility and fertility. *Data adapted from Gallant and McNiven, 1991.*

Possible causes and nature of cryoinjury

Despite advances to date at best only 30-50% of spermatozoa retain their post-thaw motility and we have little insight into when the damage occurs, the exact causes of damage and its significance for the outcome of cryopreservation. Recent studies suggest cells are fatally damaged along the entire cryopreservation process. A substantial proportion of cells are damaged prior to cooling (Lawrence, 1992, Lahnsteiner *et al.*, 1992, Linhart *et al.*, 1993, Lawrence and Rana, in press). Fluorometric assessment of *O. niloticus* sperm membrane integrity during key phases of cryopreservation revealed that up to 35% of cells are permeabilised (ie. damaged) following 30min equilibration in a 10% methanol cryodiluent prior to cryopreservation and a further 26% during the cooling/warming stages (Fig. 5).

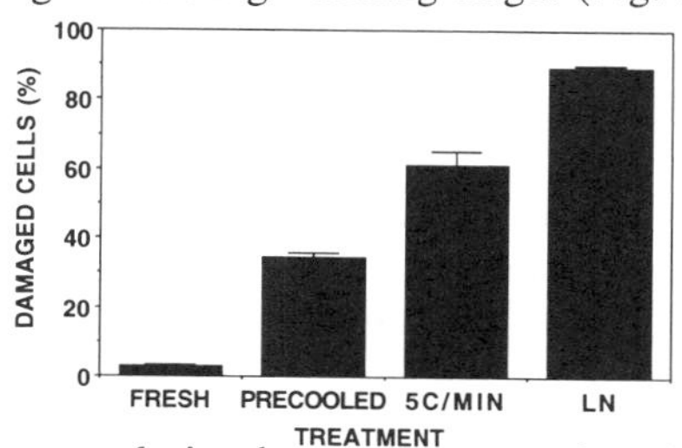

Fig 5. Damage during key cryopreservation phases.

Such damage is also observed as lower post-dilution motility following 5-20 min equilibration and this may be further reduced at higher cryoadditive concentrations (Linhart *et al.*, 1993).

Field applications

The use of the pellet technique for field cryopreservation, though convenient, is impractical for rational long term genetic resource banking. The shelf life of dry-ice is short, particularly in the tropics, and perhaps more importantly frozen samples cannot be sealed to prevent possible cross contamination of diseases in storage vessels. In addition, the thawing of large numbers of pellets is equally problematic. The fusing of pellets during thawing reduces the warming rates within aggregated pellets in an unpredictable manner.

To overcome these short comings for the tilapias a fixed rate portable cooler (FRPC) has recently been developed to generate reproducible linear cooling rates and can be used in the field for up to two weeks. Typical results from field cryopreservation trials are shown in Table 2.

Table 2 Viability of *O. niloticus* spermatozoa cryopreserved in the FRPC

Country	n	Eggs	Fert.(%)	spm.egg$\times10^6$
Sri Lanka	5	110-200	71.4 (5.1)	0.47(0.4)
Mexico	12	235-240	91.2 (3.0)	0.29 (0.01)

Large scale cryopreservation of semen

Reports on the use of large (4.5-5ml) straws and pellets for practical cryopreservation are limited. The The viability of post-thawed rainbow trout milt diluted (1:3) in 5.4% glucose and 10% egg yolk containing 10% Me_2SO and cooled on dry ice in 4.5ml straws average 84% (1.1 $\times10^7$sperm/egg) using 700 eggs (Wheeler and Thorgaard, 1991). For Atlantic salmon, milt diluted in modified Cortlands containing 10% methanol and cooled linearly at 50°C/min could fertilize around 3500 unhardened eggs (Table 2). Under controlled cooling conditions the sperm:egg ratio could be reduced from 4.2 to 1.7 $\times10^6$ without incurring any significant loss in fertility rate. At 1:8, 1500 eggs were fertilized (Table 3). In catfish (*Silurus glanis*) 6500 eggs were successfully fertilized with 5ml pellets (Linhart *et al.*, 1993).

Table 3 Mean post-thaw viability[1] of salmon milt packaged in 5ml straws at varying dilution ratios.

M:D[2]	No.eggs used	Fert. rate[1]	sperm:egg $\times10^6$
Fresh	5134(460)	92(2.8)[a]	0.35
1:1	5867(283)	61(1.9)[b]	4.2
1:2	6143(814)	59(3.5)[b]	2.7
1:4	5911(191)	59(0.6)[b]	1.7
1:8	5039(631)	31(4.0)[c]	1.1

[1] as mean [%(SEM); n=2] eyed eggs. [2] M:D= milt: extender. Means with different superscripts are significantly ($P<0.05$) different (analysis on arc-sine transformed data).

Cryopreservation of aquatic eggs and embryos

Bivalves

Among the invertebrates, bivalves are currently the most studied group. Pacific oyster embryos have been successfully cryopreserved by cooling between 0.5-2°C/min (Rana *et al.*, 1992; Lin *et al.*, 1993; Chao et al., 1993) and by vitrification (Lin *et al.*, 1993) and to date post-thaw success has been largely quantified as post-thaw ciliary movement. Post-thaw success using vitrification is low (<1%) but higher recoveries (20-50%) are reported following slow cooling (Rana *et al.*, 1992; Chao *et al.*, 1993). The ability of embryos to survive cryopreservation is reported to vary with size of eggs, embryonic stage, cryoprotectant type and concentration and cooling rate (Rana *et al.*, 1992; Gwo, 1995). Cryopreservation of unfertilized eggs and 2-8 cell stage embryos were unsuccessful and the trochophore and D-larvae were demonstrably the most tolerant to cryopreservation.

Cryobiological progress in fish eggs and embryos

Although a few reports claim successful recovery of fish embryos from LN (Zhang *et al.*, 1989; Leung and Jamieson, 1991) there is to date no reproducible cryopreservation protocols. Review of recent data suggest that while freezing contributes to fatality there is a gradual loss in viability during the entire cooling protocol (Lin *et al.*, 1993; Zhang *et al.*, 1993; Adam, 1995). Studies by Adam (1995) suggest that between 20-80% mortality may occur prior to freezing due to cryoprotectant toxicity and cold shock damage and therefore the importance of prefreezing damage may be underestimated.

Cryoprotectant-induced injury during the long equilibration and prefreezing phase can be significant. The causes of such damage remain unclear. In rosy barbs (*Puntius conchonius*) and zebra fish (*Brachydanio rerio*) embryos, high cryoprotectant concentration significantly reduced the enzymatic activity of LDH and G-6-PDH. This decrease which was highest for the blastula stage, was attributed to the rupturing of the embryo caused by hydrostatic pressure within the perivitelline space and subsequent denaturation of the leached enzymes by the cryoprotectant (Adam *et al.*, 1995).

Fish eggs and embryos are prone to cold shock damage and by 0°C up to 70% of embryos can be fatally damaged depending on cooling rate, cryoprotectant type and concentration and embryonic stage. In most studies DMSO offers the best pre freezing cryoprotection. Subzero cooling is equally traumatic and by -20 to -40°C the viability of the most resistant stage, the heart beat stage, is reduced to 10-20% (Lin et. al., 1993; Zhang et al., 1993; Adam, 1995.

To improve the low water permeability of rosy barb and zebra fish embryos (0.003-0.008μmsec^{-1} at 20°C) the feasibility of electroporation has been investigated (Adam, 1995).

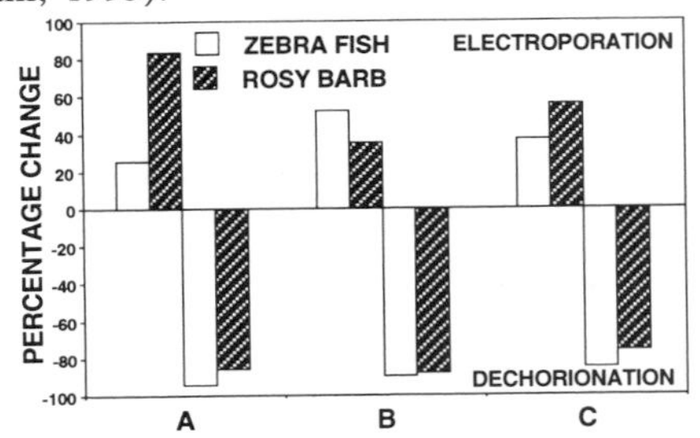

Fig.6 Effect of electroporation and dechorionation on water permeability of (A) cleavage, (B) epiboly and (C) closure of blastopore embryos.

In intact rosy barb embryos at the cleavage, epiboly and closure of blastopore stages electroporation increased permeability coefficients by 83, 34 and 56%, respectively (Adam, 1995). Much of this increase, however, may be due to improved chorion permeability. The permeability of dechorionated embryos was reduced by as much as 80% (Fig. 6) suggesting that most exchange occurs between the perivitelline space and exterior.

The removal of intracellular water is regarded as crucial for cryosuccess. Recent studies by Hagedorn *et al.* (1993) on 8h old zebra fish embryos reported a volume reduction of 8-13%. This relatively small reduction in volume probably reflects the embryonic stage used and that whole egg diameters rather than actual changes in yolk diameters were measured.

Nuclear magnetic resonance (NMR) estimates of intracellular water in epiboly and closure of the blastopore embryos suggest that 22-24% of the water could be removed (Adam, 1995). In precleavage rosy barb embryos, however, up to 36% of the water from the yolk mass (Rana *et al.*, 1995c) and around 50% from the whole egg can be removed (Fig. 7).

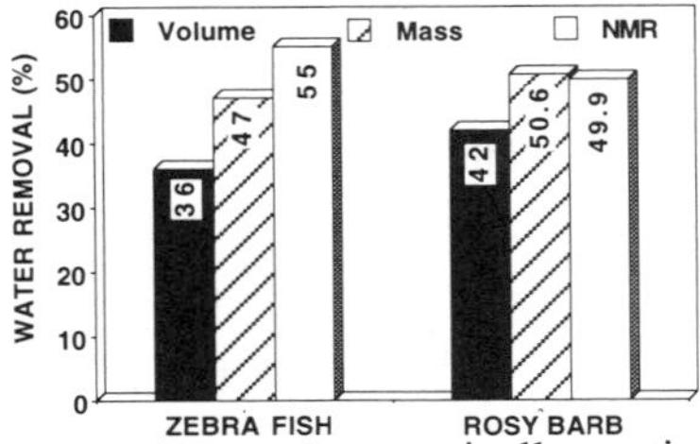

Fig. 7 Comparison of osmotically active water removed from precleavage embryos by three techniques. (*data adapted from Adam, 1995*)

The osmotically non-active volumes do not deviate markedly from other recently cryopreserved biological systems such as *Drosophila* (Steponkus *et al.*, 1990).

The requirement of adequate dehydration for cryopreservation may not be universal. *Artemia* cysts undergo natural dehydration and have a hydraulic conductivity (0.24 μmmin^{-1}.atm^{-1}) similar to mammalian embryos (0.27-1.27 μmmin^{-1}.atm^{-1}). By allowing them to hydrate in water for a fixed period of time, varying levels of internal water can be simulated and cysts cryopreserved. Moreover, the cysts can be decapsulated to improve permeability. These studies demonstrate that despite containing up to 60% water, post-thaw decapsulated and normal cysts could hatch into normal nauplii (Fig. 8).

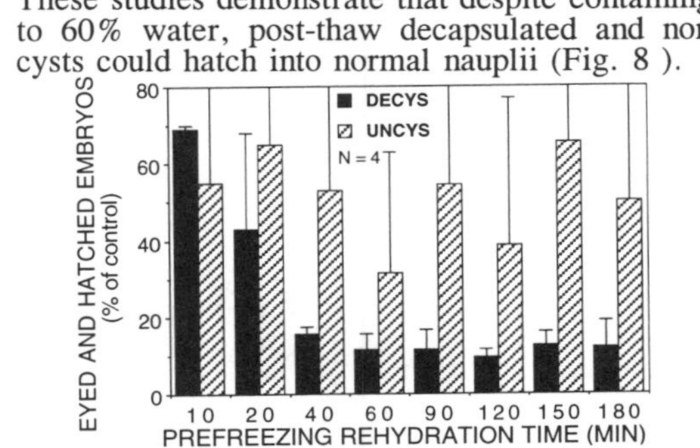

Fig. 8 Post-thaw viability of hydrated *Artemia* cysts. DECYS = decapsulated, UNCYS = encysted.

The cooling process can be observed using a cryo-microscope and internal freezing detected by flashing or blackening of the embryo. Cryomicro- scopic observations show that fully hydrated *Artemia* cysts do not show any evidence of internal freezing. Elucidation of the mechanisms for *Artemia* may provide an insight into methods for improving fish egg and embryo cryopreservation.

Acknowledgements

Studies reported here were partially funded by the British Overseas Development Administration . The assistance of Ann Gilmour and Adam Musa is greatly appreciated.

References

Adam, M.M., 1995. Pre-freezing and cryostorage problems of rosy barbs and zebra fish embryos. PhD.Thesis, Institute of Aquaculture, University of Stirling.pp 93.

Adam, M.M., Rana, K.J. and McAndrew, B.J., 1995. Effect of cryoprotectants on activity of selected enzymes in fish embryos. Cryobiology, 32:92-104.

Baynes, S.M. and Scott, A.P., 1987. Cryopreservation of rainbow trout spermatozoa: The influence of sperm quality, egg quality and extender composition on post - thaw fertility. Aquaculture, 66: 53-67.

Billard, R., Cosson, J. and Crim, L.W., 1993. Motility of fresh and aged halibut sperm. Aquatic Living Resources, 6:67-75.

Billard, R., Cosson, J., Crim, L.W. and Suquet, M., 1995a. Sperm physiology and quality. Broodstock

Management and Egg Larval Quality. N. R. Bromage and R.J.Roberts (Eds), pp. 25-52. Blackwell Science.

Billard, R., Cosson, J., Perchec, G. and Linhart, O., 1995b. Biology of sperm and artificial reproduction in carp. Aquaculture, 129:95-112.

Blaxter, J.H.S., 1953. Sperm storage and cross fertilisation of spring and autumn spawning herring. Nature, 172:1189-1190.

Chao, N., Tsai C-T., Hsu, H. and Lin, T., 1993. Selection of stepwise program parameters to cryopreserve the oyster embryo. Cryobiology, 30:615 (Abstract).

Gallant, R.K. and McNiven, M.A., 1991. Cryopreservation of rainbow trout spermatozoa. Bull. Aqua. Ass. Canada, 91:25-27.

Grout, B.W. and Morris, G.J., 1987. The Effect of Low Temperature Systems. Edward Arnold, London.

Gwo, J.C., 1995. Cryopreservation of oyster (*Crassostrea gigas*) embryos. Theriogenology, 43: 1163 - 1174.

Hagedorn, M., Westerfield, M., Wildt, D. and Rall, W.F., 1993. Preliminary studies on the cryobiological properties of dechorionated zebra fish embryos. Cryobiology, 30:604.

Holtz, W., Schmidt-Bualain, R. and Meiners-Gefkin, M., 1991. Cryopreservation of rainbow trout *(Oncorhynchus mykiss)* semen in a sucrose /glycerol extender. Fourth International Symposium on Reproductive Physiology of Fish, Abstract 63.

Lahnsteiner, F., Weisman, T. and Patzner, R.A., 1992. Fine structural changes in spermatozoa of the grayling, (*Pisces Teleostei*), during routine cryopreservation. Aquaculture, 103:73-84.

Lawrence, C. 1992. Development of computer image analysis and fluorometry to assess the fitness of cryopreserved tilapia spermatozoa. M.Sc. Thesis, Institute of Aquaculture, Stirling University. pp 79.

Leung, L. K.-P. and Jamieson, B.J.M., 1991. Live preservation of fish gametes. Fish Evolution and Systematics : Evidence from Spermatozoa. B.G.M. Jamieson(Eds), pp. 245 - 269. Cambridge University Press Cambridge.

Lin T-T., Tung H-T. and Chao H-H., 1993. Cryopreservation of oyster embryos with conventional freezing procedure and vitrification. Cryobiology, 30:614 (Abstract).

Linhart, O., Billard, R. and Proteau,J.P., 1993. Cryopreservation of European catfish (*Siluris glanis* L) spermatozoa. Aquaculture, 115:3347-3359.

Mazur, P and Cole, K. 1985. Influence of cell concentration and contribution on unfrozen fraction and salt concentration to the survival of slowly cooled human erythrocytes. Cryobiology, 22:509-536.

McAndrew, B.J., Rana, K.J. and Penman, D.J., 1993. Conservation and preservation in aquatic organisms. Recent Advances in Aquaculture. Vol IV J.F Muir & R.J.Roberts (Eds), pp. 295-336. Blackwell Science, Oxford.

Munkittrick, K.R. and Moccia, R.D., 1987. Seasonal changes in the quality of rainbow trout (*Salmo gairdneri*) semen: effect of delay in stripping on spermatocrit, motility, volume, and seminal plasma constituents. Aquaculture, 64:147-156.

Piironen, J., 1993. Cryopreservation of sperm from brown trout (*Salmo trutta*) and Arctic char (*Salvelinus alpinus* L). Aquaculture, 116:275-285.

Rana, K.J., Muiruri, R.M., McAndrew, B.J. and Gilmour, A., 1990. The influence of diluents, equilibration time and pre-freezing storage time on the viability of cryopreserved *Oreochromis niloticus* (L) spermatozoa. Aqua. and Fish. Management, 21:25-30.

Rana, K.J., McAndrew B.J. and Musa, M.A., 1992. Cryopreservation of oyster (*Crassostrea gigas*) eggs and embryos. Workshop on gamete and embryo storage and cryopreservation in aquatic organisms. Paris, p 25 (Abstract).

Rana, K.J., 1995a. Preservation of gametes. Broodstock Management and Egg and Larval Quality N.R.Bromage and R.J.R.Roberts (Eds), pp.53-75.

Rana, K.J., 1995b. Cryopreservation of fish spermatozoa. Methods in Molecular Biology Vol. 38: Cryopreservation and freeze-drying protocols. J.G. Day and M.R. McLellan (Eds), pp. 151-165.

Rana, K.J., Adam, M. and McAndrew, B.J., 1995c. Osmotic response of inseminated fish embryos exposed to various sucrose concentrations. Cryo-lett., 16:66.

Schmidt- Baulain, R. and Holtz, W., 1989. Deep freezing of rainbow trout sperm at varying intervals after collection. Theriogeneology, 32:439-443.

Scott, A.P. and Baynes, S.M., 1980. A review of the biology, handling, and storage of salmonid spermatozoa. J. of Fish Bio., 17:179-182.

Steponkus, P.L., Myer, S.P., Lynch, D.V., Gardner, L., Bronshtyen, V., Leibo, S.P., Rall, W.F., Pitt, R.E., Lin, T.T.and McIntyre, R.J., 1990. Cryopreservation of *Drosophila-melanogaster* embryos. Nature, 345:170-172.

Truscott, B., Idler, D.R., Hoyle, R.J. and Freeman H.C., 1968. Sub-zero preservation of Atlantic salmon sperm. J. Fish Res. Bd. Canada. 25(2):363-372.

Wheeler, P.A. and Thorgaard, G.H., 1991. Cryopreservation of rainbow trout semen in large straws. Aquaculture, 93:95-100.

Zhang, X.S., Zhao, L., Hua, T.C., Zhu, H.Y., 1989. Study on the cryopreservation of common carp (*Cyprinus carpio*) embryos. Cryo-lett, 10:271-278.

Zhang, T., Rawson, D., and Morris G. J. 1993. Cryopreservation of pre-hatch embryos of zebra fish (*Brachydanio rerio*) embryos. Aquat. Living Resour. 6:145-153.

CRYOPRESERVATION OF SEMEN OF SALMONID FISHES AND ITS ADAPTATION FOR PRACTICAL APPLICATION

T. Weismann[1], F. Lahnsteiner[2], and R. A. Patzner[2]

1 Bundesamt für Wasserwirtschaft, Institut für Gewässerökologie, Fischereibiologie und Seenkunde, Scharfling 18, A-5310 Mondsee, Austria
2 Institute for Zoology, University of Salzburg, Hellbrunnerstr. 34, A-5020 Salzburg, Austria[3]

Summary

For standardization of handling parameters of cryopreservation techniques the influence of storage of untreated semen, of semen dilution ratio in the extender, of equilibration period, of freezing temperature and of storage of frozen/thawed semen on the postthaw fertilization rate are investigated in *Oncorhynchus mykiss, Salmo trutta f. lacustris, Salmo trutta f. fario* and *Salvelinus fontinalis*.

Introduction

Cryopreservation techniques for Salmonidae are based on fertilization assays of small egg batches (for review see Jamieson, 1991). Freezing large semen batches or inseminating egg batches relevant in practice complicate the handling procedure (the handling at numerous straws or pellets) and may require a deviation from a standard freezing technique derived from fertilization of small amounts of eggs. Therefore handling parameters important for cryopreservation are investigated to standardize their allowable variations.

Material and methods

Semen of *Oncorhynchus mykiss, Salmo trutta f. lacustris, Salmo trutta f. fario* and *Salvelinus fontinalis* and *Coregonus sp.* was cryopreserved according to the method of Lahnsteiner *et al.* (1995): Semen was diluted in the ice cold extender (NaCl 103mM, KCl 40 mM, $MgSO_4$ 5 mM, $CaCl_2$ 1 mM, hepes 20 mM, pH 7.8, DMSO 5 %, glycerol 1%, bovine serum albumina 1.5 %, egg yolk 7 %, sucrose 0.5 %) in a ratio of 1 : 3 (semen : extender), sucked into 0.5 ml straws (Minitüb) and frozen within 5 min after dilution in an insulated box in the vapour of liquid nitrogen, after which straws were layed horizontally on a tray 1.5 cm above the surface of liquid nitrogen.

For thawing the straws were immersed in a water bath regulated by thermostat for 30 sec at a temperature of 25°C. After exactely 30 sec the straws were removed of the water bath, and the thawed semen poured onto the eggs.

For fertilization assays eggs were transfered into 4°C cold fertilization solution ($NaHCO_3$ 60 mM, glycine 20 mM, theophylline 5 mM, Tris 50 mM, pH 9 modified after Scheerer and Thorgaard, 1989) (ratio fertilization solution : eggs = 1 : 2). The desired amount of straws was thawed and the content mixed with the eggs by gentle stirring. After 2 to 3 min about 50 ml well water was added and the eggs were rinsed and incubated in flow incubators.

To exclude that differences in gamete quality might overlie the influence of methodical parameters, a series of experiments was always performed with the same pool of semen and eggs.

The influence of the following parameters on the postthaw fertility of semen was investigated:

1. Storage of untreated semen: For storage 1 ml portions of semen were incubated in reaction vessels with a diameter of 1.0 cm and a maximal volume of 12 ml at 4°C either for 10 min or for 30 min, 60 min or 120 min.
2. Dilution ratio in the extender: Semen was diluted onefold, twofold, threefold, fivefold and sevenfold in the extender and processed as described. To obtain equal sperm/egg ratios in all experiments the following number of straws was used for fertilization of 12.5 ml eggs: Dilution ratio of semen: 1:1 = ½ straw, 1 : 2 = ¾ straw, 1 : 5 = 1 ½ straws, 1 : 7 = 2 straws.
3. Equilibration period: Semen was diluted in the extender as described, sucked into the straws and equilibrated either 5 min, 10 min or 20 min at 4°C before freezing.
4. Freezing temperature: Semen was frozen 1 cm (-130 ± 1°C), 1.5 cm (-110 ± 2 °C), 2 cm (-100 ± 2°C) and 2.5 cm (-92° ± 2 °C) above the level of liquid nitrogen.
5. Storage of frozen/thawed semen: In comparison to the standard thawing procedure (25°C, 30 sec, immediate fertilization) straws were either thawed for 30 sec in 25°C water, removed out of the water bath, put onto the working table and stored for 30 sec at room temperature (10 to 12°C) or straws were thawed for only 20 sec in 25°C water (incomplete thawing) and stored for 30 sec.
6. Fertilization of large egg batches: Either 12.5 ml eggs or 100 ml eggs were fertilized according to the standard procedure and with similar sperm/egg ratios.

Control fertilization with untreated semen samples was performed in a similar way as fertilization assays with deep frozen semen and the same gamete batches and sperm/egg ratios as for cryopreservation experiments were used.

Fertilizaton rate was determined by percentage of larvae reaching the eyed stage in relation to total number of eggs. Percentage data were subjected to angular transformation and ANOVA and student *t*-test were used for data analysis. In the tables values within a row superscripted by the same letter are not significantly different ($p < 0.05$).

Results

Tab. 1. Influence of storage (4°C, 1 ml portions) of untreated semen of *Oncorhynchus mykiss* before cryopreservation on the fertilization rate of deep frozen semen. Number of repetitions of experiments (n) = 3. Sperm/egg ratio 6 x 10^6 spz./egg

duration of storage	**fertilization % absolute**
10min	60.4 ± 11.4[a]
60 min	61.4 ± 13.8[a]
120 min	46.6 ± 16.6[b]
Control	62.0 ± 11.2[a]
no. of eggs/experiment	205 ± 9

Storage up to 60 min had no influence on postthaw fertility. Longer storage periods of 120 min. significantly decreased fertilization rates.

Tab. 2. Influence of dilution ratio of semen in the extender prior to freezing on fertilization rate of deep frozen semen of *Oncorhynchus mykiss*. n = 3. Sperm/egg ratio3 x 10^6 spz./egg.

Dilution ratio (semen : extender)	**sperm concen-tration / ml extender**	**fertilization % absolute**
1 : 1	3.1 x 10^9	61.6 ± 3.6[a]
1 : 3	1.5 x 10^9	82.3 ± 2.9[b]
1 : 5	1.0 x 10^9	81.5 ± 4.2[b]
1 : 7	7.6 x 10^8	81.9 ± 3.8[b]
Control		87.9 ± 3.1[b]
no. of eggs/exper.	205 ± 9	

Significant decrease of fertilization rate at dilution ratio lower than 1:3.

Tab. 3. Influence of dilution ratio on the fertilization rate of deep frozen semen of *Salmo trutta f. lacustris*. n = 3. Sperm/egg ratio 9 x 10^6 spz./egg.

Dilution ratio	**sperm concen-tration /ml extender**	**fertilization % absolute**
1 : 3	5.3 x 10^9	50.1 ± 4.4[a]
1 : 5	3.5 x 10^9	64.9 ± 2.8[b]
1 : 7	2.6 x 10^9	73.1 ± 1.0[c]
Control		80.4 ± 4.1[d]
no. of eggs/exp.	112 ± 6	

Signifcantly higher fertilization rates at dilution ratios of 1:5 and 1:7.

Tab. 4. Influence of equilibration period in the extender on the fertilization rate of deep frozen semen of *Oncorhynchus mykiss*. n = 3. Sperm/egg ratio 1 x 10^6 spz./egg.

duration of equilibration	**fertilization % absolute**
3 min	60.7 ± 5.6[a]
10 min	60.4 ± 4.4[a]
20 min	59.4 ± 8.4[a]
Control	82.6 ± 8.0[b]
no. of eggs/exp.	210 ± 11 i

Equilibration of semen for up to 20 min.in extender did not significantly change the postthaw fertility.

Tab. 5. Influence of freezing rates on the postthaw fertility of semen of *Oncorhynchus mykiss* and *Salvelinus fontinalis*. n = 3.

distance from liquid nitrogen	**freezing temp.**	**fertilization % absolute**
Oncorhynchus mykiss		
1 cm	-130°C	63.2 ± 9.9[a]
1.5 cm	-110°C	83.8 ± 7.1[b]
2 cm	-100°C	62.8 ± 12.1[a]
Control		85.3 ± 5.5[b]
no. of eggs/experiment		196 ± 9
sperm/egg ratio	3.5 x 10^6 spz./egg	
Salvelinus fontinalis		
1.5 cm	-110°C	54.6 ± 3.3[c]
2 cm	-100°C	54.4 ± 3.4[c]
2.5 cm	-92°C	62.1 ± 2.4[d]
Control		60.3 ± 2.2[d]
no. of eggs/experiment		156 ± 7 ii
sperm/egg ratio	3.9 x 10^6 spz./egg	

Highest postthaw fertility of semen was obtained when freezing it 1.5 cm (-110°C) above N_2-level. This was similar in *Salmo trutta f. fario* and *Salmo trutta f. lacustris*. In *Salvelinus fontinalis* highest fertilization rates were obtained at 2.5 cm (-92°C) above N_2-level.

Tab. 6. Influence of storage of frozen/thawed semen on the fertilization rate in *Oncorhynchus mykiss*. Sperm/egg ratio 5 x 10^6 spz./egg. n = 3.

thawing conditions	subsequent storage in air (8-10°C)	fertilization % absolute
25°C, 30 sec	0 sec	83.8 ± 7.1 [a]
25°C, 30 sec,	30 sec	58.5 ± 20.7 [b]
25°C, 20 sec	20 sec	64.0 ± 8.1 [b]
Control		85.3 ± 5.5 [a]
no. of eggs/experiment		196 ± 9

Storage of thawed semen lowered postthaw fertility significantly in contrast to immediate use for fertilization.

Tab. 7. Fertilization of small and large egg batches of *Oncorynchus mykiss*. Sperm/egg ratio 4 x 10^6 spz./egg. n = 3.

amount of eggs	no. of eggs/ experiment	no. of straws	fertilization % absolute
12.5 ml	160 ± 7	1	60.1 ± 1.3 [a]
100 ml	1415 ± 59	8	60.5 ± 2.1 [a]
control			
12.5 ml	160 ± 7	-	62.0 ± 3.2 [a]

Similar fertzilization rates were obtained when fertilizing 12.5 ml or 100 ml egg batches with deep frozen semen.

Discussion

The described cryopreservation method is suitable for semen cryopreservation of at least five salmonid species, *Oncorhynchus mykiss*, *Salmo trutta f. lacustris*, *Salmo trutta f. fario* and *Salvelinus fontinalis*.

Influence of storage of semen before cryopreservation on its postthaw fertility: The present data are in accordance with those of Holtz (1993), that long storage of semen before cryopreservation decreases its postthaw fertility.

Influence of dilution ratio of semen in the extender on the postthaw fertility: The postthaw fertilization rates decrease at dilution ratios of less than threefold in *Oncorhynchus mykiss* and of less than five to sevenfold in *Samo trutta f. lacustris* and *Salmo trutta f. fario* (both species which higher sperm densities than *Oncorhynchus mykiss)* which demonstrates that the optimal dilution ratio must not exceed a critical value estimated to 2.0 to 2.5 x 10^9 spermatozoa/ml diluent. Higher cell concentrations in the extender significantly decrease the postthaw fertility of semen, lower sperm concentrations (tested up to 7.6 x 10^8 spermatozoa/ml extender) do not effect its postthaw fertility.

Influence of equilibration of semen in the extender on its postthaw fertility: The advantage of equilibration without a decrease of semen quality lies in facilitation of dilution and freezing of large semen portions. The present results demonstrate that equlibration of semen for maximally 20 min in the described extender does not effect the postthaw fertility. Therefore the extender has no toxic influence on spermatozoa, but longer equilibration is not necessary for freezing of salmonid spermatozoa.

Influence of freezing temperature of semen on its postthaw fertility: While *in Oncorhynchus mykiss* and also in *Salmo trutta f. lacustris, Salmo trutta f. fario* and *Coregonus sp.* (data not shown) the optimal freezing level is at 1.5 cm (approximately -110°C) above the level of liquid nitrogen, in *Salvelinus fontinalis* it is at 2.5 cm (approximately -92°C) above the level of liquid nitrogen.

The present results further demonstrate that the optimal range for freezing of salmonid spermatozoa is very narrow. Changes in the freezing distance from the level of liquid nitrogen from only 0.5 cm result in a significant decrease of postthaw fertility. This is an important factor for routine application under aspects of liquid nitrogen evaporation when using open systems.

Influence of storage of frozen/thawed semen on the fertility: Storage of frozen/thawed semen for only 30 sec (or also when incompletely thawed before storing) results in a significant decrease of postthaw fertility. This shows that the thawed semen must be used immediately for fertilization

Fertilization of large egg batches: Similar fertilization rates obtained when inseminating small egg batches of 12.5 ml and egg samples up to 100 ml indicate, that the method can be applied to large scale fertilization, too.

References

Holtz, W. 1993. Cryopreservation of rainbow trout (*Oncorhynchus mykiss*) sperm: practical recommendations. Aquaculture 110, 97 - 100.

Jamieson B. G. M., 1991. Fish evolution and systematics: Evidence from spermatozoa. Cambridge, University Press, pp. 319.

Lahnsteiner, F., T. Weismann, and R.A. Patzner, 1995. A uniform method for cryopreservation of salmonid fishes. Aquaculture Research 26, in press.

Scheerer, P.D. and G.H. Thorgaard, 1989. Improved fertilization by cryopreserved rainbow trout semen with theophylline. Prog. Fish. Cult. 51: 179 - 182.

THE PLASTICITY OF SEX DETERMINING GENOTYPES IN CHANNEL CATFISH

Kenneth B. Davis [1], Cheryl A. Goudie[2] and Bill A. Simco[1]

[1]Division of Ecology and Organismal Biology, The University of Memphis, TN 38152;
[2]USDA-ARS, Catfish Genetics Research Unit, P.O. Box 38, Stoneville, MS 38776.

Summary

An indirect method of producing all male progeny of channel catfish by hormonal sex reversal and genetic backcrossing has been developed. Oral administration of most androgens and estrogens during the phenocritical period of sex determination produced all-female populations of channel catfish. Mating sex-reversed females (XY females) with normal males (XY) produced a population with a 3:1 (male:female) phenotypic sex ratio and one-third of the males had the YY genotype. These YY males produce all-male progeny, all with the XY genotype if mated with normal females, and XY:YY genotypes if mated with XY females. Both the XY and the YY sex genotype can be feminized by hormonal treatment. Female fish with XX, XY, and YY genotypes and male fish with XY and YY genotypes are viable and fertile when mated with fish with normal sex genotypes. Prespawning concentrations of testosterone in YY-female fish were dramatically lower than XY or XX females. Mating YY males with YY females should result in all YY male fish. However, such matings resulted in less than one per cent of the eggs hatching.

Introduction

The production of channel catfish is the largest aquaculture industry of food fish in the United States. Male channel catfish may weigh from 16 to 37% more than females when they are about 500 g (Simco et al., 1989), and are, therefore, more desirable for culture.

The model of sex determination in channel catfish is female homogamety (Davis et al., 1990). This model was determined by progeny testing populations which were hormonally feminized (Goudie et al., 1983). Most hormonal preparations, including estrogens and aromatizable and non-aromatizable androgens, result in all-female populations. Sex-reversed females (genotypic male-phenotypic female) produce a population of 3:1 (males:females) with a genotypic sex ratio of YY:2XY:XX when mated with a normal (XY) male. Males with the YY sex genotype produce only male offspring when mated with a female of any genotype (Davis et al., 1991). Hormonal feminization of populations with YY genotypes should produce YY females. Identifying YY females and mating them with YY males should produce a strain of all YY, all male fish. Females for future broodstock could be produced by hormonal feminization of these known YY fish.

The presence of the unique sex genotypes in channel catfish must be detected by progeny testing. Sequences of DNA which are associated with a particular sex in amniotes, including Bkm minisatellite, human telomeric sequence, ZFY, and SRY, were detectable in channel catfish but none were sex specific (Tiersch et al., 1992). Identification of any physiological difference among these genotypes would be valuable.

Methods and Experimental Design

Male fish with the YY sex genotype were mated with females which had the XY sex genotype. The resulting populations were all male fish with XY or YY genotypes in an expected ratio of 1:1. Progeny treated orally for 21 days with 60 mg/kg 17 α ethynyltestosterone were all females.

Fish from these spawns were held in ponds until they were sexually mature. Blood samples were taken in mid-May and early June and the fish placed in spawning cages. Male fish were paired with genotypically normal females and female fish were paired with genotypically normal male fish. Male fish with the YY genotype were identified by all male progeny while normal males produced a 1:1 sex ratio. YY females produced only male offspring and females with an XY genotype were identified by a 3:1 (male:female) sex ratio of the progeny. Spawns were kept separate and the sex ratio determined by visual examination of the gonads.

Plasma concentrations of testosterone in male fish and of testosterone and estrogen in

Facilities provided by the Southeastern Fish Cultural Laboratory, USDI
Supported by funds from USDA (91-37206-6741)

female fish were determined by RIA.

Results

Pre-spawning hormone concentrations are shown in Table 1, spawning success in Table 2, and the frequency of YY animals identified in Table 3.

Table 1. Pre-spawning hormone concentrations (ng/ml) in the plasma of XY and YY males and XX, XY, and YY female channel catfish. Values are the mean ± SE. Significant differences ($p<0.05$) are shown by an * for males and different letters for females.

	Testosterone	Estrogen	n
Males			
XY	1.12±0.29	----	9
YY	4.08±1.26*	----	18
Females			
XX	19.11±1.19a	4.60±1.86a	10
XY	15.82±1.30a	2.83±0.29b	10
YY	0.92±0.54b	2.32±0.98b	3

Table 2. Spawning success of experimental males and females with XY or YY genotypes in a ratio of 1:1 mated with normal fish.

	Pairs	Spawns	Spawns
	n	n	%
Males	25	21	84
Females	48	24	50

Table 3. Frequency of YY males and YY females identified from a population with XY and YY genotypes in a 1:1 ratio mated with normal fish.

	Spawns	Expected YY		Confirmed YY	
	n	n	%	n	%
Males	9	4-5	50	3	33
Females	15	7-8	50	6	40

Spawning success of populations with expected YY females was less than that for expected YY males. However, the frequency of YY animals of both sexes was close to expected numbers.

Three pairs of known YY males and females were mated in May 1995. Each of the pairs spawned; however, each spawn had massive mortality about 36 hours after fertilization. Less than one percent of the spawn survived to hatching. Approximately 100 fish from two families and 30 fish from a third family completed development.

Discussion

Male catfish with the XY or YY genotype and female catfish with the XX, XY, or YY genotypes are viable and fertile. Sex-reversed females (XY) spawned at the same rate as did XX females and produced similar quantities of eggs and viable fry (Davis et al., 1990). Male YY catfish appear to be similar to XY male catfish in spawning success (Davis et al., 1991).

The YY genotype has been produced by spawning XY or YY male fish with XY females. Female channel catfish with the XY genotype were also spawned by gynogenesis (Goudie et al., in press). Although survival was extremely low, the sex ratio of the offspring was close to 1:1 (male:female). All of the male offspring should have the YY genotype.

Identification of genetic sex using chromosomal or molecular techniques in channel catfish is not currently possible (Tiersch et al., 1992). The parental genotype is inferred by the sex ratio of the progeny. Adequate development of the young fish to allow visual identification of the ovary requires three to four months of growth. Histological differentiation of the female gonad was detectable by 19 days after fertilization, however, no evidence of somatic or germ cell differentiation in males was apparent by 3 months of age (Schoore et al., 1995). The phenocritical period, at least for hormonal feminization, is during the first 21 days after yolk-sac absorption (Davis et al., 1992), which occurs from 12 to 14 days after fertilization.

The gonadosomatic index was higher and the testosterone and estrogen concentrations were different in hormone-treated female populations at 2.5 and 4 years than in control females (Simco et al., 1989). However, the hormonally sex-reversed female population had both XX and XY genotypes. We report here that prespawning estrogen concentrations of XY and YY females are lower than those of XX females, and that YY females have much lower testosterone concentrations than do XX and XY females. The difference in hormonal concentrations may allow selection of YY females from a mixed genotype population without progeny testing.

Reproductive success appears to be hindered when both parents have the YY genotype. The mortality may have been due to low fertilization or to a developmental problem. The sex hormone content, including testosterone, of unfertilized coho salmon eggs is equivalent to that in the ovarian fluid and must be of maternal origin (Schreck et al., 1991). The hormone concentrations decline until hatching and may participate in development. Peaks of testosterone in largemouth bass were correlated with development of successive clutches of large vitellogenic follicles (Rosenblum et al., 1991). The low testosterone concentrations in YY female catfish may have interfered with final oocyte maturation or with development.

The fish which did hatch are being presently evaluated and should all have the YY genotype. Some of these fish were feminized with hormones and can be used to establish a strain of all YY animals, which will eliminate the need for progeny testing. Recently, a direct method of masculinization has been described (J. Galvez, personal communication). If hormonally masculinized fish are fertile, this would allow the production of XX male fish, and would demonstrate that all possible sex genotypes have all the necessary information to be functional males or females. Efforts could then be focused on the regulatory pathways which dictate which sex phenotype develops.

References

Davis, K. B., B. A. Simco, C. A. Goudie, N. C. Parker, W. Cauldwell & R. Snellgrove, 1990. Hormonal sex manipulation and evidence for female homogamety in channel catfish. Gen. Comp. Endocrinol. 78:218-223.

Davis, K. B., B. A. Simco & C. A. Goudie, 1991. Genetic and hormonal control of sex determination in channel catfish. In: Proc. Fourth Internat. Symp. Reprod. Physiol. Fish. A. P. Scott, J. P. Sumpter, D. E. Kime & M. S. Rolfe (Eds.): Univ. East Anglia Printing Unit, Norwich U.K. p. 244-246.

Davis, K. B., C. A. Goudie, B. A. Simco, T. R. Tiersch & G. J. Carmichael, 1992. Influence of dihydrotestosterone on sex determination in channel catfish and blue catfish: period of developmental sensitivity. Gen. Comp. Endocrinol. 86:147-151.

Goudie, C. A., B. D. Rednor, B. A. Simco & K. B. Davis, 1983. Feminization of channel catfish by oral administration of steroid sex hormones. Trans. Am. Fish. Soc. 112:670-672.

Goudie, C. A., B. A. Simco, K. B. Davis & Q. Liu. Production of gynogenetic and polyploid catfish by pressure-induced chromosome set manipulation. Aquaculture (in press).

Rosenblum, P. M., H. Horne, N. Chatterjee & T. Brandt, 1991. Influence of diet on ovarian growth and steroidogenesis in largemouth bass. In: Proc. Fourth Internat. Symp. Reprod. Physiol. Fish. A. P. Scott, J. P. Sumpter, D. E. Kime & M. S. Rolfe (Eds.): Univ. East Anglia Printing Unit, Norwich U.K. p. 265-267.

Schoore, J. E., R. Patiño, K. B. Davis, B. A. Simco, & C. A. Goudie, 1995. Gonadal sex differentiation in channel catfish. In: Proc. Fifth Internat. Symp. Reprod. Physiol. Fish. P. Thomas and F. Goetz (Eds.). (this volume).

Schreck, C. B., M. S. Fitzpatrick, G. W. Feist & C. G. Yeoh, 1991. Steroids: developmental continuum between mother and offspring. In: Proc. Fourth Internat. Symp. Reprod. Physiol. Fish. A. P. Scott, J. P. Sumpter, D. E. Kime & M. S. Rolfe (Eds.): Univ. East Anglia Printing Unit, Norwich U.K. p. 256-258.

Simco, B. A., C. A. Goudie, G. T. Klar, N. C. Parker & K. B. Davis, 1989. Influence of sex on growth of channel catfish. Trans. Am. Fish. Soc. 118:427-434.

Tiersch, T. R., B. A. Simco, K. B. Davis & S. S. Wachtel, 1992. Molecular genetics of sex determination in channel catfish: studies on SRY, ZFY, Bkm, and human telomeric repeats. Biol. Reproduction, 47:185-192.

APPLICATION OF A NON-INVASIVE SEX TEST IN THE AQUACULTURE OF STRIPED BASS

J.L. Specker[1], L.C. Woods III[2], L. Huang[1], and M. Kishida.[1]

[1]Department of Zoology, University of Rhode Island, Kingston, RI 02881; [2]Department of Animal Sciences, University of Maryland, P.O. Box 1475, Baltimore, MD 21203, USA.

Summary

Striped bass are not sexually dimorphic. It would be beneficial in culture and management situations to discriminate mature females from immature females and from all males. A test based on the specific detection of vitellogenin (Vg), a female-specific protein, in the mucus of females led previously to perfect classification of wild males and females. This non-invasive test was used on broodstock at Crane Aquaculture Facility with limited success. In our best judgment, the hurdles to application of this test in culture situations are limited to technical problems including collection of mucus samples and the non-specific binding of mucus extracts.

Introduction

The striped bass (Morone saxatilis) is an anadromous fish indigenous to the east coast of the United States. The fish is valued in both recreational and commercial fisheries. Striped bass and its hybrids are cultured for market. Knowledge about reproduction of striped bass is important to managers of the wild stocks and to aquaculturists.

We assisted in the description of the population structure of wild striped bass by providing an updated maturity schedule for coastal females based on oocyte development (Berlinsky et al., 1995). Although males mature by age-class 3, females mature between age-classes 4 and 7. In the interests of providing a non-lethal test of female maturity, we developed a test based on detection of vitellogenin (Vg) in the surface mucus of female striped bass.

Vitellogenin is the precursor to egg yolk proteins (see Specker and Sullivan, 1994). During the annual, extended period of oocyte growth (fall-spring), female striped bass produce Vg in the liver in high amounts and transport it to the ovary (Berlinsky and Specker, 1991; Tao et al., 1993). For unknown reasons, some yolk protein, including intact Vg, appears in the surface mucus (Kishida et al., 1992). Using Vg-specific antiserum in enzyme-linked immunosorbent assay (ELISA) and Western blotting, we previously showed that the test discriminated with perfect accuracy wild female striped bass from wild male striped bass caught in the Hudson River during their spawning migration (Kishida et al., 1992; Specker and Anderson, 1994).

In the aquaculture of striped bass, choosing broodstock is made difficult in part because mature striped bass are not sexually dimorphic. We tested the utility of our non-invasive sex test for mature striped bass in an aquaculture facility. The non-invasive sex test for striped bass and probably for related species could aid the industry in broodstock selection if technical problems are solved.

Results

Striped bass are reared at Crane Aquaculture Facility (University of Maryland) as broodstock. In this study, we collected blood and surface mucus from 12 males and 24 females at four sampling times. Seven of the males and 12 females were from the 1983 yearclass, the remainder were from the 1985 yearclass. Methods for collection and analyses are described in detail elsewhere (Kishida et al., 1992). Briefly, mucus is swiped from the side of the fish. An ELISA is used to measure Vg in plasma and to approximate Vg in a mucus extract. Vg in mucus is normalized on a protein basis.

Plasma Vg concentrations are shown in Table 1. These results are comparable to those reported previously for these broodstock in another year (Tao et al. 1993). These levels are also in the range reported for wild female striped bass in the Hudson River in 1991 (Kishida et al., 1992). Plasma Vg concentrations were higher in females from the Hudson River in 1992 (Specker and Anderson, 1994). The plasma concentration of Vg in these ranges resulted in detectable Vg in the mucus of all females in the Hudson River.

Table 1. Plasma Vg concentration (ug/ml) in striped bass at Crane Aquaculture Facility and in the Hudson River. One female with 0 ug/ml plasma Vg was excluded. Mean(SE).

	Female	Male
Dec '91	165(14)	0
Jan '92	273(22)	0
Feb '92	316(19)	0
Hudson R. '91	331(43)	0
Hudson R. '92	1242(102)	0

Detection of Vg in surface mucus from broodstock striped bass was less successful than expected from our studies on wild striped bass, as shown in Table 2. In three instances, classification was correct. In the other five instances, classification was no better than random.

Table 2. Correct classification of striped bass at Crane Aquaculture Facility according to detection of Vg in mucus. Number correct/total (percent correct).

	Females	Males
Nov '91	17/24(71)	8/12(67)
Dec '91	11/24(46)	12/12(100)
Jan '92	24/24(100)	3/12(25)
Feb '92	10/24(42)	12/12(100)
Hudson R. '91	14/14(100)	3/3(100)
Hudson R. '92	14/14(100)	6/6(100)

A potentially key difference in the collection of mucus occurred between those fish sampled in the river and those sampled at the aquaculture facility. We found it easy to collect 1.5 ml mucus from the sides of the fish in the river. These fish had recently moved from seawater to fresh water. In contrast, it was difficult to collect 0.2 ml mucus from fish in the aquaculture facility where salinity during this experiment ranged from 4-8 ppt. This difference could be due either to amount of mucus produced by fish in the two locations--maybe due to salinity--or to subtleties in catching, handling, and sampling the fish.

To examine whether mucus samples collected from fish at the two locations differed, we compared the protein concentrations in mucus. Table 3 shows that mucus from fish at the aquaculture **facility had higher amounts of protein than did the fish in the river. It is possible that difficulty in collecting mucus led to deeper scraping of the skin.**

Table 3. Protein concentration (mg/ml) and apparent Vg concentration (ng/mg protein) in mucus collected from striped bass at Crane Aquaculture Facility and at the Hudson River. Mean(SE).

	Protein (mg/ml)	Vg (ng/mg protein)
Nov '91	12(0.4)	69(2)
Dec '91	57(3.7)	6(1)
Jan "92	20(1.1)	25(2)
Feb '92	26(1.2)	15(2)
Hudson R. '91	4(0.4)	176(28)
Hudson R. '92	2(0.2)	338(54)

Discussion

The application of the non-invasive sex test for mature striped bass was less successful at the aquaculture facility than in the field. A probable reason for the difference in the application of this test to wild fish and to cultured fish and possible remedies to this technical difficulty are discussed.

Mucus was more easily obtained from the surface of wild striped bass. At the aquaculture facility, the fish had to be scraped more severely to obtain surface mucus. Possibly as a result of the scraping, the protein content of the mucus from fish in the hatchery was about 10-fold higher than from fish in the Hudson River. This protein, and possibly other factors, caused interference in the ELISA; even though plasma from males did not cause any displacement in the assay, mucus extracts from these same males showed a high degree of non-specific binding and, in some cases, slight volume-dependence. Although mucus extracts from wild male striped bass cause some non-specific binding in the ELISA, the displacement curves have no slope; this is in clear contrast to the extracts of female mucus, which displace the competitor in a volume-dependent manner. In brief, serial dilutions of mucus extracts from captive males and females tended to exhibit similarly high levels of non-specific binding and similarly slight degrees of volume-related displacement.

In our best judgment, this non-invasive sex test could be technically improved to work throughout most of the year (fall through spring) in culture situations. Possible solutions to the difficulty we encountered would be to augment mucus production in captive fish and to standardize mucus collection. For example, mucus production may be increased by changing salinity for brief periods. Mucus collection could use blotting paper. We followed up this study attempting to blot mucus from the hatchery fish with nitrocellulose paper which we then used for Western blotting. The non-specific binding of unextracted mucus made this approach unworkable in our hands. Future effort needs to be directed at blocking the non-specific binding of mucus itself, developing material for collecting the mucus, and strengthening the signal associated with specific binding.

Acknowledgments

Supported by the U.S. Dept. of Commerce and Maryland Agricultural Experiment Station.

References

Berlinsky, D.L. and Specker, J.L. 1991. Changes in gonadal hormones during oocyte development in the striped bass, Morone saxatilis. Fish Physiol. Biochem. 9:51-62.

Berlinsky, D.L., Fabrizio, M.C., O'Brien, J.F., and Specker, J.L. 1995. Age-at-maturity estimates for Atlantic coast female striped bass. Transactions of the American Fisheries Society 124:207-215.

Kishida, M., Anderson, T.R. and Specker, J.L. 1992. Induction by b-estradiol of vitellogenin in striped bass (Morone saxatilis): Characterization and quantification in plasma and mucus. Gen. Comp. Endocrinol. 88:29-39.

Specker, J.L. and Anderson, T.R. 1994. Developing an ELISA for a model protein - vitellogenin. Pp. 565-576. In: Biochemistry and Molecular Biology of Fishes (P. W. Hochachka and T.P. Mommsen, eds.). Elsevier Press, Amsterdam.

Specker, J.L. and Sullivan, C.V. 1994. Vitellogenesis in fishes: Status and perspectives. Pp. 304-315. In: Perspectives in Comparative Endocrinology. (K.G. Davey, R.E. Peter, and S.S. Tobe, eds.) National Research Council, Ottawa.

Tao, Y., Hara, A., Hodson, R.G., Woods, L.C., III, and Sullivan, C.V. (1993) Purification, characterization and immunoassay of striped bass (Morone saxatilis) vitellogenin. Fish Physiol. Biochem. *12*, 31-46.

FIELD TRIALS DEMONSTRATE THE EFFICACY AND COMMERCIAL BENEFIT OF A GnRHa IMPLANT TO CONTROL OVULATION AND SPERMIATION IN SALMONIDS

A. Goren, H. Gustafson and D. Doering

AquaPharm Technologies Corp., 9110M Red Branch Rd., Columbia, MD, 21045 USA

Summary

In the trials described here we demonstrate the efficacy of a controlled release GnRHa implant to reduce the spawning period, synchronize ovulation, and advance ovulation and spermiation in three salmonid species. These trials were conducted to assess the economic benefits of the use of this implant in a large scale commercial setting. A total of 35 trials in experimental groups ranging from 30 to 500 broodfish were conducted in Chile, France, the United Kingdom, Israel, the United States and Canada between March 1994 and June 1995.

Introduction

Modern industrial aquaculture producers aim to provide a low-cost, high quality product which can be shipped fresh according to market demand. Supplying an on-demand consumer product requires a reliable and constant production system that begins with a constant supply of eggs. Egg producers must have the capabilities to breed preferred strains, maximize broodstock fecundity, and control the schedule of egg production to meet the requirements of the grow-out operation.

Many of the methods used to enhance reproductive performance, to synchronize spawning, and to advance spawning require external intervention through environmental manipulation and hormonal treatment. The hormones which are in common use include gonadotropin hormone (GtH) formulations of either fish pituitary extracts or human chorionic gonadotropins (hCG) and analogues of both mammalian gonadotropin releasing hormone (GnRH) and salmon GnRH.[1] Stable analogues of GnRH (GnRHa) have been shown to have several important advantages over gonadotropin formulations in inducing spawning in many species.[2,3] However, the rapid clearance of injected GnRHa from the circulation requires that multiple injections be given in order to maintain an effective dosage of GnRHa.[4] Multiple injections result in large fluctuations of circulating peptide concentration which may exceed recommended levels or decline below effective concentrations. The excessive handling of fish from multiple injections can lead to stress-related injuries, mortalities, and suppression of reproductive processes. It has been previously demonstrated, in both single and repetitive spawners, that the sustained administration of GnRH analogues efficiently induced sustained GtH release.[5,6] The induced endogenous GtH release stimulates the gonads more effectively than externally administered gonadotropins.

The GnRHa implant [ReproBoost™, AquaPharm Technologies Corp.] is a two millimeter diameter cylinder made from a synthetic copolymer that forms a biologically inert matrix. The matrix controls the diffusion of the hormone from the implant. The implant was designed to release the peptide for 20 days, following an initial depletion of 30% of its content during the first 3 days. The intra-muscular implantation was performed with a 12 gauge implanting device behind the dorsal fin. The [Des-Gly^{10},D-Ala^{6},Pro^{9} NHEt]-LHRH analogue was used in implants that contain 75, 150 or 250 micrograms of the peptide. In certain cases, multiple implants were used to achieve different dosages.

Synchronization of Ovulation

A population of salmonids typically reaches ovulation during a period of 6-10 weeks. The ovulation of individuals exhibits a normal, symmetric distribution centered on the peak spawning date. The length of this period over which the broodfish ovulate and the distribution of individuals' ovulation complicate the timing of egg production. The trial in Figure 1 shows that the GnRHa implants can effectively induce ovulation and can condense the final maturation period of the eggs. The treatment resulted in 93% percent ovulation within 19 days as compared to 40% for the untreated group and condensed the ovulation period from 40 days to 20 days. There was no significant difference in egg quality between the two groups.

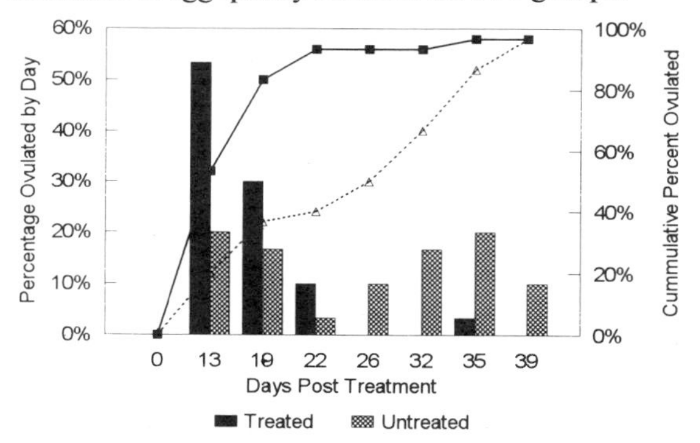

Figure 1. (See below)

Figure 1. Reduction of the Spawning Period in Coho Salmon. The trial was conducted at a commercial site in Chile during the natural spawning season. Treatment of 2SW coho salmon implanted two weeks ahead of the peak natural spawning week (■) with 150 microgram implants (average dose = 33 µg/kg) were compared to non-treated fish (group size = 40).

Advancement of Spawning

Preferential environmental conditions for accelerating growth (e.g. production of S0 smolts) and the reduction of pre-spawning mortality are two of the advantages to producing eggs 2-6 weeks ahead of the normal spawning season. 2SW Atlantic salmon of a strain which had very good growth characteristics but low fecundity even when treated with injections of GnRHa were treated with the GnRHa implants (Figure 2). By day 15 post-implantation, 83% of the implanted fish ovulated as compared to only 9% in the injected group. By day 20 post-implantation, 90% of the implanted fish had ovulated as compared to 17% of the injected fish. The average eyeing rate was 61% in the implanted group and 50% in the injected group. Thus, implantation of this low fecundity strain significantly advanced the ovulation date and increased the percentage of females that ovulated.

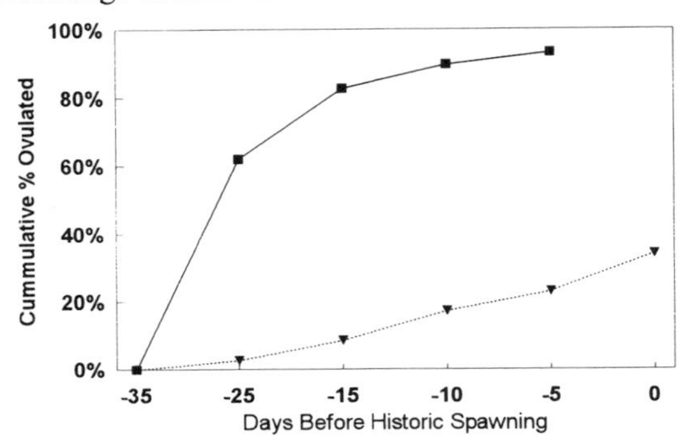

Figure 2. Advancement Of Spawning In Atlantic Salmon. Two groups of 30 and 35 females were treated five weeks before the beginning of the historical spawning period by a single implantation of two implants (■)or two separate GnRHa injections (▼). The implanted fish, each received 400 micrograms of GnRHa (average dose = 35 µg/kg). The injected fish received a total average dose of 30 µg/kg divided into two injections.

Synchronization of Spawning

In order to optimize the use of hatchery facilities and labor, it is advantageous to synchronize populations of broodfish to ovulate 'on-demand.' The trial below (Figure 3) shows the effect of discrete implant treatments to advance and synchronize groups of coho salmon broodstock. The three groups reached between 95 and 100% ovulation 19, 15 and 12 days post-implantation. As expected, the closer the treatment was to the normal spawning date, the sooner they reached ovulation.

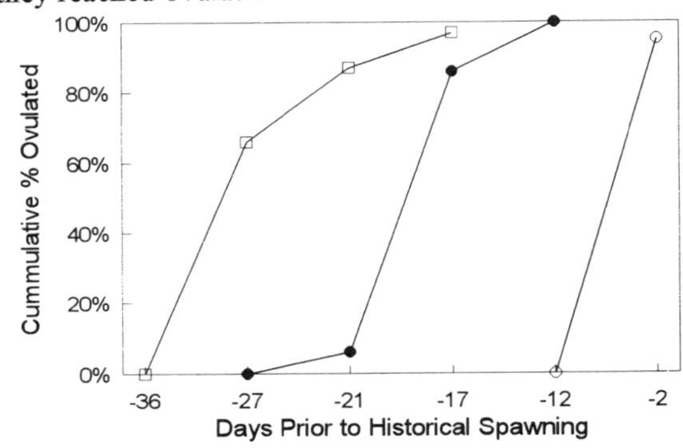

Figure 3. Synchronization of Spawning of Coho Salmon. The three lines show the ovulation of groups of 480 (□), 450 (●) and 200 (O) coho salmon that were treated with 150 microgram implants five, four, and two weeks prior to the historical spawning date.

Advancement of Photoperiod-Manipulated Fish

Production of eggs outside of the spawning season by photoperiod manipulation is common in trout and is increasingly being applied to other salmonids. The high cost of maintaining the broodstock in photo-controlled conditions makes efficient induction and timing of egg production particularly important. In the trial shown in Figure 4, photoperiod-manipulated brown trout were treated with the GnRHa implants to advance ovulation. In the first group, 93% ovulated within 11 days. In the second group, 100% ovulated within 10 days and in the last group 97% ovulated within 9 days of treatment. At the time the last group had completely ovulated, only 30% of the untreated group had ovulated (data not shown).

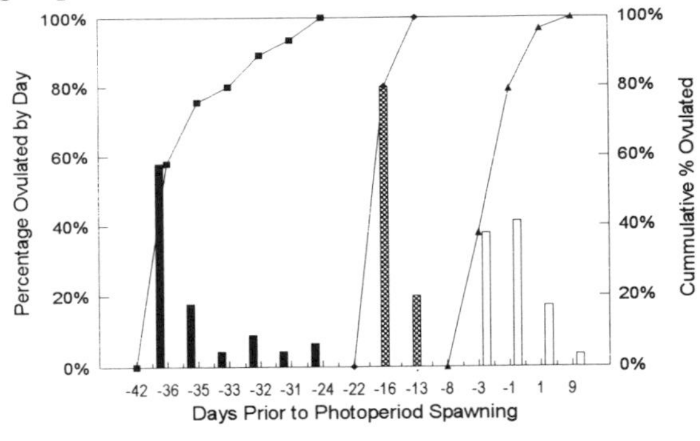

Figure 4. Synchronization and Advancement of Photoperiod-Manipulated Brown Trout. Fish were photoperiod manipulated to spawn in mid-September (day 0 on the graph). One hundred and thirty fish in 3 groups were treated 42 (■), 22 (◆), and 8 (▲) days before the planned photoperiod spawning date with 150 microgram implants (average dose = 125 µg/kg) and a control group was left untreated.

Enhancement of Atlantic Salmon Milt Production

Timing of fertilized egg production also requires controlling spermiation. Many salmonid producers encounter the problem of very low milt volumes which requires rearing large numbers of mature males. The treatment of 2SW Atlantic salmon males with the GnRHa implants resulted in significant increases in milt volume. By day 18 post-implantation, average milt volume in the treated group was over 70 ml. as compared to 12 ml. per fish in the controls (Table 1). The increased milt production in the treated fish reflects an increase in the number of motile sperm and seminal fluid volume since there was normal spermatocrit (data not shown) and sperm motility

Post-treatment	Treated	Untreated
Day 0	5 ml.	3 ml.
Day 15	18 ml.	12 ml.
Day 18	72 ml.[a]	11 ml.[b]

[a] 91% sperm motility, [b] 84% sperm motility

Table 1. Enhancement of Milt Production in Atlantic Salmon. Thirty males were treated with 250 microgram implants and compared to a similar group of untreated males (average dose = 16 μg/kg).

Discussion

The large scale trials reported here demonstrate the efficacy of administration of GnRHa with a controlled release implant to control spawning in salmonids. The prolonged exposure of the broodfish to GnRHa results in acceleration of ovulation and enables advancement of spawning ahead of the normal season. Spawning advancement is known to be possible because gamete maturation is nearly completed between 4 to 10 weeks before the normal spawning dates. During this period it is possible to induce final gamete maturation by inducing prolonged secretion of gonadotropin hormone with GnRHa.

The frequent handling and palpation to assess sexual maturity that is performed for conventional hormone injections and egg collection causes stress-related suppression of the reproductive processes. The implant effectively advances and shortens the spawning period of a population of broodstock, reducing the burden upon the egg producer to accurately select mature broodfish for hormone treatment or egg stripping. Advancement of the natural spawning season also reduces the occurrence of physiological deterioration and mortalities of the fish and eggs.

These GnRHa implants provide the broodstock manager with a valuable management tool to achieve a controllable and reliable egg production. The economic benefits to the egg producer include the reduction of egg production costs, the optimization of environmental conditions for accelerated growth, the reduction of broodstock number and the size of hatchery facilities. Similar results to those shown here have been achieved in both coldwater and warmwater freshwater food and ornamental fish and in marine species including seabass, seabream, and turbot.

ACKNOWLEDGMENTS: The authors gratefully acknowledge the assistance and participation of the salmon and trout growers that participated in these trials.

References

1. Zohar, Y. 1989. Fish reproduction: its physiology and artificial manipulation. Pages 65-119 *in* M. Shilo and S. Sarig, editors. Fish culture in warm water systems: Problem and trends. CRC Press, Florida, USA.

2. Zohar, Y. 1988. Gonadotropin releasing hormone in spawning induction in teleosts: basic and applied considerations. Pages 47-62 *in* Y. Zohar and B. Breton, editors. Reproduction in fish: basic and applied aspects in endocrinology and genetics. INRA Press, Paris, France.

3. Crim, L. W. and B. D. Glebe. 1990. Reproduction. Pages 529-553 *in* C. B. Schreck and P. B. Moyle, editors. Methods for fish biology. American Fisheries Society, Bethesda, Maryland, USA.

4. Zohar, Y., A. Goren, M. Fridkin, E. Elhanati and Y. Koch. 1990a. Degradation of native salmon GnRH, mammalian LHRH and their analogues in the pituitary, kidney and liver. General and Comparative Endocrinology 79:306-331.

5. Crim, L. W., N. M. Sherwood and C. E. Wilson. 1988. Sustained hormone release. II. Effectiveness of LHRH analog (GnRHa) administration by either single time injection or cholesterol pellet implantation on plasma gonadotropin levels in a bioassay model fish, the juvenile rainbow trout. Aquaculture 74:87-95.

6. Zohar, Y., G. Pagelson, Y. Gothilf, W. W. Dickhoff, P. Swanson, S. Duguay, W. Gombotz, J. Kost and R. Langer. 1990b. Controlled release of gonadotropin releasing hormones for the manipulation of spawning in farmed fish. proceedings of the International Congress on Controlled Release of Bioactive Materials 17:51-52.

INDUCED SPAWNING IN TELEOST FISH AFTER ORAL ADMINISTRATION OF GnRH-A

B. Breton [1], I. Roelants[2], T. Mikolajczyk[3], P. Epler[3], F. Ollevier[2]

[1] Physiologie des Poissons, INRA, Campus de Beaulieu, 35042 Rennes Cedex, France
[2] Laboratory of Aquaculture and Ecology, Catholic University, B-3000 Leuven, Belgium
[3] Dept. of Ichthobiology, Academy of Agriculture, 30149 Krakow-Mydlniki, Poland

Summary

A formulation of tween 80 (4%), oleic acid (0,6%), EDTA and tryspin inhibitor has been determined to enhance the intestinal uptake of sGnRH-a. It resulted in the stimulation of GtH2 secretion and induced-ovulation after rectal delivery in carp and African catfish, at similar dosages as after IP injection. After microencapsulation and oral delivery, these properties were preserved. The microcapsules are also efficient in rainbow trout to promote stimulation of GtH2 secretion and ovulation. Addition of nutrients did not compete with the effects of GnRH-a.

Introduction

During the reproductive season, breeders are genrally more sensitive to disease, fungus, and other hostile agents. This is generally potentiated by stress and repeated manipulations which can also block ovulation and growth, and affect the quality of gametes. Spawning induction treatments, using IP injections of GnRH-a in individuals, are also time consuming. In order to avoid these problems, we studied to what extent, GnRH-a could be absorbed after oral administration from enteric microcapsules to induce ovulation at similar dosages as those used in IP treatments, the effects of competition of GnRH-a with nutrients were investigated as well.

Although it has been reported that, because of physical and chemical barriers, the differentiated vertebrate gastro-intestinal tract is impermeable to ingested proteins (Gardner et al. 1988), there is evidence that, at least in teleost fish, these barriers are incomplete (Mc Lean and Ash 1986). It is suggested that as compared to mammals, the intestine of teleosts never reaches maturity. In this work we developed an oral delivery system for GnRH-a using microcapsules resistant to the acid pH of the stomach but soluble in the gut. The absorption of the peptide was improved by the incorporation of formulation adjuvants, such as mixed-micellar permeation enhancers and proteolytic enzyme inhibitors.

Material and methods

This work was carried out on mature females belonging to 3 species, an agastric species the carp, *Cyprinus carpio*, and 2 gastric species differing by their feeding and temperature regimes, the rainbow trout, *Oncorhynchus mykiss,* and the African catfish, *Clarias gariepinus.* DArg^6 Pro^9 salmon GnRH (sGnRH-a) was used, because the absence of the L-Arginine carboxyl group of LHRH, possibly sensitive to trypsin hydrolysis, increases resistance to degradation.

sGnRH-a was administered either orally or rectally in liquid form by intubation, or in solid form by forced-feeding, together with the other tested substances. The absolute bioavailability of sGnRH-a was determined by measurement of sGnRH-a concentrations in blood plasma at regular intervals after treatment, using a specific radioimmunoassay (Zohar et al. 1989). Bioactivity was also estimated from the ability of absorbed sGnRH-a to stimulate gonadotropin (GtH2) secretion determined by RIA, and to induce ovulation.

Results

Enhancement of intestinal sGnRH-a absorption

We studied the effects of several kinds of enhancers : medium chain fatty acids such as oleic acid; surfactants such as polysorbates (tweens); saponin adjuvants such as Quil-A-saponin; oligosaccharides such as β cyclo-dextrin and other compounds which have been proved to be efficient enhancers in mammals, (polydecanol, Arlatone, Brij 95, G 1284 and SCS 2064).

Most of these compounds had no effect. Tweens were the most potent absorption enhancers, especially as mixed micella with oleic acid (4%/0.6%). In carp, after rectal delivery of 10µg/kg of sGnRH-a, plasma levels peaked 6 hours later , being significantly increased by the tween/oleic formulation (figure 1).They remained significantly higher than in saline treated fish until 24 hours after administration. After the same delay, blood plasma levels of GtH2 were significantly stimulated by the tween-oleic sGnRH-a elevated formulation (figure 1), and were maintained at this level for more than 36

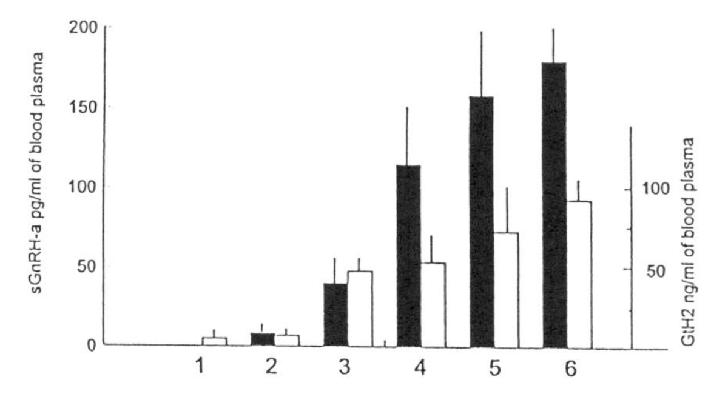

Figure 1 :Effects of rectal sGnRH-a (10µg/kg) and IP pimozide (5mg/kg) deliveries in the common carp on sGnRH-a (black bars) and GtH2 (white bars) blood plasma levels 6 hours after delivery. Results are means ± SD

1- saline, 2- sGnRH-a, 3- sGnRH-a + tween 80/oleic, 4- sGnRH-a + tween/oleic + EDTA, 5- sGnRH-a + tween/oleic + trypsin inhibitor, 6- sGnRH-a + all components

hours. It is noteworthy that, although after injection of sGnRH-a plasma levels peaked at more than 1600 pg/ml (data not shown), blood plasma levels of GtH2 were not statistically different from those measured in animals rectally treated with sGnRH-a/tween-oleic. Similar results were obtained in the catfish, using 20µg/kg of sGnRH-a. In addition 100% of catfish ovulated in the group with the enhancer as well as after IP treatment using the same dosage, while 25% of females ovulated after rectal delivery without enhancer.

In order to protect against proteolysis in the digestive tract, the effects of the bivalent ionchelator EDTA and of eggwhite trypsin inhibotor were also studied, together with tween/oleic. Both, alone or in combination, improved very significantly the effects of tween/oleic ($p<0.01$) on sGnRH-a and correlative GtH2 blood plasma levels (figure 1).

In the carp, a species devoid of a stomach, there was no significant difference between oral and rectal administration. On the contrary, in catfish a gastric species, there was no stimulation of GtH2 secretion after sGnRH-a oral delivery. This emphasizes the need of protection against pepsin and acid pH degradation in the stomach.

Any liquid formulation could stimulate either sGnRH-a uptake or GtH2 secretion in the rainbow trout.

The topical impacts of enhancer formulations on the gut mucosa were studied by histology at regular intervals following rectal intubation. With the tween/oleic mixed micellar formulation, the most important damage was observed after 2 hours, with localized detachment of enterocytes and disappearance of their goblet vacuoles.There was also an increase of cuboïdal cells, and some edema and inflammatory cells in the submucosa. The damage was totally reversed after 24 hours whereas there was no damage in saline control. Other enhancers such as polydocanol induced very drastic damage with general inflammatory processes which were not reversed after 24 hours.

The effects of dopamine antagonists, pimozide and domperidone, rectally delivered at several dosages together with tween/oleic were studied in carp and catfish receiving IP injections of sGnRH-a (20µg/kg). In both species, blood plasma levels of GtH2 were significantly higher in the presence of enhancer than in controls, the most efficient dose being 5mg/kg. It must be noted that pimozide and domperidone were as efficient in carp whereas pimozide was statistically more potent in catfish (figure 2). These results indicate intestinal uptake of dopamine antagonists and high preservation of their biological activity

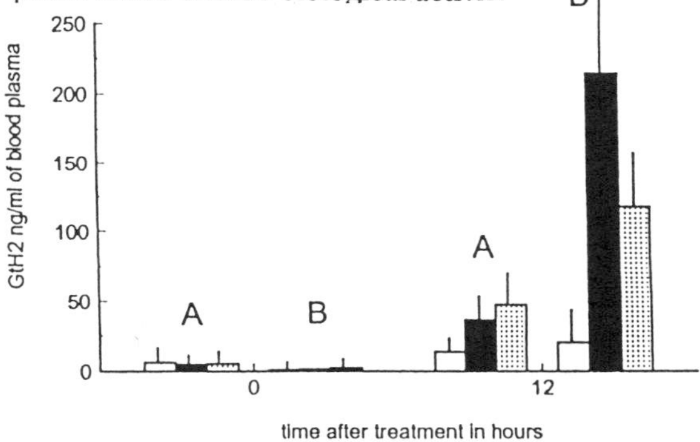

Figure 2 : effects of dopamine antagonists rectally delivered together with enhancers on the response to IP sGnRH-a treatment, values are meams ± SD :
A- in the common carp, B- in the African catfish
white bars = pimozide alone; blacks bars = pimozide with enhancer; shadowed bars = domperidone with enhancer

Oral delivery of microencapsulated sGnRH-a

In order to prepare enteric microcapsules, a powdered form of the whole formulation containing sGnRH-a, pimozide, tween/oleic, EDTA and trypsin inhibitor was prepared. The main obstacle to solve was to entrap both hydrophilic and hydrophobic substances in an enteric polymer. This was achieved by entrapping the different components in a rigid gel of kappa-carrageenan containing silicas at a ratio in agreement with their carrying capacity for high dosages of tween/oleic micella. After lyophilisation and milling, the resulting particles (70µm) were submitted to spheronisation to obtain 250µm microcapsules. These capsules were coated with Cellulose-Acetate-Phtalate (CAP) in presence of a plasticizer, this film cannot solubilized at acid pH, but dissolves in neutral and basic pH. The overall procedure is now accepted for international licence.

The effects of the entrapped sGnRH-a were studied after forced-feeding. This form stimulated GtH2 secretion in carp and catfish and also in the rainbow trout (figure 3), where it induced a dose dependent and prolonged stimulation of GtH2 secretion, Plasma GtH2 levels ahd not returned to basal level, 5 days after delivery. In all groups there were 85 to 90% ovulation within 4 days following

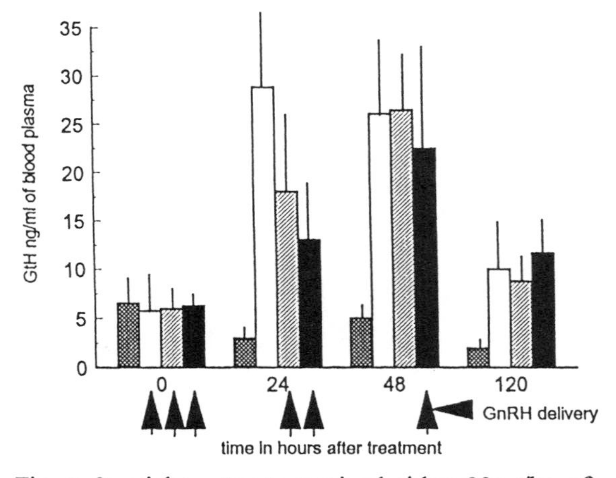

Figure 3: rainbow trouts received either 90µg/kg of microencapsulted sGnRH-a, in once time (white bars) or divided deliveries at 24 hours intervals either twice 45µg (striped bars) or 3 times 30 µg (black bars)

the treatment, while only 25% of controls ovulated. Similar results were obtained in catfish and carp. In these last species, a significant stimulation of GtH2 secretion occured with dose as low as 5 µg/kg of sGnRH-a.

The final aim of this work was to develop an "in food" delivery system. In preliminary experiments on carp and catfish, microencapsulated sGnRH-a (20 µg/k) was orally delivered together with a mixture of nutrients representative of the composition of a commercial fish diet. The ratio between the microcapsules and the nutrients varied from 100 to 33%. In both species, it was demonstrated that these nutrient ratios did not inhibit intestinal uptake of sGnRH-a, because blood plasma levels of GtH2 and rates of ovulation were not significantly different between groups fed with different ratios of microcapsules to nutrients, as in carp (figure 4).

In other experiments, the specific effects of proteins, sugars and lipids were studied. According to the species, these differents components can have either stimulatory or inhibitory effects. This is especially the case of a high lipid supplementation in the rainbow trout.

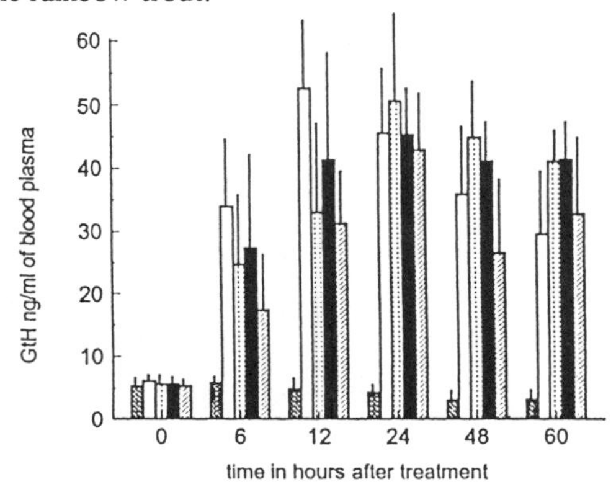

Figure 4 : common carp received 20µg/kg of microencapsulated sGnRH-a together with a mixture of nutrients at different ratios of microcapsules / nutrients: shadowed bars = blank microcapsules, white bars = 100%, pointed bars = 66/37%, Black bars = 50/50% and stripped bars = 33/67%.

Conclusion

All the results obtained in this work demonstrated that using an enhancing formulation, sGnRH-a can be delivered by the orally with the same efficiency as after IP injection or implantationas judged from stimulation of GtH2 secretion and induction of ovulation. From these results it appears possible to develop a cost effective "maturation diet" allowing induced-ovulation without stressfull manipulations of individual fish. However, more detailed studies on the composition of the food for the preservation of the stimulatory and protective effects of the procedure on GnRH-a intestinal uptake are needed.

References

Gardner, M.L.G. 1988. Gastrointestinal absorption of bioactive proteins by the gastrointestinal tract in fish : a review. J. Aquat. Health. 2; 1-11

Mc Lean, E. and Donaldson, E.M. 1990. Absorption of bioactive proteins by the gastrointestinal tract in fish: a review. J. Aquat. Health. 2:1-11

Zohar, Y., Goren, A., Tosky, M., Pagelson, G., Leeibovitz, D. and Koch, Y. 1988. The bioactivity of gonadotropin releasin hormones in its regulation in the gilthead sea bream Sparus aurata : *in vivo* and *in vitro* studies. Fish Physiol Biochem. 7; 59-67

This work was granted by EEC FAR program AQ 3.692. We are very gratefull to Pr Y. Zohar Baltimore for its generous gift of antibody against sGnRH-a, and to Dr. R. Schlutz from Utrech University for assaying catfish GtH2.

THE ROLE OF BROODSTOCK DIETARY PROTEIN IN VITELLOGENIN SYNTHESIS AND OOCYTE DEVELOPMENT, AND ITS EFFECTS ON REPRODUCTIVE PERFORMANCE AND EGG QUALITY IN GILTHEAD SEABREAM *SPARUS AURATA*

[1]M. Harel, [1]A. Tandler, [1]G.W. Kissil, [2]S.W. Applebaum

[1] National Center for Mariculture, IOLR, P.O. Box 1212 Eilat 88112, Israel
[2] The Hebrew University of Jerusalem, Faculty of Agriculture. P.O. Box 12, Rehovot 76100, Israel

Summary

Dietary Essential Amino Acids (EAA) control seabream fecundity and egg and larvae quality mainly via the synthesis and selective uptake of yolk constituents. An *in vitro* binding assay was developed and the presence of specific binding sites for vitellogenin (Vg) in oocyte membrane preparations was studied. The lowest level of plasma Vg and lowest amount of Vg binding to oocyte membrane preparation was found in fish fed wheat-gluten based diet. Supplementation of the wheat-gluten based diet with an EAA profile similar to that of seabream eggs, significantly increased by over 35% the plasma VG level and the amount of VG bound. Such a diet was associated also with doubling 15d larval survival and supported a 50% improvement in larval growth as compared to the wheat-gluten based diet.

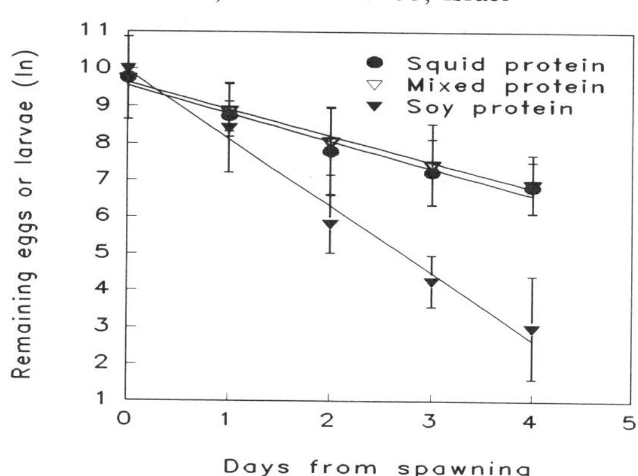

Fig. 1 The effect of broodstock diet composition on the mortality rate of eggs and larvae (n=6).

Introduction

Reproduction in many fish species generally involves a dramatic decrease in their food intake, and a substantial transfer of nutrients from various body stores into the developing oocytes (Aksnes et al., 1986; Smith et al., 1988; Nassour and Leger, 1989). In gilthead seabream, which is a continuous spawner over a period of 3-4 months in its reproductive season, this nutrient flow is mainly sustained by active feeding. Previously we have shown that the nutritional quality of seabream broodstock diet is a key factor in determining egg quality and larval growth and survival (Harel et al., 1994). We have shown also that the expression of broodstock dietary changes in egg composition and egg quality in gilthead seabream occur within 15 days. Based on these findings we continued to investigate the effect of broodstock dietary changes on vitellogenin synthesis and oocyte development.

Results

The importance of the protein fraction in seabream broodstock diet was evaluated initially by replacing the protein extract of a squid meal based diet with an equal amount of soy protein. The experimental diets were randomly assigned to six 600l spawning tanks, containing two females and three males each, and the fish were fed once a day for a period of 15 days of spawning. At the end of the experiment, eggs from each spawning tank were taken for incubation and larval rearing. The mortality rate of eggs and larvae during the first 4 days post spawning was significantly higher when the broodstock fed on soy protein based diet (Fig.1), and after 4 days post spawning very few larvae survived as compared to 25% larval survival for fish fed a squid protein or mixed proteins of squid and soy (1:1).

This result suggested that the positive effect of the squid protein fraction could be related to its balanced composition of EAA, which was very similar to that of seabream egg protein. Based on this understanding it was possible to improve the significantly lower reproductive performance of fish fed a wheat-gluten based diet by its supplementation with an appropriate EAA profile which resembles the seabream egg (Fig. 2). The wheat-gluten EAA supplemented diet was also associated with doubling the survival of 15d larvae and 50% improvement in their growth rate as compared to the gluten protein diet only.

Analysis of egg EAA composition revealed that the amino acid profile of the egg remained conservative, despite the marked changes in the protein quality of the broodstock diet and its subsequent effect on egg and

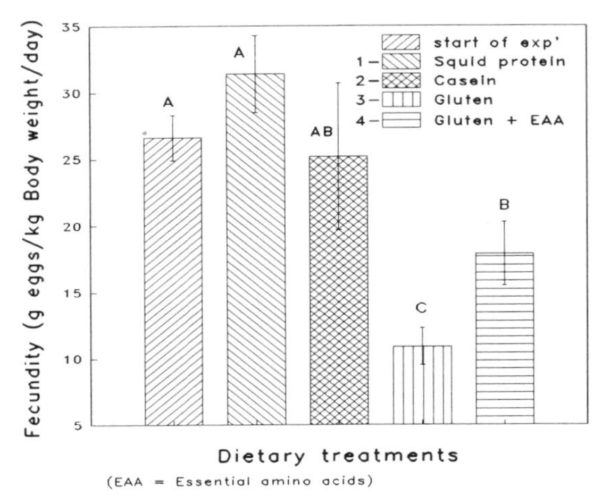

Fig.2. The effect of dietary EAA supplementation on broodstock fecundity (n=5).

larvae quality. Therefore, we investigated the dietary effects on oocyte development during vitellogenesis. An *in vitro* binding assay was developed, according to Konig and Lanzrien (1985). Blood plasma was collected from 17-ß estradiol treated male fish. Vg was purified from the plasma using a combination of gel filtration and anion exchange column chromatography. The purified Vg was labeled with ^{125}I, and the presence of specific binding sites for Vg in oocyte membrane preparations was studied. Vg binding reached an equilibrium within 90 minutes at 23°C, and was shown to be highly specific and saturable. Analysis of the saturation curves revealed that the concentration of Vg binding sites was 6.4 nmol/mg membrane protein. Scatchard analysis of these curves demonstrated high affinity (kD 4.26×10^{-8}M) of Vg to its binding sites. From all seabream plasma proteins and lipoproteins, only the Very Low Density Lipoprotein (VLDL) compete with Vg for binding to the oocyte membrane preparations, and had a 10% crossreactivity with Vg for binding (Fig. 3)

Once the parameters and the characteristics of the *in vitro* binding assay was established, it was possible to test the effect of different components in the broodstock diet including essential fatty acids and amino acids, on Vg synthesis and its binding cpacity to the growing oocyte. Another 4 diets were formulated using soy oil and wheat-gluten as a lipid and protein source. These diets were supplemented with a combination of EAA and n-3 highly un saturated fatty acids (n-3 HUFA) which we found previously as essential component in seabream broodstock diet. The squid meal based diet was used as a control diet during the experimental period.

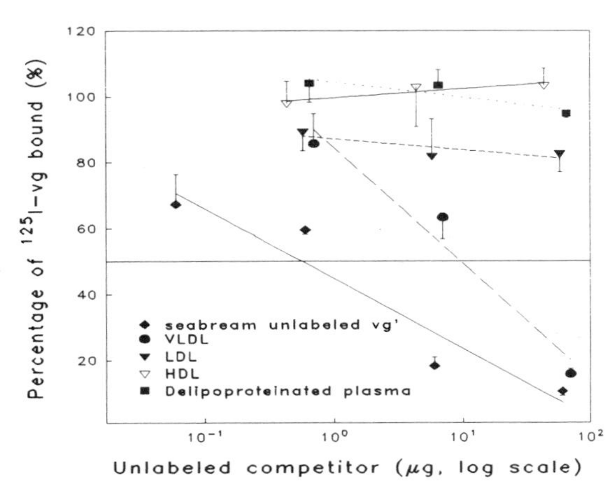

Fig.3. competition between Vg and other plasma proteins and lipoproteins on oocyte binding sites for Vg (n=3).

The EAA composition of the broodstock diet significantly effected the plasma level of Vg and its binding capacity to fish oocyte membranes. Lowest levels of plasma Vg and Vg binding sites on oocyte membrane preparations was observed when seabream broodstock were fed a wheat-gluten based diet. Supplementation of EAA to this diet significantly increased both the Vg plasma level and Vg binding capacity of fish oocyte membranes. On the other hand, the addition of n-3 HUFA did not have any dramatic effect on Vg synthesis and Vg binding capacity (Fig. 4,5)

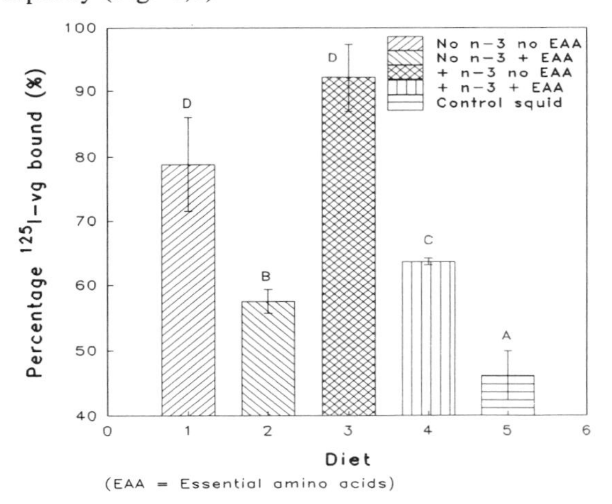

Fig.4. The effect of dietary EAA and n-3 HUFA supplementation on plasma Vg level (n=4).

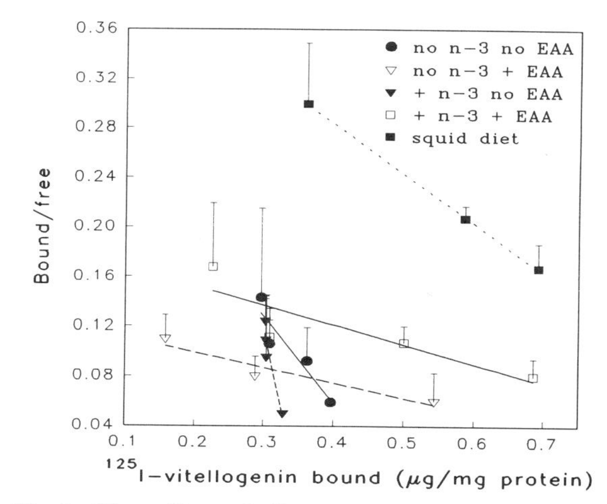

Fig.5. The effect of dietary EAA and n-3 HUFA supplementation on oocyte binding capacity to Vg (n=4).

Discussion

The binding of Vg to oocyte membrane preparations in gilthead seabream is highly specific and saturable. This finding support some other similar findings in fish, and confirm the general mechanism of selective uptake of Vg through receptor mediated endocytosis, which was previously demonstrated in chicken (Yusko et al., 1981), amphibians (Wallace et al., 1970) and insects (Ferenz, 1990).

Our results suggests that Vg is probably the major serum protein which is taken up by the developing seabream oocyte, while VLDL counts for only 10% crossreactivity with Vg for binding. The smaller uptake of VLDL probably reflects its low level in fish serum lipoprotein, where HDL is the primary transporter of serum lipids (MacFarlane et al., 1990). In contrast, chicken oocyte membranes also interact with both Vg and VLDL. But VLDL, which is the main serum lipids transporter in avian system, is much more important than Vg in contributing to the chicken egg lipids (Yusko et al. 1981; Wallace and Selman 1985).

The present study, together with our previous study (Harel et al., 1994), clearly demonstrates the different mode of action of dietary essential amino acids and fatty acids on seabream egg quality. While dietary EAA may affects egg quality through the control of Vg synthesis and its uptake, without any apparent effect on egg EAA composition, dietary EFA may affect egg quality mainly through changing the egg EFA composition without any apparent effect on Vg synthesis and its uptake.

In conclusion; the observation that dietary EAA influence the vitellogenesis process in seabream appears to have a significant potential for increasing reproductive yield and egg quality in fish, as it will be possible to improve this process by supplementation the broodstock diet with an appropriate EAA profile similar to that of fish egg protein.

References

Aksnes, A., Gjerde, B. and Rolad, S.O. 1986. Biological, chemical and organoleptic changes during maturation of farmed Atlantic salmon, *Salmo salar*. Aquaculture, 53:7-20.

Ferenz, H.J. 1990. Receptor-mediated endocytosis of insect yolk proteins. Molecular Insect Science (Hagedorn et al. Eds.). Plenum Press. New York. pp.131-138.

Konig, R. and Lanzrein, B. 1985. Binding of vitellogenin to specific receptors in oocyte membrane preparations of the ovoviviparous cockroach (*Nauphoeta cimerea*). Insect biochem., Vol.15(6), 735-747.

Harel, M., Tandler, A., Kissil, G.Wm. and Applebaum, S.W. 1994. The kinetics of nutrients incorporation into body tissues of gilthead seabream *Sparus aurata* females and the subsequent effects on egg composition and egg quality. The British J. of Nutrition, 72:45-48.

MacFarlane, R.B., Harvey, H.R., Bowers, M.J. and Patton, J.S. 1990. Serum lipoproteins in striped bass (*Morone saxatilis*): Effects of starvation, Can. J. Fish. Aquat. Sci., 47:739-745.

Nassour, I. and Leger, C.L. 1989. Deposition and mobilization of body fat during sexual maturation in female trout *Salmo gairdneri R.* Aquat. Living Resour., 2:153-159.

Smith, R.L., Paul, A.J. and Paul, J.M. 1988. Aspects of energetics of adult walleye pollock, *Theragra chalaegramma (pallas)* from Alaska. J. Fish. Biol., 33:445-454.

Wallace, R.A., Jared, D.W. and Nelson, B.L. 1970. Protein incorporation by isolated amphibian oocytes. I. Preliminary studies. J. Exp. Zool. 175:259-270.

Wallace, R.A. and Selman, K. 1985. Major protein changes during vitellogenesis maturation of fundulus oocytes. Develop. Biol., 68:172-182.

Yusko, S., Roth, T.F. and Smith, T. 1981. Receptor mediated vitellogenin binding to chicken oocytes. Biochem. J., 200:43-50.

THE EFFECT OF SEASONAL ALTERATION IN THE LIPID COMPOSITION OF BROODSTOCK DIETS ON EGG QUALITY IN THE EUROPEAN SEA BASS (*Dicentrarchus labrax*)

J. Mª Navas[1], M. Thrush[2], J. Ramos[1], M. Bruce[2], M. Carrillo[1], S. Zanuy[1] and N. Bromage[2]
[1] Instituto de Acuicultura de Torre de la Sal, Consejo Superior de Investigaciones Científicas, 12595 Torre de la Sal, Castellón-España. [2] Institute of Aquaculture, University of Stirling, Stirling, Scotland FK9 4LA, U.K.[†]

Summary

The present study investigates the assumption that broodstock lipid requirements vary seasonally depending on the state of gonad maturation. Fish were fed two pelleted lipid enriched diets (corn oil, diet 1 or high quality fish oil, diet 2) during four different periods in the annual cycle. The group fed diet 2 for twelve months and the group diet 2 only during vitellogenesis-spawning (September-February) and diet 1 during the remaining months, showed improved egg quality and higher hatch rates when compared to the remaining two groups that were fed diet 2 during pre-vitellogenesis (April-September) and post-spawning (February-April). Improved egg quality was associated with a higher total n-3 fatty acids content, including enhanced levels of both docosahexaenoic acid (DHA) and eicosapentaenoic acid (EPA).

Introduction

Economically productive aquaculture is heavily dependent upon an adequate supply of fertile eggs and juvenile fish. Thus cultivation of broodstock which produce eggs of high quality is very important. It has been generally agreed that the quality of the feed is a factor that affects greatly the spawning performance and the egg quality. The effects of broodstock diets containing different levels of protein and energy, or the effects of different rations on egg quality are conflicting (Bromage et al, 1992). However, variations in some specific broodstock dietary components like minerals, vitamins or essential fatty acids (EFA) show a clear correlation with the egg quality (Watanabe and Kiron, 1995). Recent evidence indicates that the fatty acid requirement of fish depends upon the dietary ratio of n-3 to n-6 polyunsaturated fatty acids (PUFA). This dietary intake of PUFA, particularly the n-3 series including docosahexaenoic acid (DHA), 22:6 (n-3) and eicosapentaenoic acid (EPA), 20:5 (n-3), have profound effects on oogenesis and embryonic development, both critical for reproductive success. Dietary shortage of these essential PUFA produced eggs and larvae of a poorer quality than from fish receiving more balanced diets (reviewed by March, 1993). Recently, Cerdá et al, (1995) have reported dramatic reductions in fecundity and egg viability when sea bass broodfish were fed with commercial diets deficient in n-3 PUFA. The objective of the present study is to provide information on the nutritional requirements of captive European sea bass focusing on the effects of seasonal alterations in the EFA composition of sea bass broodstock diets on egg quality.

Material and Methods

Groups of 30, three year old (wt., 500g) broodstock sea bass were reared in 8000 l concrete tanks and maintained in aerated running sea water at ambient temperature and photoperiod. During two years groups were fed to satiation with high EFA diets at various times coinciding with different phases of the reproductive cycle (table 1).

Table 1. Feeding protocol for brood stocks (Vitel=vitellogenesis, Spw=spawning,)

Groups	High EFA diet administered*	Reproductive stage
Group 1	Sep- Feb	Vitel-Spw.
Group 2	Feb-Apr	Spw-Postspw.
Group 3	Apr -Sep	Previtel.
Group 4	All Year	Control

*A diet containing a low concentration of EFA. was supplied at all other times.

Both high and low EFA diets were produced by soaking a base manufactured feed in fish oil or corn oil respectively immediately prior to feeding, which provided pellets with a total lipid content of 20%.
The fatty acid profiles of the lipids of the diets used are shown in Table 2. During the course of the spawning seasons, naturally fertilized eggs were collected each day to determine the proportion of viable (floating) to inviable (sinking) eggs as described by Carrillo et al., (1989). Further aliquots of viable eggs were transferred to incubation facilities for assessment of survival to hatch. Lipid classes were analysed using the double development high performance thin layer chromatography (TLC) method. Esterification of the extracted lipids provided fatty acid methyl esters for assessment of total lipid fatty acids, which were purified by TLC, then analysed using a Packard 436 gas chromatograph.

Table 2. Fatty acid compositions of diets. (mg/g d.y weight)

	Low EFA	High EFA
20:4 n-6	2.7	4.
20:5 n-3	44.1	79.1
22:6 n-3	51.3	72.8
Total Saturates	180.9	203.2
Total Monoenes	307.8	413.5
Total Dienes	256.5	44.1
Total PUFA	393.3	261.6
Total n-3	138.6	217.6
Total n-6	254.7	43.2
Total n-9	198.9	200.5
n-3:n-6 Ratio	0.5	5.0
DHA:EPA Ratio	1.2	0.9
Total Identified	97.9%	97.7%

Results

Long term administration of high EFA diets to broodstock groups in different stages of ovarian development did not produce consistent differences in growth. In the first spawning season, the relative fecundity was roughly the same in all groups (around 300,000 eggs / Kg♀) while in the second spawning season relative fecundity increased to 400,000-550,000 in all groups.The results of egg viability and hatching rates are represented in the figure 1 .

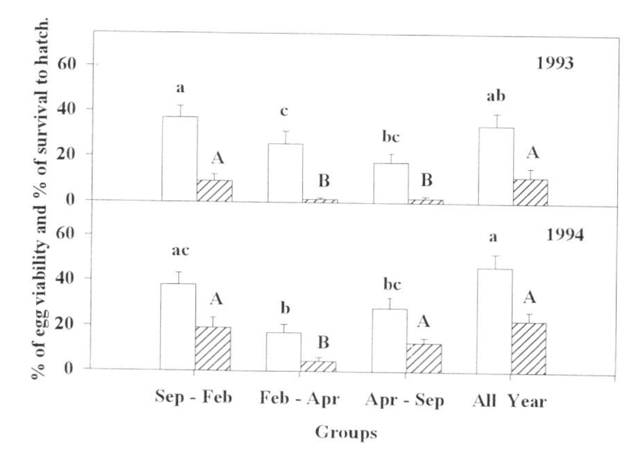

Fig.1 Viability of eggs (open bars) and survival to hatch (hatched bars) among eggs spawned by broodstock maintained on different diets during two consecutive years. Means with different letters in the same case are significantly different ($p<0.05$).

During two successive spawning seasons, egg viability and survival to hatch were higher in fish fed high EFA diet, either throughout the year or during vitellogenesis-spawning, than in those given the high EFA diet either during spawning-post-spawning or previtellogenic periods. However, during the second spawning season, a significant increase ($p<0.05$) in survival to hatch was observed in the group 3 (Apr-Sep). A general improvement either in egg viability or survival to hatch was observed in all groups during the second spawning season except in group 2 (Feb-Apr) (Fig. 1). Eggs spawned by fish maintained on the high EFA throughout the year (group 4) and during vitellogenesis-spawning (group 1) had significantly higher proportions of total n-3 ($P<0.01$), significantly lower total n-6 fatty acids ($P<0.01$) and a significantly higher n-3:n-6 ratio than those fed high levels during either the pre-vitellogenesis (group 3) or spawning post-spawning (group2) periods (Table 3). Eggs spawned by fish in group 4 had significantly higher quantities of DHA, (accounting for 20% of the total fatty acids) than all other groups ($P<0.05$) and eggs produced by group 1 had significantly higher quantities of DHA than either groups 2 or 3 ($P<0.01$). Groups 1 and 4 also had significantly higher proportions of EPA ($P<0.01$). However there were no significant differences in DHA:EPA ratios, or in 20:4n-6 (arachidonic acid) levels.

Table 3. Fatty acid composition of the total lipid extracted from fertilized eggs.

GROUPS	1	2	3	4
20:4 n-6	0.6	0.7	0.7	1.1
20:5 n-3	7.4[b]	5.4[a]	5.6[a]	7.5[b]
22:6 n-3	17.6[b]	14.5[a]	14.3[a]	20.3[c]
Total Saturates	19.1	18.0	18.3	20.2
Total Monoenes	36.9[b]	32.8[a]	33.1[a]	20.2[b]
Total Dienes	11.4[b]	23.6[c]	22.8[c]	7.2[a]
Total PUFA	41.9[a]	47.6[b]	47.1[b]	40.5[a]
Total n-3	30.1[b]	23.3[a]	23.7[a]	32.5[b]
Total n-6	11.8[b]	24.3[c]	23.4[c]	8.0[a]
Total n-9	27.1	26.3	26.0	27.0
n-3:n-6 Ratio	2.6[b]	1.0[a]	1.0[a]	4.1[c]
DHA:EPA Ratio	2.4	2.7	2.5	2.7
Total Identified	97.8	98.5	98.4	97.9

Means in the same row with different letters are significantly different ($p<0.05$).

Discussion

Results obtained from the four broodstock groups showed no consistent differences in either the growth profiles or fecundity over the course of the trial, although these parameters increased with the age of the animals. On the other hand, the diets given to sea bass broodfish have profound effects on egg and larval quality. In addition, broodstock requirements vary seasonally depending on the maturational state of the gonad. Thus, nutritional requirements of adult sea bass for growth are not as demanding as those for the attainment of good egg quality. The best spawning performance in the seasons studied correspond to groups which had received the high EFA diets either throughout the year or during vitellogenesis-spawning (Sep-Feb) as compared to those which had been reared on the high EFA either during pre-vitellogenesis (Apr-Sep) or post-spawning (Feb-Apr) periods. This indicates that the timing of administration of dietary fatty acid is important, in that the period of vitellogenesis-spawning is the critical period for incorporation of essential fatty acids into the developing oocytes. However, the supply of high EFA during five months or more during successive years can alleviate in part, this seasonal requirement of EFA as seen in group 3 (Apr-Sep) where egg survival rate to hatch improved during the second spawning season (fig. 1). This effect was not observed in group 2 (spawning-post spawning) in which the quality of the eggs was very poor even during the second spawning season. A possible explanation could be related to the natural reduced food intake of the broodfish during this period, the short duration of feeding with high EFA (two months) and the inappropriate time of high EFA supply (which coincides with the end of reproductive period). All these factors could contribute to the failure to produce eggs of good viability and the low larval survival to hatch rate of this group. The use of corn oil (as a source of low EFA) produces a number of adverse effects in the quality of the progeny when administered before or during spawning. The sea bass, like most marine fish (Sargent et al., 1993), is unable to convert C18 PUFA (provided with corn oil) to C20 and C22 homologues and as a consequence requires the preformed end products 20:4(n-6) and 22:6(n-6) available in the high EFA diets (provided by the fish oil). Watanabe and Kiron (1995) have revealed the importance of pre-spawning nutritional regimes and more specifically the levels of (n-3) PUFA (i.e., DHA and EPA) in the diets of the broodfish, on egg and larval quality. In this regard, not only the absolute levels of DHA and EPA can be considered as the most important determinants of egg quality but also the ratios of DHA:EPA and (n-3):(n-6) PUFA for marine teleosts. Sea bass eggs collected from females which had received the high EFA diet during vitellogenesis had better quality, higher total (n-3):(n-6) ratios and higher levels of DHA and EPA. These results indicate that both the levels and ratios of EFA in the diets supplied to broodfish and the time of their administration have a clear influence on the composition and viability of the eggs, stressing the importance of the transfer of fatty acid nutrients from the broodstock diet of the mother to the offspring as as reported for several marine teleosts (Zohar et al., 1995, Cerdá et al., 1995, Watanabe and Kiron, 1995). We can conclude that broodstock sea bass have a high requirement for long chain polyunsaturated n-3 fatty acid for the development of viable eggs and larvae and that the timing of the provision of dietary, fatty acid is important for the incorporation of EFA into the developing oocytes.

References

Bromage, N., J. Jones, C. Randall, M. Thrush, J. Springate, J. Duston and G. Barker. 1992. Broodstock management, fecundity, egg quality and the timing of egg production in the raibow trout (*Oncorhynchus mykiss*). Aquaculture 100: 141-166.

Carrillo, M., N. Bromage, S. Zanuy, R. Serrano and F. Prat. 1989. The effect of modifications in photoperiod on spawning time, ovarian development and egg quality in the sea bass (*Dicentrarchus labrax* L.). Aquaculture 81:351-365.

Cerdá, J., S.Zanuy, M. Carrillo, J. Ramos and R. Serrano. 1995. Short and long-term dietary effects on female sea bass (*Dicentrarchus labrax*): seasonal changes in plasma profiles of lipids in relation to reproduction. Comp. Biochem.Physiol. (In press)

March, B. E. 1993. Essential fatty acids in fish physiology. Can.J. Physiol. Pharmacol. 71:684-689.

Sargent, J.R., J.G. Bell, M.V. Bell, R.J. Henderson and D.J. Tocher. 1993. The metabolism of phospholipids and polyunsaturated fatty acids in fish. In: Aquaculture: Fundamental and applied Research. B. Lahlou and P.Vitello (Eds.): American Geophysical Union, Washington, D.C., p. 103-124.

Watanabe, T. And V. Kiron. 1995. Red sea bream.In: Broodstock management and larval quality. N.R. Bromage and R.J. Roberts (Eds.): Blackwell Science, Oxford, p. 398-414.

Zohar, Y., M. Harel, S. Hassin, A. Tandler. 1995. Gilt-head sea bream. 1995. In: Broodstock management and larval quality. N.R. Bromage and R.J. Roberts (Eds.): Blackwell Science, Oxford, p. 94-117.

Supported by an EEC FAR contract (AQ 2 406 E UK) and a fellowship from the Generalitat Valenciana to J. Navas.

EFFECTS OF CRYOPROTECTANT, SPERM DENSITY AND STRAW SIZE ON CRYOPRESERVATION OF BLUE CATFISH, *Ictalurus furcatus*, SPERM

A.N. Bart, D.F. Wolfe and R.A. Dunham
Dept. of Fisheries and Allied Aquacultures, Auburn University, Auburn, AL 36849, USA

Summary

Sperm from nine blue catfish, *Ictalurus furcatus*, were cryopreserved for 12 months. Cryopreserved sperm was then used to fertilize lots of 450 channel catfish, *Ictalurus punctatus*, eggs. Mean relative fertilization percentage for frozen sperm was 32% (26-39%) of the fresh control. Sperm frozen with the combination of an intracellular cryoprotectant, skim milk, produced no fertilization. No difference was observed between fertilization rate of sperm frozen in two straw sizes (P>0.05). The highest level of relative fertilization rate (54%) was achieved with an insemination dose of 6.00×10^9 sperm per straw (P<0.05). There was no difference (P>0.05) in fertilization rate between insemination doses of 3.75×10^8 sperm and 1.70×10^9 sperm per straw.

Introduction

Cryopreservation of catfish spermatozoa is particularly important to aquaculturists and geneticists. Domesticated catfish stocks are subjected to artificial selection pressure by breeders resulting in genetic drift. Determination of sperm collection methods, sperm fertilizing efficiency, cryopreservation method, reduction in waste of sperm during artificial fertilization, and development of long term storage of sperm requires that research establishes the best method of fertilization. The objectives of this study were to evaluate the effects of cryoprotectants, straw size, sperm concentration, and freezing and thawing procedures on viability of cryopreserved sperm.

Results and discussion

Table 1

Mean percent fertility of channel catfish, *Ictalurus punctatus*, eggs fertilized with blue catfish, *I. furcatus*, sperm cryopreserved for 12 months in two different straw sizes.*

Straw size	Reps.	**Fertility Means(SD)	%of Control
0.5(ml)	29	14.3^a(15.01)	30.2
1.0(ml)	36	19.4^a(15.22)	40.9
Control	16	47.4^b(15.39)	100.0

*A total of 450 eggs were used for each replicate. Only the dimethylsulfoxide cryoprotectant treatment was included when calculating means. Controls (fresh sperm) were extended with DMSO and Hanks solution. **Mean fertilities followed by the same letters do not differ (P>0.05)

Table 2

Mean percent fertility of blue catfish, *Ictalurus furcatus*, sperm at three different insemination doses after 12 months of cryopreservation.

Insemination dose	Sperm per egg	*Fertility Mean(SD)	%of Control
3.75×10^8	8.33×10^5	10.9^a(10.5)	23.0
1.70×10^9	3.78×10^6	16.1^a(13.8)	34.0
6.00×10^9	1.33×10^7	25.6^b(18.1)	54.0
Control	5.00×10^5	47.4 (15.4)	100.0

*Channel catfish, *I. punctatus*, egg number was 450 for all replicates. Spermatozoa were stored in two different straw sizes (0.5 and 1.0 ml) and only the dimethylsulfoxide cryoprotectant treatment was included. Controls (fresh sperm) were extended with DMSO and Hanks solution.
**Mean fertilities followed by the same letters are not different (P>0.05).

Minimally, a total of $1.25\text{-}2.50 \times 10^5$ fresh sperm/egg are needed to fertilize catfish eggs (Bart 1994). The concentrations in this study were 7, 30 and 107 times greater than 1.25×10^5 cells/egg. Highest level of post-thaw fertilization (54%) was achieved with the largest number of sperm/egg (1.3×10^7). In light of the value of sperm from blue catfish, efficiency concerning cost is an important consideration. This study indicates that one to two males are needed to provide sufficient mature sperm cells for a 32% fertilization of eggs from a single female. This would be inefficient and cost-prohibitive for production purposes. A number of variables such as larger straw sizes, various intra and extra-cellular cryoprotectants, insemination doses and, sperm and egg fitness need to be studied to achieve higher percentage of fertilization with cryopreserved sperm.

Reference

Bart, A.N. 1994. Effects of sperm concentration, egg number, fertilization efficiency with channel catfish, eggs and blue catfish sperm. Ph.D. Thesis. Auburn University

mRNA EXPRESSION AND ENZYMATIC ACTIVITY OF CATHEPSIN D RELATED WITH SEABREAM EGG QUALITY

Carnevali, O. , Centonze, F. Brooks S*., Sumpter J*. and Bromage N°.

Dept. of Biology MCA , University of Camerino, Italy; * Dept. of Biology and Biochemistry, Brunel University, UK; °Inst. of Aquaculture, University of Stirling, Scotland, UK.

SUMMARY

This study aimed to purify the seabream *Sparus aurata* and sea bass *Dicentrarchus labrax* Cathepsin D and to evaluate its enzymatic activity related with spawning period; in addition, partial cloning and sequencing of Cathepsin D gene were performed in both species.

INTRODUCTION

Cathepsin D is a member of the family of aspartic proteases and is thought to play a role in the lysosomal-mediated degradation of proteins (Tang and Wong, 1987). Its gene has been cloned and sequenced in a number of mammalian species and in chicken. As far as oviparous vertebrates are concerned, this enzyme has been recognized to be responsible for the yolk protein processing pathways during vitellogenesis. The findings obtained in salmonid and chicken provide evidence that the catalyst for the intraoocytic processing of vitellogenin is Cathepsin D. In this study the enzymatic activity of the ovary was assessed during vitellogenesis as well in good and poor quality eggs of seabream in which previous results demonstrated a second proteolytic cleavage of yolk components (Carnevali *et al.* 1992; 1993). In addition, seabream and sea bass Cathepsin D probes were developed

RESULTS

Using the two primers (20 nucleotides) corresponding to a Cathepsin D cDNA region common to mammals and chicken, fragments of 750 and of 561 bp were obtained using genomic DNA as template. Both were cloned and sequenced; the former was found to be a PCR artefact while the latter was very conserved with respect to Cathepsin D of other vertebrates, except in two regions of 97 and 68 bp, respectively.

The same primers, used on ovary cDNA as template, gave a fragment of about 400 bp, indicating the presence of intron in the 561-bp sequence.

Using the specific primer (20 nucleotides) corresponding to seabream sequence together with the previous antisense one, a fragment of 200 bp was obtained using both genomic DNA and ovary cDNA.

The evolutionary analysis of DNA sequences based on a distance matrix indicated that the Cathepsin D modifications are class specific (Fig 1).

In the affinity chromatography, enzymatic activity was found in several eluting peaks, the pool of which in denaturating electrophoresis showed only one band at 60 kDa.

The enzymatic activity of the ovary from different phases of reproductive cycle showed significant differences well related with follicle maturation (Table I).

Fig. 1 - Evolutionary tree of Catepsin D gene sequences.

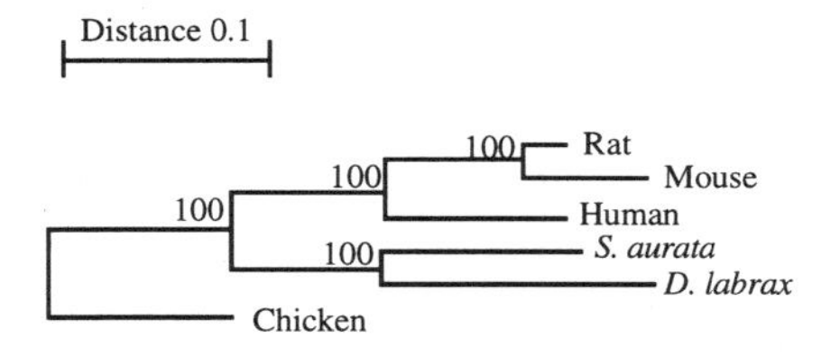

Table I - Ovarian enzymatic activity of Cathepsin D in seabream.

Pre-reproductive period	65 U/g tissue
Reproductive period	120 U/g tissue
Post-reproductive period	0.84 U/g tissue
Good quality eggs	4.5 U/g tissue
Poor quality eggs	6.2 U/g tissue

DISCUSSION

The partial cloning and sequencing of sea bass and seabream Cathepsin D, described here, represents the first Cathepsin D gene sequencing in fish. This sequencing together with the isolation of Cathepsin D protein can be considered a very productive tool in fish management, since this technology can be adopted for assessment of egg quality as well as for a better understanding of the egg maturation process.

REFERENCES

Carnevali O., Mosconi G., Roncarati A., Belvedere P., Romano M. and E. Limatola 1992. Changes in the electrophoretic pattern of yolk proteins during vitellogenesis in the gilted sea bream *Sparus aurata* L. Comp. Bochem; Physiol. 103, 955-962.

Carnevali O. Mosconi G., Roncarati A., Belvedere P., Limatola E., and A.M. Polzonetti-Magni 1993. Yolk protein changes during oocyte growth in European sea bass *Dichentrarchus labrax* L. J. Appl. Ichthyol. 9, 175-194.

Tang J. and R.N.S. Wong, 1987. Evolution in the structure and function of aspartic proteases. J. Cell Biochem. 33, 53-63.

Acknowledgments- This study was supported by AIR Project Egg Quality

MILT QUALITY AND QUANTITY PRODUCED BY YELLOWTAIL FLOUNDER (*Pleuronectes Ferrugineus*) FOLLOWING GnRH-ANALOGUE TREATMENT BY MICROSPHERES OR PELLET.

S.J. Clearwater and L.W. Crim.

Ocean Sciences Centre, Memorial University of Newfoundland, St. John's, Newfoundland, Canada A1C 5S7.

Summary

GnRH-a treatment of yellowtail flounder males results in a stimulation of milt volume without a negative effect on sperm quality. Both microspheres and 100% cholesterol pellets increase sperm volume, with a trend toward a longer term response in the males treated with a higher dosage pellet (197µg/kg).

Introduction

Collection of sperm from male yellowtail flounder is difficult because they produce small volumes of milt (0.1-1.0ml). GnRH analogue (GnRH-a) hormone treatment has been used successfully to increase milt volume in several teleost species, however the quality of the sperm so produced has not yet been examined. This study examines two methods of intramuscular administration of the GnRH-a D-Ala^6, Pro^9-NHEtLHRH in yellowtail flounder. Firstly, via the more familiar cholesterol pellet and secondly, via polymer-based biodegradable microspheres (Mylonas *et al* In Press). Sperm quality was assessed by fertilization trials in which the eggs were maintained until hatch.

Methods

The GnRH-a was implanted either by microsphere injection (30µg/kg) or in a 100% cholesterol pellet at three dosage levels (0µg/kg=control, 44µg/kg=20µg pellet, or 197µg/kg=100µg pellet) in four different groups of males (n=29) during the spawning season. All available milt was collected from each fish 2 days before hormone treatment and on 5 occasions after treatment. For logistical reasons, all treatments were divided into two groups and handled on consecutive days. Egg fertilization trials were completed when good quality eggs were available from non-treated females, 2 days before and 4 & 12 days after GnRH-a treatment of males. For each male 5µl sperm was used to fertilize 70µl eggs and the test was replicated three times. Fertilization rates were counted when the eggs had reached the 4-cell stage of development and hatch rates were counted every 2 days until all eggs had either hatched or died.

Results and Discussion

Compared to pellet control fish all groups of hormone treated fish showed an increase in milt volume, with a trend towards a longer term response in fish receiving the highest dosage pellet (Fig.1.). Although milt volume was increased by GnRH-a treatment, spermatocrit decreased throughout the duration of the experiment in all groups of fish, which may be a seasonal phenomenon or it may be due to

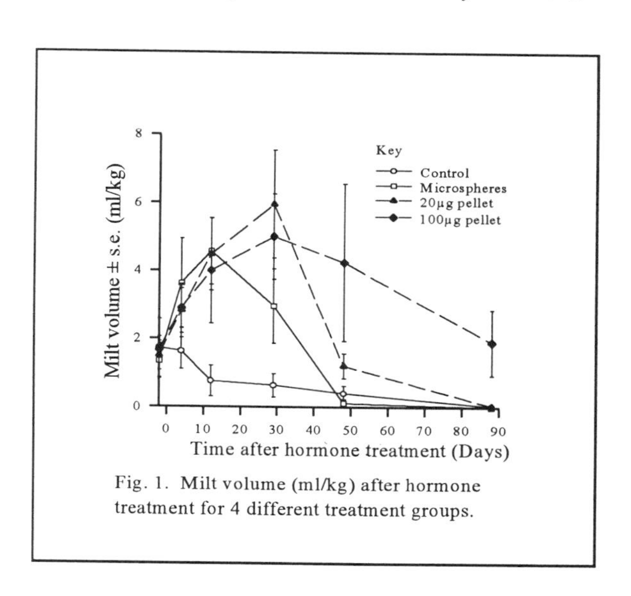

Fig. 1. Milt volume (ml/kg) after hormone treatment for 4 different treatment groups.

the GnRH-a treatments. There were no differences in fertilization rates between treatments on day -2, 4 or 12. Hatch rates showed a treatment and a group effect on day -2 (before hormone implantation), but there were no differences on day 4 after implantation. On day 12 hatch rates showed both a treatment and a group effect, and a significant interaction between these two factors, which can probably be attributed to unusually low hatch rates in group 2 of the control treatments.

References

Mylonas, C.C., Tabata, Y., Langer, R. and Y. Zohar, 1995. Preparation and evaluation of polyanhydride microspheres containing gonadotropin releasing hormone (GnRH) for inducing ovulation and spermiation in Fish. Journal of Controlled Release *In Press*.

CO_2 EFFECTS ON TURBOT (*SCOPHTHALMUS MAXIMUS*) SPERMATOZOA MOTILITY

C. Dreanno[1,4], J. Cosson[2], C. Cibert[3], F. André[4], M. Suquet[1] and R. Billard[4]

[1]IFREMER, Laboratoire de Physiologie des Poissons, BP 70, 29280 Plouzané, France; [2]CNRS URA 671, Station Marine, 06230 Villefranche/mer, France; [3]IJM, Laboratoire d'Imagerie, 2 place Jussieu, 75005 Paris, France; [4]MNHN, Laboratoire d'Ichtyologie, 43 rue Cuvier, 75231 Paris, France.

Fish spermatozoa (spz) are immotile in the genital tract and activated only when released in water. Inhibition results from high content K^+ in salmonids or low osmotic pressure in cyprinids. In turbot (*Scophthalmus maximus*), it is suggested that CO_2 plays a role in this mobility regulation. We studied the effect of CO_2 and respiration inhibitors (NaN_3, KCN and $NaHCO_3$) on the movement and the metabolism of spz.

Material and Methods

Turbots were hand stripped and the sperm collected directly from the urogenital papilla into a syringe. Spz motility was observed using a two step dilution: 1) IM (Immobilising Medium) glucose 200 mOsmol/kg, BSA (10 g/l) buffered to pH 8.2 with 20 mM Tris HCl; dilution rate (1/100); 2) AM (Activating Medium) sea water buffered at pH 8.2 with 20 mM Tris HCl, BSA (10 g/l), dilution rate (1/10). Flagellar beat frequency, wave propagation and amplitude of beating were studied and ATP concentration was measured as described in Perchec *et al.*, (1995).

Results

When exposed to a gentle CO_2 flux, spz swimming in AM stopped immediately (Fig.1a.).

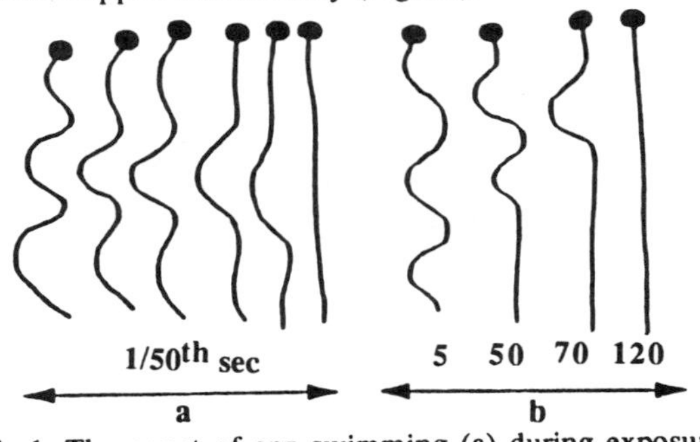

Fig.1. The arrest of spz swimming (a) during exposure to a CO_2 flux or (b) without exposure CO_2 flux.
a. Time post CO_2 flux b. Time post activation (sec)

Simultaneously, the ATP concentration decreased very rapidly from 25 to 5 nM/10^8 spz but the concentration was sufficient to sustain sperm movement. When the CO_2 flux was arrested, spz rapidly resumed a normal swimming. Simultaneously the ATP concentration increased. NaN_3, KCN induced a reversible arrest of swimming (Fig.2.) and an ATP decrease. Furthermore the addition of $NaHCO_3$ (25-100 mM) caused the arrest of motility in AM. Right after (1/50th sec) CO_2 is applied, spz showed a stiffening of the proximal part of the flagellum (Fig.1a.). On the other hand without CO_2, spz showed a stiffening of the distal part (Fig.1b.). Then for both cases, the flagellum stiffened on its whole length.

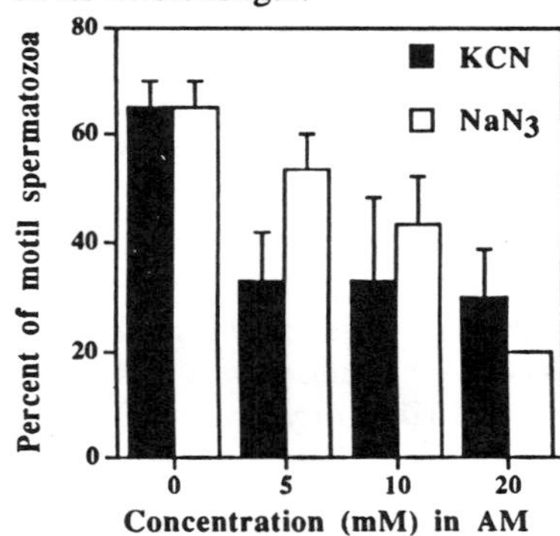

Fig.2. Effects of respiration inhibitors on spz motility. The percentage of motile spz was determined after 60 sec of movement (mean ± sem, n=3 males)

Discussion

The CO_2 inhibitory effect previously demonstrated on the sea urchin spz movement by Brokaw (1977) was also observed in turbot in our studies. $NaHCO_3$ induced likewise a reversible arrest of swimming, so the CO_2 arrest is no due to a pH shift. Sperm exposure to CO_2 induced a rapid ATP decrease probably due to a blockage of oxidative phosphorylation. However, CO_2 could have a direct action on axoneme. This suggests that CO_2 may contribute to the inhibition of turbot sperm motility in the genital tract. Spz may remain immobile by anaerobiosis and their respiration could be activated when diluted into aerobic sea water. This may explain the spontaneous motility of turbot spz observed after sperm collection that may be due to an activation by an air exposure.

References:

Brokaw, C.J. 1977. CO_2 inhibition of the amplitude of bending triton-demembranated sea urchin sperm flagella. J. Exp.Biol. 71:229-240.

Perchec, G., Jeulin, C., Cosson, J., André, F., and R. Billard, 1995. Relationship between sperm ATP content and motility of carp spermatozoa. J. Cell. Sci. 108:747-753.

OVARIAN MORPHOLOGICAL CHANGES AND PLASMA SEX STEROID PROFILES IN TWO CULTURED SALMON (***Oncorhynchus kisutch*** and ***Salmo salar***) BROODSTOCK POPULATIONS IN CHILE.

F.J. Estay[1], N.F. Díaz[1] and L. Valladares[2]

[1]Department of Ecological Sciences, Faculty of Sciences, University of Chile P.O. Box 653 Santiago, Chile. [2]Biology of Reproduction Laboratory, INTA, University of Chile.

Summary

The reproductive cycles of two cultured salmon broodstocks (*O.kisutch* and *S. salar*) were monitored for 8-9 months before ovulation. The dynamics of cytological oocyte changes, oocyte growth, gonadosomatic index and sex steroid profiles are described and discussed.

Introduction

Salmonid culture has experienced very rapid growth in Chile. However, no comparative research efforts have examined biological aspects of salmon under chilean culture conditions, including reproduction. Our purpose was to study some basic aspects of reproduction of *O. kisutch* and *S. salar* in Chile.

Results and Discussion

Ovarian tissue and blood samples, were taken from females at two farms in the south of Chile, one for coho salmon (two years old, one year sea-life) and another for Atlantic salmon (three years old, two years sea-life), between September and May (May was the ovulation time for both). Histological studies were carried out on 4 µm sections of ovaries fixed in ALFAC, stained with H+E and Alcian Blue/Van Gieson. The criteria of Bromage and Cumaranatunga (1988) were used for classification of oocyte development stages. Radioimmunoassays were performed for sex steroid determinations.

Late stage 5 oocytes were predominant in Atlantic salmon during Oct/Nov, characterized by active exogenous vitellogenesis. This is considerably more advanced maturation than which coho salmon showed a typical yolk vesicle stage at this time. Exogenous vitellogenesis in coho salmon did not start until Dec/Jan. Consistent with these developmental differences, Atlantic salmon showed a significantly greater oocyte diameter and gonadosomatic index until March. In May, however, oocytes of both species were of the same diameter (Fig. 1), reaching stage 7, and their GSI was also similar. Stage 6 (germinal vesicle migration) started earlier and lasted longer in Atlantic salmon (Jan-May), than coho salmon (Mar-May). Profiles of 17ß estradiol (E_2) were similar in both species, with high values during vitellogenesis, and very low values at ovulation. Peak E_2 levels, however, occurred two months later in coho salmon with a mean value about ten fold higher than in Atlantic salmon. (Figure 2). Ovulation values of 17α20ß DHP were very similar in both species: 20.5 ± 5.3 ng/ml and 23.5 ± 2.3 ng/mL respectively.

Dynamics of oogenesis showed obvious asynchrony between the two species, despite the fact that they share the same spawning time. It must be considered, in this respect, that the Atlantic salmon were one year older than the coho salmon. Bromage and Cumaranatunga (1988) reported that in rainbow trout exogenous vitellogenesis begins at least 8 months before spawning in 3 years old fishes but only 4-5 month before spawning in two year old fishes. Our observations suggest that the studied species could also follow this pattern, however new studies using 2 and 3 years old females of both species must be performed to test this hypothesis.

One of the posible interpretations for the large difference of the peak E_2 serum concentration between the two species, is that 2 years old females of coho salmon need higher levels of circulating E_2 than 3 year old females of Atlantic salmon in order to support a very intensive hepatic synthesis of vitellogenin, required to complete vitellogenesis in a shorter time period.

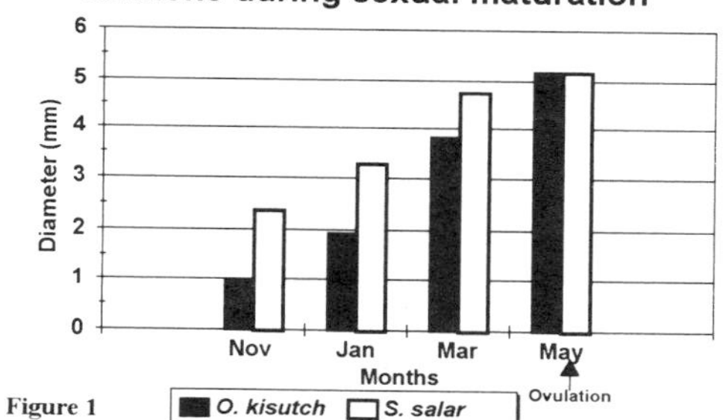

Figure 1

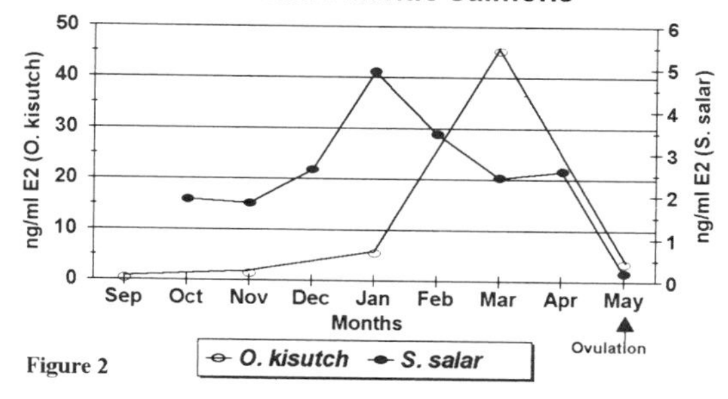

Figure 2

References

Bromage, N. R. and Cumaranatunga, R. 1988. Eggs production in the rainbow trout. Chapter 2 In: Recent Advances in Aquaculture, volume 3. J. F. Muir and R. J. Robert, editors. Timber Press, Portland, Oregon.

FONDEF PI-10 and FONDECYT 1940259-94

PLASMA STEROID HORMONES IN ADULT TRIPLOID TILAPIA (*OREOCHROMIS NILOTICUS*)

F. Freund, G. Hörstgen-Schwark , W. Holtz

Institute for Animal Husbandry and Genetics, University of Göttingen, Albrecht-Thaer-Weg 3, D-37075 Göttingen, Germany

Summary

Hormone profiles were examined in adult diploid and triploid *Oreochromis niloticus*, kept in pond culture, over a 60 day period. There were no significant differences between plasma levels of testosterone in diploid and triploid males. In triploid females with retarded ovaries no endocrine signs of maturation were found. Triploid females showing advanced stages of gonadal development exhibited plasma concentrations of testosterone (T) and estradiol-17ß (E2) comparable to mature diploid females.

Introduction

In spite of being sterile, testes of triploid tilapia males may be fully developed (Penman *et al.,* 1987). In triploid females gametogenesis may be either completely blocked (Puckhaber *et al.*, 1991; Mol *et al.*, 1994), or there may be considerable gonadal development (Brämick *et al.*, 1995). The present study presents profiles of gonadal steroids in adult triploid *O. niloticus* of different developmental stages in comparison to their diploid sibs.

Materials and methods

Diploid and triploid males and females of an average age of 360 days were kept together in tropical pond culture (Nigeria) at 28°C and optimal breeding conditions. All fish were tagged to follow them individually during the 60 day experiment. Every five days fish were bled by caudal puncture. The levels of testosterone and estradiol-17ß were determined in plasma extracts by specific ELISAs.

Results and discussion

Males of both ploidy groups showed secondary sexual characteristics. There was no difference between testosterone profiles in diploid and triploid male tilapia, with triploids having a secretion of 26.4±16.5 ng/ml (mean ± SD; n=3) and diploids of 17.0±10.4 ng/ml (n=5).

In triploid females with retarded ovaries T and E2 were low (Fig.1B). Levels of 2.3±1.3 ng T/ml and 2.0±1.6 ng E2/ml (n=3) were comparable to the basal secretion of sexually active diploid females.

Maturing triploid females with advanced gonadal development and postvitellogenic oocytes exhibited typical cyclical endocrine profiles, as did the diploid females (Fig. 1A,C,D). Mean secretion of T and E2 (4.7±1.4 and 9.8±4.2 ng/ml, resp., n=9) as well as maximum concentrations reached were not distinct from diploids. While the average duration of a breeding cycle was 15-20 days in diploid females, the ovarian cycles of maturing triploid females were significantly longer (20-40 d).

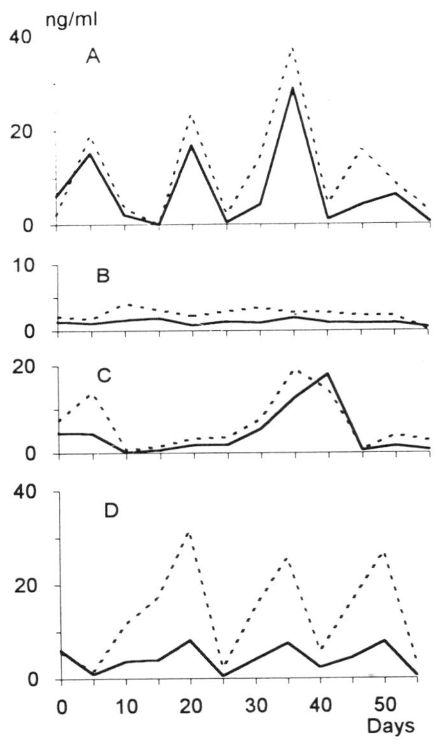

Figure 1: Hormone profiles (——— T, ·········· E2) of individual female *O.niloticus*
(A Mature diploid, B Immature triploid, C,D Maturing triploids)

Our data demonstrate, that steroid hormone concentrations depend on the respective stage of gonadal development. As is the case in males, triploid females do reach full maturity. In contrast to salmonids (Benfey *et al.*, 1989) *O. niloticus* obviously generate enough follicle cells to produce the estrogenic stimulus for liver vitellogenin production. These triploid female *O. niloticus* may be considered endocrinologically fully competent.

References

Benfey et al. (1989). Gen. Comp. Endocrinol. 75, 83-87.

Brämick et al. (1995). Aquaculture, in press.

Mol et al. (1994). Fish Physiol. Biochem. 13, 209-218.

Penman et al. (1987). Proc. World Symp. Sel., Hybrid., Genet. Eng. Aquacult., Vol. II, 277-288.

Puckhaber et al. (1991). Abstracts III. Int. Symp. Tilapia Aquacult.

IMMERSION OF NILE TILAPIA (OREOCHROMIS NILOTICUS) IN 17α-METHYLTESTOSTERONE AND MESTANOLONE FOR THE PRODUCTION OF ALL-MALE POPULATIONS

W.L. Gale, M.S. Fitzpatrick, and C.B. Schreck

Oregon Cooperative Fishery Research Unit, Oregon State University, Corvallis, OR. 97331

Summary

Immersion of Nile tilapia (Oreochromis niloticus) fry in 17α-methyltestosterone or in mestanolone was evaluated for masculinization. Three hr exposures of fry at 10 and 13 days post fertilization in mestanolone at 500 μg/L produced sex ratios greater than 93 percent male.

Introduction

The production of mono-sex populations has greatly improved the efficiency of aquaculture operations. The common method for producing mono-sex populations of tilapia involves dietary treatment with 17α-methyltestosterone (MT). We describe an alternative method for the masculinization of Nile tilapia, which involves immersion of fry in mestanolone (methyldihydrotestosterone; MDHT).

Methods

At 10 days post fertilization (dpf), fry (n=100/group) were placed in separate jars (3L). Treatments were given in two separate 3 hr immersions at 10 and 13 dpf. Steroid was delivered in 0.5 ml of ethanol. Fry were immersed in MT or MDHT at 100 or 500 μg/L. Control groups included: immersion in water and ethanol vehicle (ETH), an immersion in water alone (CTL), and dietary MT (FED; 60 mg/kg from 10 dpf - 30 dpf). The first experiment (EX I) was replicated (EX II) with omission of the FED group. Grow-out occurred in 20 L tanks. At about 100 dpf, sex ratios were determined by examination of in-situ (40X) and squash (100X) preparations. Iron aceto-hematoxylin was employed for staining purposes.

Sex ratio data were analyzed using chi-square ($\alpha<0.05$). The CTL and ETH groups were not significantly different, and were pooled for comparison to other groups. The final weight of the sampled fish in EX II were analyzed for differences between groups using one-way ANOVA.

Results

Immersion in MDHT at 500 μg/L resulted in 100 (EX I) and 94 (EX II) percent male populations (Fig.1). Other immersion treatments resulted in significant skewing of the sex ratio towards males, but these results were not consistent between replicates. There was no significant difference among the groups (EX II) in terms of final weight.

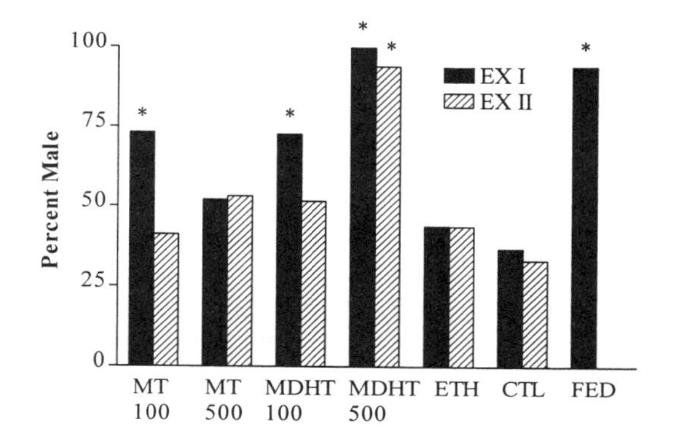

Fig. 1. Percent males in each group for experiments 1 and 2. Asterisks indicate significant differences in sex ratios from the pooled control group. Sample size ranged from 19 to 50 with an average of 35 fish.

Discussion

Although dietary administration of androgen is effective in masculinizing tilapia, alternative techniques may allow for a reduction in the amount of steroid needed, and a decrease in worker and environmental exposure. Immersion of fry in 500 μg/L of MDHT is effective for masculinizing tilapia and may be safer and more efficient than use of steroid-treated feeds. Further research is needed to determine the minimum effective dose, and minimum effective fish density, before immersion technology can be adapted to production scale operations.

Funded by the Pond Dynamics/Aquaculture CRSP Egypt Project

FAILURE OF GYNOGENETICALLY-DERIVED MALE CHANNEL CATFISH TO PRODUCE ALL-MALE OFFSPRING

Cheryl A. Goudie[1], Bill A. Simco[2] and Kenneth B. Davis[2]

[1]USDA-ARS, Catfish Genetics Research Unit, P.O. Box 38, Stoneville, MS 38776
[2]Division of Ecology and Organismal Biology, The University of Memphis, TN 38152

Summary

Sex ratios of offspring from matings of gynogenetically-derived male channel catfish Ictalurus punctatus (presumed YY sex genotype) with normal XX females produced the expected 100% male offspring in only 7 of 18 males tested. Aberrant sex ratios (<100% males) were not expected, suggesting some disturbance in the meiotic process or that instability in the sex determination system occurs as a result of induced gynogenesis of XY females.

Introduction

Hormonal manipulations and selective breeding techniques have been used to produce channel catfish males with YY sex genotype and females with XY or YY sex genotype (Davis et al., 1990). Induced gynogenesis of XX females resulted in only female offspring (Goudie et al., 1995), confirming female homogamety. Sex reversed (XY) females were used in gynogenesis for development of presumptive YY male broodstock that would produce all-male channel catfish for commercial aquaculture.

Results

Meiotic gynogens were produced by fertilizing eggs from XY females with UV-irradiated sperm from blue catfish I. furcatus, then subjecting the eggs to a pressure shock. Offspring from three females were maintained separately until sex could be positively identified. Sex ratios were nearly equal (48, 50 and 53% males), apparently reflecting genotypes that were XX in females and YY in males. Mating of these males with unrelated females resulted in 18 families, seven (39%) with all-male offspring, five (28%) with 62 to 98% males, and six families (33%) with 1:1 male:female sex ratios.

Discussion

Varadaraj and Pandian (1989) successfully used the strategy of gynogenesis of XY female tilapia Oreochromis mossambicus to obtain YY male broodstock that produced all-male progeny when mated with normal females. Similarly, the populations of all-male channel catfish were expected and confirmed the YY genotype of the male parent. However, populations of channel catfish with <100% male offspring were unexpected. The males which produced offspring with equal sex ratios apparently had an XY genotype. Recombination during meiosis I of the XY females could result in gynogenetic XY males. However, the frequency was much higher than expected based on low rates of recombination of the sex-determining alleles (Liu et al., 1995). Sex ratios intermediate between 50 and 100% males are difficult to explain. The sex determining mechanism may have been disturbed, either as a result of the gynogenetic procedure itself or possibly due to interaction with blue catfish genome from the irradiated sperm. We continue to study the cause of the sex ratios observed in this study. Nevertheless, as a large percentage of gynogenetically-derived males did not possess the YY sex genotype, gynogenesis may not be a suitable method for production of definitive YY male channel catfish.

References

Davis, K.B., B.A. Simco, C.A. Goudie, N.C. Parker, W. Cauldwell and R. Snellgrove, 1990. Hormonal sex manipulation and evidence for female homogamety in channel catfish. General and Comparative Endocrinology 78:218-223.

Goudie, C.A., B.A. Simco, K.B. Davis and Q. Liu. 1995. Production of gynogenetic and polyploid catfish by pressure-induced chromosome set manipulation. Aquaculture 133:185-198.

Liu, Q., C.A. Goudie, B.A. Simco and K.B. Davis. 1995. Sex-linkage of glucose phosphate isomerase-B and mapping of the sex-determining gene in channel catfish. Cytogenetics and Cell Genetics *accepted for publication*.

Varadaraj, K. and T.J. Pandian. 1989. First report on production of supermale tilapia by integrating endocrine sex reversal with gynogenetic technique. Current Science 58:434-441.

ULTRASTRUCTURAL STUDY OF OSMOLALITY EFFECT ON SPERM OF THREE MARINE TELEOSTS

Jin-Chywan Gwo

Department of Aquaculture, National Taiwan Ocean University, Keelung 20224, Taiwan

Summary

The morphological changes of the sperm of 3 marine teleosts: black porgy (Acanthopagrus schlegelli),black grouper (Epinephelus malabaricus),and Atlantic croaker (Micropogonias undulatus), were compared either after activation in artificial sea water (ASW) or when immersed in media of different osmotic pressure. Following activation with ASW, sperm became motile and both the size and number of mitochondria decrease and then totally disappear. The present study strongly suggests that an energy source(s), responsible for motility, is located within the mitochondria in the midpiece of these 3 marine teleost sperm.

Introduction

Low resistance of sperm to hypotonic osmotic shock is considered to be one of the limiting factors for the duration of sperm motility in fresh water fish. However, very little information is presently available on marine teleost sperm morphological changes following either exposure to hypotonic solution or activation in hypertonic solution and the relation of these conditions to the duration of motility. To investigate the effects of osmolality on marine teleost sperm structure, the sperm ultrastructure of 3 species of Perciformes; black porgy, black grouper and Atlantic croaker was studied.

Results

Sperm were motile in 1,000 mM glucose and in ASW 1,100 mOsm/kg, which was hypertonic to the seminal plasma (about 340 mOsm/-kg). The percentage of motile sperm and the duration of sperm motility were comparable in electrolyte (ASW) and nonelectrolyte (glucose) solutions. Drastic change in sperm structure was observed when the osmolality of diluents increased or decreased from 300 mOsm/kg. The morphological changes induced by the osmolality effect in black porgy sperm (Figs. 1-6) were also observed in black grouper and Atlantic croaker (data not shown). Severe morphological distortion, including the swelling of the nucleus, mitochondria completely disappeared and bursting of the plasma membrane, occurred following exposure of the fresh semen

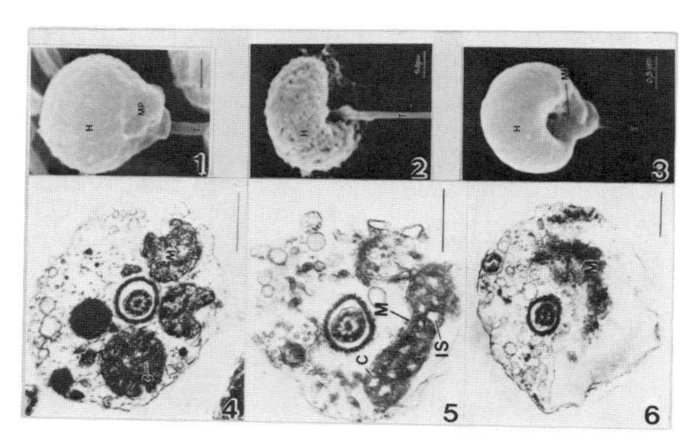

Figures 1-6. Electron micrograph of black porgy sperm showing typical head, midpiece and tail. Fig. 4. Normal fresh black porgy sperm with mitochondria (M) showing typical ultrastructure including well-defined cristae (C). Scale = 0.5μm.

to hypotonic solution (distilled water; Fig. 2). At 1 min after activation in ASW, the intracristal spaces were grossly dilated, the matrix space reduced to a minimum, and bud-like evaginations of the cristae membranes grew into the intracristal spaces (Fig. 5). After activation for 5 min in ASW, mitochondria shrunk and eventually disappeared (Figs. 3,5 and 6).

Discussion

Osmolality is an external trigger for the initiation of sperm motility in marine teleosts. Following the activation of the 3 marine fish sperm in artificial sea water in the present study, the mitochondria of the sperm were shrunken and completely disappeared at the end of motility duration, with no obvious change in other sperm ultrastructures. In spite of the diverse taxonomic positions of the 3 fish studied, the sperm demonstrated common ultrastructural changes upon hypoosmotic shock and after activation in ASW. This observation suggests that marine fish sperm motility is closely related to the existence of mitochondria in the midpiece. Probably the exhaustion of the energy supply contained in mitochondria restricted the motility of marine fish sperm motility in the present study.

This research was supported by grant from National Science Council, Taiwan, to J.-C. Gwo, under project number NSC 81-0409-B-019-17.

THE EFFECTS OF SUSTAINED ADMINISTRATION OF TESTOSTERONE AND GnRHa ON GTH-II LEVELS AND GAMETOGENESIS IN IMMATURE STRIPED BASS, MORONE SAXATILIS

M.C.H. Holland, S. Hassin, E.L. Mañanos, C.C. Mylonas and Y. Zohar

Center of Marine Biotechnology and Agricultural Experiment Station, University of Maryland, Columbus Center, 701 E. Pratt St., Baltimore, MD 21202, USA

Summary

Two-year-old immature striped bass were treated with testosterone (T) and/or GnRHa (G) for either 50 or 150 days. In females, T alone or in combination with G (T+G) elevated pituitary GtH-II levels, while in immature and precocious males, G alone or in combination with T increased pituitary GtH-II levels. GnRHa also had a stimulatory effect on GSI in immature males but decreased GSI after 150 days of treatment in precocious males. No treatment effect on gonadal development could be observed in the females.

Introduction

The onset of sexual maturity in fish is accompanied by increases in pituitary and plasma GtH levels. Treatment of immature fish with estrogens or aromatizable androgens increases pituitary GtH accumulation, but does not always result in GtH release or initiation of gametogenesis (Dufour et al., 1983). However, a combination of GnRH and steroid treatment can enhance GtH release and induce precocious gametogenesis (Crim and Evans, 1983). In the present study, we used delivery systems for the continuous administration of testosterone (T) and/or GnRHa (G) to study the effects of long-term T and G treatment on pituitary GtH-II levels and gametogenesis in immature striped bass.

Results and Discussion

A delivery system for testosterone was developed which, when administered to striped bass at a dose of 1.6 mg T/kg, results in elevated plasma testosterone levels (>1.5 ng/ml) for up to 7 weeks at 16 °C. This system was used alone or in combination with an 8-week-GnRHa delivery system (levels > 0.2 ng/ml plasma).

In December, when gonadal recrudescence occurs in maturing striped bass, 2-year-old immature striped bass, which were maintained under simulated natural photo- and thermoperiod, were treated with T, G or a combination of both (T+G). Controls (C) received a similar delivery device devoid of any hormone. After 50 days, half of the fish (35/treatment) were sacrificed. The remaining fish were given the same treatment at days 50 and 100, and were sacrificed on day 150. Gonadosomatic index (GSI) was determined and gonadal development was assessed macroscopically and histologically. Pituitaries were analyzed for GtH-II levels using a homologous ELISA (Fig. 1).

Females: No treatment effects on GSI and gonadal morphology could be observed in the females, all of which had ovaries with oocytes in the primary growth stage. Average GSI at both sampling times was

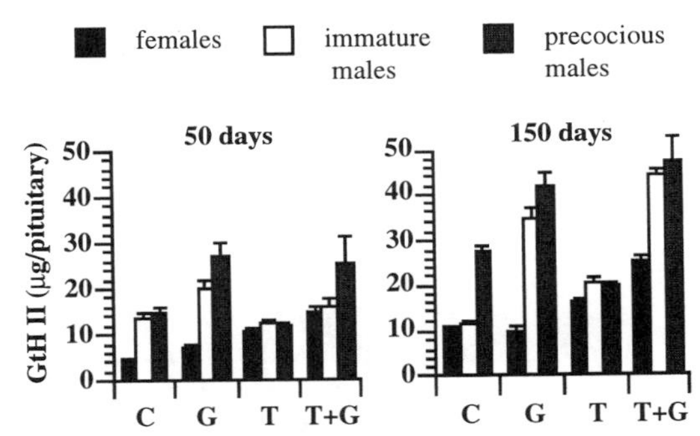

Figure 1 : The effect of a 50-day and 150-day testosterone and/or GnRHa treatment on pituitary GtH-II levels (Mean levels ± SEM).

0.38 ± 0.03%. Pituitary GtH-II levels in all treatments were higher after 150 days of treatment than after 50 days. T+G treatment elevated pituitary GtH-II levels significantly ($p<0.05$) at the 50-day sampling, while both T alone and T+G treatment elevated pituitary GtH-II levels at the 150-day sampling. This confirms the finding that T stimulates GtH-II accumulation in the pituitary of immature fish (Swanson and Dickhoff, 1988).

Males: In immature males, G alone increased GSI after 150 days of treatment, compard to controls (0.78±0.13% and 0.12±0.01% for the G and C, respectively). However, testis morphology did not seem to be affected. Both G and T+G elevated pituitary GtH-II levels significantly ($p<0.05$) after 150 days of treatment.

Precocious males could be found in all groups. After 150 days, GSI was reduced significantly in all precocious males ($p<0.05$), indicating that at this time the gonads were already regressing. G and T+G had an inhibiting effect on GSI at this time (1.6±0.4% and 1.5±0.2%, respectively, versus 3.7±0.4% in the controls). However, males from these two groups, but not the T group, had significantly elevated GtH-II levels at both sampling times. This observation indicates that in males, G and not T has a stimulatory effect on pituitary GtH-II levels.

References

Crim, L.W., and D.M.Evans. 1983. Influence of testosterone and/or LHRHa on precocious sexual development in the juvenile rainbow trout. Biol. Reprod. 29:137-142.

Dufour, S., N. Delerue-LeBell, and Y-A. Fontaine. 1983. Effects of steroid hormones on pituitary immunoreactive GtH in European freshwater eel. Gen.Comp.Endo. 52: 190-197.

Swanson, P. and W.W. Dickhoff. 1988. Effects of exogenous gonadal steroids on GtH I and II. Amer. Zool. 28:55A (Abstract)

BREEDING AUSTRALIAN LUNGFISH IN CAPTIVITY.

Jean Joss and Greg Joss

School of Biological Sciences, Macquarie University, Sydney, NSW 2109, Australia.

Summary

A lungfish breeding facility has been established at Macquarie University, Sydney, Australia. It comprises two large earthen ponds, one of which contains nine adult lungfish, obtained from the three river systems where lungfish occur in the wild, and an aquarium room for hatching and rearing juvenile lungfish. The lungfish have spawned during both breeding seasons since their introduction to the pond and eggs have been collected, hatched and successfully reared in the laboratory.

Introduction

The Dipnoi (lungfish) are a group of fish which arose in the Silurian, at about the same time as the other groups of jawed fish and some 50 million years prior to the first tetrapods (Joss *et al*, 1991). The fossil record indicates that they were a diverse and successful group of fish during the Devonian. Today however, they are represented by only three genera in two families. The Australian lungfish, *Neoceratodus forsteri*, is the sole living member of the oldest of these two families, dating back to the Carboniferous. As a species, it appears to have survived unchanged since the Cretaceous. Thus in its own right it is of great interest as a "living fossil" but recent molecular evidence (Meyer, 1995) strongly suggests that it is the closest living species to the ancestor of the tetrapods which makes it a key species in the study of the evolution from fish to amphibian.

The natural distribution of *Neoceratodus* is confined to a few coastal river systems in south-eastern Queensland which makes it vulnerable to extinction. Because of this and its very great interest it has been totally protected for nearly 100 years. Over the last 100 years. several attempts have been made to breed *Neoceratodus* in captivity with only one previous record of success by an amateur who kept a single pair which spawned for three years in an outdoors pool. He did not successfully rear any of the progeny.

Methods

In 1992, work began at Macquarie University to construct two large earthen ponds (15 x 50 m each). The ponds have a shallow (0.5 m depth) end planted with *Vallisneria* (used for spawning in the rivers) and a deep (2.5 m depth) end to simulate the pools in the rivers where they occur naturally. Paddle wheel aerators provide water movement and aeration. Floating islands of *Eichornia* remove excess nitrogen and provide shady areas for the lungfish during daylight hours. The ponds attract a large variety of insects, frogs, birds, etc. and native shrimp and bivalve molluscs have been introduced. No additional food is provided. The ponds were completed in September, 1992 and one of these was stocked progressively over the ensuing 12 months with 9 adult lungfish (4 male, 5 female), obtained from several sources and representing fish derived from the Brisbane, Mary and Burnett Rivers.

Results

The lungfish spawned on the *Eichornia* roots in December, 1993 and again in October-December, 1994. Eggs have been collected from both spawning seasons and reared according to the method described by Kemp (1981). Each egg, and after hatching, each fish is kept separately in shallow, sterilised pond water (changed every 3-4 days). From two weeks to 6 months of age, they eat only live food which is treated with antibiotic to prevent infection. After 6 months, they will take fish pellets and become increasingly hardy. I have lost none through natural causes after this age.

Growth rates are slower in the aquarium-reared fish than in the ponds. In mid-1994, the ponds were completely enclosed in bird-proof netting to reduce losses to fishing birds of juvenile lungfish growing out in the ponds.

The embryos and hatchlings are providing material for much-needed ontogenetic studies and the fish growing out in the ponds will provide the numbers of lungfish necessary to carry out physiological studies of statistical significance which have been impossible hitherto.

References

Joss, JMP, Cramp, N, Baverstock, PR and Johnson, AM., 1991. A phylogenetic comparison of 18S ribosomal RNA sequences of lungfish with those of other chordates. *Aust J Zool* **39**:509-18.

Kemp, A., 1981. Rearing of embryos and larvae of the Australian lungfish, *Neoceratodus forsteri* under laboratory conditions. *Copeia* **1981**:776-84.

Meyer, A., 1995. Molecular evidence on the origin of tetrapods and the relationships of the coelacanth. *TREE* **10**:111-16.

REPRODUCTIVE VARIATION IN HATCHERY-RELEASED SOCKEYE SALMON, ONCORHYNCHUS NERKA

M. Kaeriyama[1], H. Ueda[2], S. Urawa[1], and M. Fukuwaka[1].

[1]Research Division, Hokkaido Salmon Hatchery, Fisheries Agency of Japan, 2-2 Nakanoshima, Toyohira-ku, Sapporo 062, Japan.
[2]Faculty of Fisheries, Hokkaido University, 122 Tsukiura Abuta-cho, Hokkaido 049-57, Japan.

Summary

In a study on reproductive variation of hatchery-released sockeye salmon (Oncorhynchus nerka), larger adult females had higher fecundity. There was no difference of egg size between lake-resident and age-1.1 anadromous sockeye salmon. In hatchery-released sockeye salmon having a constant gametic effort without breeding competition and parental care, egg size may be stable within a cohort or a population regardless of body-size variation, although fecundity is expressed by the function of body size affected by environmental factors.

Introduction

Anadromous sockeye salmon (Oncorhynchus nerka) is not naturally found in Japan, although lake-resident sockeye salmon (the resident salmon) are distributed in several oligotrophic lakes. The Lake Shikotsu resident salmon maintained by hatchery releases has been geographically landlocked in this lake for more than 15 generations. Anadoromous sockeye salmon have been produced from the Lake Shikotsu resident salmon by smolt release technology in the Bibi River, central Hokkaido (Kaeriyama et al. 1995).

To study factors affecting reproductive characters of hatchery-released sockeye salmon, we examined body size, gonad somatic index (GSI), fecundity and egg size of the resident adult salmon in Lake Shikotsu during 1988-1994 and anadromous adult sockeye salmon returning to the Bibi River during 1990-1993.

Results and Discussion

Body size of the resident adult salmon in Lake Shikotsu has shown an increasing trend since 1989. Fecundity also increased with body size. However, neither GSI nor egg size showed annual changes (Fig. 1).

There were significant differences of fork length and fecundity among the resident salmon, age-1.1 and age-1.2 anadromous sockeye salmon (Kruskal-Wallis test, $P < 0.001$). There was, however, no difference between any groups in GSI (Kruskal-Wallis test, $P > 0.05$). There was not any difference of egg size between the resident and age-1.1 sockeye salmon (LSD test, $P = 0.708$), although age-1.2 sockeye salmon had eggs approximately 12% larger than the resident and age-1.1 sockeye salmon (LSD test, $P < 0.001$; Fig. 2).

Allometric regressions by each population were observed between fork length and fecundity ($r^2 > 0.18$, $P < 0.05$), but were not observed between fork length and egg size ($r^2 < 0.003$, $P > 0.05$). Allometric regressions for all populations combined were observed between fork length and fecundity ($r^2 = 0.947$, $P < 0.001$), between fork length and egg size ($r^2 = 0.315$, $P < 0.001$), and between fecundity and egg size ($r^2 = 0.508$, $P < 0.001$).

In wild Pacific salmon, gametic effort is influenced by breeding competition, parental care, and migration cost. There appear to be a trade-off between fecundity and egg size (Fleming and Gross 1990). Sockeye salmon examined in this study, have been excluded from breeding competition and parental care for more than 15 generations because they have been reproduced in hatchery. For hatchery-released sockeye salmon having a constant gametic effort (GSI) without breeding competition and parental care, fecundity may be determined by the function of body size affected by environmental factors such as food and population density. However, their egg size may be stable within a cohort or a population regardless of body-size variation.

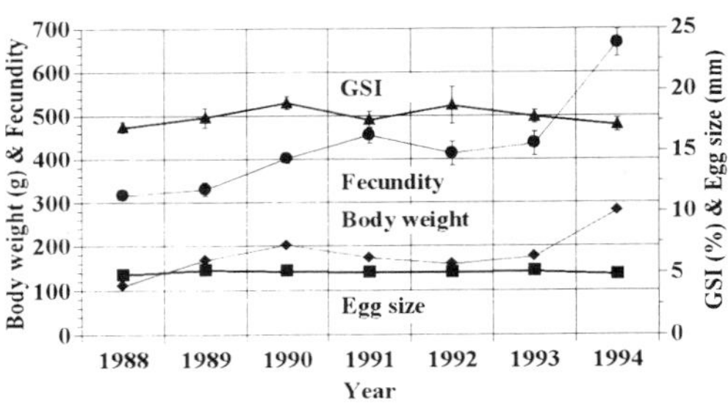

Fig. 1. Annual changes in average and standard errors of reproductive characters for adult female resident sockeye salmon in Lake Shikotsu during 1988-1994.

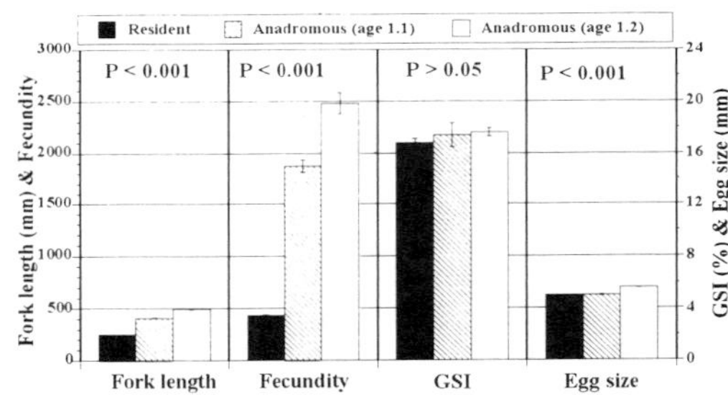

Fig. 2. Average and standard errors of reproductive characters for resident and anadromous sockeye salmons released from hatchery. Probability level of significance between populations is represented by P.

References

Fleming, I. A. and M. R. Gross. 1990. Latitudinal clines: a trade-off between egg number and size in Pacific salmon. Ecology, 71: 1-11.

Kaeriyama, M., S. Urawa, and M. Fukuwaka. 1995. Variation in body size, fecundity, and egg size of sockeye and kokanee salmon, *Oncorhynchus nerka*, released from hatchery. Sci. Rep. Hokkaido Salmon Hatchery, (49); 1-9.

INCREASED MILT PRODUCTION BY GONADOTROPIN RELEASING HORMONE ANALOG (GnRHa)-TREATED ATLANTIC SALMON (*Salmo salar*) AFTER INJECTION OF 17α-HYDROXYPROGESTERONE.

H. King[1] and G. Young[2]

[1]Salmon Enterprises of Tasmania Pty. Ltd., P.O. Box 1, Wayatinah, Tasmania, Australia 7140.
[2]Department of Zoology, University of Otago, P.O. Box 56, Dunedin, New Zealand.

Summary

Three trials over two years of study showed that injection of 17α-hydroxyprogesterone (HP) reduced 11-ketotestosterone (KT), increased 17α,20β-dihydroxy-4-pregnen-3-one (DHP) levels and enhanced milt production in GnRHa-treated Atlantic salmon. HP appears to be a cost-effective method of further promoting spermiation in GnRHa-treated salmon.

Introduction

GnRHa and gonadotropins have been used to stimulate milt production in a variety of teleosts. Evidence suggests that some of the effects of these hormones are mediated through gonadal progestogens, in particular DHP (see Pankhurst, 1994). However, the high cost of DHP precludes its routine use in broodstock management and our past use of a commercial salmon GnRHa preparation in Atlantic salmon has given equivocal results. The objective of the present study was to further investigate the effects of GnRHa as well as DHP and its precursor HP.

Materials & Methods

Non-spermiating male Atlantic salmon received an i.m. injection of a commercial salmon GnRHa (10μg/kg) preparation or vehicle on day 0. On days 6 and 7, they received an i.m. injection of HP, DHP (1mg/kg) or vehicle. On day 8, milt was collected by catheterising the animals. Blood samples were also collected on days 0 and 8 for steroid analysis by radioimmunoassay.

Results

In all three trials, HP and DHP treatment markedly increased plasma progestogens, particularly DHP. The effect was even greater when GnRHa and steroids were combined.

GnRHa increased milt volume 2-4 fold compared to controls. HP or DHP alone also tended to cause a 2 fold increase in milt volume. Combination of GnRHa with HP, but not with DHP, caused a further 2-fold increase in milt production.

The level of milt production tended to display a negative correlation with day 8 plasma KT which, in turn, tended to be negatively correlated with day 8 plasma DHP.

The results of one trial are displayed in Table 1.

Discussion

The results reveal significant 20β-hydroxysteroid-dehydrogenase (20β-HSD) activity and tend to indicate enhanced activity in response to GnRHa. The apparent negative correlation between plasma DHP and KT is consistent with the observed negative *in vitro* effect of DHP on androgen production in male carp (*Cyprinus carpio*; Barry *et al.*, 1990).

The effect of GnRHa+HP, HP and DHP on milt volume supports the involvement of progestogens, in milt production. The reason for the failure of DHP to further increase milt production when combined with GnRHa is unclear, particularly as this treatment resulted in steroid profiles similar to those in GnRHa+HP treated males.

Table 1. Typical effect of GnRHa, HP and DHP on Day 8 Milt Volume and Plasma Steroid Levels.

Treatment	Milt	Plasma Steroid (ng/ml)		
	(ml/kg)	KT	HP	DHP
Control	0.8 ± 0.1	18.8 ± 4.0	3.3 ± 0.6	1.9 ± 0.3
GnRHa	1.6 ± 0.3	13.8 ± 2.1	3.2 ± 1.5	1.7 ± 0.4
HP	1.7 ± 0.2	8.1 ± 1.4	24.0 ± 3.0	24.1 ± 3.8
DHP	1.0 ± 0.2	12.2 ± 3.5	15.6 ± 8.7	47.8 ± 16.6
GnRHa+HP	2.6 ± 0.7	7.3 ± 1.7	13.5 ± 3.4	36.2 ± 11.1
GnRHa+DHP	1.5 ± 0.3	6.7 ± 2.0	1.9 ± 0.2	67.9 ± 20.1

References

Barry, T.P., Aida, K. and Hanyu, I., 1990. Effects of 17α,20β-dihydroxy-4-pregnen-3-one on the *in vitro* production of 11-ketotestosterone by testicular fragments from the common carp, *Cyprinus carpio*. J. Exp. Zool. 251: 117-120.

Pankhurst, N.W., 1994. Effects of gonadotropin releasing hormone analogue, human chorionic gonadotropin and gonadal steroids on milt volume in the New Zealand snapper, *Pagrus auratus* (Sparidae). Aquaculture, 125: 185-197.

INTERACTION OF CRYOPROTECTANTS WITH RAINBOW TROUT (*Oncorhynchus mykiss*) SPERM PLASMA MEMBRANE DURING FREEZE-THAWING : A BIOPHYSICAL STUDY

C. Labbé[1,2], G. Maisse[1] and J.H. Crowe[2]

[1]Laboratoire de Physiologie des Poissons, INRA, Campus de Beaulieù, 35042 Rennes cedex, France
[2]Section of Molecular and Cellular Biology, Storer Hall, UCDavis, Davis CA 95616, USA

Summary

Stability of the phospholipid (PL) fraction extracted from rainbow trout spermatozoa plasma membrane was studied after freeze-thawing. As temperature dropped below physiological levels, PLs underwent a phase transition. We tested the stabilizing effect of different molecules incorporated or not to the bilayer and observed negative interactions between some of them.

Introduction

The lipid transition from the liquid crystalline to gel state as temperature drops is known to induce phase separation of biological and artificial membranes. We wanted to know if lipid phase transition can explain the instability of trout spermatozoa plasma membrane (PM) during freeze-thawing (FT) and how cryoprotectants interact with the lipid fraction of this membrane and stabilize it.

Results

Trout spermatozoa plasma membrane or artificial membranes made with its phospholipid (PL) fraction were analyzed by Fourier Transform Infrared spectroscopy (FTIR). A high CH_2 stretch frequency of the acyl chain corresponds to the liquid phase of the lipids while a low one corresponds to the gel phase.

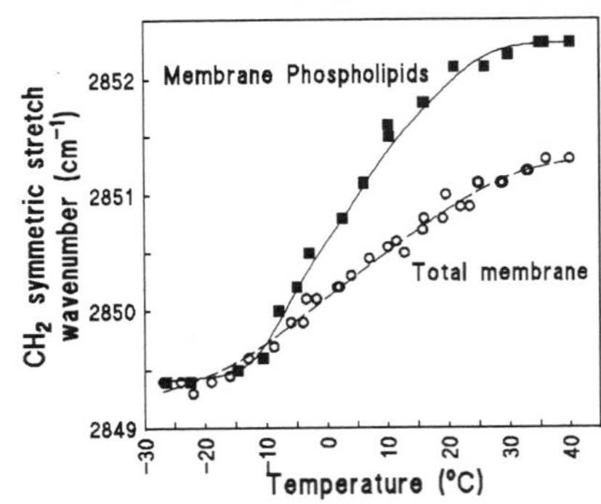

Fig.1 Lipid phase transition in trout spermatozoa PM

The total membrane showed an even increase in acyl chain order without any abrupt lipid phase transition. The isolated PL reconstituted in liposomes showed a sharper phase transition that could partly explain local instability of the PM (Fig.1).

The ability of PL liposome to retain their internal content was assessed using carboxyfluorescein entrapped at a self quenching concentration inside these vesicles.

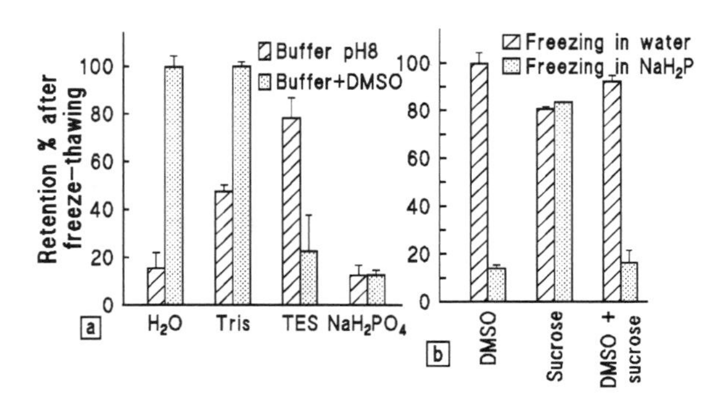

Fig.2 Stability of PL liposomes during freeze-thawing in different solutions

The Tris and TES (tris aminoethane sulfonic acid) buffer (10mM pH8) used alone stabilized the liposomes after FT, as does DMSO 10% (Fig. 2A). However, this stabilizing effect disappeared when DMSO and TES or NaH_2PO_4 were used together. A high sugar concentration (100mM) overcame the NaH_2PO_4-induced instability, but the addition of DMSO is still deleterious (Fig. 2B).

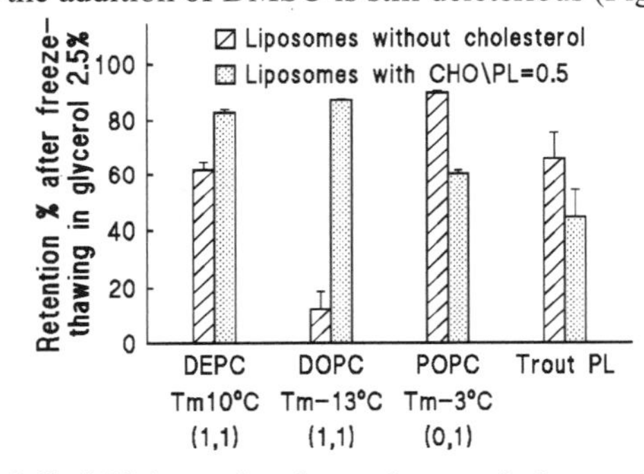

Fig.3 Stabilizing role of membrane cholesterol

The incorporation of cholesterol in dielaidoyl and dioleoyl phosphatidylcholine (DEPC, DOPC) liposomes increased their FT stability although it destabilized palmitoleoyl phosphatidylcholine (POPC) liposomes. The effect of cholesterol on trout PL liposomes was less clear, but it tended to destabilize the bilayer too.

Conclusion

We showed that the PL fraction of trout spermatozoa PM is very stable during FT provided that negative interactions between cryoprotectants are avoided. This discrepancy with the whole membrane instability indicates that lipid phase transition is not the only source of membrane damage.

We gratefully thank Ms Micheline Heydorff for the purification of trout spermatozoa plasma membrane.

EVALUATION OF SEMEN FITNESS OF THE RAINBOW TROUT (*ONCORHYNCHUS MYKISS*) FOR CRYOPRESERVATION BY PHYSIOLOGICAL AND BIOCHEMICAL PARAMETERS

F. Lahnsteiner[1], B. Berger[2], T. Weismann[3] and R. A. Patzner[1]

[1] Institute for Zoology, University of Salzburg, Hellbrunnerstr. 34, A-5020 Salzburg, Austria; [2] Bundesanstalt für Fortpflanzung und künstliche Befruchtung von Haustieren, Austr. 10, A-4601 Thalheim, Austria; [3] Bundesamt für Wasserwirtschaft, Institut für Gewässerökologie, Fischereibiologie und Seenkunde, Scharfling 18, A-5310 Mondsee, Austria.

Summary:

Correlations between the fertilization rate of deep frozen semen and motility parameters and seminal plasma composition of semen were investigated to determine the suitability of semen for cryopreservation and the quality of frozen/thawed semen.

Introduction:

From experience it is known that fish semen reveals wide variations in its tolerance for cryopreservation. So indicators are necessary to determine the suitability of semen for cryopreservation and the quality of frozen/thawed semen. Until now no statistical reliable correlations have been reported.

Material and methods

In 50 individual semen samples of the rainbow trout obtained by catheterisation the following parameters were investigated: Motility of fresh and deep frozen semen (% immotile, % local motile and % motile; % circular, % non linear, % linear swimming types; swimming velocity), seminal plasma composition (pH, osmolality, Na^+, K^+, Ca^{++}, protein, sperm density, lactate dehydrogenase, aspartate aminotransferase, ß-D-glucuronidase, alkaline and acid phosphatase, adenosine triphosphatase, pyruvate, lactate, creatine phosphate, choline, cholesterol, triglycerides, glucose) and postthaw fertility.

Sperm motility was evaluated by computer assisted cell motility analysis (Stroemberg, Mika): Semen prediluted 1 : 3 in physiological saline or in the extender was activated in fertilization solution to a final dilution ratio of 1 : 300 whereby homogenous motility was obtained with the used activator medium. Motility was recorded on a videotape and 100 spermatozoa were analyzed in each sample 10 ± 1 sec after the onset of motility.

For analysis of seminal plasma semen was centrifuged (350 g, 10 min, 4°C). Sperm density, pH and osmolality were measured with routine methods and Na^+, K^+ and Ca^{++} with ion sensitive membranes. Routine spectrophotometrical assays were used for the determination of enzyme activities and enzymatic assays were standardized for rainbow trout semen by definition of maximal reaction velocity, K_m and optimal pH.

For cryopreservation the method of Lahnsteiner et al. (1995) was used.

Results:

A statistical significant positive correlation was found between the sperm motility rate of fresh semen and its postthaw fertility rate ($r = 0.365$, $y = 0.23x + 28.44$, $p < 0.02$) and the sperm velocity rate of fresh semen and its postthaw fertility rate ($r = 0.340$, $y = 0.356x + 17.38$, $p < 0.05$) as expressed by linear regression.

The sperm motility rate of deep frozen semen was also positively correlated with the postthaw fertility rate. The correlation was significantly higher than in fresh semen ($r = 0.458$, $y = 1.44x - 34.20$, $p < 0.01$). The correlation between the velocity of frozen/thawed spermatozoa and postthaw fertility rate was also positive ($p< 0.02$) and significantly higher than in fresh semen. This relationship was best characterized by a quadratic function ($r = 0.400$, $y = 0.84x - 0.0054 x^2 + 20.88$, $p < 0.02$).

Of the seminal plasma compounds investigated, osmolality ($r = 0.632$, $y = 0.42x - 91.14$, $p < 0.005$) and pH ($r = -0.604$, $y = -74.06x + 655.30$, $p < 0.005$) were very significantly correlated with the semen postthaw fertility rate. The correlation between osmolality and postthaw fertility rate was positive while pH and postthaw fertility rate were negatively correlated. Both relationships were linear.

Discussion

1. The motility rate of semen is a predictor for its postthaw fertilization rate. For fresh semen it is a rough estimator to exclude immotile or low motile semen batches. The motility of frozen/thawed semen is a reliable parameter for quality determination, since there is a statistically greater correlation with postthaw fertilization rate.
2. High sperm velocities are quality parameters of fresh semen since they are positively correlated with the postthaw fertility rate. In frozen/thawed semen, sperm velocities of 70 to 120 μm x sec^{-1} characterize a high postthaw fertility of semen, while sperm samples with lower and higher sperm velocities have a reduced postthaw fertility rate.
3. Of all parameters investigated, osmolality and pH of the seminal fluid are most significantly correlated to the semen posthaw fertilization rate. A high osmolality of more than 320 mosmol/kg and a pH less than 8.2 characterize semen highly suitable for cryopreservation.

References

Lahnsteiner, F., T. Weismann, and R.A. Patzner, 1995. A uniform method for cryopreservation of salmonid fishes. Aquaculture Research, 26, in press.

EXTENDED SPERMIATION BY REPEATED INJECTION WITH CARP PITUITARY GLAND IN THE EUROPEAN CATFISH (*SILURUS GLANIS* L.): O.LINHART[1-2], M.PROD'HOMME[2], R.BILLARD[3], J.KOUŘIL[1] & J.HAMÁČKOVÁ[1]; [1]RIFCH, 38925 Vodňany, Czech Republic; [2]Angers Univ., CHRU, 49033 Angers, France; [3]NMNH, 43 Rue Cuvier, F-75231 Paris, France

INTRODUCTION

In artificial reproduction of the European catfish (*Silurus glanis L.)* a single injection of carp pituitary glande (CP) at doses from 3 to 5 $mg \cdot kg^{-1}$ of fish can increase sperm release (Linhart *et al.*, 1986; Saad & Billard 1995), but the response lasts only 3 days (Linhart & Billard, 1995). In the present work it was attempted to expose the males to continuous stimulation of spermiation by repeated injection of CP homogenates.

MATERIAL AND METHODS

At the time of reproduction the males four-eight years old (the average weight 6.3 kg), were kept separately each in 200 l tank with recirculating water of 0.1 $l \cdot s^{-1}$ at 23.5±1.2 °C and 7.0-10.0 mg $O_2 \cdot l^{-1}$. Two groups, each of 4 males were treated with carp pituitary homogenate (CP) dissolved in Ringer solution and injected intramuscularly at a dose of 5 $mg \cdot kg^{-1}$ fish in Ringer solution (0.5 $ml \cdot kg^{-1}$ bw). One control group was injected only with Ringer solution. The males were injected 4 times: 0, 6, 12 and 18 days of experiment with collection of sperm 2 days after each injection. After each stripping the males were feed with 1 or 2 forage fish (100 g) per male. After each manipulation, males were treated with a bath of 1 g potassium hypermanganate per 100 l of water for dessinfection. Sperm was sucked in a serynge at the level of the genital papila into the immobilizing solution (NaCl: 200 mM, Tris: 30 mM, pH 7, (Saad & Billard, 1995) in order to prevent spontaneous initiation of motility. Percentage of motile spermatozoa was assessed microscopically after final dilution 1:200 using a 17 mM NaCl, 5 mM Tris-HCl pH 8 activating solution at 20-22 °C. Significance was assessed using ANOVA, followed by multiple range test ($P<0.05$).

RESULTS AND DISCUSSION

The volume of sperm collected was in the range of 0.23-0.48 $ml \cdot kg^{-1}$ bw (fig. 1A). During entire time of CP injecting the males produced significantly with the best result after third injection with 0.12-$0.13 \cdot 10^9$ spermatozoa$\cdot kg^{-1}$ bw and fourth injection with 0.12-$0.14 \cdot 10^9$ spermatozoa$\cdot kg^{-1}$bw, compared to 1th and 2nd injections with 0.03-$0.07 \cdot 10^9$ spermatozoa$\cdot kg^{-1}$ bw (fig. 1B). The motility during the forth stripping of spermatozoa shows that the motility were better after 2nd-4th injection with motility from 85-100 %, compare to 1th injection with 75-95 % of motility (fig. 1C), but these diferences are not statistically significant. The quantity and quality of spermatozoa per one male after repeated injections were sufficient for repeated inseminating of the eggs of as many as 3-4 females during the 20 days period.

This work has been supported by CEC grant no CIPA-CT93-0274 and PROTAGORAS, Beacouze, France.

REFERENCES

Linhart O., Kouřil J. & Hamáčková J., 1986. The motile spermatozoa of wels, *Silurus glanis* L. and tench, *Tinca tinca* L. after sperm collection without water activation. Práce VÚRH Vodňany, 15, 28-41.

Linhart O. & Billard R., 1995. Spermiation and sperm quality of European catfish (*Silurus glanis* L.) after GnRH implantation and injection of carp pituitary extracts. J.Appl.Ichtyol., 10, 182-188.

Saad A. & Billard R., 1995. Production et gestion des spermatozoides chez le silure European (*Silurus glanis* L.). Aquatic. Living. Res., 8: (in press).

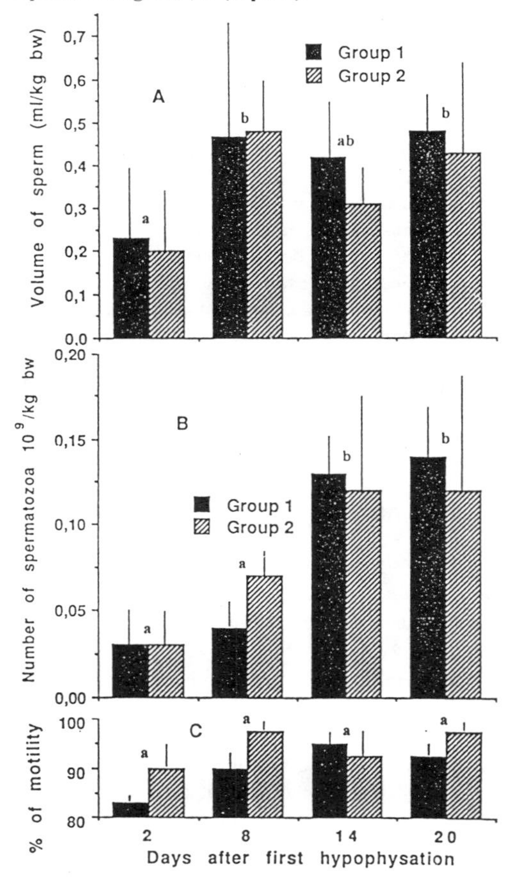

Fig. 1. Changes the volume of sperm (A) and number of spermatozoa (B) per kg bw and percentage of motil spermatozoa (C) in the replicate groups after 4 repeated injection of CP. Groups with a common superscrip do not different significantly ($P<0.05$).

REPRODUCTION OF DELTA SMELT (*Hypomesus transpacificus*) IN CAPTIVITY

R.C. Mager, S.I. Doroshov, J.P. Van Eenennaam

Department of Animal Science, University of California, Davis, CA 95616

Summary

Captive rearing, spawning, embryonic development and larval rearing of delta smelt are described.

Introduction

Delta smelt (family Osmeridae) is a threatened species endemic to the Sacramento-San Joaquin estuary, California. This is a small and short lived fish dependent on the estuarine zooplankton (Moyle et al., 1992). The effects of environmental factors on the abundance of delta smelt are not clearly understood due to the inadequate knowledge of its reproduction.

Published observations on gonadal development, breeding habits, and early life stages of delta smelt are practically absent. To obtain information about the reproductive biology of this species, we have initiated rearing and breeding of delta smelt in captivity.

Methods

Juvenile wild fish were collected in late summer-fall from the Sacramento and San Joaquin Rivers and raised in recirculation systems with controlled temperature, photoperiod, and feeding. Captive populations were sampled regularly and gametogenesis was monitored by histological observations.

Spawning was allowed to occur naturally in tanks. The adhesive eggs were scraped from the tank bottom and incubated in jars in fresh water at 14.5-16.5°C. Hatched larvae were reared in aerated aquaria (6-8 ppt salinity) with upflow current. Larvae were fed rotifers, *Brachionus plicatilis*, and brine shrimp, *Artemia franciscus*, starting 5 and 15 days after hatch, respectively. In addition, in vitro insemination was conducted to study early development.

Results

From August to May, fish grew in mean fork length from 47 to 70 mm, body weight from 0.8 to 2.6 g, and their gonadosomatic index reached 1.9% and 10.9% in males and females, respectively. The paired gonads of delta smelt are asymmetrical: the left ovary or testis is hypertrophied whereas the right gonad is reduced in size and is positioned posterior to the left gonad. However, development of germ cells in both gonads is synchronous.

Vitellogenesis and spermatogenesis occurred within 2-3 months starting in January-February. Spawning occurred at night during April-June, and embryos collected from the tank bottom in morning hours were at morula stage.

The translucent ova range 0.75-0.85 mm in diameter, and contain granular yolk and several oil droplets. Upon activation the outer chorion forms a highly adhesive "foot" that attaches to the substrate and zygote diameter increases to 0.80-1.00 mm due to formation of the perivitelline space. By 10 hours the embryo is at late morula, and epiboly and formation of the embryonic axis are completed at 50 hours post fertilization. Differentiation of eyes, auditory vesicles, and heart occur within a period 80-140 hours after fertilization. Hatching starts at the end of the tenth day, peaks on day 12 and is completed by day 14.

Newly hatched larvae have an average length of 5.3 mm and possess an ovoid-shaped yolk sac containing a single oil globule. The prolarvae have an inferior mouth opening but the jaws are not fully developed. Two rows of melanophores are present on both ventral sides of the body along the gut.

By 6-9 days posthatch, at mean length 6.8 mm, the yolk is fully resorbed, though a reduced oil globule is still present during the next 2-3 days. The mouth is fully developed with a lower jaw protruding forward. The ingestion of algae and rotifers was observed at 5 days posthatch, followed by brine shrimp nauplii at 15 days posthatch. By 21 days posthatch, at mean length 9.02 mm, the differentiation of unpaired fins starts, though the larvae do not possess a functional swimbladder.

Discussion

Delta smelt were successfully captured, transported, and raised until full sexual maturity in laboratory conditions. They spawned naturally in tanks. Embryos develop and hatch in 10-14 days at temperatures of 14.5-16.5°C. Larvae hatch at an average of 5.3 mm and resorb the yolk within 6-9 days and the oil globule within 10-12 days. Feeding begins 4-5 days posthatch with rotifers, and artemia nauplii within 14-18 days after hatching. Differentiation of fins begins 21 days after hatch, though the functional swimbladder is still absent. The preliminary observations on larval development of delta smelt suggest adaptations to life in the freshwater-saltwater mixing zone and retention of larvae in the tidal estuary as it was described for the anadromous rainbow smelt (Ouellet and Dodson, 1985).

References

Ouellet, P. and J.J. Dodson, 1985. Dispersion and retention of anadromous rainbow smelt (*Osmerus mordax*) in the middle estuary of the St. Lawrence River. Can. J. Fish. Aquat. Sci. 42:332-341.

Moyle, P.B. and B. Herbold, 1992. Life history and status of delta smelt in the Sacramento-San Joaquin estuary, California. Trans. Amer. Fish. Soc. 121:67-77.

Funding was provided by the California Department of Water Resources.

PERMEABILITY AND TOXICITY OF ^3H DMSO TO DEVELOPING ORNAMENTAL CARP EGGS

Magnus, Y.[1,2], A. Ar[1] and E. Lubzens[2]

[1]Department of Zoology, George S. Wise Faculty of Life Sciences, Tel Aviv University, Tel Aviv 69978,Israel; [2]Israel Oceanographic & Limnolological Research, P.O.Box 8030, Haifa 31080, Israel.

Introduction

Specific problems arise in attempting to preserve fish ova or embryos: 1) Fish eggs are relatively large and are surrounded by a relatively impermeable chorion; 2) The eggs or embryos contain yolk; 3) Fish eggs are easily activated after ovulation and quickly lose their viability during storage. Attempts to preserve fish embryonic cells and isolated blastomeres (Harvey, 1983; Nilsson and Cloud, 1992, 1993; Lin et al., 1992) were made to overcome partially,these problems. These techniques, however, are far from adequate to serve in routine fish culture.

We aim to establish methodologies for cryopreservation of fertilized eggs of the Japanese ornamental carp (*Cyprinus carpio* L.), which is an important cultured fish in Israel. We determined the sensitivity of embryos to cold thermic shock, the permeability of ^3H DMSO into embryos, the effect of temperature on uptake of DMSO and its toxicity to embryos.

Results and Discussion

Embryos that reached 8 different developmental stages at 24°C, were exposed to a 2 h cold thermic shock at 4°C. Embryos that were exposed at early stages of development, before reaching the gastrulation stage, did not survive after the exposure to 4°C. Variable results were obtained for embryos at the gastrulation stage, which lasts for more than 8 h at 24°C. Most embryos (range 69-86%) that were exposed at later stages of development, were not affected by this treatment.These results support the accumulated evidence that embryos of yolk containing eggs are more tolerant to cold thermic shock at late developmental stages. Roubaud et al., (1985) suggested that the progressive acquisition of tolerance to cold shock in carp embryos maybe related to desynchronisation and the slowing down of cell division cycles at later stages of embryonic development. Zhang et al.,(1989), however, reported 62% survival of 8-cell carp embryos that were exposed to -5°C for 10 h.

Of the total amount of cryoprotectant permeating fertilized eggs that were incubated in 0.5 M ^3H DMSO, the yolk fraction contained 32% (~0.05 M DMSO) and 36% (~0.1 M DMSO), after 15 min or 2 h of incubation, respectively. The relative amount retained in the non-soluble membrane fraction decreased from 35% to 22% during the same periods of incubation. The remainder was found in the supernatant fluid obtained after washing the membrane fraction. The amount of ^3H DMSO that permeated these fertilized eggs, increased with the concentration of DMSO in the incubation media, duration and temperature of incubation. About 0.2 M of DMSO (10%) entered fertilized eggs after 2 or 4 h of incubation in 2 M DMSO at 24°C, which is higher than the amount reported for salmonid or zebra fish (Harvey and Ashwood-Smith, 1982; Harvey et al.,1983). All the tested concentrations of DMSO (0.5 - 2.0 M) reduced the survival of embryos exposed to them for more than 1 h. Zhang et al., (1989) showed, however, that 80% of embryos hatched after exposure to 2 M DMSO.

Conclusions

Only embryos past a specific developmental stage have the potential to survive cold exposure and therefore, potentially, cryopreservation. DMSO permeates rapidly and reaches the yolk within 15 min of incubation, but its concentration within the surviving embryos, is relatively low. Future work should consider the use of non-permeating cryoprotectants.

References

Harvey, B., 1983. Cryobiology, 20: 440-447.
Harvey, B. and Ashwood-Smith, M.J., 1982. Cryobiology, 19: 29-40.
Harvey, B., Kelley, R.N. and Ashwood-Smith, M.J., 1983. Cryobiology, 20: 432-439.
Lin, S., Long, W., Chen, J. and Hopkins, N., 1992.Proc. Natl. Acad. Sci. USA, 89: 4519-4523.
Nilsson, E.E. and Cloud, J.G., 1992. Proc. Natl. Acad. Sci. USA, 89: 9425-9428.
Nilsson, E.E. and Cloud, 1993. Aquat. Living Resour., 6: 77-80.
Roubaud, P., Chaille, C. And Sjafei, D., 1985. Can. J. Zool., 63: 657-663.
Zhang, X.S., Zhao, L., Hua, T.C. and Zhu, H.Y., 1989. Cryo Lett. 10: 271-278.

STEROID INDUCED SEX REVERSAL OF PADDLEFISH

S.D. Mims[1], W.L Shelton[2] and J.A. Clark[1]

[1]Aquaculture Research Center, Kentucky State University, Frankfort, KY 40601 USA
[2]Zoology Department, University of Oklahoma, Norman, OK 73019 USA

Summary

Sex reversal of paddlefish, Polyodon spathula, was evaluated to determine the proper dosage of 17-α methyltestosterone (MT) in order to direct the phenotypic sex of juvenile paddlefish. Level of testosterone was significantly higher in the blood plasma of fish implanted with one or two MT-filled capsule(s) compared with those implanted with an empty capsule. Fish implanted with two capsules gave best sex reversal results of 90% males, 10% hermaphrodite, and 0% females.

Introduction

Female paddlefish are highly valued for their roe as a commercial caviar. Development of broodstock to produce all-female progeny would be desirable in maintaining a viable caviar industry in the US. The protocol involves steroid-induced sex reversal of genotypic females, and then using these phenotype-altered broodstock to produce monosex population (Shelton, 1986). Androgenic hormones must be provided at a proper dosage and administered throughout a critical period of gonadal differentiation in order to direct the phenotypic sex of juvenile paddlefish. Since gonadal differentiation of paddlefish occurs 14 to 16 months after hatch (Mims et al., 1995) and the fish does not readily accept feed, silicone elastomer capsules filled with MT and implanted intraperitoneally might be an effective method to induce sex reversal. The objective of this study was to evaluate effects of MT-filled capsule(s) implanted in mixed-sex juvenile paddlefish.

Methods

Nine-week old paddlefish with PIT tags weighing 30 to 40 g were implanted with a PIT tag and with an empty capsule (Control), one 5 mg MT capsule or two 5 mg MT capsules. A total of 270 fish were stocked into three 400 m^2 ponds (30 fish/treatment/pond). Blood samples were taken monthly and analyzed for plasma testosterone by RIA. When the fish were 70 weeks old, the gonadal tissue were removed and examined histologically for sex determination.

Results and Discussion

Level of testosterone measured in the plasma responded quadratically to the amount of MT implanted in the fish. There was a significant increase in testosterone level from fish implanted with one or two capsule(s) compared to those with an empty capsule (Fig. 1.) Fish implanted with two capsules resulted in the highest percentage of males (Fig. 2).

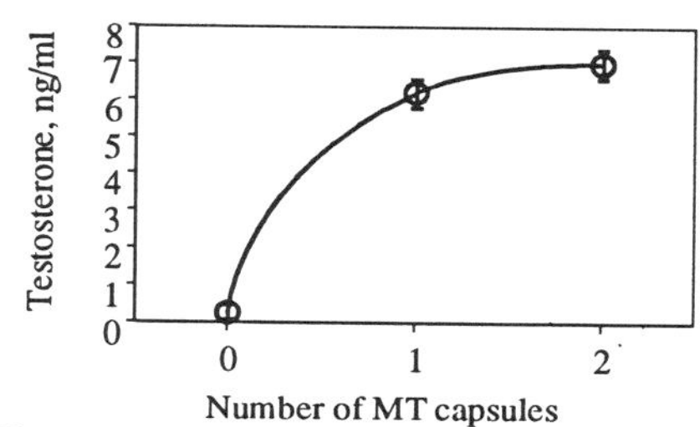

Fig.1 Blood testosterone in paddlefish

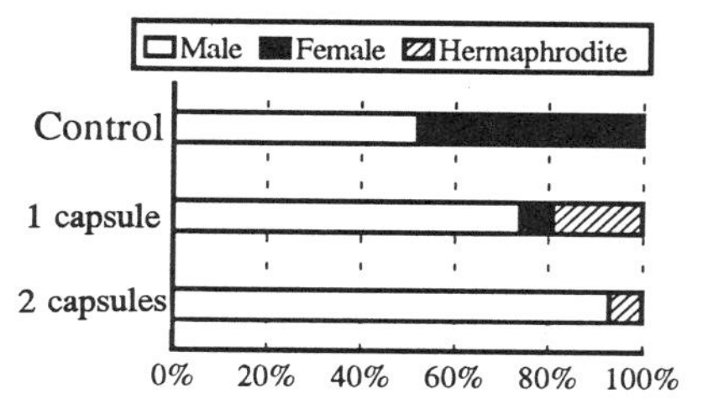

Fig 2. Sex ratios of implanted paddlefish

Paddlefish implanted with two MT capsules is an effective method for producing phenotypic males. Currently, presumptive females produced by gynogenesis are being tested with MT capsules.

References

Mims, S. D., J.A. Clark, T.E. Curry, and C. Moore. 1995. Gonadal differentiation of paddlefish, Polyodon spathula. Book of Abstracts, Aquaculture '95, San Diego, CA.

Shelton, W.L. 1986. Broodstock development for monosex production of grass carp. Aquaculture 57:311-319.

PREPARATION AND EVALUATION OF GnRHa-LOADED, POLYMERIC DELIVERY SYSTEMS FOR THE INDUCTION OF OVULATION AND SPERMIATION IN CULTURED FISH

Costadinos C. Mylonas and Yonathan Zohar

Center of Marine Biotechnology and Agricultural Experiment Station, University of Maryland, Columbus Center, 701 E. Pratt St., Baltimore, MD 21202, USA

Summary

We have developed various GnRHa-delivery systems able to achieve elevated plasma GnRHa levels for more than 30 days at 18°C. Both non-degradable and biodegradable systems have proven successful in inducing 100 % ovulation and enhancing spermiation up to 7-fold in various marine and freshwater fishes in captivity, including striped bass (*Morone saxatilis*), white bass (*M. chrysops*), Atlantic salmon (*Salmo salar*) and American shad (*Alosa sapidissima*).

Introduction

Most commercially-cultured fishes exhibit some degree of reproductive dysfunction. In females it is usually manifested as asynchrony or absence of final oocyte maturation, whereas in males the problem is usually diminished sperm production. Multiple injections of gonadotropin-releasing hormone analogs (GnRHa) have been successful in inducing ovulation, spermiation and spawning in a variety of species (Zohar, 1989). Our objective was to develop sustained-administration delivery-systems for GnRHa, which can obviate the need for multiple handling and treating of sensitive and valuable broodstock.

Results and Discussion

GnRHa-loaded delivery systems (DAla^6 , Pro^9 [Net] LHRH) were prepared with Ethylene-Vinyl Acetate in the form of implantable, non-degradable disks (implants), and with poly[fatty acid dimer-sebasic acid] in the form of injectable, biodegradable microspheres (microspheres) (Mylonas et al. 1995). Both delivery systems maintained elevated plasma GnRHa concentrations for at least 30 days (Fig. 1).

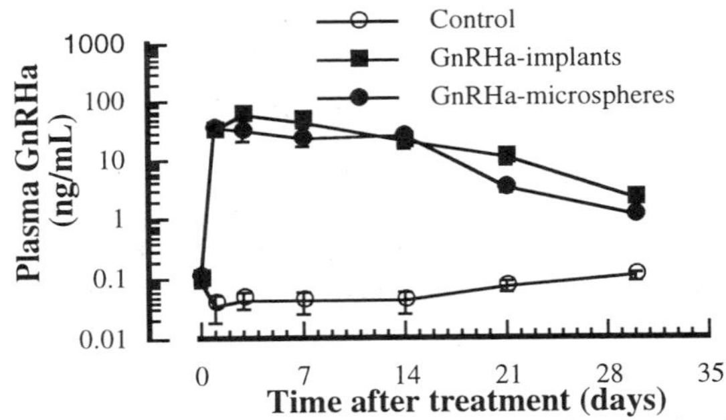

Figure 1. Mean (+ s.e.m) plasma GnRHa of striped bass treated with GnRHa-delivery systems.

American shad (an asynchronous spawner) treated with GnRHa-implants begun spawning after 2 days (Fig. 2) and continued producing fertile eggs for 12 days. The implants induced 100% ovulation of Atlantic salmon within 14 days, with better than 80% fertilization, compared to only 10% ovulation of control fish (data not shown). The implants also induced a 2.5-fold increase in sperm production.

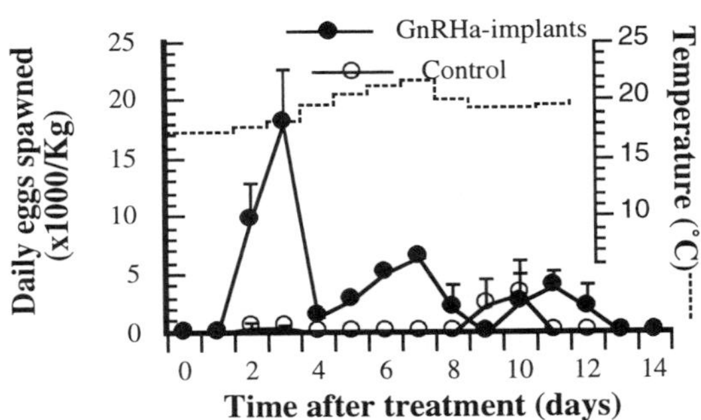

Figure 2. Mean (+ s.e.m.) cumulative spawning of American shad (n = 9) given GnRHa -implants.

Both GnRHa implants and microspheres induced 100% ovulation of striped bass and white bass, with egg fertilization averaging 45 and 81%, respectively (data not shown). In male striped bass, both delivery systems enhanced milt production many-fold (Fig. 3) without any reduction in sperm cell concentration (data not shown).

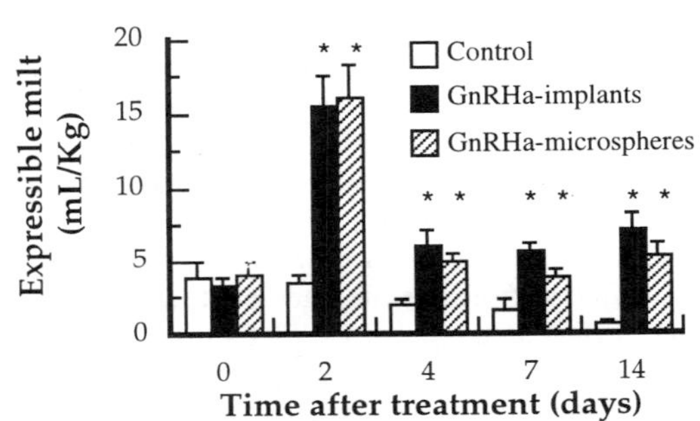

Figure 3. Mean (+ s.e.m.) expressible milt of striped bass treated with GnRHa delivery systems. "*" $P<0.05$

Sustained-administration of GnRHa via delivery-systems can be a very effective alternative to multiple GnRHa injections. The long-term GnRHa release from such systems makes them applicable to both one-time spawners, as well as sequential spawners.

References

Mylonas, C.C., Tabata, Y., Langer, R. and Y. Zohar. 1995. Preparation and evaluation of polyanhydride microspheres containing gonadotropin-releasing hormone (GnRH), for inducing ovulation and spermiation in fish, J. Controlled Release, 35:23-34.

Zohar, Y. 1989. Fish reproduction: its physiology and artificial manipulation. In: Fish Culture in Warm Water Systems: Problems and Trends. M. Shilo and S. Sarig (Eds.), CRC Press, Florida, p. 61-119.

EFFECT OF THE LIPID COMPOSITION OF THE DIET ON HORMONAL LEVELS AND SPAWNING PERFORMANCE OF SEA BASS (*Dicentrarchus labrax*)

J.Mª. Navas[1], E. Mañanós[2], M. Thrush[3], J. Ramos[1], S. Zanuy[1‡], M. Carrillo[1], Y. Zohar[2] and N. Bromage[3]

[1]Instituto de Acuicultura de Torre de la Sal, CSIC, Torre de la Sal, 12595 Castellón, Spain; [2]Center of Marine Biotechnology, University of Maryland, Baltimore, MD 21202, USA; [3]Institute of Aquaculture, University of Stirling, FK94LA, Scotland, UK.

Summary

In this study, GTHII was measured for the first time in sea bass (*Dicentrarchus labrax*). The results obtained suggest that dietary deficiencies may affect GTHII release and the quality of progeny.

Introduction

In mammals, inadequate nutrition is associated with dysfunctional LH release and a decline in the rate of ovulation. Dietary manipulation may modify reproductive function in fish. In the sea bass, changes in dietary components affect the number, viability and quality of eggs. As in other teleosts, dietary lipids are very important for reproductive success. Long-term dietary deficiencies in PUFA (polyunsaturated fatty acids) may account for altered patterns of testosterone (T) and estradiol (E_2,) and for the early appearance of atresia in the ovary. On the other hand, when sea bass were fed a low protein (P)-high carbohydrate(CH) diet, GnRH release was affected at the time of spawning. Recently, it has been shown in fish that GTHII, structurally homologous to tetrapod LH, is associated with final oocyte maturation. Our aim in this study was to examine whether dietary changes in the sea bass may affect GTHII plasma levels during the reproductive cycle and lead to an alteration in the ovarian function and a decline in the quality of the progeny.

Materials and Methods

Six year old fish were hand-fed daily for one year with three pelleted diets of different composition. A: 55% P; 9% oil; 17% CH; B: 45% P, 20% oil (9% enriched with 11% marine oil rich in PUFA mainly 22:6-n-3), 16% CH and C: 46% P, 28% oil; 17% CH. The control group was fed sliced fish (*Boops boops*). Females were bled monthly and plasma GTHII, VTG and E_2 were measured by ELISA and RIA, respectively. During the spawning season, quality of eggs, fecundity and percent hatching were determined.

Results and Discussion

Sea bass is a multiple spawner which spawns 3 to 4 discrete clutches in quick succesion. Thus, the ovary contains at least 3 or 4 populations of oocytes in different stages of vitellogenesis. Fig 1. shows the seasonal profile of GTHII plasma levels. In general, levels of GTHII were low from October to December. In January, a modest rise occurred followed by a sharp

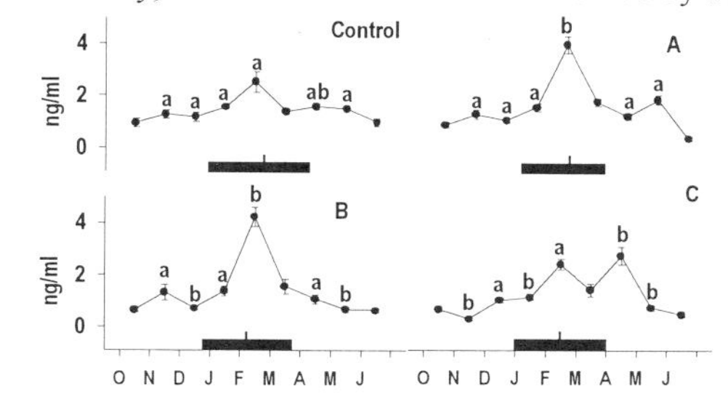

Fig. 1. Monthly seasonal plasma GTHII levels in sea bass fed different diets. Horizontal black bars and vertical lines indicate duration of spawning and the maximum mean spawning time, respectively. Different letters show differences between dietary treatments.

significant increase in Febuary coinciding with the maximum rate of oocyte maturation, ovulation and spawning. Levels were low after March (post spawning) except in group C where a second significant peak was observed in April ($p<0.05$). In all groups, plasma E_2 exhibited a steady increase from December to February, corresponding with active vitellogenesis, which returned to basal levels in March. There were no differences in plasma levels of E_2 between groups except in the control group, where significantly lower values occurred in January ($P<0.05$). In all cases, VTG increased steadily from November to January, when the maximum percentage of vitellogenic oocytes was observed in the ovary. Control group exhibited higher values of egg quality, hatching rate ($p<0.05$) and fecundity compared to groups A, B and C. Thus, dietary treatment significantly affected reproductive performance of sea bass.

From the present data it can be concluded that, as in other teleost species, GTHII may be associated with final maturation and ovulation although no relationship was found with vitellogenesis. The present results suggest that dietary deficiencies may account for a dysfunction in the GTHII release.

Supported by an EEC Contract AIR2-CT93-1005 and a fellowship from the Generalitat Valenciana to J.Navas.

CHARACTERIZATION OF A TWO YEAR REPRODUCTIVE CYCLE FOR COHO SALMON (*O. kisutch*) IN CHILE[1].

R. Neira[a], F.J. Estay[b], N.F. Díaz[b] and X. García[a]

[a]Dept. of Animal Production. Faculty of Agricultural & Forest Sciences, University of Chile. P.O.Box 1004. Santiago. Chile.
[b]Dept. of Ecological Sciences. Faculty of Sciences, University of Chile. P.O.Box 653. Santiago. Chile.

Summary

Favorable fresh water temperatures in the south of Chile allow fast growth and sexual maturation in two years in coho salmon, with high fecundity. Spawning distribution and reproductive variables from 1,230 individual females are analyzed and compared with respect to the populations of coho salmons in the northern hemisphere (N.H).

Introduction

During the seventies, several ocean ranching projects in the south of Chile contributed to improve the knowledge of salmon culture, and served as the interphase to intensive culture in net pens, which has subsequently shown an impressive growth. The availability of warm freshwater in lakes and springs in this geographic area allowed the generalized use of 8-9 month old smolts (S0) enabling to complete the reproductive cycle in two years. Two year old mature females, coming from S0 smolts, were sampled in three farms from Chiloé Island (42°30'S; 73°45'W) and one farm from Aysén fiord (45°47'S, 72°45'); 431 individuals were sampled in 1993 and 799 in 1994. Management of artificial reproduction was standardized in all farms, using the same procedures during spawning, fertilization and incubations. Fertilization was performed by the dry method, 2-3 ml of pooled semen of 5-6 males was used. Eggs from individual females were incubated separately. Well or spring water of 8-12°C with a flow of 10-15 l/min was available for incubation. Eggs were shocked after 250-300 degree days. Fecundity and egg survival were obtained gravimetrically. Egg size was measured with a Von Bayer trough.

Results

Spawning of parental stock occurred in May and 8-9 moths old smolts were obtained in next January-February. Sexual maturation and spawning occurred in May, after 16 months of growth in the sea (Fig 1). Spawning season periods were 39 and 41 days in 1993 and 1994 respectively, starting the last week of April. Most spawning occurred during the last 15 days of May (63%). Table 1 shows important reproductive variables from 1993-94 seasons.

Table 1. Reproductive Variables of farmed coho salmon under a two year cycle in the south of Chile

VARIABLE	AVE. ± SD	N
FEMALE WEIGHT (g)	3,996 ± 855	1212
CONDITION FACTOR (K)	1.32 ± 0.2	971
TOTAL FECUDITY (eggs)	3,802 ± 956	1188
RELATIVE FECUNDITY (eggs/kg)	986 ± 274	1170
EGG DIAMETER (mm)	7.11 ± 0.5	1185

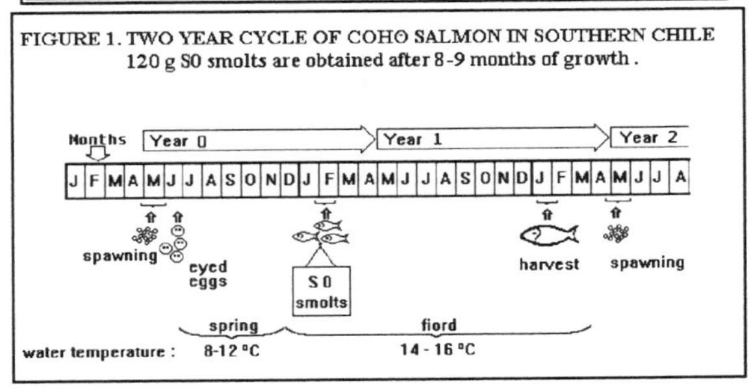

FIGURE 1. TWO YEAR CYCLE OF COHO SALMON IN SOUTHERN CHILE 120 g S0 smolts are obtained after 8-9 months of growth.

Discussion

The reproductive cycle of coho salmon in culture has normally been described as a three year cycle (range: 2-4 years, Gordon et al., 1987). Two year cycles have been described using S0 smolts, produced as the result of accelerated growth under warm water temperature (Donaldson and Brannon, 1976).

S0 smolts are now extensively used in Chile because of the advantages of a two year cycle, and coexist with a three years cycle, with 18 month old smolts produced in November-December. Spawning season is short, with 90% of females spawning within a month and no late spawning populations have been observed, as those mentioned by Gordon et al., (1987) in the N.H.

Coho salmon under an S0 cycle showed better fecundity and bigger eggs, compared with wild cohos from the NH (2100-2800 eggs; 6.6 mm diam., Gordon et al.,1987). Mature 2 years old females showed similar growth than 3 years old females (2.7-5.4 kg, ibid.) from the N.H.

References

Gordon, M. R., Klotin, K. C.; Campbell, V. M. and Cooper, M. M. 1987. Farmed salmon broodstock management. Ministry of Environment, Victoria, British Columbia, Canada. 194 pp.
Donaldson, L.R. and Brannon, E.L. 1976. The use of warmed water to accelerate the production of coho salmon. Fisheries, 1:12-16.

[1] FONDEF PI-10 and FONDECYT 1940259/94 Grants

COOL STORAGE OF THE JAPANESE EEL (*ANGUILLA JAPONICA*) SPERMATOZOA

H. Ohta[1] and T. Izawa[2]

[1] National Research Institute of Aquaculture, Nansei, Mie 516-01, Japan.
[2] Hokkaido Fish Hatchery, Eniwa, Hokkaido 061-14, Japan.

Introduction

This study was designed to consider the prerequisite constituents for Japanese eel sperm diluent and to examine the possibility of its application for cool storage.

Materials and methods

Milt was obtained from the males which had been given over 9 times weekly injections of HCG (1 IU/ gBW / wk). After the storage of the milt with 50 times dilution with various isotonic solutions (from 3 hours to 28 days), sperm motility (%motility) was measured at 15 sec after dilution with 450 mM NaCl (final dilution ratio was x1,000).

Results and Discussion

Spermatozoa diluted with artificial seminal plasma (ASP), consisting of 149.3 mM NaCl+15.2 mM KCl +1.3 mM $CaCl_2$+1.6 mM $MgCl_2$ buffered with 5 mM $NaHCO_3$-NaOH at pH 8.2, or with isotonic NaCl+KCl solution showed high percentage motility indices (percentage motility / initial percentage motility x100) at 3 hours after cool storage at 5 °C (Fig. 1). However, those diluted with isotonic NaCl solution, mannitol solution, NaCl+$CaCl_2$ solution, or NaCl+$MgCl_2$ solution showed significantly lower motility compared to the above two solutions.

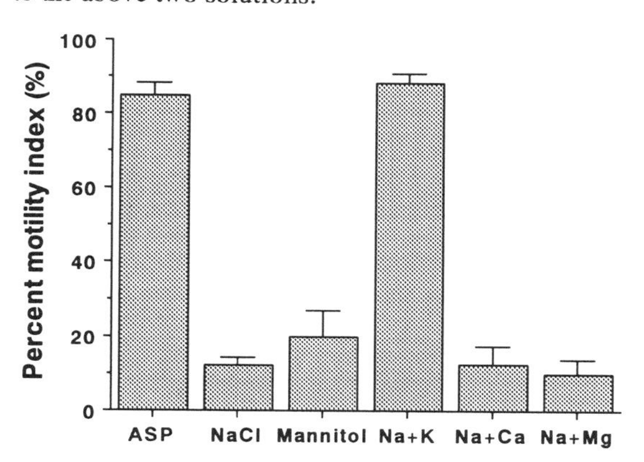

Fig. 1. Changes of % motility after 3-hour storage in the isotonic solutions.

Spermatozoa were then diluted with ASP at various potassium concentrations (0, 5, 15, 25, 35, 45, 75, and 164.5 mM), and stored at 3 °C for up to 28 days (Fig. 2). Spermatozoa diluted with the 5 - 45 mM KCl solutions showed a sharp decrease of percentage motility at 24 hours, however, they showed a marked recovery of the motility by the 7th day. The severity of the decrease at 24 hours became smaller with an increase of potassium concentration in the isotonic diluent. Spermatozoa diluted with the 75 or 164.5 mM KCl solutions did not show the temporal decrease of motility at 24 hours, but their motility at 14 - 28 days were considerably lower than those in lower concentration KCl solutions. Spermatozoa diluted with the 15, 25, 35, and 45 mM KCl isotonic solutions showed high motility indices of 47.9 - 64.8% at 21 days, and of 29.7 - 35.1% at 28 days, respectively.

These findings indicate that potassium ion is an essential constituent of isotonic diluent for cool storage of Japanese eel spermatozoa. The optimum KCl concentration for short term storage (up to 7 days) is 45-75 mM, and that for longer storage is 15-45 mM.

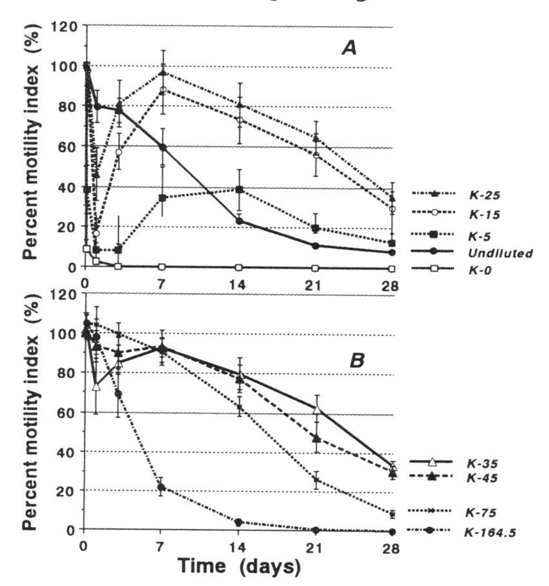

Fig. 2. Changes of % motility during 28-day storage in ASP with various concentrations of KCl.

Supported by a grant from the Ministry of Agriculture, Forestry and Fisheries, Japan (BMP 95-II-2-3).

EFFECTS OF METHYLTESTOSTERONE CONCENTRATION ON SEX RATIO, GROWTH AND DEVELOPMENT OF NILE TILAPIA

M. Okoko and R.P. Phelps

Department of Fisheries & Allied Aquacultures, Auburn University, AL 36849

Summary

Sex ratios of nile tilapia Oreochromis niloticus populations were highly skewed to males (97% male) when fry were fed methyltestosterone at concentration of 15 to 60 mg/kg of diet for 28 days. Higher concentrations increased the frequency of females and intersex fish.

Introduction

Male tilapia can be obtained by feeding a feed containing methyltestosterone (MT) or other androgens to fry with sexually undifferentiated gonads until gonadal differentiation is complete. Sex direction has been obtained in O. niloticus over a range of MT concentrations but a minimum effective dose has not been well defined. High concentrations may result in abnormalities or possible reduced masculinization. Intersex fish have been observed when tilapia have been treated with MT but the relationship of their occurrence and dose rate has not been examined.

Materials and Methods

Nile tilapia fry with an initial length of 11 mm TL were fed a commercial trout ration containing 0, 3.75, 7.5, 15, 30, 60, 120, 240, 480, 600, or 1200 mg MT/kg of diet for 28-d. Three hundred fish were stocked in 45 L aquaria with three replicates per treatment. After 28-d of treatment all fish were harvested and average length, weight and survival determined. Fish were restocked into 20m^2 outdoor tanks and fed a non-hormone treated feed for an additional 30-d a minimum size of 5 cm was reached. Fish were harvested and preserved in formalin.

Gonads were removed and examined microscopically, the sex of the fish classified as male, female, or intersex. Gonads of intersex fish were further classified as to the relative abundance of ovarian tissue in the gonad.

Results

Sex ratios are given in the table. MT at concentrations up to 1,200 mg/kg di not effect fish growth, survival, or feed conversion during the 28-d treatment period. Hormone concentrations of 15, 30 and 60 mg/kg of ration resulted in 98, 99 and 97% male populations. The percent females in the non-hormone treated population was 55% and progressively decreased to 0.67% at 60 mg/kg. At dose rates greater than 60 mg/kg the percent males decreased and was not significantly different from non-treated fish at dose rates of 240 to 1200 mg/kg.

Intersex fish were 1 to 25% of the fish at a specific dose rate with intersex fish most common at dose rates of 120 to 600 mg/kg. Intersex gonads were predominantly composed of testicular tissue at doses of 3.75 to 60 mg/kg and predominantly ovarian tissue at doses of 120 mg/kg and greater.

Table 1. Percent of population (Mean ± S.D) as male, female or intersex when Nile tilapia were fed various rates of methyltestosterone for 28 days.

Dose	% Male	% Female	% Intersex
0	41.0±7.9	58.0±7.9	1.0±0.6
3.75	80.0±7.8	1.0±1	19.0±7.8
7.5	91.7±3.8	0	8.3±3.8
15	98.3±1.1	0	1.7±0.6
30	99.3±1.1	0	0.7±0.6
60	97.3±1.0	0.7±1.1	2.3±0.6
120	71.9±7.4	9.7±4.2	18.0±7.2
240	50.7±2.1	24.0±7.2	25.3±5.5
480	48.3±4.0	44.0±5.2	7.7±1.1
600	55.0±7.0	19.7±3.8	25.3±3.8
1200	52.0±1.7	48.0±1.7	0

Discussion

A minimum effective dose of methyltestosterone to obtain a 95% male population (acceptable in most production settings) would be 10.2 mg/kg when given indoors in a flowing water environment. A dose of 7.5 mg/kg resulted in no females in the population but 19% of the fish were intersex having predominantly testicular tissue. These reduced rates of MT are significantly below the dose rate commonly used (30-60 mg/kg). Use of such rates would meet the goal of controlling tilapia reproduction while reducing even further any potential MT residue when such fish are cultured for food.

ALTERATION OF CARP (*CYPRINUS CARPIO*) SPERMATOZOA MOTILITY BY URINE CONTAMINATION DURING SAMPLING

G. Perchec[1], J. Cosson[2], C. Jeulin[3], C. Paxion[1], F André[1] and R. Billard[1]

[1]Ichtyologie, MNHN, 43, rue Cuvier, 75231 Paris Cedex 05, France; [2]CNRS URA 671 Marine Station F, 06230 Villefranche/Mer, France; [3]Centre Hospitalier Universitaire, 94275 Kremlin-Bicêtre Cedex, France.

The aim of this study was to investigate the impact of urine contamination occurring at sampling on sperm quality.

Methods

After gonadotropic stimulation, carp sperm was collected with a catheter. Samples without spontaneous motility were mixed with urine or glucose solution added directly to undiluted milt. After one hour incubation, sperm was diluted 1/100 into: IM (Immobilizating Medium), 200 mM KCl, 30 mM Tris, pH 8, op >400 mOsm.kg^{-1}, in which no flagellar movement occurred. Motility of spermatozoa was immediately triggered in AM (Activating Medium, 1/2000), 45 mM NaCl, 5 mM KCl, 30 mM Tris, pH 8, op <160 mOsm.kg^{-1} and measured by a microscopic estimation. Change in sperm ATP content was assayed by bioluminescence.

Results

After activation in AM, the populations of spermatozoa exposed to increasing amounts of urine showed a progressive inversion of the ratio between spermatozoa with "high" and "low" velocity (Fig. 1). The ATP content also declined when the percentage of added urine increased (Fig. 2).

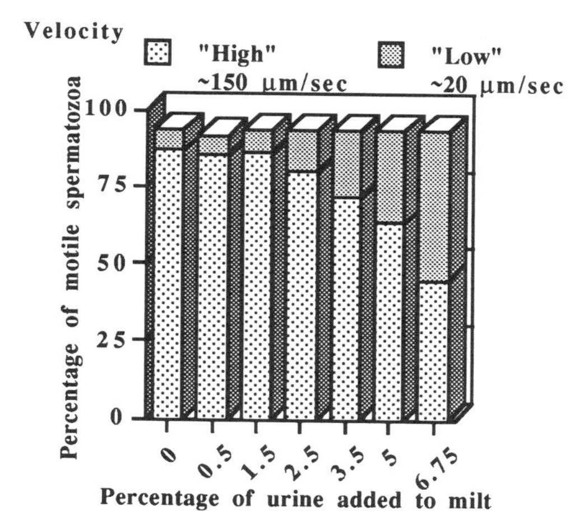

Fig. 1: Velocity and percentage of motile spermatozoa mixed with urine. (mean±sem, n=3 males).

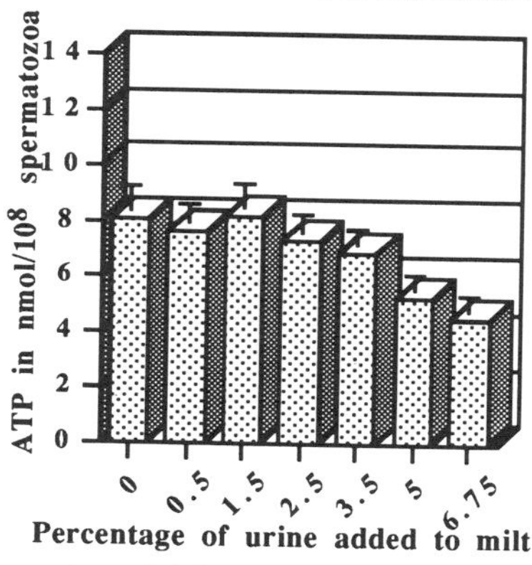

Fig 2: Dynamics of ATP content in sperm following exposure during one hour to various quantities of urine. (mean±sem, n=3 males).

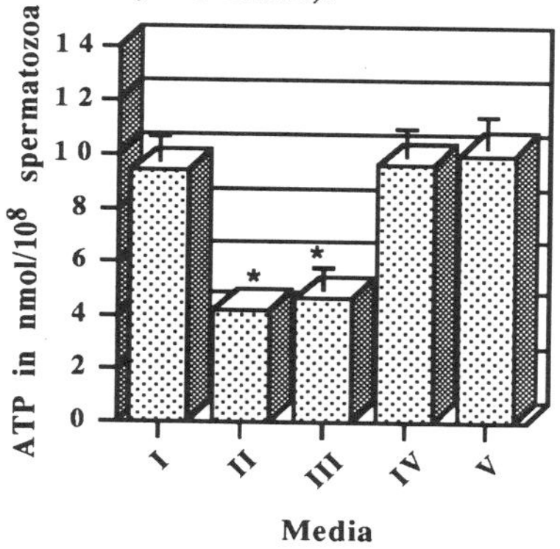

Fig 3: Changes in ATP content of intact sperm (I) after addition of 7.5% urine 18 mOsm.kg^{-1} (II); 7.5% glucose 20 mOsm.kg^{-1} (III); 7.5% urine with glucose 340 mOsm.kg^{-1} (IV); 7.5% glucose 340 mOsm.kg^{-1} (V). (*, $p< 0.05$, Student test) (mean±SD, n=2 males).

Conclusion

Low osmotic pressure conditions which trigger carp spermatozoa motility induce a rapid decrease in ATP content concomitantly with velocity and flagellar beat frequency (Perchec et al., 1995. J. Cell Sci. 108: 743-753). When sperm was collected, the exposure of carp milt to urine results in the occurrence of a spermatozoa population with low motility due to a partial sperm activation and hydrolysis of ATP. These results suggest that the "poor" quality of some carp sperm sample (Redondo et al, 1991. Mol. Reprod. Dev. 29, 259-270) may be due to such a spontaneous activation after stripping.

CRYOPRESERVATION OF YELLOWTAIL FLOUNDER (*PLEURONECTES FERRUGINEUS*) SEMEN

G.F. Richardson,[1] L.W. Crim,[2] Z. Yao,[2,3] and C. Short[2]

[1]Department of Health Management, Atlantic Veterinary College, University of Prince Edward Island, Charlottetown, PE, Canada C1A 4P3; [2]Ocean Sciences Centre, Department of Biology, Memorial University of Newfoundland, St. John's, NF, Canada A1C 5S7.

Summary

Semen from yellowtail flounder was diluted 1:3 with each of four extenders. The semen was frozen over liquid nitrogen in 0.25 mL straws. All extenders were effective for semen cryopreservation.

Introduction

Semen from various species of fish has been successfully frozen but much less is known about the cryopreservation of flatfish semen. The ability to cryopreserve flatfish semen would be a convenience and may also be of commercial value. The objective of this study was to determine the effectiveness of four extenders for the cryopreservation of yellowtail flounder (*Pleuronectes ferrugineus*) semen.

Materials and Methods

The four extenders used in this study consisted of diluent A (diluent 1 without lecithin, Gallant et al.) and diluent B (plaice ringer, Cobb et al.), each with 10% (v/v) dimethyl sulfoxide (DMSO) or propylene glycol (PG). In two trials, fresh pooled yellowtail flounder semen was diluted 1:3 (semen:extender) and drawn into 0.25 mL straws (I.M.V., L'Aigle, France). The sealed straws were suspended 5.5 cm above liquid nitrogen on a floating styrofoam rack. After about 12 minutes, the temperature of the extended semen reached -95°C and the straws were plunged into liquid nitrogen.

Batches of approximately 200 eggs (trial 1) or 120 eggs (trial 2) per replicate, with three replicates per treatment, were used for fertility trials. Semen frozen in each extender, fresh pooled semen diluted 1:3 with each diluent and each extender, and fresh undiluted pooled semen were compared. The straws were thawed for 7 seconds in a 30°C water bath. Five μL of undiluted semen or 20 μL of diluted semen were added to each batch of eggs. The fertilized eggs were incubated at 5°C until hatching.

Results and Discussion

The results from both trials were pooled. There were no significant differences between the means for fertilization rate, hatch rate (percentage of fertilized eggs that hatched), and percentage of deformed larvae for the undiluted semen as compared with all other treatments, except that the means for frozen semen were lower for fertilization rate ($p<0.05$). The fertilization rate was not reduced for unfrozen extended semen and therefore the reduction in fertilization rate was due to freezing and not extender toxicity.

When results involving extended semen were compared, the fertilization rate for frozen semen was lower than for unfrozen semen ($p<0.001$). In contrast, the hatch rate was higher ($p<0.05$) and the percentage of deformed larvae was lower ($p<0.05$) for frozen than for unfrozen semen. These results also showed that the fertilization rate was higher when A was used with DMSO than with PG ($p<0.05$), but this difference between cryoprotectants was not observed with B. Semen diluted with extender containing A had a higher hatch rate than did semen diluted with extender containing B ($p<0.05$). This difference was also evident when just the means for frozen semen were compared ($p<0.05$).

The use of cryopreserved semen resulted in an average fertilization rate, hatch rate and percentage deformed larvae of 70%, 66% and 8%, respectively. This is in comparison to results of 82%, 63% and 14%, respectively, for fresh undiluted semen. All extenders were effective for the cryopreservation of yellowtail flounder semen.

References

Gallant, R.K., G.F. Richardson and M.A. McNiven, 1993. Comparison of different extenders for the cryopreservation of Atlantic salmon spermatozoa. Theriogenology 40:479-486.

Cobb, J.L.S., N.C. Fox and R.M. Santer, 1973. A specific ringer solution for the plaice (*Pleuronectes platessa* L.). J. Fish Biol. 5:587-591.

[3]Present address: Department of Zoology, University of Guelph, Guelph, ON, Canada N1G 2W1.

CRYOPRESERVATION OF SPERM FROM STRIPED TRUMPETER *LATRIS LINEATA*

Arthur Ritar and Marc Campet

Dept. Primary Industry and Fisheries, Marine Research Laboratories, Taroona Tasmania 7053 Australia

Summary

Semen was diluted with a saline (300 mOsm) medium containing 2 M glycerol and frozen as pellets. The optimum dilution rate was 1:5. Motility after dry thawing was 40.5± 4.18% but declined when thawed in sea water. Thawing temperatures of 10°, 20° or 30°C were equally effective. There was no effect of pellet volume (0.2-0.6 ml) or number of pellets thawed (1-6 per tube) on motility. For fresh and thawed semen, egg fertilisation rates were 82.7 ± 3.07% and 71.2 ± 7.22% and larval hatch rates were 45.1 ± 4.93% and 31.5 ± 5.30%. Fresh sperm were immotile within 9 minutes after activation with sea water. Chilled sperm were still motile 8 days after collection.

Introduction

Spermiating males may not be available for the fertilization of eggs, especially early in the spawning season (which lasts only 8 weeks, August to October) or for fish held under alternative photothermal environments to induce out-of-season spawning. A bank of frozen semen which may be thawed as required would assist hatchery management. This study examined the freezing of semen and fertilization of eggs and other aspects of semen handling.

Results

From Table 1, the post-thawing recovery was higher for the modified glycerol diluent (Dil. A: Thorogood and Blackshaw, 1992) than the DMSO diluent (Dil. B: Gwo *et al.*, 1993) although both media were previously found suitable for frozen storage of flounder sperm. Motility was higher for pellets thawed Dry (in dry test tubes) than Wet (in 0.5 ml sea water).

Table 1. Effect of diluent and thawing on motility.

Diluent	Thawed	Mean±s.e.m.
Dil. A	Dry	40.5±4.18%
	Wet	34.5±4.31%
Dil. B	Dry	13.5±3.40%
	Wet	5.0±0.75%

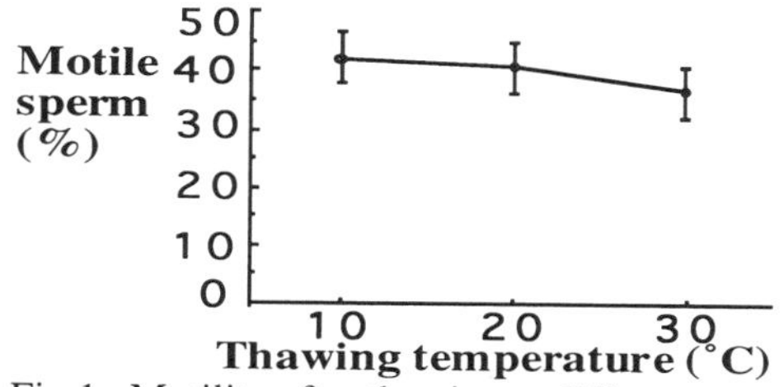

Fig 1. Motility after thawing at different temperatures.

For 0.2 ml pellets, there was no effect of thawing temperature on sperm motility (Fig. 1). The post-thawing recovery was highest for semen diluted 1:5 (Table 2). There was no effect on motility of the number of 0.4 ml pellets thawed in individual test tubes. Another study also found that the motilities after thawing of 1, 2 or 4 pellets per test tube were similar for pellet sizes of 0.2, 0.4 or 0.6 ml.

Table 2. Effect of dilution rate and number of pellets thawed on sperm motility.

Dilution rate (semen:diluent):		
1:2	1:5	1:11
24.9±3.81%	30.3±4.17%	27.2±3.75%
Number of pellets thawed:		
1	3	6
28.1±3.90%	27.9±4.02%	26.5±3.91%

Table 3. Fertilisation and larval hatch after inseminating eggs with fresh or frozen-thawed semen.

	Fresh	Frozen	P value
% motile sperm	> 80	43.3±6.15	-
% eggs fert.	82.7±3.07	71.2±7.22	0.32NS
% dead eggs-Day 0	29.8±4.53	47.8±4.21	0.03*
Day 2	17.9±1.23	19.5±1.33	0.48NS
Day 4	1.2±0.14	0.6±0.14	0.03*
Day 6	6.0±1.98	0.7±0.18	0.01**
No. eggs & larvae	392±21.5	363±26.6	0.51NS
% larval hatch	45.1±4.93	31.5±5.30	0.15NS

Fertilisation and larval hatch rates were similar for eggs inseminated with fresh or thawed semen even though motility differed by more than 30% (Table 3). Egg mortality for frozen semen was higher on Day 0 but lower on Days 4 and 6. After activation of fresh semen at 18°C, motility declined asymptotically and sperm were immotile within 9 min. For undiluted semen held at 4°C, motility declined sigmoidally from >70% on Day 0 to <10% by Day 8 after collection.

Discussion

Semen diluted at 1:5 with the simple glycerol medium and frozen by the pellet method maximised post-thawing motility resulting in egg fertilisation and larval hatch rates similar to fresh semen. The thawing temperature, volume of individual pellets and number of pellets thawed were not critical to post-thawing motility. Motility of fresh sperm declined rapidly after activation but sperm remained viable for up to 8 days when chilled and stored neat.

References

Gwo, J. C. Kurokura, H. and Hirano, R. 1993. Cryopreservation of spermatozoa from rainbow trout, common carp, and marine puffer. *Nippon Suisan Gakkaishi*, 59, 777-782.

Thorogood, J. and Blackshaw, A. 1992. Factors affecting the activation, motility and cryopreservation of the spermatozoa of the yellowfin bream, *Acanthopagrus australis (Gunther). Aquaculture and Fisheries Management*, **23**, 337-344.

DELAYED OVARIAN DEVELOPMENT AND REDUCED FECUNDITY IN LARGEMOUTH BASS RAISED ON A PELLETED FEED CONTAINING HIGH LEVELS OF STEROIDS AND LOW LEVELS OF ARCHIDONIC ACID.

P. Rosenblum,[1] H. Horne,[1] G. Garwood,[2] T. Brandt,[2] and B. Villarreal[3]

[1]Southwest Texas State University, [2]National Biological Service, and [3]Texas Parks and Wildlife Department, San Marcos, Texas, USA

Summary

Female largemouth bass, *Micropterus salmoides*, raised on a pelleted feed containing high levels of testosterone (T) and 17β-estradiol (E2), and a low level of arachidonic acid (AA), exhibited delayed ovarian recrudescence and low serum levels of T and E2. Ovaries from pellet-fed fish contained low levels of AA. Pellet-fed bass spawned less frequently, and produced fewer eggs and fry per spawn than forage-fed controls. These data indicate that dietary steroid and fatty acid levels can affect ovarian development and spawning in this species, perhaps accounting for previously observed reproductive impairment of pellet-fed largemouth bass.

Introduction

Largemouth bass fed pelleted feeds have been reported to have reduced reproductive success. Previous studies in our laboratory have shown that pellet-fed bass had larger gonads than bass fed a forage diet of goldfish, *Carassius auratus*, but that gonadal growth was delayed in the pellet-fed fish (Rosenblum *et al*., 1994). Since these studies were done in ponds, no information on egg or fry production was obtained. The present studies were designed to examine indices of fecundity in female largemouth bass raised on pelleted and forage diets, and to examine whether dietary steroid and/or arachidonic acid levels might underlie any observed differences in reproductive success.

Results

Feed Steroid and Arachidonic Acid Levels. The pelleted feed contained significantly higher concentrations (ng/g dry wt) of both T (pellet, 4.47 ± 0.43 ; forage, 0.27 ± 0.08) and E2 (pellet, 2.21 ± 0.08; forage, 0.05 ± 0.01), and a lower amount (% wt) of AA (pellet, 0.21 ± 0.05; forage, 6.33 ± 0.20).

Patterns of Gonadosomatic Index (GSI) and Circulating Gonadal Steroids. The seasonal pattern of GSI differed between the pellet-fed and forage-fed fish. Forage-fed bass exhibited a steady increase in GSI from December, reaching a peak in March (13.21 ± 1.13). In pellet-fed bass, GSI increased slightly between December and January, but no further increases were seen until April, when maximum GSI (11.44 ± 0.04) was reached. Patterns of circulating T and E2 reflected patterns of GSI. In forage-fed fish, levels of both steroids rose in parallel with GSI through March (T, 0.79 ± 0.29 ng/ml; E2, 0.72 ± 0.10 ng/ml). During the period of suppressed ovarian growth, pellet-fed fish had significantly lower serum levels of both T (0.18 ± 0.03 ng/ml) and E2 (0.20 ± 0.03 ng/ml) than did forage-fed fish. Levels of both steroids rose as GSI increased in April (T, 0.54 ± 0.12 ng/ml; E2, 0.98 ± 0.26 ng/ml).

Ovarian Arachidonic Acid Levels. AA levels in mature ovaries from pellet-fed bass (0.87 ± 0.08 % wt) were significantly lower than in mature ovaries from forage-fed fish (2.96 ± 0.05 % wt).

Raceway Spawning Performance. A total of four spawns were obtained from pellet-fed bass; forage-fed fish produced eight spawns. Pellet-fed fish spawned significantly fewer eggs/spawn (6,023.23 ± 1,284.46) and produced significantly fewer fry/spawn (5,417.28 ± 921.05) than did forage-fed fish (eggs, 13,538.10 ± 2,293.19; fry, 12,970.86 ± 2,344.63). Fertilization and hatching rates were not affected by diet.

Discussion

Pellet-fed largemouth bass had delayed ovarian recrudescence. During this time serum levels of T and E2 were low. These data suggest that during early and mid-recrudescence, dietary steroids inhibit gonadal development, possibly through negative feedback effects on gonadotropic hormone secretion. The negative effects of dietary steroids are reversed during the later stages of recrudescence, and ovarian growth is resumed.

Pellet-fed bass spawned fewer times, and produced fewer eggs and fry per spawn than did forage-fed bass. Fertilization and hatching rates were not affected by diet. Together, these data suggest that reduced fecundity in pellet-fed bass may be due to altered spawning behavior and a failure to ovulate and release adequate numbers of eggs during spawning.

Ovaries from pellet-fed bass contained low levels of the prostaglandin precursor arachidonic acid. Since prostaglandins are known to be involved in spawning behavior and ovulation in fish, a reduced capacity to synthesize prostaglandins might underlie the observed reductions in spawning behavior and ovulation in pellet-fed largemouth bass.

References

Rosenblum, P. M., Brandt, T. M., Mayes, K. B., and Hutson, P. (1994) Annual cycles of growth and reproduction in hatchery-reared Florida largemouth bass, *Micropterus salmoides floridanus*, raised on forage or pelleted diets. *J. Fish Biol.* **44**: 1045-1059.

GYNOGENESIS IN ALBINO GRASS CARP, Ctenopharyngodon idella (Val.)

S. Rothbard[1], Y. Hagani[1], B. Moav[2], W.L. Shelton[3] and I. Rubinshtein[1]

[1]YAFIT Lab., Fish Breeding Centre, Gan Shmuel 38810; [2]Zoology Dept., Tel Aviv Univ., Tel Aviv 69987, Israel; [3]Zoology Dept., Univ. of Oklahoma, Norman OK 73019, USA.

Summary

Gynogenesis was induced in albino grass carp (AGC) by insemination of eggs with gamma- or UV-irradiated common carp (CC), koi carp (KC) or wild-type colored grass carp (WGC) sperm. Activated eggs were diploidized by application of 1-2 minutes early heat or pressure shocks, 4-6 minutes post-activation (at the developmental duration of $\approx 0.2\tau_o$). In spite of the high rate of abnormal embryos, better survival of larvae was recorded in batches of eggs inseminated with UV- as compared to gamma-treated sperm.

Introduction

The study was carried out on eggs obtained from albino grass carp (AGC), Ctenopharyngodon idella, carrying a single recessive gene for albinism at an autosomal locus (Rothbard and Wohlfarth, 1993). The objectives of the study were (i) to produce diploid gynogenotes by exposing eggs to early shock at the developmental duration $\approx 0.2\tau_o$, according to Shelton and Rothbard (1993), (ii) to examine assessment of successful gynogenesis in the hatch-out larvae, in AGC eggs inseminated with irradiated sperm of donor koi and common carp (Cyprinus carpio), carrying dominant color markers, (iii) to stimulate embryogenesis of artificially diploidized AGC eggs, activated with sperm of heterologous species and (iv) to examine integration and incorporation of foreign pigment genes and their expression in fish.

Materials and Methods

1. Ovulated AGC eggs were inseminated at 22°C with UV-irradiated (800 J/m²) CC sperm and heat- (40.0±0.5 °C) or pressure-shocked (7500 psi) for 120 or 40 seconds, respectively, 4-6 minutes (0.2-$0.3\tau_o$) post-activation. Controls were fertilized with non-irradiated normal CC or AGC sperm. Eggs were incubated in 50L zugers.
2. Sperm of tricolored Japanese ornamental (koi) carp (KC), common carp (CC) and wild-type colored grass carp (WGC) was gamma-irradiated (Co^{60}) with dosages of 60, 75 and 90 Krad. Irradiated sperm was used to inseminate AGC eggs. Batches of eggs, activated at 22-24°C prior to shock, were diploidized either with early heat- (42±1°C/1 min) or pressure-shock (8,000 psi/2 min), 4 minutes (at 0.2-$0.25\tau_o$) post-activation.

Results and Conclusions

Survival of gynogenotes (79-95%) produced from AGC eggs inseminated with UV-irradiated sperm, was indicated by diploid albino larvae (**Table** 1) and did not differ significantly from their respective AGCxAGC diploid controls (77-83%; t-test <0.87; P=0.5). Although high survival (about 70%) was found in AGCxCC (AGC eggs fertilized with intact CC sperm) embryos, examined 10h post-fertilization, no embryos were hatched.

Table 1. Survival of diploidized AGC gynogenotes*, shocked 6 minutes post-activation (at $0.3\tau_o$) and their respective controls, AGCxAGC embryos (Ctrl).

Female number	Eggs vol. (ml)	Conditions of shock	Surv. (%) **
1. Gyn	60	40.0±0.5°C/2 min	93.9
Ctrl	1000	no-shock	77.2
2. Gyn	150	7500 psi/40 sec	78.5
Ctrl	750	no-shock	83.6
3. Gyn	60	40.0±0.5°C/2 min	94.6
Ctrl	500	no-shock	85.3

* Inseminated with UV-irradiated sperm.
** Embryos survival examined in samples of $150<N<320$ eggs, at blastula or gastrula stage.

The crosses AGCxKC treated with gamma-irradiated sperm, yielded a few pigmented (wild-type) fish plus some (10) albinos, or large numbers of albinos. We presume that the genes responsible for the black pigment (e.g. tyrosinase) from the KC, were transferred to the AGC eggs via chromosome fragments and were expressed in the dark-colored individuals (to be further examined by DNA-fingerprintings). Other crosses, AGCxCC and AGCxWGC, yielded all albino offspring. Survival of albino fish in the latter combinations was about 10% (total of about 2,500 hatch-out larvae).

References

Rothbard, S. & Wohlfarth, G.W., 1993. Inheritance of albinism in the grass carp, Ctenopharyngodon idella. **Aquaculture**, 115:13-17.

Shelton, W.L. & Rothbard, S., 1993. Determination of the developmental duration (τ_o) for ploidy manipulation in carps. **Israeli J. Aquacult. - Bamidgeh**, 45:73-81.

CHARACTERIZATION OF EMULSION PREPARED WITH LIPOPHILIZED GELATIN AND ITS APPLICATION FOR THE INDUCTION OF VITELLOGENESIS IN JAPANESE EEL *Anguilla japonica*

N.Sato[1], I.Kawazoe[2], Y.Suzuki[1] and K.Aida[1]

[1]Dept. Fisheries, Fac. Agriculture, The University of Tokyo,Bunkyo, Tokyo, 113, Japan
[2]Central Research Institute, MARUHA Corporation, Tsukuba, 300-42, Japan.

Summary

A new water in oil in water (W/O/W) type emulsion using lipophilized gelatin (LG) and cotton-seed oil was developed for the administration of hormones. Plasma profiles of salmon gonadotropin (sGtH II) in eels showed gradual changes when LG emulsion containing salmon pituitary extract was administered to fish. The immature Japanese eel (BW 566-1017g) received weekly intramuscular injections of LG emulsion, water in oil (W/O) emulsion prepared with Freund incomplete adjuvant (FIA), or saline solution each of which contained a salmon pituitary GtH fractions. LG emulsion was found to be more effective than the other treatments in inducing vitellogenesis in the eels.

Introduction

It is of importance to develop an efficient system of hormonal delivery which can be applied to improve fish reproduction in aquaculture. The most commonly employed method has been hormonal injection. However, the effects of hormones are not usually maintained for long enough periods using this methods. For example, in order to induce maturation and spawning of eels, the repeated administration of the pituitary extract is necessary. In this study, we developed a new method of hormone administration using LG to minimize the damage to fish due to the repeated hormone administration and rapid clearance. LG was prepared by attaching palmitic anhydride to gelatin.

Firstly, we investigated of hormone release properties from the best formed emulsions by *in vitro* and *in vivo* experiments. Secondly we administered sGtH-loaded LG emulsion to immature adult eels in order to compare the effectiveness of LG emulsion with that of the other types of hormone solutions.

Methods

The W/O/W type emulsion was prepared with LG solution and cotton-seed oil and its structure was observed microscopically. *In vitro* release experiments were done by measuring the amount of glucose (used as an indicator) released from the emulsion. In the *in vivo* release experiments, plasma profiles of sGtH II were monitored in cultured young eels (BW about 200g) treated with LG emulsion, FIA emulsion or saline solution containing salmon pituitary extract in sea water at 20°C. Cultured immature Japanese eels (BW 566-1017g) received weekly intramuscular injections of LG emulsion, FIA emulsion or saline solution, each of which contained 2mg of a salmon pituitary GtH fraction. Injections were repeated ten times. Fish were kept in sea water at 20°C and their body weight (BW) were measured every week. At the end of the experiment, all the fish were dissected and gonadosomatic index (GSI%) was determined.

Results and Discussion

The final concentration of LG in water and the volume ratio of cotton-seed oil to LG solution adequate for preparing the best-formed and the most stable W/O/W emulsion were determined to be 2% and 2:1, respectively, by the microscopic observation and the *in vitro* release experiment.

In the *in vivo* release experiments, plasma profiles of sGtH II in young eels showed gradual changes in the LG group, while sGtH II levels were constantly low in the FIA group. Plasma profiles of sGtH II in the saline solution group showed acute changes. These results indicate LG emulsion does not cause acute and massive changes, and has a mild hormone releasing property different from saline solution and FIA emulsion.

When immature Japanese eels received weekly intramuscular injections of the salmon pituitary GtH fraction, in the group treated with LG emulsion, BW increased significantly after 9 weeks and GSIs were 33.6-45.2. In the group treated with FIA emulsion, BW increased after 7 weeks and GSIs were 6.9-55.9. In the group treated with saline solution, BW increased after 9 weeks and GSIs were 11.7-40.3. In the LG emulsion group, all the fish matured with small variances of GSI and percent increase in BW. In the FIA emulsion group, fish showed great variances of GSI and percent increase in BW. In the saline solution group, it took longer period of time to complete vitellogenesis. LG emulsion was thus more effective than the other treatments for inducing vitellogenesis in the eel.

GONADAL SEX DIFFERENTIATION IN CHANNEL CATFISH

J.E. Schoore[1], R. Patiño[1], K.B. Davis[2], B.A. Simco[2], and C.A. Goudie[3]

Texas Coop. Fish & Wildl. Res. Unit, Texas Tech Univ., Lubbock, TX 79409-2125;[1] Dept. Biology, Univ. of Memphis, Memphis, TN 38152;[2] USDA Agricult. Res. Service, Catfish Genetics Res. Unit, Stoneville, MS 38776[3]

Summary

Channel catfish gonads were histologically examined from Day 7 to Day 90 after fertilization. No differences in gonadal organization were seen in Day 7 to Day 16 individuals. However, somatic differentiation (onset of cavity formation) was observed in about half the individuals sampled at Day 19. Meiosis in these gonads was detected at Day 22. These gonads later developed into typical young ovaries containing an ovocoel and growing oocytes. The gonads of the rest of the individuals remained indifferent for the duration of the study. Therefore, in channel catfish, ovaries differentiate histologically much earlier than testes.

Introduction

It has been shown that male catfish grow faster than females, and therefore the culture of male monosex or male-biased populations offers an economic advantage to catfish growers (Goudie et al. 1994). In many teleosts, the manipulation of gonadal sex by steroid treatment has become a fairly simple procedure. Estrogen treatment yields female or mostly female populations whereas androgen treatment produces male or male-biased populations (Patiño, submitted). However, in channel catfish the production of phenotypic males by steroid treatment of genotypic females has been difficult. Namely, treatment of catfish fry with a variety of androgens failed to masculinize genetic females and instead produced paradoxical feminization of genetic males (Goudie et al., 1983). (There is recent evidence that some synthetic androgens can masculinize genetic female catfish; J. Galvez, personal communication.)

The timing and pattern of normal gonadal sex differentiation in channel catfish has not been established. This knowledge could have practical applications for the development of effective techniques of hormone-induced masculinization, and is also of scientific value to understand the basis of paradoxical feminization in this species.

Methods, Results, and Discussion

Eggs were collected on the day of spawning, and eggs and fish were reared at 27-29° C. Fish were collected from Day 7 to Day 90 postfertilization. Samples were processed for histological analyses using routine procedures (paraffin sections, H&E). Table 1 summarizes our main findings:

Table 1. Chronology of relevant events during early ontogeny of channel catfish gonads

Days after fertilization	Incidence of Appearance	
	Indifferent	Ovary-like
7-16	All gonads	
19	4/10	6/10 (onset of ovocoel formation)
22	6/10	4/10 (onset of germ cell meiosis)
48-90	about 50%	about 50% (onset of previtellogenic growth)

This study showed that channel catfish ovaries differentiate much earlier than testis. Histologically, somatic differentiation of the ovaries seemed to start earlier than germ cell differentiation. Also, the onset of ovarian differentiation correlated with the time period for effective steroid-induced feminization in this species (Davis et al., 1990).

References

Davis, K.B., Simco, B.A., Goudie, C.A., Parker, N.C., Cauldwell, W., Snellgrove, R. 1990. Hormonal sex manipulation and evidence for female homogamety in channel catfish. Gen. Comp. Endocrinol. 58:218-223.

Goudie, C.A., Redner, B.D., Simco, B.A., Davis, K.B. 1983. Feminization of channel catfish by oral administration of steroid hormones. Trans. Am. Fish. Soc. 112:670-672.

Goudie, C.A., Simco, B.A., Davis, K.B., Carmichael, G.J. 1994. Growth of channel catfish in mixed sex and monosex pond culture. Aquaculture 128:97-104.

Patiño, R. Submitted. Manipulations of the reproductive system of fishes by means of exogenous chemicals.

Study supported in part by Texas Higher Education Coordinating Board Grant No. 003644-060

The Effect of Stripping Frequency on Sperm Quantity and Quality in Winter Flounder (*Pleuronectes americanus* Walbaum)

B. Shangguan and L. W. Crim

Ocean Sciences Centre, Memorial University of Newfoundland, St. John's, Nfld, Canada A1C 5S7

Summary

In the winter flounder, increased sampling frequency significantly raised the total sperm production collected in the first eight weeks and did not influence sperm motility and fertility. In contrast, delaying collection of milt reduced the quantity and quality of sperm in this species.

Introduction

Fish held in captivity experience abnormal conditions and may be frequently handled increasing stress and potentially leading to a reduction in sperm production and quality. To test the effects of sampling frequency on sperm production and quality in winter flounder, males were stripped at different intervals, either weekly or biweekly, during the summer spermiation season. The study would provide valuable information for the proper choice of sampling protocol in research and aquaculture practice.

Results and Discussion

Frequent collections of milt from males demonstrated that variations occur in both sperm production and quality (% motility) throughout the period of spermiation. However similar seasonal patterns in these parameters were observed in males irrespective of the sampling frequency.

Despite similarities in seasonal variation, significant differences between the two groups of males were observed in sperm quantity and quality. Clearly, males stripped less frequently had lower total sperm production during the first eight weeks and higher mortalities by week nine (Table 1). Interestingly, this study suggests that the increased handling of males did not decrease sperm quality in this flatfish. Milt obtained with less sampling frequency (biweekly) contained sperm of lower quality, e.g. reduced motility and egg fertilization rate (Table 2) in contrast to previous work showing that increased stripping frequency detrimentally influences sperm quality (motility), as well as milt volume and sperm concentration, in the male turbot (Suquet *et al.* 1992). One difference may be that at low water temperature (0-5°C), winter flounder are quiescent and may be relatively tolerant of handling stress. On the other hand, since spermatogenesis and spermiogenesis occur in this species from October to December, well before the spawning season, sperm may be more sensitive to the aging process. Extended residence in the testis or spermduct might intensify the aging process in sperm cells.

Table 1. Effects of stripping frequency on milt volume, sperm production, spermatocrit and mortality of winter flounder

Group	Milt V (ml/kg BW)	Sp N (10^{10}/kg BW)	SCT (%)	Mort (%)	GSI (%)
Biwk	15.5 ±3.4**	21.1 ±4.8**	67.7 ±8.0	83.3	8.3*
Wk	38.6 ±5.9	73.0 ±13.4	64.6 ±3.1	33.3	3.7

Note: Milt Volume and Sperm Number: total values in 0-8 week; SCT: average of spermatocrit in 0-8 week; Mort: mortality by the week 9; GSI: gonadsomatic index when fish stopped spermiation or died; Data are mean(±sem), *P<0.05, **P<0.01.

Table 2. Effects of stripping frequency on sperm motility, egg fertilization rate and larval hatching rates in winter flounder

Group	Motility (%)	Fert Rate (%)	Hatch Rate (%)
Biwk	20.2 ±4.3**	33.8 ±6.8*	78.4 ± 4.1
Wk	41.9 ± 4.3	53.0 ± 8.6	79.9 ± 7.9

Note: values are mean±sem, sperm motility evaluated by five replicates at biweekly intervals from 0-8 week, fertilization and hatching rates determined twice at week 4 and 6.

Reference

Suquet, et al., (1992) Influence of photoperiod, frequency of stripping and presence of females on sperm output in turbot, *Scophthalmus maximus* (L.). Aquac. & Fish Manag., 23 (2):217-225.

GONADAL DIFFERENTIATION OF BLACK CARP

W.L. Shelton[1] and S. Rothbard[2]

[1]Zoology Department, University of Oklahoma, Norman, Oklahoma 73019 USA
[2]YAFIT Laboratory, Fish Breeding Centre, Gan Shmuel 38810 ISRAEL

Summary

Gonadal differentiation of black carp, Mylopharyngodon piceus, was examined with reference to age and size, to estimate the appropriate schedule for steroid treating juveniles to induce functional sex reversal. Gonads were cytologically undifferentiated throughout 1-1.5 years of life at sizes smaller than 395 g (340 mm TL). Gametogenesis was initiated between 18-24 months of age in fish larger than 210 g (300 mm).

Introduction

Black carp have potential for biological control of nuisance mollusks, which may include combating disease-bearing snails (Ling 1977) or mollusca which interfere with water utilization (Leventer and Teltsch 1990). Development of broodstock to produce monosex populations is a practical management system to test or utilize non-native fishes in an ecologically sensitive manner. The protocol involves steroid-induced sex reversal of genotypic females, and then using these phenotype-altered broodstock to produce monosex populations (Shelton 1986). System development begins by characterizing the pattern of gonadal differentiation to identify the labile period for steroid treatment (Jensen and Shelton 1983), then developing androgen-inducing protocols which are applied to functionally sex reverse genetic females. All-female populations are initially developed through gynogenesis. The sex-reversed males (genetic females) will sire only female progeny which can be stocked or can be used for direct production of next-generation sex-reversed male broodstock.

Methods

Artificial propagation and fry nursing were conducted in Israel. Juveniles were reared in Israel and Egypt in earthen ponds. Gonads were examined from fish between 9 and 26 months of age, and 5 to 1180 g (70-560 mm TL) using the gonadal-squash technique (Guerrero and Shelton 1974).

Results and Discussion

No development of gametocytes was documented during the first 1-1.5 years of age for fish as large as 395 g (340 mm). Initial cytological differentiation of gonads occurred in 18-26 month-old fish as small as 210 g (295 mm). Some slower growing 24-26 month-old fish still had undifferentiated gonads at sizes less than 190 g (285 mm). Androgen-induced sex reversal in black carp should be effective if initiated at ages of 18-24 months between 150 and 250 g (250-300 mm). Extended release methyltestosterone implants (Shelton 1986) are being tested in mixed-sex juvěniles, as well as in presumtive females which were produced by gynogenesis (Rothbard and Shelton 1993).

References

Guerrero, R.D. and W.L. Shelton. 1974. An aceto-carmine squash technique for sexing juvenile fishes. Prog. Fish-Cult. 36:56

Jensen, G.L. and W.L. Shelton. 1983. Gonadal differentiation in relation to sex control of grass carp, Ctenopharyngodon idella (Pices: Cyprinidae). Copeia 1983:749-755.

Leventer, H. and B. Teltsch. 1990. The contribution of silver carp (Hypophthalmichthys molitrix) to the biological control of Netofa reservoirs. Hydrobiologia 191:47-55.

Ling, S.-W. 1977. Aquaculture in Southeast Asia, a historical overview. Univ. Wash. Press, Seattle.

Rothbard, S. and W.L. Shelton. 1993. Gynogenesis in the black carp, Mylopharyngodon piceus. Israeli J. Aquacult. - Bamidgeh 45:82-88

Shelton, W.L. 1986. Broodstock development for monosex production of grass carp. Aquaculture 57:311-319.

Partially funded by USAID PD/A CRSP 263-0152-G-00 and NSF INT 9114675

INDUCED OVULATION OF CHINOOK SALMON USING A GnRHa IMPLANT: EFFECT ON SPAWNING, EGG VIABILITY AND HORMONE LEVELS

I.I. Solar[1], J. Smith[1], H.M. Dye[1], D. D. MacKinlay[2], Y. Zohar[3] and E.M. Donaldson[1]

[1] West Vancouver Laboratory, Department of Fisheries and Oceans, 4160 Marine Dr., W. Vancouver, B.C. Canada, V7V 1N6; [2] Salmonid Enhancement Program, DFO, 555 W. Hastings St., Vancouver, B.C. Canada, V6B 5G3; [3] Center of Marine Biotechnology, University of Maryland, Baltimore, MD 21202, U.S.A.

Introduction

In the present study we compared the effect of similar doses of a GnRH analogue in solution injected intraperitoneally or as an implant inserted into the dorsal muscle of adult maturing female chinook salmon (*Oncorhynchus tshawytscha*). The fish had been caught at their return to a coastal site in British Columbia and transported by truck to the Tenderfoot River Hatchery, a DFO salmonid enhancement facility near Squamish, B.C.

Procedures

An implant (ReproBoost™ AquaPharm Tech. Corp.) containing 250 μg of GnRHa ([D-Ala^6,Pro^9 NEt]-LHRH) was inserted at a shallow angle under the skin into the dorsal muscle of 30 fish (mean weight 7.2 Kg). The same amount of GnRHa in saline solution was injected intraperitoneally into a second group of 30 fish (mean weight 7.9 Kg). A third group (same number, mean weight 7.7 Kg) received an injection of 0.7% saline. The fish were held in raceways with freshwater at 7°C and checked for maturation at 3-day intervals. During the course of the study blood samples were collected from 10 fish from each group for measurement of cortisol, estradiol, 17α-hydroxy,20ß-dihydroxyprogesterone (17,20ß-P), and GnRHa. Random egg samples from 5 fish from each group were incubated to determine the effect of the treatment on viability and survival to hatch.

Results and Conclusions

Within 18 days post-treatment 93% of the implant group, 60% of the injection group and 13% of the control group had spawned. Final cumulative spawning in each group was as follows: 97% in the implant group at day 22, 93% in the injection group at day 35, and 87% in the control group at day 37 (Fig 1). Mean egg viability to hatch of the egg samples from each group was not significantly different. Cortisol levels were consistently high in all three groups during the course of the experiment. Plasma estradiol levels averaged over 100 ng/ml at the time

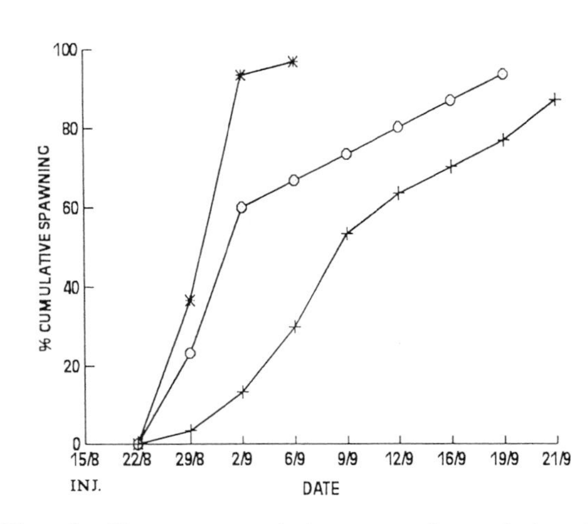

Fig. 1. Percent cumulative spawning of female chinook salmon.

of injection; after 21 days levels in controls had reached an average of 14.7 ng/ml, while in the experimental groups estradiol levels had declined earlier in the experiment, indicating that maturation had been accelerated in the experimental groups. Levels of 17,20ß-P were near the detection limit of the radioimmunoassay at the beginning of the experiment. However 17,20ß-P increased dramatically after 7 days of treatment, while the controls did not exhibit the same increase in the same period of time. This increase in 17,20ß-P also indicates that final maturation had been initiated earlier than in the control group. Measurement of GnRHa levels in plasma showed that while very low levels of GnRHa were present in the injected females at day 7 post treatment, elevated levels of GnRHa persisted in the blood of the implanted females for at least 18 days. We conclude that the administration of the GnRHa in the form of an implant, resulting in a sustained exposure to the peptide over several days, is more effective than a single injection of the same dose of GnRHa in aqueous solution. In this experiment egg viability was not affected by the acceleration of final maturation.

Post-smolt maturation in Atlantic salmon: Timing of decision and effects of ration levels

Sigurd O. Stefansson[1] and Richard L. Saunders

DFO, Biological Station, St. Andrews, N.B., E0G 2X0, Canada, [1]Present address: Department of Fisheries and Marine Biology, University of Bergen, N-5020 Bergen, Norway

Summary

Atlantic salmon (*Salmo salar* L.) post-smolts were maintained in freshwater and fed 100%, 50%, 25% and 0% of recommended rations for 6 weeks. From 22 to 43% of males matured, with no trend among rations, while 16% of females matured in the 25% ration group.

Introduction

If salmon smolts are maintained in fresh water, mechanisms for hyper-osmoregulation will be re-established. Salmonid reproduction is restricted to freshwater, hence smolting and maturation are regarded as mutually exclusive processes (Thorpe, 1987). Salmonid maturation is dependent on energy status during a critical decision period (late winter/spring), i.e. concurrent with the final stages of smolting. The present study focuses on effects of ration level during early post-smolt stages for re-adaptation to freshwater and early maturation.

Experimental design

Atlantic salmon 1+ smolts were maintained in freshwater and fed four ration levels from 22 May: 100%, 50%, 25% and 0%, according to temperature and fish size. From 9 July onwards all groups were fed full ration. Maturation status was determined on 1 October.

Results

Ration level significantly influenced growth rate and condition factor (Table 1, Fig.1), and loss of gill Na^+,K^+-ATPase activity (Fig.2). When fed in excess, groups from restricted rations grew at a higher rate than previously full fed fish (Table 1). Condition factor increased in late summer, with the most rapid increase in the restricted ration groups (Fig.1). From 22% to 43% of males matured as post-smolts, with no trend among previous ration levels. Female maturation was found only in the 25% ration group (Table 1).

Table 1. Mean length (sem) at initiation (22 May) and termination (9 July) of different ration levels. Growth rate (mm/day), final mean length (1 October) and % maturation of Atlantic salmon fed different ration levels (MM/MF - Mature males/females, n indicates number of mature fish).

	Ration level			
Date	0%	25%	50%	100%
22 May	21.4 (.18)	21.1 (.20)	21.3 (.21)	21.2 (.18)
9 July	22.3 (.28)	22.8 (.34)	23.8 (.41)	25.8 (.41)
1 October	31.8 (.45)	32.8 (.47)	32.7 (.46)	34.3 (.53)
Growth rate	1.14	1.20	1.07	1.02
MM/MF	43%/0%	22%/16%	42%/0%	30%/0%
(n)	3/0	2/3	5/0	3/0

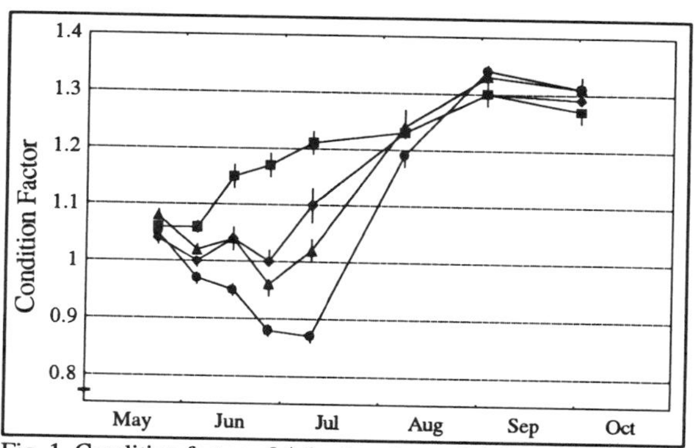

Fig. 1. Condition factor of Atlantic salmon fed different ration levels. Error bars indicate sem. ●=0%, △=25%, ◆=50%, ■=100%.

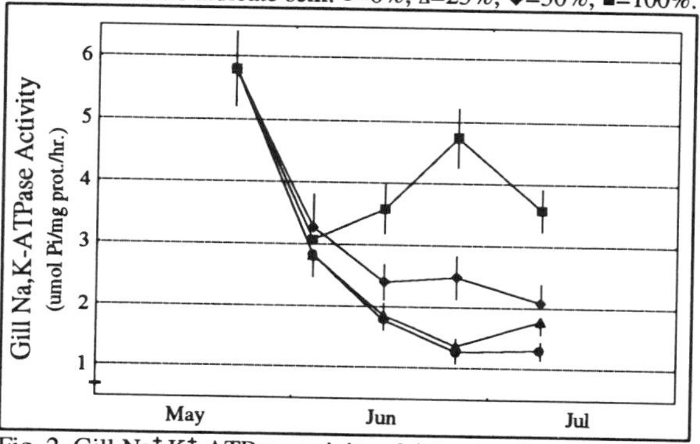

Fig. 2. Gill Na^+,K^+-ATPase activity of Atlantic salmon fed different ration levels. Error bars indicate sem. Symbols as in Fig. 1.

Discussion

Low gill Na^+,K^+-ATP-ase activity suggests that groups on reduced ration quickly lost the ability for ion excretion, with re-establishment of osmoregulation characteristic of freshwater. Maturation was high in male post-smolts, and even some females matured. The results, which are in contrast with current views on the relationship between smolting and maturation (Hoar, 1988), suggest that the decision to commence maturation may be made during the completion of smolting. Gonadal development then proceeds soon after the completion of smolting. Alternatively, given favourable growing conditions, the decision may be taken during late spring or early summer, later than previously believed.

References

Hoar, W. S., 1988. The physiology of smolting salmonids. *In:* Hoar, W. S. and D. J. Randall [eds.]: *Fish physiology, Vol. XIB*. Academic Press, New York, NY. Pp. 275-343.

Thorpe, J.E., 1987. Smolting versus residency: Developmental conflict in salmonids. American Fisheries Society Symposium 1: 244-252.

Cryopreservation of Rainbow Trout (*Oncorhynchus mykiss*) Semen in Straws

H. Steinberg, A. Hedder, R. Baulain, W. Holtz
Institute of Animal Husbandry and Genetics, University of Göttingen, Albrecht-Thaer-Weg 3, 37075 Göttingen, Germany

Summary

Pooled trout semen was frozen in 5ml-straws after being extended with a sucrose/DMSO extender. Maximum fertilization rates of 83.8 ± 2.8% (SEM) of the controls were achieved when fertilizing batches of 2000 eggs with 1.5ml semen : 2.5ml extender (0.6M sucrose with 20% DMSO). Freezing occurred in N_2-vapor at -180°C and thawing at +25°C for 35s.

Introduction

Rainbow trout semen is commonly cryopreserved in pelleted form. The technique is tedious and only suitable for small batches of eggs. The present investigation follows up attempts by WHEELER and THORGAARD (1991) to freeze larger volumes in straws.

Materials and Methods

Five cryopreservation experiments were conducted. Varying relationships of semen (pooled from > 5 males) and 0.6M sucrose solution (HOLTZ, 1993) with either 10 or 20% DMSO (Table 1) were compared. A volume of 4ml was drawn into 5-ml straws immediately after extension and frozen in crushed dry ice at -79°C (Exp. 1) or nitrogen vapor at -180°C. Straws were stored in LN_2 (-196°C). In Exp. 1, 2, 4 and 5 straws were thawed at +25°C/35sec, in Exp. 3 at +40°C/15sec. The partially thawed extended semen was added to 20ml of 0.119M $NaHCO_3$, swirled and poured onto egg batches of 1800-2000 eggs. Fertilized eggs were incubated 100 d° and the % fertilization was assessed from a random sample of 500 eggs. As a control, batches of 100 eggs were fertilized with semen frozen as pellets (HOLTZ, 1993) at a similiar sperm-egg-ratio.

Results

Mean fertilization rates of the experimental groups were not significantly different ($p > 0.05$). However, it appeared that 1.5ml semen diluted with 2.5ml of a 0.6M sucrose solution containing 20% DMSO, frozen at -180°C and thawed at 25°C/35s gave the best fertilization rates of 83.8 ± 2.6 % of pellet-frozen semen (Exp. 5).

Discussion

Cryopreservation of rainbow trout semen as 0.1ml-pellets gives good fertilization results with small batches of eggs (HOLTZ, 1993). In this investigation pellet-frozen semen gave an average fertilization rate of 82% which is close to results obtained with fresh semen. The pellet-technique therefore, served as a control for Experiments 1 to 5, which were conducted with the purpose to establish a suitable method for the freezing of larger volumes of semen. The technique devised permits fertilization of all the eggs of one stripping (about 2000) with the semen contained in one straw. The most likely reason for the good fertilization rate, as compared to the results of WHEELER and THORGAARD (1991), is the high freezing rate. The straws we used were thinner (4.8 mm vs. 5.4mm) and the freezing temperature was lower (-180 vs. -79°C). Freezing in N_2-vapor turned out to be much more practical than in crushed dry ice. Rapid freezing brought about by dropping straws directly into LN_2, killed the spermatozoa. Though differences among groups were just beyond the level of significance, the groups with a closer relationship of semen to extender (Exp. 4 and 5) gave better results than the others. More research along those lines seems indicated, although the present results justify application in the field.

References

HOLTZ, W., 1993, Aquaculture 110, 97-100

WHEELER, P.A. and G.H. THORGAARD, 1991, Aquaculture, 93, 95-100

Table 1:
Fertilization rates (% of pellet-frozen controls, 81.7 ± 0.9% fertilization, n = 130) of trout semen cryopreserved in a 0.6M sucrose extender with different concentrations of DMSO, and different freezing and thawing temperatures

Experiment No.	Extension rate Semen:Extender (ml)	DMSO-conc. (%)	Freezing temperature (°C)	Thawing temp. and time (°C/s)	No. of Replicates	Fertilized eggs $\bar{x}$	SEM
1	1 : 3	10	-79	25/35	26	**68.3**	**4.5**
2	1 : 3	10	-180	25/35	30	**70.7**	**2.1**
3	1 : 3	10	-180	40/15	26	**68.6**	**3.6**
4	1.5 : 2.5	10	-180	25/35	23	**81.0**	**3.7**
5	1.5 : 2.5	20	-180	25/35	25	**83.8**	**2.6**

($p > 0.05$, Scheffé-test)

CRYOPRESERVATION OF FISH SPERM: LABORATORY, HATCHERY AND FIELD STUDIES OF TWENTY SPECIES

Terrence R. Tiersch

Forestry, Wildlife, and Fisheries, Louisiana Agricultural Experiment Station, Louisiana State University Agricultural Center, Baton Rouge, LA 70803

Summary

We have developed an integrated system for rapid identification of conditions useful for refrigerated storage and cryopreservation of fish sperm. Best results have been obtained with use of standard straws (0.25 to 5.0-ml), permeating cryoprotectants such as dimethyl sulfoxide or methanol (without egg yolk), rapid freezing (-45°C/min) in nitrogen vapor, and rapid thawing (~20°C/sec) in a water bath. Shipping dewars have proven invaluable for cryopreservation studies in the field.

Introduction

During the past 4 years we have studied sperm of 20 species representing 9 families of teleost fishes. This has involved development of techniques for use with previously unstudied species in a variety of settings, including wilderness areas. The objectives of this report are to overview our procedures, to indicate typical results, and to provide examples of applications for cryopreserved fish sperm.

Procedures

Contamination of sperm with water, urine or feces is avoided during collection. The intensity and duration of motility is estimated by dilution of sperm with water, and morphology is evaluated. Samples are mixed at different dilutions in extender solutions, and at various osmotic pressures to optimize refrigerated storage, activation and fertilization. Sperm are exposed to cryoprotectants in concentrations of 5% to 15%. Acute toxicity (~1 hr) is evaluated to choose the concentration used for cryoprotection. In the field, vapor shipping dewars are used to freeze samples. All equipment necessary for fieldwork is contained in 3 cases (2 shipping dewars and a waterproof case with portable microscope, temperature recorder, pipettors, etc.). We use a computer-controlled freezer in the laboratory, and evaluate 2 freezing rates for new species (-45°C/min and -4°C/min). When the samples reach -80°C, they are plunged into liquid nitrogen for long-term storage. We use rapid thawing (~20°C/sec) to minimize ice crystal formation. Because the relationship of sperm motility, viability, and fertilization ability is unresolved, fertilization is assessed whenever possible. We use zipper-lock plastic bags to provide a portable "micro-hatchery" for field use.

Results

We have performed hundreds of studies addressing various aspects of refrigerated and frozen storage. Hanks' balanced salt solution (HBSS) and calcium-free HBSS (C-F H) have been useful extenders for sperm of many species when osmotic pressure was optimized (Fig. 1). Methanol and dimethyl sulfoxide have proven most satisfactory as cryoprotectants, often yielding post-thaw motility values of >40%.

Discussion

Many applications for cryopreserved sperm have been identified. Fisheries can be augmented and monitored by use of genetic markers and cryopreserved sperm. Cryopreserved sperm can be used for selective breeding in ornamental fish (such as *Cyprinus carpio*) to improve coloration or markings. The advantages of cryopreservation are available for application in commercial aquaculture. Sperm cryopreservation has become increasingly important for endangered species. We have assisted in hatchery production and germplasm storage of *Pangasius gigas* the giant Mekong catfish (reaching sizes of >400 kg) which is considered to be endangered in Thailand. We have developed laboratory, hatchery and field techniques for use with the endangered razorback sucker *Xyrauchen texanus*, Colorado squawfish *Ptychocheilus lucius*, bonytail chub *Gila elegans* and humpback chub *Gila cypha*.

Family	Species studied	N	Extender (mOsm/kg)	Storage (days)	Cryoprotectant	Post-thaw motility
Ictaluridae	5	>350	HBSS (300)	15	5% methanol	40%
Pangasiidae	2	12	C-F H (300)	7	10% dimethyl sulfoxide	15%
Cyprinidae	4	108	C-F H (300)	7	10% dimethyl sulfoxide	45%
Catastomidae	1	70	C-F H (300)	10	10% methanol	15%
Mugilidae	1	1	HBSS (200)	2	10% dimethyl sulfoxide	50%
Percichthydae	2	15	HBSS (300)	1	10% glycerol	<1%
Sparidae	1	5	HBSS (203)	10	10% dimethyl sulfoxide	40%
Sciaenidae	3	43	HBSS (300)	6	10% dimethyl sulfoxide	45%
Cichlidae	1	45	HBSS (310)	6	15% dimethyl acetamide	20%

Fig. 1. Summary of studies on refrigerated and frozen storage of sperm of twenty species of fish.

Supported in part by funding from the National Sea Grant College Program, U. S. D. A., U. S. Fish and Wildlife Service, U. S. Bureau of Reclamation, U. S. A. I. D. and Louisiana Catfish Promotion and Research Board.

EFFECT OF CATIONS, pH AND OSMOLALITY ON SPERM MOTILITY OF MALE WHITE CROAKER, MICROPOGONIAS FURNIERI

D. Vizziano[1], J.R. García Alonso[1] and D. Carnevia[2]
[1]Facultad de Ciencias. T. Narvaja 1674. Montevideo, Uruguay; [2] Facultad de Veterinaria, T. Basañez 1160. Montevideo, Uruguay.

Summary

Factors affecting the sperm motility of eurhyhaline *M. furnieri* were studied *in vitro*. High levels of NaCl (250 mM) associated with a basic pH (8.7) induced the optimum sperm motility.

Introduction

The aim of the present study was to establish the optimum conditions for artificial insemination of *M. furnieri*. This euryhaline species reproduces in the estuary of the Río de la Plata (Uruguay). In freshwater and marine fishes sperm are usually activated by hypo and hyperosmotic media, respectively. While optimum media for sperm activation are known for salmonids (Maisse, 1990), in marine fishes the role of different cations and pH is less studied.

Materials and methods

Three spermiating males of *M. furnieri* were injected with hCG (750 IU kg-1 b.w.). Following 24 h they were sacrified and the testes were removed. The sperm was collected from the vas deferens and maintained in the cold (near 0ºC). In this condition it remained mobile for 4 to 5 h. The motility was evaluated using a scale of 0 to 5 according to Sánchez-Rodríguez and Billard (1977) and taking into account the time of activation. One drop of sperm was diluted in solutions of buffer Tris 0.02 M: a) including NaCl (31.2 to 500 mM), KCl (200 mM), $CaCl_2$ (30 to 240 mM) or $MgCl_2$ (5 to 40 mM) at pH 8.9; b) mixed with buffer Tris-HCl 0.02 M, resulting in pH varying between 5.2 and 9.9; c) at pH 8.9 with NaCl (osmotic pressure in mOsm kg^{-1}, OP = 82 to 817), KCl (OP = 57), $CaCl_2$ (OP = 80), $MgCl_2$ (OP = 25); d) a sodium free medium at pH 9 including varying concentrations of a salt mixture: $CaCl_2$, $MgCl_2 x6H_2O$, KCl, $MgSO_4 x7H_2O$, $NaHCO_3$ (OP = 560, 832, 1700).

Results

Among the cations tested, only the Na+ activated the spermatozoa (spz) of white croaker. High molarities of NaCl (250 and 500 mM) immediately induced the maximal response in 100 % of spz. Compared with the higher concentration, at 250 mM NaCl, the sperm activation lasted longer. When lower concentrations of NaCl were used (175 mM to 31.2 mM), the sperm was not activated immediatley and the level of activation fell (4 to 0) while the period of activation increased (7 to 19 min). Changes in pH (5.3 to 8.9) did not activate spz. However, when NaCl (250 mM) was added, the optimum sperm activation was reached at pH 8.7. Lower pH (8.1 to 5.5) decreased but not inhibited the sperm motility. Hyperosmotic solutions (OP = 560, 832, 1690) containing a mixture of different sea salts without NaCl activated spz, but optimum motility was reached when NaCl (OP = 480, 817) was used. Solutions containing Na+ and lower OP (254, 168, 82) but neither those with K+ (OP = 57), Ca++ (OP = 80) nor Mg++ (OP = 25) induced the sperm activation.

Conclusions

Our results show that sperm activation in white croaker can be induced by a hyperosmotic medium, like in other marine fish. The best condition to activate spz of this species is the combination of elevated concentrations of NaCl (250 mM) and basic pH (8.7). However, hyposmotic media containing NaCl and acid (5.2) pH also induced the sperm motility.

References

Sánchez-Rodríguez, M. and R. Billard, 1977. Conservation de la motilité et du pouvoir fécondant du sperme de la truite arc-en-ciel maintenu à des températures voisines de 0ºC. Bull. Fr. Piscic. 265:144-152.

Maisse, G., 1990. Le sperme des salmonidés: le point sur les connaissances. Applications à la salmoniculture. INRA Prod. Anim. 3:223-228.

Cryopreservation, motility and ultrastructure of sperm from the ocean pout (Macrozoarces americanus L.), an internally fertilizing marine teleost.

Z. Yao[1*], L.W. Crim[1], G.F. Richardson[2] and C.J. Emerson[1]

1.Ocean Science Centre and Dept.of Biology, Memorial University of Newfoundland, St.John's, NF, Canada A1C 5S7. 2.Dept.of Health Management, Atlantic Veterinary College, University of Prince Edward Island, Charlottetown, PEI, Canada C1A 4P3

Summary

The male ocean pout produces motile mammalian-like sperm in a thin-watery milt. Cryopreservation of the sperm requires extension of the sperm in a diluent of similar ionic composition to the seminal plasma and a high level of cryoprotectant, e.g., 20% dimethyl sulfoxide. Successful cryopreservation of ocean pout sperm also depends on the freezing-thawing rates and the milt dilution ratio. Cryopreservation caused sperm ultrastructural changes including the shrinkage of cell membrane leading to the exposure of mitochondria and the death of sperm.

Introduction

The ocean pout is a candidate for cold-water aquaculture. However, it uses internal fertilization which makes artificial propogation of the fish difficult (Crim et al., 1995). The present study describes sperm cryopreservation of this species and the use of the frozen sperm for artificial insemination of eggs.

Results and discussion

Semen sampled from the spermduct of mature males contained 50-60% motile sperm. Dilution of the semen 1:3 in diluent C formulated according to the ionic composition of the ocean pout seminal plasma (Yao and Crim, 1995) improved the sperm motility (%), which however dropped to 0% after freezing in liquid N_2 (Fig.1). This indicates that cryopreservation of the ocean pout sperm requires cryoprotectant in the diluent.

Addition of dimethyl sulfoxide (DMSO) to the diluent C improved the post-thaw sperm motility and 20% DMSO produced the highest motility (25%) although it was also affected by freezing rate (Fig.2).

Semen dilution and thawing rate also affected the motility of post-thaw sperm (Fig.3). Increasing dilution ratio or increasing the thawing rate both reduced the motility of post-thawed sperm.

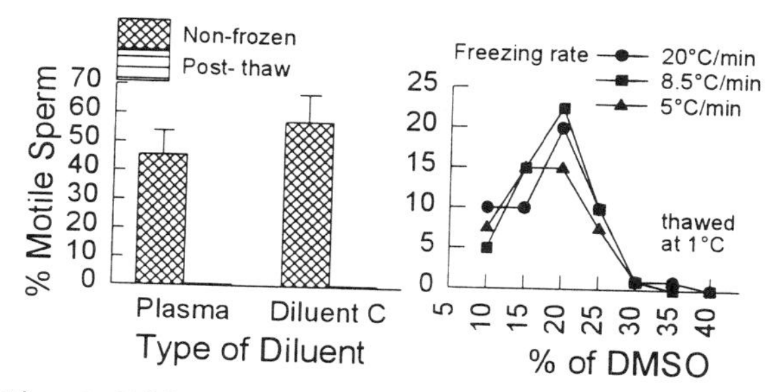

Fig.1 Effect of diluent C on sperm motility before freezing (left).
Fig.2 Effect of freezing rate & DMSO on post-thaw sperm motility (right)

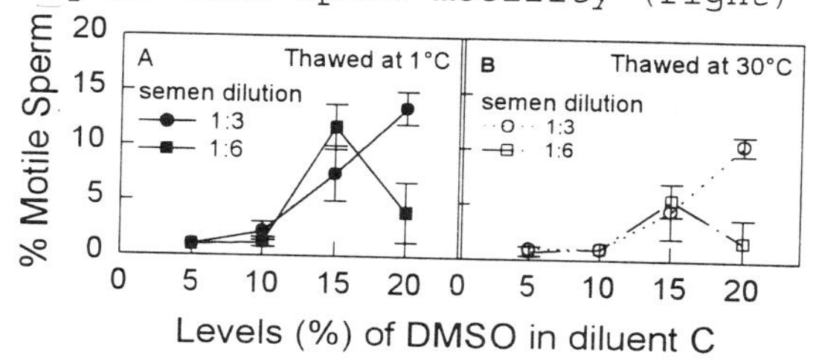

Fig.3 Effect of semen dilution ratio, thawing rate and % DMSO on post-thaw sperm motility.

The decrease of sperm motility after freeze and thaw might result from the ultrastructural changes of cell membrane at the midpiece leading to the exposure of mitochondria. In vitro fertilization of eggs using post-thaw sperm yielded up to 32% fertilization rates. This suggests that post-thaw sperm were viable and may be used for artificial fertilization.

References

Crim, L.W., Yao, Z. and Wang, Z., 1995. Reproductive mechanism for the ocean pout (Macrozoarces americanus L.), a marine temperate fish, including internal fertilization and parental care of eggs. Reported elsewhere in this Proceedings

Yao, Z.and Crim, L.W., 1995. Spawning of the ocean pout (Macrozoarces americanus):Evidence in favour of internal fertilization of eggs. Aquaculture 130:361-372.

SUSTAINED ADMINISTRATION OF GnRHa INCREASES SPERM VOLUME WITHOUT ALTERING SPERM COUNTS IN THE SEA BASS (*Dicentrarchus labrax*)

L.A. Sorbera, C.C. Mylonas*, S. Zanuy, M. Carrillo and Y. Zohar*

Instituto de Acuicultura (CSIC), Torre de la Sal, 12595 Castellón, Spain; * Center of Marine Biotechnology, University of Maryland, Baltimore, MD 21202, USA

Summary

Spermiation was stimulated in the European sea bass (*Dicentrarchus labrax*) by administration of a GnRH analog using different delivery systems. Results revealed that slow, sustained delivery of GnRHa significantly increased the total volume of sperm produced without altering counts or motility of sperm.

Introduction

Although GnRH regulation in female reproduction has been widely investigated, the role of GnRH and its analogs in spermiogenesis is not well understood. Previous studies have shown that a mammalian GnRH analog ([D-Ala^6,Pro^{9}NEt]-mGnRH;GnRHa) can induce spawning in preovulatory, mature sea bream (*Sparus aurata*) by stimulating gonadotropin activity (Zohar *et al.*, 1989). Here, the ability of GnRHa to stimulate spermiation was investigated in the sea bass (*Dicentrarchus labrax*). Several different GnRHa delivery systems were implemented and total expressible sperm over a 44 day period was analyzed for volume, motility, and counts.

Methods and Results

Male sea bass (≥5 yr), maintained in natural photoperiod and fed a commercial diet, were separated into 5 groups (*n*=8). Groups were treated during the reproductive season as follows: (**C**) injected with dH_20, 250 µl/kg; (**I**) injected with GnRHa in dH_20, 25 µg/kg; (**M**) injected with biodegradable GnRHa microspheres, 50 µg/kg; (**EVSL**) implanted with slow releasing or fast releasing (**EVAc**) polymeric GnRHa implants, 100 µg/fish. All injections and implants were given *i.m.* on Day 0 of the experimental period. Total expressible sperm was collected at approximate 7 day intervals beginning at Day -2. Expressible sperm volume was noted in ml and the density (spermatozoa/ml) as well as the motility of spermatozoa were microscopically assessed. Analysis of variance followed by Tukey's HSD multiple range test were used to compare groups. Differences were accepted when $P<0.05$. Body weight had no significant effect on the amount of sperm collected. Fig. 1 shows the amount of expressible sperm obtained for each treatment group over the 44 day period. All GnRHa treatments induced a significant increase in collected volume over controls. However, while sperm volumes in group I returned to control levels by Day 14, EVAc exhibited prolonged spermiation until Day 28 and both M and EVSL groups

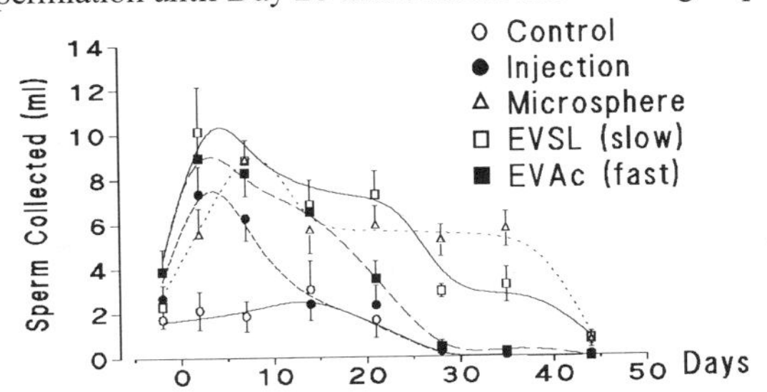

Figure 1. Total expressible sperm in sea bass after treatment with GnRHa via different delivery systems.

maintained high levels of spermiation through Day 35, well past the end of the natural spawning period. Sperm motility was found to be consistently good to excellent (75-100% vigorously motile) in all groups even when spermiation was prolonged. There were no significant differences in sperm counts between groups (mean spermatozoa density for all groups was $5.5 \times 10^9 \pm 1.2 \times 10^8$ spermatozoa /ml).

Discussion

All modes of GnRHa delivery resulted in stimulation of spermiation in the sea bass, possibly indicating induction of gonadotropin activity. The slow delivery systems of GnRHa (M and EVSL) were capable of augmenting the quantity of sperm produced as well as prolonging the period of spermiation past the end of the natural spawning season. Although GnRHa increased spermiation, it did not affect the quality or concentration of sperm suggesting that spermiogenesis, although super-activated, proceeds in a physiological manner. Further studies are necessary to determine the mechanism by which GnRHa stimulates, augments, and prolongs spermiation in the sea bass.

References

Zohar, Y., Goren, A., Tosky, M., Pagelson, G., Leibovitz, D. and Koch Y. 1989. The bioactivity of gonadotropin-releasing hormones and its regulation in the gilthead seabream, *Sparus aurata*: *in vivo* and *in vitro* studies. Fish Physiol. Biochem.7:59-67.

Supported by NATO CRG 940889 and a postdoctoral fellowship from the Spanish government to L.A.S.

APPLICATION OF SEX PHEROMONES TO ENHANCE FERTILITY IN MALE CYPRINIDS: STUDIES IN GOLDFISH (*Carassius auratus*) AND COMMON CARP (*Cyprinus carpio*)

W. Zheng, J.R. Cardwell, N. E. Stacey, and C. Strobeck.

Department of Biological Sciences, University of Alberta, Edmonton, T6G 2E9, Canada.

Summary

In common carp, the stimulatory pheromonal effect of 17α, 20ß-dihydroxy-4-pregnen-3-one (17,20ß-P) on milt volume does not occur if fish are exposed and stripped at frequent intervals (3 - 7 days). In goldfish, 17,20ß-P exposure increases fertilization rate at the onset of spawning.

Introduction

17,20ß-P is a potent pheromonal stimulator of milt volume in goldfish (Sorensen *et al.*, 1990) and common carp (Stacey *et al.*, 1994), and can be detected by other cultured carps such as *Mylopharyngodon piceus*, *Ctenopharyngodon idella* and *Aristichthys nobilis* (Stacey, unpub.). In these studies, we examined the effects of repetitive 17,20ß-P exposure on milt production in common carp, and 17,20ß-P exposure on goldfish fertility, aspects relevant to aquaculture application.

Methods

In carp, size-matched precocious males (35-120 g) were held in groups of 4 in 70 l flow-through aquaria at 20 °C and stripped of milt after overnight exposure to 17,20ß-P ($5x10^{-10}$M) or ethanol control, given at 3-, 7- or 30-day intervals. In goldfish, ovulated females were allowed to perform two spawning acts with a male from one treatment group (either 17,20ß-P-exposed or control), and then two acts with a male from the other treatment group. Eggs developing to the eyed stage were used to calculate fertility.

Results and Discussion

In carp, on the first day of stripping, controls in the three exposure-schedule groups had similar milt volumes , and 17,20ß-P carp in the three groups exhibited similar significant increases in milt volume (Fig. 1). Although on subsequent strippings milt volumes of 17,20ß-P-exposed fish remained numerically larger than controls, a significant effect was observed only early in the schedule of the 3-day group, and at the second sample of the 30-day group. The cumulative milt volumes of 17,20ß-P-exposed fish were similar in the 3 groups, but significantly greater than controls only in the 30-day group. Results indicated that repeated exposures and/or strippings reduce milt response to 17,20ß-P since the 30-day group under similar holding conditions markedly increased milt volume when exposed to 17,20ß-P. These data suggest that application of pheromones in aquaculture may be more effective in cases of natural spawning, which do not require frequent handling.

In goldfish, males exposed to 17,20ß-P achieved significantly higher fertilization rate than the controls (Fig.1). This is consistent with our recent studies using microsatellite DNA fingerprinting in which 17,20ß-P exposure dramatically increases fertility during competitive spawning. These goldfish data suggest that pheromonal 17,20ß-P can be applied to improve male fertility in cultured cyprinids.

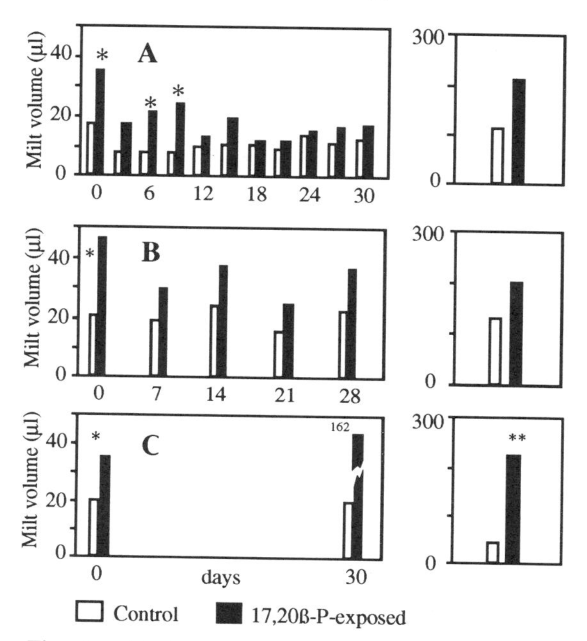

Fig. 1. Median milt volume (left panels) and median cumulative milt volume (right panels) of carp stripped every 3 days (A), every 7 days (B) or every 30 days (C). N=8 - 16 for each group. *p<0.05; **p<0.01 (U test).

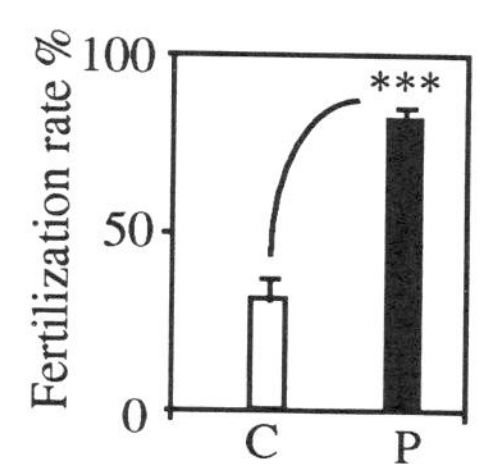

Fig. 2. Fertilization rates (mean+SEM) of control (C) and 17,20ß–P exposed male goldfish (P) at the initiation of natural spawning. N=6. *** p<0.001 (t test)

Supported by NSERC grant to N.E.S.

References

Sorensen, P.W., T.J. Hara, N.E. Stacey, and J.G. Dulka, 1990. Extreme olfactory specificity of male goldfish to the preovulatory steroidal pheromone 17α, 20ß-dihydroxy-4-pregnen-3-one. J. Comp. Physiol. A, 166: 373-383.

Stacey, N.E., W. Zheng, and J.R. Cardwell, 1994. Milt production in common carp (*Cyprinus carpio*): stimulation by a goldfish steroid pheromone. Aquaculture, 127: 265-276.

Environmental Influences on Reproduction

TEMPERATURE MANIPULATION OF SEX DIFFERENTIATION IN FISH

C.A. Strüssmann[1] and R. Patiño[2]

[1]Department of Aquatic Biosciences, Tokyo University of Fisheries, Konan 4-5-7, Minato, Tokyo 108, Japan;
[2]Texas Cooperative Fish & Wildlife Research Unit, Texas Tech University, Lubbock, TX 79409, USA.

Summary

This paper reviews the effects of the thermal environment on the primary sex differentiation of fishes, with special regard to the evidence of thermolabile sex-determination (TSD). TSD is common among Atherinids, and temperature effects on gonadogenesis in this group range from functional sex change to virtual sterilization. Within this group, exposure to extreme (sub-lethal) temperatures seems to allow sex manipulation even in species with otherwise strong genetic sex determination (GSD). In other taxa, including several economically important species, there are instances in which (phenotypical) gonadal sex appears at least partly modulated by environmental temperature. Practical considerations for the demonstration of TSD and its prospects for the control of fish sex in aquaculture are discussed.

Introduction

Environmental factors define the reproductive strategies and the output of fish reproductive activity by setting the timing and the amount of energy available for reproduction (Lam, 1983). In a number of species, environmental factors also influence the relative contribution of each sex to reproduction by temporally or permanently inducing functional sex-change (environmental sex determination, ESD). The occurrence of the various forms of ESD among fishes (temperature-, behavioral-, salinity-, light-, water quality-, pH-, or nutrition- dependent sex determination) has been dealt with in a number of reviews (Chan and Yeung, 1983; Adkins-Reagan, 1987; Korpelainen, 1990; Shapiro, 1990; Francis, 1992). Thermolabile sex determination (TSD) seems to be the prevalent form of ESD among fishes and other animals in general (Korpelainen, 1990). Although most authors agree that TSD is likely widespread among teleosts (Conover, 1984, 1992; Schultz, 1993; Adkins-Reagan, 1987; Korpelainen, 1990; Francis, 1992), reports of TSD are still scarce (Schultz, 1993). One reason is that demonstration of TSD in fish imposes practical difficulties such as poor survival (Schultz, 1993; see also below). Another reason may be that TSD is generally viewed in an ecological or evolutionary context, with much emphasis being placed in demonstrating its adaptive significance. Only recently have scientists realized the broader implications and potential of TSD for aquaculture and hence studies in this area are gradually being undertaken. This review deals with TSD and other effects of temperature on the gonadogenesis of gonochoristic fishes, particularly in atherinids where most studies have been conducted. We also discuss the practical use of TSD in aquaculture.

Effects of temperature on sex differentiation of atherinids

Perhaps the most thoroughly studied and best known example of TSD in fishes occurs in one member of the family Atherinidae, the Atlantic silverside, *Menidia menidia*. Individuals of this species become male or female in response to the temperature regime experienced during a specific period in early life (Conover and Kynard, 1981; Conover and Fleisher, 1986). Presumably, TSD confers an adaptive advantage for *M. menidia* by the creation of sex-linked size dimorphism. Females are born earlier than males as a result of incubation under the low temperatures prevailing in the beginning of the spawning season and thus have a longer growing season, becoming larger than males at the time of sexual maturation. Size has a relatively greater effect on the fecundity of females than that of males (Conover, 1984, 1992). Interestingly, it was shown that environmental sex- determination is itself under genetic control (Conover and Kynard, 1981; Conover and Heins, 1987a). Also, the degree of environmental versus genetic sex-determination and the magnitude of the response to temperature varied among local populations of *M. menidia*, reflecting an adaptive compensation to local differences in the thermal environment (Conover, 1992; Conover and Heins, 1987b). The gonadal sex of Northern populations of *M. menidia*, for instance, was completely unresponsive to temperature (Fig. 1).

TSD has been subsequently demonstrated in three other atherinids, the tidewater silverside, *M. peninsulae* (Middaugh and Hemmer, 1987), and two South American silversides from mean latitudes, *Odontesthes bonariensis* (Strüssmann et al., 1995a), a freshwater and estuarine species, and *O. argentinensis* (Strüssmann et al., unpublished data), a marine species. A fifth species, *Patagonina hatcheri*, an inhabitant of the cold freshwaters of Patagonia, initially appeared unresponsive to temperature and was therefore considered strongly genetically sex determined (Strüssmann et al., unpublished data). However, recent studies have succeeded in manipulating the sex of *P. hatcheri* by use of even lower temperatures (Saitoh and Strüssmann, unpublished data). The adaptive significance of TSD for these species, if any, remains unclear (Strüssmann et al., 1995a) or conjectural at best (Middaugh and Hemmer,

1987).

In all these species, temperature effectively modifies gonadal sex only when applied at a specific ontogenetic stage. This stage has been determined to be around metamorphosis and it culminates with the period of histological sex differentiation of the gonads (Conover and Fleisher, 1986; Strüssmann et al., 1995a, b). All atherinid species examined seem to be gonochorists and intersexes are rare. Thus, temperature treatments before or after the phenocritical stage of sex determination do not alter sex.

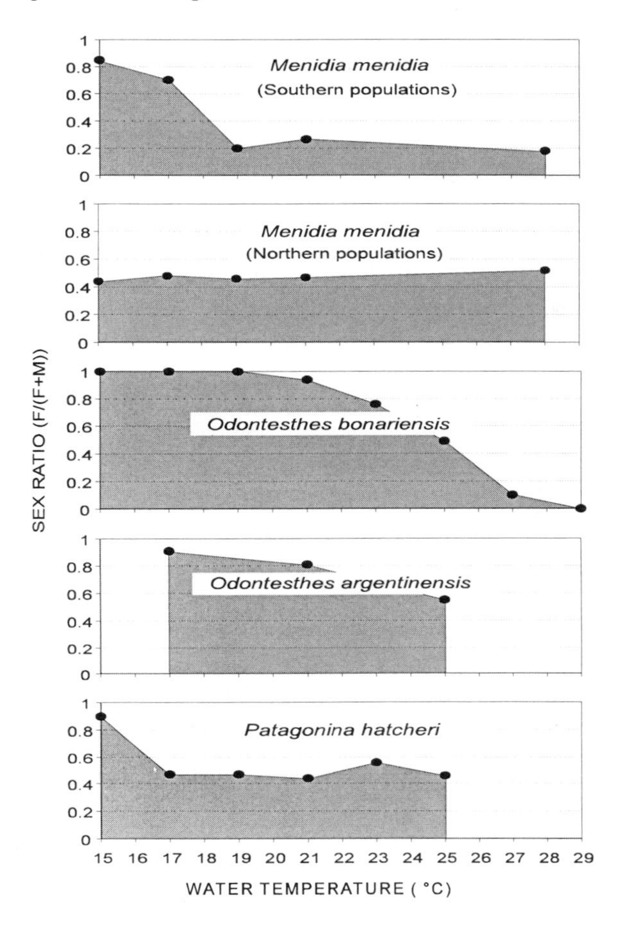

Fig. 1. Relation of water temperature during the period of sex determination and sex-ratio in some atherinid species. Means from data of Conover and Heins (1987b; *M. menidia*) and Strüssmann et al. (1995a, unpublished data).

The direction of the response to temperature is markedly constant among atherinids: higher temperatures favor the formation of males whereas lower temperatures favor the formation of females (Conover and Kynard, 1981; Middaugh and Hemmer, 1987; Strüssmann et al., 1995a; Fig. 1). Transition from female-producing to male-producing temperatures appears monotonic and gradual (Conover and Heins, 1987a; Fig. 1), in contrast to the steep threshold or sometimes "U-shaped" response of some reptiles (Korpelainen, 1990). In *Patagonina hatcheri*, however, it appears that threshold temperatures occur at the lower range of viable temperatures. In this species, sex-ratios are remarkably invariable from 1:1 between 25 and 17°C but nearly 90% females can be produced at 15°C (Strüssmann et al., unpublished data).

Effects of temperature on sex differentiation in other species

Evidence suggesting environmentally-related modifications of phenotypical sex among other fish groups are available from a number of sources. Circumstantial evidence can be drawn from some studies aimed at the detection of sex chromosomes as well as from those attempting the production of monosex fish by mating of sex-reversed parents or by gynogenesis induction. In short, the majority of these studies have produced sex ratios which conformed most of the time to simple genetic sex inheritance systems (XX-XY or ZW-ZZ) but occasionally with a few inexplicable exceptions (*Betta splendens*, Lowe and Larkin, 1975; *Brachydanio rerio*, Streisinger et al., 1981; *Carassius auratus*, Oshiro, 1987; *Gnathopogon caerulescens*, Fujioka, 1993; *Carassius carassius grandoculis*, Fujioka et al., 1995). Some of these studies even suggested a possible link between water temperature and variable sex ratios. For instance, Winge (1934) and Aida (1936) noted the formation of higher proportions of males in *Poecilia reticulata* and *Oryzias latipes*, respectively, during the warmer months. It is unclear whether all these exceptional cases are evidence of thermal (or other form of environmental) sex reversal, but some were subsequently shown to be (see below).

More direct evidence of TSD has since being produced by direct comparison of the sex ratios of broods reared under controlled thermal conditions. Although limited in scope, Table 1 shows that trials with either normal or putative monosex progeny of various species sometimes produced significantly different sex ratios at different temperatures. It must be noted that several of these species have been previously assumed to possess stable genetic sex determining mechanisms. In fact, the mostly biased sex ratios of putative monosex (Table 1) seem to support a strong role for genotypic sex determination (GSD) in these species. Even in such species, however, temperature appears able to override genetic sex determination. It is not known whether all the unsuccessful trials reflect parental genotypes that are unresponsive to temperature, such as in the case of *Poeciliopsis lucida* (Sullivan and Schultz, 1986; Schultz, 1993), or simply those with different thermal thresholds for TSD. Low temperatures favor the production of females in some species, including all atherinids previously described, whereas in others the reverse is the case (Table 1).

Table 1 Effects of rearing temperature on sex ratios of fishes other than atherinids.

Species	Successful trials[*1]	% females at low	mean temperatures[*2]	high	Effective temp. to produce[*3] females	males	Type of brood used[*4]	Reference
Gasterosteus aculeatus	1/1	55.9	-	80.2		low	normal bisexual	Lindsey, 1962
		(16-20)	-	(22-26)				
Oreochromis mossambicus	2/2	22.0	40.9	42.4		low	normal bisexual	Mair et al., 1990
		(20)	(26)	(32)				
		11.1	85.1	100.0		low	probably putative all-female brood	
		(19)	(25.5)	(31)				
O. aureus	1/7	4.8	2.9	20.0		low	putative all-male brood	Mair et. al., 1990
		(26)	(29)	(32)				
Poeciliopsis lucida	1/2	62.0	46.3	8.0	low	high	normal bisexual	Sullivan and Schultz, 1986
		(24)	(25.5)	(30)				
Carassius auratus	4/6	80-100	60-70	-		high	putative all-female brood	Oshiro et al., 1988[*5]
		(20)	(25)	-				
C. carassius grandoculis	2/6	50.0	-	21.4		high	normal bisexual	Fujioka, 1995
		(19-21)	-	(28-31)				
		100.0	-	46.6		high	putative all-female brood	
		(19-21)	-	(28-31)				
Paralichthys olivaceus	2/2	-	50.0	10.0		high	normal bisexual	Yamamoto and Masutani, 1990
		-	(20)	(25)				
		-	100.0	60.0		high	putative all-female brood	
		-	(20)	(25)				
Misgurnus anguillicaudatus	3/3	-	50.0	NS		high	normal bisexual	Nomura et al., 1994[*6]
		-	(20)	(27-30)				
		-	100.0	NS			putative all-female brood	
		-	(20)	(30)				
Oryzias latipes	1/1	-	100.0	NS		high	putative all-female brood	Ochiai et al., 1994[*7]
		-	(26)	(30)				
Gnathopogon caurulescens	1/1	-	93.7	70.0		high	putative all-female brood	Fujioka, 1993
		-	(20-30)	(25-32)				

[*1] Trials that produced significantly different sex ratios at variable temperatures over the total number of trials performed. Each trial used progeny from different parents.
[*2] Results of successful trial(s). Actual temperatures are given within parenthesis. NS: not specified.
[*3] As concluded by the authors of these studies.
[*4] Some studies used putative monosex broods produced by gynogenesis or from sex-reversed broodstock.
[*5] Oshiro, T., Zhang, F., and Takashima, F. Presented at the Annual Meeting of the Japanese Society of Scientific Fisheries, Spring 1988.
[*6] Nomura, T., Arai, K., and Suzuki, R. Presented at the Annual Meeting of the Japanese Society of Scientific Fisheries, Autumn 1994.
[*7] Ochiai, T., Oshiro, T., and Aida, K. Presented at the Annual Meeting of the Japanese Society of Scientific Fisheries, Autumn 1994.

Temperature-dependent sterilization

Thermally sex-reversed fish generally have normal gonadal histology and are presumed to be functional. However, Strüssmann et al. (1995a, unpublished data) noted the occasional occurrence of individuals of *O. bonariensis* and *P. hatcheri* whose gonads were poor in or completely devoid of germ cells in groups reared at the high temperatures. Current research has since examined this problem and established that prolonged exposure to temperatures around 29°C causes the disappearance of germ cells in both male and female gonads of *O. bonariensis* (Strüssmann and Saitoh, unpublished data; Fig. 2). Since at 29°C only males are formed, sterilization of female-gonads involves previous exposure to lower temperatures during sex-differentiation. Conversely, formation of normal testes in all-male groups requires short-term exposure to high temperatures. Insufficiently high temperatures or short exposures to 29°C seem to cause two types of partial germ cell disappearance: either their number is greatly reduced or they disappear completely from only one of the two lobes of the gonads (Strüssmann and Saitoh, unpublished data).

Warm temperature-sterilization likely occurs in other fish groups. For instance, Nakamura and Nagahama[*] recently reported the nearly complete elimination of germ cells in the gonads of *Oreochromis niloticus* exposed to 36-37°C between 7-50 days after hatching. They also showed that these effects are permanent and resulted in sterile adults.

[*] Nakamura, M. and Nagahama, Y. Presented at the Annual Meeting of the Japanese Society of Scientific Fisheries, Spring 1995.

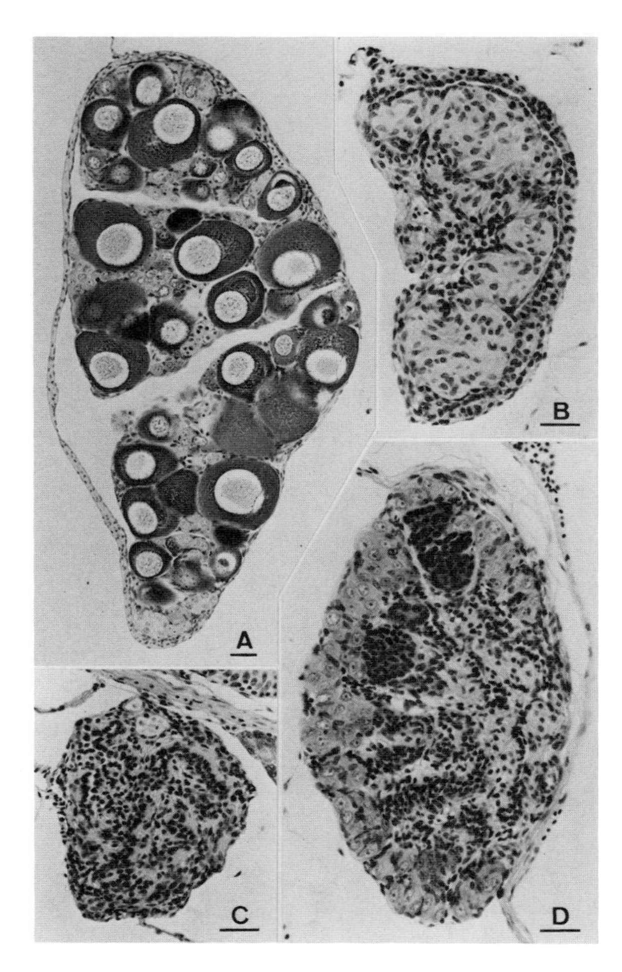

Fig. 2. Histological appearance of normal (A, D) and thermally-sterilized (B, C) ovaries and testes of *O. bonariensis*. Bars under letters represent 25 µm.

Thermal manipulation of sex in aquaculture

From what has been described above, i.e. that thermal lability of sex appears widespread and that thermal manipulation of sex may allow both functional sex reversal and production of neutered fish, the possibilities of application of TSD for aquaculture appear extraordinary. In many species, uncontrolled reproduction impairs growth and survival during grow-out or poses environmental problems upon stocking in open waters. Present methods for sex control of fish involve either hormonal treatment or chromosome manipulation (Dunham, 1990). Both methods, though, present problems for practical application such as consumer reaction to hormone-treated or chromosome-manipulated fish, or require maintenance and periodical replenishment of sex-reversed broodstock. Thermal sex manipulation is a promising alternative to the methods above.

For practical application of TSD in aquaculture, combinations of temperature and duration of thermal manipulation for reliable production of specific sex ratios (all-female, all-male) will have to be established. As shown for *O. bonariensis* (Strüssmann and Saitoh, unpublished data), it is a narrow step between functional all-male producing temperatures and sterilization. Thus, temperatures and treatment duration for selective functional sex reversal or sterilization must be carefully chosen. It will be necessary to optimize other conditions that promote survival during thermal stress and to reduce thermal treatment duration. Wherever the desirable sex is obtained at the lower temperatures, it will be necessary to minimize the duration of thermal treatments also because low temperatures cause depression of growth rates. Finally, growth rates are likely to differ between hatcheries even when temperature is kept constant, and thus criteria to determine the end-point of treatments must be worked out.

Temperature control, even if not sufficient by itself, may still be needed for successful manipulation of sex by other methods. As shown in Table 1, putative monosex broods produced by gynogenesis or from sex reversed parents often present varied proportions of the opposite sex, which may depend on rearing temperature as well as on parental genotype. Temperature also seems to act synergistically with exogenous steroid hormones during hormonal sex manipulation. Not surprisingly, the temperatures that potentiated the action of androgen hormones coincided with those that favored the formation of males during thermal sex manipulation, e.g. high temperature in *Gnathopogon* (Fujioka, 1993; conf. Table 1). Also, complete masculinization of carp was only possible under high temperature (Nagy et al., 1981), a condition that favors male formation in other cyprinids (Table 1).

Concluding remarks

Information reviewed herein indicates that thermolability of gonadal sex and sensitivity of germ cells to temperature are not unique to a certain group of fishes, nor do they seem to be exclusive attributes of species which adaptively take advantage of environmental sex determination. Many cases of unexpected sex ratios from crosses with otherwise genotypically sex-determined species reported in the literature may have an environmental basis, either through direct thermally induced-sex reversal of part of the progeny or indirectly through use of naturally sex-reversed parents.

The seemingly widespread occurrence of TSD and perhaps temperature-dependent sterilization among fishes ought to have important implications for the production of monosex and sterile fish in aquaculture. Practical use will require that exposure to extreme temperatures is limited to a minimum, e.g. the period encompassing sex differentiation, in order to maintain adequate survival and growth rates. Externally visualized developmental events that are concomitant with gonadal sex differentiation may constitute cues to determine the end-point of thermal treatments, and warrant further study. These external indicators may prove helpful especially in circumstances where substantial differences in growth rates between hatcheries (or trials) are expected. Temperature control, if not sufficient by itself for sex manipulation in some species, may be complementary to other sex control techniques such as hormone treatment and chromosome manipulation.

Acknowledgements

Original research cited in this review was supported by grants from Moritani Scholarship Foundation, Ministry of Education, Science, and Culture of Japan, and Tokyo University of Fisheries to C.A.S. and from the Texas Higher Education Coordinating Board (#003644-060) to R.P. Provision of travel expenses for C.A.S. by Tokyo University of Fisheries is greatfully acknowledged.

References

Adkins-Regan, E., 1987. Hormones and sexual differentiation. In: Hormones and reproduction in fishes, amphibians, and reptiles. D.O. Norris and R.E. Jones (Eds.): Plenum Press, New York, p. 1-29.

Aida, T., 1936. Sex reversal in *Aplocheilus latipes* and a new explanation of sex differentiation. Genetics 21:136-153.

Chan, S.T.H. and W.S.B. Yeung, 1983. Sex control and sex reversal in fish under natural conditions. In: Fish physiology, Vol. 9B, Reproduction: behavior and fertility control. W.S. Hoar, D.J. Randall, and E.M. Donaldson (Eds.): Academic Press, New York, p. 171-222.

Conover, D.O., 1984. Adaptive significance of temperature-dependent sex determination in a fish. Amer. Nat. 123:297-313.

Conover, D.O., 1992. Seasonality and the scheduling of life history at different latitudes. J. Fish Biol. 41(Supplement B):161-178.

Conover, D.O. and M.H. Fleisher, 1986. Temperature-sensitive period of sex determination in the Atlantic silverside, *Menidia menidia*. Can. J. Fish. Aquat. Sci. 43:514-520.

Conover, D.O. and S.W. Heins, 1987a. The environmental and genetic components of sex ratio in *Menidia menidia* (Pisces: Atherinidae). Copeia 1987(3):732-743.

Conover, D.O. and S.W. Heins, 1987b. Adaptive variation in environmental and genetic sex determination in a fish. Nature 326:496-498.

Conover, D.O. and B.E. Kynard, 1981. Environmental sex determination: interaction of temperature and genotype in a fish. Science 213:577-579.

Dunham, R.A., 1990. Production and use of monosex or sterile fishes in aquaculture. Rev. Aquat. Sci. 2:1-17.

Francis, R.C., 1992. Sexual lability in teleosts: developmental factors. Quart. Rev. Biol. 67(1):1-18.

Fujioka, Y., 1993. Sex reversal in honmoroko, *Gnathopogon caurulescens* by immersion in 17-methyltestosterone and an attempt to produce all-female progeny. Suisan Zoshoku 41(3):409-416 (in Japanese).

Fujioka, Y., 1995. Production and some properties of gynogenetic diploids in nigorobuna *Carassius carassius grandoculis*. Rep. Shiga Prefec. Fish. Exp. Station (in press; in Japanese).

Fujioka, Y., H. Tanaka, and N. Sawada, 1995. Monosex female production by the use of sex-reversed males of nigorobuna, *Carassius carassius grandoculis*. Suisan Zoshoku (in press; in Japanese).

Korpelainen, H., 1990. Sex ratios and conditions required for environmental sex determination in animals. Biol. Rev. 65:147-184.

Lam, T.J., 1983. Environmental influences on gonadal activity in fish. In: Fish physiology, Vol. 9B, Reproduction: behavior and fertility control. W.S. Hoar, D.J. Randall, and E.M. Donaldson (Eds.): Academic Press, New York, p. 65-116.

Lindsey, C.C., 1962. Experimental study of meristic variation in a population of threespine stickleback, *Gasterosteus aculeatus*. Can. J. Zool. 40:271-312.

Lowe, T.P. and J.R. Larkin, 1975. Sex reversal in *Betta splendens* Regan with emphasis on the problem of sex differentiation. J. Exp. Zool. 191:25-32.

Mair, G.C., J.A. Beardmore, and D.O.F. Skibinski, 1990. Experimental evidence for environmental sex determination in *Oreochromis* species. In: Proceedings of the. Second Asian Fisheries Forum. R. Hirano and I. Hanyu (Eds.): Tokyo, p. 555-558.

Middaugh, D.P. and M.J. Hemmer, 1987. Influence of environmental temperature on sex-ratios in the tidewater silverside, *Menidia peninsulae* (Pisces: Atherinidae). Copeia 1987(4):958-964.

Nagy, A., M. Bercsenyi, and V. Csanyi, 1981. Sex reversal in carp (*Cyprinus carpio*) by oral administration of methyltestosterone. Can. J. Fish. Aquat. Sci. 38:725-728.

Oshiro, T., 1987. Sex ratios of diploid gynogenetic progeny derived from five different females of goldfish. Nippon Suisan Gakkaishi 53:1899.

Schultz, R.J., 1993. Genetic regulation of temperature-mediated sex ratios in the livebearing fish *Poeciliopsis lucida*. Copeia 1993(4):1148-1151.

Shapiro, D.Y., 1990. Sex-changing fish as a manipulable system for the study of the determination, differentiation, and stability of sex in vertebrates. J. Exp. Zool. (Sup.) 4S:132-136.

Sullivan, J.A. and R.J. Schultz, 1986. Genetic and environmental basis of variable sex ratios in laboratory strains of *Poeciliopsis lucida*. Evolution 40(1):152-158.

Streisinger, G., C. Walker, N. Dower, D. Knauber, and F. Singer, 1981. Production of clones of homozygous diploid zebra fish (*Brachydanio rerio*). Nature 291:293-296.

Strüssmann, C.A., S. Moriyama, E.F. Hanke, J.C. Calsina Cota, and F. Takashima, 1995a. Evidence of thermolabile sex determination in pejerrey, *Odontesthes bonariensis*. J. Fish Biol. (in press).

Strüssmann, C.A., F. Takashima, and K. Toda, 1995b. Sex differentiation and hormonal feminization in pejerrey *Odontesthes bonariensis*. Aquaculture (in press).

Winge, O., 1934. The experimental alteration of sex chromosomes into autosomes and vice versa, as illustrated by *Lebistes*. Compt.-rend. Trav. Lab. Carlsberg, Ser. physiol 21(1):1-51.

Yamamoto, E. and R. Masutani, 1990. Production of gynogenetic olive flounder. Yoshoku 27:80-85 (in Japanese).

TEMPERATURE SEX DETERMINATION IN TWO TILAPIA, OREOCHROMIS NILOTICUS AND THE RED TILAPIA (RED FLORIDA STRAIN): EFFECT OF HIGH OR LOW TEMPERATURE.

[1,2]J.F. Baroiller, [2,3]F. Clota and [2,4]E. Geraz

[1]Laboratoire de Physiologie des Poissons, INRA, 35042 Rennes Cédex France; [2]CIRAD-EMVT, GAMET, BP 5095, 34033 Montpellier Cédex 1, France.[3]Département Hydrobiologie et Faune Sauvage, INRA, 35042 Rennes Cédex France. [4]ENITA-Bordeaux, Productions Animales, BP 201, 33175 GRADIGNAN Cédex France.

Summary

Very few studies have evaluated the role of external factors on sex determination in gonochoristic fish species. In tilapia, our recent results clearly demonstrate that high temperature has a strong effect on the Oreochromis niloticus sex-ratio. In reptiles and amphibians, 3 different patterns of temperature sex determination (TSD) have been described according to the sex produced at high and low temperature. In order to define this pattern in the tilapia, Oreochromis niloticus, we compared the effects of low and high temperature on the sex-ratio of standard or genetically female progenies. High temperatures (34-36°C) significantly increased the proportion of males (extreme percentages = 69-91% males) but low temperatures (19-23°C) did not affect the sex-ratio. Furthermore, we also investigated TSD in red tilapia of the Red Florida strain (taken as a gene pool of 2 MH and 2 FH species). In this red tilapia, treatments at 36°C also significantly increased the percentage of males in most progenies (extreme percentages = 60-97%). These results suggest that TSD could exist in several tilapia species, with strong genotype-temperature interactions. The pattern of this TSD resembles that of the other thermosensitive fish, Menidia menidia.

Introduction

Since the first evidence of Temperature Sex Determination (TSD) in a gonochoristic fish, Menidia menidia (Conover and Kynard, 1981), very few studies have focused on other species (Sullivan and Schultz, 1986; Baroiller et al., 1993; 1995). In Oreochromis niloticus, sex is determined by genetic factors, temperature, and genotype/temperature interactions: at standard temperatures, sex-ratios generally tally with the sexual genotype, whereas high temperatures applied during a critical period may affect the gonadal sex differentiation (Baroiller et al., 1993; 1995). In the fish, M. menidia, as in many thermosensitive reptile and amphibian species, low and high temperatures oppositely affect the sex-ratio (Conover and Kynard, 1981; Spotila et al., 1994). As wild tilapia may experience a wide range of temperatures, we have tested the effects of low and high temperature on the Oreochromis niloticus sex-ratio. Moreover, as different Oreochromis niloticus progenies tend to vary as to the the level of TSD (Baroiller et al., 1993; 1995), we have examined some characteristics of these variations.

Materials and methods

Standard O. niloticus ("Bouaké strain") and red tilapia (Red Florida strain) or genetically all-female O. niloticus progenies were subjected for 28 days to various temperature treatments during the thermosensitive period (Baroiller et al., 1993; 1995). Variation in thermosensitivity was examined in 2-3 successive progenies from 5 different O. niloticus breeding pairs. The inheritance of thermosensitivity was examined in 2 temperature sex-reversed males. In both experiments, progenies were divided in 2 groups and exposed to 27 and 36°C. Five red tilapia progenies from different breeding pairs were temperature-treated as described above for O. niloticus. Sex-ratios were determined on all fish by microscopic examination of gonad squashes, 90-120 days post-fertilization (PF) when the histological characteristics of male or female differentiation are already in place (Baroiller, 1988). Survival rates were not significantly affected by high or low temperature treatment. The Chi-square test ($p<0.05$) was used to compare the sex-ratios.

Results

Pattern of temperature sex determination

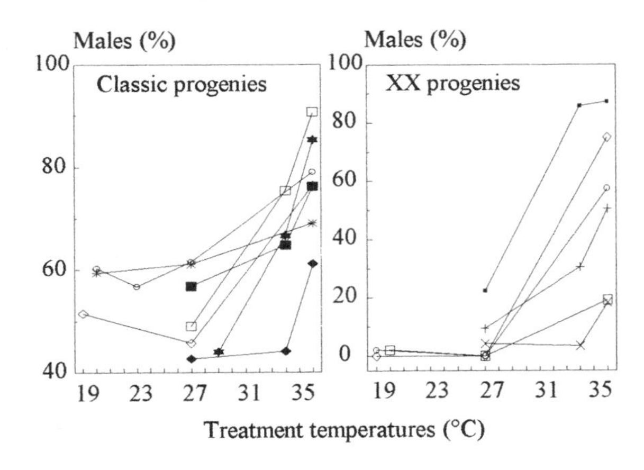

Figure 1: Effect of various temperatures on O. niloticus sex-ratio.

Author's present address: Lab. de Physiologie des Poissons, INRA,35042 RENNES Cédex, FRANCE

In both standard and genetically female O. niloticus progenies, high temperatures (34-36°C) significantly increased the proportion of males whereas low temperatures (19-23°C) did not affect the sex-ratio.

Variations and inheritance of thermosensitivity:

There was no significant difference in the response to treatment at a given temperature between replicate samples or among successive progenies issued from a same breeding pair. The progenies of 3 breeding pairs (I, II and III) responded to high temperature treatment by a significant increase in the proportion of males. In contrast, the progenies of the breeding pairs IV and V were insensitive to temperature treatment.

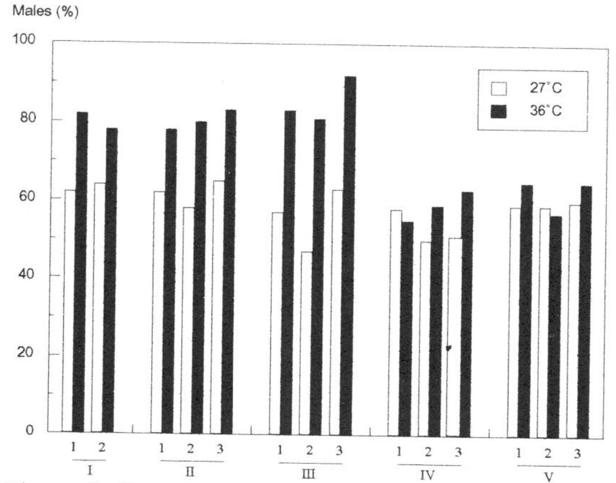

Figure 2: Response of the sex-ratio to treatment at 27 or 36°C in several successive progenies (1-3) from 5 different O. niloticus breeding pairs (I-V).

Temperature sex-reversed males for the inheritance experiment were obtained by the following procedure.

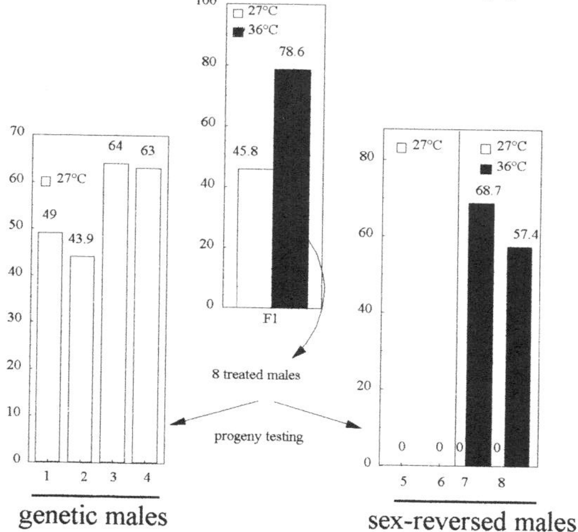

Figure 3: Inheritance of thermosensitivity in the progenies of 2 temperature sex-reversed males of O. niloticus. Progenies 7 and 8 have been subjected to treatments at 36°C.

Temperature treatment was applied to a standard progeny. In the 36°C treated group, the percentage of males obtained was 78% (Fig. 3a). Eight of these males were raised to maturity . By analysing the sex-ratio in the progenies of these 8 males it was possible to identify the temperature sex-reversed males: 4 males sired all-female progenies (Fig. 3c), whereas balanced sex-ratios were obtained from the other 4 males (Fig. 3b).

Progenies of 2 temperature sex-reversed males which were subjected to 36°C yielded significantly higher proportion of males (57.4-68.7%) than in the groups at 27°C (Fig. 3c).

Effect of temperature in red tilapia

Here again, treatment at 36°C significantly increased the percentage of males in most of the progenies. As in O. niloticus, there was a strong parental influence on the response to high temperature-treatment.

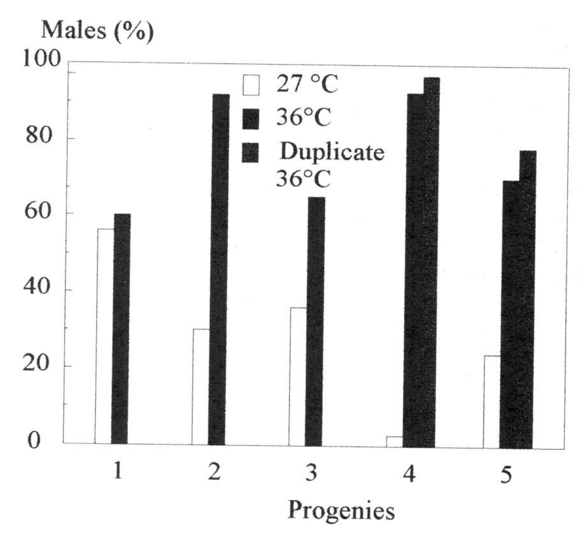

Figure 4: Response of the sex-ratio to treatment at 36°C among the progenies of different red tilapia breeding pairs (with replicate samples).

Discussion

O. niloticus sex-ratios are governed by a female homogamety and modified by high temperatures: at low temperatures, the sex-ratios usually tallies with the sexual genotype Elevated temperatures result in predominantly male sex-ratios but never in all-male populations: the highest percentage of males recorded for O. niloticus was 92%; for the red tilapia, it was 98.3%.

In both tilapia species, strong genotype-temperature interactions are suggested:

- progenies from different breeding pairs vary as to their thermosensitivity (from 0 to 98.3% males following a same 36°C treatment).
- similar sex-ratios are observed among successive progenies of a same breeding pair when treated at a given temperature
- temperature sex-reversed males can sire thermosensitive progenies

Red tilapia of the Red Florida strain was originally obtained by crossing a mutant red O. mossambicus with 3 other species, O. hornorum, , O. niloticus, O. aureus. The existence of TSD in red tilapia may simply reflect the O. niloticus contribution to the genome of this tilapia. It might also mean that such an effect exists in all four species from which it derives. This possibility deserves further study.

The O. niloticus TSD resembles that of another thermosensitive gonochoristic fish, the Atlantic sides Menidia menidia; however, in the latter species which has major sex-determining genes but no detected sex chromosomes, low temperatures produce mostly female offsprings (Conover et al., 1992).
In thermosensitive reptiles and amphibians, genotype-temperature interactions are generally not reported (Dournon et al., 1990).

In nature, wild tilapia may well encounter masculinizing temperatures during the critical thermosensitive period. However, the biological significance of TSD and the underlying physiological mechanisms remain to be elucidated. TSD may be more widespread than recognized.

References

Baroiller, J.F., 1988. Etude corrélée de l'apparition des critères morphologiques de la différenciation de la gonade et de ses potentialités stéroïdogènes chez Oreochromis niloticus. Thèse dr., Univ. Pierre et Marie Curie, Paris, 70 p.

Baroiller, J.F., Fostier, A., Cauty, C. and B., Jalabert, 1993. Significant effects of high temperatures on sex-ratio of progenies from Oreochromis niloticus with sibling sex-reversed males broodstock. In R.S.V., Pullin, J. Lazard, M. Legendre, J.B. Amon Kothias and D. Pauly, (eds) "Third International Symposium on Tilapia in Aquaculture". 11-16 Nov. ICLARM Conf. Proc. 41. In press

Baroiller, J.F., Chourrout, D., Fostier, A. and B., Jalabert, 1995. Temperature and sex chromosomes govern sex-ratios of the mouthbrooding cichlid fish, Oreochromis niloticus. J. Exp. Zool. In press.

Conover, D.O. and B.E., Kynard, 1981. Environmental sex determination: interaction of temperature and genotype in a fish. Science 213: 577-579.

Conover, D.O., Van Voorhees, D.A. and A., Ehtisham, 1992. Sex-ratio selection and the evolution of environmental sex determination in laboratory populations of Menidia. Evolution 46: 1722-1730.

Dournon, C., Houillon, C. and C., Pieau, 1990. Temperature sex reversal in amphibians and reptiles. Int. J. Dev. Biol., 34: 81-92.

Spotila, J.R., Spotila, L.D. and N.F., Kaufer, 1994. Molecular mechanisms of TSD in reptiles: a search for the magic bullet. J. Exp. Zool., 270: 117-127.

Sullivan, J. D. and R. J., Schultz, 1986. Genetic and environmental basis of variable sex ratios in laboratory strains of Poeciliopsis lucida. *Evolution,* 40(1), 152-158.

The authors wish to thank D. CHOURROUT for comments on the manuscript

SALMON GnRH GENE EXPRESSION FOLLOWING PHOTOPERIOD MANIPULATION IN PRECOCIOUS MALE MASU SALMON

K. Aida[1] and M. Amano[2]

[1] Department of Fisheries, Faculty of Agriculture, The University of Tokyo, Bunkyo, Tokyo, Japan
[2] Nikko Branch, National Research Institute of Aquaculture, Nikko, Tochigi, Japan

Summary

In spring, a large individual variation appeared in the number of sGnRH neurons expressing sGnRH mRNA in the ventral telencephalon (VT) and the preoptic area (POA) of underyearling male masu salmon. Numbers showed positive correlation with GSI, although GSI was very low. This suggests that development of sGnRH neurons in the VT and POA is involved in the determination of future precocity.

In early summer, precocious maturation of underyearling male masu salomon was accelerated when fish were transferred from natural photoperiod to 8L16D, compared to fish transferred to 16L8D: spermiation was observed in August in the former, and in September in the latter. The number of sGnRH neurons expressing sGnRH mRNA in the VT and POA increased in August in the 8L16D group, whereas numbers increased in September in the 16L8D group. These results indicate that sGnRH synthesis is accelerated by decreasing daylength, and activated sGnRH synthesis is involved in the acceleration of precocious maturation. This may be the first demonstration that photoperiod regulates sGnRH neuronal activity in teleost fish.

Introduction

It has been known that male salmonid fishes precociously mature. Despite their interesting life history, mechanisms of precocious maturation have not yet been clarified. In this paper, we used masu salmon *Oncorhynchus masou* as an experimental fish, and examined how precocious maturation is regulated. Precocious maturation occurs in autumn in underyearling male masu salmon which show rapid growth, suggesting that precocious maturation is controlled by somatic growth and photoperiod changes.

Since gonadal maturation of salmonids is controlled by salmon type gonadotropin-releasing hormone (sGnRH) via stimulation of pituitary gonadotropin synthesis and release, sGnRH is considered to be involved in the regulation of precocious maturation (Amano et al. 1991). In a previous paper (Amano et al. 1994), we reported that when underyearling males are transferred from natural photoperiod to 8L16D or 16L8D in June, precocious maturation is accelerated in the 8L16D group with rapid increase in pituitary sGnRH content, suggesting that decreasing daylength accelerates sGnRH gene expression. Recent analysis of the nucleotide sequence of the masu salmon sGnRH cDNA (Suzuki et al. 1992) has made it possible to investigate sGnRH synthetic activity by using *in situ* hybridization techniques.

In a previous experiment using underyearling males in June (Amano et al. 1993), there was no difference in the appearance rates of precocious males between the 8L16D and 16L8D groups. This result suggests that future precocity is determined before June. In the experiment, sGnRH was also suggested to be involved in the determination of precocious maturation.

Therefore, in this paper we aimed to clarify the relationship between brain sGnRH gene expression and the regulation of precocious maturation. Two experiments were carried out at the Nikko Branch, National Research Instutute of Aquaculture. Masu salmon used were offspring of wild fish which had migrated to the Shiribetu River in Hokkaido.

Determination of future precocity and sGnRH gene expression in spring

Underyearling males which were reared under natural photoperiod and 9-10℃ were sampled on April 16, May 7 and June 11. On the day of autopsy, fish were randomly selected and anesthetized in ethyl-p-aminobenzoate (0.05%). For *in situ* hybridization, brains were removed rapidly and were fixed with 4% paraformaldehyde and 1 % picric acid in 50 mM phosphate buffer (pH 7.3) at 4 ℃. *In situ* hybridization (ISH) was carried out according to the method previously reported (Amano et al. 1994). Testes were fixed with Bouin's solution and weight was measured for the calculation of GSI.

Body length and weight gradually increased from April through June. Mean GSIs were 0.060%, 0.043% and 0.143 % in April, May and June, respectively: they increased significantly in

June. In June, gonadal weight and GSI showed wider distribution than in April and May. Proliferation of spermatogonia was not observed in all individuals in April and May. In June, proliferation occurred in fish which had relatively high GSI in June.

The number of neurons expressing sGnRH mRNA in the VT + POA increased in May and remained at the same level in June. Changes in total silver grains in the VT + POA showed a similar pattern as observed in the number of neurons expressing sGnRH mRNA. The number of neurons and total silver grains per individual fish showed wide distribution in May and June. There were positive correlations between GSI and the number of neurons expressing sGnRH mRNA in the VT + POA in April and June. Positive correlations were also observed between GSI and total silver grains in the VT + POA.

Effects of photoperiod on acceleration of precocious maturation and sGnRH gene expression in summer

In June, 320 underyearling fish, including males and females, were randomly selected from a stock reared under natural photoperiod and 9-10°C. They were then transferred to short photoperiod (8L16D) or long photoperiod (16L8D) conditions (160 fish each).Tisues were sampled on June 23 (initial), July 23, August 19 and September 28. Sampling procedures were as described above. *In situ* hybridization was carried out according to the method previously reported (Amano et al. 1994). Sex and gonadal stages were checked and only precocious males were used as material. Precocious males were easily distinguishable by increased GSI at all periods except at the beginning in June.

In the 8L16D group, GSI increased in July, remained at high levels in August, and then decreased in September. Spermiation was observed in August. In the 16L8D group, GSI increased from August and attained a peak in September, when spermiation was observed. In July, GSI was significantly higher in the 8L16D group than in the 16L8D group. On the other hand, GSI was significantly higher in the 16L8D group than in the 8L16D group in September (Fig. 1).

In the 8L16D group, the number of neurons expressing sGnRH mRNA rapidly increased in the VT and POA until August. In the 16L8D group, numbers were unchanged until August, increasing in September. In August, the number of ISH-positive neurons in both areas were significantly greater in the 8L16D group than in the 16L8D group (Fig. 1). There were no marked changes in the number of neurons expressing sGnRH mRNA in the terminal nerve ganglion and the olfactory bulbs in either photoperiodic groups during gonadal maturation. There were no significant changes in the number of silver grains per neuron in all brain regions, although numbers tended to decrease in the olfactory bulbs in both photoperiodic groups.

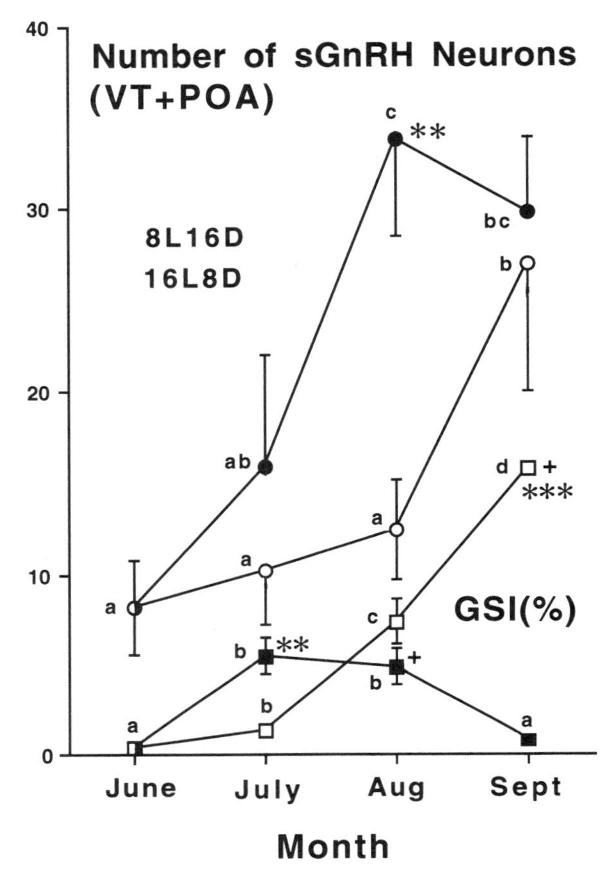

Fig. 1 Changes in GSI and the number of neurons expressing sGnRH mRNA in the VT+POA in underyearling male masu salmon following photoperiod changes.

+ indicates that spermiation was observed. In each group, means indicated by differing letters, for example a and b, differ significantly ($p<0.05$). ** ($p<0.01$) and *** ($p<0.001$) indicate significant difference between the short and long photoperiod groups in each month.

Discussion

The number of sGnRH neurons in the VT + POA increased in May, one month earlier than the increase in GSI. Positive correlations between GSI and the number of sGnRH neurons were observed in April and June. In June, precocious males became distinguishable by their increased GSI. Taken together, it is possible that the development of sGnRH neurons in the VT and POA is directly related to the determination of future precocity of male masu salmon, suggesting that the brain-pituitary-gonadal axis is established.

Precocious mauration and the activation of sGnRH synsthesis in the VT and POA were coincidently accelerated by changing photoperiod from natural conditions to 8L16D in June, suggesting that the following accelerating process exists. Decreased daylength accelerated an increase in the number of sGnRH neurons expressing sGnRH mRNA in both areas, and synthesized sGnRH was transported to the pituitary. Accumulated sGnRH in the pituitary (Amano et al, 1994) stimulated GTH synthesis and release, and secreted GTH accelerated precocious matuation. This may be the first demonstration that photoperiod regulates sGnRH neuronal activity in teleost fish.

Long photoperiod delayed, but did not suppress precocious maturation of underyearling males. Precocious males were expected to be destined for sexual maturity in May prior to the commencement of the experiment (Amano et al., 1993). Once the brain-pituitary-gonadal axis becomes active, gonadal maturation may proceed irrespective of photoperiod. Thus, changes in photoperiod from long to short daylength accelerates gonadal maturation via activation of sGnRH neurons in the VT and POA possibly through a photoperiodic signal input system; long photoperiod only delays the onset of sGnRH neuron activation and hence gonadal maturation.

It is unlikely that sGnRH neurons in the terminal nerve ganglion are involved in testicular maturation through GTH synthesis and release. sGnRH neurons in the terminal nerve ganglion are considered to have a neuromodulatory function in the brain of dwarf gourami, *Colisa lalia* (Oka and Ichikawa, 1990; Oka 1992), and goldfish, *Carassius auratus* (Kobayashi et al., 1992; 1994). Such may also be the case in masu salmon. sGnRH neurons in the olfactory bulbs also showed no remarkable changes during photoperiod manipulation, indicating that they are not involved in gonadal maturation.

References

Amano, M., Y. Oka, K. Aida, N. Okumoto, S. Kawashima and Y. Hasegawa, 1991. Immunocytochemical demonstration of salmon GnRH and chicken GnRH-II in the brain of masu salmon *Oncorhynchus masou*. J. Comp. Neurol. 314:587-597.

Amano, M., K. Aida, N. Okumoto, Y. Hasegawa 1993. Changes in levels of GnRH in the brain and pituitary and GTH in the pituitary in male masu salmon, *Oncorhynchus masou*, from hatching to maturation. Fish. Physiol. Biochem. 11, 233-240.

Amano, M., N. Okumoto, S. Kitamura, K. Ikuta Y. Suzuki and K. Aida, 1994. Salmon GnRH and GTH are involved in precocious maturation induced by photoperiod manipulation in underyearling male masu salmon, *Oncorhynchus masou*. Gen. Comp.Endocrinol. 95, 368-373.

Kobayashi, M., M. Amano, Y. Hasegawa K. Okuzawa and K. Aida, 1992. Effects of olfactory tract section on brain GnRH distribution, plasma gonadotropin levels, and gonadal stage in goldfish. Zool. Sci. 9, 765-773.

Kobayashi, M., M. Amano, M. Kim, K. Furukawa Y. Hasegawa and K. Aida, 1994. Gonadotropin-releasing hormones of termional nerve origin are not essential to ovarian development and ovulation in goldfish. Gen. Comp. Endocrinol. 95, 192-200.

Oka, Y. and M. Ichikawa, 1990. Gonadotropin-releasing hormone (GnRH) immunoreactive system in the brain of the dwarf gourami (*Colisa lalia*) as revealed by light microscopic immunocytochemistry using a monoclonal antibody to common amino acid sequence of GnRH. J. Comp. Neurol. 300, 511-522.

Oka, Y., 1992. Gonadotropin-releasing hormone (GnRH) cells of the terminal nerve as a model neuromodulator system. Neurosci. Lett. 142, 119-122.

Suzuki, M., S. Hyodo, M. Kobayashi, K. Aida and A. Urano, 1992. Characterization and localization of mRNA encoding the salmon-type gonadotropin releasing hormone precursor of the masu salmon. J. Mol. Endocrinol. 9, 73-82.

HOW DO PHOTOPERIOD, THE PINEAL GLAND, MELATONIN, AND CIRCANNUAL RHYTHMS INTERACT TO CO-ORDINATE SEASONAL REPRODUCTION IN SALMONID FISH?

N.R. Bromage, C.F. Randall, M.J.R. Porter, and B. Davies

Institute of Aquaculture, University of Stirling, Stirling FK9 4LA, Scotland

Introduction

It is now well established that the principal environmental determinant of the timing of seasonal breeding in rainbow trout is the ambient photoperiod with only minimal influence imparted by temperature (Bromage *et al*, 1993). Advances in spawning time are elicited by long days early in the yearly cycle of reproduction particularly if followed later by short days; by contrast delays in spawning are brought about by constant short or by short and then long daylengths.

Although the influences of photoperiod are known to be mediated by the hypothalamic-pituitary-gonadal axis, with alterations in daylength being paralleled by changes in the timing of onset and patterns of GTH and other hormones (Bromage *et al*, 1982), many questions remain regarding the mechanisms by which information on daily and seasonally-changing daylength is perceived and transmitted by the fish and in turn how this influences the neuroendocrine axis. In mammals the pineal hormone melatonin has been shown to be involved. However, it is not known whether melatonin has a similar role in fish.

A further complication is that the overall process of reproduction in salmonid fish is now thought to be controlled by an endogenous circannual rhythm or "clock" (Bromage *et al*, 1993). Although this rhythm exhibits many of the features of other biological clocks, its level of influence on the photoperiodic response and in particular with the putative functions of melatonin remains unclear. Aspects of these questions are further considered in the present study.

Results

(i) Photoperiod and Melatonin Secretion: Two series of experiments were conducted. In the first, a group of potential S2 Atlantic salmon parr were maintained from hatching under a natural photoperiod (Lat. 52°N). On 15 Sept. (LD 12.5:11.5) serum samples were taken at 3hr intervals, over a 30hr period, for RIA of melatonin (Randall *et al*, 1995). Fig 1 shows that as light levels fell at dusk, melatonin values increased and remained elevated (>250pg/ml) throughout the dark period. By contrast at dawn, melatonin values rapidly decreased and were maintained at basal levels (approx 90pg/ml) throughout the light period.

In the second experiment, 3 groups of 2+ year old Atlantic salmon post-smolts were acclimated for 2-3 months to constant photoperiods of LD 12:12, LD 16:8 or LD 20:4 and serum samples taken at 2hr intervals over 30hr periods for melatonin assay. Under each regime, melatonin levels were significantly ($p<0.001$) elevated throughout the dark periods (Fig 2); the duration of the raised melatonin values exactly reflecting night-length.

(ii) Seasonal Patterns of Melatonin Secretion: Plasma samples were taken at 3hr intervals over 27hr sampling periods, at approximately monthly intervals, from groups of potential S2 salmon parr exposed to a natural photoperiod (56°N). Measurements of melatonin showed that as daylength increased in the Spring, up to the Summer Solstice, then the duration of raised melatonin levels each day gradually decreased and was shortest on the longest day (Fig 3). Conversely, as daylength decreased in the Autumn/Fall up to the shortest day in December, the duration of raised melatonin progressively increased.

(iii) Effects of Pinealectomy and Melatonin Implantation on Spawning Time: Several pinealectomy (PNX) experiments have been carried out with pre-pubertal trout and salmon. Blood samples taken mid-way through light and dark periods in PNX and intact and sham-PNX control fish, at regular intervals starting 3 weeks after surgery and continuing until after spawning, confirmed that on all occasions controls had normal diel patterns of melatonin, high at night and basal during the day, whereas PNX fish had basal (daytime) levels

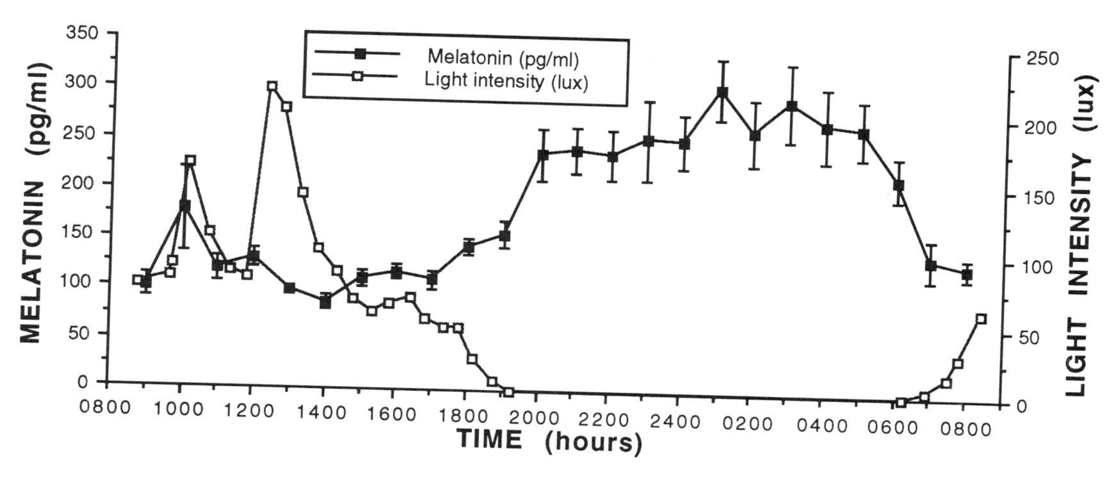

Fig 1: Diel melatonin profile (LD12.5:11.5)

throughout the day and night. Weekly examinations of fish to determine the times of ovulation revealed significant delays in spawning in PNX fish compared to controls. Fig 4 shows the delay in spawning following PNX at the Summer Solstice of 2 year old rainbow trout.

For the implantation experiment two groups of salmon smolts were transferred to sea water on 15 March. On 18 April the following year, half of the fish received intramuscular implants of melatonin, which had previously been shown to provide supraphysiological levels of melatonin (approx 1500pg/ml) for periods up to a year (Randall & Bromage, unpublished). Weekly examinations of the fish showed no significant differences in spawning times of the implanted and control fish although there appeared to be a loss in synchrony in the implanted group, possibly indicative of a "free-running" rhythm. The absence of a clear response to continuous melatonin infusion may be due to the requirement for periods of low or minimal melatonin levels for proper transmission of the photoperiodic message.

<u>(iv) Endogenous Rhythms of Melatonin Secretion</u>: Groups of salmon S1 post-smolts were subjected to the following photoperiod regimes: (1) a simulated natural seasonal light cycle, 6 months out-of-phase with the ambient photoperiod, (2) a simulated natural seasonal cycle until 21 December with the daylength on that day (ie short day) then held constant (Winter Solstice-hold) and (3) a stimulated natural seasonal cycle until 21 June and then

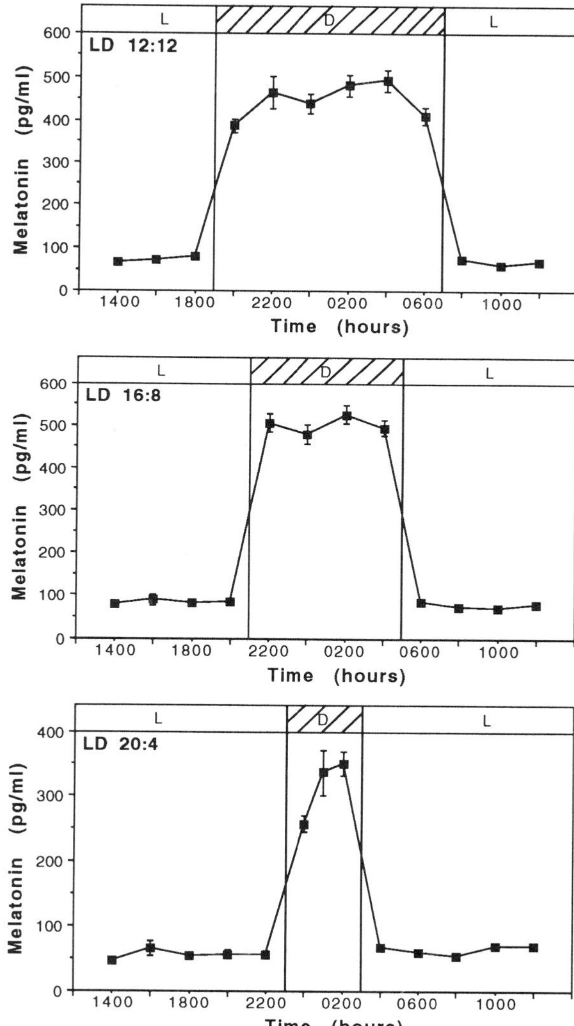

Fig 2: Diel serum melatonin under LD12:12, LD16:8 and LD20:4.

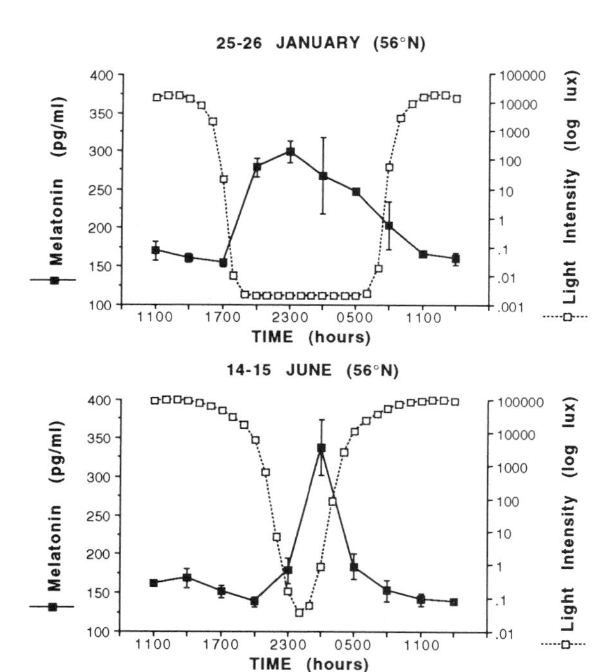

Fig 3: Seasonal profiles of diel melatonin

daylength held constant (Summer Solstice-hold). Plasma samples were collected at 2hr intervals over 24hr periods at randomly selected times over the following year for Group 1 and for 2-3 months following the Solstice holds for Groups 2 and 3.

In all fish, the diel patterns of melatonin accurately reflected the lengths of the dark periods of the photoperiods to which the fish were being exposed. On no occasion was there any indication that the patterns of melatonin secretion were under the influence of any circadian or circannual rhythm.

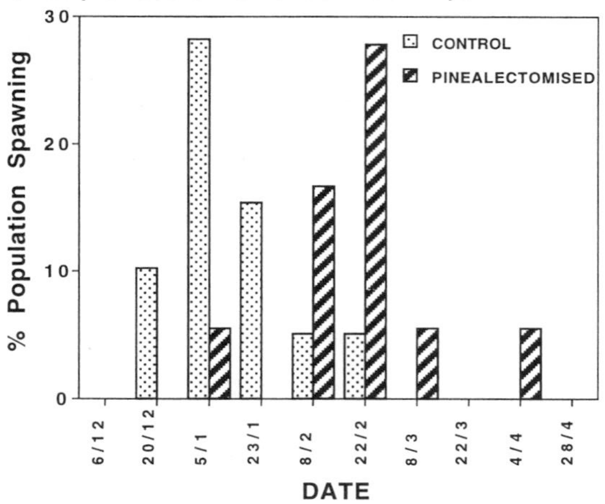

Fig 4: Effects of Summer Solstice pinealectomy on spawning time.

Conclusions

(1) The results show clearly that under a range of seasonally-changing and constant photoperiods, melatonin levels accurately reflect the length of the period of darkness. It is suggested that the diel and seasonally-changing patterns of circulating melatonin encodes information on both daily and calendar time which is then used to time daily and seasonal events.

(2) The ability of alterations in photoperiod to bring about advances and delays in spawning and in parallel to produce modifications in profile of melatonin secretion suggests that melatonin is a mediator of the photoperiodic control of reproduction.

(3) Pinealectomised fish did not exhibit any diel changes in circulating melatonin. Pinealectomy at the time of the Summer Solstice produced a delay in spawning, compared to controls suggesting that the pineal (and melatonin) is necessary for the entrainment of reproduction which, in part, is normally brought about by the decreasing daylengths (ie. increased daily duration of melatonin secretion) of Autumn.

(4) Implants of melatonin failed to modify spawning time possibly because alternating periods of high and low melatonin are required for successful transmission of information on daylength.

(5) The consistent representation of night length by the duration of melatonin increase under both the out-of-phase seasonal and Solstice-hold photoperiods indicates that there is no endogenous circannual influence on melatonin secretion.

(6) It is concluded that the endogenous rhythm(s) which control reproduction in salmonid fish must be acting "down-line" from the information provided by melatonin on daily and calendar time.

Bromage, N. *et al* (1982) Relationships between serum levels of GTH, E2 and vitellogenin in the control of reproductive development in the rainbow trout. General and Comparative Endocrinology 47, 366-376.

Bromage, N. *et al* (1993) Environmental control of reproduction in salmonids. Recent Advances in Aquaculture. Vol.IV, pp55-65.

Randall, C. *et al* (1995) Melatonin rhythms in Atlantic salmon maintained under natural and out-of-phase photoperiods. General Comparative Endocrinology 98,73-86.

(Supported by a NERC grant to N.B)

PHOTOPERIOD CONTROLS THE TIMING OF REPRODUCTION IN ATLANTIC COD (GADUS MORHUA).

Birgitta Norberg[1], Björn Thrandur Björnsson[2] and Carl Haux[2]

[1]Institute of Marine Research, Austevoll Aquaculture Research Station, N-5392 Storebö, Norway
[2]Department of Zoophysiology, University of Göteborg, Medicinaregatan 18, S-413 90 Göteborg, Sweden

Summary

Juvenile Atlantic cod (Gadus morhua) were submitted to four different photoperiod treatments, designed to advance or delay spawning time. The results clearly show that photoperiod manipulation can be used to alter and control the time of first and subsequent spawnings in Atlantic cod of both sexes.

Introduction

The Atlantic cod is a teleost species of great commercial importance to the fishing nations surrounding the North Atlantic. Methods for successful cultivation of cod exist, and the depletion of several wild cod stocks has created a renewed interest for aquaculture of this species. A stable, continous supply of juveniles is crucial in order to achieve a commercially viable cod production. Since the cod is a seasonal breeder, manipulation of spawning time is an important tool for obtaining a year-round production of fry.

The present study describes the successful manipulation of spawning time in male and female Atlantic cod, by treatment with compressed or extended annual photoperiod cycles. Plasma levels of the sex steroids estradiol-17β and testosterone were monitored in females, and plasma levels of testosterone were monitored in males.

Materials and methods

Experimental design

Juvenile, one year old, Atlantic cod of both sexes, with an average weight of 310 g, were randomly divided in four groups of 150 individuals. The groups were kept in circular indoor tanks (3m diameter), surrounded by light-proof black plastic and with a fluorescent daylight tube, connected to a time switch, as light source. On July 1 1992, the experiment was started, by submitting the four groups to four different annual photoperiod cycles:

1. ADV6 - first annual photoperiod cycle compressed to six months, second annual photoperiod cycle twelve months.
2. ADV3 - first annual photoperiod cycle compressed to nine months, second annual photoperiod cycle twelve months.
3. DEL - first annual photoperiod cycle extended to eighteen months, second annual photoperiod cycle twelve months.
4, NORM - first and second annual photoperiod cycles twelve months.

Each month, the fish were checked for signs of sexual maturation, i. e. running eggs and milt and blood samples were taken from twenty randomly chosen individuals. To reduce sampling stress, no individual was sampled two months in a row. The blood samples were immediately centrifuged and the plasma frozen on N_2(l). The frozen plasma samples were stored at -20°C until analysis.

Steroid analysis

Plasma levels of estradiol-17β and testosterone were analysed by previously validated radioimmunoassays (Methven et al., 1992). Antisera against estradiol-17β and testosterone were purchased from ICN Immunobiologicals Inc., reference standards were purchased from Sigma Chemical Co. and tritium labelled steroids were obtained from Amersham.

Results

Spawning time

In group ADV6, fish with running eggs and milt were first detected in January and February 1993. In July 1993, running males were again detected. The following month, spawning behaviour was observed in the tank and at sampling, all females had running eggs. Eggs and milt was detected until October, while only mature males were found in November.

In group ADV3, running males were first detected in January 1993. In February and March, both male and female fish were mature. In April, only mature males were found. Running males were again detected in this group in November and December 1993, while both females and males were mature in January 1994.

In group DEL, running males were found from February to August 1993, while mature fish of both sexes were found from March until July this year. In 1994, the fish spawned from September to November, and mature males were found until December.

In the control group, NORM, mature males were found from January to May both in 1993 and in 1994. Females had running eggs from February to May in 1993 and from February to April in 1994.

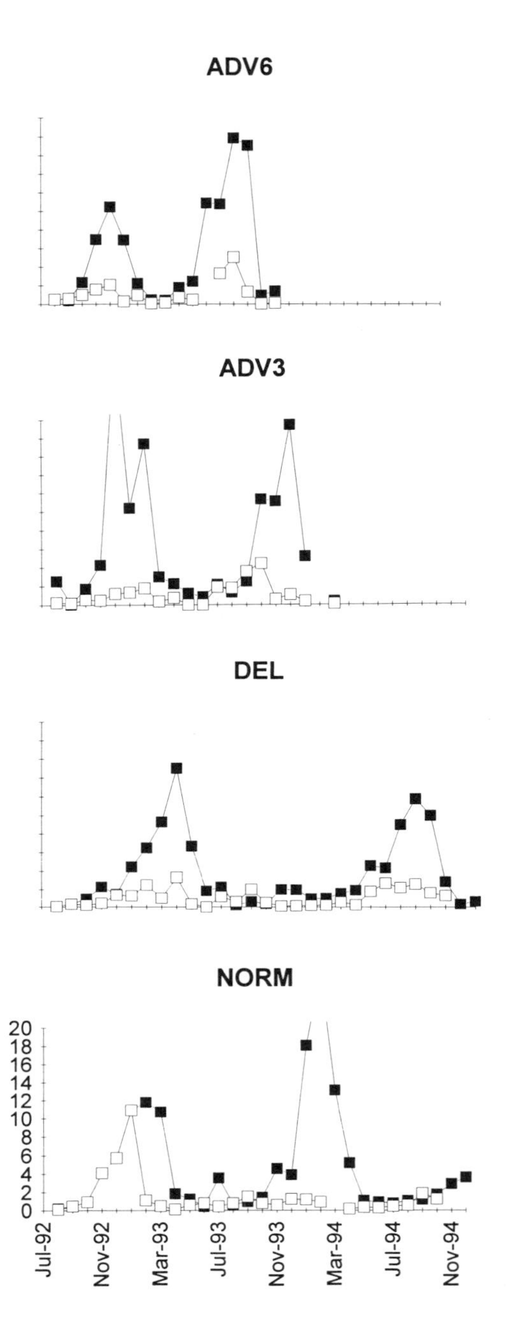

Fig. 1. Plasma levels of estradiol-17β and testosterone in female Atlantic cod, held at four different annual photoperiod cycles. Filled squares: estradiol-17β; open squares: testosterone

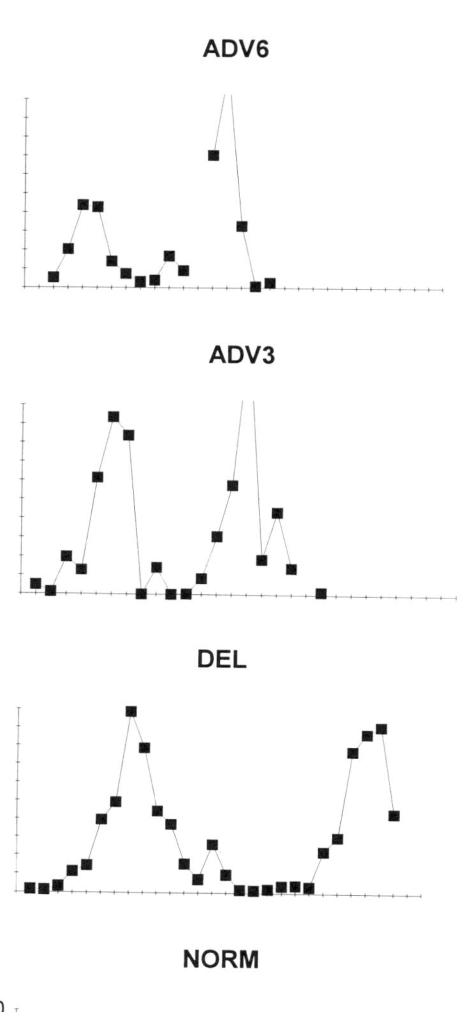

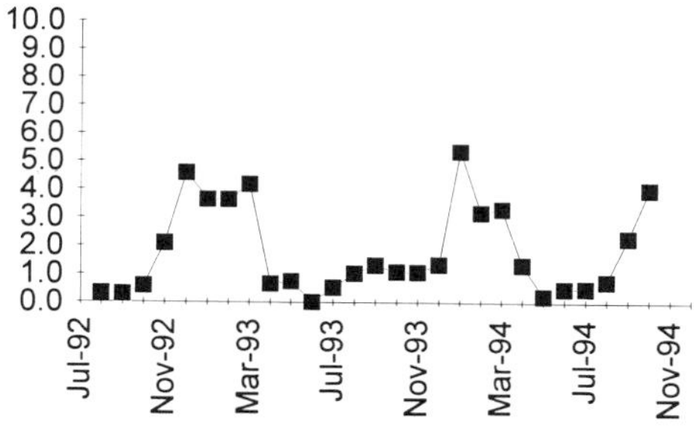

Fig. 2. Plasma levels of testosterone in male Atlantic cod held at four different annual photoperiod cycles.

Plasma steroid hormone levels

Plasma levels of estradiol-17β and testosterone in female cod, and of testosterone in male cod, are shown in Figs. 1 and 2, respectively. The highest plasma steroid levels were detected shortly before or during spawning in both female and male cod.

Discussion

Photoperiod manipulation is a powerful tool for controlling the timing of reproduction in Atlantic cod, as shown by the results of the present study. Steroid hormone levels were similar and in accordance with previous investigations (Kjesbu, Kryvi and Norberg, unpublished results) in the various groups. The underlying physiological mecanisms enabling different groups of cod to mature and spawn twice in one year, or extend their spawning period from three to five months in response to different photoperiod treatment, remain largely unknown and merit further investigation.

References

Methven, D. A., Crim, L. W., Norberg, B., Brown, J. A., Goff, G. P. and Huse, I. (1992). Seasonal reproduction and plasma levels of sex steroids and vitellogenin in Atlantic halibut (Hippoglossus hippoglossus) Can. J. Fish. Aquat. Sci. 49:754-759.

EFFECTS OF SPRING TEMPERATURE AND FEEDING REGIME ON SEXUAL MATURATION IN ATLANTIC SALMON (*SALMO SALAR* L.) MALE PARR

Ingemar Berglund

Department of Aquaculture, Swedish University of Agricultural Sciences, S-901 83 Umeå, Sweden

Summary

Food restriction in spring had a significant negative effect on the incidence of maturation in male salmon parr, but maturation seems to be suppressed in only a small part of the experimental populations. Even males with very low or negative growth rate in June matured. Water temperature in spring seems, however, to have a more general effect on the onset of rapid gonadal growth in late June; low water temperature during May and June inhibited sexual maturation in most males.

Introduction

Sexual maturation in male Atlantic salmon (*Salmo salar* L.) parr is common under both natural and hatchery conditions. The probability of maturation in male parr aged 1+ yrs is positively related to its size at the end of the first summer under both wild (Myers et al. 1986) and experimental conditions (Berglund 1992; Simpson 1992). Furthermore, it has been found that the opportunity for growth, in spring, before the start of rapid gonadal growth in early summer, affects the proportion of mature parr (Rowe and Thorpe 1990; Berglund 1992; Prevost et al. 1992).

The preliminary "decision" of a male parr to mature at age 1+ seems to be based on its size after the first summer (Myers et al. 1986; Berglund 1992; Simpson 1992). The accomplishment of this decision seems, however, to depend on the growth rate in spring the second summer. It has been suggested that rapid spring growth and accumulation of fat are necessary for the onset of rapid gonadal growth in early summer. The present paper reports on the effects feeding opportunity and water temperature in spring on the probability of sexual maturation in experimental groups of 1+ male salmon parr.

One-year-old Atlantic salmon parr of Baltic origin, belonging to the Skellefteälven (Skellefte River stock, 64°45'N, 21°E), were used in this study. Fish were raised from eggs under standard hatchery conditions with through-flowing river water at ambient temperature and under natural photoperiod.

Experiments were performed in 1992, 1993, and 1994. During the experiments groups of fish were subjected to a treatment, feeding or temperature regime, from early May (5-9, depending on year) to late June (23-27), after that they were reared at ambient temperature and fed at excess until the incidence of sexual maturation were determined in September. In all experiments 150-220 fish, selected at random to be representative of the stock, were used in each treatment group. Experimental groups were kept in 4-m^2 tangential-flow tanks and fed commercial fish-feed pellets dispensed by automatic feeders. All fish were individually tagged with passive integrated transponder (PIT) tags.

In 1992, effects of individual size and feeding opportunity in spring on the size-specific probability of sexual maturation were studied (see Berglund 1995 for details). Fish were subjected to three different feeding regimes: (1) controls feed at excess, (2) alternate starvation (three days)/excess feeding (two days), and (3) restricted (ca. 40% of recommended) daily ration, in May and June. Fish were reared at ambient temperature, which increased rapidly from 3° C in early May to about 16° C in mid June. There were two replicate tanks with 220 fish in each regime. In 1993, a preliminary temperature experiment with two temperature regimes was performed. In the ambient regime (two tanks) the temperature rose from 5° C in mid May to 12° C in mid June, and in the accelerated regime (one tank) from 4° C to 16° C. In 1994, three temperature regimes (two tanks in each) were applied. All regimes started at 3 ° C on 9 May, and reached 11, 13, and 16° C, respectively, by mid June.

All fish were measured at the start and at the end of the treatment periods, and their individual increase in weight and condition factor ($100(\text{weight}/\text{length}^3)$) were calculated.

Results

In 1992, the proportion of mature males decreased with decreasing feeding opportunity in May-June (Table 1, Berglund 1995). The mean proportions of mature males in the restricted regimes were significantly reduced compared with the control regime.

Generally, the probability of maturation increased with increasing body length, reached a maximum, and then decreased for the largest males (Fig. 1). By logistic regression it was established that the effect of length of yearling parr on the probability of maturation differed among treatments (Log Likelihood ratio test, $p < 0.001$). The decrease in maturation rate among fish larger than 9 cm, was due to an increasing frequency of one-year-old smolts, which not mature sexually, with increasing size (Berglund 1995). Food restriction during June or May and June shifted the peak in maturation rate towards larger fish and suppressed maturation in males smaller than ca. 10 cm in length; males larger than 10 cm showed similar or higher maturation rate compared with controls (Fig. 1).

Table 1. Proportion of mature males and growth in weight in groups of salmon parr subjected to different feeding and temperature regimes in May and June.

Year/Regime	Proportion mature (%)[a]	Growth[b]
1992 Feeding		
Control	61	0.88
Starvation	46	0.10
Low ration	35	0.19
1993 Temperature		
High	54	1.01
Low	13	0.81
1994 Temperature		
High	37	1.07
Medium	15	0.87
Low	5	0.69

a) Within year, control or high temperature treatments differed from the other treatments by χ^2-test, $p<0.01$.
b) ln(weight at end of treatment) - ln(weight at start of treatment). Data represent fish with a length of 7 to10 cm in early May.

Maturing males grew faster than non maturing males in the low ration regime, (ANCOVA, $F_{1,196} = 9.69$, $p < 0.01$), but growth rates were similar in the other regimes. It should be noted that 10 of the 71 maturing males in the low ration regime decreased in weight during June.

In 1993 and 1994, the proportion of mature males was positively related to the temperature increase during May and June (Table 1, Berglund unpublished). Furthermore, maturation rate seems to decline more rapidly with decreasing growth in May and June if the water temperature is kept at a low level than if food ration is restricted.

Discussion

Food restriction in spring had a significant negative effect on the incidence of maturation in male salmon parr in this study, but maturation seems to be suppressed in only a small part of the experimental populations. Water temperature in spring seems, however, to have a more general effect on the onset of rapid gonadal growth in late June; low water temperature in May and June inhibited sexual maturation in most males.

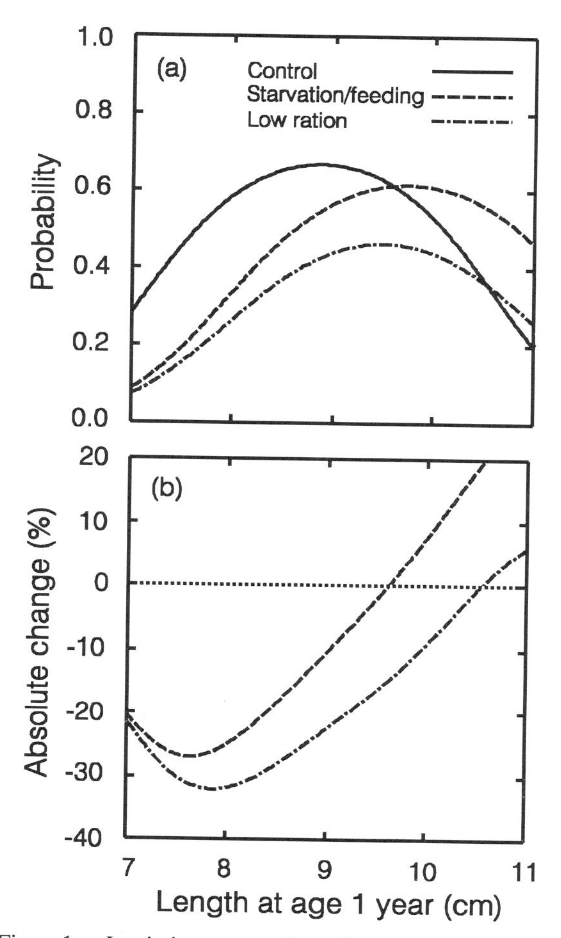

Fig. 1. Logistic regression lines showing the relationship between the probability of sexual maturation in male salmon parr at age 1+ in relation to size at age 1 year. Fish were subjected to three different feeding regimes in May and June: controls at excess, alternate starvation /feeding, and low (maintenance) ration. The general form of the regression equation was: Probability of maturation = $1/(1+\exp(\alpha - \beta_1(\text{length}) + \beta_2(\text{length})^2))$. (b) Absolute change in maturation rate for groups on restricted food. The change was calculated as: logistic regression equation for restricted group - regression equation of control. Reference line at change = 0 is given.

The occurrence of maturing males among starved fish with very low growth rate in June, and in poor condition by the end of June, implies that high growth rate in June is not necessary for the onset of rapid gonadal growth in early summer. This is further supported by the finding that slow temperature development in spring depressed maturation rate despite comparatively high growth rates.

Rowe et al. (1991), suggested that aromatization of testosterone in fat stores should be the physiological link between fat accumulation during spring and the initiation of gonadal growth in early summer. The above results, and the finding that no estrogens were formed from radiolabelled androstenedione in incubations of mesenteric fat tissue from male parr in mid June (Borg, Mayer, Berglund, and Lambert, unpublished data), makes the suggested mechanism unlikely.

The present findings suggest that water temperature in spring and early summer has stronger influence on maturation rate in male salmon parr than food availability and growth rate *per se*. It seems likely that the long-term growth, affecting size-at-age, of individuals, and the water temperature in spring and early summer are the most important factors explaining between-year variation in the proportion of mature male parr within salmon populations.

Acknowledgements

I thank the staff at the Kvistforsen hatchery for their help and hospitality. Financial support was given by the Swedish Forestry and Agricultural Research Council. I wish to thank Dr. H. Lundqvist, H. Königsson, and B.-S. Wiklund for valuable support and technical assistance.

References

BERGLUND, I. 1992. Growth and early sexual maturation in Baltic salmon (*Salmo salar*) parr. Can. J. Zool. 70: 205-211.

BERGLUND, I. 1995. Effects of size and spring growth on sexual maturation in 1+ Atlantic salmon (*Salmo salar* L.) male parr: interactions with smoltification. Can. J. Fish. Aquat. Sci. (In press)

MYERS, R.A., J.A. HUTCHINGS, AND R.J. GIBSON. 1986. Variation in precocious maturation within and among populations of Atlantic salmon. Can. J. Fish. Aquat. Sci. 43: 1242-1248.

PREVOST, E., E.M.P CHADWICK, AND R.R CLAYTOR. 1992. Influence of size, winter duration and density on sexual maturation of Atlantic salmon (*Salmo salar*) juveniles in Little Codroy River (southwest Newfoundland) J. Fish Biol. 41: 1013-1019.

ROWE, D.K. AND J.E. THORPE. 1990. Suppression of maturation in male Atlantic Salmon (*Salmo salar* L.) parr by reduction in feeding and growth during spring months. Aquaculture 86: 291-313.

ROWE, D.K., J. E. THORPE, AND A.M. SHANKS. 1991. The role of fat stores in the maturation of male Atlantic salmon (*Salmo salar*) parr. Can. J. Fish. Aquat. Sci. 48: 405-413.

SIMPSON, A.L. 1992. Differences in body size and lipid reserves between maturing and nonmaturing Atlantic salmon parr, *Salmo salar* L. J. Zool. 70: 1737-1742.

APPLICATION OF REPRODUCTIVE PHYSIOLOGICAL TESTING TO UNDERSTAND THE MECHANISMS OF ENVIRONMENTAL ENDOCRINE DISRUPTORS

G. Van Der Kraak[1], M.E. McMaster[1] and K.R. Munkittrick[2]

[1]Dept. Zoology, University of Guelph, Guelph, Ontario, Canada, N1G 2W1; [2]Dept. Fisheries and Oceans, Great Lakes Laboratory for Fisheries and Aquatic Sciences, Burlington, Ontario Canada, L7R 4A6

Abstract

Assessment of the reproductive performance of fish is increasingly being used to evaluate the impacts of environmental disturbances. Reproductive physiologists are being called upon to provide quick answers to questions related to the identity and actions of compounds in complex mixtures, and the effectiveness of chemical process changes and remedial treatment regimes. At this time, there are no surrogates for long term experiments, although physiological testing of wild fish populations is providing a means of focusing attention to key components of the reproductive axis sensitive to environmental disturbances.

Introduction

There is growing awareness and concern that environmental contaminants can influence the endocrine system and thereby play a role in the developmental and reproductive problems observed in wildlife and human populations. The term endocrine disrupting chemicals (EDCs) has been used to categorize the broad suite of environmental compounds which affect endocrine function (Colborn et al. 1993). There is ongoing debate as to whether EDCs contribute to the increased incidence of breast cancer in women, or the increase in reproductive tract disorders in men and proposed progressive declines in human sperm counts (Colborn et al., 1993; Safe 1995).

A large number of EDCs have been shown to alter reproductive endocrine function in fish in field and laboratory studies (Table 1). For example, depressed levels of circulating steroids and altered reproductive function have been observed in fish collected downstream of pulp mills and sewage treatment plants, at sites contaminated with polynuclear aromatic hydrocarbons (PAHs) and polychlorinated biphenyls (PCBs) (Johnson et. al. 1988; Munkittrick et al. 1992; Purdom et al. 1994; see review in McMaster et al. 1995a). Other alterations in reproductive function include smaller gonads and egg size, delayed sexual maturity, reduction or sex-reversal of secondary sexual characteristics, changes in sex ratios, and increased embryonic and early life stage mortality.

Table 1. Some of the EDCs that affect reproductive function in fish.

Pesticides
Herbicides, Insecticides, Fungicides
Industrial and Synthetic Chemicals
Metals (Cd, Pb, Hg)
Persistent organochlorines (Dioxins, PCBs)
PAHs
Alkylphenols
Phthalates
Phytochemicals
Isoflavones, Sterols

While monitoring wild fish populations may provide the strongest measure of ecological relevance, rapid laboratory testing will be required to evaluate both the increasing number of suspected EDCs and the effectiveness of remediation techniques once problems are identified. There is also a growing awareness that existing, standard, toxicological test methods have not been effective in detecting the actions of EDCs. Standardized test methods to screen for the activity of EDCs, have to be established for fish. *In vitro* (receptor binding, steroid and vitellogenin production) and short term whole animal tests (plasma hormone levels, functional testing of the hypothalamo-pituitary gonad axis) have been regularly used to assess the physiological mechanisms mediating reproductive processes. It is necessary to adapt the existing technology for:

a) use in evaluating the effects of EDCs,
b) identification of EDCs present in complex environmental mixtures, and
c) developing functional linkages between physiological changes and whole organism

effects.

This paper reviews some of the limitations of current *in vivo* and *in vitro* testing protocols for application in evaluating the effects of EDCs.

In Vivo Endocrine Responses

Many studies evaluating wild fish populations have included measurement of reproductive hormone levels. Our work downstream of a bleached kraft pulp mill showed that white sucker (*Catostomus commersoni*) and lake whitefish (*Coregonus clupeaformis*) had depressed plasma sex steroid and GTH II levels which correlated with whole organism responses (e.g. gonad size, sexual maturity, secondary sexual characteristics) (Munkittrick et al. 1992; Van Der Kraak et al. 1992). By comparison, longnose sucker (*Catostomus catostomus*) from this site had depressed hormone levels but no consequences at the whole organism level. These findings illustrate the difficulty in interpreting the effects of EDCs on reproductive function. Although the endocrine events which mediate reproductive processes are well understood, we have a limited understanding of which endocrine responses offer the most potential for prediction of whole animal and population level effects.

There is a lot of current interest in the measurement of vitellogenin levels as an indicator of exposure to EDCs with estrogenic activity (Purdom et al, 1994). While there is great public concern over the presence of female specific proteins in male fish, the consequences of vitellogenin production to the long term reproductive performance is unknown. The relevance of these changes needs to be carefully assessed in wild fish.

Laboratory exposures are useful for screening compounds and elucidating mechanisms, but often do not integrate all of the conditions and stressors present in a field environment. Although indications of changes in wild fish exposed to contaminants is the best measure of ecological relevance, artificial exposures of naive fish to effluents in the laboratory or at field locations can be invaluable. These tests can provide critical information to aid in establishing cause and effect linkages and separating the confounding effects of environmental variables (altered habitat, temperature changes, etc.) in field studies. Field caging studies using goldfish (*Carassius auratus*) have been successfully applied to our work on pulp mills (McMaster et al. 1995b) and by Sumpter's group in England using rainbow trout (*Oncorhynchus mykiss*) downstream of sewage treatment lagoons (Purdom et al. 1994). Generally, these studies have been of short duration and the long term consequences of exposure, including multi-generational effects, have not been addressed. In mammals, the developing embryo tends to be the most sensitive to EDCs exposure (Colborn et al. 1993).

Testing the functional integrity of the pituitary gonadal axis in wild or laboratory exposed fish can be used to characterize the actions of EDCs (Van Der Kraak et al. 1992). Monitoring the response of fish to exogenous GnRH adminstration and *in vitro* incubation of gonadal tissue served to identify sites within the pituitary-gonadal axis altered by EDCs. Knowledge of the mechanisms underlying reproductive changes can serve to identify the impact points, which can focus follow up studies on relevant endpoints to identify or characterize the responsible stressors.

In Vitro Endocrine Responses

EDCs impact the reproductive process through a wide variety of mechanisms (Table 2). However, most of the work to date has focused on interactions with the estrogen receptor and induction of estrogen dependent responses. A diverse array of environmentally relevant chemicals including the phytosterols, non-ionic surfactants (nonylphenol), pesticides (o,p'-DDT and DDE) and hydroxylated PCBs (White et al., 1994; Tremblay et al., this volume, P.Thomas unpublished) bind to the hepatic estrogen receptors of fish. Compared to 17β-estradiol, the receptor affinity of xenobiotics with estrogenic activity are often 4 to 6 orders of magnitude lower in activity. This has led some investigators to question whether interactions of xenoestrogens with the estrogen receptor account for the biological effects of these compounds (ie. Safe, 1995). EDCs with estrogenic activity show significant differences in the relative potencies between *in vitro* (receptor binding) and *in vivo* bioassays (plasma steroid levels) which questions the predictive value of the *in vitro* test methods. Understanding the actions of EDCs with estrogenic activity is further complicated by 17β-estradiol having multiple sites of action which may differ in sensitivity (Table 3). This calls into question which endpoint would be most predictive.

Table 2. Some of the mechanisms of action of EDCs.

1. Analogues of endogenous hormones
   a. Steroids
   b. Thyroid hormones
   c. Fatty acids/peroxisomal proliferators
2. Aryl hydrocarbon (Ah) Receptor agonists
   a. Induction of P450 enzymes
   b. Growth factor/receptor expression
3. Modulation of the immune system
4. Modulation of vitamin metabolism

Table 3. The diverse actions of 17β-estradiol

Target	Action
Brain/Hypothalamus	Differentiation
	GnRH secretion
Pituitary	**GTH Secretion**
Liver	Vitellogeninsecretion
	Estradiol receptors
Ovary	Steroid biosynthesis
	Steroid metabolism
	DNA synthesis
	Sex differentiation
Plasma	Carrier Proteins

Many *in vitro* test methods could serve as surrogates for whole animal testing for reproductive toxicity (e.g. receptor binding, mitogenic responses, hormone biosynthesis and metabolism). EDCs may exert their effects through more than one mechanism. For example, hydroxylated PCBs are estrogenic, but they also interact with the Ah receptor, which is responsible for turning on the synthesis of a variety of proteins, including hepatic P450 detoxification enzymes. The potential endocrine consequences of the interactions of the various disruptions is poorly understood.

Conclusion

There are no surrogates for whole animal testing, and laboratory testing lacks the integration provided by multi-generational exposures of wild organisms. Existing physiological testing protocols must be carefully integrated into studies of EDCs to provide insight into the mechanisms of disruption before predictive test methods can be developed.

References

Colborn, T., F.S. vom Saal and A.M. Soto. 1993. Developmental effects of endocrine-disrupting chemicals in wildlife and humans. Environ. Health Perspect. 101: 378-384.

Johnson, L.L., E. Casillas, T. Collier, B.B. McCain and U. Varanasi. 1988. Contaminant effects on ovarian development in English sole (Parophrys vetulus) from Puget Sound, Washington) Can. J. Fish Aquat. Sci. 45:2133-2146.

McMaster, M.E., G.J. Van Der Kraak and K.R. Munkittrick. 1995a. An epidemiological evaluation of the biochemical basis for steroid hormonal depressions in fish exposed to industrial wastes. J. Great Lakes Res. IN PRESS.

McMaster, M.E., K.R. Munkittrick, G.J. Van Der Kraak and M.R. Servos. 1995b. Recovery of steroid function in fish exposed to pulp mill effluent at Jackfish Bay. In: M.R. Servos, et al. (eds.), Environmental Fate and Effects of Pulp and Paper Mill Effluents, St. Lucie Press, Boca Raton, FL. IN PRESS.

Munkittrick, K.R. G.J. Van Der Kraak, M.E. McMaster and C.B. Portt. 1992. Reproductive dysfunction in three species of fish exposed to bleached kraft mill effluent at Jackfish Bay, Lake Superior. Water Poll. Res. J. Canada 27:439-446.

Purdom, C.E., P.A. Hardiman, V.J.Bye, N.C. Eno, C.R. Tyler and J.P.Sumpter. 1994. Estrogenic effects of effluents from sewage treatment works. Chem. Ecol. 8: 275-285.

Safe, S. 1995. Do environmental estrogens play a role in development of breast cancer in women and male reproductive problems? Human Ecol. Risk Assess. 1:17-24.

Van Der Kraak, G.J., K.R. Munkittrick, M.E. McMaster, C.B. Portt and J.P. Chang. 1992. Exposure to bleached kraft pulp mill effluent disrupts the pituitary-gonadal axis of white sucker at multiple sites. Toxicol. Appl. Pharmacol. 115: 224-233.

White, R., S. Jobling, S.A. Hoare, J.P. Sumpter, and M.G. Parker, 1994. Environmentally persistent alkylphenolic compounds are estrogenic. Endocrinology **135**:175-182.

A LONG TERM STUDY OF THE EFFECTS OF POLLUTED SEDIMENT ON THE ANNUAL REPRODUCTIVE CYCLE OF THE FEMALE FLOUNDER, PLATICHTHYS FLESUS

J.G.D. Lambert and P.A.H. Janssen

Department of Experimental Zoology, Research Group Comparative Endocrinology, University of Utrecht, Padualaan 8, 3584 CH Utrecht, The Netherlands.

Summary.

The euryhaline flatfish *Platichthys flesus* inhabits coastal/estuarine waters and is therefore suitable for monitoring organic chemical pollutants. Compounds such as polycyclic hydrocarbons, polychlorinated biphenyls and pesticides are easily incorporated and often metabolized by enzyme systems which are also involved in steroid metabolism. Disturbances of reproduction may therefore occur through interference of these compounds with the endocrine system.
Fish were kept for three years in mesocosm systems containing polluted sediment from Rotterdam harbour and sampled twice a year in May and November. The gonadal development and steroidogenesis were studied. Moreover, plasma levels of estradiol and vitellogenin were determined. The November females all contained ovaries with vitellogenic oocytes, comparable with those captured in nature. In May, however, all the control animals were in previtellogenic phase while the ovaries of the "polluted" animals contained in addition to previtellogenic oocytes large numbers of yolk containing oocytes. The levels of estradiol and vitellogenin were elevated in "polluted" females, but the ovarian capacity to synthesize estrogens was comparable to that in the controls. In "polluted" males, however, no trace of vitellogenin could be detected. This indicates that a direct estrogenic effect of the pollutants can be excluded. Steroids, as well as xenobiotics, will be metabolized in the liver by mixed function oxygenase (cytochrome-P450) enzymes. An interference of both compounds is possible resulting in a decreased clearing of steroids. This might explain the increased level of estradiol and consequently the presence of vitellogenin.

Introduction.

The Dutch coastal waters contain a variety of pollutants, including polycyclic aromatic hydrocarbons (PAHs) and polychlorinated biphenyls (PCBs). These organic chemical pollutants are easily taken up in fish after which compounds such as PCBs are known to be stored in adipose tissue, liver, muscle and gonads (Von Westernhagen *et al*. 1987). In fish these organic compounds (xenobiotics) can reduce reproductive success by interacting directly with gonads and germ cells (Armstrong 1986). Indirect effects on reproduction, via interference with the regulating hormonal system have also been reported (Freeman *et al.* 1980; Thomas 1989). For monitoring the quality of the Dutch coastal waters the reproductive success of fish might be a good indication.
The flounder, *Platichthys flesus*, has been selected as a model species because this euryhaline flatfish lives on and in the sediment of coastal waters and estuaries, which can result in a high exposure to pollutants that enter these areas from the rivers.
As a basis, a detailed description of the developmental phases of the ovary of the flounder was made (Janssen *et al*. 1995), as well as an extensive description of the steroids identified in the blood of females during the annual cycle (Asahina *et al*. in prep). To study the influence of pollution, flounder were kept for about three years in a large mesocosm system filled with polluted sediment. In May (spring) and November (fall), ovarian morphology and steroidogenesis were studied. Moreover steroid metabolism in the liver was studied to investigate whether PAHs (benzo[a]-pyrene) (BaP)) can influence the hydroxylations (cytochrome P450 activity) of steroids.

Material and Methods.

Animals.

The effects of pollution were studied in fish held in two large self-supporting mesocosm systems of 40 x 40 x 3 m. The reference mesocosm contained relatively clean Wadden Sea sediment and water. The polluted mesocosm contained Rotterdam harbour sediment, contaminated with a variety of organic chemical pollutants and heavy metals. One year old fish captured in an area of the Dutch Wadden Sea, known to have a relatively low level of pollution, were divided at random over the two mesocosms.

To induce cytochrome P450 enzyme activity in the liver, animals captured in the Wadden Sea were injected intraperitoneally with BaP (2 mg/kg) dissolved in DMSO (Eggens et al.). After two days the livers were used for steroid metabolism studies.

Histology.

Ovarian tissue was fixed in Smith's formalin-dichromate (6-7 hrs) and after rapid dehydration embedded in paraffin. Sections (7 µm) were stained with Mayer's hematoxylin-eosin.

Steroid metabolism.

Minced ovarian tissue was incubated in L15 medium with ^{3}H-androstenedione to study the aromatase activity. To study hepatic steroid catabolism (hydroxylations), homogenized liver tissue was incubated separately with tritiated pregnenolone, testosterone and estradiol in the presence of NADPH in a phosphate buffer. The produced steroids were isolated and identified by TLC as described previously by Schoonen and Lambert (1986).
Competition studies were carried out by incubating liver tissue with steroid precursor in combination with a 1000 fold higher BaP concentration

Results and discussion.

Ovarian morphology

The ovaries of the females collected in May from the reference mesocosm were morphological identical to the ones from the Wadden Sea. The ovary was characterized by numerous oocytes in the late perinucleolus stage. Also the younger stages were present such as oogonia and oocytes in chromatin nucleolus and early perinucleolus stage. The ovaries were previtellogenic. The ovaries of females from the polluted mesocosm, however, contained besides the smaller oocytes, oocytes in yolk granule stage. Exogenous vitellogenesis had been started with deposition of acidophilic yolk granules at the periphery of the oocyte. Moreover, the oocytes seemed to develop less synchronous than oocytes in ovaries from the reference mesocosm. No difference could be observed between the control and polluted ovaries of the November animals. Vitellogenesis was advancing and the ovary consisted mainly of large oocytes, densely packed with yolk granules.

From these results it is obvious that pollution can alter the morphology of the ovary; vitellogenesis is induced. Normally, estradiol is responsible for this induction. The premature vitellogenesis in the polluted fish might be explained by a change in steroidogenesis, resulting in a higher level of estradiol or by a decrease in the catabolism of estradiol as a result of a change in liver cytochrome P450 activity or by a competitive inhibition by organic pollutants such as PAHs and PCBs (Waxman,1988). An alternative explanation might be that the estrogenic effect is not accomplished by endogenous estrogens but by exogenous compounds (xeno-estrogens) that act directly on the liver (Purdom et al., 1994). Males, however, living for three years in the polluted mesocosm do not show any vitellogenin in the plasma, indicating that an exogenous estrogenic effect is not very plausible.

Steroid levels in plasma.

Estradiol (E2) and testosterone (T) levels were measured by RIA in the plasma of control and "polluted" females collected from the mesocosms in May. The levels of both steroids in the control animals (E2: 0.12 ng/ml; T: 0.08 ng/ml) is even lower than in the previtellogenic animals from the Wadden Sea (Asahina et al. in prep.). In the "polluted" flounder, however, a 100 per cent increase could be demonstrated (E2: 0.3 ng/ml; T: 0.15 ng/ml). Although the E2 level is not nearly as high as during natural vitellogenesis, a long term small increase might be sufficient to induce vitellogenin synthesis. The increased level of estradiol can be explained by an increased ovarian synthesis or by a decreased hepatic catabolism. An increase in steroid levels is contradictory to the results of Thomas (1989) and Yano and Matsuyama (1986) who found respectively a decrease of steroids in the Atlantic croaker and carp after a treatment with a mixture of PCBs

Ovarian estrogen synthesis.

Minced ovarian tissue of control and "polluted" animals were incubated with ^{3}H-androstenedione as precursor. The main products were testosterone, estrone and estradiol. No difference could be observed between the two incubations. This is an argument to hypothesize a decreased hepatic catabolism to explain the increased estradiol level.

Hepatic steroid catabolism.

In the intermediary metabolism of steroids the hepatic cytochromes P450 play an important role. In addition the P450s participate in the oxidation of many xenobiotics. It is known that both steroids and xenobiotics can serve as substrates for the same enzyme. In the presence of an excess amount of xenobiotics, a competitive enzyme inhibition is likely, resulting in a decreased hydroxylation of steroids. This was studied by incubating liver homogenates with ^{3}H-testosterone in the presence of BaP (1000 fold). The main products were 5ß-reduced androgens such as 5ß-dihydrotestosterone, 5ß-androstane-3α,17ß-diol and its water soluble conjugate. Only a minor percentage of testosterone was converted into cytochrome P450 catalyzed products 2ß(OH)-testosterone and 6ß(OH)-testosterone and androstenedione (besides 17ß-HSD, also a P450 isozyme can catalyze the conversion to androstenedione). Therefore a difference in conversion could hardly be demonstrated. To overcome this problem an incubation was set up with pregnenolone as substrate. This steroid was converted by the liver into a non identified polar compound: "hydroxylated pregnenolone". A competitive inhibition, however, could not be demonstrated; an incubation in combination with BaP did not alter the conversion rate of pregnenolone. It is therefore not likely that the increased steroid levels in the plasma were due to substrate-enzyme competition. Another cause for the increased steroid level could be the changes that occur in cytochrome P450 activity after induction by xenobiotics. A single i.p. injection of BaP induces in the liver a 5-fold increase of the EROD activity and a 50 % increase of the total P450 amount (Eggens et al. in prep). To study the influence of an increased hepatic P450 activity on the hydroxylation of steroids, flounder were treated once with 2 mg/kg BaP. In this pilot study, the control and induced livers of four animals were pooled separately, homogenized and incubated with ^{3}H-pregnenolone. It appeared that under the experimental conditions, the hepatic conversion of pregnenolone into the hydroxylated product was about 50 % reduced in the P450 stimulated animals. In vivo this can lead to a decreased clearing of steroids resulting in an increased steroid level. Apparently, it is a contradiction: a decrease in pregnenolone conversion while the cytochrome P450 has been increased. This can be explained, however, by a competition at the level of the enzyme, NADPH-cytochrome P450 reductase (Cawley et al., 1995). The steroids and the xenobiotics will be metabolized by different P450 isozyms and since NADPH-cytochrome P450 reductase is the limiting component, different P450 isozyms must compete for the available reductase molecules.

References

Armstrong,D.T.(1986) Environmental stress and ovarian function. Biology of Reproduction **3**, 29-39.

Asahina,K., Janssen,P.A.H., Vermeulen,G.J., Lambert,J.G.D. and Goos,H.J.Th. (in prep.) Changes in plasma steroid profiles during the annual cycle of the female flounder, Platichthys flesus: Identification and quantification by gaschromatography - mass spectrometry.

Cawley,G.F., Batie,C.J. and Backes,W.L. (1995) Substrate dependent competition of different P450 isozymes for limiting NADPH-cytochrome P450 reductase. Biochemistry **34**, 1244-1247.

Eggens,M.L., Vethaak,A.D., Leaver,M.J., Horbach,G,J,M.J., Boon,J.P.and Seinen,W. (in prep). Differences in CYP1A response between flounder and plaice after long term exposure to harbour dredged spoil in a mesocosm study

Freeman,H.C., Uthe,J.F. and Sangalang,G.(1980) The use of steroid hormone metabolism studies in assessing the sublethal effects of marine pollution. Rapp. P.-v.Reun. Cons.t. Explor. Mer. **179**, 16-22

Janssen,P.A.H., Lambert,J.G.D. and Goos, H.J.Th.(1995). The annual ovarian cycle and the influence of pollution on vitellogenesis in the flounder Platichthys flesus. Journal of Fish Biology, in press.

Schoonen,W.G.E.J. and Lambert,J.G.D. (1986) Steroid metabolism in the testes of the African catfish, Clarias gariepinus, during the spawning season, under natural conditions and kept in ponds. Gen. comp. Endocr. **61**, 40-52.

Purdom,C.E., Hardiman,P.A., Bye,V.J., Eno, N.C., Tyler,C.R. and Sumpter,J.P. (1994). Estrogenic effects of effluents from sewage treatment works. Chemistry and Ecology **8**, 275-285.

Thomas,P. (1989). Effects of Arochlor 1254 and cadmium on reproductive endocrine function and ovarian growth in Atlantic croaker. Mar.env.Res. **28**,499-503.

Von Westernhagen,H., Dethlefsen,V.,Cameron, P. and Janssen,D. (1987). Chlorinated hydrocarbons residues in gonads of marine fish and effects on reproduction. Sarsia **72**, 419-422.

Waxman, D.J. (1988). Interactions of hepatic cytochrome P-450 with steroid hormones. Regioselectivity and stereospecificity of steroid metabolism and hormonal regulation of rat P-450 enzyme expression. Biochem. Pharm. **37**, 71-84

Yano,T. and Matsuyama,H. (1986). Stimulatory effect of PCB on the metabolism of sex hormones in carp hepatopancreas. Bull. Jap. Soc. scient. Fish. **52**, 1847-1852

EFFECTS OF AROMATASE INHIBITORS ON SEXUAL MATURATION IN THREE-SPINED STICKLEBACK, GASTEROSTEUS ACULEATUS.

C. Bornestaf, E. Antonopoulou, I. Mayer and B. Borg.

Department of Zoology, University of Stockholm, S-106 91 Stockholm, Sweden.

Introduction

Sticklebacks mature when exposed to long but not to short photoperiod in winter. Maturation in long photoperiod involves the activation of gonadotropic (GTH) cells, via a positive feedback from the gonads on the (hypothalamus)-pituitary-gonad axis (Borg et al. 1989). Androgens are aromatized to estrogens in the brain in teleosts. Aromatization has been indicated to be of critical importance for feedback(s) on the hypothalamus-pituitary-gonad axis in fish, including positive feedbacks in salmonids (Crim et al. 1981, Antonopoulou et al. 1995). The aim of the present experiment was to investigate the role of aromatization in the photoperiodic response in the stickleback.

Experimental procedure

Male sticklebacks were implanted with Silastic capsules filled with one of two aromatase-inhibitors, 1,4,6-androstatrien-3,17-dione (ATD) or the non-steroidal CGS16949 A, 4-benzonitrile (CGS), or empty capsules and exposed to one of two different photoperiods; LD 8:16 or LD 16:8 at 18°C. After six weeks the fish were dissected and organs were excised, fixed and sectioned. In male sticklebacks the kidney hypertrophies at maturation and produces a glue for the building of the nest. As a maturation parameter, kidney development was studied by measuring the proximal tubule epithelium height. Immunocytochemistry was performed on the pituitaries using antibodies for coho salmon β-GTH II kindly provided by Dr. Penny Swanson.

Results

All fish kept in LD 16:8 matured, i.e. displayed a high kidney epithelium height. Controls in LD 8:16 showed little maturation, whereas both groups implanted with aromatase-inhibitors had higher kidney epithelium heights.

Spermatogenesis is quiescent in the stickleback during the natural breeding season, when the seminiferous tubules contain little but spermatozoa. This condition was found in most animals in all groups except in the control fish in LD 8:16, who showed active spermatogenesis.

In most LD 8:16 control fish the GTH-II-immunoreactive cells were small and contained little immunoreactive material. In the ATD and CGS-treated LD 8:16 fish, as well as in all LD 16:8 groups the GTH-II-ir cells were large, numerous and heavily stained, indicating a higher activity than in the LD 8:16 controls.

Discussion

As expected control fish in LD 16:8 had stimulated kidneys, quiescent spermatogenesis and GTH-II-ir cells with an active appearance, whereas the control fish in LD 8:16 displayed unstimulated kidneys, active spermatogenesis and GTH-II-ir cells with an inactive appearance. These parameters were not affected by aromatase-inhibitors in long photoperiod. This suggests that the previously observed positive feedback from the testes on the (hypothalamus and) pituitary is not dependent on aromatization in the stickleback, unlike what has been found in salmonids. In short photoperiod, the treatment with aromatase-inhibitors stimulated the kidneys and GTH-II-ir cells. Thus, it seems likely that aromatization is important for some (feedback?) mechanism that inhibits breeding under short photoperiod.

References

Antonopoulou, E., Mayer, I., Berglund, I. and Borg, B. (1995). Effects of aromatase inhibitors on sexual maturation in Atlantic salmon, Salmo salar, male parr. Fish Physiol. Biochem., 14: 15-24.

Borg, B., Andersson, E., Mayer, I., Zandbergen, M.A. and Peute, J. (1989). Effects of castration on pituitary gonadotropic cells of the male three-spined stickleback, Gasterosteus aculeatus L., under long photoperiod in winter: indications for a positive feedback. Gen. Comp. Endocrinol. 76: 12-18.

Crim, L.W., Peter, R.E. and Billard, R. (1981). Onset of gonadotropic pituitary gland in response to estrogen or aromatizable androgen steroid hormones. Gen. Comp. Endocrinol. 44: 374-381.

DOES CORTISOL INFLUENCE EGG QUALITY IN THE RAINBOW TROUT, *ONCORHYNCHUS MYKISS*?

S. Brooks[1], T. G. Pottinger[2], C. R. Tyler[1] and J. P. Sumpter[1].

[1.]Fish Physiology Research Group, Brunel University, Uxbridge, Middlesex, U.K;
[2.]Institute of Freshwater Ecology, Ferry House, Far Sawrey, Ambleside, Cumbria, U.K.

Summary

Female rainbow trout, *Oncorhynchus mykiss*, were treated with cortisol implants, 6 weeks prior to ovulation. The eggs produced from both cortisol treated and control fish were fertilised with a single pool of milt and one half dosed with 500 µg/l cortisol for 2 hours. Survival to hatching data indicated that cortisol does not directly affect egg quality.

Introduction

Various studies on human and mammalian reproduction have shown that cortisol may affect ovarian physiology (Hillier 1994). High levels of cortisol are associated with several reproductive disorders and have been shown to reduce ovarian weight. Stress has been clearly shown to decrease gamete quality in fish (Campbell *et al.* 1994) and elevation of plasma cortisol is a major component of the stress reaction. Cortisol implantation in trout has previously been shown to have deleterious effects on ovarian development (Carragher *et al.* 1989). This study investigated the relationship between cortisol and egg quality in rainbow trout.

Materials and Methods

This experiment was conducted on 2+, virgin, female rainbow trout. Six weeks before ovulation fish received one of three treatments:- Group 1 - maintained as controls, Group 2 - received a subcutaneous implant of 1ml cocoa butter (shams) and Group 3 - received a subcutaneous implant of 45 mg cortisol in 1 ml cocoa butter. Blood samples were taken at 3 weeks after implantation and at ovulation. Eggs were stripped manually from ovulated fish and fertilised with milt pooled from 5 males. Half of the fertilised eggs were dipped in a solution of 500 µg/l cortisol for 2 hours, the remaining eggs were kept as controls. The eggs were left to develop and the number surviving monitored throughout. Cortisol levels in the plasma, unfertilised and fertilised eggs were measured by RIA.

Results

Plasma levels of cortisol appeared to be elevated in implanted fish (40.4 ± 13.8ng/ml, mean ± SE), but this rise was not statistically significant (control fish 21.5 ± 7.6 ng/ml). Fertilised eggs had significantly lower levels of cortisol than freshly ovulated eggs in all groups (Fig. 1). Dosing the fertilised eggs with cortisol caused a 30 fold increase in egg cortisol content (Fig. 1). Survival to hatching data (Fig. 2) showed that there was no difference in egg survival between the three groups of fish and that dosing the eggs with cortisol after fertilisation did not influence egg survival.

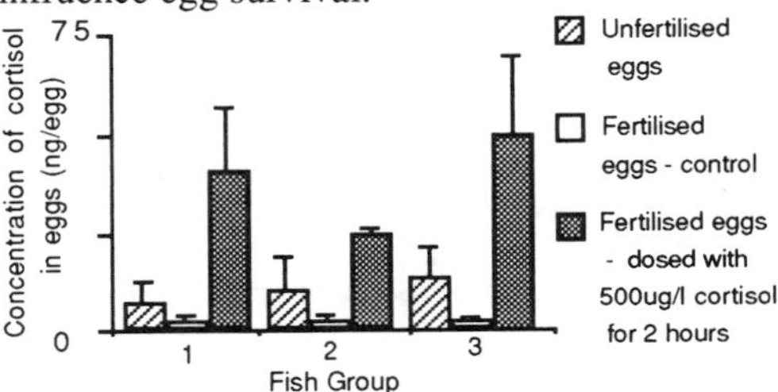

Fig. 1. Concentration of cortisol in eggs. Cortisol levels were measured in 10 eggs from each female.
1 - Controls (n = 6), 2 - Shams (n = 2), 3 - Cortisol implanted (n = 4). Values represent mean ± SE.

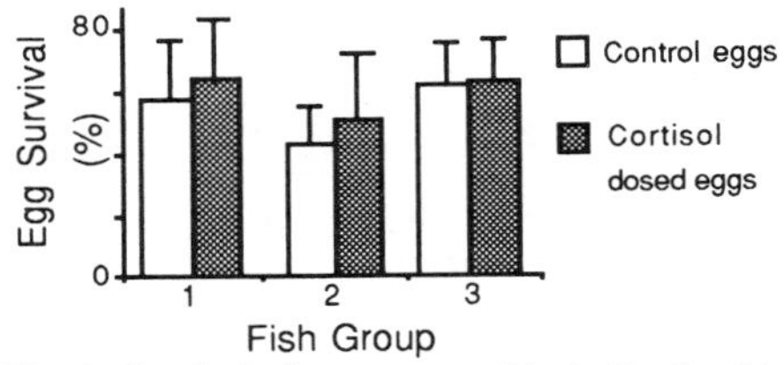

Fig. 2. Survival of eggs upto and including hatching. Fish groups etc. as in Fig. 1.

Discussion

Previous experiments have shown that chronic confinement stress during the final stages of reproductive development leads to significantly reduced levels of plasma vitellogenin, reduced egg size and higher rates of mortality in eggs and larvae in both rainbow and brown trout (Campbell *et al.* 1994). In this study plasma cortisol was raised, by implants, to levels similar to those in chronically stressed fish (Campbell *et al.* 1994). The survival data obtained showed that there was no relationship between plasma cortisol levels and egg mortality, plasma cortisol and levels of cortisol in the eggs were not linked at any time, before or after fertilisation, and dosing the eggs with cortisol had no effect on embryonic development and survival.

References

Campbell, P., Pottinger, T. and Sumpter, J. 1994. Preliminary evidence that chronic confinement stress reduces the quality of gametes produced by brown and rainbow trout. Aquaculture, 120:151-169.
Carragher, J., Sumpter, J., Pottinger, T. and Pickering, A. 1989. The deleterious effects of cortisol implantation on reproductive function in two species of trout, *Salmo trutta* L. and *Salmo gairdneri* Richardson. Gen. Comp. Endocrinol. 76:310-321.
Hillier, S. 1994. Cortisol and oocyte quality. Clin. Endocrinol. 40:19-20.

THE EFFECT OF SPAWNING TEMPERATURE ON EGG VIABILITY IN THE ATLANTIC HALIBUT, (HIPPOGLOSSUS HIPPOGLOSSUS).

N.P.Brown[1], N.R. Bromage[1], and R.J.Shields[2].

[1]Institute of Aquaculture, University of Stirling, Stirling, FK9 4LA Scotland; [2] Marine Farming Unit, Seafish Industry Authority, Ardtoe, Acharacle, Argyll, PH36 4LD Scotland.

Introduction

The cue for the onset of spawning in Atlantic halibut is taken from photoperiod and water temperature (Holmefjord et al, 1993). In the wild, spawning takes place in deep water characterised by relatively stable water temperatures of 5 to 7°C, (Haug, 1990). Variations in egg quality leading to variable hatch rates are well demonstrated in Atlantic halibut (Norberg et al, 1991). Temperature fluctuations in land-based hatcheries are common. The aim of this work was to determine whether temperature control through water chilling, is effective in improving egg viability and a necessary feature of halibut broodstock management in Scotland.

Materials and Methods

Wild caught broodstock from Iceland were used for the experiments, which were carried out at the Seafish Industry Authority, Ardtoe. Two broodstock groups were used, both held under ambient photoperiod in sea water of 33-34°/₀₀. The 'chilled stock' were maintained at 6°C during the spawning season and the 'ambient stock' received water at ambient temperature. Manually stripped, fertilised eggs were examined at the 8-cell stage and the following characteristics were recorded:

- Total annual egg production.
- Batch fertilisation rate.
- Survival to hatch for each batch, assessed from eggs incubated in microtitre plates.

Results and Discussion.

Table 1. Spawning characteristics of 'chilled' and 'ambient' halibut broodstock groups in 1994 & 1995. Data are pooled for all females in each group. Values are means (+/- s.d).

	Chilled stock		Ambient stock	
	1994	1995	1994	1995
# Females spawned	6	5	7	6
Spawning period (wks)	16	14	10	11
Total egg prod'n ($x10^6$)	4.8	4.2	3.2	2.5
Fertilisation rate (%)	71.0 (26.5)	65.9 (30.9)	54.7 (33.6)	26.7 (22.3)
Hatch rate (%)	64.9 (16.7)	69.9 (13.9)	27.9 (27.5)	2.7 (3.6)

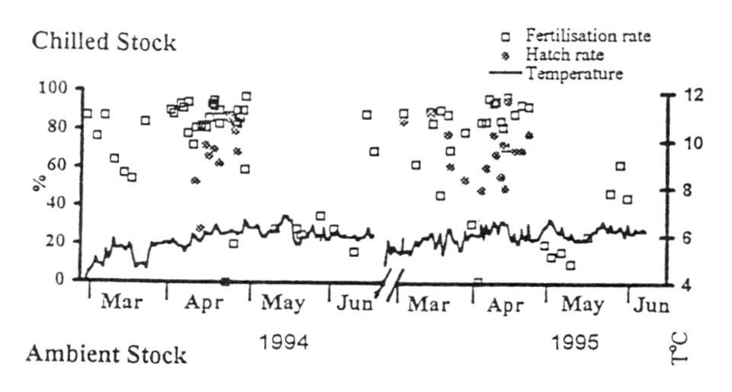

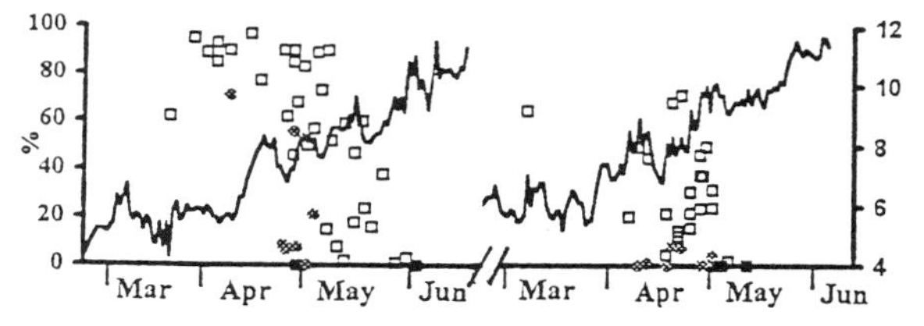

Figure 1. Spawning performance of 'chilled' and 'ambient' broodstock groups in 1994-5.

Seasonal spawning characteristics are presented in Figure 1. The spawning performance of the 'ambient' stock was generally poorer than that of the 'chilled' stock. This was shown by:

- A shorter and delayed spawning season.
- Lower fertilisation and hatch rates, which fell dramatically once the water temperature exceeded 8°C.
- Lower total egg number.
- The incidence of eggs exhibiting abnormal cell cleavage patterns was also found to be higher in batches from the 'ambient' stock (unpublished data).

References

Haug, T. (1990). Biology of the Atlantic halibut, Hippoglossus hippoglossus (L. 1758). Advances in Marine Biology **26**, 1-69.

Holmefjord, I., Gulbrandsen, J., Lein, I. and Reftsie, T. (1993). An intensive approach to Atlantic halibut fry production. Journal of the World Aquaculture Society **24**, (2) 275-284.

Norberg, B., Valkner, V., Huse, J., Karlsen, I. and Grung, G.L. (1991). Ovulatory rhythms and egg viability in the Atlantic halibut Hippoglossus hippoglossus. Aquaculture **97**, 365-371.

Project funded by a grant from M.A.F.F., U.K. N.P.Brown was supported by Trouw Ltd, U.K.

LINKS BETWEEN NUTRITION AND REPRODUCTION IN FISH

M.P.M. Burton

Department of Biology and Ocean Sciences Centre, St.John's, Newfoundland, Canada A1B 3X9

Introduction

Although it has usually been assumed that iteroparous (repeat) reproduction in fish involves regular repetition, some fish species, such as winter flounder, *Pleuronectes americanus* have irregular reproductive omission related to nutritional status (Burton 1994). Winter flounder inhabiting inshore waters off Newfoundland have a prolonged winter fast and will not initiate or maintain the following year's spawn if they have poor nutritional status in the critical period which occurs for females close to the normal spawning season (Burton 1994). There must be factors linking nutritional status to reproduction: the present study investigated plasmatic levels of potential transmitters particularly amino acids to check whether any could be capable of signalling good or poor nutritional status to the hypothalamus, pituitary or gonad (HPG) thereby affecting reproduction. Fish sampled were adult female *P.americanus* which either had a 5 week starvation period in the early summer shortly after the winter fast, or were satiation fed, with sampling occurring 24 hours after feeding or after a 7 day fast. Analyses of ninhydrin positive plasma components, after deproteination, were done through the Amino Acid Analysis Facility of the MUN Biochemistry Department.

Results

Recently fed fish had, as expected, high levels of some amino acids such as alanine, leucine and valine. Potential inhibitors such as taurine, glycine and GABA for the satiation fed (7 day fast) and the starved (5 week fast) fish are shown in Fig.1.

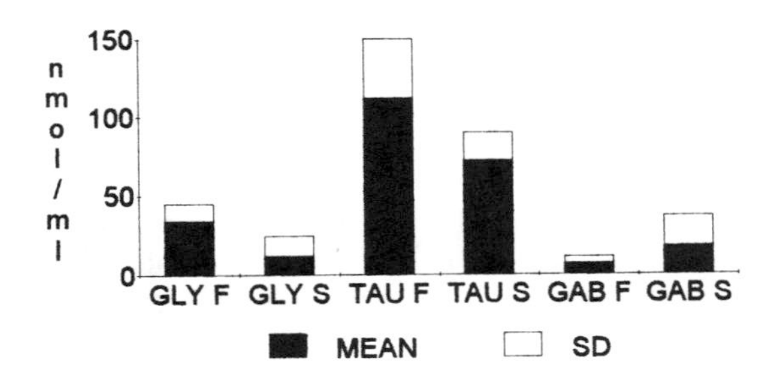

Fig. 1 A comparison of plasmatic "inhibitors" in fed and starved female flounder. F = fed, S = starved.

Transmitter precursors tryptophan (W), tyrosine (Y) and phenyl alanine (F), and the excitatory amino acids (EAA), aspartic (D) and glutamic acid (E), are shown in Fig 2, (same fish as Fig.1)

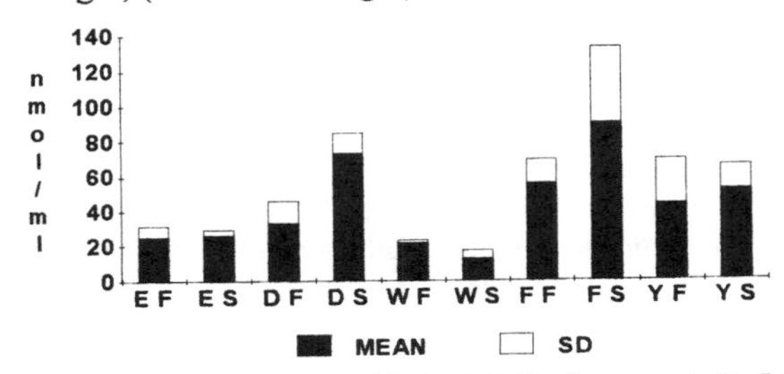

Fig.2 Plasmatic amino acids in fed, F, & starved, S, fish

Discussion

In order for a plasmatic message to convey either good or bad nutritional status to the HPG and thus provide the link between condition and reproduction the messenger must be consistent and not confounded by a recent meal. Some of the differences shown above may provide the basis for a reliable message. The levels of tryptophan were significantly different between satiation fed and starved fish, which could indicate involvement of 5HT. The relationship between the EAAs may also be important in signalling condition, the message could be sustained against the background noise of an occasional meal or short fast. Glycine was significantly higher in satiation fed fish than in starved fish and this would be biologically significant if glycine was acting synergistically with glutamate at an NMDA receptor. However Donoso et al (1990) reported non-NMDA receptors involved with LHRH release in mammals. In fish it is possible that transmitters conveying either nutritional excitation or inhibition function presynaptically close to the pituitary; which would provide amplification.

References

Burton, M.P.M. 1994 A critical period for control of early gametogenesis in female winter flounder *Pleuronectes americanus*. J.Zool.Lond.233: 405-415.

Donoso, A.O, Lopez, F.J. & Negro-Vilar, A. 1990 Glutamate receptors of the non-N-methyl-D-aspartic acid type mediate the increase in luteinizing hormone release by excitatory amino acids *in vitro*. Endocrinology, 126: 414-420.

EFFECT OF STRESS ON THE REPRODUCTIVE PHYSIOLOGY OF RAINBOW TROUT, Oncorhynchus mykiss.

W.M. Contreras-Sanchez, C.B. Schreck, and M.S. Fitzpatrick.

Oregon Cooperative Fishery Research Unit, Oregon State University, Corvallis, OR. 97331

Summary

The effect of stress over the final stages of sexual maturation of female rainbow trout Oncorhynchus mykiss was studied. Significant differences were observed in terms of spawning time, size of the eggs and fry, and mortality at different stages of development. Levels of cortisol in plasma were high in all groups, but significantly less cortisol was observed in the ovarian fluid and eggs.

Introduction

The environment under which broodstock are maintained before reproduction is often stressful; however, the impact of stress on broodstock and gamete quality is not well known. We investigated the effects of stress applied at various times during maturation on the reproductive physiology of rainbow trout females and their progeny.

Methods

Stress was administered over the period of early vitellogenesis (one and a half months), late vitellogenesis-final maturation (one and a half months), or during both periods (three months). Each stress treatment and control was triplicated with eight females in each replicate (n=24 fish per treatment). The eggs and progeny of each female were kept separate and observations were made until two months after ponding. Cortisol levels were measured in plasma, ovarian fluid and eggs by RIA.

Results

Fish that experienced stress during final maturation and those that were under stress during the whole experiment spawned earlier than the control group. In contrast, fish stressed during the period of early vitellogenesis spawned at the same time as the controls (Fig. 1). Fecundity and fertilization were not significantly affected in any group. Stress applied early in vitellogenesis resulted in smaller eggs ($p > 0.01$); however, no differences were found in the size of fry 10 days after hatching or juveniles 8 weeks after hatching.

Circulating levels of cortisol were high at ovulation in all groups, but significantly less cortisol was observed in the ovarian fluid and eggs (Fig. 2).

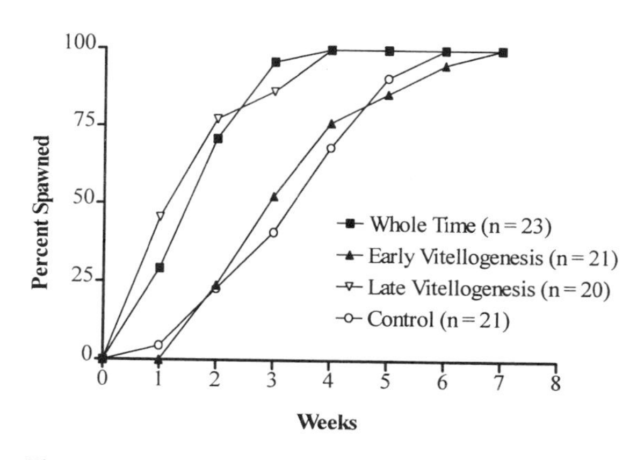

Fig. 1. Cumulative frequency of females spawned

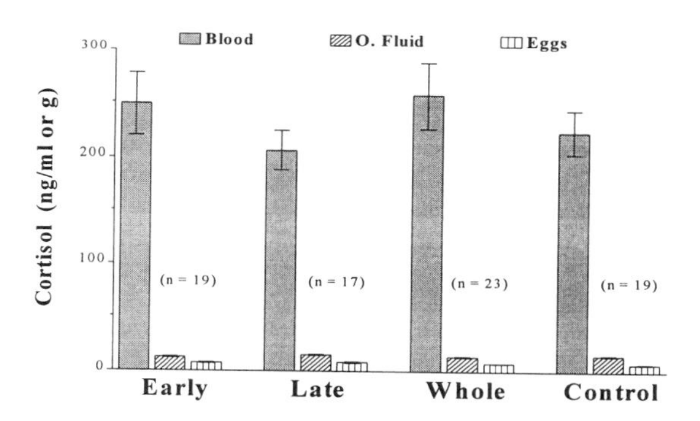

Fig. 2. Cortisol levels in blood, ovarian fluid and eggs.

Discussion

Effects of stress on the reproduction, quality of the eggs and progeny of rainbow trout varies depending on when the female is stressed during maturation. Time of spawning is significantly affected, and the lower levels of cortisol in ovarian fluid and eggs compared with plasma suggests that there is a mechanism by which the female protects the eggs from potentially deleterious effects of prolonged exposure to elevated concentrations of corticosteroids.

Density-dependent inhibition of spawning in the substrate-spawning cichlid *Tilapia tholloni* (Sauvage).

K. Coward* & N.R. Bromage.
Institute of Aquaculture, University of Stirling, Stirling. Scotland. FK9 4LA. U.K.
*Present address: Fish Physiology Research Group, Department of Biology and Biochemistry, Brunel University, Uxbridge, Middlesex. UB8 3PH. U.K.

SUMMARY.

In female *Tilapia tholloni* (Sauvage), confined conditions caused a suppression in the levels of serum 17ß-oestradiol and testosterone. Levels of both steroids rose sharply following transfer to individual aquaria and were maintained consistently higher than those of fish remaining under confinement. Following return of the fish to confined conditions, blood steroid levels fell dramatically and were reduced to levels seen in fish which had remained under confined conditions throughout.

INTRODUCTION.

The regulation of spawning periodicity in tilapias is poorly understood but is known to be influenced by latitude, temperature, photoperiod, diet and stocking density. Earlier work in this laboratory found that female *T. tholloni* subjected to crowding failed to spawn despite possessing ovaries dominated by late-vitellogenic oocytes. Furthermore, such fish displayed a marked tendency to spawn within several days of transfer to individual aquaria. The present study investigated sex steroid profiles and ovarian physiology associated with these earlier observations.

MATERIALS AND METHODS.

14 female *T. tholloni* were maintained in a 1 x 1 x 0.5m stock tank along with 20 males. Blood samples were taken from each female at the time of confinement (day 0) and approximately 5 day intervals thereafter for a total period of 30 days. Fish were then divided into two equal groups; group-1 (conditions unchanged) and group-2 (moved into partitioned glass aquaria). Blood sampling continued for a further 30 days. Group-2 fish were subsequently returned to the stock tank. Final blood samples were taken from each group after a further 5 days. Serum testosterone and 17ß-oestradiol were measured by radioimmunoassay.

RESULTS.

During 30 days of crowding, fish in both group-1 and group-2 exhibited consistently low levels of both serum 17ß-oestradiol and testosterone (Figs. 1a and 1b). Following transfer to individual aquaria, levels of both steroids in group-2 fish rose sharply and were maintained consistently higher than in the confined fish. Two fish in group-2 spawned; one after just 4 days and one after 30 days. No spawning activity was observed in group-1 fish. Just five days after return to crowded conditions levels of both 17ß-oestradiol and testosterone in group-2 fish were much reduced and very similar to those observed in group-1.

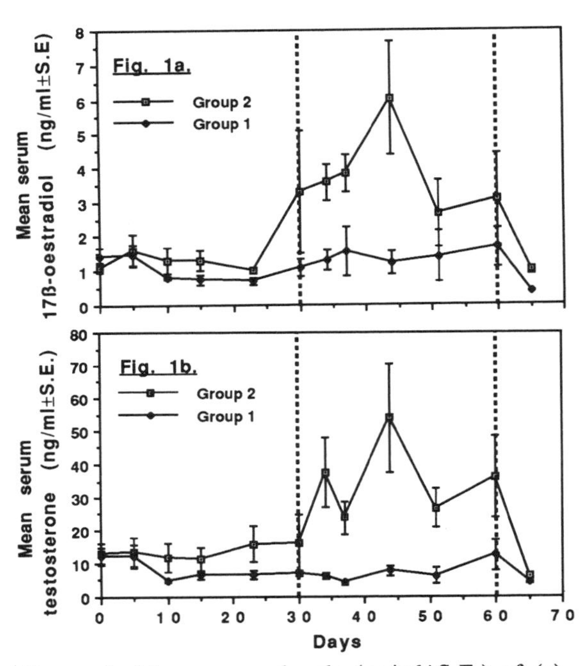

Figure 1. Mean serum levels (ng/ml±S.E.) of (a) 17ß-oestradiol and (b) testosterone in 2 groups of female *T. tholloni* over a 65-day period. Days 0 - 30: both groups crowded. Days 30 - 60: fish in group-1 crowded and fish in group-2 in individual aquaria. Days 60 - 65: both groups crowded. Broken lines represent the times of transfer.

DISCUSSION.

Under holding conditions, the levels of 17ß-oestradiol and testosterone in female *T. tholloni* appeared suppressed and exhibited little change throughout confinement. Both steroids rose sharply following transfer to individual aquaria and were maintained at consistently higher levels than those of confined fish. The present study demonstrates that under confined conditions 17ß-oestradiol and testosterone are maintained at consistently low levels in *T. tholloni*. It is suggested that such levels may remain insufficient for oocytes to complete ovarian growth thus preventing ovulation and oviposition. Following transfer to individual aquaria however, levels of both steroids rise sharply, thus inducing a resumption of ovarian growth (data not shown) and renewed potential for spawning activity.

This study was supported by a N.E.R.C. award to N.R.B.

THE EFFECTS OF PHOTOPERIOD AND TEMPERATURE ON SERUM GTH I AND GTH II AND THE TIMING OF MATURATION IN THE FEMALE RAINBOW TROUT.

B. Davies, P. Swanson* and N. Bromage

Institute of Aquaculture, University of Stirling, Stirling, FK9 4LA, UK. *Northwest Fisheries Science Centre, Seattle, WA, USA.

Introduction

The reproductive cycle of rainbow trout appears to be controlled by an endogenous circannual clock which is entrained primarily by the yearly cycle of photoperiod with temperature appearing to perform a modulatory role (Bromage *et al.*, 1993). The following research has been conducted to determine the mechanisms by which such environmental factors affect the reproductive axis.

Method

Four groups of post-spawned 2 year old female rainbow trout, with a natural spawning time of January-February, were exposed to the following experimental regimes commencing on the 1st February;

Group A- Ambient photoperiod (56°N), River water (0-21°C).
Group B- Stimulatory long-short photoperiod (18L:6D up to May 10, then 6L:18D), River water (0-21°C).
Group C- Ambient photoperiod, Borehole water (8.5±1°C).
Group D- Stimulatory long-short photoperiod (18L:6D up to May 10, then 6L:18D), Borehole water (8.5±1°C).

Results

The stimulatory long-short photoperiod produced a 4 month advance in spawning compared to the groups exposed to the ambient photoperiod and maintained on similar temperatures (Figure 1). The borehole water temperatures produced 3-4 week advances in spawning compared to the groups on river water and exposed to similar photoperiods.

In fish exposed to the stimulatory long-short photoperiod, GTH I levels rose to reach a peak during May-June and then fell prior to spawning, levels then subsequently increased after the spawning period.

The fish exposed to the ambient photoperiod initially showed high levels of GTH I which then fell to low levels from June onwards, with a rise occurring after spawning for the fish maintained on borehole water.

GTH II levels were only detectable close to and during the respective spawning periods of the four experimental groups.

17ß-oestradiol and testosterone levels (not shown) increased relative to each group's spawning time with 17ß-oestradiol and testosterone falling prior and during spawning respectively.

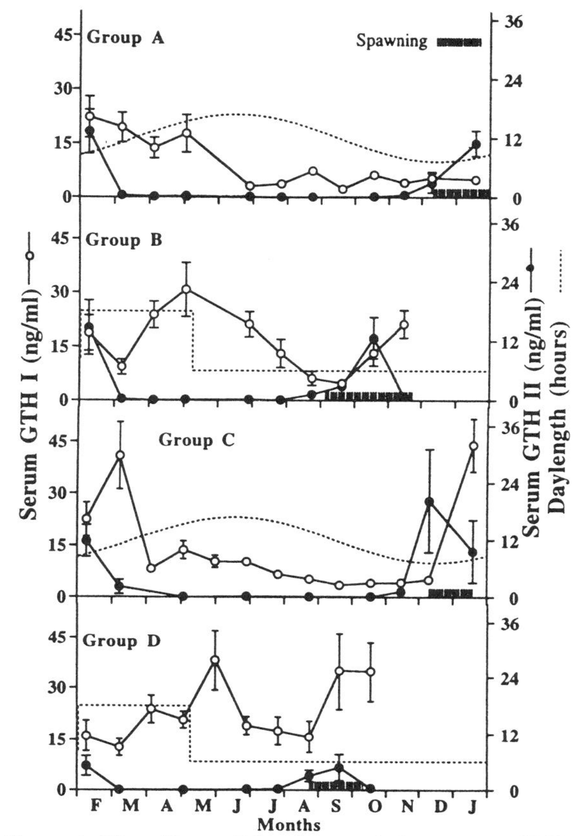

Figure 1. The effects of the photoperiod on serum GTH I and GTH II levels (n=6-10) and spawning time for the fish in Groups A-D. Values shown are mean±sem. IIIIIIIII represents the spawning period.

Conclusions

An abrupt increase in daylength after spawning resulted in an increase in serum GTH I that coincided with an advance in the timing of vitellogenesis.

Stimulatory long-short photoperiods increased GTH I levels during vitellogenesis compared to the fish exposed to an ambient photoperiod.

GTH II levels rose just prior to and during the spawning period whilst GTH I levels rose during and after spawning.

References

Bromage, N., Randall, C., Duston, J., Thrush, M. and Jones, J. (1993). Environmental control of reproduction in salmonids. In "Recent Advances in Aquaculture IV", pp. 55-65.

Acknowledgements: This work was funded by grants from Scott Trout, BP (nutrition) Ltd. and NERC to NRB.

GROWTH, GONADAL DEVELOPMENT AND SPAWNING TIME OF ATLANTIC COD (*Gadus morhua*) REARED UNDER DIFFERENT PHOTOPERIODS

T. Hansen[1], O.S. Kjesbu[2], J.C. Holm[3] and Ø. Karlsen[3]

[1]Institute of Marine Research, Matre Aquaculture Research Station, N-5198 Matredal;
[2]IMR, P.box 1870 Nordnes, N-5024 Bergen; [3]IMR, Austevoll Aquaculture Research Station, N-5392 Storebø

Atlantic cod (*Gadus morhua*) were reared in four 30m^3 tanks supplied with running sea water for 36 months. In the period between mid-summer 93 and mid-summer 94, the four groups were subjected to four different photoperiod regimes (Figure 1).

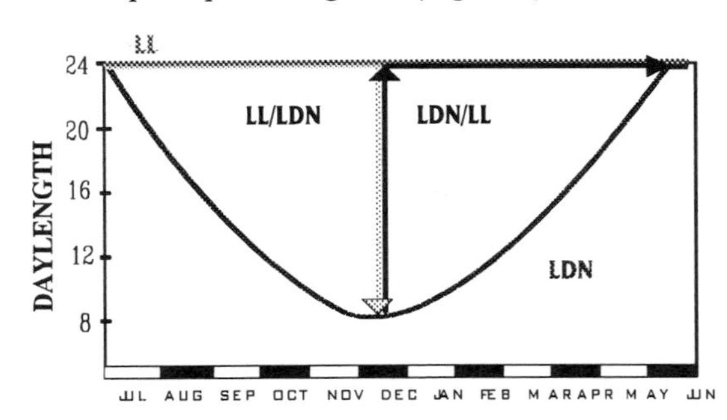

Figure 1: Photoperiod regimes: LL-continuous light; LDN-natural photoperiod

Sexual maturation reduced feed intake and growth. At an age of 26 months the weights of the cod reared under natural photoperiod and continuous light were 1.5 and 2.5 kg respectively (Figure 2).

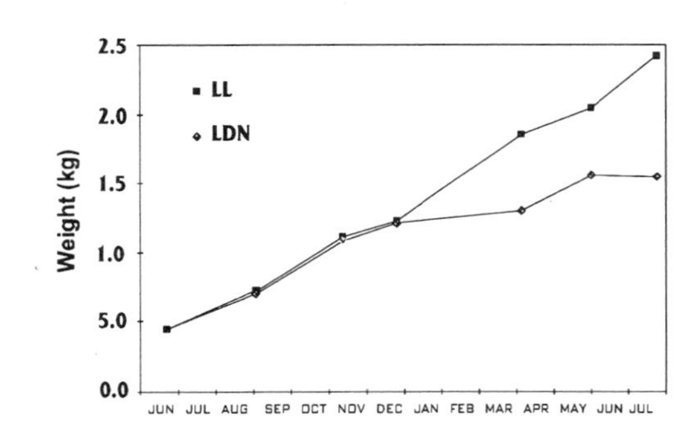

Figure 2: Growth of cod reared under natural photoperiod (LDN) and continuous light (LL).

Cod reared under natural photoperiod spawned in the period between January and April at an age of two years (Figure 3). Oocytes of females reared under continuous light were arrested in the cortical alveoli stage. When transferred to natural photoperiod in December, females ovulated within 4-5 months.

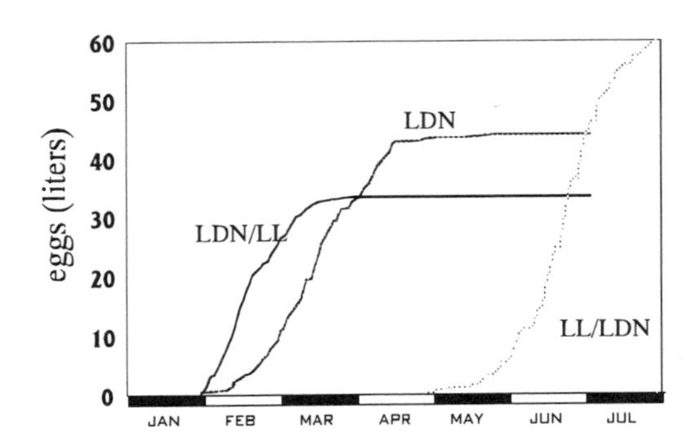

Figure 3: Spawning of female cod under the different photoperiod regimes. The LL group did not spawn

Between mid-summer 94 and mid-summer 95 the experiment was continued with only two experimental groups. The LL group was still kept under continuous light and the LL/LDN group continued on natural photoperiod.

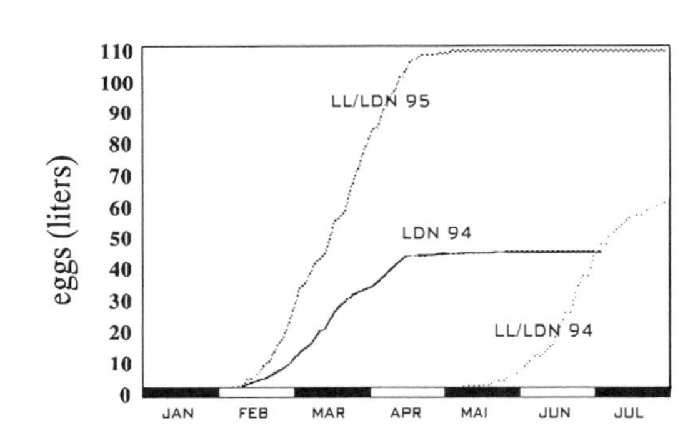

Figure 4: Spawning of the LL/LDN female cod during the 94 and 95 spawning seasons. The LDN 94 spawning is included for comparison. The values of the LL group are covered by the x-axis.

Two year old cod that spawned three months delayed (May-July) spawned again at their normal spawning time (as three year olds) when reared under a natural photoperiod (Figure 4). A few of the females reared under continuous light ovulated at an age of three years.

DEPRESSION OF BLOOD LEVELS OF REPRODUCTIVE STEROIDS AND GLUCURONIDES IN MALE WINTER FLOUNDER (*PLEURONECTES AMERICANUS*) EXPOSED TO SMALL QUANTITIES OF HIBERNIA CRUDE, USED CRANKCASE OIL, OILY DRILLING MUD AND HARBOUR SEDIMENTS IN THE 4 MONTHS PRIOR TO SPAWNING IN LATE MAY-JUNE.

D.R.Idler[1], Y.P.So[1], G.L.Fletcher[1] and J.F.Payne[2]

[1]Memorial University of Newfoundland, Ocean Sciences Centre, St. John's, NF. A1C 5S7 and [2]Department of Fisheries and Oceans, St. John's NF. A1C 5X1

Introduction

Oily fractions were added to 45 kg of clean dry sand exposed to flowing seawater for one week before flounder were introduced (Truscott et al.,1992). Hibernia crude and crankcase oil were tested at 0, 5, 25, 50, 100 and 250 ml per 45 kg sand and oily drilling mud at 0, 33, 110, 1000 and 4000 ml. Bottom sediments were tested at 0, 1, 2, 5, 10 and 20% of the sand.

Radioimmunoassays (Truscott et al.,1992) were done using 3H-11ketotestosterone (11-KT) and 125I 3-0-carboxy methyl oxime testosterone. Specificity was such that paper chromatography was not necessary.

UDP-glucuronyl transferase was assayed in the supernatant from 30 mg of testes homogenate, p-nitrophenol (PNP), 18 nmole, 14C-PNP, 0.05μCi and UDP glucuronic acid, 3.4 nmole incubated for 15min. After precipitation with methanol the supernatant was applied to an ion exchange HPLC C18 column (Table 1).

Results

For complete data on flounder exposed to harbour sediments see Table 2. The following is a summary of the flounder exposed to the highest concentrations of the various xenobiotics used. Plasma levels of the principal male sex steroid, 11-KT glucuronide, was greatly reduced following the 4-month exposure to all xenobiotics. Reduction was 72% using Hibernia crude oil, 40% with crankcase oil, 43% with oily drilling mud, and 30% using sediments from St. John's harbour. Testosterone glucuronide was reduced 43% by harbour sediments, 68% by drilling mud and 29% by Hibernia crude. Testosterone was reduced significantly by harbour sediments, 20% and crankcase oil, 50%. 11-KT was diminished by harbour sediments, 26%, oily mud, 40% and Hibernia crude, 29%.

TABLE 1. URIDINE DIPHOSPHATE GLUCURONYL TRANSFERASE IN TESTIS OF WINTER FLOUNDER EXPOSED TO HARBOUR SEDIMENTS

Treatment	N	% Transformation / 100 mg tissue	
CONTROL	14	40.4 +/- 4.8	
20% H.S.	15	24.9 +/- 4.1	P<0.02

TABLE 2.

	HARBOUR SEDIMENT					
LEVELS	CONTROL	E1	E2	E5	E10	E20
N =	16	15	16	14	13	15
FREE STEROIDS (ng/mL)						
T	34* ±3	38 ±4 A	25 ±3 B	26 ±2 B	25* ±2 B	27* ±2 B
11-KT	308* ±26	202 ±17 D	213 ±22 C	221 ±17 C	213* ±21 C	227* ±19 C
STEROID GLUCURONIDES (ng/mL)						
T	183** ±16	213 ±27 A	133 ±12 B	110 ±13 C	109* ±5 D	105* ±9 D
11-KT	429** ±57	515 ±59 A	244 ±32 C	257 ±30 B	268* ±29 B	302* ±34 B

A,NSD; B,P<0.05; C,P<0.01; D,P<0.001.
*Assayed 2X in duplicate.
**Assayed 3X in duplicate.

Discussion

Mixed function oxygenases are recognized as very sensitive indicators of petroleum hydrocarbons in biological monitoring (Payne et al.,1987). It is of interest that the steroids measured in this study are even more sensitive indicators of these xenobiotics.

The correlation of glucuronyl transferase activity in the testes with blood levels of the steroids suggest that these xenobiotics may act by suppressing this enzyme system.

References

Truscott, B., Idler, D.R., and G.L. Fletcher, 1992. Alteration of reproductive steroids of male winter flounder chronically exposed to low levels of crude oil in sediments. Can J Fisheries Aquat Sci., 49:2190-2195.

Payne, J.F., Fancy, L.L., Ramitula, A.D., and E.L. Porter, 1987. Review and perspective on the use of mixed function oxygenase enzymes in biological monitoring. Comp. Biochem. Physiol., 86C: 233-245.

EFFECTS OF ENDOCRINE-DISRUPTING CHEMICALS ON MARINE FLATFISH: AN APPROACH TO ENVIRONMENTAL RISK ASSESSMENT

Lyndal L. Johnson, Sean Y. Sol, Daniel P. Lomax, and Tracy K. Collier

Northwest Fisheries Science Center, NMFS, 2725 Montlake Blvd. E. Seattle, WA 98102

Introduction

The potential risk to marine fish populations of endocrine-disrupting chemicals in the environment is an issue of increasing concern both to the public and to resource managers. In field studies and laboratory studies we have investigated the impact of chemical contaminants on reproductive function in several Pleuronectid species. From these field data we are developing dose-response models to aid in identification of specific chemical risk factors and in quantifying threshold sediment or tissue contaminant levels where signs of reproductive pathology initially occur, and have incorporated the data into population models to predict impacts on fish abundance. This integrated multidisciplinary approach has enabled us to examine effects of endocrine-disrupting chemicals on marine fish not only at the biochemical or individual level, but at the population level as well, and to make progress toward developing methodologies for assessing the environmental risk of endocrine-disrupting chemicals. This paper presents sample results for two flatfish species from Puget Sound, WA: rock sole *(Pleuronectes bilineata)* and English sole *(Pleuronectes vetulus)* which differ in habitat preference and in reproductive traits, and possibly in sensitivity to contaminant exposure, on the basis of their susceptibility to toxicopathic lesions (Myers et al. 1995). Consequently, comparison of their responses may provide insight into how ecological factors and physiological factors might influence species sensitivity to the impacts of endocrine-disrupting chemicals.

Methods

Fish for these studies were by collected from by otter trawl from five sites in Puget Sound: Eagle Harbor, a creosote-contaminated embayment, Sinclair Inlet, a moderately contaminated urban waterway, and Port Susan and Pilot Point, non-urban reference areas. Concentrations of dioxin-like PCBs and other congeners were determined in liver and ovary samples, and exposure to aromatic hydrocarbons was assessed by measuring levels of fluorescent aromatic hydrocarbons in the bile. Gonadal development, fecundity, and spawning success were determined and these data were then incorporated into a simple, deterministic, age-classified Leslie matrix model to estimate the potential consequences of contaminant-associated alterations in reproduction function on fish population growth rates *(r)*. Methods and English sole data are reported in detail in Johnson et al. 1988, 1995; Casillas et al. 1991; and Landahl and Johnson 1993).

Results and Discussion

Reproductive potential was significantly reduced in both rock sole and English sole from polluted sites (Table 1). English sole exhibited both impaired gonadal development (for which exposure to AHs was a major risk factor), and reduced spawning success (which was associated with exposure to both AHs and PCBs). In rock sole, only spawning success was affected. Model results suggest that contaminant-associated reductions in reproductive potential may be sufficient to reduce the intrinsic rate of increase *(r)* in both English sole and rock sole populations from Eagle Harbor, provided their impact is not offset by mitigating factors such as immigration of juveniles from other sites. While on the individual level, rock sole appeared to be less sensitive to the impacts of pollutants on reproductive function than English sole, the impact of the alterations in fertility that were observed was equally severe because of the relatively small egg number and late age of maturation in this species. These results highlight the importance of both toxicant sensitivity and life history traits in assessing health risks in fish and other wildlife. The information from these studies is now being used as part of an ongoing program to monitor the success of sediment remediation efforts at the Eagle Harbor site.

Table 1. Flatfish reproductive success.

	Reference Site	Sinclair Inlet	Eagle Harbor
English sole			
% maturing	80 (50)	90 (96)	57 (97)
% spawning	90 (60)	75 (23)	35 (21)
% fertilization	52 (54)	35 (17)	24 (7)
% norm. larvae	74 (53)	54 (14)	68 (4)
r	0.0	-0.10	-0.26
Rock sole			
% maturing	84 (43)	86 (37)	91 (77)
% spawning	89 (9)	71 (7)	20 (25)
% fertilization	79 (8)	74 (5)	35 (5)
% normal larvae	37 (8)	38 (4)	13 (4)
r	0.0	-0.03	-0.31

References

Casillas. E. *et al.* 1991. *Mar. Environ. Res.* 31:99-122.

Johnson, L. L. *et al.* 1988. *Can. J. Fish. Aquat. Sci.* 45:2133-2146.

Johnson, L. L., *et al.* 1995. *Proc. Puget Sound Research '95* (in press)

Landahl, J. T. and L. L. Johnson. 1993. *Amer. Fish Soc . Symp. 14:117-123.*

Myers, M. S. *et al.* 1995. *Mar. Env. Res.* (in press).

THE HORMONE MIMIC ß-SITOSTEROL ALTERS REPRODUCTIVE STATUS IN GOLDFISH (*Carassius auratus*).

D.L. MacLatchy[1], Z. Yao, L. Tremblay and G.J. Van Der Kraak

Dept. of Zoology, University of Guelph, Guelph, ON, Canada N1G 2W1

Summary

This study demonstrates ß-sitosterol (ß-sit) is potentially responsible for some reproductive dysfunctions in fish exposed to bleached pulp and paper kraft mill effluent (BKME).

Introduction

Hormone mimics in the environment can alter reproductive status. Previous studies have shown that fish downstream of pulp and paper mills have reduced gonad size, delayed maturation, and decreased plasma sex steroids. This experiment investigated the possible role of the dominant BKME plant sterol (ß-sit) as a modifier of reproductive status.

Results

Male goldfish were exposed in the laboratory for 12 days to environmentally-relevant ß-sit concentrations [0 (controls), 75, 300, 600 or 1200µg/L] in two separate experiments. At all ß-sit exposure levels, plasma testosterone (T) and 11-ketotestosterone were significantly decreased (T only shown; Fig. 1).

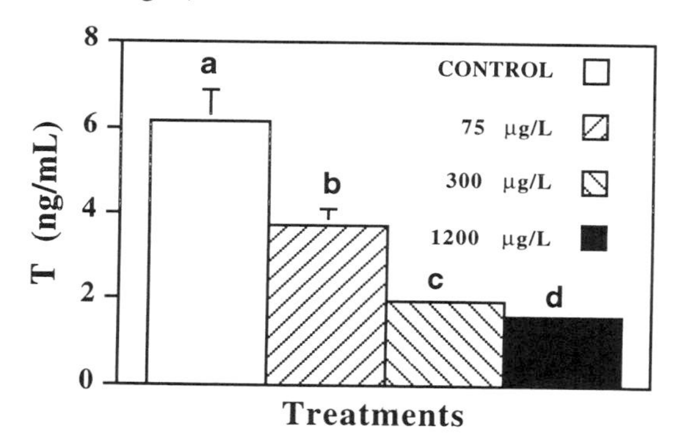

Fig. 1. Plasma T levels in goldfish exposed to ß-sit.

This effect appears to be mediated, at least partially, at the level of the testes, as basal and hCG-stimulated *in vitro* gonadal incubations show decreased T and pregnenolone (preg) production (Fig. 2). Preg was measured as it is the first post-cholesterol steroid in the biosynthetic pathway to T. Reduced preg content suggests that chol availability/metabolism might be affected. However, elevated preg production in the absence of concomitant increases in T production in incubates with exogenous 25-OH chol addition suggests effects of ß-sit downstream of preg production. Free and esterified tissue chol levels are also decreased in testes from fish exposed to 1200µg/L ß-sit (data not shown).

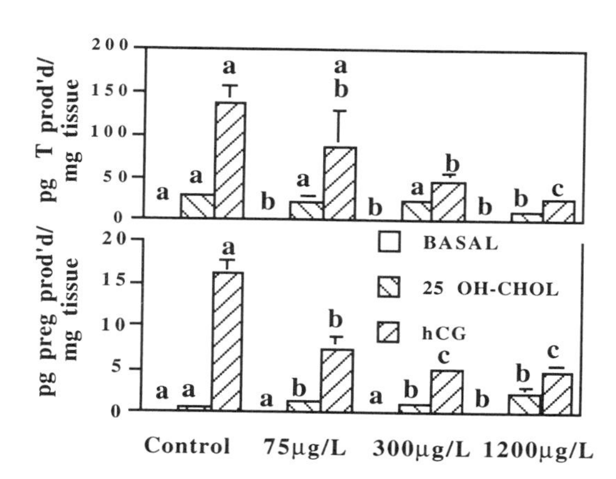

Fig. 2. Effect of ß-sit on *in vitro* testicular steroid production. Letters indicate significant differences amongst exposures within each *in vitro* treatment.

ß-sit might also be acting as an estrogen mimic, as plasma vitellogenin levels increased with ß-sit treatment, at levels similar to estradiol (E_2) treatment. Vitellogenin levels (as measured by ELISA) were not detectable in control fish, while they ranged from 9.4±1.0 (75µg/L ß-sit) to 18.1±1.6 mg/mL (1200µg/L ß-sit) in exposed fish. Water-borne estradiol (75µg/L) causes plasma vitellogenin to increase to 15.6±2.6mg/mL.

Binding studies using KCl extracts of trout hepatic nuclear estrogen receptors suggest that ß-sit binds to the estrogen receptor with low affinity (Tremblay and Van Der Kraak, 1995).

Discussion

Treatment with ß-sit leads to decreased steroid levels which might be caused by a reduction in the activity of the sidechain cleavage enzyme P_{450scc} due to decreased chol availability. ß-Sit also has a second site of action downstream of preg formation. In addition, ß-sit is a potential estrogen mimic as it increases plasma vitellogenin levels in male goldfish and weakly binds to trout hepatic estrogen receptors.

References

Tremblay, L. and Van Der Kraak, G. (1995). Interactions of the environmental estrogens nonylphenol and ß-sit with liver estrogen receptors. *Proceedings of the 5th International Symposium on the Reproductive Physiology of Fish*, Austin, Texas.

[1]present address: Dept. of Biology, University of New Brunswick, Saint John, NB, Canada E2L 4L5

TRANSPORT AND OVARIAN ACCUMULATION OF o,p'-DDT IN THE ATLANTIC CROAKER (MICROPOGONIAS UNDULATUS) DURING GONADAL RECRUDESCENCE

Janet R. Ungerer and Peter Thomas

The University of Texas, Marine Science Institute, Port Aransas, TX 78373, USA

Introduction

Exposure of adult fish to organochlorine xenobiotics, such as o,p'-DDT (DDT), critically impacts the growth and development of the resulting larvae. These environmentally persistent, endocrine-disrupting compounds accumulate in the ovaries of exposed fish, thus providing a means of transfer to the offspring. Organochlorines have also been shown to bind to plasma lipoproteins in a number of species. However, no investigations have focused on the ability of lipoproteins to bind xenobiotics and transport them into the growing oocytes. In the present study the binding of DDT to plasma lipoproteins and its subsequent accumulation in tissues of fish fed DDT during gonadal recrudescence were investigated.

Results

Females in all treatment groups (1.8, 10.8, or 54.6μg DDT/100g fish/day for 21 days) accumulated significantly more DDT in their gonads than males (Table 1). Sex differences in accumulation in other tissues were not as pronounced. DDT concentrations in ovaries increased as GSI increased in the females at all doses (r^2=0.52, 0.85 and 0.76 for the low, middle and high doses, respectively), but there was very little change in DDT concentration with gonad size in males. DDT analysis of the oocyte components revealed that the oil globule contained most of the accumulated DDT (>95%), while only 4.5% was present in the yolk.

Table 1. o,p'-DDT concentration in the gonads (μg/g tissue)

Dose	Male	Female
Low	0.0015±0.0005	0.237±0.046*
Middle	0.006±0.002	0.709±0.047*
High	0.207+0.017	4.94+0.47*

*significantly different from males ($p \leq 0.001$)

Lipoproteins in plasma from fish dosed with 54.6μg DDT/100g fish/day were separated by density ultracentrifugation. Densities were adjusted with a NaCl/KBr solution (d=1.34 g/ml). VLDL contained 39% of the total plasma DDT in females, whereas the very high density lipoprotein fraction (VHDL) contained only 7%. No DDT was found in the VHDL fraction of males, which contained no vitellogenin as evidenced by the lack of protein-bound phosphorus. Ovarian uptake of DDT following injection of DDT as a DDT/VLDL complex was six-fold higher than that following injection of a DDT/vitellogenin complex or DDT in corn oil.

Discussion

These results provide the first demonstration that lipoproteins are involved in the binding of organochlorine xenobiotics and their subsequent uptake into the oocytes of an oviparous vertebrate. The increase in ovarian DDT concentration with increasing GSI in Atlantic croaker suggests a continuous uptake of DDT as the ovary grows. This growth is primarily due to lipoprotein uptake during vitellogenesis. Indeed, Niimi showed that organochlorine accumulation in the eggs of environmentally exposed fish was correlated with their percent lipid content and the percent of total lipid deposited in the eggs (Niimi, 1983).

DDT is compartmentalized in the oocyte, being primarily associated with the triglyceride-rich oil globule. Leger and coworkers have found that VLDL and LDL are the primary sources of triglycerides in the eggs of rainbow trout (1981). Interestingly, several workers observed increased rates of mortality and deformities in larvae from organochlorine-exposed females during the period when the oil globule lipids are utilized. (Burdick et al., 1972)

Furthermore, VLDL carries 40% of the plasma DDT in exposed croaker females. In additon, VLDL complexation enhances the ovarian uptake of DDT. Our results from the analysis of tissues from DDT-dosed fish reveal a mechanism of transport of xenobiotics into the ovary involving transport in plasma via VLDL, uptake of VLDL by the growing oocyte and ultimate deposition of VLDL-sequestered toxicants into the oil globule.

References

Burdick, G.E., Dean, H.J., Harris, E.J., Skea, J., Karcher, R. and Frisa, C. (1972) Effect of rate and duration of feeding DDT on the reproduction of salmonid fishes reared and held under controlled conditions. *N.Y. Fish and Game J.*, **19**, 97-115.

Leger, C., Fremont, L., Marion, D., Nassour, I. and Desfarges, M.-F. (1981) Transfer of very low density lipoprotein from hen plasma into egg yolk. *Lipids*, **16**, 593-600.

Niimi, A.J. (1983) Biological and toxicological effects of environmental contaminants in fish and their eggs. *Can. J. Fish. Aquat. Sci.*, **40**, 306-312.

DIFFERENTIAL MODES OF HORMONAL DISRUPTION IN FISH EXPOSED TO VARIOUS ORGANIC CONTAMINANTS

M.E. McMaster[1], G.J. Van Der Kraak[1], and K.R. Munkittrick[2].

[1]Dept. Zoology, University of Guelph, Guelph, Ontario, Canada, N1G 2W1; [2] Dept. of Fisheries and Oceans, Great Lakes Laboratory for Fisheries and Aquatic Sciences, Burlington, Ontario, Canada, L7R 4A6.

Summary

Our field studies indicate that a) reproductive alterations in fish exposed to organic wastes are widespread, b) different chemical stressors are capable of causing reproductive changes by similar mechanisms, but that more than one mechanism may exist, and c) species differences in reproductive responsiveness to contaminants exist and must be considered when evaluating effluent effects in a field situation.

Introduction

A number of environmental contaminants are known to alter reproductive function in fish through depression of circulating steroid levels. However few studies have examined the mechanisms of reproductive toxicity. The *in vitro* isolation and examination of specific sections of the hypothalamic-pituitary-gonadal axis may permit identification of contaminant action on reproductive function. In this study we use an *in vitro* gonadal incubation procedure to examine reproductive function in two species of fish, white sucker *Catostomus commersoni* and brown bullhead *Ictalurus nebulosus*. Fish were collected at sites exposed to bleached kraft pulp mill effluent (Jackfish Bay, Ontario; Androscoggin River, Maine), effluent from a sulphite pulp mill (Pine Falls, Manitoba), and effluent from a steel mill (Black River, Ohio) and comparisons of the reproductive responses were made between contaminants and species.

Results and Discussion

In depth studies at Jackfish Bay demonstrated that the reduced production of steroid hormones by ovarian follicles paralleled the reductions found in circulating steroid levels. These reductions also corresponded to other reproductive alterations such as decreased gonad size, reduced expression of secondary sexual characteristics and increased age to maturation in the white sucker population. Studies into the mechanisms of BKME action on steroid production indicate that multiple sites within the steroid biosynthetic pathway are disrupted (McMaster et al. 1995ab). Studies on the same species that exhibit similar reductions in circulating steroid levels at a site receiving non-chlorinated pulp mill effluent (Friesen et al. 1994) illustrate similar biochemical lesions within the steroid biosynthetic pathway resulting in reduced steroid capacity of the follicles. Brown bullheads collected at a site heavily contaminated with PAH's demonstrated similar reductions in gonad size, circulating steroid levels and *in vitro* steroid production as pulp mill effluent sites. However, brown bullhead exposed to pulp mill effluent (Androscoggin) failed to show signs of reproductive alteration due to effluent exposure. White sucker collected from this site showed reduced gonadal size, but failed to show a consistent correlation between reduced circulating steroids and *in vitro* steroid production. These studies indicate that reproductive alterations in feral fish exposed to organic contaminants are widespread. Although some sites exhibit identical reproductive responses, differences within species between sites suggest that that different mechanisms may be responsible for the reproductive changes. The lack of reproductive responses in brown bullhead exposed to BKME in Maine indicate species differences in responsiveness to effluent exposure. Similar species differences in terms of reproductive responses to contaminant exposure have been described by Munkittrick et al. (1992). This indicates that multiple species should be evaluated for reproductive alterations when conducting field assessments.

References

Friesen, C., W.L. Lockhart, and S.B. Brown. 1994. Results from the analysis of Winnipeg River water, sediment and fish. Report to Dept. Indian Affairs and Northern Development, Winnipeg. 10p.

McMaster, M.E., G.J. Van Der Kraak, and K.R. Munkittrick. 1995a. Exposure to bleached kraft pulp mill effluent reduces the steroid biosynthetic capacity of white sucker ovarian follicles. Accepted Comp. Biochem. Physiol. June 1995.

McMaster, M.E., G.J. Van Der Kraak and K.R. Munkittrick. 1995b. An epidemiological evaluation of the biochemical basis for steroid hormonal depressions in fish exposed to industrial wastes. Submitted J. Great Lakes Res.

Munkittrick, K.R., G.J. Van Der Kraak, M.E. McMaster, and C.B. Portt. 1992. Reproductive dysfunction and MFO activity in three species of fish exposed to bleached kraft mill effluent at Jackfish Bay, Lake Superior. Water Poll. Res. J. Canada. 27(3): 439-446.

STRESS-RELATED CHANGES OF PITUTARY-INTERRENAL AXIS IN MALE GILTHEAD SEABREAM

G. Mosconi[1], A. Gallinelli[2], F. Facchinetti[2], and A.M. Polzonetti-Magni[1]

[1]Dipartimento di Biologia M.C.A., Università di Camerino, Via Camerini, 62032 Camerino (MC), Italy; [2]Dipartimento di Ostetricia e Ginecologia, Università di Modena, Via del Pozzo, 71; 41100 Modena, Italy.

SUMMARY

The effects of confinement and crowding on plasma cortisol levels were assessed in seabream together with the involvement of endorphin in the stress response.

INTRODUCTION

It has been well demonstrated that when teleost fish are exposed to stressful stimuli, cortisol output from interrenal cells dramatically increases, and that high blood levels of cortisol for extended periods of time have deleterious effects on growth, reproductive function, and immune system. The possibility of habituation of stress response in the marine teleost fish evidenced by a decline of initially elevated plasma cortisol levels, was investigated in marine teleost, *Sparus aurata*. Also, the involvement of proopiomelanocortin-derived endorphin in the regulation of interrenal function by treatment with antaxone, the long-acting opiate receptor antagonist, was studied.

In male seabream of average body weight 400 g, the most common aquaculture stressors such as confinement, crowding and manipulation were applied during spring months.

Plasma cortisol levels were measured by coated-tube RIA; plasma and pituitary N-terminal acetylated endorphin (N-ac-ßEP) was measured with salmon antibody used previously for this species (Facchinetti *et al.*, in press). The N-ac-ßEP antibody, was a gift from Dr. C.H. Kawauchi (Kitesato University, Japan).

RESULTS

Experiment I - One month confinement plus crowding produced a permanent increase in plasma cortisol levels (Table I), since cortisol levels rose from a baseline of 12.5 ± 1.5 to 164 ± 14.3 ng/ml ($P < 0.001$). Plasma N-ac-ßEP significantly rose from 3.5 ± 0.4 to 19 ± 1.6 fmol/ml ($P < 0.001$), while pituitary N-ac-ßEP significantly ($P < 0.001$) declined from 97.5 ± 8.6 to 29 ± 3.0 pmol/organ.

Table I. **Ambient condition** = fish maintained in outdoor basins; **Confinement and crowding** = ten fish confined in 50-liter tanks for 1 month. Results are expressed as means ± SE (n=10)

	Ambient condition	Confinement and crowding
Plasma Cortisol (ng/ml)	12.5±1.5	164±14.3
Plasma N-ac-ßEP (pmol/ml)	3.5±0.4	19±1.6
Pituitary N-ac-ßEP (pmol/organ)	97.5±8.6	29±3.0

Experiment II - In the experiments in which crowding and manipulation was added for 30 min to long-term (one month) confinement (Table II), the significant ($P < 0.001$) elevation of cortisol was found in the crowded, manipulated and confined (group C-nT) fish compared with the confined ones (group B). But, in the group (C-T) injected with 1 µg of antaxone, the rise in cortisol was eliminated. Moreover, when the conditions were restored to those of group B (i.e. group D), the plasma cortisol levels significantly decreased coming close to baseline. In contrast with the decrease of pituitary N-ac-ßEP found in Experiment I, confinement, crowding and manipulation together increased pituitary N-ac-ß-EP content (Table II).

Table II. **A**=fish maintained in outdoor basins; **B**=fish confined in 10,000-liter tanks for 1 month; **C**=fish previously confined in 10,000-liter tanks for 1 month, then crowded (ten fish confined in 50-liter tanks) and manipulated for 30 min. (nT, not treated; T, injected with 1 µg of antaxone); **D**= fish previously subjected to the conditions of group C, and then confined again in 10,000-liter tanks for 20 h. Results are expressed as means ± SE (n=10)

	Plasma Cortisol (ng/ml)	Plasma N-ac-ßEP (pmol/ml)	Pituitary N-ac-ßEP (pmol/organ)
A	12.5±1.5	3.5±0.4	97.5±8.6
B	142±9.5	4±0.4	9±0.8
C-nT	400±38.3	5±0.6	57.3±6.0
C-T	150±12.9	5±0.4	- -
D	50.5±4.9	- -	47.6±5.0

DISCUSSION

These results suggest the involvment of opioid peptides in seabream stress-response.

Plasma N-ac-ßEP titers significantly increased together with concomitant decrease of pituitary N-ac-ßEP content, suggesting the discharge of pituitary peptide in the blood of one-month confined fish. On the contrary, an increased content of N-ac-ßEP in the pituitary of stressed fish in Experiment II was found. It seems that, as in salmonids (Sumpter *et al.*, 1985), the melanotrope response to stress is essentially stressor-specific. Interestingly, in this stress paradigm antaxone treatment reversed the activation of the pituitary-interrenal axis, suggesting that endogenous opioid activation accounts for the stress effects.

REFERENCES

Sumpter, J.P., Pickering, A.D. and Pottinger, T.G. (1985) Stress-induced elevation of plasma α-MSH and endorphin in brown trout, *Salmo trutta* L. Gen. Comp. Endocrinol. 59: 257-265.

Facchinetti, F., Radi D., Mosconi M., Carnevali O., Pestarino M. and Polzonetti-Magni AM.(1995) Acetyl salmon endorphin in the ovary of two teleostean species: changes with environmental conditions. J. Reprod. Fertil. *In press.*

Acknowledgments- This study was supported by Ministero per le Risorse Agricole Alimentari e Forestali of Italy

HAS MELATONIN A ROLE IN REPRODUCTIVE SEASONALITY IN THE FEMALE RAINBOW TROUT, *Oncorhynchus mykiss* ?

J. Nash[1], D. E. Kime[1], W. Holtz[2] and H. Steinberg[2]

[1] Animal and Plant Sciences, The University of Sheffield, Sheffield, S10 2TN, United Kingdom.
[2] Institut für Tierzucht und Haustiergenetik, Universität Göttingen, D-37075, Göttingen. Germany.

Summary

Slow release melatonin implants elevated daytime plasma melatonin but did not affect the reproductive seasonality of the female rainbow trout.

Introduction

The role of melatonin in the transduction of photoperiodic information to the reproductive axis has been well established in mammals. Melatonin increases at night and acts within the brain as a signal of daylength status. The function of melatonin in fish however remains unclear even though it shows a diurnal rhythm of nocturnal secretion (Randall *et al.*, 1991). By introducing an exogenous source of melatonin to interrupt the natural signal it might be possible to ascertain whether melatonin is involved in reproductive seasonality.

Materials and Methods

Groups (n=14) of two year old adult female trout were intramuscularly implanted with 0.5 and 3.5 mg/kg b.w. of slow-release melatonin microspheres or placebos at four different times over one year under natural photoperiod. In a further trial, treatment and control groups (n = 35) were implanted both into the peritoneal cavity and muscle (7 mg/kg b.w.) continuously at six week intervals over a period of 9 months. In this trial a positive control group (n = 35) was subjected to continuous darkness. Vitellogenin was quantified by ELISA and melatonin by RIA. All statistical significances were tested to $P < 0.05$.

Results

The implants significantly elevated plasma melatonin above normal daytime levels for up to eight weeks post implant (Fig. 1).

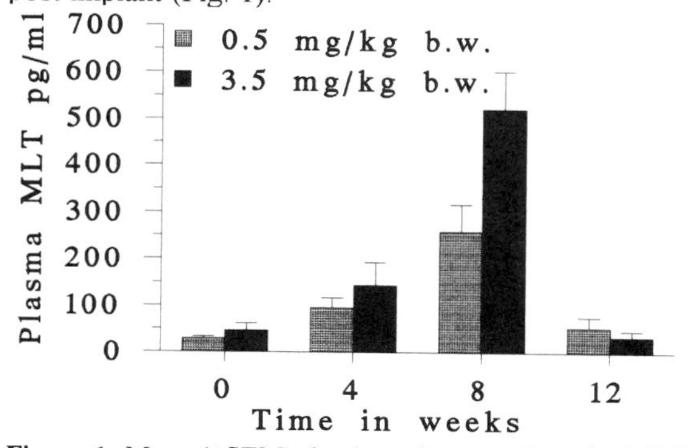

Figure 1. Mean (±SEM) daytime plasma melatonin (MLT) levels before and up to 12 weeks after implantation.

In the first trial none of the treatments altered the spawning time outside the normal season. In the second trial the group in continuous darkness spawned significantly earlier than the group in natural photoperiod but there was no difference in spawning time in the melatonin implanted group (Fig. 2).

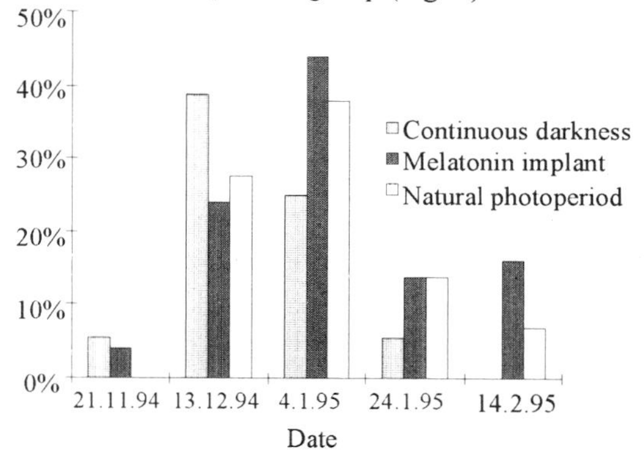

Figure 2. Proportion of fish which spawned in the second implant trial over five sampling dates.

At the end of both trials, gonad size and vitellogenin levels were not significantly different between any of the treatments including the group in continuous darkness. Vitellogenin profiles for individual fish show the expected pattern of a sustained peak prior to spawning with no differences between treatments.

Discussion

The role of melatonin in reproductive seasonality in the female rainbow trout remains unclear. Daytime melatonin was successfully manipulated in the plasma using slow release implants but this caused no effect on seasonality. Several possible explanations of these results can be proposed, for example: 1) the experiments were too short in duration or the levels of melatonin released were too small, 2) melatonin is not the transductional hormone of photoperiodism in trout, 3) manipulation of melatonin in the plasma may not interrupt the signal at the receptor sites inside the brain.

References

Randall, C. F., Bromage, N. R., Thrush, M. A., Davies, B. 1991. Photoperiodism and melatonin rhythms in salmonid fish. In " Proceedings of the Fourth International Symposium on the Reproductive Physiology of Fish", pp. 136-138, FishSymp 91, Sheffield.

This research has been financed by the Commission of the European Communities within the frame of the EEC Research Programme in the Fisheries Sector ("FAR").

EFFECT OF LIGHT ON MELATONIN SECRETION *IN VITRO* FROM THE PINEAL OF THE HAMMERHEAD SHARK, *SPHYRNA LEWINI*

D. K. Okimoto and M. H. Stetson

School of Life and Health Sciences, University of Delaware, Newark, Delaware 19716

Introduction

The influence of light on melatonin secretion was investigated in the pineal of the hammerhead shark, *Sphyrna lewini*, maintained in organ culture.

Methods

Individual pineals were incubated in 1 ml of culture medium (Eagle's MEM, Sigma, M-4144) at 23°C under a normal (12L:12D) and a reversed (12D:12L) photoperiod for 6 days (Fig. 1), continuous light (LL) for 2 days followed by a normal (12L:12D) or a reversed (12D:12L) photoperiod for 4 days (Fig. 2), continuous dark (DD) for 4 days and DD for 2 days followed by a normal (12L:12D) or a reversed (12D:12L) photoperiod for 2 days (Fig. 3). Samples were collected every 12 hr and frozen at -20°C until melatonin determination by radioimmunoassay. Samples under DD were taken under dim red light.

Results

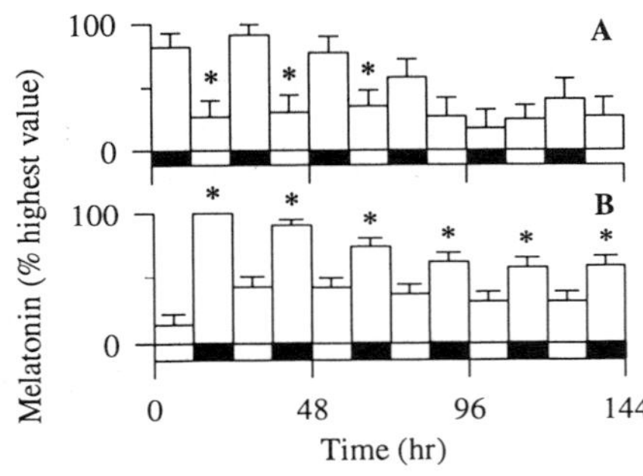

Fig. 1. Effect of A: LD 12:12 or B: DL 12:12 on melatonin release from shark pineals.

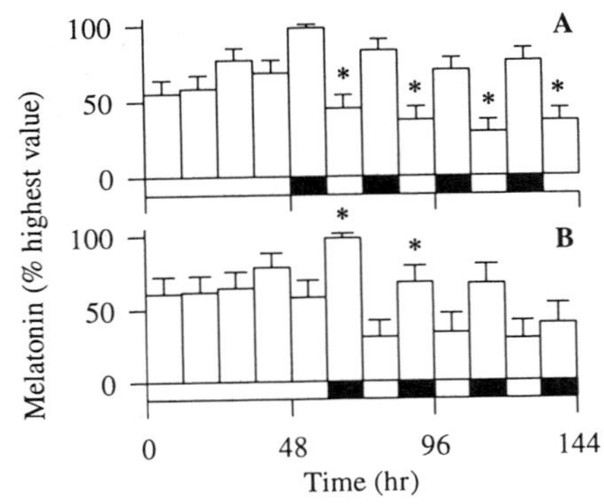

Fig. 2. Effect of LL followed by A: LD 12:12 or B: DL 12:12 on melatonin release from shark pineals.

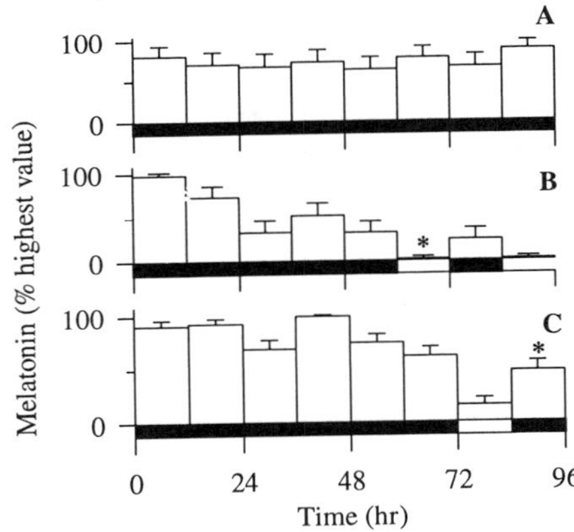

Fig. 3. Effect of A: DD, B: DD followed by LD 12:12 and C: DD followed by DL 12:12 on melatonin release from shark pineals.

Melatonin secretion was high during the dark and low during the light under normal (12L:12D; $P<0.0027$) and reversed (12D:12L; $P<0.001$) photoperiodic conditions. Moreover, melatonin release was suppressed under constant light (LL). Under constant dark (DD), melatonin release was constantly elevated and no circadian rhythmicity in secretion was found. In Fig. 1-3, an asterisk (*) denotes significance at $P<0.05$ compared with the preceding 12-hr incubation period.

Discussion

This is the first report demonstrating that light affects melatonin release *in vitro* from the pineal of an elasmobranch.

Melatonin release from cultured shark pineals occurs in direct response to ambient photoperiod. Under 12L:12D or 12D:12L, melatonin release is rhythmic with greater release in the dark than in the light. Moreover, melatonin release from shark pineals appears to be suppressed by light; rhythmic release was not detected under LL, but was reestablished after changing from LL to 12L:12D or 12D:12L. These data suggest that the isolated gland contains photoreceptors which are directly responsive to light.

Under DD, melatonin release was constantly elevated with no detectable rhythmicity, suggesting that the pineal of the hammerhead shark does not contain an intrapineal circadian oscillator (=clock).

In summary, the pineal of the hammerhead shark acts as a photoneuroendocrine transducer which may play a role in regulating seasonal aspects of reproduction in sharks. Supported by NSF grants DCB87-14638 and IBN93-18203.

A REASSESSMENT OF THE INHIBITORY EFFECTS OF CORTISOL ON OVARIAN STEROIDOGENESIS

N.W. Pankhurst, [1]G. Van Der Kraak and [2]R.E. Peter

Department of Aquaculture, University of Tasmania, Launceston Tasmania 7250, Australia. [1]Department of Zoology, University of Guelph, Guelph, Ontario N1G 2W1, Canada. [2]Department of Biological Sciences, University of Alberta, Edmonton, Alberta T6G 2E9, Canada.

Summary

Incubations of trout ovarian follicles with cortisol (F) resulted in inconsistent suppression of basal 17β-estradiol (E_2) production. F had no effect on follicles from goldfish, carp and snapper.

Introduction

Stress has an inhibitory effect on reproduction in every teleost species so far examined, and this effect is generally assumed to be mediated by F. In particular, work on trout suggests that F inhibits basal steroidogenesis by ovarian follicles (Carragher and Sumpter, 1990). Recently, using identical protocols, we were unable to show that F had any effect on either basal, GtH or steroid precursor-stimulated production of testosterone or E_2 by isolated ovarian follicles of carp, goldfish or the sparid *Pagrus auratus* (Pankhurst et al. 1995). In view of the difference between these outcomes and those from the earlier work on salmonids, we re-examined the effects of F on *in vitro* production of E_2 by ovarian follicles from rainbow trout.

Methods and Results

Isolated ovarian follicles were incubated for 21h at 12°C in modified Cortland solution containing F at concentrations of 0-1000 ng.ml^{-1}. Levels of E_2 in the media were measured by RIA. Basal E_2 production was inhibited by F at 1000 ng.ml^{-1} in follicles from 2 fish, but unaffected in a further 3 fish (Fig. 1). There was no effect of F on E_2 production by follicles from 3 other fish incubated at 18°C (the incubation temperature used by Carragher and Sumpter, 1990).

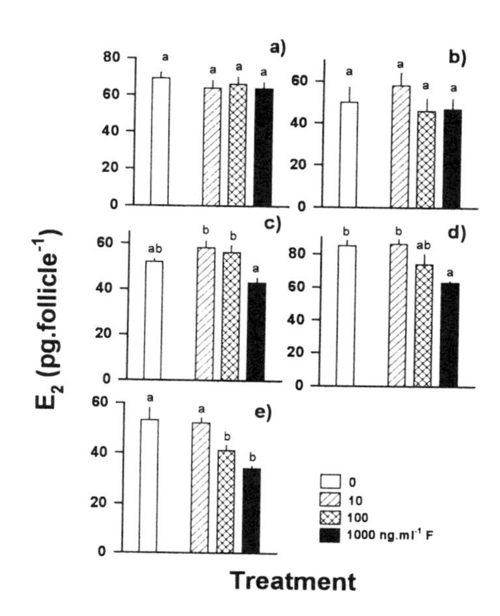

Fig. 1. *In vitro* E_2 production by follicles from 5 rainbow trout incubated for 21h at 12°C with 0-1000 ng.ml^{-1} F. Values are mean + SE (n=4). Different superscript = significantly different (P<0.05).

Discussion

Our results indicate that in contrast to non-salmonid species, F can inhibit E_2 production by trout follicles, albeit not consistently. It is not clear to us why we found suppression of E_2 production so much less frequently than the previous workers (2/8 compared with 7/8 fish tested), or why salmonids should behave differently with respect to F than the non-salmonids we have tested. The difficulties that we have had in detecting a consistent effect of F at the ovarian level, and the high doses of F generally required to produce the effect suggest to us that direct inhibition of ovarian steroidogenesis by F is not the principal mechanism whereby stress affects reproduction.

References

Carragher, J.F. and Sumpter, J.P. 1990. The effect of cortisol on the secretion of sex steroids from cultured ovarian follicles of rainbow trout. Gen. Comp. Endocrinol. 77: 403-407.

Pankhurst, N.W., Van Der Kraak, G. and Peter, R.E. 1995. Evidence that the inhibitory effects of stress on reproduction are not mediated by the action of cortisol on ovarian steroidogenesis. Gen. Comp. Endocrinol. In press.

PRELIMINARY OBSERVATIONS ON THE EFFECTS OF MELATONIN IMPLANTS AND PINEALECTOMY ON THE TIMING OF REPRODUCTION IN RAINBOW TROUT

C.F. Randall, M.J.R. Porter, N.R. Bromage and B. Davies

Institute of Aquaculture, University of Stirling, Stirling FK9 4LA, Scotland

Introduction

It is not known how information on daylength is transmitted to the reproductive axis of salmonid fish. It has been suggested that the pineal gland transduces photic information into an endocrine signal, via release of the hormone melatonin, or a neural signal (see Bromage *et al.*, this volume). This paper reports preliminary observations on the effects of melatonin implants and pinealectomy on the timing of reproduction in the female rainbow trout.

Results and Discussion

To determine whether administration of constant-release melatonin implants could advance spawning of female rainbow trout to the same extent as a reduction from a long to a short daylength 3 groups of fish were subjected to the treatments shown in Fig. 1. Spawning was significantly advanced (P≤0.001; Dunn's test) in fish exposed to a reduction from a long to a short day in early May (Gp A) compared to fish maintained on a constant long day throughout the experiment (Grps B and C); there was no significant difference in spawning time between sham-operated and melatonin-implanted fish maintained on constant long days (Grps B and C), and both these groups exhibited a desynchronized spawning profile relative to Grp A (Fig. 1). Thus, melatonin administration did not mimic the phase-advancing and synchronizing effects on spawning of a reduction from a long to a short day, as would have been expected if constantly elevated melatonin levels were perceived as a short-day signal.

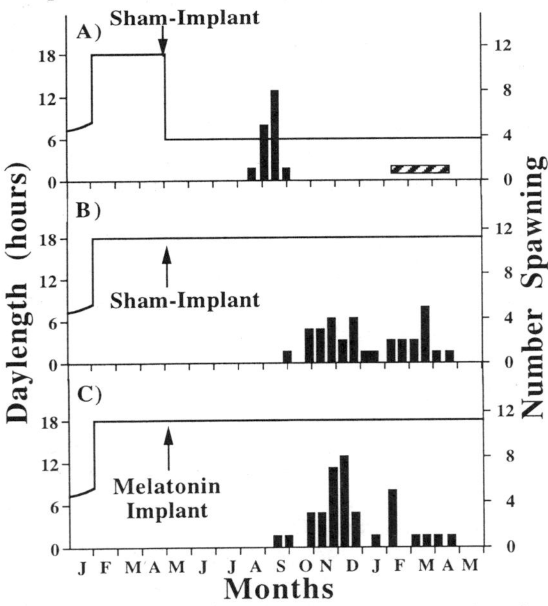

Fig. 1. Effect of melatonin implants on spawning time (histograms). Daylength is indicated by line graphs. The natural spawning period is indicated by the hatched horizontal bar on the top graph.

To determine whether removal of the pineal gland could prevent female rainbow trout responding to a change from a long to a short daylength with an advance in spawning time 3 groups of fish were exposed to the same changes in photoperiod as shown in Fig. 1A, except that the increase to a long day occurred in late March and the reduction to a short day occurred in late May. Just prior to the reduction in daylength one group was sham-pinealectomized and another pinealectomized (see Porter *et al.*, this volume); the remaining group was left intact. A further two groups were maintained under constant long days (from late March) or under ambient daylength. Sham-pinealectomized fish spawned significantly earlier (P≤0.001; Dunn's test) than pinealectomized fish (Fig. 2). Although spawning also occurred earlier in intact than pinealectomized fish the difference was not statistically significant. However, plasma calcium levels (measured as an index of vitellogenin) began to increase earlier in both intact and sham-pinealectomized fish (late Aug) than in the pinealectomized group (mid Nov). Calcium levels also began to increase in Nov in fish maintained under constant long days or ambient daylength, but these groups did not spawn until Mar-Apr and Feb-Apr, respectively. Thus, pinealectomized fish did not respond to a reduction from a long to a short daylength in the same manner as sham-operated or intact controls, but spawning was still advanced when compared with controls maintained under constant 'long' days or ambient daylength.

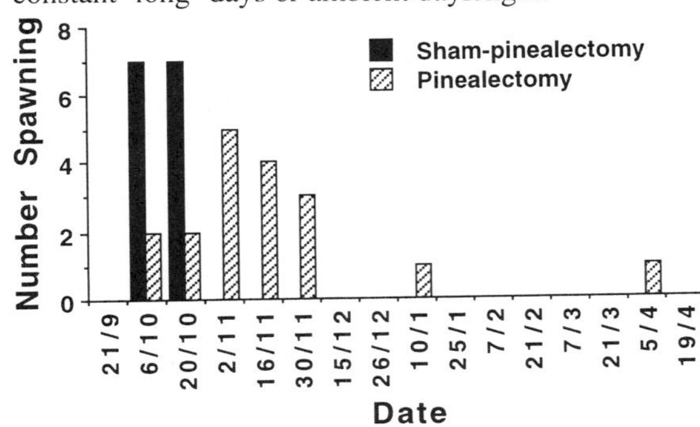

Fig. 2. Effect of pinealectomy on spawning time.

Summary

The reproductive response of pinealectomized rainbow trout to a reduction from a long to a short daylength was different to that observed in sham-operated and intact controls. This suggests that the pineal gland is involved in conveying photoperiodic information to the reproductive axis of the rainbow trout. Melatonin implants were not perceived as a short-day signal, possibly indicating that a 'melatonin-free' interval is required for the correct interpretation of the melatonin signal or perhaps suggesting a neural, rather than a hormonal, mechanism for the pineal-mediated transduction of photoperiodic information. However, further work is required to distinguish between these possibilities.

This work was supported by grants from the NERC and SERC

EFFECT OF PHOTOPERIOD AND TEMPERATURE ON REPRODUCTIVE ACTIVITY OF THE MUMMICHOG *FUNDULUS HETEROCLITUS* DURING VARIOUS SEASONS

A. Shimizu

Arasaki Marine Biological Station, National Research Institute of Fisheries Science, Nagai 6-31-1, Yokosuka, Kanagawa 238-03, Japan

Summary

Environmental control of the annual reproductive cycle in the mummichog (a reared strain; spawns during spring and summer) was examined. Both photoperiod and temperature are involved in regulation of the annual reproductive cycle, although their relative importance varies among the different reproductive phases. Some internal factor such as an endogenous circannual rhythm may also be involved in the mechanism regulating the cycle.

Introduction

One strain of the mummichog ('Arasaki Strain'; originating from the Chesapeake Bay, U.S.A., introduced to Arasaki Marine Biological Station, NRIFS Japan, in 1985) which has been reared at NRIFS shows a distinct and permanent annual reproductive cycle. This fish spawns from late March until late July in the underyearlings, and from late March until late August in the yearlings or older fish in Arasaki. This fish has also shown a regular daily spawning cycle during the spawning season, without showing a lunar or semilunar rhythm. Fish with these distinct and simple rhythms would be ideal models for studying the mechanism regulating the reproductive cycles in estuarine or marine teleosts.

Material and Methods

Mummichog were reared in outdoor tanks supplied with running natural sea water before the experiments. Experimental fish were kept under specific regimes of photoperiod and temperature for several weeks in 60 *l* glass aquarium, and the GSI and gonadal histology were examined at the time of sampling.

Results and Discussion

In the experiment started in early February, fish kept under warm temperature (16°C) showed marked gonadal development and attained functional maturity regardless of photoperiod (11 or 16L), while fish kept under low temperature (7°C) did not (Table 1). Gonadal development towards the spawning season is therefore accelerated by the elevation of water temperature during early spring. In the experiment started from middle June, yearling fish kept under long day length maintained gonadal maturity regardless of temperature (22 or 30°C), while the fish kept under moderate temperature with shorter day length (22°C-13L) showed marked gonadal regression (Table 2). Termination of the spawning season is therefore mainly caused by short daylength during late summer, although high temperature is also involved in the underyearlings (from results of another experiment). In the experiment started from the middle of October, fish kept under moderate temperature (16 or 22°C) did not show gonadal development regardless of photoperiod (11 or 16L) indicating that this species becomes photorefractory after the spawning season. Females kept under constant conditions of 22°C-16L for more than 1 year showed high GSI values (8~15%) during spring and summer and low values (1~5%) during autumn and winter, suggesting the existence of an endogenous circannual rhythm. Cooperation of environmental factors and internal factors may be important for regulation of the reproductive cycle in this species.

Table 1. Effects of photoperiod and temperature on the attainment of functional maturity (ovulation in females, milt production in males) of the mummichog during early spring.

	Fully Matured Individuals (%)			
	7°C-11L	7°C-16L	16°C-11L	16°C-16L
Female	0	0	86	100
Male	0	0	100	86

Table 2. Effects of photoperiod and temperature on the maintenance of gonadal maturity of the mummichog during summer.

	Fully Matured Individuals (%)			
	-initial	22°C-13L	22°C-16L	30°C-16L
Female	100	0	100	83
Male	100	0	100	100

THE EFFECTS OF STRESS ON SPAWNING PERFORMANCE AND LARVAL DEVELOPMENT IN THE ATLANTIC COD, *GADUS MORHUA* L.

C.E. WILSON[1], L.W. CRIM[1] AND M.J. MORGAN[2]

1. Ocean Sciences Centre, Memorial University of Newfoundland, St.John's, NF A1C 5S7
2. Department of Fisheries and Oceans, NWAFC, St.John's, NF A1C 5X1

Summary

Stress, as a result of capture and confinement elevated plasma cortisol in Atlantic cod. No differences occurred in the spawning performance of the control and stressed fish with regard to egg fertilization and hatching rates. However, there was an increase in the number of egg batches producing abnormal larvae from stressed fish as compared to the control fish.

Introduction

With the decline of the natural stocks, cod (*Gadus morhua* L.) is rapidly becoming an important species for the aquaculture industry. More basic and applied research must be performed to increase knowledge of broodstock husbandry and reproductive biology. The aim of this study was to evaluate the impact of stress on the reproductive success of spawning cod.

Methods

Four groups of male and female cod were held in separate 10600 L tanks and allowed to spontaneously spawn. Two tanks of fish were exposed to stress by netting and identifying individuals for ½ hr period 3x/wk and the remaining two tanks were undisturbed controls. Spawning activity and success was measured by daily egg volumes, egg fertilization and hatch rates as well as observations of larval development.

Results

There was no significant difference between the egg fertilization rates of stressed fish as compared to the undisturbed controls (Fig.1).

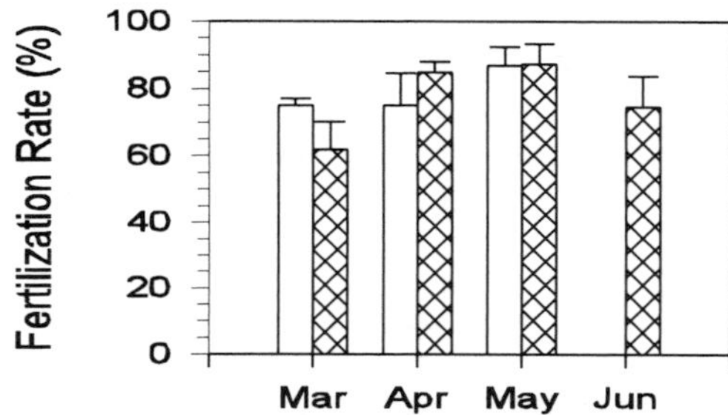

Fig.1. Fertilization rates of eggs from control (clear box) and stressed (hatched) fish.

The most obvious effect of stress was noted during larval development where higher frequencies of abnormal larvae ("twirlers") were produced by stressed cod compared with the control broodstock (Table 1.).

Table 1: Egg batches containing >75% abnormal* larvae.

Date	Control		Stressed	
	#[1]	%[2]	#	%
March	10	0	11	9
April	17	17	22	45
May	15	26	17	59
June	0	0	13	100

* Abnormal larvae were curved and swam in circles
1 = total number of batches of eggs
2 = percent abnormal batches

Discussion

By virtue of their age and size, broodstock are far more tolerant of stress than fry and juveniles; however, broodstock reproductive performance is sensitive to stress (Campbell *et al.*, 1992).

Exposure to stress can result in irregular spawning intervals, low egg fertilization rates and increased occurrence of abnormal embryos (Kjesbu, 1989). However, this study showed that fertilization and hatching rates were not affected when the broodfish were exposed to stress from chasing, capture and confinement. The only visible indication of the effects of stress was the increased percentage of abnormal larvae. According to Kjörsvik (1994) parameters such as egg survival and hatching rates are very crude measures of egg quality whereas morphological parameters of larvae and rates of "normal" viable larvae are considered more reliable indicators of gamete quality.

References

Campbell, P.M., *et al.* (1992). Stress reduces the quality of gametes produced by rainbow trout. Biol. of Repro., 47: 1140-1150.

Kjesbu, O.S. (1989). The spawning activity of cod, *Gadus morhua* L.. J. Fish Biol., 34: 195-206.

Kjörsvik, E. (1994). Egg quality in wild and brood-stock cod *Gadus morhua* L., J. World Aquaculture Society. Vol.25, No.1, p.22-29.

SEASONAL SHIFTS IN DIEL CHANGES IN FREE T4 AND T3 AND ROLE OF SEX STEROIDS IN BINDING ABILITY OF T4 AND T3 TO THEIR THYROID BINDING PROTEINS IN *CLARIAS BATRACHUS*

T.P. Singh, N. Sinha, B. Lal and K. Acharia

Fish Endocrinology Laboratory, Department of Zoology,
Banaras Hindu University, Varanasi- 221 005, India.

Summary

Plasma levels of free T4 and T3 changed daily with two peaks being found, one in the photophase and other in the scotophase; their timings and magnitude varied with reproductive status. 17α, 20β-DHP reduced the ability of T4 and T3 to bind to their transport proteins.

Introduction

Thyroid hormones in the circulation are bound to transport proteins. It is their free forms only which enter the target cells and bring characteristic changes. T3 synergizes the effects of other hormones and maintains diel rhythms of tissue sensitivity to hormones. Therefore, plasma levels of free T4 and T3 were measured by RIA at 4 h interval over a daily cycle during different reproductive phases in *C. batrachus* and also the *in vitro* effect of E2 and 17α, 20β-DHP on the binding ability of T4 and T3 to their transport proteins.

Results

Daily variations were observed in circulating levels of free T4 and T3 with a shift in their peak times and the magnitude of the peak during the annual reproductive cycle. They showed two peaks one in photophase and other in scotophase. The results are summarized in the tables. Values of peak 1 of total and free T4 an annual scale revealed similar trend of increase and decrease, but peak 2 presented opposite trend. Comparison of peak values of free T3 with total T3 showed similar trends. In a preliminary in vitro experiment, 17α, 20β-DHP decreased binding ability of T4 and T3 to their transport proteins, while E2 failed to do so.

Discussion

Though significance of such diel changes are yet to be determined, these could be a long term adaptation to many macro- and micro-environmental factors and are perhaps entrained through many other circadianally changing hormones. These entraining factors might be influencing concentration, capacity and affinity of their transport proteins also. The increase in free T4 and T3 during oocyte maturation and ovulation, provides evidence of their involvement in these reproductive events. It was supported by *in vitro* findings also where 17α, 20β-DHP reduced the ability of T4 and T3 to bind to their transport proteins.

Table 1 : Daily changes in plasma FT4 and FT3 (p mol/l) during different phases. (Values are Mean ± SEM; n=5)

Phase		0400 h	0800 h	1200 h	1600 h	2000 h	2400 h
PREPARATORY	FT4	8.01 ± 0.33*	2.75 ± 0.30	3.23 ± 0.24	4.12 ± 0.30*	1.03 ± 0.08	2.91 ± 0.22
	FT3	0.27 ± 0.01*	0.18 ± 0.01	0.38 ± 0.01*	0.15 ± 0.01	0.13 ± 0.01	0.11 ± 0.05
PRESPAWNING	FT4	2.64 ± 0.20	1.84 ± 0.10	9.42 ± 0.37*	2.18 ± 0.18	6.71 ± 0.22*	3.44 ± 0.22
	FT3	0.18 ± 0.01	0.75 ± 0.03	1.89 ± 0.07*	0.17 ± 0.01	1.76 ± 0.11*	0.19 ± 0.01
SPAWNING	FT4	4.08 ± 0.29	3.81 ± 0.30	5.55 ± 0.26*	4.70 ± 0.28	5.97 ± 0.22*	5.29 ± 0.17
	FT3	0.18 ± 0.01	0.23 ± 0.02	0.60 ± 0.01	1.38 ± 0.11*	0.16 ± 0.01	1.12 ± 0.08*
RESTING	FT4	2.27 ± 0.33	3.18 ± 0.25	8.56 ± 0.24*	3.91 ± 0.23	9.27 ± 0.28*	4.23 ± 0.26
	FT3	0.10 ±0.01	0.13 ± 0.01	0.29 ± 0.01*	0.11 ± 0.01	0.21 ± 0.01*	0.13 ± 0.01

* Denotes peak values

Table 2 : Seasonal changes in peak values of total (ng/ml) and free T4 and T3 (p mol/l). (Values are Mean ± SEM; n=5)

Phase	Peak 1		Peak 2		Peak 1		Peak 2	
	FT4	TT4	FT4	TT4	FT3	TT3	FT3	TT3
Preparatory	4.12 ± 0.30	2.16 ± 0.12	8.01 ± 0.33	4.84 ± 0.16	0.38 ± 0.01	2.37 ± 0.16	0.27 ± 0.01	2.90 ± 0.10
Prespawning	9.42 ± 0.37*	8.22 ± 0.29	6.71 ± 0.22	8.42 ± 0.30	1.89 ± 0.07*	3.03 ± 0.15	1.76 ± 0.11*	2.95 ± 0.20
Spawning	5.55 ± 0.26	4.35 ± 0.16	5.97 ± 0.22	4.63 ± 0.17	1.38 ± 0.11	0.31 ± 0.01	1.12 ± 0.08	0.32 ± 0.01
Resting	8.56 ± 0.24*	3.40 ± 0.12	9.27 ± 0.28*	3.14 ± 0.19	0.29 ± 0.01	1.26 ± 0.15	0.21 ± 0.01	0.96 ± 0.11

* Denotes maximum values on seasonal scale of peak values of free T4 and T3

Table 3 : Effect of E2 and 17α, 20β-DHP on the binding of T4 and T3 to their transport proteins. (Values are Mean ± SEM; n=5)

	Control	E2	17α, 20β-DHP
T4	100 ± 5	95 ± 4	76 ± 5
T3	100 ± 5	95 ± 5	73 ± 5

Values are normalized to % to their respective controls.

EFFECTS OF CONTINUOUS LIGHT ON GROWTH AND SEXUAL MATURATION IN SEA WATER REARED ATLANTIC SALMON

G.L. Taranger[1], H. Daae[2], K.O. Jørgensen[2] and T. Hansen[3]

[1]Institute of Marine Research, Aquaculture Centre, P.O. Box 1870 N-5020 Bergen, Norway. [2]Stolt Sea Farm AS, P.O. Box 1798, N-5020 Bergen, Norway. [3]Institute of Marine Research, Matre Aquaculture Research Station, N-5198 Matredal, Norway.

Summary

Exposure of post-smolts to continuous additional light from January until May increased growth and decreased the proportion of sexually maturing Atlantic salmon (*Salmo salar* L.).

Introduction

Sexual maturation after 1.5 years in sea water (i.e. grilsing) is a major problem in Atlantic salmon farming. Maturation leads to reduced flesh quality and loss of growth. Mature fish can also experience high mortality if kept in sea water throughout maturation. The problem may be reduced by 1) applying techniques which reduce or prevent early maturation or 2) by enhancing growth rate in order to achieve market size before maturity imposes any negative effects. The present study was conducted to test if use of continuous additional light (LL) on Atlantic salmon post-smolts from Jan. until May could enhance growth and reduce grilsing. Salmon post-smolts (N = 30,000) were distributed randomly among two conventional 2250m^3 sea cages, and two 800m^3 tanks with pumped sea water on a commercial fish farm in Norway on Nov. 20. Mean body weight was 1.1±0.3 kg (±SD). One tank (LL-TANK) and one sea cage (LL-PEN) were exposed to LL from Jan. 1 to May 1 [6 x 1000 W halogen lamps over each unit]. Two control groups received natural light only (NL-TANK and NL-PEN). The fish were fed *ad libitum* during the hours of natural day light. Weight and length were recorded in Feb. and Apr. Weight and sexual maturity were recorded at the time of harvest in Aug. and Sep., in tanks and sea cages, respectively.

Results and discussion

Appetite decreased in the LL groups compared with the NL groups during a period of approx. two months following onset of LL (Fig. 1). Subsequently, appetite increased in the LL groups, and remained higher than in the NL groups during March-June. Growth followed the same pattern as appetite, and mean body weights were higher in the LL groups than in the corresponding NL groups at time of harvest (Table 1). Appetite and growth was generally higher in the tanks compared with the sea cages during winter and spring, partly due to higher water temperature. The proportions of grilse were 21 and 26% in the NL-TANK and NL-PEN groups, but only 11 and 9% in the LL-TANK and LL-PEN groups, respectively.

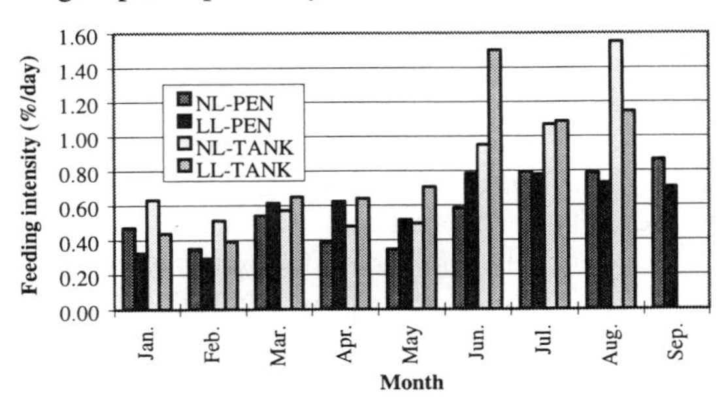

Fig. 1. Feeding intensity (feed amount * biomass^{-1} * day^{-1} * 100%) in Atlantic salmon fed *ad libitum*.

Table 1. Mean weights (±SD) of Atlantic salmon exposed to natural (NL) or continuous additional light (LL) from January until May. Same superscript letter indicate groups which are not significantly different.

	Group			
Date	NL-PEN	LL-PEN	NL-TANK	LL-TANK
26/02	1.6±0.5^{ab}	1.4±0.4^{a}	2.0±0.5^{c}	1.8±0.6^{bc}
29/04	2.1±0.6^{a}	2.2±0.6^{a}	2.2±0.6^{a}	2.4±0.6^{b}
14/08	-	-	2.9	3.7
08/09	3.3	3.9	-	-

The results suggest that an abrupt change from short to long photoperiod in January both increases growth rate and reduces the proportion of maturing salmon. This is in accordance with previous studies on 2 sea-winter Atlantic salmon (Taranger 1993), and offers a simple and effective method to enhance growth and reduce grilsing in Atlantic salmon farming.

References

Taranger, G.L. 1993. Sexual maturation in Atlantic salmon, *Salmo salar* L.; aspects of environmental and hormonal control. *Dr. scient.* thesis. University of Bergen, Norway. ISBN 82-7744-006-5.

PHASE-SHIFTED PHOTOTHERMAL CYCLES ADVANCE SEXUAL MATURATION OF MORONE HYBRIDS

A.E. Tate[1] and L.A. Helfrich

Department of Fisheries and Wildlife Sciences, Virginia Polytechnic Institute and State University, Blacksburg, VA

Summary

Six and nine month photothermal regimes, formed by excising summer/fall or summer periods of the annual light and temperature cycle, advanced sexual maturation of striped bass hybrids (Morone chrysops x M. saxatilis). Early maturity allowed spawning in December and January, respectively, without significantly reducing reproductive performance.

Introduction

Expansion of striped bass aquaculture is dependent on off-season fry and fingerling production. Seed fish are currently produced from an unpredictable supply of wild stocks captured and spawned during their natural reproductive season (April - June).

This study examined the efficacy of six and nine month phase-shifted cycles, formed by excising summer/fall or summer photothermal conditions coinciding with dormant phases of the annual reproductive cycle, for advancing maturation of Morone broodstock.

Results

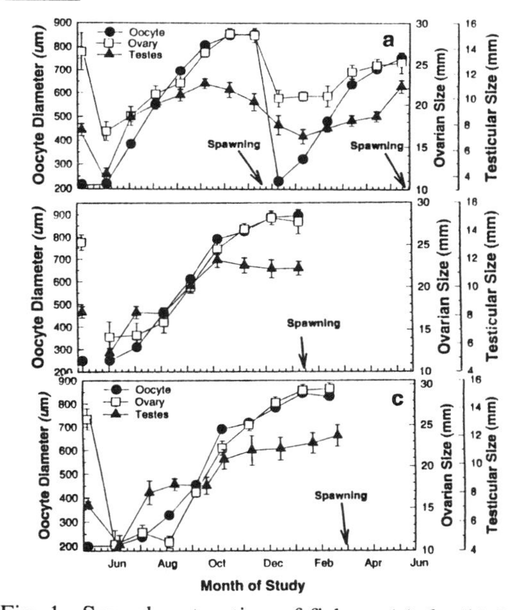

Fig. 1. Sexual maturation of fish on (a) 6-, (b) 9- and (c) 12-mo cycles.

Maximum oocyte and ovarian width occurred in December, January and March, and did not differ significantly ($p>0.05$) among females on 6-, 9- and 12-mo cycles (Fig. 1). Testicular width peaked in October for males on 6- and 9-mo cycles, and March for 12-mo fish (Fig. 1).

Spawning was induced with hormone injection in Dec. (day 200), Jan. (day 240) and March (day 283). All females (n=37) matured and ovulated, and most (>90%) males (n=36) spermiated (Table 1). Fertility did not differ ($p>0.05$) among fish on 6- (69%), 9- (59%) and 12- (68%) mo cycles, and survival was high (>85%) in all cycles. All fish exposed to a second, consecutive 6-mo cycle matured and spawned, but exhibited decreased ($p<0.05$) fertility (29%).

Table 1. Reproductive performance of hybrid striped bass on 6-, 9- and 12-mo cycles.

Reproductive Parameter	Photothermal Cycle		
	6-Month	9-Month	12-Month
Ovulating (%)	100	100	100
Spermiating (%)	92	100	100
Fertility (%)	69	59	68
Latent Period (h)	33	33	34
Survival (%)	96	92	86

Discussion

Phase-shifted photothermal cycles advanced gonadal recrudescence, shortened the reproductive cycle, and permitted off-season spawning of hybrid bass, without compromising reproductive performance. This technique can be used to spawn hybrids more frequently (i.e. biannually) and can provide culturists with an option to compressed cycles, delayed spawning, or hormone treatment for producing fingerlings and market-size hybrid striped bass year-round.

Results suggest that a lengthy refractory period between spawning and recrudescence is unnecessary, and that photoperiod and temperature are strong maturational cues, although other unconfirmed factors (e.g. endogenous rhythms) may modify maturational rates. Early maturity (1-2 mo) of fish on control cycles (12 mo) may be due to a premature increase in spring temperatures. Future research should examine interactions between endogenous rhythms and photothermal cues.

Current Address: University of Delaware, College of Marine Studies, Lewes, DE.

INTERACTIONS OF THE ENVIRONMENTAL ESTROGENS NONYLPHENOL AND β-SITOSTEROL WITH LIVER ESTROGEN RECEPTORS IN FISH

L. Tremblay, X. Yao and G. Van Der Kraak

Department of Zoology, University of Guelph, Guelph, Ontario, Canada, N1G 2W1

Summary

Two environmentally relevant chemical contaminants nonylphenol (Np) and β-sitosterol were shown to bind to rainbow trout and goldfish hepatic estrogen receptors (ER). They function as estrogen agonists by affecting estrogen dependent process including ER induction and vitellogenin (Vg) synthesis.

Introduction

17-β estradiol (E_2) stimulates the production of Vg and induces the synthesis of ER in fish hepatocytes (Mommsen and Lazier, 1986). There are numerous reports of environmental chemicals that may act as estrogen mimics. Two of the most abundant estrogenic compounds found in water are Np a widely used nonionic surfactant (White *et al.*, 1994) and β-sitosterol a phytoestrogen found in pulp and paper mill effluents. Np and β-sitosterol were evaluated for their ability to compete with E_2 for the ER and to induce estrogen dependent responses including Vg production and up-regulation of ER.

Methods

Specific, high affinity and low capacity E_2 binding sites were isolated and characterized in KCl extract of rainbow trout liver nuclei and in goldfish liver cytosol. Isolated ER were used to evaluate the potency of Np and β-sitosterol relative to E_2 using a competition assay. Enzyme linked immunosorbant assays (ELISA) were developed and validated to evaluate Vg in both species.

Results and Discussion

Figure 1 shows the competitive binding of NP and sitosterol to rainbow trout hepatic ER. Np had a relative potency of 0.001 while sitosterol had a somewhat lower potency but solubility problems prevented testing higher doses. Similar results were obtained with isolated goldfish hepatic ER. Genistein is another phytoestrogen but is not found in pulp mill effluents.

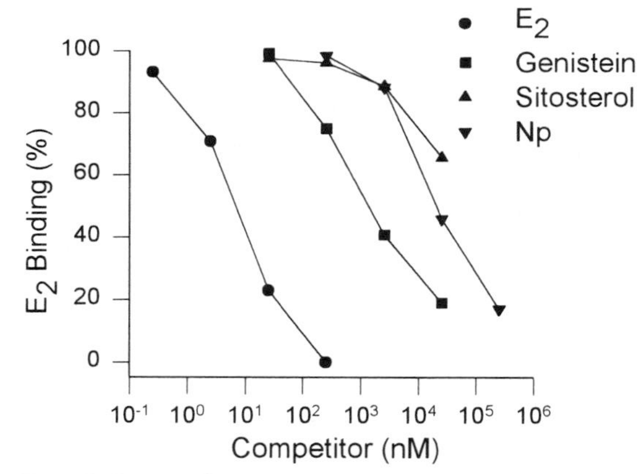

Fig. 1 Competitive displacement of tritiated E_2 from the ER by Np and β-sitosterol.

Further tests to evaluate whether these compounds act as E_2 agonists were based on their ability to induce ER and plasma Vg. Plasma concentrations of Vg were not induced in juvenile rainbow trout exposed to Np (25 μg/L) for 8 days. However, there was a 20% increase in hepatic E_2 binding sites. No significant induction of ER was observed in mature male goldfish exposed to Np or sitosterol (12 days) above the high levels of hepatic ER found in control fish. In male goldfish, both Np and sitosterol induced the production of Vg after 11 days.

In summary, these studies demonstrate that NP and sitosterol bind weakly to ER in fish and in some tests function as E_2 agonists.

References

Mommsen T.P., and C. B. Lazier, 1986. Stimulation of estrogen receptor accumulation by estradiol in primary cultures of salmon hepatocytes. FEBS Lett. **195**:269-271.

White, R., S. Jobling, S.A. Hoare, J.P. Sumpter, and M.G. Parker, 1994. Environmentally persistent alkylphenolic compounds are estrogenic. Endocrinology **135**:175-182.

Reproductive Life History

ELASMOBRANCH REPRODUCTIVE LIFE-HISTORIES: ENDOCRINE CORRELATES AND EVOLUTION

Ian P. Callard, Oliver Putz, Marina Paolucci and Thomas J. Koob

Mount Desert Island Biological Laboratory, Salsbury Cove, ME 04672, USA

Summary

All elasmobranch strategies are similar in that the group has evolved slow growth, large adult size, late reproduction, reduced fecundity, and large well developed offspring. Here we review the basic categories of reproductive strategy in the context of reproductive biology and endocrinology. We suggest that the gonadal steroid hormone progesterone controls progression of the cycle through specific "gates" by inhibitory actions at various functional levels. These include the rate of vitellogenin synthesis, as emphasized here. At the molecular level we believe that the early evolution of two progesterone receptor isoforms and their differential tissue-specific expression provides a molecular basis for both known inhibitory and stimulatory actions of this hormone.

Introduction

The cartilaginous fishes, the oldest group of jawed vertebrates, include a minor group, the Holocephali, and the Elasmobranchii, which diverged early in vertebrate evolution about 400 - 350 mybp. Elasmobranchs include the sharks (Selachii) and their sister group the Batoidea (Skates and Rays) which diverged in the early Jurassic (150 mybp) from a neoselachian ancestor. The group comprises 8 orders of sharks with 375 to 500 extant species and 5 orders of batoids representing 472 - 494 species. Compagno (1990) has brought some taxonomic order to the diversity of elasmobranch species on the basis of alternative life styles, and he has divided them into at least 18 types based on habitat and morphology. In Comagno's system, the coastal (littoral) "ecomorphotype" is seen as most plesiomorphic.

During the past 30 years, reproduction, the central life-history trait (Stearns 1976), has been subjected to analysis by life-history theory. This combines ecological and classical life history studies with evolutionary aspects as they emerge from population genetics. This type of analysis brings a powerful interdisciplinary approach and a variety of theoretical and experimental tools to the study of life-history traits (such as the number of offspring) in which the trait is treated as a gene and its survival over generations can be modeled.

Life-history theory defines a "strategy" (e.g. reproductive strategy) as a complex of co-adapted traits designed by natural selection to solve ecological problems (Stearns, 1976). Reproductive traits, such as brood size, age at first reproduction, parity, seasonal reproduction and reproductive behavior are related to one another by trade-offs where one trait can only be increased at the expense of another (Calow, 1981). Hence, a reproducing organism has to "evaluate" the

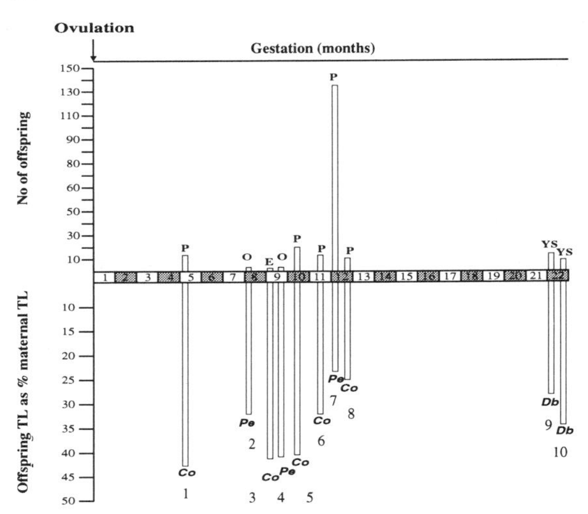

Fig.1 Differential fecundities, relative embryo sizes and gestation periods in 10 viviparous species. **O** = oophagous, **YS** = yolk sac dependent, **P** = placental, ***Co*** = coastal, ***Pe*** = pelagic, ***Db*** = deep benthic. 1 = *Sphyrna tiburo*, 2 = *Lamna nasus*, 3 = *Carcharias taurus*, 4 = *Alopias vulpinus*, 5 = *Mustelus canis*, 6 = *Carcharhinus leucas*, 7 = *Prionace glauca*, 8 = *Negaprion brevirostris*, 9 = *Squalus acanthias*, 10 = *S. blainviellei*.

resources it will spend on its reproduction instead of on itself. Once the resources available to offspring are determined, they have to be divided among the offspring. Litter sizes will decrease when the energy spent on individual offspring increases; further, when the energy expended on the individual offspring increases, the fitness of that offspring will increase. Individual fitness is dependent upon the fecundity of an organism measured over its life span. In elasmobranchs, differential fecundities are observable (Fig. 1) and indicate different reproductive strategies; endocrine regulation is necessary for their evolution and maintenance.

All elasmobranch reproductive strategies are similar in that the group has evolved slow growth, large adult size, late reproduction, reduced fecundity and large well-developed offspring (Holden, 1977). They all employ internal fertilization by paired intromittent organs and viviparity is the predominant strategy in more than 400 chondrichthyan species (Wourms, 1981). Otake (1990) divided elasmobranch viviparity into four basic types, where offspring can be provided with (a) nutrients stored in yolk sacs only (b) by uptake of maternal nutrients via trophonemata (c) by a placenta or (d) by oophagy /embryophagy. In all cases of viviparity, the embryos hatch within the uterus and subsequently develop until parturition swimming freely in the uterus. Oviparity is considered the plesiomorphic character, whereas viviparity has evolved subsequently. Viviparity is a trait which has evolved independently in various vertebrate groups (Wourms, 1981; Callard et al., 1992). The "cladogram" (Fig. 2) shows our analysis and the hypothetical relationship of reproductive traits as observed in the elasmobranchs. No conclusions should be drawn about the systematic relationship between the different species which are indicated. However, assuming that oviparity is plesiomorphic, viviparity may have evolved on as many as 18 different occasions within the group (Compagno, 1990) and placental reproduction is the most recent (apomorphic) character. Based on the diversity of reproductive strategies exemplified in Fig. 2, the question of regulation of these different strategies is important to our understanding of their evolution.

In the past ten years, a number of reviews and products of conferences on elasmobranch reproductive biology have integrated morphological, developmental, genetic, behavioural, physiological / endocrinological, ecological and evolutionary information available on elasmobranchs (see Dodd and Sumpter, 1984; Hamlett, 1989; Hamlett et al., 1993; Wourms et al., 1988; Demski and Wourms, 1993). Although the putative endocrine function of the elasmobranch gonad, and specifically the corpus luteum, has been of interest since the early work of Hisaw (Hisaw and Abramowitz, 1939), only beginning with the work of Dodd and colleagues spanning the period 1950 to 1985, and Chieffi and colleagues (1955 to present) has the endocrinology of the group become a serious pursuit. Since the mid-60's, the effort has been increasingly directed towards the physiology and biochemistry of elasmobranch reproductive systems with an emphasis on ovarian regulation, steroid synthesis and action as primary components. In the field of male elasmobranch reproductive physiology, the work of G.V. Callard on testicular function and regulation of spermatogenesis is notable (Callard et al., 1994).

Our intent in the current presentation is to examine for the first time the endocrine aspects of elasmobranch reproductive life-history strategy in the context of their general reproductive biology. Several species have been chosen on the basis of current knowledge of endocrine correlates of their reproductive cycles. Figure 2 allows us to infer probable evolutionary direction from the plesiomorphic character of oviparity through various intermediate strategies to placental viviparity, considered to be the most apomorphic in the type species *Scoliodon laticaudus* (spadenosed shark, Wourms, 1993). In this species eggs are virtually devoid of yolk and are the smallest of any known shark (65 µg dry weight); here placental viviparity is very close to the eutherian condition.

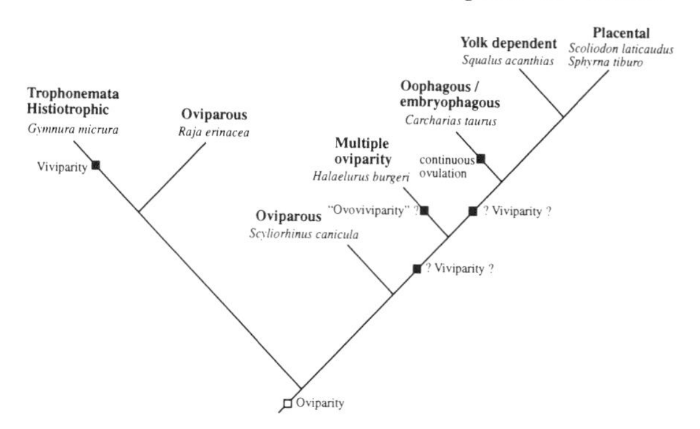

Fig. 2 Hypothetical cladogram for the evolution of elasmobranch reproductive strategies. Oviparity is the assumed plesiomorphic trait, whereas viviparity might have evolved more than once. The tree does not resemble actual relationships in recent elasmobranch species.

Although knowledge of the endocrinology of reproduction is by no means complete for any of these species, enough information is available, when combined with other observations to suggest the adaptive modifications in the blue print of vertebrate viviparity. It should be noted that elasmobranch reproductive adaptations are more similar to amphibian and amniote patterns, and antedate them, than those of teleosts.

A. Oviparous cycle. *Raja erinacea* (Batoidea) / *Scyliorhinus canicula* (Squalomorpha) Figure 3

Although a considerable amount of information is available on *S. canicula*, sex steroid hormones have not been measured throughout the cycle in this species. Nevertheless, based on the descriptions of its ovulatory cycle and interval we think it likely that it follows that depicted for *Raja* although different in its temporal aspects. We have thus taken the liberty of using the cycle of the little skate as the plesiomorphic type from which others are evolved. As can be seen it is characterized by high levels of progesterone during the preovulatory period, with titer falling during the luteal phase; nevertheless, corpora lutea of this species are functional on the basis of progesterone production (Callard et al., 1993). When actively producing eggs, this species always has a hierarchy of large yolked eggs in the ovary, and plasma estrogen shows a modest peak associated with the follicular phase. Follicular growth may be discontinous as we have reported significantly lower levels of vitellogenin around ovulation (Perez and Callard, 1993) and the associated formation of the corpus luteum. Plasma testosterone levels (not shown here) peak in advance of progesterone.

As shown in the "cladogram" of reproductive strategies, we suggest that multiple oviparity, as practiced by *Halaelurus burgeri* (Wourms et al., 1988) in which 4 - 12 encapsulated eggs may be found in the oviduct with each egg at a stage of development consistent with the temporal sequence of ovulation; eggs are retained for varying lengths of time (in *H. burgeri* 6 - 8 months) and may hatch *in utero* or *ex utero*. Bass (1975) has suggested that *H. lutarius* may be truly ovoviviparous. If so, it may have much in common with *S. acanthias*. A similar strategy has been described for *Galeus* species by Compagno (1984). During the actively reproducing period, these species, as well as *Raja* and *Scyliorhinus* are characterised by continous ovulatory cycles. This characteristic ap-

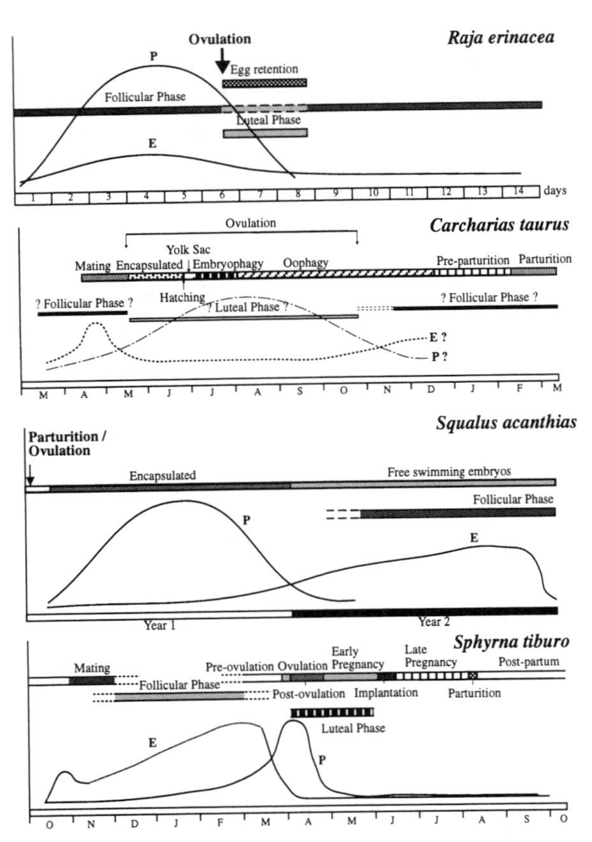

Fig. 3 Reproductive and endocrine events associated with selected elasmobranch cycles. For explanations see text.

pears to have been retained and used to considerable advantage in the postulated next stage in the evolution of viviparity, as exemplified by *C. taurus*.

B. Oophagous / embryophagous viviparity. *Carcharias taurus*

Plasma hormone levels are not yet available for this species; their presumed pattern, based on what is known of follicular growth, ovulation and pregnancy, is indicated by dotted lines. In this species gestation is 9 - 10 months the onset of which follows a defined mating period documented by Gilmore (1993). The first egg capsule contains only one ovum, followed by capsules with multiple ova (Gilmore et al., 1983). Up to three embryos have been observed to develop within a single capsule (Gilmore, 1993). The embryo that is first to hatch *in utero* utilises the other embryos (embryophagy) and subsequently feeds on ova that continue to be ovulated at intervals during the succeeding 5 months of gestation (oophagy). It is estimated that the embryos consume approx. 19000 ova throughout gestation (Putz, unpubl.). Figuring an average daily cohort of 15 eggs/ovulation (Gilmore, 1993), one can predict exhaustion of ova after 5 months (19000 total ova / 15 ova per cohort = 150 days), which in fact is true (Putz, unpubl.).

Ingested yolk is stored in an expanded yolk stomach and used for continued nutrient supply after ovulation ceases until parturition. Based on Gilmore et al. 1983, and unpublished data (Putz and Gilmore), it appears that the follicular phase occupies late gestation (see also *Squalus*) and the non-pregnant mating period. Ovaries show highest ova numbers around mating and are largest at the onset of oophagy; ovarian size and egg number is progressively depleted by the ovulation of cohorts of eggs necessary to sustain embryo growth in the absence of new follicle development. We suggest that continued ovulation generates successive waves of corpora lutea which maintain plasma progesterone until the ovary is depleted of follicles. If this interpretation is correct, this type of viviparity may be plesiomorphic in as much as it retains an oviparous ovulatory pattern (and its hypothalamic pituitary correlates) while progressing to viviparity. It is worth noting that the oophagous strategy of viviparity produces the largest embryos as percentage of maternal body length, and from these large and advantaged embryos develop the largest sharks known (see Fig. 1 and 4).

C. Yolk-dependent aplacental strategy. *Squalus acanthias*

This well-known cycle is characterized by the longest gestation period of any vertebrate; once sexually mature, females are likely to be pregnant continuously with the exception of a short period after parturition. Eggs are very large (5 - 6 cm) and are packaged together in a thin membrane (from 2 - 8 per uterine horn), depending on adult size. After ovulation corpora lutea are formed and are functional (Tsang and Callard, 1987) and no doubt contribute to the elevated plasma progesterone levels during the first year of gestation. During this period, follicular growth is slow, and vitellogenesis cannot be induced by estrogen. After embryos break out of the capsule and become free in the uterus, progesterone levels fall as estrogen levels rise associated with the follicular phase which occupies the second year of gestation. The strategy of advancing the follicular phase into the latter half of gestation ensures that eggs are ready to ovulate at the end of the long gestation period, thus enhancing life-time fecundity. This is probably made possible by the decline in luteal function by mid-term, thus removing negative feedback on vitellogenesis and hypothalamic-pituitary function. Embryos utilize egg yolk as the sole source of nutrition during the two year gestation. It can be seen from Fig. 1 that based on number and size of offspring the aplacental strategy is as effective in generating offspring as the placental mode; however, in these aplacental species the time for completion of development is twice as long suggesting improved efficiency of nutrient delivery in placental species.

D. Placental viviparity. *Sphyrna tiburo*

The complete cycle of this species has only recently been described by Manire et al. (1995). This cycle is closer to the "mammalian pattern" than those described above, consisting of sequential mating, follicular, ovulatory, pregnant and postpartum phases. Mating coincides with a rise in estrogen level, which

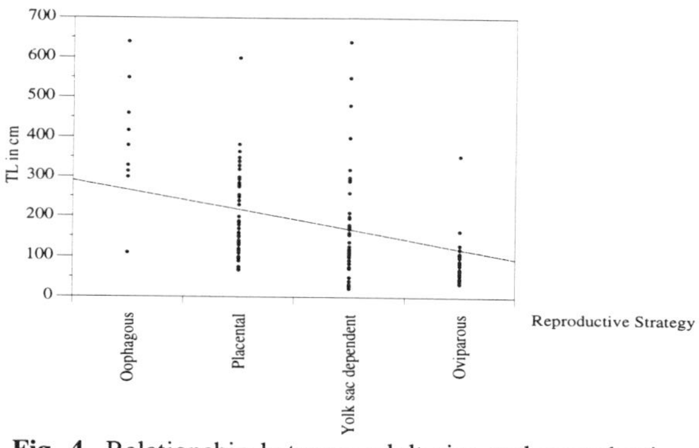

Fig. 4 Relationship between adult size and reproductive strategy in 138 viviparous elasmobranch species. A Kruskal-Wallis nonparametric ANOVA revealed significant differences between median sizes of the four groups (KW = 38.014; $P < 0.0001$).

must derive from small or medium follicles at this stage of the cycle. It should be noted that small and medium sized follicles from *Raja* and *Squalus* are most active in synthesizing estrogen *in vitro* (Callard et al., 1993; Tsang and Callard, 1987). In *Sphyrna*, estrogen levels peak at the end of the follicular phase and progesterone peaks post-ovulation during early pregnancy. Both plasma progesterone and estrogen remain low for the remaining 3 - 4 months of the 5 month gestation period, and the role of peripheral progesterone in the maintenance of gestation in this species, as well as in *S. acanthias*, is in question. Nevertheless, the potential for local placental production of progesterone in *Sphyrna* must be considered. Further, as we have suggested for *Squalus*, peptide hormones such as relaxin may be important for maintenance of gestation and uterine function after mid-term or earlier (Sorbera and Callard, 1995). Testosterone titers (not shown) follow those of estrogen, suggesting the precursor / product relationship; levels of testosterone are however higher than either of the other two steroids; further, they remain quite high during the luteal phase. As can be seen from Fig. 1, placental viviparity in this species is very efficient in producing large off-spring in a short time.

Discussion

Examination of the reproductive patterns in any vertebrate species suggests key control points or gates at which "choices" are made with regard to cycle progression. These are (a) mate selection and mating (b) follicular development / atresia (c) steroidogenesis (d) vitellogenesis / reproductive tract preparation (e) ovulation (f) corpus luteum formation, function and maintenance (g) reproductive tract growth and maintenance (h) parturition. This sequence depends to a large degree upon the temporal relationships of a number of hormones which interact with target tissue genes. Chief among the hormones are the steroids estradiol (E) and progesterone (P).

Using the mammalian "pattern" of hormonal control, derived largely from studies of a few rodents, primates and domestic animals, which undoubtedly evolved from a common ancestor with the plesiomorphic chondricthyan traits, early estradiol synthesis is a primary trigger from which other events flow once environmental factors have triggered the central nervous system. While E exerts regulatory influence by both positive and negative feedback at the central control mechanism (hypothalamus-pituitary), its primary effect on peripheral reproductive targets (liver, gonad, reproductive tract and accessories) via its specific transcription factor, the estrogen receptor (ER), is positive and provides an "on" signal. Progesterone may act as both a synergist and an antagonist (brake) on estrogen action, first on the progression of the cycle and subsequently to maintain *in utero* development and all of its correlates. In mammals, the latter actions are most remarkable and well documented; further, in as much as the E/ER interaction induces the PR gene, and P/PR appears to exert regulatory effects on ER transcription, the effects of both steroids are precisely controlled at the cellular level to orchestrate a successful reproductive outcome.

In almost all feral non-mammals, with the exception of a few apomorphic viviparous species (e.g., *S. laticaudus*, see above), follicular development cannot progress without estrogenic induction of hepatic vitellogenesis. The time course of yolk deposition in oocytes determines when eggs are large enough for ovulation and to sustain developing oviparous or viviparous young. Vitellogenesis must therefore be regulated precisely with regard to environmental conditions so as to ensure the well-being of both the maternal organism and the future brood. It is our hypothesis that this brake on the process of folliculogenesis is provided by P/PR to ensure appropriate rates of follicular development. This effect may be observed at various levels (a) inhibition the hypothalamic / pituitary unit and thus gonadotropin (GTH) output and both ovarian E synthesis and vitellogenin production and uptake (b) directly on the ovary to inhibit E synthesis and vitellogenin uptake (c) directly on the liver to inhibit E induced vitellogenin gene transcription. Although all of these effects are likely to operate in the regulation of elasmobranch reproduction cycles discussed above, experimental evidence only exists for the last so far (see Perez and Callard, 1993).

The action of both E and P on the liver requires specific receptors and both ER and PR have been demonstrated by us in elasmobranch liver; however, the mode of action of P/PR, whether by direct interaction with the 5′ regulatory elements of the vtg-gene, as for E/ER, or indirectly, remains to be determined (see Callard et al., 1993 for review). The nature of the putative PR of elasmobranchs and its relationship to other transcription factors of the steroid hormone receptor family is of great interest from the point of view of the evolution of reproductive control mechanisms, steroid receptors presumably evolving from a common ancestral molecule (Evans, 1989). A PR with a biphasic Scatchard plot, as distinct from an androgen receptor (AR) and glucocorticoid receptor (GR) has been described in the testis of *S. acanthias* by Cuevas and G.V. Callard (1992). PR appear to exist as two isoforms (a smaller, 80 - 90 kD PR-A, and a larger, 115 - 120 kD PR-B) in representative apomorphic vertebrates. In recent studies progesterone binding activity has been demonstrated in the skate liver and oviduct (Paolucci and Callard, unpubl.). A single binding moiety, which elutes from DNA-cellulose at ≈ 0.065 M in liver and 0.14 M in oviduct has many characterictics of a true receptor. Its k*d* values are ≈ 10^{-9}, it binds to DNA-cellulose and cross-reacts on Western blots with antibodies against PR-A (oviduct) and PR-B (liver). It is of great importance to complete elucidation of the characteristics of the elasmobranch PR. Although the physiologic importance of two PR isoforms is not yet understood, we have contended that different tissue specific ratios of PR-A and PR-B, or the expression of one or other only may determine whether P effects are synergistic or antagonistic with regard to E action (see Callard et al., 1993).

Recently, Vegeto et al. (1993) have shown that PR-A may have a dual role as both an activator or a repressor of ligand mediated transcription, allowing dissimilar responses to a single hormone and providing a molecular explanation for the co-existence of two forms of PR in the human and other species. We believe that the vertebrate liver is at least one target at which inhibitory effects of PR are readily seen on vitellogenin gene transcription. The presence of this mechanism in the earliest jawed vertebrates indicates it is a plesiomorphic characteristic of the endocrine control system for reproduction.

References

Bass, A.J.; J.D. D'Aubrey and N. Kistnasamy. 1975. Sharks of the east coast of southern Africa. II. The families Scyliorhinidae and Pseudotriakidae. *Invest. Rep. Oceanogr. Res. Inst.* **37**: 1 - 63.

Callard, G.V., M. Betka and J. Jorgensen. 1994. Stage-related functions of Sertoli cells: Lessons from lower vertebrates. In A. Bartke (ed.): Function of Somatic Cells in the Testis. *Springer Verlag*: 27 - 54.

Callard, I.P.; L.A. Fileti; L.L. Perez; L.A. Sorbera; G. Giannoukos; L.L. Klosterman; P. Tsang and J.A. McCracken. 1992. Role of the corpus luteum and progesterone in the evolution of vertebrate viviparity. *Amer. Zool.* **32**: 264 - 275.

—, L.A. Fileti and T.J. Koob. 1993. Ovarian steroid synthesis and the hormonla control of the elasmobranch reproductive tract. *Env. Biol. Fish.* **38**: 175 - 185.

Calow, P. 1981. Resource utilization and reproduction. In C.R. Townsend and P. Calow (eds.): Physiological Ecology: An Evolutionary Approach to Resource Use. *Sinauer Associates.*

Compagno, L.J.V. 1984. Sharks of the World: An Annotted and Illustrated Catalogue of Shark Species Known to Date. FAO Species Catalogue. *U.N. Dev. Prog., FAO, Rome.*

—. 1990. Alternative life-history styles of cartilaginous fishes in time and space. *Env. Biol. Fish.* **28**: 33 - 75.

Cuevas, and G.V. Callard. 1992. Androgen and progesterone receptors in shark (*Squalus*) testis: Characteristics and stage-related distribution. *Endocrinology* **130**: 2173 - 2182.

Demski, L.S. and J.P. Wourms (eds). 1993. The Reproduction and Development of Sharks, Skates, Rays and Ratfishes. *Kluwer Academic Publishers*. Repr. *Env. Biol. Fish.* **38**.

Dodd, J.M. and J.P. Sumpter. 1984. Fishes. In G.E. Lamming (ed.): Marshalls' Physiology of Reproduction. Vol. 1 Reproductive Cycles of Vertebrates. *Chuchill Livingstone*. 1 - 126.

Evans, R.M. 1988. Molecular characterization of the glucocorticoid receptor. In J.H. Clark (ed.): Recent Progress in Hormone Research. *Academic Press.* 1 - 21.

Gilmore, R.G. 1993. Reproductive biology of lamnoid sharks. *Env. Biol. Fish.* **38**: 95 - 114.

—, J.W. Dodrill and P.A. Linley. 1983. Reproduction and embryonic development of the sand tiger shark, *Odontaspis taurus* (Rafinesque). *Fish. Bull.* **81**: 201 - 225.

Hamlett, W.C. 1989. Evolution and morphogenesis of the placenta in sharks. *J. Exp. Zool.* Suppl. 2: 35 - 52.

—, A.M. Eulitt; R.L. Jarrell and M.A. Kelly. 1993. Uterogestation and placentation in elasmobranchs. *J. Exp. Zool.* **5**: 347 - 367.

Hisaw, F.L. and A.A. Abramowitz. 1939. Physiology of reproduction in the dogfishes, *Mustelus canis* and *Squalus acanthias*. *Ret. Woods Hole Oceanog. Inst.* **1938**: 22.

Holden, M.J. 1977. Elasmobranchs. In J.A. Gulland (ed.): Fish Population Dynamics. *John Wiley & Sons.*

Manire, C.A; L.E.L. Rasmussen; D.L. Hess and R.E. Hueter. 1995. Serum steroid hormones and the reproductive cycle of the female bonnethead shark, *Sphyrna tiburo*. *Gen. Comp. Endocrinol.* **97**: 366 - 376.

Otake, T. 1990. Classification of reproductive modes in sharks with comments on female reproductive tissues and structures. In H.L. Pratt, S.H. Gruber and T. Taniuchi (eds.): Elasmobranchs as Living Resources: Advances in the Biology, Ecology, Systematics, and the Status of the Fisheries. *NOAA Tech. Rep. NMFS circ.* **90**: 111 - 130.

Perez, L.E. and I.P. Callard. 1993. Regulation of hepatic vitellogenin synthesis in the little skate (*Raja erinacea*): Use of homologous enzyme-linked immunosorbent assay. *J. Exp. Zool.* **266**: 31 - 39.

Sorbera, L.A. and I.P. Callard. 1995. Myometrium of the spiny dogfish *Squalus acanthias*: peptide and steroid regulation. *Am. J. Physiol.* **269** (*Regulatory Integrative Comp. Physiol.* **38**): R389 - R379.

Stearns, S.C. 1976. Life-history tactics: A review of the ideas. *Q. Rev. Biol.* **51**: 3 - 47.

Tsang, P. and I.P. Callard. 1987. Luteal progesterone production and regulation in the viviparous dogfish, *Squalus acanthias*. *Gen. Comp. Endocrinol.* **70**: 164 - 168.

Vegeto, E; M. Manonchehr; D.W. Shabaz; M.E. Goldman; B.W. O'Malley and D.P. McDonnell. 1993. Human PR-A form is a cell- and promoter-specific repressor of human PR-B function. *Molec. Endocrinol.* **7:** 1244 - 1255.

Wourms, J.P. 1981. Viviparity: the maternal-fetal relationship in fishes. *Amer. Zool.* **21**: 473 - 515.

—. 1993. Maximization of evolutionary trends for placental viviparity in the spadenose shark, *Scoliodon laticaudus*. *Env. Biol. Fish.* 269 - 294.

—, B.D. Grove and J. Lombardi. 1988. The maternal-embryonic relationship in viviparous fishes. In W.S. Hoar and D.J. Randall (eds.): Fish Physiology, Vol. XI, Part B. 1 - 134.

Supported by NSF-Grants (I.P.C.) and Mount Desert Island Biological Laboratory (T.J.K.).

EVOLUTION OF GnRH IN FISH OF ANCIENT ORIGINS

Stacia A. Sower

Dept. of Biochemistry and Molecular Biology, Univ. of New Hampshire, Durham, 03824, USA.

Summary

A key neuroendocrine function of the hypothalamus is the release of the decapeptide gonadotropin-releasing hormone (GnRH) which in turn acts on the pituitary regulating the pituitary-gonadal axis for all vertebrates. Agnathans are of particular importance in understanding hypothalamic-pituitary relationships since they represent the oldest lineage of vertebrates which evolved over 550 million years ago. The agnathans are classified into two groups, myxinoids (hagfish) and petromyzonids (lamprey). Lampreys are the first vertebrates to clearly demonstrate roles for multiple GnRH molecules as neurohormones involved in reproduction. In addition, we suggest from structure-activity and receptor studies that lamprey GnRH receptor requirements for GnRH are different in the lamprey from those of all other vertebrates. This paper summarizes our current studies on the structure, function, distribution and embryonic origin of lamprey GnRH-I and -III and the evolution of GnRH systems in vertebrates.

Reproductive Cycle of the Lampreys

There are approximately 32 species of living lampreys that are classified as parasitic or nonparasitic. Lamprey spawn only once in their lifetime after which they die. Sexual maturation is a seasonal, synchronized process. The sea lampreys, *Petromyzon marinus*, begin their lives as freshwater, filter feeding larvae which burrow in the bottoms of streams (Hardisty and Potter, 1972). After approximately five to seven years in freshwater streams, metamorphosis occurs and the larvae become free swimming, sexually immature lampreys, which migrate to the sea or lakes. While the lampreys are in the parasitic sea phase, gametogenesis progresses. In males, spermatogonia proliferate and develop into primary and secondary spermatocytes and in females, vitellogenesis occurs. After approximately 15 months at sea, lampreys return to freshwater streams to spawn and undergo the final maturational processes resulting in mature eggs and sperm.

Gonadotropin-Releasing Hormone: Structure

A key neuroendocrine function of the hypothalamus is the release of the decapeptide, GnRH, which in turn acts on the pituitary regulating the pituitary-gonadal axis for all vertebrates. Currently, nine primary structures of GnRH have been determined in various representatives of vertebrates. Included in this family are the structures of GnRHs of three fish species of ancient origin, an agnathan, the sea lamprey (lamprey GnRH-I and III) (Sherwood et al., 1986; Sower et al., 1993); an elasmobranch, the spiny dogfish shark, *Squalus acanthias*, (dogfish GnRH and chicken GnRH-II) (Lovejoy et al., 1992); and a holocephalan, the ratfish, *Hydrolagus colliei*, (chicken GnRH-II) (Lovejoy et al., 1991).

Previous studies have led to the identification of two molecular forms of gonadotropin-releasing hormone (GnRH-I and II) in the brain of the sea lamprey. From analysis of these two forms, the primary structure of GnRH-I and the amino acid composition of GnRH-II were determined (Sherwood et al., 1986). We have now isolated a third molecular form of GnRH (lamprey GnRH-III) from the brain of this species that is different from GnRH-I and -II. We determined the primary structure of lamprey GnRH-III as pGlu-His-Trp-Ser-His-Asp-Trp-Lys-Pro-Gly-NH_2 (Sower et al., 1993). The primary structure of lamprey GnRH-III differs in three amino acids compared with lamprey GnRH-I. Lamprey GnRH-III is more closely related to the other members of the GnRH family than lamprey GnRH-I. Lamprey GnRH-III has 80% identity with chicken GnRH-II and dogfish GnRH; 70% identity with catfish GnRH-I, lamprey GnRH-I, and salmon GnRH; and 60% identity with mammal GnRH and chicken GnRH-I (Sower *et al*., 1993). In all GnRH peptides, certain regions of the molecule have been highly conserved including the NH_2-terminal, $pGlu^1$- His^2 and Ser^4 and the COOH- terminal. These regions and the length of the molecule have remained unchanged during 500 million years of evolution.

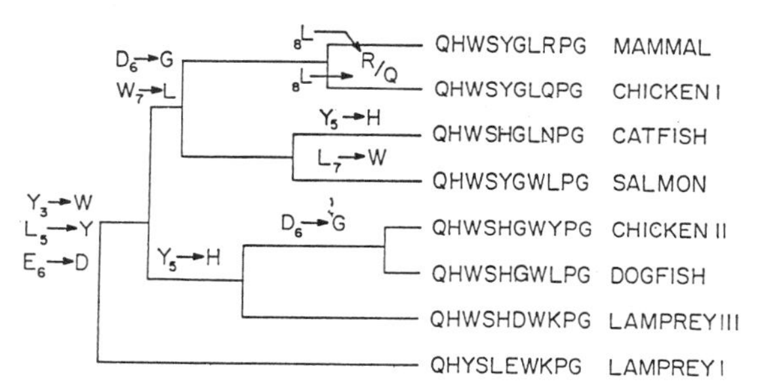

Fig. 1. Phylogenetic tree based on eight sequences of the primary structures of GnRH. One arrow signifies one base change. (Sower et al., 1993)

Gonadotropin-Releasing Hormone: Function

Until the past few years, there was little evidence for a regulatory influence of the hypothalamus on the pituitary-gonadal axis in Agnathans. Using synthetic lamprey GnRH-I and analogs in our earlier studies, these experiments provided the first evidence of neuroendocrine control of reproduction in lampreys. The biological activity of lamprey GnRH-I or -III has been assessed by steroidogenesis or gametogenesis in *in vitro* and *in vivo* studies (Review: Sower, 1990; Sower et al., 1993; Deragon and Sower, 1994). Other studies showed seasonal correlations between changes in brain GnRH and gametogenic and steroidogenic activity of the gonads in adult male and female sea lampreys

(Supported by NSF and the Great Lakes Fisheries Commission)

(Fahien and Sower, 1990; Bolduc and Sower, 1992). Our recent studies indicate that lamprey GnRH-III is also a neurohormone involved in reproduction based on its ability to stimulate steroidogenesis and gametogenesis in adult sea lampreys (Sower et al., 1993; Deragon and Sower, 1994) and of the occurrence of this peptide in lampreys undergoing different stages of metamorphosis coinciding with the acceleration of gonad maturation (Youson and Sower, 1991).

The purification of lamprey gonadotropin(s) has been a very difficult project due to the size of the pituitary and difficulties associated with purification. However, all available information strongly suggests that lamprey pituitaries have a reasonably typical pituitary-gonadal axis. (see Review, Sower, 1990).

Lamprey GnRH-I and lamprey GnRH-III are the only two members of the GnRH family to have substitutions in the sixth position, Glu^6 and Asp^6, respectively; all other GnRH peptides have Gly in the sixth position suggesting a different conformational structure. Thus, a structure-activity study of lamprey GnRH-I or analogs that were cyclized or with sixth position substitutions were determined *in vivo* in adult female sea lamprey (Sower, et al., In Press). The following analogs which were tested, ([D-Glu^6]-GnRH-I; cyclo-[D-Glu^6-Trp^7-Lys^8]-GnRH-I; or cyclo-[Glu^6-Trp^7-Lys^8]-GnRH-I), significantly elevated plasma estradiol compared to controls. However, [D-Glu^6]-lamprey GnRH-I was the only analog to significantly stimulate ovulation while another analog [Gly^6]-lamprey GnRH-I significantly delayed ovulation. These data suggest that the sixth position of lamprey GnRH is critical for function.

Pituitary GnRH Receptor:

Quantitative *in vitro* autoradiography was used to characterize and localize putative gonadotropin-releasing hormone (GnRH) receptors in the anterior pituitary of the adult female sea lamprey (Knox et al., 1994). Scatchard analysis revealed two classes of high affinity binding sites with K_d's of 1.5×10^{-12} M and 5×10^{-9}M and B_{max}'s of 8.4×10^{-14} M and 5×10^{-11}M, respectively. Binding to the GnRH receptors was saturable, reversible, tissue specific and time- and temperature-dependent. Displacement studies showed that labeled peptide could be displaced by chicken GnRH-I, chicken GnRH-II, synthetic mammal, salmon lamprey GnRH-I, lamprey GnRH-III, $DAla^6$,Pro^9 NEt mammalian GnRH and $DPhe^{2,6}$,Pro^3 lamprey GnRH. The proximal pars distalis region of the anterior pituitary contained most of the GnRH binding sites with slight binding in the rostral pars distalis. These data provide the first direct evidence of GnRH activity in the pituitary of an Agnathan and are the first to demonstrate that a vertebrate pituitary contains two high affinity binding sites for GnRH.

Brain-Pituitary Relations and Distribution of GnRH:

Unlike most vertebrates, a distinct vascular or neural link between the hypothalamus and the adenohypophysis has not been observed in either lamprey or hagfish (Gorbman, 1965). In the lamprey, the neurohypophysis and adenohypophysis are separated by avascular connective tissue (Gorbman, 1965). However, there is anatomical evidence to support the concept of hypothalamic control of adenohypophysial function by diffusion of the neurohormones from the neurohypophysis to the pars distalis of the adenohypophysis (Nozaki et al., 1984; King et al., 1988). In the lamprey, GnRH-like neurons identified by immunocytochemistry project their fibers primarily into the neurohypophysis from the preoptic region (Nozaki et al., 1984; King et al. 1988). In these same studies, GnRH was not demonstrated to be widely distributed in extra hypothalamic regions as noted for other neuropeptides within the lamprey brain. We propose in this latter study that an additional route of GnRH is via secretion into the third ventricle and transported by tanycytes to the adenohypophysis (King et al., 1988). We experimentally examined the functional anatomical relationship between the hypothalamus and adenohypophysis in sea lamprey. Horseradish peroxidase (HRP) was injected into the third ventricle of the brain of adult lampreys (Nozaki et al., 1994). Within 5 to 15 minutes HRP had passed through the neurohypophysis, which forms the floor of the third ventricle, and diffused throughout the connective tissue separating the adenohypophysial follicles from the neurohypophysis and into intracellular spaces in the adenohypophysis. We conclude that neurosecretory peptides are able to diffuse from the brain to the adenohypophysis, and thus regulate its secretory activity in lampreys. Thus, there is evidence of normal occurrence of GnRH in a part of the lamprey brain homologous with that brain region in higher vertebrates in which GnRH localization forms part of a neuroendocrine mechanism for gonadotropin secretion.

Using antibodies to the lamprey GnRH-III molecule, several immunocytochemistry studies have been completed. Both lamprey GnRH-I and -III immunoreaction were found in the cell bodies in the rostral hypothalamus and preoptic area in larval (Wright et al., 1994; Tobet et al., 1995a) and adult sea lamprey (Nozaki, Gorbman and Sower, unpublished). We have suggested that in the larval stage, the majority of the irGnRH is lamprey GnRH-III indicating that GnRH-III perhaps is the more active form during reproductive maturation (Tobet et al., 1995a).

Origin and Evolution of GnRH During Development

Chromatographic and immunological studies of vertebrate brain extracts have shown two or more GnRH-like peptides in representative species of all vertebrate classes (Muske, 1993). The functional significance of multiple forms of GnRH within the brain and in extrahypothalamic locations within the same species has not been established with the possible exception of lampreys. However, the GnRHs have apparently multiple actions on reproductive physiology and behavior either through pituitary or non-pituitary stimulation depending on the origin of the GnRH system during development. Muske (1993) has proposed that gnathostome vertebrates have two principle GnRH systems each with different embryonic origins expressing different molecular forms of GnRH and projecting to different targets. In the vertebrates examined, neurons which contain forms of GnRH which are considered to regulate pituitary-gonadal functional are thought to be derived from progenitor cells in the olfactory placode which then will migrate to its final position in the preoptic /hypothalamic areas. With some exceptions, the other GnRH system arises from non-placodal origin and involved in non-pituitary-gonadal function. Our most recent experiments in lampreys were conducted to characterize the earliest development of GnRH neurons and determine the probable pathway of their migration (Tobet et al., 1995b). GnRH neurons were first visualized at day 22 after fertilization in the preoptic area and hypothalamus. GnRH neurons were not seen within the olfactory system. In contrast to all other vertebrates, we propose that GnRH neurons in developing lampreys originate within proliferative zones of the diencephalon and not in the olfactory system. Thus, we propose the following hypothesis as shown in Fig. 2.

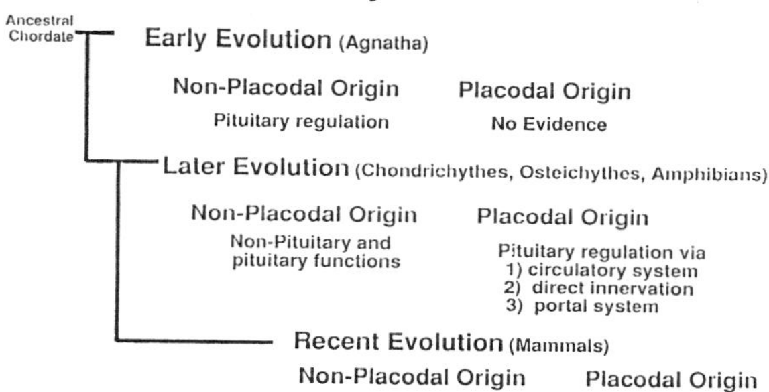

References

Bolduc, T.G. and S.A. Sower, 1992. Changes in brain gonadotropin-releasing hormone, plasma estradiol 17-ß and progesterone during the final reproductive cycle of the female sea lamprey, *Petromyzon marinus*. J. Exp. Zool. 264:55-63.

Deragon, K. and S.A. Sower., 1994. Effects of lamprey gonadotropin-releasing hormone-III on steroidogenesis and spermiation in male sea lampreys. Gen. Comp. Endocrinol. 95:363-367.

Fahien, C.M. and S.A. Sower, 1990. Relationship between brain gonadotropin-releasing hormone and final reproductive period of the adult male sea lamprey, *Petromyzon marinus*. Gen. Comp. Endocrinol. 80:427-437.

Gorbman, A., 1965. Vascular relations between the neurohypophysis and adenohypophysis of cyclostomes and the problem of evolution of hypothalamic neuroendocrine control. Arch. Anat. Micros. Morphol. Exp. 54:163-194.

Hardisty, M.W. and I.C. Potter, 1972. The general biology of adult lampreys. In: The Biology of Lampreys. Vol. 1, Chapt. 4. Hardisty, M.W. and I.C. Potter (eds), London: Academic Press.

King, J.C., S.A. Sower, and E.L.P. Anthony, 1988. Lamprey gonadotropin releasing hormone neurons in the brain of the sea lamprey, *Petromyzon marinus*. Cell Tiss Research. 253:1-8.

Knox, C.J., S.K. Boyd and S.A. Sower, 1994. Characterization and localization of gonadotropin-releasing hormone receptors in the adult female sea lamprey, *Petromyzon marinus*. Endocrinol. 134: 492-498.

Lovejoy D.A., N.M. Sherwood, W.H. Fischer, B.C. Jackson, J.E. Rivier and T. Lee, 1991. Primary structure of gonadotropin-releasing hormone from the brain of a holocephalan (Ratfish: *Hydrolagus colliei*) Gen. Comp. Endocrinol. 82: 152-161.

Lovejoy D.A., W.H. Fischer, S. Ngamvongchon, A.G. Craig, C.S.Nahorniak, R.E. Peter, J.E. Rivier and N.M. Sherwood, 1992. Distinct sequence of gonadotropin-releasing hormone (GnRH) in dogfish brain provides insight into GnRH evolution. Proc. Natl. Acad. Sci.USA 89:6373-6377.

Muske, L., 1993. Evolution of gonadotropin-releasing hormone (GnRH) neuronal systems. Brain Behav. Evol. 42: 215-230.

Nozaki, M. ,T. Tsukahara, and H. Kobayashi, 1984. An immunocytochemical study on the distribution of neuropeptides in the brain of certain species of fish. Biomedical Res., Suppl. 36:135-184.

Nozaki, M., A. Gorbman, and S.A. Sower, 1994. Diffusion between the neurohypophysis and the adenohypophysis of lampreys, *Petromyzon marinus* Gen. Comp. Endocrinol.96:385-391.

Sherwood, N.M., S.A. Sower, D.R. Marshak, B.A. Fraser and M.J. Brownstein, 1986. Primary structure of gonadotropin-releasing hormone from lamprey brain. J. Biol. Chem. 261:4812-4819.

Sower, S.A., 1990. Neuroendocrine control of reproduction in lampreys. Fish Physiol. and Biochem. 8:365-374.

Sower, S.A., Y.C. Chiang, S. Lovas and J. M. Conlon, 1993. Primary structure and biological activity of a third gonadotropin-releasing hormone from lamprey brain. Endocrinol. 132:1125-1131.

Sower, S.A., M. Goodman, and Y. Nishiuchi. *In vivo* effects of lamprey GnRH-I and cyclized analogs: A structure-activity study. Neuropeptides. (In Press)

Tobet, S.A., M. Nozaki, J. H. Youson, and S.A. Sower, 1995a. Distribution of lamprey gonado-tropin hormone-releasing hormone-III (GnRH-III) in brains in larval lampreys (*Petromyzon marinus*). Cell Tissue Research 279:261-267.

Tobet, S.A., T.W. Chickering, and S.A. Sower, 1995b. Relationship of gonadotropin-releasing hormone (GnRH) neurons to the olfactory system in developing lamprey.Endocrinol.77 Abst.:554.

Wright, G.M., K.M. McBurney, J.H. Youson and S.A. Sower, 1994. Distribution of lamprey gonadotropin-releasing hormone in the brain and pituitary of larval, metamorphic and adult sea lamprey, *Petromyzon marinus*. Can J. Zool. 72:48-53.

Youson, J.H. and S.A. Sower, 1991. Concentration of brain gonadotropin-releasing hormone during metamorphosis in the lamprey, *Petromyzon marinus*. J. Exp. Zool. 259: 399-404.

DIURNAL RHYTHM IN TESTICULAR ACTIVITY IN THE SECONDARY MALE OF A PROTOGYNOUS WRASSE, *Pseudolabrus japonicus*

M. Matsuyama, S. Morita,[1] N. Hamaji,[1] M. Kashiwagi,[1] and Y. Nagahama[2]

Department of Fisheries, Faculty of Agriculture, Kyushu University, Fukuoka 812, Japan

Summary

Under the captive condition, the secondary male of a protogynous wrasse *Pseudolabrus japonicus* spawned daily over one month between 6:00 and 9:00 from October to November. Spermatogonial proliferation and meiosis occurred between 0:00 and 15:00. Spermiation occurred between 18:00 and 6:00. At around 15:00 the activity of 11β-hydroxylase, 11β-HSD, and 21-hydroxylase was high, leading to the producing of 11-ketotestoterone (11-KT) and $17\alpha,20\beta$,21-trihydroxy-4-pregnen-3-one (20β-S). The serum level of 11-KT showed a positive correlation with the number of B-type spermatogonia and spermatocytes. Thus, the secondary male of *P. japonicus* exhibits a diurnal rhythm in both spermatogenesis and steroidogenesis.

Introduction

Many marine teleost species undergo multiple cycles of gamate maturation and spawning within a single spawning season. However, the short-term cyclical change in testicular activity of male fish during the spawning season has not been investigated. The wrasse, *Pseudolabrus japonicus*, exhibits diandric protogyny, with populations consisting of small initial-phase (IP) males (primary males), IP females (primary females), and large terminal-phase (TP) males. TP males may be derived from either females which have undergone sex change to become males (secondary males), or from IP males (TP primary males). In this species, the spawning behavior is mainly pair spawning performed by a TP male and a female. TP males defend individual spawning territories, and spawn exclusively with one female in the morning time. After the pair spawning with one female, the pair dissolvs and the TP male begins to show spawning behavior again with another female. Based on the underwater observation, it is suggested that a single TP male performs daily spawning (Nakazono 1979).

In this paper, we ask the question: whether the sperm are produced and released daily in the TP males or TP males store sperm for future release ? In other words, whether TP males exhibit a diurnal rhythm of spermatogenesis and spermiation or not ? In the present study, we used TP secondary males as the experimental fish because the appearance rate of TP primary male is very low.

Results and Discussion

Daily spawning of a single secondary male

Captive population consisted of one secondary male and two females, and was established in two round tanks of 200 *l* capacity (tank A and B). Spawning was monitored everyday (Fig. 1). The

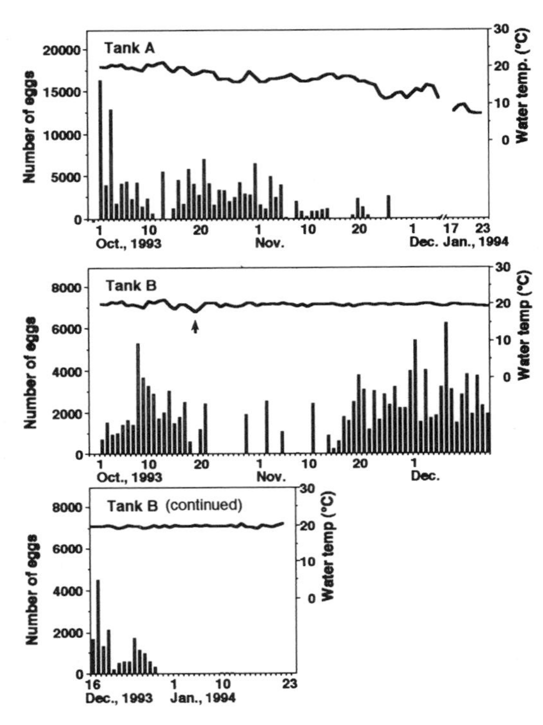

Fig. 1. Number of eggs in the presence of a single TP secondary male and two females of *P. japonicus*. Tank A was kept under natural conditions. In tank B, the water temperature was maintained at 20 °C from October 19, 1993. Open and solid bars show floating and sinking eggs, respectively.

first spawning occurred in both tanks on October 1, 1993, one day after the collection of experimental fish. In tank A (natural photoperiod and water temperature) daily spawning continued up to mid November, and the last spawning was observed on November 26. Colder water temperature tended to result in decreased egg release during spawning. In

1, Faculty of Bioresources, Mie University, Tsu 514, Japan, and 2, National Institute for Basic Biology, Okazaki 444, Japan.

tank B (natural photoperiod and water temperature until October 19, after which temperature was maintained at 20 °C, Fig. 1) two series of daily spawning were observed. The first one from October 1 to October 18, and the second from November 14 to December 28. Daily spawning in both tanks occurred between 6:00 and 9:00. Thus, it appears that a single secondary male of *P. japonicus* spawns daily during the spawning season.

Germ cell composition

TP males were collected in the field at 9:00, 12:00, 15:00 and 18:00; five fish each were sacrificed for histological study. Remaining fish samples from each time were stored in the outdoor tanks under the natural conditions. After confirming the daily spawning, TP males (five each) were sampled at 21:00, 24:00, 3:00 and 6:00. A total of 39 secondary males were used. Tissue for the quantitative study of germ cells was sampled from site 5 (Fig. 2) of the left lobe. Ten cross sections of different seminal lobules were selected randomly from each sample. The number of three types of testicular germ cell – *(i)* B-type spermatogonia and spermatocytes, *(ii)* spermatids and *(iii)* spermatozoa – per cross section of a seminal lobule was counted (Fig. 3).

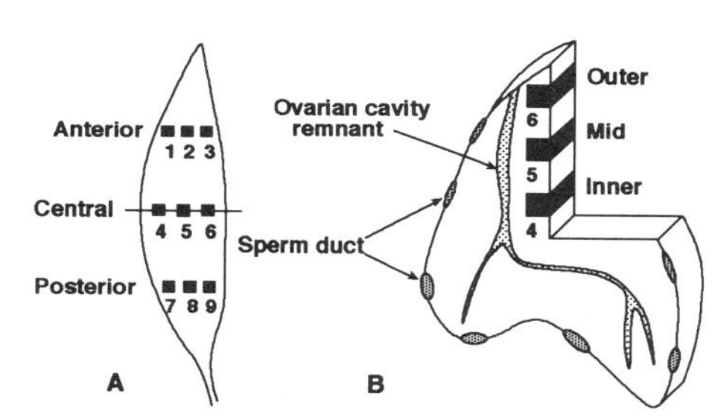

Fig. 2. **A**, Left testicular lobe of a TP secondary male *P. japonicus*, and nine subsamples of testicular tissue for quantitative analysis of germ cells, inner (sites 1, 4, and 7), mid (sites 2, 5, and 8), and peripheral sites (sites 3, 6, and 9). **B**, Cross section of the central part of the left testicular lobe taken at the level indicated by the cross-line in Fig. 2A.

The number of B-type spermatogonia and spermatocytes were lowest at 0:00 ($P<0.05$ by Duncan's multiple range test, vs 15:00 and 21:00), increased gradually thereafter, peaked at 15:00 ($P<0.05$, vs 0:00), and decreased rapidly from 21:00 to 0:00. The number of spermatids exhibited a pattern similar to that of B-type spermatogonia and spermatocytes. However, there was no significant change throughout a day ($P>0.05$). At 6:00, just prior to the spawning, the lobular lumens were occupied with a large amount of spermatozoa which had been released from the cysts. At 9:00, during or after spawning, spermatozoa showed a marked

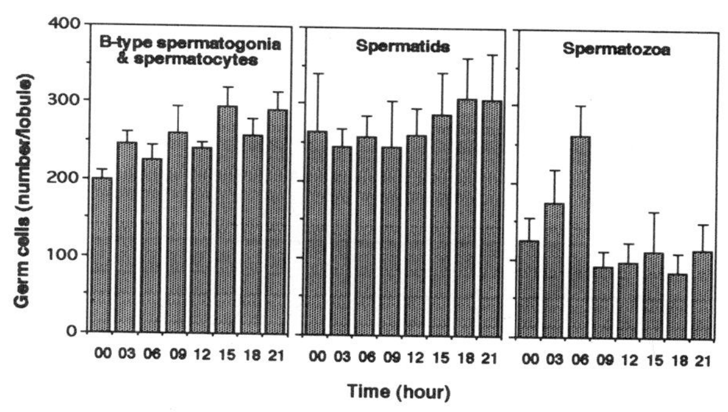

Fig. 3. Diurnal changes in testicular germ cell compositions in TP secondary male *P. japonicus*.

decrease in number ($P<0.01$, vs 6:00), with a relatively small amount of spermatozoa in the lobular lumen. Thereafter spermatozoa displayed constant low levels of occurrence, and began to increase in number after 18:00. Thus, spermatogonial proliferation and meiosis occurred between 0:00 and 15:00, followed by spermiation (release of spermatozoa into the lobular lumen from the cysts) between 18:00 and 6:00, immediately prior to the spawning.

Serum steroid hormone levels

Serum 11-KT levels were lowest (406 pg/ml) at 21:00 ($P<0.01$, vs 12:00), increased at 3:00, peaked at 12:00 (1785 pg/ml, $P<0.01$, vs 0:00 and 21:00), and returned back to those seen previously at 21:00 (Fig. 4). 11-KT levels showed a positive correlation with the numbers of B-type spermatogonia and spermatocytes. There was no significant difference in the serum levels of DHP at the various times ($P>0.05$).

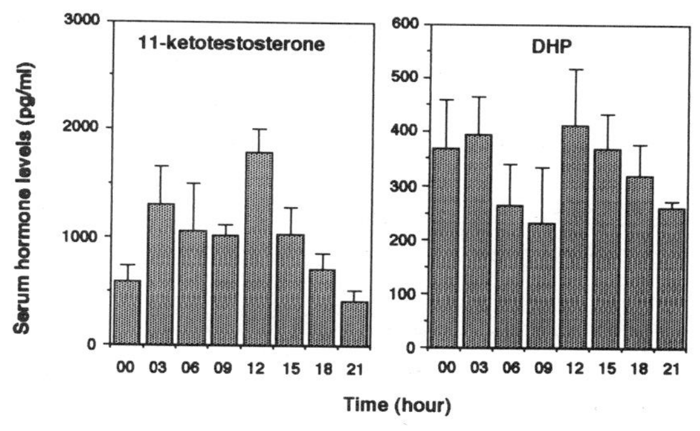

Fig. 4. Diurnal changes in serum steroid hormone levels in TP secondary male *P. japonicus*.

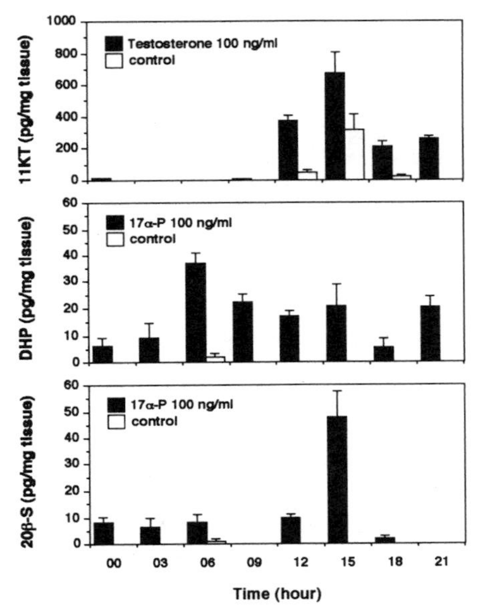

Fig. 5. *In vitro* steroid production by testicular fragment of TP secondary male *P. japonicus* incubated with or without 100 ng of exogenous steroid precursor.

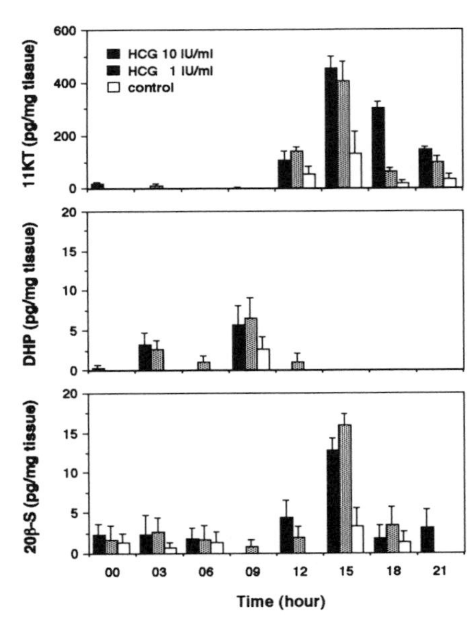

Fig. 6. *In vitro* production of steroids by testicular fragments of TP secondary male *P. japonicus* incubated with 0, 1, 10 IU/ml HCG.

Conversion of unlabeled precursors

Testicular fragments (20 to 40 mg) were incubated in 1 ml of L-15 medium with or without 100 ng/ml of radioinert exogenous precursors, 17α-hydroxyprogesterone (17α-P) or testosterone (T), for 18 hr at 20 °C. Converted steroids (11-KT, DHP, and 20β-S) were measured by enzyme immunoassay.

Exogenous T was converted to 11-KT between 12:00 and 21:00 (Fig. 5), indicating that both 11β-hydroxylase and 11β-hydroxysteroid dehydrogenase (11β-HSD) activity is high during this period. Exogenous 17α-P was converted to DHP throughout a day. By contrast, exogenous 17α-P was largely converted to 20β-S at 15:00. These results suggest that 20β-hydroxysteroid dehydrogenase (20β-HSD) activity is present throughout the daily cycle and 21-hydroxylase is high activity at 15:00.

In vitro responsiveness of testicular steroidogenesis to gonadotropin

In vitro production of 11-KT, DHP, and 20β-S in response to gonadotropin (HCG) was investigated using testicular fragments from secondary males (Fig. 6). Testicular fragments were incubated in 1 ml of medium with 0, 1, or 10 IU/ml HCG for 18 hr at 20 °C.

The HCG-enhanced 11-KT production increased rapidly at 15:00 ($P<0.01$, vs 12:00), then decreased gradually to 0:00. There were no significant differences in HCG-enhanced DHP production at the various times ($P>0.05$). HCG increased 20β-S production only at 15:00 ($P<0.01$). These results suggest that the steroidogenic enzyme system involved in the synthesis of 11-KT and 20β-S was activated by gonadotropin at around 15:00. These findings, together with those from experiments with unlabeled precursors, suggest that at around 15:00 gonadotropin increases the activity of 11β-hydroxylase, 11β-HSD, and 21-hydroxylase, leading to the production of 11-KT and 20β-S. The functions of 11-KT and 20β-S in the reproductive cycle of the secondary male *P. japonicus*, however, remain largely unknown.

References

Nakazono, A., 1979. Studies on the sex reversal and spawning behavior of five species of Japanese labrid fishes. Rep. Fish. Res. Lab., Kyushu Univ. 4: 1-64.

A COMPARATIVE STUDY OF THE VITELLOGENESIS DYNAMIC AND REPRODUCTIVE ECOLOGY IN SINGLE AND MULTISPAWNER CYPRINIDS

P. Kestemont, J. Rinchard, and R. Heine

Unité d'Ecologie des Eaux Douces, Facultés Universitaires N.D. de la Paix, 61, rue de Bruxelles, B-5000 Namur, Belgium.

Summary

Dynamics of vitellogenesis correlated with seasonal profiles of oocyte growth, vitellogenin and estradiol-17ß has been compared in a single-spawner (the roach *Rutilus rutilus*) and two multispawner cyprinids (the white bream *Blicca bjoerkna* and the bleak *Alburnus alburnus*) collected in the river Meuse (Belgium). Different patterns of gonadosomatic index (GSI), oocyte growth, vitellogenin (expressed by plasma protein phosphorus) and estradiol-17ß levels have been observed between the single and multispawner fish, and also among the two multi-spawners. Compared to the rapid decline of GSI in the roach population, the GSI of the multispawners decreased progressively during the spawning season. However, the different parameters involved in the vitellogenesis process and oocyte growth indicated that, in white bream, gonadal development was mainly oriented to the first batch of ova, the ovaries being fully mature and vitellogenesis markedly reduced at the onset of the spawning period whereas in the bleak vitellogenic activity remained high throughout the spawning season. These results are discussed in term of reproductive ecology of the mono and multispawner fish.

Introduction

Different fish reproductive strategies have been reviewed by Wootton (1984), Lambert and Ware (1984), Mann et al. (1984), Kamler (1992) and Balon (1990). Among the iteroparous species, dynamics of vitellogenesis and spawning frequency have been used to classify the species as group-synchronous or asynchronous (Wallace and Selman, 1981, Kestemont and Philippart, 1991) and as single, multiple or continuous spawners (McEvoy and McEvoy, 1992). The aim of this paper was to compare the different patterns of oocyte recruitement and the endocrine regulation of vitellogenesis in three cyprinids exhibiting similar characteristics of reproduction (spring spawners, high fecundity of small ova) but differing by their spawning frequency : the roach *Rutilus rutilus* as single spawner and the white bream *Blicca bjoerkna* and the bleak *Alburnus alburnus* as multispawners.

Material and methods

The study was conducted between April and October 1994 in the river Meuse. Fish were sampled bi-weekly during vitellogenesis, weekly during the spawning period and monthly after that period. Five females of each species were collected at each sampling date. Capture techniques and sampling sites were choosen according to the spawning habitat of each species. White bream were mainly collected by gill-nets installed in the river Meuse whereas roach and bleak were sampled in a fish pass allowing the upstream migration of fish through the dam of Tailfer (Belgium). All samplings were done around 10.00 a.m. in order to avoid nychtemeral effects.

Immediately after capture, fish were transferred to the laboratory, anesthetized with ethylen glycol mono-phenyl ether, measured (±0.1cm) and weighed (±0.1g). Blood samples were taken from the caudal vessel into a heparinized syringe, centrifuged for 15 min. at 4000 g and the plasma was stored at -20°C until radio-immunoassay or vitellogenin dosage. The ovaries were removed, weighed (±0.001g) and fixed in Bouin's solution for histological examination. Tissues embedded in paraplast were prepared into 6 µm sections and stained with trichrome : hemaluin, phloxine and light green (Langeron, 1942).

Estradiol-17ß and vitellogenin dosages

Plasma estradiol-17ß (E2) was measured by radioimmunoassay. Samples of 25, 30 or 50 µl of plasma were extracted with 1ml cyclohexane/ethyl acetate (v/v), dried down and redissolved in 300 µl gelatinized phosphate buffer (0.1 %, pH 7.25). Plasma concentrations of E2 were measured according to the methods of Breton et al (1983). Cross-reactivities of various steroids with the antisera used in the radioimmunoassays have been described by Fostier and Jalabert (1986). Indirect method of vitellogenin dosage has been used, according to the technique of Martin and Doty (1949) modified by Jarred and Wallace (1968). Plasma protein phosphorus (µg/ml plasma) were measured colorimetrically after protein extraction from 100 µl plasma and denaturation to release phosphorus.

Histological examination

Each gonad was classified according to the most advanced type of oocyte present (Table 1). Ovarian development was examined by a histomorphometric analysis modified from Kestemont (1987). Two parameters were examined: 1) distribution of oocyte size, assessed by measuring 50 profiles for each stage; 2) relative proportion (%) of each stage, i.e. by counting 200 cells per ovary and then dividing the percentage of a given stage by the corresponding mean diameter.

Stage	Ovarian maturity
1	Previtellogenesis (protoplasmic oocytes)
2	Endogenous vitellogenesis
3	Exogenous vitellogenesis
4	Final maturation
5	Intermediate multispawning stage
6	Post-spawning

Data analysis

All data are expressed as the mean ± SEM. Data were statistically analysed by a variance analysis

(ANOVA 1) followed by comparison of means using Scheffe F's test. A Hartley test was used to verify the homogeneity of variance.

Results

Profiles of gonadosomatic index and corresponding stages of ovarian maturity are shown in fig. 1. After a sudden increase early May, GSI of roach females reached 21.6% whereas the maximum GSI of bleak and white bream attained 17.7 and 14.5%, respectively. The duration of the spawning period (defined as the period during which ovulating females and/or in stage 5 were collected) was clearly shorter in roach than in bleak and white bream. Compared to the rapid decline of GSI in the roach population, the GSI of the multispawners decreased progressively during the spawning season. Despite a GSI decrease since mid-June, the bleak remained mature and ovulated until the end of July (stage 5*).

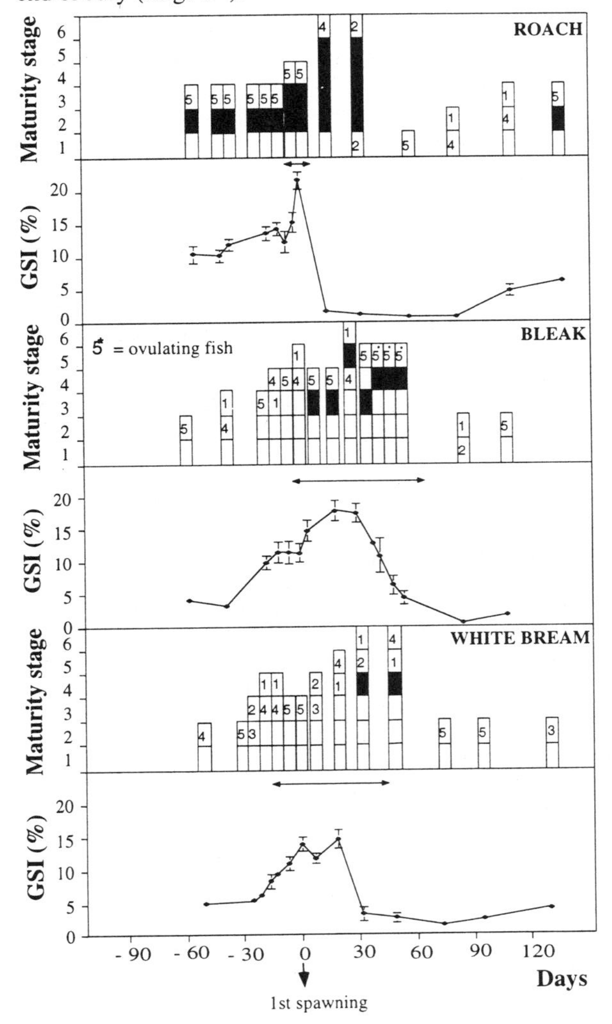

Fig. 1. Seasonal profiles in gonadosomatic index and corresponding ovarian stages in roach, white bream and bleak females.

Different patterns of E2 and PPP levels have been observed between the single and the multiple spawners but also among the two multiple spawners (Fig. 2).

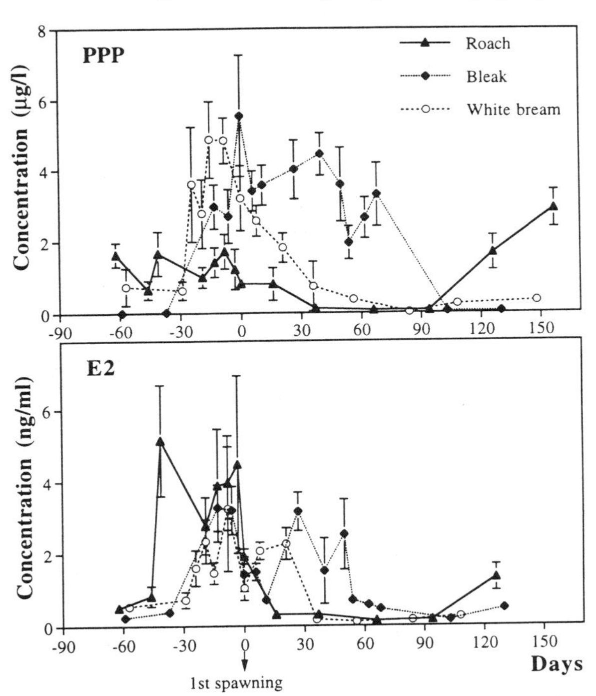

Fig. 2. Variations of plasma protein phosphorus (PPP) and estradiol-17β (E2) levels in roach, bleak and white bream. Time is calculated according to the onset of their respective spawning period.

Maximum levels of E2 were recorded during late exogenous vitellogenesis in roach and white breams and decreased (but not significantly in white bream) from the spawning period onwards, whereas the plasma E2 contents measured in bleak remained high during the entire spawning season. During the months preceding the respective spawning season of these species, the PPP profiles recorded in roach were opposed to those observed in white bream and bleak (Fig. 2). From day -30 to day 0 (day 0 being calculated as the onset of their respective spawning period), a marked increase of PPP levels was measured in these latter species whereas the PPP content of roach females remained low. Significant differences ($P<0.01$) in PPP levels occurred also after day 0, roach and white bream showing a progressive decrease of PPP concentrations while high levels of PPP were measured in bleak along the spawning season. In roach, the recrudescence of gametogenesis observed from September onwards was clearly supported by an increase of E2 and PPP content.

Discussion

According to these results, ovarian growth of roach differs clearly from those of bleak and white bream. As already reported by Mattheeuws et al. (1981), gonadosomatic index of roach females increased early after the spawning season, oocytes in exogenous vitellogenesis being present in the ovary from September onwards, whereas the ovarian development of the two multispawners was limited to the endogenous vitellogenesis stage until March. Moreover, the levels of plasma protein phosphorus measured in roach in September and October suggest that exogenous vitellogenesis would be highly advanced before winter. Vitellogenesis dynamics of bleak and white bream appeared similar during the pre-spawning period. However, during the spawning period, the patterns of vitellogenesis differed markedly. Supported by histological observations and fecundity assessment (see Rinchard et al., this volume), profiles of estradiol-white bream, gonadal development appeared mainly oriented to the first batch of ova in which the ovaries were fully mature and vitellogenesis markedly reduced at the onset of the spawning period whereas in the bleak vitellogenic activity remained high throughout the spawning season.

Based on the present results, these two multi-spawner cyprinids exhibit two different reproductive strategies in relation with the energy allocated to the ovary in the onset of the spawning season. In this respect, although white bream develops its ovaries asynchronously and spawn several times into one breeding season, this species might be associated to the single spawners whereas bleak limits the energy allocated to the ovary before the onset of the spawning season but maintains a production of mature oocytes during that period. According to the definition proposed by Hunter et al. (1985), white bream can be considered as a determinate and bleak as an indeterminate multi-spawner fish.

References

Balon, E.K., 1990. Epigenesis of an epigeneticits: the development of some alternative concepts on the early ontogeny and evolution of fishes. Guelph Ichthyol. Rev., 1:1-48.

Breton, B., Fostier, A., Zohar, Y., Le Bail, P.Y. and Billard, R., 1983. Gonadotropine glycoprotéique maturante et oestradiol-17β pendant le cycle reproducteur chez la truite (*Salmo trutta*) femelle. Gen. Comp. Endocrinol., 49:220-231.

Fostier, A. and Jalabert, B., 1986. Steroidogenesis in rainbow trout (*Salmo gairdneri*) at various preovulatory stages: changes in plasma hormone levels and *in vivo* and *in vitro* responses of the ovary to salmon gonadotropin. Fish Phys. Bioch., 2:87-99.

Hunter, J.R., Lo, N.C.H. and Leong, R.J.H., 1985. Batch fecundity in batch spawning fishes. In: An Egg Production Method for Estimating Spawning Biomass of Pelagic Fish: Applications to the Northern Anchovy, *Engraulix mordax*. R. Lasker (Ed): NOAA Tech. Rep., NMFS, 36:66-77.

Jarred, D.W. and Wallace, R.A., 1968. Comparative chromatography of the yolk proteins of teleosts. Comp. Biochem. Physiol., 24:437-443.

Kamler, E., 1992. Early Life History of Fish. An Energetics Approach. Chapman et Hall, London, 267 p.

Kestemont, P., 1987. Etude du cycle reproducteur du goujon *Gobio gobio* L. 1. Variations saisonnières dans l'histologie de l'ovaire. J. Appl. Ichthyol., 4:145-157.

Kestemont, P. and Philippart, J.C., 1991. Considèrations sur la croissance ovocytaire chez les poissons à ovogenèse synchrone et asynchrone. Belg. J. Zool., 120:263-274.

Lambert, T.C. and Ware, D.M., 1984. Reproductive strategies of demersal and pelagic spawning fish. Can. J. Fish Aquat. Sci., 41:1565-1569.

Langeron, M., 1984. Précis de Microscopie Technique- Ex-périmentation-Diagnostic. Masson et Cie, Paris, 1340 p.

Mann, R.H.K., Mills, C.A. and Crisp, D.T., 1984. Geographical variation in the life-history tactics of some species of freshwater fish. In: Fish Reproduction: Strategies and Tactics. G.W. Potts and R.J. Wootton (Eds): Academic Press, London, p. 171-186.

Martin, J.B. and Doty, D.M., 1949. Determination of inorganic phosphate. Analyt. Chem., 21:965-967.

Mattheeuws, A., Genin, M., Detollenaere, A. and Micha, J.C., 1981. Etude de la reproduction du gardon (*Rutilus rutilus*) et des effets d'une élévation provoquée de la température en Meuse sur cette reproduction. Hydrobiologia, 85:271-282.

McEvoy, L.A. and McEvoy, J., 1992. Multiple spawning in several commercial fish species and its consequences for fisheries management, cultivation and experimentation. J. Fish. Biol., 41:125-136.

Wallace, R.A. and Selman, K., 1981. Cellular and dynamic aspects of oocyte growth in teleosts. Amer. Zool., 21:325-343.

Wootton, R.J., 1984. Introduction: strategies and tactics in fish reproduction. In: Fish Reproduction: Stra-tegies and Tactics. G.W. Potts and R.J. Wootton (Eds): Academic Press, London, p. 1-12.

HOMING MECHANISMS IN SALMON: ROLES OF VISION AND OLFACTION

H. Ueda[1], M. Kaeriyama[2], A. Urano[3], K. Kurihara[4] and K. Yamauchi[5]

[1]Toya Lake Station for Environmental Biology, Faculty of Fisheries, Hokkaido University, Abuta 049-57, [2]Hokkaido Salmon Hatchery, Fisheries Agency of Japan, Sapporo 062, [3]Division of Biological Sciences, Graduate School of Science, Hokkaido University, Sapporo 060, [4]Faculty of Pharmaceutical Sciences, Hokkaido University, Sapporo 060, [5]Department of Biology, Faculty of Fisheries, Hokkaido University, Hakodate 041, Hokkaido, Japan

Summary

The mechanisms underlying the amazing ability of salmon to migrate a long distance from open water to natal streams for spawning are still unknown. Kokanee salmon (*Oncorhynchus nerka*) in Lake Toya offer an excellent model system for studying the salmonid homing mechanisms. We telemetrically tracked mature males, and found that the fish released at a long distace from the natal area returned directly to the natal area mainly using visual cues. In addition, we examined biochemical and cytophysiological changes in a salmonid olfactory system-specific protein (N24) and salmon gonadotropin-releasing hormone (sGnRH). Quantitative changes in olfactory N24 and sGnRH were apparently correlated to the homing behavior in salmonids. These findings suggest that kokanee salmon return first to the vicinity of the natal area by visual orientation mechanisms, and then recognize the natal stream odorants using olfactory discrimination.

Introduction

One of the most interesting misteries in the "reproductive life history of fish" is the homing mechanism of salmon to the natal stream. Salmon have an amazing ability to migrate thousands of kilometers from the ocean to natal streams for spawning after a few years of oceanic life (Fig. 1). The contribution of vision, rheotaxis to water current, and magnetic sense for open water orientation, has been discussed (Smith, 1985). Tracking of salmon in the ocean is, however, rather difficult, so that few behavioral experiments on open water orientation have been carried out. Moreover, it is well established that salmon return from the coast to the natal branch of rivers using their olfactory sense (Stabell, 1992). However, few attempts have been made to investigate biochemical aspects of any molecules specific to the olfactory system (olfactory epithelium, olfactory nerve and olfactory bulb) during salmonid homing migration.

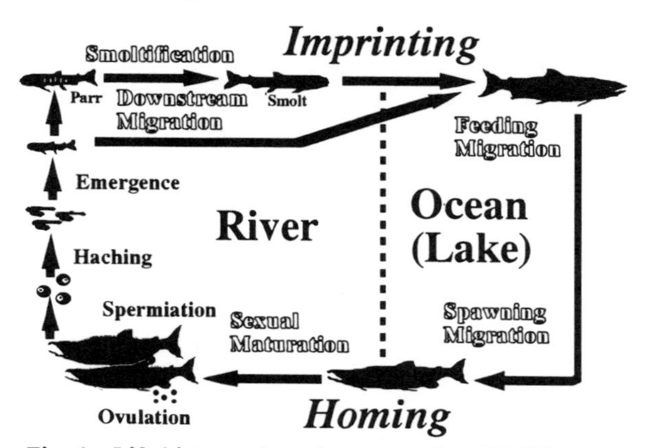

Fig. 1. Life history of anadromous salmonid fish.

Land-locked sockeye salmon (kokanee salmon; *Oncorhynchus nerka*) in Lake Toya can be a good model fish for studying the salmonid homing mechanisms, because mature fish return to the natal area with high accuracy. We thus attempted to investigate homing mechanisms mainly in kokanee salmon using the telemetric tracking of homing behavior as well as the biochemical and cytophysiological analyses of salmonid olfactory system-specific molecules.

Telemetric Tracking of Homing Behavior

In Lake Toya (surface area 70 km^2, average diameter 9.4 km and average depth 116 m), four mature males (average fork length, 33.5 cm; average body weight, 474.0 g) captured at the natal area during the spawning season were subjected to one of the following treatments: (A) attachment of a brass ring on the lateral head for control, (B) attachment of a NdFe magnet ring on the lateral head to interfere with magnetic cue, (C) injection of a mixture of carbon toner and corn oil into the eyeball to detach retinas, (D) detached retinas plus attachment of magnetic ring. An ultrasonic transmitter was placed in the stomach of individual fish. Each fish

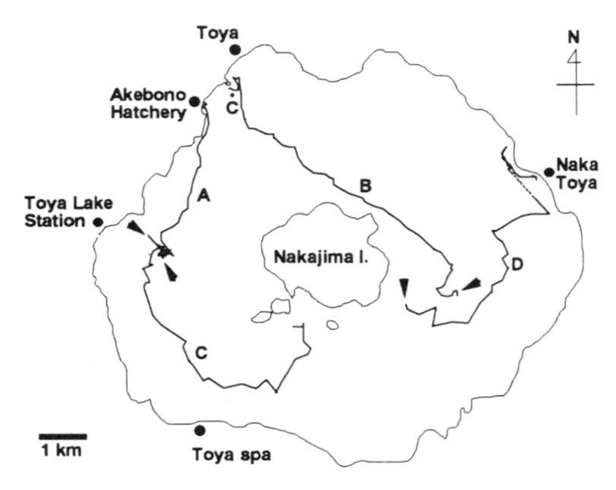

Fig. 2. Tracks of four mature male kokanee salmon in Lake Toya during the spawning season. Arrowhead indicates the releasing point of each fish. A, control fish; B, magnetic cue-interfered fish; C, visual cue-interfered fish; D, visual and magnetic cues-interfered fish.

was released in the middle of the lake, and then telemetrically tracked by an ultrasonic receiver on a boat. The position of the boat was recorded by the global positioning system with signals from artificial satellites. The weather was clear on all days when fish were released.

The control fish released at a point 3.5 km distant from the natal area returned directly to the natal area (Fig. 2A). Similar to the control fish, when the magnetic cue-interfered fish was released 6.8 km southeast of the natal area, it returned directly to the vicinity of the natal area (Fig. 2B). In contrast, the blinded fish moved to a direction opposite to the natal area and reached Nakajima Island in the evening, and he was rediscovered in the natal area on the next evening (Fig. 2C). The fish whose visual and magnetic cues were blocked also moved randomly and finally reached the shore far from the natal area (Fig. 2D).

The present results demonstrate for the first time the direct return of kokanee salmon using visual cues, and suggests that kokanee salmon have an amazing ability to precisely identify their position in open water and the direction of the natal area. Rheotaxis to the water current does not contribute to the open water orientation in Lake Toya, because both mature fish released either south or southeast of their natal area, could return directly. Kokanee salmon do not primarily use magnetic cues in the selection of the natal area, but it is still likely that fish supplementally use magnetic cues for their orientation (Quinn and Groot, 1983). The random movement of the blinded kokanee salmon suggests that fish return to the vicinity of the natal area using visual cues. It is possible that the fish use a "sun-compass" (Hasler and Schwassmann, 1960), polarized skylight patterns (Waterman and Forward, 1970), or memory of landmarks (Guilford, 1993) for identification of position and direction.

Biochemical and Cytophysiological Analyses of N24 and sGnRH

Salmonid olfactory system-specific protein (N24)

Using sodium dodecyl sulfate-polyacrylamide gel electrophoresis (SDS-PAGE), we have identified an olfactory system-specific 24 kDa protein (N24) in kokanee salmon by electrophoretic comparison of proteins restricted to the olfactory system with those found in other parts of the brain (Shimizu *et al.*, 1993). A specific polyclonal antiserum to N24 recognized only the 24 kDa protein in the olfactory system as determined by Western blotting analysis (Fig. 3). In various species of teleosts, N24 immunoreactivity was found in the olfactory system of species migrating between sea and river, such as Japanese eel (*Anguilla japonica*). However N24 immunoreactivity was not observed in carp (*Cyprinus carpio*) and tilapia (*Oreochoromis nilotics*), which exhibit no migratory behavior (Ueda *et al.*, 1994). Both at the time of imprinting of the natal stream odorants in masu salmon (*O. masou*) and at the time of homing to the natal stream in chum salmon (*O. keta*), the immunoreactivity of N24 in fish in the natal stream was stronger than that in fish in seawater. Immunocytochemical and immunoelectron microscopic observations revealed that N24 positive immunoreactivity occurred in ciliated and microvillus olfactory receptor cells and in the olfactory bulb innervated by the olfactory nerve. These results suggest that N24 has a possible role in neuromodulation in the olfactory system, and may be important to both olfactory imprinting and homing mechanisms in salmonids.

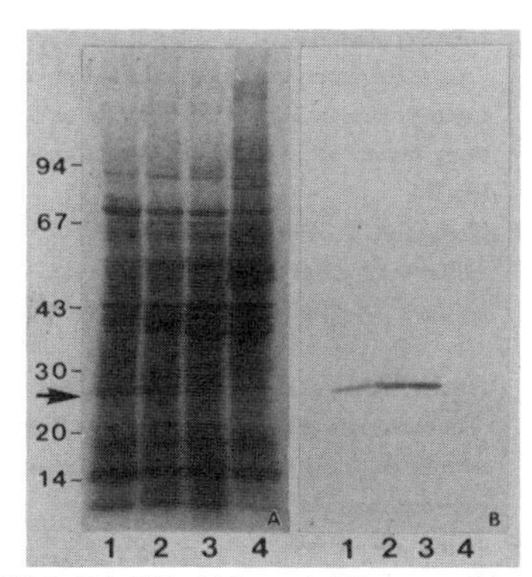

Fig. 3. SDS-PAGE (A) and Western blotting (B) of soluble extracts of the olfactory system (1, olfactory epithelium; 2, olfactory nerve; 3, olfactory bulb) and the telencephalon (4). Arrow indicates N24.

Salmon gonadotropin-releasing hormone (sGnRH)

Cytophysiological changes of sGnRH producing neurons in chum salmon were examined during homing migration (Ueda et al., 1984) by immunocytochistry with a specific antiserum to sGnRH (Okuzawa *et al.*, 1990) and *in situ* hybridization with an oligonucleotide encoding the sGnRH precursor (pro-sGnRH, Suzuki *et al.*, 1992). In the forebrain (olfactory nerve, ON; olfactory bulb, OB; telencephalon, T; preoptic area, POA), sGnRH immunoreactive neurons and neurons showing signals for pro-sGnRH mRNA were compared between fish in the coastal sea and those on the spawning ground. Neurons in the ON and between the ON and OB (ON-OB) exhibited strong sGnRH immunoreactivity and strong hybridization signals in fish in the coastal sea, whereas these activities and

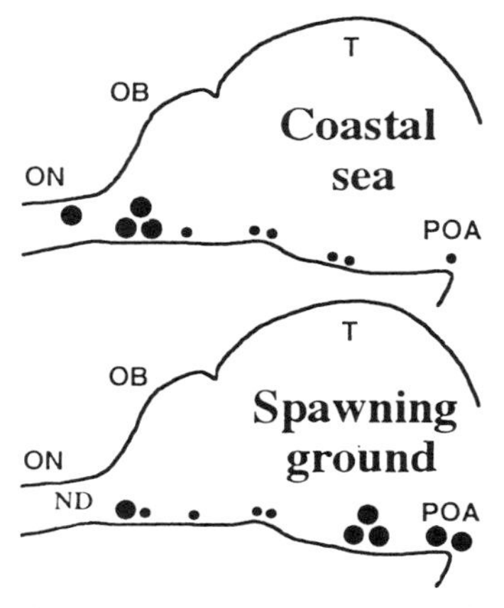

Fig. 4. Distribution of neurons showing sGnRH immunoreactivity and hybridization signals for pro-sGnRH mRNA in the forebrain of chum salmon during homing migration. Small and large circles indicate weak and strong hybridization signals, respectively. ND, non-detectable.

signals disappeared or decreased in animals on the spawning ground. In contrast, neurons in the OB, T and POA showed sGnRH immunoreactivity and hybridization signals of sGnRH during homing migration, and the hybridization signals of sGnRH in the T and POA were stronger in fish on the spawning ground than those in the coastal sea (Fig. 4). Moreover, sGnRH immunoreactive electron-dense granule-like structures of 50 nm in diameter were observed in fibers of the ON close to the olfactory epithelium and to the OB in masu salmon (Kudo *et al.*, 1994). These findings suggest that sGnRH may participate in neurotransmission and/or neuro-modulation in the olfactory system, and that sGnRH neurons in the ON and ON-OB may be involved in olfactory functions. SGnRH neurons in the T and POA appear to participate in gonadal maturation during the early and final phases of homing migration.

Conclusion

The present study provides several new findings on the visual and olfactory homing mechanisms mainly in kokanee salmon (Fig. 5). But further studies are needed to more precisely clarify both the visual mechanisms of the open water orientation and the olfactory mechanisms of the natal stream discrimination. Furthermore, little is known about the visual and olfactory imprinting mechanisms in juvenile salmon. Several intensive behavioral and molecular biological analyses are under investigation using kokanee salmon in Lake Toya.

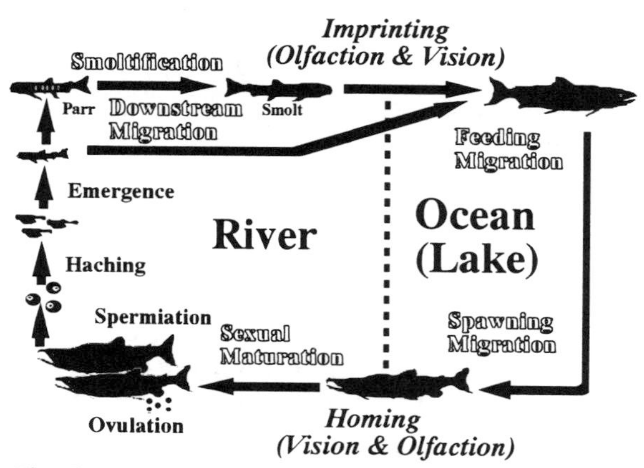

Fig. 5. Model for the visual and olfactory imprinting and homing mechanisms in kokanee salmon.

Acknowledgements

The authors thank Drs. O. Hiroi, A. Hara, T. Shoji, S, Hyodo, and Messrs. H. Kawamura, M. Fukuwaka, H. Kudo for their contribution to this project. Studies in this paper were partly supported by research funds from the Ministry of Education, Sciences, and Culture of Japan, and the Fisheries Agency of Japan.

References

Guilford, T. 1993. Homing mechanisms in sight. Nature 363:112-113.

Hasler, A.D. and H.O. Schwassmann. 1960. Sun orientation of fish at different latitudes. Cold Spring Harbor Symp. on Quant. Biol. 25:429-441.

Kudo, H., H. Ueda, H. Kawamura, K. Aida and K. Yamauchi. 1994. Ultrastructural demonstration of salmon-type gonadotropin-releasing hormone (sGnRH) in the olfactory system of masu salmon (*Oncorhynchus masou*). Neurosci. Lett. 166:187-190.

Okuzawa, K., M. Amano, M. Kobayashi, K. Aida, I. Hanyu, Y. Hasegawa and K. Miyamoto. 1990. Differences in salmon GnRH and chicken GnRH-II contents in discrete brain areas of male and female rainbow trout according to age and stage of maturity. Gen. Comp. Endocrinol. 80:116-126.

Quinn, T.P. and C. Groot. 1983. Orientation of chum salmon (*Oncorhynchus keta*) after internal and external magnetic field alternation. Can. J. Fish. Aquat. Sci. 40:1598-1606.

Shimizu, M., H. Kudo, H. Ueda, A. Hara, K. Shimazaki and K. Yamauchi. 1993. Identification and immunological properties of an olfactory system-specific protein in kokanee salmon (*Oncorhynchus nerka*). Zool. Sci. 10:287-294.

Smith, R.J.F., 1985. The control of fish migration. Springer-Verlag, Berlin Heidelberg, p. 1-243.

Stabell, O.B., 1992. Olfactory control of homing behavior in salmonids. T.J. Hara (Ed.): Fish Chemoreception, Chapman & Hall, London, p. 249-270.

Suzuki, M., S. Hyodo, M. Kobayashi, K. Aida and A. Urano. 1992. Characterization and localization of mRNA encoding the salmon-type gonadotropin-releasing hormone precursor of the masu salmon. J. Mol. Endocrinol. 9:73-82.

Ueda, H., O. Hiroi, A. Hara, K. Yamauchi, Y. Nagahama. 1984. Changes in serum concentrations of steroid hormones, thyroxine, and vitellogenin during spawning migration of the chum salmon,*Oncorhynchus keta*. Gen. Comp. Endocrinol. 53:203-211.

Ueda, H., M. Shimizu, H. Kudo, A. Hara, O. Hiroi, M. Kaeriyama, H. Tanaka, H. Kawamura and K. Yamauchi. 1994. Species-specificity of an olfactory system-specific protein in various species of teleosts. Fish. Sci. 60:239-240.

Waterman, T.H. and R.B. Forward. 1970. Field demonstration of polarized light sensitivity in the fish *Zenarchopterus*. Nature 228:85-87.

PLASMA SEX STEROIDS IN FEMALE NEW ZEALAND FRESHWATER EELS (*ANGUILLA* SPP.) BEFORE AND AT THE ONSET OF THE SPAWNING MIGRATION

P. Mark Lokman and Graham Young

Department of Zoology, University of Otago, PO Box 56, Dunedin, New Zealand

Summary

In a field study, we compared plasma levels of sex steroids in migratory and non-migratory female eels in order to determine which steroids may play a role in gonadal development in these fishes. Levels of sex steroids were generally low in non-migratory, but elevated in migratory eels. In the latter group, estradiol-17β was found at high levels in longfinned eels (2.5 ng/ml), whereas levels of its precursor testosterone were higher in shortfins (2.5-3 ng/ml). 17α,20β-dihydroxy-4-pregnen-3-one levels were low in all groups (less than 0.2 ng/ml). Surprisingly, very high levels of 11-KT were found in migrating eels, averaging 2.7 ng/ml in longfins and about 20 ng/ml in shortfins. The possible role of this compound in female eels is discussed.

Introduction

Unlike Northern Hemisphere eel species, New Zealand longfinned eels (*Anguilla dieffenbachii*) at the onset of the spawning migration are commonly found in vitellogenesis (Todd, 1981). As a result, early stages of gonadal development may be studied in wild longfins, rather than in hormone-treated eels from other species.

The present study has exploited that advantage and describes for the first time which plasma sex steroids are found in vitellogenic female eels in the wild and how they relate to ovarian development. The New Zealand shortfin, *A. australis schmidtii* was included in the study for comparative purposes.

Materials and Methods

Female eels were caught with fyke nets which were emptied and reset daily during the migratory season. The fish were blood sampled immediately upon removal from the net. Blood samples were centrifuged and plasma collected and stored frozen until analyses for estradiol-17β (E2), testosterone (T), 11-ketotestosterone (11-KT) and 17α,20β-dihydroxy-4-pregnen-3-one (17,20-DHP) by radioimmunoassay. Body and gonad weights were determined and the gonadosomatic index (GSI) was calculated as the ratio of gonad weight over total body weight. Gonad tissue was fixed in calcium-formalin for microscopy.

Values are expressed as means ± standard errors. Data were log-transformed and analysed by General Linear Models to test for differences between groups. Main or interaction effects were subjected to Scheffé's multiple comparisons test and considered significant for $p<0.05$.

Results

The GSI was significantly higher in migratory than in non-migratory eels of both species ($p<0.001$). Among non-migrants, ovarian development had progressed further in shortfins (SF; GSI=0.21 ± 0.06%, n=15) than in longfins (LF; 0.08 ± 0.02%, n=11), while the opposite was found among migratory eels (GSI=3.32 ± 0.13%, n=20, and 7.29 ± 0.34%, n=16, for SF and LF, respectively; $p<0.001$).

Microscopically, ovaries of non-migratory eels of both species contained previtellogenic oocytes. However, early to midvitellogenic oocytes, characterised by yolk granules and platelets in the cytoplasm, predominated in ovaries of migratory LF, whereas in migratory SF, oocytes in early vitellogenesis or in the oil droplet stage were most common (data not shown).

Plasma E2 was significantly higher in migratory LF (2.52 ± 0.26 ng/ml, n=16) than in migratory SF ($p<0.002$). In non-migratory eels, this trend was reversed, with higher levels found in SF (0.82 ± 0.21 ng/ml, n=15) than in LF females (less than 0.1 ng/ml). Plasma T levels were significantly higher in SF than in LF, both in migratory and in non-migratory eels. Levels averaged 2.42 ± 0.17 ng/ml in migratory SF, 0.88 ± 0.10 ng/ml in migratory LF and less than 0.3 ng/ml in non-migratory animals. Plasma 11-KT followed a trend similar to plasma T, but the differences between the groups were greatly accentuated and highly significant ($p<0.001$; Fig. 1). Concentrations of 17,20-DHP in plasma were low, at near-detectable limits (around 0.1 ng/ml), in all groups.

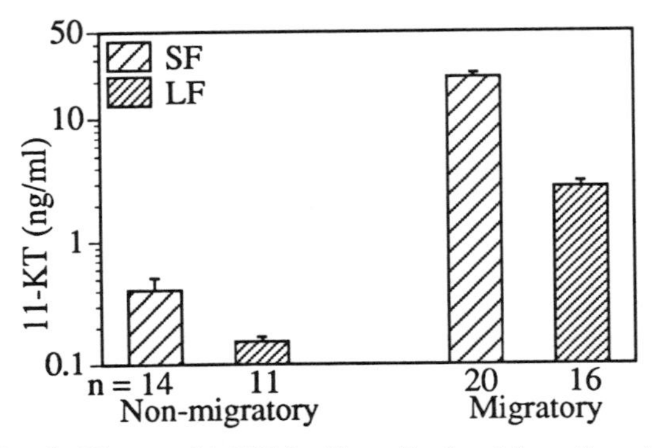

Fig. 1. Plasma 11-KT in New Zealand longfinned (LF) and shortfinned (SF) eels before and at the onset of the spawning migration.

Discussion

The results from our study are in good agreement with those from Todd (1981) and clearly demonstrate that gonadal development is more advanced in migratory LF than in migratory SF eels. The higher GSI and the predominant yolky oocytes in the ovary of LF indicate that the animals were undergoing vitellogenesis at the time of capture. In contrast, yolk granules were less often seen in SF eels and, if present, they were always small and confined to the periphery of the oocyte. In other eel species, the GSI at the onset of the spawning migration varies from around 6.0% in *A. rostrata* (American eel; experimental controls) to 1.7% in *A. anguilla* (Boëtius *et al.*, 1962) and 2% in *A. japonica* (experimental controls; Yamamoto *et al.*, 1974). This implies that at the onset of the spawning migration vitellogenesis may have started in migratory female American eels also, although we have not been able to verify this in the published literature.

In migratory female LF, a great elevation in plasma E2 levels was observed compared to non-migratory eels, whereas the elevation in SF, although significant, was considerably less. Given that E2 has been implicated in vitellogenesis in oviparous vertebrates, the differences between E2 in migratory LF (2.5 ng/ml) and SF (1.3 ng/ml) may reflect the developmental stage of the ovary. High levels of T in migratory SF agree with this scenario and could indicate that aromatase activity in SF is not as high as that in LF. However, we have not been able to demonstrate a relationship between plasma E2 and GSI in migratory LF (R^2=0.10; p=0.22). The difference in E2 between LF and SF could therefore be based on species differences rather than on developmental stage .

Unfortunately, our data do not provide information on changes in plasma E2 with time. The use of the GSI as a measure of time does not alter this situation, given the absence of a relationship between the GSI and plasma E2. Hence, it is difficult to compare the results of our study with those from induced spawning studies in *A. anguilla* and *A. japonica* (e.g., Ijiri *et al.*, in press).

Plasma levels of 17,20-DHP were low in all groups. This steroid has been tentatively identified as the maturation-inducing steroid in *A. japonica* (Yamauchi, 1990) and low levels before and during vitellogenesis are physiologically appropriate.

Similar to T and E2, plasma 11-KT levels were low (generally less than 0.5 ng/ml) in non-migratory female eels. However, surprisingly high levels of this steroid were found in migratory females, especially in SF (over 20 ng/ml). To our knowledge, the presence of 11-KT, generally considered a male-specific androgen in teleosts, in females has only been reported for several salmonid species (e.g., Leatherland *et al.*, 1982; Slater *et al.*, 1994). Preliminary *in vitro* studies suggest that 11-KT may be produced in the interrenal gland by enzymatic 11-hydroxylation and subsequent 11-oxidation of testosterone. However, it will have to be established whether these enzymes can act on androgen (i.e., testosterone), rather than progestogen substrates. Although our preliminary results favour the interrenal gland over the ovary as source for 11-KT, the presence of 11-hydroxylated and 11-ketosteroids in ovarian incubations has been repeatedly reported in *A. anguilla* (e.g., Colombo and Colombo Belvedere, 1976).

The function of 11-KT in female eels remains unknown. Perhaps levels increase during stress, a link that could be easily made if the interrenal gland is the source of this steroid. However, we have found high levels of 11-KT in female SF within 10 minutes of capture (only slightly elevated cortisol levels; data not shown) which would not support this theory. In a recent study on rainbow trout (*Oncorhynchus mykiss*), 11-KT stimulated growth of the heart and red muscle (Thorarensen *et al.*, submitted), while earlier studies by Idler *et al* (1961; cited by Slater *et al.*, 1994) and Leatherland *et al.* (1982) related 11-KT to increased skin thickness in sockeye salmon (*O. nerka*) and secondary sexual characteristics in coho salmon (*O. kisutch*), respectively. In eels, increases in red muscle and skin thickness have been reported in association with the transformation from non-migratory into migratory animals (Pankhurst, 1982a,b). Hence, we postulate that 11-KT may have anabolic functions that prepare the animals for migration.

Acknowledgments

We are grateful to Lake Ellesmere eel fishermen for their help and advice in obtaining the eels and to Trevor and Noeline Gould for their hospitality during field work. We extend our gratitude to Professor Yoshitaka Nagahama, National Institute of Basic Biology, Okazaki, Japan, and Professors Kohei Yamauchi and Shinji Adachi, Hokkaido University, Hakodate, Japan, for their generous gift of specific antibodies. We would also like to acknowledge financial support from the Royal Society of New Zealand and the Division of Sciences, University of Otago.

References

Boëtius, J., Boëtius, I., Hemmingsen, A.M., Bruun, A.F. and E. MØller-Christensen, 1962. Studies of ovarial growth induced by hormone injections in the European and American eel (*Anguilla anguilla* L. and *Anguilla rostrata* Le Sueur). Medd. Danm. Fisk. Havund. 3: 183-198.

Colombo, L., and P. Colombo Belvedere, 1976. Steroid biosynthesis by the ovary of the European eel, *Anguilla anguilla* L., at the silver stage. Gen. Comp. Endocrinol. 28: 371-385.

Ijiri, S., Kazeto, Y., Takeda, N., Chiba, H., Adachi, S., and K. Yamauchi. Changes in serum steroid hormones and steroidogenic ability of ovarian follicles during artificial maturation of cultivated Japanese eel, *Anguilla japonica*. Aquaculture (in press).

Leatherland, J.F., Copeland, P., Sumpter, J.P., and R.A. Sonstegard, 1982. Hormonal control of gonadal maturation and development of secondary sexual characteristics in coho salmon, *Oncorhynchus kisutch*, from Lakes Ontario, Erie, and Michigan. Gen. Comp. Endocrinol. 48: 196-204.

Pankhurst, N.W., 1982a. Changes in body musculature with sexual maturation in the European eel, *Anguilla anguilla* (L.). J. Fish Biol. 21: 417-428.

Pankhurst, N.W., 1982b. Changes in the skin-scale complex with sexual maturation in the European eel, *Anguilla anguilla* (L.). J. Fish Biol. 21: 549-561.

Slater, C.H., Schreck, C.B., and P. Swanson, 1994. Plasma profiles of the sex steroids and gonadotropins in maturing female spring chinook salmon (*Oncorhynchus tshawytscha*). Comp. Biochem. Physiol. 109A: 167-175.

Thorarensen, H., Young, G., and P.S. Davie. 11-ketotestosterone stimulates growth of heart and red muscle in rainbow trout. Can. J. Zool. (submitted).

Todd, P.R., 1981. Morphometric changes, gonad histology, and fecundity estimates in migrating freshwater eels (*Anguilla* spp.). NZ J. Mar. Freshw. Res. 15: 155-170.

Yamauchi, K., 1990. Studies of gonadal steroids involved in final gonadal maturation in the Japanese eel, *Anguilla japonica*, a review. Int. Revue ges. Hydrobiol. 75: 859-860.

Yamamoto, K., Morioka, T., Hiroi, O., and M. Omori, 1974. Artificial maturation of female Japanese eels by the injection of salmonid pituitary. Bull. Jap. Soc. Sci. Fish. 40: 1-7.

Reproductive Mechanisms for the Ocean Pout (*Macrozoarces americanus* L.), a Marine Temperate Fish, Include Internal Fertilization and Parental Care of the Eggs

L.W. Crim, Z. Yao and Z. Wang

Ocean Sciences Centre, Memorial University, St. John's, Newfoundland, A1C 5S7, Canada.

Summary

Past reports and more recent investigations of the reproductive biology of the ocean pout indicate that after copulation in the summer, the female ocean pout spawns a single mass of eggs for which she provides parental care for 2-3 months. Studies of sperm physiology and egg fertilization suggest that these reproductive mechanisms including internal fertilization, oviparity and parental care positively contribute to evolution of a successful reproductive strategy in ocean pout based upon relatively low male and female fecundity.

Introduction

The ocean pout occupies a benthic marine habitat stretching across both sides of the Atlantic from the North American to the European Continents (Scott & Scott. 1988). In the colder western North Atlantic ocean environment, the ocean pout is protected from below zero temperatures by antifreeze proteins (Fletcher *et al.*, 1985) although it is reported to migrate to warmer off-shore waters for the winter and return in-shore for breeding during the warm summer and fall months (Keats *et al.*, 1985).

Results

The Seasonal Reproductive Cycle

Laboratory studies of the ocean pout demonstrated that the seasonal reproductive cycle for both sexes can be divided into 4 phases including periods of 1) slow gonadal recrudescence during the winter 2) rapid continued gonad growth in early summer 3) spawning in late summer and finally 4) post-spawned gonadal quiescence.

In males the inactive reproductive period is short since spermatocytes were already found in the testes by October. Although the testes contained spermatids by March, the major increase in GSI remained delayed until July-September together with the period of spermiation (Fig 1).

For females, changes in the GSI, VTG and oocyte size can be used to track seasonal reproductive activity. After spawning, the ovary was populated only by small diameter oocytes (~1 mm) and the GSI declined to approximately 2% in the fall (Fig. 1). Recrudescence of the ovaries began in January and oocyte growth was well established by April based upon an increased GSI (~5%), elevated plasma VTG (~20 mg/ml, unpublished results) and oocyte diameters which ranged from 3.5-5.0 mm (Fig. 2). By July and August, oocyte diameters were larger, ranging from 5.5-6.5 and 6.5-8.0 mm, respectively and the GSI of prespawning females exceeded 15% indicating that oocyte development was nearing completion.

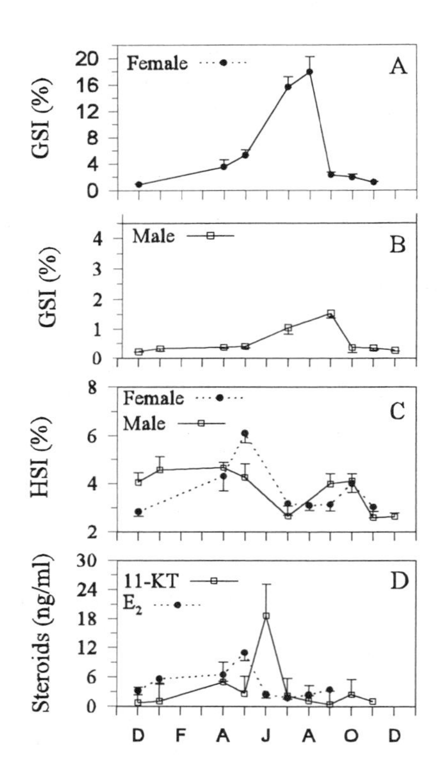

Figure 1. Seasonal changes in (A) the gonadosomatic index for females (GSI), (B) the GSI for males, (C) hepatsomatic index (HSI) and (D) plasma sex specific steroid levels (♀ estradiol; ♂ 11-ketotestosterone). Values are mean ± SEM (N=3-8).

Changes in the plasma sex steroid levels and the hepatosomatic index (HSI) also accompany the seasonal reproductive cycle. For example, in females parallel increases in plasma estradiol and significant liver enlargement (HSI) coincided with oocyte development evident in April and May. Afterwards, the HSI dropped for the balance of the reproductive season (Fig. 1). No such HSI changes occurred for the males in different stages of the reproductive cycle although peak levels of 11-ketotestosterone (11-KT) were found in sexually mature males at the beginning of spermiation.

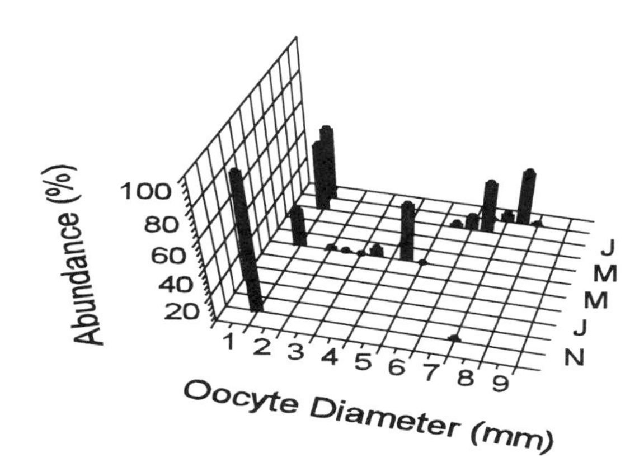

Figure 2. Seasonal changes in oocyte size-class abundance in the ovaries of the female ocean pout.

Sperm Physiology

Male ocean pout produce a "thin-watery" milt containing comparatively low sperm concentrations reflected in low spermatocrit values ranging from approximately 1-2%. The sperm are suspended in a seminal fluid containing an assortment of ions including Na^{+}, K^{+}, Ca^{++}, Mg^{++} and Cl^{-} and is characterized by pH and osmolarity values of near 7.4 and 380 mOsM, respectively (Yao & Crim, 1995). Interestingly, in June early in the male spawning season, ocean pout milt contains two types of germ cells, non-motile spermatids and mature, motile spermatozoa (Wang & Crim, 1995). Furthermore, the sperm are actively motile during their residence in the male reproductive tract and this continuous motility is very long-lasting *in vitro* e.g. several days of storage at 4°C. Although the spermiation response of male ocean pout lasts 3 or 4 months, approximately from June-September, the highest sperm motility is evident for a shorter 6 week period during July and August overlapping the female spawning season (Wang & Crim, unpublished). Exposure of ocean pout sperm to seawater immediately terminates motility yielding an explanation for the unsuccessful attempts to fertilize ocean pout eggs *in vitro* in a seawater medium. Not surprisingly, the morphological features of ocean pout sperm include a well developed mid-piece containing numerous mitochondria and a biflagellated tail both of which perhaps account for the characteristics of extended periods of motility (Yao *et al.*, 1995a).

Egg Fertilization and Spawning

Prior to spawning, the body colouration of prespawning females darkens, the girth greatly expands and finally the ovipore opens in preparation for copulation (Yao 1994). For males, external signs of seasonal reproductive development are less obvious although enlargement of the genital papilla does occur. However, despite occasional instances of courtship behaviour being observed between captive males and females, the egg fertilization rates always remained very low (6%). As an alternative, Yao & Crim, 1995 reported much better results when females were artificially inseminated *in vivo* (AI) after opening of their ovipore. In this case, 67% of the AI females spawned fertilized eggs as opposed to 0% for non-inseminated controls.

Methods for artificial insemination of ocean pout eggs *in vitro* have also been developed beginning with an examination of the effects of semen dilution with different extenders on sperm motility. A synthetic diluent based upon the ionic composition of ocean pout seminal plasma was shown to be effective for cool- temperature preservation of ocean pout sperm motility (Yao, Richardson and Crim, unpublished) and for frozen storage of viable sperm when combined with DMSO (Yao *et al.*, 1995b). Further studies demonstrated that artificial insemination of freshly stripped ocean pout eggs can produce high egg fertilization rates of approximately 85% by suspending the eggs and sperm in the diluent for a period of 5 hr. with stirring before adding and continuing the incubation of the eggs in seawater.

Discussion

After migrating into shallow waters during the spring, pairing of the sexes occurs in both the ocean pout and the Atlantic wolffish (Anarhichas lupus) prior to spawning of their egg masses within rocky crevices in the autumn (Keats *et al.* 1985). Although these two marine species share many reproductive adaptations, namely internal egg fertilization, oviparity and single-time seasonal spawning, parental care is provided the egg mass by the male wolffish and not the female as in the ocean pout. Interestingly, both of these species have low egg and sperm fecundity which seems linked with internal fertilization and post-spawning parental care of the eggs. In a related viviparous species, the eel pout (*Zoarces viviparous* L.), parental care of a limited number of developing embryos (up to 300) is provided by the mother until birth (Korsgaard 1986).

Laboratory studies confirm the seasonality of reproduction in ocean pout which develop their gonads throughout winter and into the summer prior to the autumn spawning. Seasonal increases in plasma estradiol (E2) and VTG in females and 11-KT in males can be used to identify the sex of individuals and also indicate seasonal reproductive development prior to the spawning season. Since infertile spontaneous reproduction has been reported for males and females housed together under captive conditions, collection of viable eggs and sperm from both ocean pout (Yao & Crim, 1995) and the wolffish (Pavlov & Moksness, 1994) represents a significant break-through. Opening of the ovipore provides the best clue to when ripe females should be injected with sperm for *in vivo* artificial insemination and also indicates the time for stripping eggs from females for **artifical insemination *in vitro*.**

Two reproductive mechanisms displayed by ocean pout, namely internal fertilization and sperm with long life-span characteristics, presumably are related to the need for intra-ovarian sperm dispersal following insemination of the female. Although fertilization of large eggs may be facilitated by the presence of more than a single micropyle, as reported in wolffish (Dzerzhinskiy *et al.*, 1992) and ocean pout (Yao *et al.*, 1995), the capability for extended periods of motility demonstrated by ocean pout sperm may be equally important.

Comparing other teleosts, ocean pout seminal plasma contains low levels of K^+ and high levels of Mg^{++} and motility is preserved over a considerable range of K^+ (Yao *et al.*, unpublished). Further studies of various cyroprotectants and freezing regimes have shown that ocean pout sperm can be frozen and stored in liquid nitrogen retaining their ability to fertilize eggs after thawing.

References

Dzerzhinskiy, K.F., D.A. Pavlov and E.K. Radzikhvskaya, 1992. Features of the structure of the egg shell of the White Sea Wolffish, *Anarhichas lupus* marisalb: Discovery of several micropyles. UDC.597.08.591.81.

Fletcher, G.L., C.L. Hew, X. Li, K. Haya and M.H. Kao, 1985. Year-round presence of high levels of plasma antifreeze peptides in a temperate fish, ocean pout (*Macrozoarces americanus*). Can. J. Zool. 63:488-493.

Keats, D.W., G.R. South and D.H. Steele, 1985. Reproduction and egg guarding by Atlantic wolffish (*Anarhichas lupus*: Anarhichidae) and ocean pout (*Macrozoarces americanus*: Zoarcidae) in Newfoundland waters. Can. J. Zool. 63:2565-2568.

Korsgaard, B., 1986. Trophic adaptations during early intraovarian development of embryos of *Zoarces viviparous* (L). J. Exp. Mar. Biol. Ecol. 98:141-152.

Scott, W.B. and M.G. Scott, 1988. Atlantic fishes of Canada. Can. Bul. Fish. Aquat. Sci. 291:412-414.

Pavlov, D.A. and E. Moksness, 1994. Production and quality of eggs obtained from wolffish (*Anarhichas lupus* L.) reared in captivity. Aquaculture 122:295-312.

Wang, Z. and L.W. Crim, 1995. Aspects of spermatogenesis and spermiogenesis in the ocean pout *Macrozoarces americanus*. In: Proceedings of the Fifth Int. Symp. on the Reproductive Physiology of Fish, University of Texas at Austin, July 2-7, 1995.

Yao, Z., 1994. A study of reproductive biology of the ocean pout (*Macrozoarces americanus* L.) in captivity. Memorial University of Newfoundland, Ph.D. Thesis, pp. 1-189.

Yao, Z. and L.W. Crim, 1995. Spawning of ocean pout (*Macrozoarces americanus*L.): evidence in favour of internal fertilization of eggs. Aquaculture 130:361-372.

Yao, Z. and L.W. Crim, 1995. Copulation, spawning and parental care in captive ocean pout. Jour. Fish Biol. (in press).

Yao, Z., C.J. Emerson and L.W. Crim, 1995a. Ultrastructure of the spermatozoa and eggs of the ocean pout (*Macrozoarces americanus* L.), an internally fertilizing marine fish. Mol. Reprod. Dev. (in press).

Yao, Z., L.W. Crim, G.F.Richardson and C.J. Emerson. 1995b. Cryopreservation, motility and ultrastructure of sperm of ocean pout (*Macrozoarces americanus* L.), an internally fertilizing marine teleost. In: Proceedings of the Fifth Int. Symp. on the Reproductive Physiology of Fish. University of Texas at Austin, July 2-7, 1995.

POPULATION DENSITY INFLUENCES PLASMA LEVELS OF GONADAL STEROIDS IN BOTH TERRITORIAL AND NON-TERRITORIAL MALE DAMSELFISH, *CHROMIS DISPILUS*

C.W. Barnett, and N.W. Pankhurst

Department of Aquaculture, University of Tasmania, P.O. Box 1214, Launceston, Tasmania 7250, Australia

Introduction

During spawning of the temperate damselfish *Chromis dispilus*, some males set up territories so as to gain access to ovulated females, however, most males do not occupy territories, but hold station in the water column above nesting sites. Synchronised spawning cycles consist of 1-2 days of spawning followed by 4-5 days of brooding. Plasma testosterone (T) and 11-ketotestosterone (11 KT) are consistently elevated in association with spawning activity, and are significantly higher in territorial males than non-territorial males. Plasma levels of 17α,20β-dihydroxy-4-pregnen-3-one (17,20βP) are also elevated during spawning activity but only in territorial males (Pankhurst, 1990; Barnett and Pankhurst, 1994).

Density of territorial males is correlated with population load of non-territorial males and females in the water column above. During periods of spawning activity, territorial males from areas of high population density have higher levels of gonadal steroids than territorial males from lower population density. This is correlated with higher frequency of spawning and territorial interaction at high density (Pankhurst and Barnett, 1993). In this study we investigated whether or not plasma levels of gonadal steroids in non-territorial males also show density effects.

Methods

Blood samples were taken from non-territorial males, captured by SCUBA diving on reef areas of low (< 1.25 nests.m^2) or high (>1.50 nests. m^2) population density. Blood samples were assayed for the gonadal steroids T, 11 KT and 17,20βP by RIA.

Results and Discussion

Plasma levels of T and 11 KT, but not 17,20βP were elevated at high relative to low density sites during spawning but not brooding (Fig.1). This indicates, that as in territorial males, plasma androgens are elevated in areas of high population density, suggesting that the social influences that appear to act on territorial fish (numbers of females, frequency of spawning and territorial interactions; all of which are higher at high density sites) also influence the endocrine status of the non-territorial fish. In contrast, plasma 17,20βP levels showed no differences with respect to density, consistent with our earlier findings that 17,20βP levels do not change in non-territorial males during spawning periods; further evidence that changes in 17,20βP levels result from direct participation in spawning activity by territory holders.

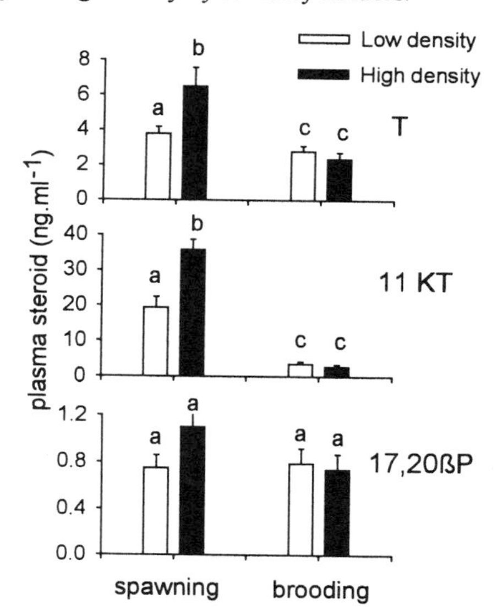

Fig. 1. Plasma levels of gonadal steroids in non-territorial males, sampled from areas of low or high population density during the spawning or brooding phase of the reproductive cycle. Values are means ± s.e. (n values range from 13-40). Different letters show significantly different means.

References

Barnett, C.W. & N.W. Pankhurst. 1994. Changes in plasma levels of gonadal steroids and gonad morphology during the spawning cycle of male and female demoiselles *Chromis dispilus*. Gen. Comp. Endocrinol. 93: 260-274.

Pankhurst, N.W. 1990. Changes in plasma levels of gonadal steroids during spawning behaviour in territorial male demoiselles *Chromis dispilus* (Pisces: Pomacentridae) sampled underwater. Gen. Comp. Endocrinol. 79: 215-225.

Pankhurst, N.W. & C.W. Barnett. 1993. Relationship of population density, territorial interaction and plasma levels of gonadal steroids in spawning male demoiselles *Chromis dispilus* (Pisces: Pomacentridae). Gen. Comp. Endocrinol. 90: 168-76.

AN EVALUATION OF ROCKFISH (*Sebastes spp.*) AS MODELS FOR THE STUDY OF REPRODUCTION AND DEVELOPMENT IN VIVIPAROUS MARINE FISH

B.R. Barron, W.N. Tsang, S. Larsen, and P.M. Collins

Department of Biological Sciences, University of California, Santa Barbara, CA, 93106

Introduction

The present work evaluates two species of Pacific rockfish as models for the study of reproductive seasonality, larval development (*Sebastes rastrelliger*, grass rockfish) and growth (*Sebastes auriculatus*, brown rockfish) in viviparous marine fish. These studies provide the first analysis of seasonal variations in reproductive status in grass rockfish, chronicle critical events during larval development, and establish growth rates of cultured brown rockfish from fingerlings to maturity.

Results

Captured grass rockfish exhibited testicular recrudescence during September and all specimens were in full spermatogenic condition by December. Progressive regression of the testis occurred during March and April. In females, ovaries in various stages of vitellogenesis were observed in December. Fertilization and gestation were recorded between January and March and by April all female fish were in a post-partum condition (fig. 1).

Larvae obtained from grass rockfish spawned in captivity were cultured on a combination of live and microencapsulated marine microorganisms. Linear growth rates were evident from birth to day 40 ($y = 0.133x + 4.534$, $r = 0.97$). Direct correlations were also established between age and both the degree of caudal fin development ($r = 0.96$) and the number of fin rays ($r = 0.97$; table 1).

Captured brown rockfish grown in culture showed a progressive acceleration in incremental growth rate from the young fingerling stage (0.23 g/day at 7 cm S.L.) until they reached a maximum growth rate (0.38 g/day at 12 cm S.L.) which was sustained for the remainder of juvenile development. Under subsistence conditions, brown rockfish are estimated to grow to sexual maturity and marketable size (400 g) in 3-3.5 years.

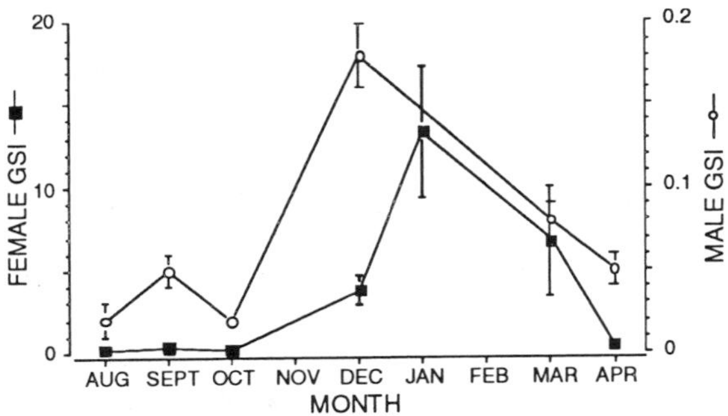

Fig. 1. Seasonal changes in gonadosomatic index (GSI) in grass rockfish (*Sebastes rastrelliger*).

Table 1. Parameters of development in cultured grass rockfish (*Sebastes rastrelliger*) larvae.

DAY	LENGTH (mm)	CAUDAL FIN DEVELOPMENT % DEVELOPED	NUMBER OF RAYS
2	5.33 ± 0.29	13.49	0
6	4.82 ± 0.43	14.72	0
11	6.05 ± 0.18	19.17	0
15	6.51 ± 0.31	25.59	5
18	6.56 ± 0.38	25.97	5
22	7.28 ± 0.26	36.46	11
30	9.33 ± 0.73	65.27	15
40	9.58 ± 0.40	97.32	19

* Flexion began at 22 days

Discussion

As in other species of rockfish (Echeverria, 1987; Love et al., 1990), the grass rockfish displays a restricted breeding season in both sexes with maximum gonadal activity and fertilization occurring in the winter months. The larval rearing component of the work establishes well defined parameters of timed development which will allow more direct aging of wild rockfish larvae. This information can be incorporated into a variety of studies on the effects of diet, environmental pollutants and other factors on larval recruitment. Our growth rate data for the brown rockfish indicate that selected rockfish species are potential candidates for commercial culture and replenishment programs. The ability to raise rockfish larvae, fingerlings, and juveniles in culture will allow the continued exploration of the central mechanisms regulating reproductive seasonality, early development, and growth in viviparous marine fish (unpublished data).

REFERENCES

Echeverria, T.W. 1987. Thirty-four species of California rockfishes: Maturity and seasonality of reproduction. Fishery Bulletin 85(2): 229-250.

Love, M., Morris, P., McCrae, M., and Collins, R. 1990. Life history aspects of 19 rockfish species (Scorpaenidae: *Sebastes*) from the Southern California Bight. NOAA Technical Report NMFS 87.

SEASONAL CHANGES IN WEAKFISH SONIC MUSCLE FIBER MORPHOLOGY AND METABOLIC SUBSTRATE CONCENTRATIONS

M.A. Connaughton,[1] M.H. Taylor.[2]

[1]UT-Houston Medical School, Lab. for Neuroendocrinology, P.O. Box 20708, Houston, TX 77225, U.S.A.; [2]School of Life and Health Sciences and College of Marine Studies, University of Delaware, Newark, DE 19716, U.S.A.

Introduction

The paired sonic muscles of the male weakfish Cynoscion regalis are used to produce a drumming sound during the spawning season and undergo a tripling in mass during this period (Connaughton and Taylor, 1995A). The purpose of the present study was to examine the morphological and biochemical changes which take place in the sonic muscles during the course of their seasonal change in condition.

Methods

Male weakfish were collected from the Delaware Bay from May through August 1992. One sonic muscle was used for analysis of protein, glycogen, and lipid contents, and the other for analysis of fiber cross-sectional area (CSA). Two CSA measurements were taken for each muscle fiber examined: that of the myofibrillar (contractile) tissue, and that of both the myofibrillar tissue and the ring of mitochondria rich (Ono and Poss, 1982) sarcoplasm surrounding it.

Results and Discussion

Both myofibrillar and sarcoplasmic CSA increased significantly early in the spawning season (Fig. 1), and peak fiber CSA coincided with maximal sonic muscle mass. This observed hypertrophy is likely due both to increasing testosterone titers and increased use of the muscle, since sonic muscle hypertrophy induced by exogenous testosterone administration without use of the muscle resulted in increased myofibrillar, but not sarcoplasmic, CSA (Connaughton and Taylor, 1995B). Increased sonic muscle mass results in an increase in sound pressure level of the produced drumming sounds (unpublished data). An increase in the area of the mitochondria-rich sarcoplasm might improve the capacity of the muscle to produce sound for longer periods.

Sonic muscle protein concentration also increased significantly during the spawning season, rising from 28 to 37 mg·g dry tissue^{-1}, before decreasing to 28 mg·g dry tissue^{-1} at the end of the summer. This increase likely reflects rises in both contractile and mitochondrial proteins. Glycogen and lipids began the spawning season high, then decreased rapidly during the period of maximal drumming activity (Fig. 2), suggesting that both of these energy substrates may be used by the sonic muscles.

Given the apparent importance of sound production in weakfish reproductive behavior (Connaughton and Taylor, 1995A; C) these seasonal modifications of the sonic muscle may play an important role in the reproductive success of the species.

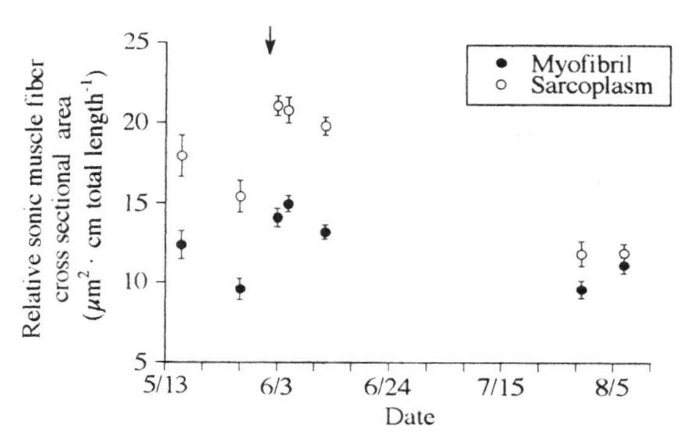

Fig. 1. Seasonal changes in mean relative myofibrillar and sarcoplasmic CSA (standardized by fish total length). The arrow represents the time of maximal sonic muscle mass (Connaughton and Taylor, 1995A).

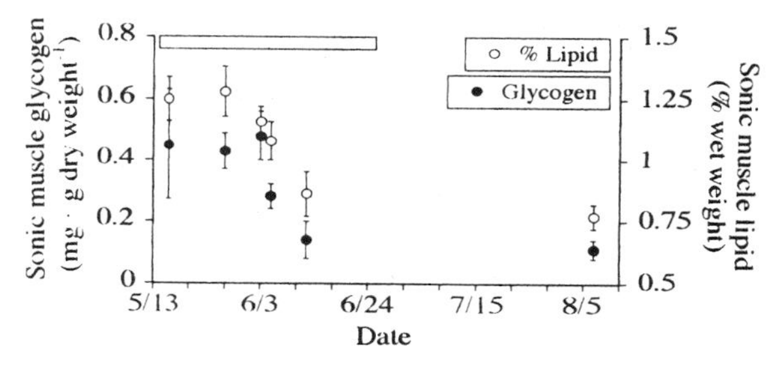

Fig. 2. Seasonal changes in mean glycogen concentration and lipid content. The open bar represents the period of maximal seasonal drumming activity in the field (Connaughton and Taylor, 1995A).

References Cited

Connaughton, M.A., and M.H. Taylor. 1995A. Seasonal and daily cycles in sound production associated with spawning in the weakfish, Cynoscion regalis. Environ. Biol. Fish. 42:233-240.

Connaughton, M.A., and M.H. Taylor. 1995B. Effects of exogenous testosterone on sonic muscle mass in the weakfish, Cynoscion regalis. Gen. Comp. Endocr., In press.

Connaughton, M.A., M.H. Taylor. 1995C. Drumming, courtship and spawning behavior in captive weakfish, Cynoscion regalis. In press.

Ono, R.D., and S.G. Poss. 1982. Structure and innervation of the swim bladder musculature in the weakfish, Cynoscion regalis (Teleosti:Sciaenidae). Can. J. Zool. 60:1955-1967.

CONTROL OF REPRODUCTION IN *TRICHOGASTER TRICHOPTERUS*

Gad Degani, MIGAL - Galilee Technol. Center, Kiryat Shmona 10200, Israel

Introduction

This study examines the control mechanism over the female reproductory cycle in the blue gourami, *Trichogaster trichopterus* L. a male-dependent, multi-spawning fish with asynchronic ovaries, which spawns all year round.

Results

There are two main stages of oogenesis: vitellogenesis, which is not male-dependent, and maturation and ovulation (Degani, 1994), which are controlled by the male. The female is ready for maturation when the ovarian oocytes are in an advanced stage of vitellogenesis (Degan and Boker, 1992a). This stage is controlled by gonadotropin I (GtH I) and 17β-estradiol (E).

Gonadotropins (GtH) control vitellogenesis, maturation and ovulation in the female. When she is in advanced vitellogenesis, she swims into the territory of the male, who secretes pheromones (steroid glucoronides) into the water, triggering oocyte maturation in the female (Degani and Schreibman, 1993). Male sexual behavior (courtship and nets-building) and pheromones together then trigger the process of maturation and ovulation (Degani, 1993). Our unpublished studies suggest that the GtH synthesized in the pituitary at the end of vitellogenesis is GtH II, and that testosterone (T) may be involved in the transition from GtH I to GtH II synthesis. Pheromones and male behavior stimulate the production of a GtH-releasing hormone (GnRH), which triggers the secretion of GtH from the pituitary to the plasma (Degani et al., in press). The effect of GtH on the thecal cells of the ovary is to convert the 17α-hydroxy-progesterone (17-P) to 17α,20β-dihydroxy-4-pregnen-3-one (17,20-P), which is the steroid that induces maturation and ovulation (Degani and Boker, 1992b).

Once the female has spawned, the number of her oocytes in vitellogenesis decreases, with the result that she is no longer in reproductive condition. She is unaffected by male pheromones and courting behavior, and swims out of his territory to begin the cycle of vitellogenesis again.

Five distinct sub-stages of maturation have been described: 1. The nucleus migrates to the periphery of the oocytes; 2. Nucleus is near oocyte envelope, surrounded by homogenous ooplasma; 3. Oocyte envelope invaginates slightly; 4. Micropylar cell appears; 5. The nuclear membrane breaks down (Jackson et al., 1994).

Discussion

These results, arising from research conducted continually over the past five years, have enabled us to suggest a model for the control mechanism governing reproduction in *T. trichopterus*, as shown below.

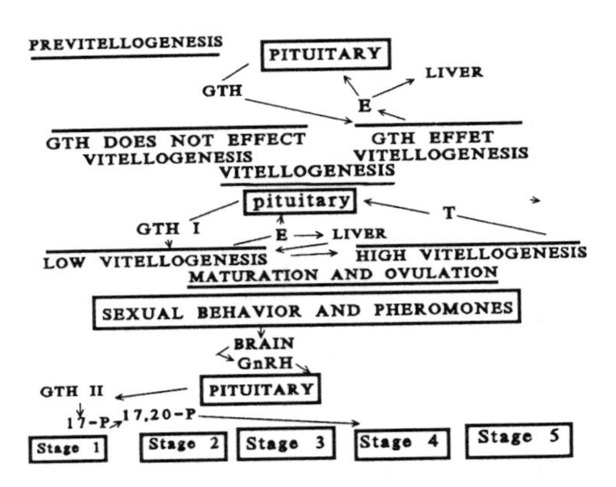

References

Degani, G. 1990. Comp. Biochem. Physiol. 96A:525-528.

Degani, G. 1993. Copeia (4):1091-1096.

Degani, G. 1994. J. Aquacult. Trop. 9:15-23.

Degani, G. & Boker, R. 1992a. Gen. Comp. Endocrinol. 85:430-439.

Degani, G. & Boker, R. 1992b. J. Exp. Zool. 263:330-337.

Degani, G. & Schreibman, M.P. 1993. J. Fish Biol. 43:475-485.

Degani, G., Jackson, K. & Mermelstein, G. 1995. J. Aquacult. Trop. (in press).

Jackson, K., Abraham, M. & Degani, G. 1994. J. Morphol. 220:1-9.

REPRODUCTIVE CYCLES OF COMMON SNOOK, _CENTROPOMUS UNDECIMALIS_[1]

Harry J. Grier, Ronald G. Taylor, and Ruth O. Reese

Florida Department of Environmental Protection, Stock Enhancement Research Facility, 14495 Harllee Road, Port Manatee, Florida 34221

Summary

The common snook spawns between May and September in the early evening hours.

Introduction

The common snook is an inshore centropomid that forms reproductive schools near passes and beaches; however, it is also commonly found in grass flats near mangroves. The annual reproductive data is not complete, although studies have been done on both coasts (west - Marshall, 1958; Volpe, 1959. east - Gilmore, et al., 1983; Tucker and Campbell, 1988).

Results

To study the annual reproductive cycle, snook were collected monthly in south Tampa Bay during January 1988 - December 1989. Gonadosomatic indices plotted by month (Fig. 1) demonstrated that the breeding season extended from May through September. During the breeding season, all of the females caught had ripe oocytes within the ovarian lamellae.

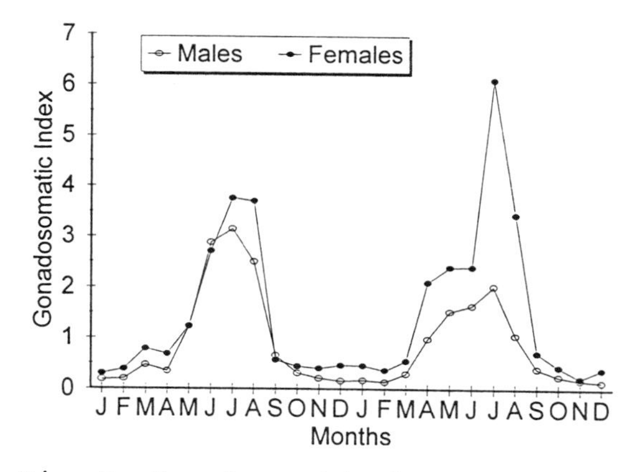

Fig. 1. Gonadosomatic index, 1988-89.

To study the daily cycle during the 1986 spawning season, snook collections were made at spawning sites during three daily intervals: 0700-0900, 1100-1300, and 1700-1900 hrs. For investigation of the daily cycle, the staging criterion used was determination of the most advanced oocyte in the sample. During the 0700-0900 period, about 35% of the females had oocytes with nuclear migration and early fusion of lipid droplets. During the 1100-1300 period, 85% of the females had oocytes with advanced lipid fusion and/or yolk coalescence. During the 1700-1900 period, 80% of the females had cleared oocytes.

Discussion

Snook have a daily, early evening spawning cycle during their annual breeding season of May to September, which is when day lengths are longest and water temperatures are highest. It is possible that daylength, temperature, and tidal cycles all play some role in determining the reproductive cycle in this fish.

References

Gilmore, R. G., C. J. Donahoe and D. W. Cooke. 1983. Observations of the distribution and biology of the common snook, _Centropomus undecimalis_ (Bloch). Fla. Sci. 46:313-336.

Marshall, A. R., 1958. A survey of the snook fishery of Florida, with studies of the biology of the principal species, _Centropomus undecimalis_ (Bloch). Fla. St. Brd. Consrv. Tech. Ser. No. 22.

Tucker, J. W., Jr. and S. W. Campbell, 1988. Spawning season of common snook along the east central Florida coast. Fla. Sci. 51:1-6.

Volpe, A.V., 1959. Aspects of the biology of the common snook, _Centropomus undecimalis_ (Bloch) of southwest Florida. Fla. St. Brd. Consrv. Tech. Ser. No. 31.

[1]Contribution No. 1 of the Snook Project, sponsored jointly by the Florida Department of Environmental Protection and Mote Marine Laboratory.

TESTICULAR STEROIDOGENESIS IN RELATION TO SOCIAL INTERACTIONS IN THE NILE TILAPIA *OREOCHROMIS NILOTICUS*

Gene A. Hines, Kristina M. Wasson and Stephen A. Watts

Department of Biology, University of Alabama at Birmingham, Birmingham, AL 35294-1170, USA

Summary

The social environment exerts a limited effect on testicular androgen and estrogen steroidogenesis, however, the physiological significance of these differences has not been determined. The principal androgen produced in the testes is 5β-androstane-3α,17β-diol, not 11-ketotestosterone.

Introduction

Endocrine correlates of social environment with sexual status have indicated the importance of hormones in reproductive behavior and physiology of teleost fish (Stacey et al., 1994). Cichlids exhibit relatively complex social interactions, especially during reproduction, and may serve as useful models to understand psychoendocrine phenomena of teleosts. This study examines the relationship between the social environment and steroidogenesis in the testis of the Nile tilapia, *Oreochromis niloticus*.

Results

In this study, adult *Oreochromis niloticus* were exposed to various social treatments in individual recirculating aquaria including single males (n=3, held for 2wk), grouped males (n=5/group, held together for 2 wk), or paired males with females (n=3 pairs held for 2 d). Testes were excised, minced and incubated with ^{3}H-androgen precursors for up to 3 hr. Organic-soluble steroidal metabolites were identified by TLC, derivatizations and recrystalizations. Generally, the precursor androstenedione was converted into 7-10 organic-soluble metabolites (Table 1). Less than 5 % of the total metabolites were synthesized as aqueous-soluble compounds.

Table 1. Percentage (Mean and SE) of organically-extracted C19 and C18 steroids in relation to total metabolites synthesized in 3 hr from the androstenedione precursor.

Steroid	Single	Paired	Grouped
5β-androstane-3α,17β-diol	26±8	32±8	55±6
Etiocholanolone	26±3	23±7	24±3
Epietiocholanolone	19±8	9±3	1.5±0.3
5β-DHT	5±2	10±4	4±2
5β-androstanedione	11±6	9±5	0
Testosterone	2±2	7±3	8±4
Estradiol	0	0	2±0.3
Unidentified	11	10	6

Discussion

Since Idler's (1969) review, most reports indicate that the 11-oxyandrogens, particularly 11-ketotestosterone, are the principal androgens synthesized by the testes of teleosts. 11-keto-testosterone was not synthesized by the tilapia testis. A major metabolite synthesized at all social treatments was 5β-androstane-3α,17β-diol, and was the principal metabolite in grouped and paired males. Whether or not this androstanediol serves as the principal, functional androgen in the tilapia testis is not known. Single and paired males synthesized and accumulated 5β-androstanedione. Only grouped males synthesized estradiol. Little or no water-soluble compounds were produced. These studies indicate that the social environment can have a limited influence on testicular steroidogenesis in *O. niloticus*. However, due to a small sample size and potential variations in reproductive status of the gonads in the individuals used in this study, it is difficult to relate the observed differences in steroid metabolism to socially-induced reproductive behavior and physiology. These limited differences in socially-induced differential steroidogenesis, as well as the lack of gonadally-derived steroid conjugates, suggest that extra-gonadal tissues may also be involved.

References

Idler, D., B. Truscott and H. Stewart. 1969. Some distinctive aspects of steroidogenesis in fish. In: Progress in Endocrinology: International Congress Series, Exerpta Medica Foundation, Vol. 184, p. 724-729.

Stacey, N., J. Cardwell, N. Liley, A. Scott and P. Sorenson. 1994. Hormones as sex pheromones in fish. In: Perspectives in Comparative Endocrinology. K. Davey, R. Peter and S. Tobe (Eds.): National Research Council of Canada, Ottawa, p. 438-448.

THE WHITE PERCH, *MORONE AMERICANA* : A LABORATORY MODEL FOR REPRODUCTION OF TEMPERATE BASSES

L.F. Jackson,[1] W. King V,[1] E. Monosson,[2] and C.V. Sullivan.[1]

[1]Department of Zoology, North Carolina State University, Raleigh, NC 27695-7617; [2]Marine Science Research Center, State University of New York, Stony Brook, NY 11794-5000.

Summary

The gametogenic cycle of white perch was characterized in detail and effectively controlled by manipulating water temperature and photoperiod to initiate maturation, and administering GnRHa implants and hCG injections to induce spawning.

Introduction

The white perch Morone americana matures rapidly at a small size and is easily kept in captivity. In order to evaluate its utility as a laboratory model for reproductive physiology, three studies were performed: 1) the annual gametogenic cycle was characterized in detail with respect to circulating levels of the gonadal steroid hormones estradiol-17β (E_2), testosterone (T) and 11-ketotestosterone (11-KT), and of the yolk protein precursor, vitellogenin (Vg); 2) hormonal control of final oocyte maturation (FOM) was investigated using in vivo and in vitro assays; 3) white perch broodstock were induced to mature in the laboratory with photoperiod manipulation and spawned with implanted GnRH analogue (GnRHa) and injected hCG.

Materials and Methods

Adult white perch were sampled from ponds at monthly intervals for one year and gonadal steroids and Vg were measured in their blood plasma by RIA and radial immunodiffusion, respectively (Woods and Sullivan 1993). Gonadal tissue was embedded in glycomethacrylate for histological examination. Circulating levels of 17α,20β-dihydroxy-4-pregnen-3-one (DHP) and 17α,20β,21-trihydroxy-4-pregnen-3-one (20β-S) were evaluated for females undergoing FOM, and for culture of ovarian incubates undergoing FOM in vitro (King et al. 1995). Alternating light cycles of constant length were used to manipulate the reproductive cycle. GnRHa implants (150 μg/kg) and hCG injections (100 IU/kg) were administered to induce spawning (Hodson and Sullivan 1993).

Results

Increased levels of E_2, T and Vg accompanied oocyte growth, while increased T and 11-KT levels coincided with the progressive maturation of the testes. Histochemical analyses of ovarian tissue indicated that recruitment into vitellogenesis began in November and that the spawning period extended from late March to mid-May. Spermatogenesis was underway by October, and the spermiation period extended from early January to mid-May. At the peak of spawning (April) ovaries contained oocytes at all stages of development (multiple-clutch, group-synchronous maturation). Plasma levels of DHP and 20β-S were highest in females during FOM, and follicles undergoing GVBD produced DHP and 20β-S in vitro. Under compressed photoperiod schedules the reproductive cycle could be shortened to ≤ 6 months. Ovarian growth was initiated under short days (9L:15D) and rapid follicle growth commenced under long days (15L:9D). Broodstock maturing under natural light cycles were reliably spawned in the laboratory.

Discussion

Endocrine correlates of gonadal growth in white perch and its associated changes in vitellogenesis and FOM in females are virtually identical to those observed in striped bass M. saxatilis and other teleosts (Jackson and Sullivan 1995). White perch respond very well to experimental manipulation of their reproductive cycle. Maturation and spawning can be induced as needed using environmental and pharmacological treatments. These observations indicate that white perch can serve as a valid model for studying reproductive physiology in Morone species like striped bass or other perciformes that are too rare, valuable, large or sensitive for routine laboratory investigation.

References

Hodson, R.G. and C.V. Sullivan. 1993. Aquaculture and Fisheries Management 24:389-398.

Jackson, L.F. and C.V. Sullivan. 1995. Transactions of the American Fisheries Society. in press.

King V, W., Berlinsky, D.L. and C.V. Sullivan. 1995. Fish Physiology and Biochemistry in press.

Woods, C. and C.V. Sullivan. 1993. Aquaculture and Fisheries Management 24:211-222.

This work was supported by grants from the National Coastal Resources Research and Development Institute (#AQ119.87S-5628-2-09-1) and the UNC Sea Grant College Program (#NA90AA-D-SG-062).

SEASONAL CHANGES IN GONADAL HISTOLOGY AND SEX STEROID HORMONE LEVELS IN THE PROTOGYNOUS HERMAPHRODITE, *EPINEPHELUS MORIO*

A.K. Johnson[1] and P. Thomas[2].

[1] The Department of Marine Science, University of South Florida, St. Petersburg, FL 33705.
[2] The University of Texas at Austin, Marine Science Institute, Port Aransas, Texas 78373.

Introduction

The red grouper, *Epinephelus morio,* is a protogynous hermaphrodite belonging to the family serranidae. Although the reproductive cycle and sex change have been described previously (Moe, 1969), no information is available on the changes in sex steroid hormone levels during the reproductive cycle and sex-reversal. Therefore, in the present study, seasonal changes in gonadal histology and circulating sex steroid hormones of estradiol-17β (E_2), testosterone (T), and 11-ketotestosterone (11-KT) were examined in a field population of red grouper.

Methods

Red grouper were caught by hook-and-line in the eastern Gulf of Mexico between February 1993 and August 1994. Gonads and blood samples were collected immediately after capture. Gonads were fixed histologically and classified according to Moe (1969). Plasma levels of E_2, T, and 11-KT were determined by RIA.

Results and Discussion

Females. Females with perinucleolar stage (previtellogenic) oocytes had low levels of E_2, T, and a low GSI. The ovaries of females caught during the spawning season (March-May) contained oocytes at all stages of development (previtellogenic, cortical alveoli, vitellogenic, and final stage of oocyte maturation (FOM)). Ovaries of individuals undergoing FOM also contained several cohorts of vitellogenic oocytes. The mean GSI increased rapidly between January and March, reaching a peak in April. Plasma levels of both E_2 and T rose during ovarian recrudescence and remained elevated during FOM. The maintenance of high levels of E_2 and T in fish undergoing FOM has been observed in other multiple spawning species, e.g., *Epinephelus striatus* (Smith, 1972), presumably because vitellogenesis and the recruitment of new oocytes is continuing in these individuals.

Transitionals. Sexual succession occurred in a small proportion (approximately 2%) of female red grouper (4-7 years old) which had previtellogenic and/or cortical alveoli stage oocytes during the breeding and nonbreeding seasons. Similarly, in the graysby (*Epinephelus cruentatus*), sexual transition is not restricted to any particular time of the reproductive cycle (Nagelkerken, 1979). Cysts containing male tissue at the same developmental stage formed "pockets" within the predominantly female ovary (Figure 1). These fish become functional males when the sperm duct is formed. Transitionals had low levels of all three steroids (E_2, T, and 11-KT). The ratio of E_2 to androgen declined dramatically during sex change. However, the low levels of E_2 and T were similar to those in previtellogenic females, but not to females with cortical alveoli stage oocytes.

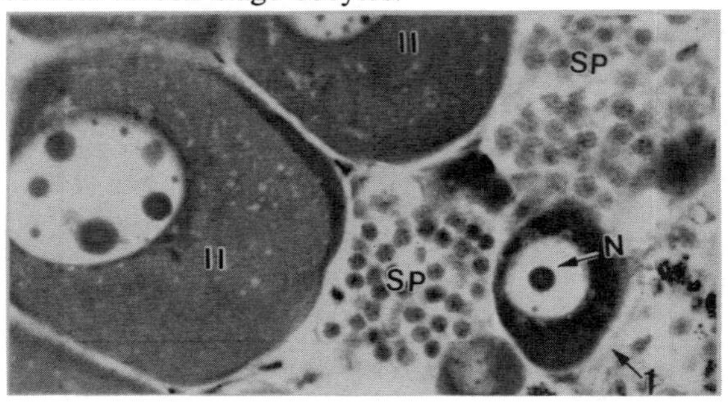

Figure 1. Cross-section of a transitional gonad showing propagation of male germ cells (SP) with female oocytes (1 and 11).

Males. All the males examined had remnant previtellogenic oocytes in their gonads which confirms that this species is protogynous. Plasma steroid levels (T and 11-KT) were only elevated during active spermatogenesis (March-May) in fully developed males and declined in postspawning individuals. Some males retained mature gametes in their gonads up to August, three months longer than females.

References

Moe, M.A. Jr. 1969. Biology of the red grouper (*Epinephelus morio* Valenciennes) from the eastern Gulf of Mexico. Prof. Pap. Ser. Mar. Lab. Fla. No. 10, 95pp.

Nagelkerken, W.P. 1979. Biology of the graysby *Epinephelus cruentatus*, of the coral reef of Curacao. *In* Studies on the fauna of Curacao and other Caribbean Islands (P.W. Hummelinck and L. J. van Der Steen, eds.). 60:1-118.

Smith, C.L. 1972. A spawning aggregation of nassau grouper, *Epinephelus striatus*. Trans. Amer. Fish. Soc. 101:257-261.

PLASMA LEVELS OF STEROID AND THYROID HORMONES IN THE INDIAN MAJOR CARP, *Labeo rohita*, DURING SEXUAL MATURATION

S.F. Khanam, S. Ali and S. Khan

Department of Zoology, Dhaka University, Dhaka, Bangladesh

Summary

A monthly hormonal analysis of captive *L. rohita* showed that males mature earlier than the females. Sex steroid levels were undetectable until the age 18 months in females and 9 months in males. In both sexes, these hormones increased during the preparatory to the spawning period. The levels of cortisol, T_4 and T_3 in both sexes also showed an increase during the same period.

Introduction

The fast growing carp, *Labeo rohita,* is an important cultivated fish in the Indian subcontinent and a thorough knowledge of factors involved in its sexual maturation is important. Since we know that there is a direct and indirect involvement of a variety of hormones in sexual maturation, we chose to examine plasma levels of steroids (sex and adrenal) and thyroid hormones.

Methods

A total of 810 fish, ages 5 months old, were reared over the next 23 months. Eight fish were anesthetized monthly and blood samples were taken from the caudal artery. T_4, T_3, oestriol and cortisol were measured in the plasma by radioimmunoassay using commercial kits (Amerlex RIA). Plasma testosterone (tes) was also assayed by RIA as described by Stupnicki and Kula (1982).

Results

Table. 1. Seasonal variations in plasma T_4 and T_3 (mean±S.E.) of *L. rohita*.

PERIOD	SEX	T4 (ng/ml)	T3 (ng/ml)
Preparatory	F	8.20±0.19	1.84±0.05
(Feb - Mar)	M	7.55±0.43	1.77±0.06
Prespawning	F	4.58±0.003	0.49±0.03*
(Apr - May)	M	5.58±0.44*	1.52±0.18
Spawning	F	11.71±1.02*	2.40±0.20*
(Jun - Aug)	M	12.14±1.06*	1.70±0.25
Postspawning	F	5.79±0.16*	0.62±0.03*
(Sep - Dec)	M	5.58±0.43*	0.96±0.11*

* significant (P < 0.05) from the previous period.

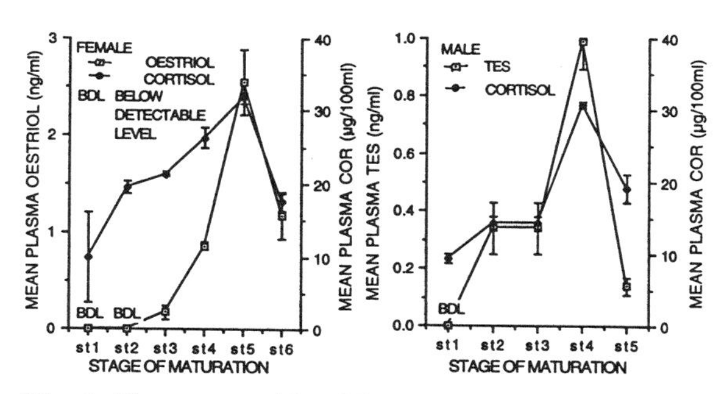

Fig. 1. Plasma steroids of *L. rohita* at different stages of maturation. (st1 immature; st2 maturing; st3 advanced maturing; st4 mature; st5 ripe female, resorption in male; st6 resorption).

The maximum levels of oestriol in females, tes in males, and cortisol in both sexes, were found during the spawning period when fish were at a fully mature stage and low at a resorption stage (Fig.1). The levels of T_4 and T_3 showed an increase as fish became mature; highest levels were found during the spawning period (Tab.1).

Discussion

The high levels of hormones during the reproductive period (Summer-Monsoon) of *L. rohita* coincide with the first phase of gonadal maturation. The seasonal pattern of sex steroid and thyroid hormones, and gonadal growth of this species are comparable with that found in plasma thyroid and steroid hormones of other teleosts (Billard *et al.* 1982; Flett *et al.* 1994).

References

Billard, R., A. Fostier, C. Weil and B. Breton, 1982. Endocrine control of spermatogenesis in teleost fish. Can. J. Fish. Aquat. Sci., 39:65-79.

Flett, P.A., J.F. Leatherland, and G. Van der Kraak, 1994. Endocrine correlates of seasonal reproduction in fish: thyroid hormones and gonadal growth, steroidogenesis, and energy partitioning in Great Lakes salmon. In "Perspective in Comparative Endocrinology" (K. Davey, R. Peter, and S. Tobe Eds), Nat. Res. Coun. Canada, Ottawa. 602-610.

Stupnicki, R. and E. Kula, 1982. Direct RIA of steroids in human plasma. Endokrinologie, 80:1-7.

Ackmts: We appreciate the help of Dr. J. Youson, U of Toronto, in the preparation of this manuscript.

CHANGES IN PLASMA ESTRADIOL AND TESTOSTERONE CONCENTRATIONS DURING A BROODING CYCLE OF FEMALE MOUTHBROODING TILAPIA, OREOCHROMIS MOSSAMBICUS, AND MALE MOUTHBROODING TILAPIA, SAROTHERODON MELANOTHERON

M. Kishida, W. A. Tyler III[1], and J. L. Specker

Department of Zoology, University of Rhode Island, Kingston, RI 02881, U. S. A.

Summary

Concentrations of estradiol (E2) and testosterone (T) in plasma were measured in female O. mossambicus during fry-brooding periods and in male S. melanotheron during brooding cycles. In female O. mossambicus, E2 concentrations were elevated during the fry-brooding period compared to those after separation of the fry, while T concentrations did not change. In male S. melanotheron, both E2 and T concentrations increased towards the day of fry release. These data suggest a possible involvement of E2 and T with changes in mouthbrooding behavior in tilapia.

Introduction

Although parental behavior is widespread among fishes, its endocrine control is not well understood. Most studies have been done on male substrate spawners (Kindler et al., 1989; Sikkel, 1993). There is only one study on the steroid profiles in a female mouthbrooder, O. mossambicus, during ovarian cycles (Smith and Haley, 1988). Examining both sexes which exhibit similar behavior will help understand endocrine controls of parental behavior.

In this study, we used two species of tilapia, O. mossambicus (female mouthbrooders) and S. melanotheron (male mouthbrooders) to investigate changes in plasma E2 and T concentrations during brooding cycles.

Results

In female O. mossambicus, E2 concentrations were 6.7 ± 1.5 and 6.1 ± 1.2 ng/ml at 2 days and 5-6 days after the first release of the fry, respectively, and decreased to 2.5 ± 0.9 ng/ml 7 days after separation from the fry. T concentrations did not change significantly at these sampling times (17.8 ± 2.0, 32.6 ± 16.0, 15.3 ± 6.6 ng/ml). In male S. melanotheron, E2 concentrations were 0.4 ± 0.1 ng/ml at the day of spawning, 0.3 ± 0.1 ng/ml 5 days after spawning, increased to 0.7 ± 0.1 and 0.8 ± 0.1 ng/ml 10 days after spawning and the day of releasing fry, respectively, and decreased to 0.5 ± 0.2 ng/ml 10 days after the fry release. T concentrations were low during the early brooding period (5.7 ± 1.7 ng/ml, the day of spawning; 5.7 ± 1.8 ng/ml, 5 days after spawning), increased to 17.9 ± 5.4 ng/ml at 10 days after spawning and to 69.2 ± 33.1 ng/ml at the day of fry release, and then decreased to 13.1 ± 4.5 ng/ml 10 days after the release.

Discussion

The results in this study suggest that E2 and T concentrations in plasma change in association with the changes in behavior over brooding cycles in mouthbrooding tilapia, except that T concentrations in O. mossambicus did not change over the sampling time.

The decrease in T concentrations during the early brooding period in S. melanotheron is similar to those reported for other male substrate spawners (Kindler et al., 1989; Sikkel, 1993).

Both E2 and T increased at the day of fry release in S. melanotheron. It is possible that the elevated steroid levels induce the release of fry rather than result from the behavior, because the plasma levels appeared to increase before the behavior occurred.

Further studies are required to elucidate causal relationships between hormones and parental behavior.

References

Kindler, P. M., D. P. Philipp, M. R. Gross and J. M. Bahr, 1989. Serum 11-ketotestosterone and testosterone concentrations associated with reproduction in male bluegill (Lepomis macrochirus: Centrachidae). Gen. Comp. Endocrinol. 75:446-453.

Sikkel, P. C., 1993. Changes in plasma androgen levels associated with changes in male reproductive behavior in a brood cycling marine fish. Gen. Comp. Endocrinol. 89:229-237.

Smith, C. J. and S. R. Haley, 1988. Steroid profiles of the female tilapia, Oreochromis mossambicus, and correlation with oocyte growth and mouthbrooding behavior. Gen. Comp. Endocrinol. 69:88-98.

[1]Present address: Dept. of Biology, Adams State College, Alamosa, CO 81102, U. S. A.
Supported by NSF-IBN 9420268.

REPRODUCTIVE CYCLE AND EMBRYONIC GROWTH DURING GESTATION OF THE VIVIPAROUS TELEOST, *ZOARCES ELONGATUS*

Y. Koya[1*], T. Matsubara[1], T. Ikeuchi[2], N. Okubo[1], S. Adachi[2], and K. Yamauchi.[2]

[1]Hokkaido Natl. Fish. Res. Inst., Kushiro, Hokkaido, 085, Japan; [2]Dep. Biology, Fac. Fish. Hokkaido Univ., Hakodate, Hokkaido, 041, Japan.

Introduction

Within the Zoarcidae, three species of the genus *Zoarces* have evolved viviparity. Heretofore, almost all the information concerning such Zoarcid viviparity has been centered on the European species, *Z. viviparus* (e.g. Korsgaard, 1986). We have thus initiated several investigations to elucidate the mechanisms in the maintenance of gestation, including endocrine control, in Japanese species, *Z. elongatus*.

Results

The annual reproductive cycle of female *Z. elongatus* can be divided into the following five periods; 1) early vitellogenesis (Dec to Apr): pre-vitellogenic oocytes in which gradual yolk accumulation begins, 2) late vitellogenesis (May to Aug): yolk accumulation proceeds rapidly, 3) early gestation (Sep to Oct): ovulated eggs fertilize and develop in the ovarian lumen, 4) mid-gestation (Oct to Nov): hatched larvae absorb their yolk-sac, 5) late gestation (Dec to Apr): embryos continue to grow further until parturition, and oocytes for the next clutches begin vitellogenensis.

The ripe eggs of *Z. elongatus* were 4.4 mm in diameter. Their wet and dry weights were 45 mg and 7 mg, respectively (Fig. 1). Within one month after hatching, the embryos grew to 170 mg and 20 mg in wet and dry weights, respectively. Finally, they reached 540 mg and 65 mg in wet and dry weights, respectively in February.

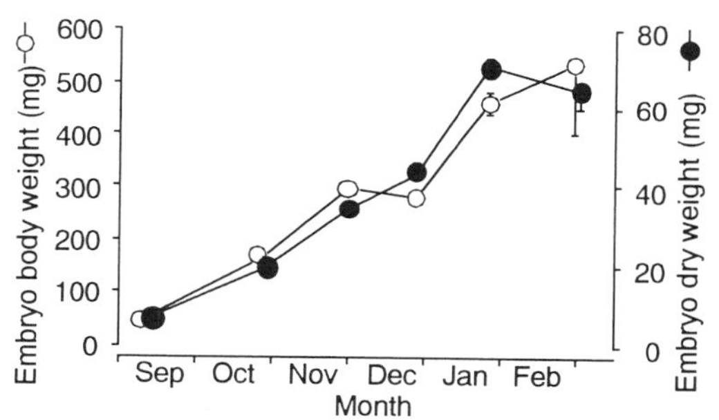

Fig. 1. Changes in wet body weight and dry weight of embryo in *Zoarces elongatus* during gestation (mean ± SE).

Vitellogenin concentrations in female serum were measured by quantitative immunodiffusion using an antiserum against estradiol-17β (E2) treated male serum absorbed with immature male serum. The serum vitellogenin levels (Fig. 2) increased from May to a peak (6 mg/ml on average) in September, then decreased to a low level (0.1 mg/ml on average) in October. The levels remained low (<0.3 mg/ml) throughout the gestation period.

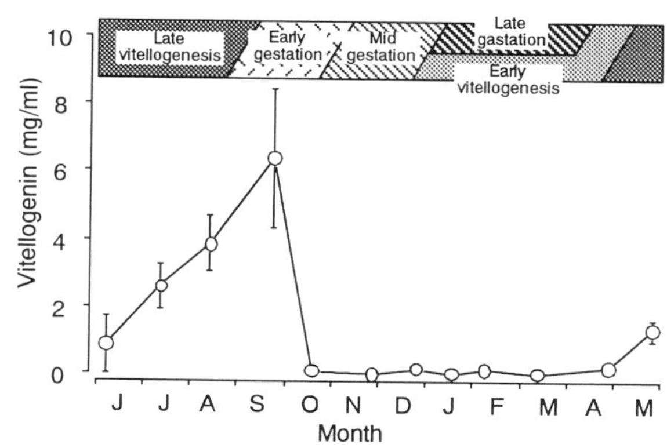

Fig. 2. Annual profile of vitellogenin levels in the serum of female *Zoarces elongatus* (mean ± SEM).

Serum E2 levels in females, analyzed by RIA, were high during the late vitellogenesis and declined to a low level during the early gestation (Table 1).

Table 1. Serum concentrations of estradiol-17β in female *Zoarces elongatus*.

Date	Reproductive period	N	Serum estradiol-17β (ng/ml) (Mean ± SE)
14 Jun	Late vitellogenesis	9	1.31 ±0.16
5 Aug	Late vitellogenesis	6	2.95 ±0.56
13 Sep	Early gestation	5	0.43 ±0.06
26 Oct	Early gestation	4	0.33 ±0.05

Discussion

The present study demonstrated that *Z. elongatus* is an intraluminal gestation type. The embryos of this species increased nine times in dry weight during gestation, indicating that this species is a matrotrophic type in which embryos depend on a continual supply of maternal nutrients (Wourms, 1981). Serum levels of vitellogenin and E2 in females were high during the late vitellogenesis, showing good correlation with the advance of oocyte development, then a rapid decline during the early gestation period. These results combined with the matrotrophic growth of embryos during gestation suggest that there is a shift in the consumption of nutrients from the yolk to other maternal nutritional products during the mid-gestation period.

References

Korsgaard, B., 1986. Trophic adaptations during early intraovarian development of embryos of *Zoarces viviparus*. J. Exp. Mar. Ecol., 98: 141-152.

Wourms, J. P., 1981. Viviparity: The maternal-fetal relationship in fishes. Am. Zool., 21: 473-515.

*Present adress: Dept. Biol. Fac. Educ. Gifu Univ., Gifu, Gifu, 502, Japan.

VARIABILITY IN EGG QUALITY AND PRODUCTION IN A BATCH-SPAWNING FLOUNDER, *Pleuronectes ferrugineus* .

A.J. Manning and L.W. Crim

Ocean Sciences Centre, Memorial University of Newfoundland, St. John's, NF, A1C 5S7 CANADA

Summary

Female yellowtail flounder frequently have a daily ovulatory rhythm. Females of an average size of 37cm will produce over a half a million eggs. The production can be portioned out in as many as 27 batches . Strong individual variation in egg quality is apparent.

Introduction

Yellowtail flounder is a batch-spawning flatfish species which has its northern distribution in the waters off Newfoundland, Canada. The aims of the study were to determine the inter-ovulatory period for the females, as well as the rates of seasonal egg production for individuals. In Zamarro's (1991) histological study of yellowtail ovaries, it was proposed that a daily spawning pattern occurs and a female of 42cm length could produce 1 456 000 eggs in 7 ±1 batches .

In addition, the project was designed to determine the variability in egg quality among batches within individuals and the variability in egg quality among the different females. A daily stripping protocol was employed and egg quality was assessed by fertilization and hatching rates.

Results & Discussion

Egg production for this group of 11 captive females was estimated at over 6 million eggs, female A having the best production (Table 1). The mean female production was ~550 000. The average female weight was 694g and the mean total length was 37cm. The duration of the spawning season for an average female lasted 48 days. Production was portioned out in an average of 14 batches, ranging up to 27 batches for female I measuring 33.5cm (Table 1). This data indicates that a female can portion out her production in smaller amounts than suggested by Zamarro (1991).

Inter-ovulatory periods were pooled for the group, the pooled data indicates that 65% of the time a female will ovulate a batch one or two days apart. A one day interval between batches was more frequent as Zamarro (1991) had predicted. Females with interrupted spawning, as indicated by large average inter-ovulatory periods (e.g. ♀ L), were generally associated with poor egg quality (Table 2).

Statistically significant differences in egg quality were found among the different females and among batches within individuals. These results lead to the conclusion that female yellowtail in a captive situation show strong individual differences in reproductive performance. Hence, there is no typical female in terms of egg quality. The strong batch variability within females indicates that females are not consistent in their performance during their spawning season (Table 2). As to the performance on the group level, the egg quality data suggests that one third of the eggs produced can be expected to hatch.

Table 1. Data from three representative females illustrating differences in prespawning total length and weight, the number of batches ovulated and the total seasonal egg production.

♀ id	Prespawning L (cm)	Prespawning Wt (g)	Batches Produced	Total egg production
A	37	739	19	1 324 762
I	33.5	468	27	723 217
L	41.5	864	13	345 135

Table 2. Data from three representative females illustrating differences in spawning duration, average inter-ovulatory period, average fertilization rate (FR) and average hatching rate (HR) with standard errors.

♀ id	Duration (days)	Avg Period	Avg FR (%)	Avg HR (%)
A	31	1.72	46 ±5.6	78 ±4.9
I	51	1.96	53 ±4.3	68 ±5.3
L	95	7.92	4.5 ±2.3	38 ±11

References

Zamarro, J., 1991. Batch fecundity and spawning frequency of yellowtail flounder (*Limanda ferruginea*) on the Grand Bank. NAFO Sci. Coun. Studies. 15:43-51.

SEASONAL CHANGES IN PLASMA LEVELS OF SEXUAL HORMONES IN THE TROPICAL FRESHWATER TELEOST, *Pygocentrus cariba* (formely *P. notatus*, TELEOSTEI: Characidae)

Hilda Y. Guerrero, Gisela Cáceres-Dittmar and Dayssi Marcano

Laboratorio de Neuroendocrinología, Departamento de Fisiología, Escuela de Medicina J.M.Vargas and Instituto de Medicina Experimental. Universidad Central de Venezuela. Apartado 47633. Caracas 1041-A. Venezuela.

Introduction

In previous studies we have shown the changes that occur in gonadotropin releasing hormone (GnRH) (Gentile et al., 1986) and catecholamines content (Guerrero et al., 1990) in the brain of *Pygocentrus cariba* during the reproductive cycle. In the present study, annual variations in the plasma levels of 17ß-estradiol (17β-E) and testosterone (T) in the male and female *P. cariba* have been determined. Steroid levels have been correlated with the gonadosomatic index.

Results

In females, plasma 17β-E concentrations changed significantly during the year (Fig. 1).

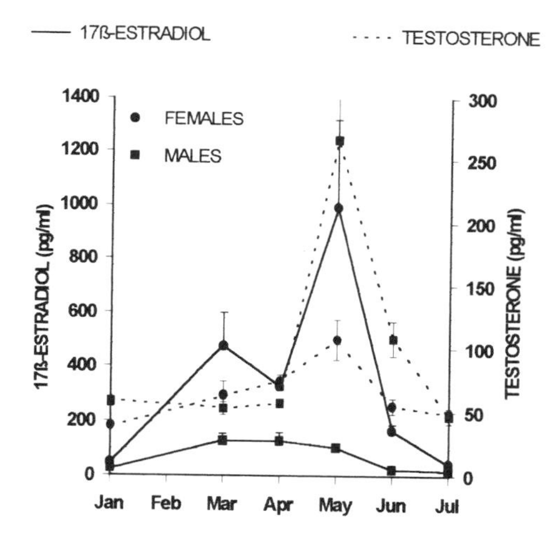

Figure 1. Plasma levels of T and 17β-E in male and female *P. cariba* during the annual reproductive cycle. Values represent means ±SEM (n=4-20).

Lower levels were found in January (post-spawning), June and July (pre-spawning and spawning respectively). Highest 17β-E levels were recorded in May. Testosterone was detectable in plasma at all sample times and the highest concentrations were found in females sampled in May. In males (Fig. 1), T was low in fish that were spent, regressed, or in early recrudescence and elevated at prespawning period. The maximun T concentration was two times higher than the maximum 17β-E concentration. Sex differences in T concentration were observed with a three-fold increase in the T levels in males compared to female levels. In contrast, plasma 17β-E levels were lower in males than in females. Low levels of 17β-E were found in the plasma of male fish captured in January, June and spent fish captured in July.

Discussion

The present study demonstrated significant differences in plasma concentrations of sexual steroids related to the sex or the physiological condition of the fish. In males, elevated testosterone prior to spawning may be related to spermatogenesis and spermiogenesis and could be responsible for the release of mature spermatozoa. The exact time of the gonadotropin (GTH) surge in *P. cariba* is unknown. However, we have shown that the content of both GnRH and GTH, was greatest the month before the maximal GSI was attained (Gentile et al., 1986). This suggests that, in this species, the GTH surge may occur just before spawning. The increase of 17β-E during the preparatory period suggests that this steroid may prime the pituitary and produce these changes in GnRH content (Gentile et al., 1986), resulting in successful spawning in this species. Positive effects of sex steroids on GTH have been reported for several fish species.

References

Gentile, F.,O. Lira, and D. Marcano-de Cotte. 1986. Relationship between brain gonadotropin releasing hormone (GnRH) and seasonal reproductive cycle of the caribe colorado, *Pygocentrus notatus*. Gen. Comp. Endocrinol. 64:239-245.

Guerrero, H., G. Cáceres, C.L. Paiva., and D. Marcano. 1990. Hypothalamic and telencephalic catecholamine content in the brain of the teleost fish, *Pygocentrus notatus*, during the annual reproductive cycle. Gen. Comp. Endocrinol. 80, 257-263.

Acknowledgments This work was supported by grants from the Consejo de Desarrollo Científico y Humanístico de la Universidad Central de Venezuela (CDCH 10-12-2770/94) and CONCIT-2427 to D.M.

ALLOZYME VARIATION IN TWO MORPHOLOGICAL TYPES OF THE SWAMP EEL, *Synbranchus marmoratus*, BLOCH.

G. Rey Vázquez, M.C. Abel, J.C. Vilardi and M.C. Maggese.

Dept. Cs. Biológicas, Fac. Cs. Exactas y Naturales, Universidad de Buenos Aires, (1428) Buenos Aires, República Argentina.

Summary

The swamp eel, *Synbranchus marmoratus*, from Corrientes Province (Argentina) exhibits two distinctive defined types with respect to head shape. With the aim of determining whether these phenotypes correspond to different populations, isoenzymes were used as genetic markers. EST, ADH, LDH and GOT were analysed in three tissues: liver, skeletal muscle and kidney. Morphological types were differentiated by both allelic frequencies and the number of alleles in some loci.

Introduction

Protein electrophoresis allows one to investigate genetic variation within populations by examining different genetic loci as expressed by their enzymatic gene products (Allendorf and Phelps, 1980). They also allow one to correlate biochemical traits with morphological and physiological effects.

The present paper describes a study in a sample of 40 adult fish from Corrientes, Argentina, where half individuals exhibited any of two different head shapes: round and flat. Four isoenzimatic systems esterase (EST), alcohol dehydrogenase (ADH), lactate dehydrogenase (LDH) and glutamate-oxaloacetate transaminase (GOT) were analysed in liver (L), skeletal muscle (M) and kidney (K) by horizontal electrophoresis on polyacrylamide gels. The loci detected were used as genetic markers to evaluate the degree of reproductive isolation between morphological types.

Results

Table 1 summarizes for each locus in wich tissues they were expressed, the number of alleles detected in each morphological type and the significance level (SL) of homogeneity chi square tests to compare allelic frequencies.

Discussion

Recently there has been some disagreement concerning the specific status of *S. marmoratus* as a valid species. Nakamoto et al (1986) proposed that at least to distinct species or population groups occur in the State of Sao Paulo associated with different geomorphological provinces.

We found two different head shapes (flat and round). Two measures, the maximum head width (MHW) and the body diameter estimated over the spiracle (BD), were employed to describe these morphological types (Abel et al, in press). The present allozymic study indicates a substantial genetic divergence between the two morphs.

The demonstration of genetic differences between the morphological types in *S. marmoratus* suggests a certain degree of reproductive isolation. The possible causes of such isolation are currently under study.

Table 1. Tissue specificities, number of alleles of each locus and significance level for the comparison of allelic frequencies between morphs. (20 individuals of each head shape were studied)

Locus	Tissues	N° alleles		SL
		Flat	Round	
Est-I	L,K	4	3	**
Est-II	L,K	3	4	**
Est-III	L,K	4	4	**
Ldh	M,L,K	1	1	--
Adh-I	M	1	1	--
Adh-II	L	2	2	**
Got-I	K	1	1	--
Got-II	L,K	3	3	NS

** highly significant; NS non significant; M: skeletal muscle L:liver K:kidney

References

Abel, M., G.Rey Vázquez, M.Maggese & J.Vilardi 1995. Esterase variation in two morphological types of the swamp eel, *Synbranchus marmoratus*, Bloch. Evolución Biológica, in press.

Allendorf, F. and S.R. Phelps, 1980. Loss of genetic variation in a hatchery stock of cut-throat trout. Trans. Am. Fish. Soc. 119: 537-543.

Nakamoto, W., P.Machado & F.Foresti 1986. Hemoglobin patterns in different populations of *Synbranchus marmoratus* Bloch, 1795 (Pisces, Synbranchidae). Comp. Biochem. Physiol. 84B, N°3: 377-381.

This work was carried out thanks to the finantial support of Universidad de Buenos Aires (Grant EX-208)

A MULTIDISCIPLINARY APPROACH TO THE STUDY OF THE REPRODUCTIVE ECOLOGY OF SINGLE AND MULTIPLE SPAWNING CYPRINIDS

J. Rinchard, P. Kestemont, R. Heine, J.C. Micha

Unité d'Ecologie des Eaux Douces, Facultés Universitaires N.-D. de la Paix, 61, rue de Bruxelles, B-5000 Namur, Belgium

Summary

The dynamics and regulation of oogenesis in single and multispawner fish have been investigated by a multidisciplinary approach including the seasonal profiles of gonadosomatic index, steroid hormones (estradiol-17ß, testosterone and 17,20ß-dihydroxy-4-pregnen-3-one) and vitellogenin levels correlated with the most advanced ovarian stages, fecundity and ultrastructure of the hepatocytes.

Introduction

To compare the dynamics and regulation of oogenesis in single and multispawner cyprinids, a multidisciplinary approach to reproduction has been undertaken with a special emphasis on three species from the river Meuse (Belgium) : the roach, *Rutilus rutilus*, a single spawner, and the bleak, *Alburnus alburnus*, and the white bream, *Blicca bjoerkna*, multiple spawners.

From 1992 to 1994, fish were collected in the River Meuse by electrofishing, gillnets or by regular control of a fish pass. Gonads were removed, weighed and submitted to histological routine (see Kestemont et al., this volume). GSI and histomorphometric analysis of oocyte growth were used to assess seasonal changes in ovarian development. Each gonad was classified according to the most advanced type of oocyte present (previtellogenesis, early endogenous vitellogenesis, late endogenous vitellogenesis, exogenous vitellogenesis, final maturation, intermediate multispawning stage and post-spawning). Serum levels of testosterone (T), estradiol-17ß (E2) and 17,20ß-dihydroxy-4-pregnen-3-one (17,20ßP) were measured by radioimmunoassay. Vitellogenin levels were assessed using an indirect method based on the content of plasma protein phosphorus (PPP). Seasonal changes of hepatocyte ultrastructure (abundance and status of mitochondria and rough endoplasmic reticulum) were studied by electron microscopy. Total and fractional fecundities were determined by counting the mature oocytes in single and multiple spawner fish respectively.

Results and discussion

A short spawning season, represented by a steep GSI curve with a single peak and highest value (from 18.4 to 21.6%) is characteristic of roach. On the other hand, depending on sampling years, maximum GSI of multispawners reached 12.8 to 17% and 14.5 to 17.7% for white bream and bleak respectively and decreased progressively during the protracted spawning season. Before the spawning period, levels of PPP increased in multispawners but remained low in roach. After the first spawning, the PPP and E2 contents decreased in white bream and roach while high levels were found in bleak during the whole spawning season. This latter pattern has also been observed in other multispawning fish as *Carassius auratus* (Kagawa et al., 1983) and *Dicentrarchus labrax* (Prat et al., 1990). The absolute value concentrations of 17,20ßP observed during final maturation were very low in the three species compared with those in salmonid fish. As reported in *Carassius auratus* (Kagawa et al., 1983), the follicle might have a lower capacity to produce this steroid. On the other hand, 17,20ßP would be rapidly metabolized and present only for a brief period and rapidly deactivated. A peak is unlikely to be found during weekly sampling unless it coincides closely with spawning. Although liver changes can not be related to vitellogenin synthesis only, ultrastructural observations of hepatocytes showed that vitellogenic activity seemed limited to the prespawning period in roach and white bream but remained intense during the whole spawning season in bleak. The hepatocytes contained extensive areas of rough endoplasmic reticulum and numerous mitochondria. As opposed to bleak which maintained a high production of eggs during the whole reproductive period, the white bream concentrated most of its enery into first spawning act (Table 1).

Table 1 : Relative fecundity (number of mature oocytes/kg fish ± SEM) of roach, white bream and bleak

Species	Year	1st spawning	Next spawnings
roach	92	138108±5083	-
	93	114350±11356	-
white bream	92	149802±26625	60472±9940
	93	136667±4564	60262±11023
bleak	92	135396±8735	82215±9452
	93	81033±18224	66642±17499

In conclusion, different patterns in the reproductive indicators were observed between single and multispawners but also among multispawners. In white bream, differentiation of vitellogenic follicles from smaller oocytes was completed before the onset of spawning season. Oogenesis dynamics of this species appears relatively similar to the one of roach while in bleak, recruitment of vitellogenic follicles occurs during the whole reproductive period.

References

Kagawa, H., Young, G. and Nagahama, Y., 1983. Changes in plasma steroid hormone levels during gonadal maturation in female goldfish (*Carassius auratus*). Bull. Jpn. Soc. Fish., 49: 1783-1787

Prat, F., Zanuy, S., Carillo, M., De Mones, A. and Fostier, A., 1990. Seasonal changes in the plasma levels of gonadal steroids of sea bass *Dicentrarchus labrax* L. Gen Comp. Endocrinol., 78: 361-373.

DEPLETION OF FATTY ACID RESERVE DURING EMBRYONIC DEVELOPMENT IN THE MEDAKA (ORYZIAS LATIPES)

Karen Charleson and Murray D. Wiegand †

Department of Biology, University of Winnipeg, Winnipeg, MB, CANADA R3B 2E9

Introduction

Yolk lipid provides energy, structural fatty acid and other essential nutrients to fish embryos. The medaka is a useful model for the study of fish embryonic and reproductive physiology but its embryonic lipid metabolism has not been extensively examined.

Materials and Methods

A colony of medaka was held at 26°C and 16L/8D photoperiod. Spawning occurred shortly after lights on. Eggs were collected within an hour and were incubated at 25°C and 16L/8D. Samples were collected at fertilization, 96 hr (eyed stage) and hatching. Neutral lipids (NL) and neutral phospholipids (phosphatidyl choline + phosphatidyl ethanolamine, PC+PE) were isolated from lipid extracts on aminopropyl columns (Kim & Salem 1990). Fatty acid methyl esters were analyzed by gas chromatography (Moodie et al. 1989). Fatty acid mass was quantified by internal standard.

Results and Dicussion

1096 eggs were collected from 77 spawnings. Fertility averaged 90.2%. Total mortality was 13% between fertilization and hatch. Hatching occurred between 196 and 338 hr post fertilization with a mean of 237 hr and a mode of 216 hr.

There was a slight, significant decline in PC+PE fatty acid during development and a major decline in NL fatty acid after 96 hr. (Table 1).

Table 1. Mass of total lipid (TL), NL and PC+PE fatty acids in developing medaka.

Sample	TL	NL	PC+PE
eggs	43±3[1]	27±1	11±1
96 hr	42±3	26±2	9±1
hatch	22±2	11±1	7±0.5

[1] μg fatty acid/embryo or larva ± SD.

Medaka eggs were unusual in their high content of 18:2(n-6) and low levels of 20:5(n-3) (Table 2). PC+PE was richer in (n-3) polyunsaturated fatty acids and saturated fatty acids (SFA) and poorer in monounsaturated fatty acids (MUFA) than the NL.

Table 2. Major fatty acids in medaka egg TL, NL and PC+PE.

Fatty Acid	TL	NL	PC+PE
16:0	16.6	15.1	20.6
18:0	6.1	4.9	8.1
Σ SFA	24.7	22.0	30.5
18:1(n-9)	16.0	17.9	10.3
18:1(n-7)	4.7	5.0	3.8
Σ MUFA	25.8	28.8	16.7
18:2(n-6)	18.0	20.5	12.0
20:4(n-6)	3.6	2.3	4.5
Σ (n-6)	26.5	27.7	21.9
20:5(n-3)	0.5	0.5	0.5
22:6(n-3)	13.9	12.6	21.8
Σ (n-3)	18.1	17.0	25.4

Values are mean weight percent for four assays. Unknowns averaged 5%.

By hatching, the largest changes in fatty acid profile were in PC+PE, where 16:0 increased to 25.3% and 18:2(n-6) decreased to 7.2%. In the total lipids, 22:6(n-3) increased slightly to 15.3%. Unlike most other fish embryos studied to date, fatty acid consumption in medaka was largely non-specific. MUFA were not preferentially depleted and selective retention of 22:6(n-3) was slight.

References

Kim, H-Y. & Salem, N. (1990) Separation of lipid classes by solid phase extraction. J. Lipid Res. **31**, 2285-2289.

Moodie, G.E.E., Loadman, N.L., Wiegand, M.D. & Mathias, J.A. (1989) Influence of egg characteristics on survival, growth and feeding in larval walleye (Stizostedion vitreum). Can. J. Fish. Aquat. Sci. **46**, 516-512.

† Author to whom correspondence should be addressed

Behavior

HORMONES AS SEX PHEROMONES IN FISH: WIDESPREAD DISTRIBUTION AMONG FRESHWATER SPECIES

N. E. Stacey and J. R. Cardwell

Department of Biological Sciences, University of Alberta, Edmonton, Alberta, CANADA T6G 2E9

Summary

Hormonal pheromones have evolved in seven orders of freshwater fishes and are widely distributed among at least two of these (Cypriniformes, Characiformes). Fish hormonal pheromone systems are diverse, electro-olfactogram surveys having detected species responsive to combinations of free, glucuronidated, and sulphated C21, C19 and C18 steroids as well as prostaglandins. Surprisingly, the compounds detected vary little within most lower taxa (genera, tribes) examined, suggesting that knowledge of hormonal pheromone systems in a few key species can be transferred directly to basic and applied research in a great number of species

Introduction

Studies of hormones and reproductive behavior in tetrapods have examined the mechanisms by which gonadal steroids not only regulate development and activation of adult behaviors, but also respond to socio-sexual stimuli (Becker et al., 1992). Although similar phenomena have been demonstrated in fish (Mayer et al., 1994; Olsen & Liley, 1993), studies over the past 15 years show that fish have evolved more than these classical hormone-behavior interactions. In many fish, gonadal hormones and their metabolites, act not only as internal chemical signals, but also are released to act as external chemical signals (*hormonal pheromones*) with actions on the physiology and behavior of conspecifics (Sorensen, 1992; Stacey et al., 1994a; Bjerselius et al., 1995a,b). In revealing this new dimension of fish physiology, hormonal pheromones influence basic concepts of how endocrine and neural systems interact to organize and synchronize reproductive function.

First, hormonal pheromones illustrate the speed with which change in individual hormonal status can be signalled to conspecifics. Typically, steroid hormones act indirectly through slowly responding somatic effectors to generate sexually dimorphic visual, chemical, auditory (Brantley & Bass, 1991), or electrical (Zakon, this volume) signals reflecting the signaller's *past* hormonal status. With hormonal pheromones, on the other hand, the delay between synthesis and release is such ongoing hormonal change can be directly monitored by conspecifics via release of the hormone or its metabolites..

Second, Cardwell et al. (1995) show that in a cyprinid (Poropuntius schwanenfeldi) androgen treatment of juveniles increases olfactory receptor response to pheromonal prostaglandin (PG). This finding illustrates a novel mechanism whereby steroids induce sexually dimorphic CNS function, through sensory gating of a chemical signal that can in turn change steroid levels.

Third, hormonal pheromones might cause us to reconsider what we define as the "hormonal system" regulating reproduction. Traditionally the unit of a species' reproductive function has been the individual's hypothalamo-pituitary-gonadal (HPG) axis, capable of internally-generated tonic and cyclic function, but also responsive to exogenous stimuli. But because released conspecific hormones can be effective exogenous stimuli, the reproductive unit might be expanded to include pheromonal feedback actions of the individual's released hormones, either directly via autostimulation, or indirectly by stimulating hormonal pheromone release by conspecifics (Fig. 1).

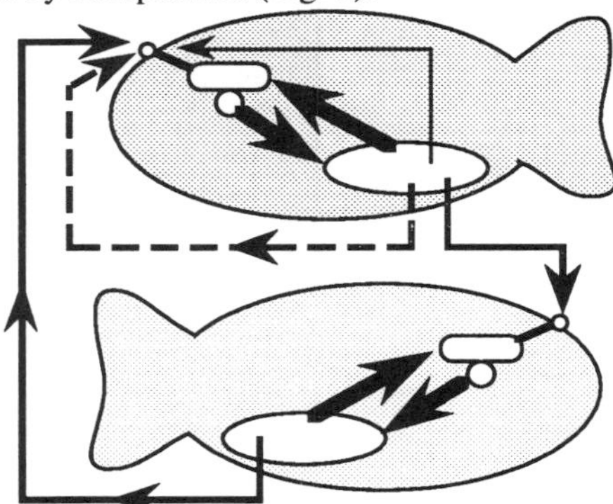

Fig. 1. Hormonal control of reproduction in a fish (top) can involve both *endogenous* hormonal mechanisms (HPG interactions - thick lines; steroid regulation of olfactory receptors - thin line), and *exogenous* olfactory actions of released hormones that can operate either directly (dashed line) or indirectly by changing hormone release from conspecifics.

Although no fish is known either to change its hormone levels by olfactory monitoring of their release, or to change a conspecific's hormonal pheromone release by changing its own, studies of pheromonal 17a,20ß-dihydroxy-4-pregnen-3-one (17,20ß-P) in goldfish (Carassius auratus) suggest this can occur. Goldfish increase ovulatory rate if they are exposed to 17,20ß-P, demonstrating the ability to respond to a pheromone they release (see Stacey et al., 1994a). As well, 17,20ß-P-exposed males increase blood 17,20ß-P (Dulka et al., 1987) and, presumably, 17,20ß-P release. Therefore, to understand reproductive function of species using hormonal pheromones, the individual should be viewed not in isolation, but as one portion of a 'hormonal system' comprised of those conspecifics chemically linked by water-borne hormones.

Known hormonal pheromone systems

Currently, there are only nine fish species for which there is more than preliminary evidence that released hormones and their metabolites function as sex pheromones: black goby (Gobius jozo; Colombo et al., 1982), zebrafish (Danio rerio; Van Den Hurk & Resink, 1992), African catfish (Clarias gariepinus; Van Den Hurk & Resink, 1992), Atlantic salmon (Salmo salar; Moore and Scott, 1992), goldfish (Stacey et al., 1994a), crucian carp (Carassius carassius; Bjerzelius et al., 1995a), common carp (Cyprinus carpio; Irvine & Sorensen, 1993; Stacey et al., 1995), Japanese weatherloach (Misgurnus anguillicaudatus; Kitamura et al., 1994a; Ogata et al., 1994), and tinfoil barb (Poropuntius; Cardwell et al., 1995. Studies of these species provide evidence for at least two of the following criteria: pheromone synthesis, pheromone release, olfactory detection, biological response. Only for goldfish have all four criteria been met. Although few in number, the species in this list are diverse both taxonomically (Cypriniformes: Danio, Carassius, Cyprinus, Poropuntius and Misgurnus; Siluriformes: Clarias; Perciformes: Gobius; Salmoniformes: Salmo), and in the chemistry of their proposed pheromones (prostaglandins [PGs]; free, glucuronidated and sulphated C21 steroids; free and glucuronated C19 steroids, and glucuronated C18 steroids). This group of species also has diversity in pheromonal function, exhibiting not only cases in which females stimulate males (Carassius, Cyprinus, Poropuntius, Misgurnus, Salmo) or males stimulate females (Gobio, Clarias), but also in Danio apparent production and response by both sexes. And in goldfish, as noted above, evidence that 17,20ß-P increases both ovulation rate and milt production suggests that this steroid has both intra- and intersexual pheromonal functions (Stacey, 1991). The proposed pheromones in these nine species have been reviewed extensively (Sorensen, 1992; Sorensen & Goetz, 1993; Stacey & Sorensen, 1991; Stacey et al., 1994a) and therefore will not be covered further here.

Two factors should determine the potential impact and contribution of hormonal pheromones to fish reproduction research:

1. the likelihood that they are practical culture and management tools for commercially important species:
   -although there is evidence for hormonal pheromones in species of commercial interest (Salmo, Cyprinus, and recently Oncorhynchus - Dittman & Quinn, 1994) and a report of pheromone-induced milt increase in Cyprinus (Stacey et al., 1994b), no attempt has been made to manipulate reproduction with pheromones under culture conditions.
2. the degree to which research on a few model species can reveal principles of direct relevance to related taxa:
   -as yet, there is good evidence for hormonal pheromones in the nine species described above, but only preliminary reports comparing pheromone systems in related taxa (e.g. Stacey et al., 1994a).

Distribution and diversity of hormonal pheromones systems: an electro-olfactogram survey

To assess the prevalence and nature of hormonal pheromones, we are using electro-olfactogram (EOG) recording to determine olfactory response of fresh-water fish to approximately 10 PGs and 150 steroids, assuming for simplicity that olfactory responsiveness is indicative of pheromonal function. Although the close correlations between EOG responses and biological responses in goldfish and common carp (Irvine and Sorensen, 1993; Sorensen et al., 1990; Stacey et al., 1994b) support this assumption, we acknowledge that some of our reported responses may indicate non-pheromonal functions. Limited space prevents listing all compounds tested in our EOG survey; however, they include the F-series PGs and many of their 15-keto and 13,14-dihydro metabolites (Cayman), and available (Sigma, Steraloids) free, glucuronated and sulphated C21 (n = 105), C19 (n = 34) and C18 (n = 9) steroids that might be likely candidates for pheromonal activity. This approach will generate 'false negatives' for any species with hormonal pheromones entirely different from our test compounds, and thus should provide a conservative estimate of the extent of hormonal pheromone evolution.

When complete, EOG recording (Cardwell et al., 1995) for each species will include screening with all PGs (10^{-8} M) and steroids (10^{-9} or 10^{-8} M), threshold determination for the most potent compounds, and cross-adaptation experiments (Cardwell et al., 1995) to determine the number of olfactory receptor mechanisms present. As this analysis is complete for only a few species, this preliminary report simply presents a summary of detected compounds arranged by chemical structure (Tables 1 and 2). It must be kept in mind that there is no simple relationship between number of chemical categories detected and number of receptor mechanisms: e.g. in goldfish, free and glucuronated 17,20ß-P are detected by the same receptor mechanism (Sorensen et al., in press). In other cases, there is evidence for more than one receptot mechanism within one of our arbitrary chemical categories.

Hormonal pheromones at the order level

Although EOG studies have begun on ten orders (Table 1), in all but two (Cypriniformes, Characiformes) very few species have been examined. In particular, the three orders in which no EOG responses were observed are represented by only a single species, although absence of EOG response to PG in Acipenseriformes (sturgeon) has been reported recently by Kitamura et al. (1994b). Despite the small number of groups examined, however, EOG responses indicate a wide distribution of hormonal pheromone systems, from relatively primitive (Elopiformes, tarpons) to advanced groups (Perciformes).

In the largely marine Elopiformes and Perciformes, EOG responses in two euryhaline species (Megalops cyprinoides; Neogobius melanostoma; see Murphy et

Table 1. Orders which have been tested for electro-olfactogram (EOG) responses to prostaglandins and steroids [a].

ORDERS [b]	No. species	PG	C21			C19			C18		
			F	G	S	F	G	S	F	G	S
Acipenseriformes	1[c]	- [g]	-	-	-	-	-	-	-	-	-
Osteoglossiformes	1	-	-	-	-	-	-	-	-	-	-
Elopiformes	1	+[h]	+	-	+	-	-	-	-	-	-
Ostariophysi											
Gymnotiformes [d]	1	+	-	-	-	-	-	-	-	-	-
Siluriformes [e]	3	+	+	+	-	-	+	-	-	-	-
Cypriniformes	70	+	+	+	+	+	+	-	+	+	-
Characiformes	25	+	+	+	+	+	+	-	+	-	+
Salmoniformes [f]	5	+	-	-	+	+	-	-	-	-	-
Gadiformes	1	-	-	-	-	-	-	-	-	-	-
Perciformes	6	-	+	-	-	-	+	+	+	+	+

PG = prostaglandins; C21, C19 and C18 = 21-, 19-, and 18-carbon steroids respectively; F = unconjugated; G = glucuronated; S = sulphated; [a] = not all species were tested with all categories of compounds; [b] = after Nelson, 1994; [c] = number of species tested shown in parentheses; [d] = J.G. Dulka, unpublished; [e] = includes data reviewed by Van Den Hurk and Resink, 1992; [f] = includes data from Moore & Scott, 1992 and Kitamura et al., 1994b; [g] = dash indicates compounds in this category not detected; [h] = plus indicates compounds in this category detected by at least one species.

Table 2. Electro-olfactogram responsiveness to prostaglandins and steroids in order Cypriniformes [a].

Superfamily / FAMILY / Sub-family *Tribe* Sub-tribe	No. species	PG	C21			C19			C18		
			F	G	S	F	G	S	F	G	S
Cyprinoidea											
CYPRINIDAE[b]											
Acheilognathinae	1	+[d]	-[e]	-	-	-	-	-	-	-	-
Danioninae	17	+	(+)[f]	(+)	(+)	-	(+)	-	-	-	-
Leuciscinae	8	+	+	+	+	-	-	-	-	-	-
Cyprininae *Squaliobarbini*	2	+	+	+	+	-	-	-	-	-	-
Cyprinini											
Cyprini	2	+	+	+	+	+	-	-	-	-	-
Tor	2	+	+	+	+	-	-	-	-	-	-
Systomini	20	+	+	+	+	(+)	(+)	-	(+)	-	-
Labeonini	15	+	(+)	-	+	(+)	-	-	-	-	-
Cobitoidea											
GYRINOCHEILIDAE	1	+	-	-	-	-	-	-	-	-	-
CATOSTOMIDAE	5	+	-	-	-	-	-	-	-	-	-
COBITIDAE[c]	1	+									

Prostaglandin and steroid abbreviations as in Table 1. [a] = Cypriniformes classification after Nelson, 1994; [b] = Cyprinidae classification after Rainboth, 1991; [c] = from Kitamura et al., 1994b - responses to steroids not examined in this study; [d] = compounds in this category detected by all species in taxon; [e] = compounds in this category not detected by any species in taxon; [f] = compounds in this category detected by only some species in taxon.

al., this volume) suggest hormonal pheromones are not restricted to freshwater species. Interestingly, etiocholanolone glucuronide, detected by Neogobius, also was proposed as a pheromone in the related Gobius jozo (Colombo et al., 1982).

All four Ostariophysin orders detected PGs and three detected at least some steroids (Table 1). We presently are focussing on Cypriniformes and Characiformes because these orders contain many commercially important species, and because a diversity of species is readily available. Although this paper presents only our results on Cypriniformes, a similar picture is developing in Characiformes (Cardwell et al., this volume and unpublished).

Hormonal pheromones within order Cypriniformes

Of the five familes recognized in order Cypriniformes (Nelson, 1994; Table 2), we have examined species from Cyprinidae (carps and minnows), Gyrinocheilidae (algae eaters) and Catostomidae (suckers), and Kitamura et al. (1994a) have recorded from one species of family Cobitidae (loaches). Family Balitoridae (river loaches) remains unexplored.

Every cypriniform species examined responds to PGs, PGF2α, 15-keto-PGF2α, or 13,14-15-keto-PGF2α being the most potent. So far, responses to steroids have been observed only in superfamily Cyprinoidea, although few species from superfamily Cobitoidea have been tested. At least for family Catostomidae (suckers), however, the lack of response to steroids is likely meaningful, since the five species examined represent the three recognized subfamiles (Cycleptinae; Myxocyprinus: Ictiobinae; Ictiobius: Catostominae; Catostomus, Moxostomus).

Only four of the the seven recognized cyprinid subfamilies (Rainboth, 1991) have been examined, and only two (Danioninae and Cyprininae) in any detail. Leuciscins examined include both Asian (bighead carp, Hypophthalmichthyes nobilis) and North American species (e.g. Compostoma, Notropis), but much of the North American diversity remains unexplored.

Two Cyprinid taxa noteworthy for the specificity of their steroid receptors are subfamily Danioninae and tribe Labeonini . Of 17 Danioninae tested, 9 (Rasbora and Luciosoma spp) detected no steroids, 2 (Rasbora) detected only 17,20ß-P, and 4 (Leptobarbus and Danio) detected only 17,20ß-P-sulphate. Of the remaining 2 species, one (Leptobarbus) detected 17,20α-P in addition to 17,20ß-P-sulphate, and the other (Rasbora) detected glucuronated C21, C19 and C18 steroids, all of which evidently act through one receptor mechanism that is most sensitive to 17,20ß-P-glucuronide. In contrast to earlier reports (Van Den Hurk & Resink, 1992) that C19 and C18 glucuronides have pheromonal activity in Danio rerio, we observed responses only to 17,20ß-P-sulphate in this and two congeners (D. albolineatus and malabaricus.

In tribe Labeonini, all 15 species tested responded to 17,20ß-P-sulphate, which for 11 of these species (4 genera) was the only steroid detected. All 4 of the remaining species (Labeo) also detected free C21 steroids, through a broadly-tuned receptor that also detects 17a-androstendione.

Unlike Danionini and Labeonini, tribes Cyprinini (Carassius, Cyprinus, Probarbus, Tor) and Systomini (Hampala, Poropuntius, Puntius), responded not only to a wide variety of C21 and C19 steroids, but also to many of their reduced and/or conjugated metabolites. Within tribe Systomini is the only cyprinid species (Puntius lateristriga) known to have a receptor mechanism specifically responsicve to C18 steroids. It is not yet known if the pheromonal systems of Cyprinini and Systomini are more complex than those of Danioninae and Labeonini, or whether their olfactory receptor mechanisms are simply less specific.

Summary and prospectus

Although our EOG survey has only begun, patterns are emerging that provide insight into hormonal pheromone evolution. The most obvious is that olfactory responsiveness to water-borne hormones and metabolites is not distributed randomly among the species examined, but is highly correlated with accepted phylogenies. There is considerable diversity among and within higher taxa (order, family), but at lower taxonomic levels (tribe, subtribe, genus), this virtually disappears: e.g. common carp and goldfish (Irvine and Sorensen, 1993); Danioninae and Labeonini (Table 2). The most parsimonius interpretation is that olfactory responses to water-borne hormones have evolved relatively few times and diversified slowly.

A second pattern is that, except for Perciformes, responsiveness to PGs is far more common than is responsiveness to steroids. This prevalence of PG pheromones might simply reflect the fact that released PGs, signalling ovulation, should be important cues in any mating system, whereas released steroids should be important only in certain situations (Stacey & Sorensen, 1991). However, leuciscin and cyprinin species with vastly different mating systems (broadcast spawning vs. male nesting and parental behavior) appear to use similar steroidal pheromones.

A third generalization emerging from EOG studies of hormonal pheromones is that if these systems have species specificity within lower taxa, specificity is likely to be achieved through subtle variations in pheromonal mixtures rather than gross differences in the compounds used. No study has yet tackled this difficult and important problem directly, though the techniques are now available and many appropriate species groups have been identified.

Finally, our findings emphasize the degree to which understanding of hormonal pheromones depends on knowledge of released hormones. Because our current views have been so heavily influenced by the recent discovery of sulphated steroid pheromones (Scott & Vermeirssen, 1994), it is expected that further research in hormone metabolism will lead to equally exciting advances in the study of hormones as pheromones.

References

Becker, J.B., S.M. Breedlove and D. Crews, 1992. Behavioral Endocrinology, MIT Press, Cambridge MA, 574 p.

Brantley, R.K. and A.H. Bass, 1991. Secondary sex characters in a vocalizing fish: intra- and intersexual dimorphism and role of androgens. In: Proc. Fourth Intl. Symp. Reprod. Physiol. Fish, A.P. Scott, J.P. Sumpter, D.E. Kime, and M.S. Rolfe (Eds.), FishSymp 91, Sheffield, p. 197-199.

Bjerselius, R., K.H. Olsen and W. Zheng, 1995a. Endocrine, gonadal and behavioral responses of male crucian carp to the hormonal pheromone 17α,20ß-dihydroxy-4-pregnen-3-one. Chem. Senses 20: 221-230.

Bjerselius, R., K.H. Olsen and W. Zheng, 1995b. Behavioral and endocrinological responses of mature male goldfish to the sex pheromone 17α,20ß-dihydroxy-4-pregnen-3-one in the water. J. Exp. Biol. 198: 747-754.

Cardwell, J.R., N.E. Stacey, E.S.P. Tan, D.S.O. McAdam and S.L.C. Lang, 1995. Androgen increases olfactory receptor response to a vertebrate sex pheromone. J. Comp. Physiol. A 176: 55-61.

Colombo, L., P.C. Belvedere, A. Marconato and F. Bentivegna, 1982. Pheromones in teleost fish. In: Proc. Second Intl. Symp. Reprod. Physiol. Fish, C.J.J. Richter and H.J.Th. Goos (Eds.), Pudoc, Wageningen, p. 84-94.

Dittman, A.H. and T.P. Quinn, 1994. Avoidance of a putative pheromone, 17α,20ß-dihydroxy-4-pregnen-3-one, by precociously mature spring salmon (Oncorhynchus tshawytscha). Can. J. Zool. 72: 215-219.

Dulka, J.G., N.E. Stacey, P.W. Sorensen and G.J. Van Der Kraak, 1987. A sex steroid pheromone synchronizes male-female spawning readiness in goldfish. Nature 325: 251-253.

Irvine, I.A.S. and P.W. Sorensen, 1993. Acute olfactory sensitivity of wild common carp, Cyprinus carpio, to goldfish sex pheromones is influenced by gonadal maturity. Can. J. Zool. 71: 2199-2210.

Keller, C.H., H.H. Zakon and D.Y. Sanchez, 1986. Evidence for a direct effect of androgens upon electroreceptor tuning. J. Comp. Physiol. A 158: 301-310.

Kitamura, S., H. Ogata, and F. Takashima, 1994a. Activities of F-type prostaglandins as releaser sex pheromones in cobitid loach, Misgurnus anguillicaudatus. Comp. Biochem. Physiol. A 107A: 161-169.

Kitamura, S., H. Ogata and F. Takashima, 1994b. Olfactory responses of several species of teleost to F-prostaglandins. Comp. Biochem. Physiol. A 107A: 463-467.

Mayer, I., N.R. Liley and B. Borg, 1994. Stimulation of spawning behavior in castrated rainbow trout (Oncorhynchus mykiss) by 17α,20ß-dihydroxy-4-pregnen-3-one, but not by 11-ketoandrostenedione. Horm. Behav. 28: 181-190.

Moore, A and A.P. Scott, 1992. 17α,20ß-dihydroxy-4-pregnen-3-one-20-sulphate is a potent odorant in precocious male Atlantic salmon parr which have been pre-exposed to the urine of ovulated females. Proc. Roy. Soc. Lond. Ser. B Biol. Sci. 249: 205-209.

Nelson, J.S., 1994. Fishes of the World (3rd ed.), Wiley, New York, 600 p.

Ogata, H., S. Kitamura and F. Takashima, 1994. Release of 13,14-dihydro-15-keto-prostaglandin F2α, a sex pheromone, to water by cobitid loach following ovulatory stimulation. Fish. Sci. 60: 143-148.

Olsen, K.H. and N.R. Liley, 1993. The significance of olfaction and social cues in milt availability, sexual hormone status, and spawning behavior of male rainbow trout (Oncorhynchus mykiss). Gen. Comp. Endocrinol. 89: 107-118.

Rainboth, W.J., 1991. Cyprinids of South East Asia. In: Cyprinid Fishes: Systematics, Biology and Exploitation, I.J. Winfield and J.S. Nelson (Eds.): Chapman and Hall, London, p. 156-210.

Scott, A.P. and E.L.M. Vermeirssen, 1994. Production of conjugated steroids by teleost gonads and their role as pheromones. In: Perspectives in Comparative Endocrinology, K.G. Davey, R.E. Peter, and S.S. Tobe (Eds.): National Research Council, Ottawa, p.645-654.

Sorensen, P.W., 1992. Hormones, pheromones, and chemoreception. In: Fish Chemoreception, T.J. Hara (Ed.), Chapman and Hall, London, p. 199-221.

Sorensen, P.W. and F.W. Goetz, 1993. Pheromonal and reproductive function of F-prostaglandins and their metabolites in teleost fish. J. Lipid Mediators 6: 385-393.

Sorensen, P.W. and A.P. Scott, 1994. The evolution of hormonal sex pheromone systems in teleost fish: poor correlation between the pattern of steroid release by goldfish and olfactory sensitivity suggests that these cues evolved as a result of chemical spying rather than signal specialization. Acta Physiol. Scand. 152: 191-205.

Sorensen, P.W., T.J. Hara, N.E. Stacey and J.G. Dulka, 1990. Extreme olfactory specificity of male goldfish to the preovulatory steroidal pheromone 17α,20ß-dihydroxy-4-pregnen-3-one. J. Comp. Physiol. A 166: 373-383.

Sorensen, P.W., A.P. Scott, N.E. Stacey and L. Bowdin. Sulfated 17α,20ß-dihydroxy-4-pregnen-3-one functions as a potent and specific olfactory stimulant with pheromonal actions in the goldfish. Gen. Comp. Endocrinol. (in press).

Stacey, N.E., 1991. Hormonal pheromones in fish: status and prospects. In: Proc. Fourth Intl. Symp. Reprod. Physiol. Fish, A.P. Scott, J.P. Sumpter, D.E. Kime, and M.S. Rolfe (Eds.), FishSymp 91, Sheffield, p. 177-181.

Stacey, N.E. and P.W. Sorensen, 1991. Function and evolution of fish hormonal pheromones. In: The Biochemistry and Molecular Biology of Fishes, Vol. 1, P.W. Hochachka and T.P. Mommsen (Eds.), Elsevier, Amsterdam, p. 109-135.

Stacey, N.E., J.R. Cardwell, N.R. Liley, A.P. Scott and P.W. Sorensen, 1994a. Hormonal sex pheromones in fish. In: Perspectives in Comparative Endocrinology, K.G. Davey, R.E. Peter, and S.S. Tobe (Eds.), National Research Council, Ottawa, p. 438-448.

Stacey, N.E., W. Zheng and J.R. Cardwell, 1994b. Milt production in common carp (Cyprinus carpio): stimulation by a goldfish steroid pheromone. Aquaculture 127: 265-276.

Van Den Hurk, R. and J.W. Resink, 1992. Male reproductive system as pheromone producer in teleost fish. J. Exp. Zool. 261: 204-213.

A PHEROMONE IN FEMALE TROUT URINE.

Etiënne L. M. Vermeirssen[1], Alexander P. Scott[1], John P. Sumpter[2], and Francisco Prat[2]

[1]MAFF, Fisheries Laboratory, Pakefield Road, Lowestoft, Suffolk, NR33 0HT, and [2]Department of Biology and Biochemistry, Brunel University, Uxbridge, Middlesex, UB8 3PH, United Kingdom

Summary

When male rainbow trout (*Oncorhynchus mykiss*) are exposed to female urine, plasma levels of GtH II rise within 5 min and levels of 17,20β-P between 20 and 30 min. However, 30% of the males show no response. This lack of response cannot be correlated with initial levels (i.e. pre-urine exposure) of cortisol, 17,20β-P, GtH II or GSI.

Introduction

Several studies have shown that female trout release an attractant (releaser pheromone) via the ovarian fluid.

Recently, Olsén and Liley (1993) also showed that female trout produce a priming pheromone, stimulating milt availability and levels of sex steroids in males. A yet more recent study showed that this priming pheromone is likely to be released by the females in their urine (Scott *et al.*, 1994).

In this study, we address two questions posed by our previous study: is GtH II involved in stimulating the rise in 17α,20β-dihydroxy-4-pregnen-3-one (17,20β-P) levels which occurs when males are exposed to female urine; what causes the apparent non-responsiveness of approximately 30% of the males to female urine?

We sought to answer the first question by collecting a series of blood samples from males as soon as possible after exposing them to urine and then radioimmunoassaying them for GtH II and 17,20β-P. We sought to answer the second question by correlating levels of 17,20β-P with levels of cortisol and the size of the gonads (GSI).

Materials and Methods

Urine collection

Preovulatory females were cannulated via the urinary bladder and urine was collected for 3 days. It was stored at -20° C until used in the experiment (for urine collection methods see: Scott *et al.*, 1994).

Experiment

Spermiating male trout were brought from a trout farm and left to recover for 2 weeks in outdoor tanks. Ten males were cannulated via the dorsal aorta and left to recover for 18 h. Each fish was held in a net compartment (0.5 x 0.4 x 0.2 m) in a separate tank with a separate water supply. A blood sample was taken at 0 min, after which a fish was either exposed to 10 ml tank water (controls) or 1 ml of female urine in 9 ml tank water. Solutions were squirted under water at the head of the fish. Further blood samples were taken at: 5, 10, 20, 30, 45, 60, 90, 120, 180 and 300 min. The fish were killed after the last blood sampling, and body weight and gonad weight were measured.

The procedure was repeated three times and 19 series of blood samples were obtained from fish squirted with female urine, and eight from fish squirted with water.

Radioimmunoassays

17α,20β-dihydroxy-4-pregnen-3-one was assayed as described by Scott *et al.* (1994); cortisol was assayed using antiserum C001 from Steranti Research Ltd., St. Albans, Herts., UK, and [1,2,6,7-^{3}H]-cortisol from Amersham Int. plc, Little Chalfort, Bucks., UK; GtH II was measured as described by Sumpter and Scott (1989), with the modification that an antibody was used raised specifically against the β-subunit of GtH II.

Statistics

Hormone and GSI data were transformed logarithmically. Treatments were compared with control levels using Student's *t*-test.

Results

17α,20β-dihydroxy-4-pregnen-3-one

Data from the control fish (n = 8) were pooled. Data from the urine-treated fish (n = 19) were split into two groups: Group 1 - fish whose 17,20β-P level rose to more than double the initial level (responders; n=12); Group 2 - fish with 17,20β-P levels that never rose above double the initial level (non-responders; n=7).

Moreover, two fish were excluded from Group 2 on the grounds of outlying initial levels of 17,20β-P (23 rising to 30 ng/ml and 44 rising to 73 ng/ml; initial mean of non-responders ± S.E.M.: 5.2 ± 2.5 ng/ml). Data are shown in Figure 1.

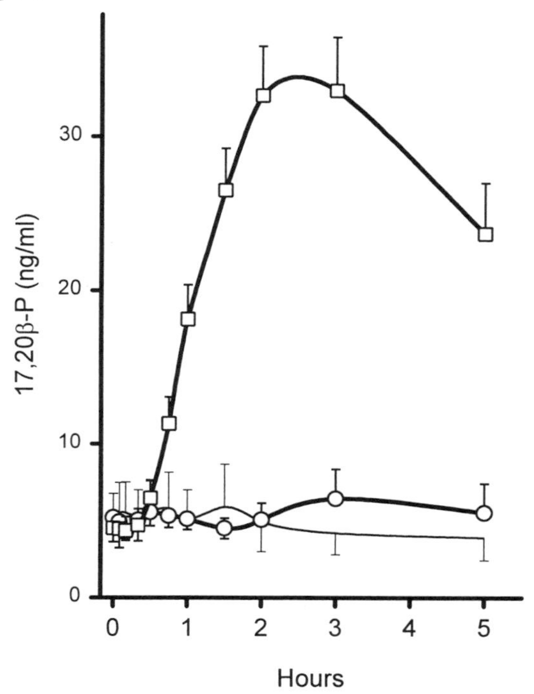

Fig. 1. The effect of a single squirt of 1 ml female urine or tank water on levels of 17,20β-P in males (mean ± S.E.M.; □ = responders, n=12; ○ = controls; 'no symbol' = non-responders, n=5).

Levels of 17,20β-P started to rise between 20 and 30 min and were significantly elevated over control levels at 45 min ($P<0.005$). Levels peaked at *c.* 3 h after the exposure to female urine. Levels in control and non-responding fish were unchanged over 5 h.

Gonadotrophin II

GtH II levels in the same three groups of fish are shown in Figure 2; note that, the two fish with outlying 17,20β-P levels have still been excluded from these plots; the GtH II levels in these males were 0.82 rising to 1.31 ng/ml and 2.46 rising to 3.12 ng/ml.

Following exposure to female urine, GtH II levels rose rapidly in responding males and peaked at 10 min. GtH II levels in responding males were significantly elevated over levels in non-responding and control males at all times between 5 and 60 min. There were no significant differences between control and non-responding males.

The lowest GtH II level found in *c.* 300 samples was 0.45 ng/ml. The 'basal level' of GtH II in males was *c.* 0.8 ng/ml. The minimum

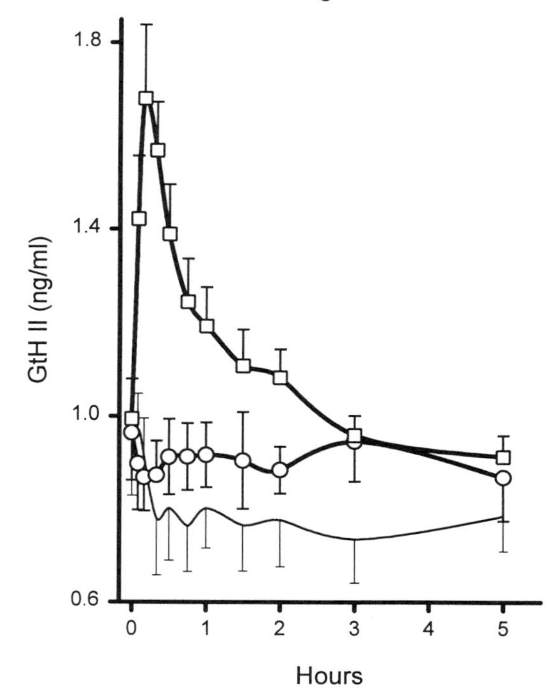

Fig. 2. The effect of a single squirt with 1 ml female urine or tank water on levels of GtH II in males (for legend see Fig. 1).

assayable amount of GtH II in trout plasma with the assay we used was probably between 0.4 and 0.8 ng/ml (the midpoint of the GtH II standard curve was at 1.6 ng/ml), and this apparent GtH II level can probably be explained by non-specific binding.

Cortisol

Cortisol levels in the same three groups of fish are shown in Figure 3. There were no initial differences in cortisol levels between the three groups. However, cortisol levels in non-responding males were significantly higher ($P<0.05$) than those in responding males at 10 min.

Body weight and GSI

There were no significant differences in body weight and GSI between control and urine-treated fish (mean ± S.E.M.; 399 ± 18 g *vs.* 390

± 12 g; 3.4 ± 0.4% *vs.* 3.5 ± 0.2%). Also, no differences were found in body weight and GSI between responding and non-responding males (mean ± S.E.M.; 401 ± 14 g *vs.* 373 ± 21 g; 3.4 ± 0.2% *vs.* 3.5 ± 0.3%).

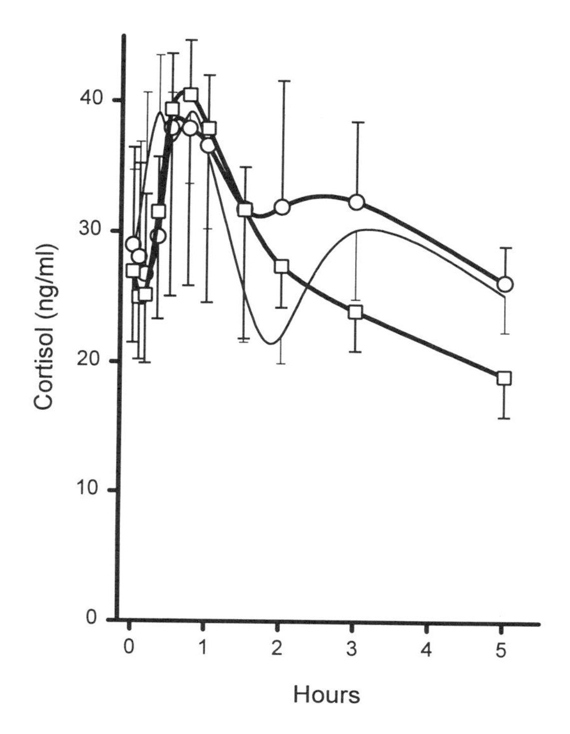

Fig. 3. Profile of cortisol levels in males following a single squirt with female urine or tap water (for legend see Fig. 1).

Discussion

One aim of this study was to establish if the rise in levels of 17,20β-P in males exposed to female urine, observed by Scott *et al.* (1994), was preceded by a rise in levels of GtH II. The data show that GtH II is released very rapidly into the blood following the exposure of males to female urine. Furthermore, the rise in GtH II levels is correlated with the subsequent rise in levels of 17,20β-P; Zohar *et al.* (1986) also observed a close correlation between fluctuations in both GtH and 17,20β-P in periovulatory female rainbow trout.

In this study, an apparently small (two-fold) elevation of GtH II levels, for a short time, led to a long and substantial (seven-fold) elevation of 17,20β-P levels. However, data published in Scott *et al.* (1994) indicated that basal levels of GtH II in isolated male trout were lower than those found in the present study (0.2 ng/ml *vs.* 0.8 ng/ml). Two different antisera have been used in these studies. The present one is not as sensitive as the previous one, and thus the 'basal levels' of GtH II of 0.8 ng/ml are probably an overestimate of what is actually present. Much more sensitive GtH II assays are still needed to assess 'true basal GtH II levels' in male trout (see also discussion in Sumpter and Scott, 1989).

Scott *et al.* (1994) noted that there was a large variation in levels of sex hormones in isolated males in response to female urine. Results of the present study reveal two distinct pools of males: responders and non-responders. Although 'stress' and more specifically cortisol, have been implicated in the suppression of sex steroid levels (see discussion by Scott *et al.*, 1994), non-responders did not have different levels of cortisol to responders.

The correlation between GtH II and 17,20β-P levels shows that non-responsiveness to the pheromone is unlikely to be due to non-responsiveness at the level of the testis. It remains to be established at which level the problem occurs - whether in the pituitary, the hypothalamus or at the olfactory epithelium.

References

Olsén, K. H., and Liley, N. R. (1993). The significance of olfaction and social cues in milt availability, sexual hormone status, and spawning behavior of male rainbow trout (*Oncorhynchus mykiss*). Gen. Comp. Endocrinol. 89, 107-118.

Scott, A. P., Liley, N. R., and Vermeirssen, E. L. M. (1994). Uring of reproductively mature female rainbow trout, *Oncorhynchus mykiss* (Walbaum), contains a priming pheromone which enhances plasma levels of sex steroids and gonadotrophin II in males. J. Fish Biol. 44, 131-147.

Sumpter, J. P., and Scott, A. P. (1989). Seasonal variations in plasma and pituitary levels of gonadotrophin in males and females of two strains of rainbow trout (*Salmo gairdneri*). Gen. Comp. Endocrinol. 75, 376-388.

Zohar, Y., Breton, B., and Fostier, A. (1986). Short-term profiles of plasma gonadotropin and 17α-hydroxy,20β-dihydroprogesterone levels in the female rainbow trout at the periovulatory period. Gen. Comp. Endocrinol. 64: 189-198

ORIGINS AND FUNCTIONS OF F PROSTAGLANDINS AS HORMONES AND PHEROMONES IN THE GOLDFISH

P.W. Sorensen[1], A.R. Brash[2], F.W. Goetz[3], R.G. Kellner[1], L. Bowdin[1], and L.A. Vrieze[1]

[1] Dept. Fisheries & Wildlife, University of Minnesota, 1980 Folwell Ave., St. Paul, MN 55108 USA,
[2] Dept. Pharmacology, Vanderbilt University, Nashville, TN 37232 USA,
[3] Dept. Biological Sciences, University of Notre Dame, Notre Dame, IN 46556 USA

Summary

Using gas chromatography-mass spectrometry we have definitively identified prostaglandin F2α (PGF2α) in the circulation and water of ovulatory goldfish to demonstrate that it is both a behavioral hormone and the precursor of this species' post-ovulatory sex pheromone. Plasma levels of PGF2α increased approximately 100-fold coincident with ovulation and the onset of spawning activity in female goldfish. Tracing the fate of circulating PGF2α by injecting a radio-label, we discovered that this compound is cleared to the water as 4 metabolites and that levels of circulating PGF2α peak within 45 min, the time at which female behavior is also most intense. Of the PGF metabolites released to the water, 15-keto-PGF2α, a potent olfactory stimulant, was also released in the greatest quantities. Thus, it is now clear that PGF2α plays an essential role synchronizing ovulation with the expression of both female and male behavior in the goldfish, and likely other species of teleost fish.

General Introduction

It is becoming increasingly evident that F prostaglandin(s) (PGFs) function as blood-borne signals in many species of fish, synchronizing ovulation with the expression of female sexual behavior and then acting as a precursor for sex pheromone production. However, the precise nature of this role has yet to be defined. Not only have there been few studies of the origins and fates of PGFs in fish, but all of these studies have relied exclusively upon radioimmunoassay (RIA) and/or radio-tracers. Furthermore, none have examined the relationship between circulating PGFs and either female behavior or pheromone production. Lastly, PGFs have not been definitively identified using biochemical means in the gonad, blood, or holding water of any fish. This is a particular concern because these compounds, which are traditionally thought to have autocrine and paracrine activity, are notoriously diverse and labile, and antibodies to them are often non-specific.

Where examined in fish (three species), the production of PGFs has been related to ovulation or the resumption of meiosis in the oocyte (Goetz et al., 1991). In all cases, an increase in ovarian production of immunoreactive PGF (IR-PGF) occurs at the time of follicular rupture, suggesting that PGF synthesis is associated with ovulation. The possibility of a link between ovarian PGFs and ovulation is strengthened by the observation that in several species, including the goldfish (Carassius auratus), ovulation is blocked by indomethacin (ID, a prostaglandin synthetase inhibitor) (Stacey & Pandey, 1975) and exogenous PGF2α specifically stimulates ovulation in vitro in the goldfish as well as at least 6 other species of fish (Goetz, 1991). Although the fate of ovarian IR-PGF is unclear, it seems possible that it finds its way into the plasma because small (several ng/ml) increases in circulating IR-PGF have been described in ovulated brook trout (Goetz & Cetta, 1983). An unpublished study (Bouffard, 1979) describes similar changes in the goldfish but the levels reported are questionable because of problems with the radioimmunnossay used.

A variety of circumstantial evidence suggests that circulating PGFs may function to stimulate female sexual behavior in recently-ovulated oviparous fish (Stacey & Goetz, 1982; Sorensen & Goetz, 1993). For instance, ovulated eggs and inanimate egg-like objects placed into the oviduct of goldfish trigger female sexual receptivity (Stacey & Liley, 1974), and ID blocks this response. Additionally, PGF2α–injection stimulates female sexual behavior in non-ovulated females within minutes and, of several PGs tested, PGF2α is the most potent (Stacey, 1981; Sorensen et al., 1989). PGF2α-injection is now known to stimulate female receptivity in at least a dozen other fish species (Stacey, personal communication).

It has also been hypothesized that circulating PGFs are cleared to the water to function as a pheromone(s), thereby synchronizing male and female sexual behavior (Sorensen et al., 1988). The data supporting this hypothesis are also indirect -- only two studies have measured PGF release and biochemical identification of water-borne PGFs has not been described (Sorensen et al., 1988; Ogata et al., 1993). The clearest data for this hypothesis is from the goldfish. When ovulated, goldfish, like many oviparous fish (Stacey et al., 1986), release a chemical cue which stimulates male sexual behavior. Not only does the odor of PGF2α-injected fish elicit sexual arousal in males, but electro-olfactogram (EOG) recording conducted from the male goldfish olfactory epithelium finds it to be acutely and specifically sensitive to water-borne PGFs (Sorensen et al., 1988). PGF2α and its metabolite 15-keto prostaglandin F2α (15K-PGF2α) are the most potent, the former has a detection threshold of 10^{-10}Molar (M), and the latter 10^{-12}M (Sorensen et al., 1988). EOG recording has also now established that several dozen species of fish, principally the Cypriniformes and Salmoniformes, are equally sensitive to commercially available PGFs (Stacey et al., this symposium), although there is no direct information on whether these fish naturally release PGFs to the water. Over the past four years we have sought to clarify the identities, roles, and fates of PGFs in the goldfish to establish whether these compounds function as a blood-borne hormone which is released as a sex pheromone. Here we review some preliminary data which address several questions related to this central objective: 1) Does the level of circulating PGF2α increase in female goldfish at the time of ovulation?

2) Are PGF metabolites with olfactory activity released by ovulated and PGF2α-injected fish? 3) What are the fates and functions of injected PGF2α, and do they support a behavior role for this compound?

Does circulating prostaglandin F2α in female goldfish increase after ovulation when goldfish are sexually active and then decrease after spawning?

Vitellogenic female goldfish were placed into aquaria and the temperature of these aquaria increased to 18°C. Five of these fish were then placed into 1 L of water for 1 h to collect a water sample from pre-ovulatory fish. Afterwards, their water was extracted by passing it through an activated C18 Sep-Pak (Waters) and a sample of blood drawn using a heparinized syringe rinsed with ID. Blood was placed into a chilled tube containing ID, and centrifuged to separate the plasma which was immediately diluted with buffer and extracted by Sep-Pak. That evening, the fish were injected with human chorionic gonadotropin and the next morning they were checked for ovulation. Approximately 75% of these fish had ovulated, including all those used in the pre-sample. Several ovulated fish were then individually placed into 1L of water and later bled for PGF determination. Another group of ovulated fish was placed with mature males and allowed to spawn. Several hours later these fish were found to be 'spawned out' (i.e. all ovulated eggs had been released), and their water and blood sampled as described above.

For analysis, aliquots of extracted plasma and water were spiked with tetradeuterated PGF2α and 15K-PGF2α as internal standards. Samples were then derivatized to the pentofluorbenzyl ester, purified by thin layer chromatography, and converted to the methoxime (15K-PGF2α only) and trimethylsilyl derivatives (Parsons & Roberts, 1988). Samples were then analyzed by stable isotope dilution and combined gas chromatography-mass spectrometry (GC-MS) using negative ion/chemical ionization detection of the pentaflurobenzyl ester trimethylsilyl ether derivatives (Blair, 1990). PGF2α and 15K-PGF2α were quantified from the ratios of peak heights of the unlabeled and deuterated internal standards in relation to known mixtures of authentic compounds.

Circulating levels of PGF2α in pre-ovulatory goldfish were relatively low (0.2 - 0.8 ng/ml) but increased nearly 100-fold at the time of ovulation and ranged between 15 - 143 ng/ml. All ovulated fish spawned with males and after they had released their eggs their plasma PGF2α levels were once again low (0.3 - 1.9 ng/ml). Pre-ovulatory fish also released little PGF2α and 15K-PGF2α to the water (less than 1.0 ng/hr); however, when ovulated, the release rate of 15K-PGF2α increased nearly 1000-fold, ranging between 240 - 1,975 ng/hr.

What is the fate of PGF2α-injected into female goldfish and how does it correspond with the expression of female sexual activity?

Female goldfish were injected with radiolabeled PGF2α at a dosage of 1μg/g body weight and placed into 18°C water. These individuals were then bled as quickly as possible at 15-min time intervals over a 2 h period. Water samples were also collected. Blood and water samples were extracted by Sep-Pak, and their radiation levels monitored by scintillation counting. Samples were then pooled by time period and analyzed by radiochromatography. This was accomplished by injecting the samples onto a reverse-phase C18 column using a mobile phase of H_3PO_4: acetonitrile with a flow of 1.5 ml/min and a gradient going from 66:33 to 55:45 over 25 min. Radiometabolites were detected by an inflow radionucleotide detector and identified by comparing retention times of both cold (190 nm and 230 nm wavelengths) and tritiated labels. The amount of PGF2α in the plasma was quantified using total plasma radiation and percent attributable to each metabolite. Next, to determine the relationship between PGF content in the plasma of PGF2α-injected fish and the expression of female spawning behavior, female fish were injected with 1μg/g body weight PGF2α and placed into an aquarium with spawning substrate. Sexually-active mature males were then removed from aquarium with a spawning female, and placed with the test female whose spawning activity was then noted for a 2 h period.

PGF2α-injected female goldfish consistently released small quantities of PGF2α as well as 4 metabolites of this compound (Fig. 1). One of these metabolites was 15K-PGF2α but the identities of the others were unknown. Occasionally, small quantities of 13,14-dihydro-15keto-PGF2α were also released as was another unknown compound (not shown). Of these compounds only three, P1, P3 and PGF2α, were found in the plasma of PGF2α injected fish (Fig. 1).

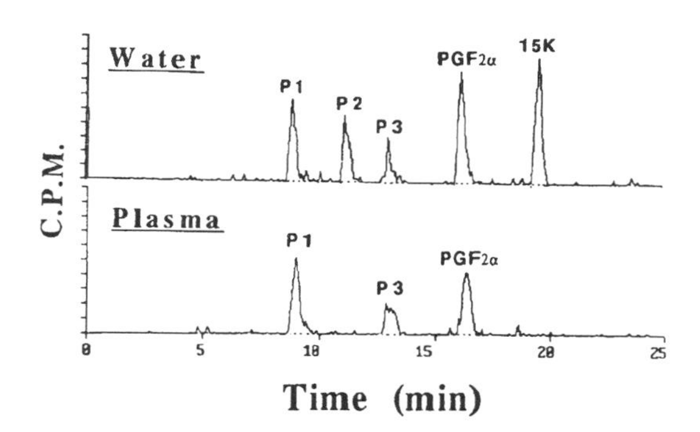

Fig.1. Radiochromatographs of PGF-Ms released to water and found in the plasma of a typical PGF2α-injected fish. Peaks labeled P1, P2, and P3 are unknowns. Peak F elutes at the same time as PGF2α and peak 15K at the same time as 15K-PGF2α.

When PGF2α metabolites were traced in the plasma of PGF2α-injected female goldfish over time, PGF2α was seen to peak between 30-60 min and then decline (Fig. 2). P1 and P3 showed different time courses and no 15K-PGF2α was measured in the circulation (data not shown). Fish injected with PGF2α exhibited sexual behavior with an intensity which approximately paralleled that of circulating PGF2α in the plasma, peaking at 30-60 min and then dropping (Fig. 2).

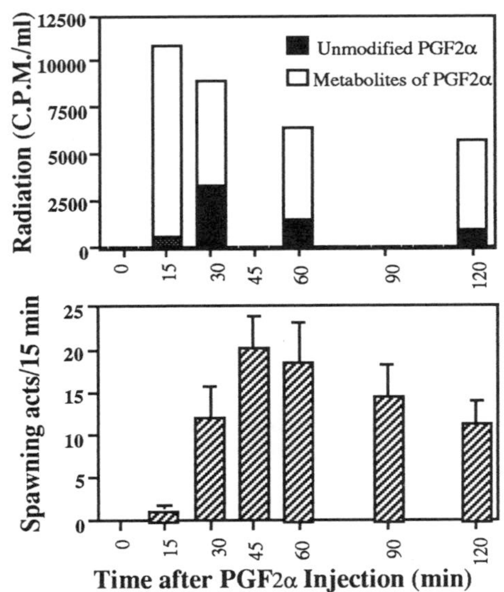

Fig. 2 Fate of injected PGF2α in the blood over time, and the behavior of PGF2α-injected females over time.

General Discussion and Conclusions

This study establishes that prostaglandin PGF2α is a behavioral hormone and pheromone in the goldfish. Not only is this compound produced in great quantities at the time of ovulation when goldfish are sexually active, but its levels drop dramatically after spawning. Other studies conducted with greater temporal resolution have shown that circulating PGF2α increases coincident with ovulation (Sorensen et al., unpublished). It is also now clear that PGF2α-injection produces a physiological increase in circulating PGF2α. Furthermore, the decline of PGF2α in the circulation matched a decline in female sexual activity. Thus, it seems reasonable to conclude that PGF2α functions as a hormone to directly stimulate female sexual activity.

The levels of PGF2α which we have now conclusively measured in ovulated goldfish are likely the highest ever measured in a vertebrate. Interestingly, circulating levels of PGF2α measured in goldfish exceed those measured in the trout (Goetz & Cetta, 1983), a species which lacks an oviduct and which has been suggested not to use PGF2α as a behavioral hormone (Stacey, 1987). It seems possible that the rather impressive ability of goldfish to produce PGF2α may be associated with a specialized physiological mechanism. Preliminary studies injecting egg substitutes into the reproductive tracts of non-ovulated goldfish demonstrate that this stimulates PGF2α production: ovulated eggs can not be the sole source of PGF2α (Sorensen et al., unpublished data).

Finally, just as significant as our elucidation of the presence of circulating PGF2α in ovulated goldfish was our discovery that circulating PGF2α is rapidly cleared and released to the water, primarily as 15K-PGF2α. The olfactory potency of this compound and its effects on male behavior are already well established (Sorensen et al., 1988): it can now be concluded that it is a sex pheromone. Because 15K-PGF2α is not found in the blood and is released in the urine (Appelt et al., this symposium), the kidney is likely involved in its production. On the basis of the present findings, it seems likely that many other species of fish, particularly the Cypriniformes, use PGFs as both hormones and pheromones. However, the possibility of species-specific differences in production remains. It will be exciting to extend our studies to other species and to examine the specialized mechanisms which must underlie PGF production and release in the goldfish.

References

Blair, I.A., 1990. Electron-capture negative ion chemical ionization mass spectrometry of lipid mediators. Methods Enzymol. 187:13-23.

Bouffard, R.E., 1979. The role of prostaglandins during sexual maturation, ovulation, and spermiation in the goldfish, Carassius auratus. unpublished MSc. Thesis, University of British Columbia, Canada.

Goetz, F.W., 1991. The compartmentalization of prostaglandin synthesis within the fish ovary. Amer. J. Physiol. 260: R862-865.

Goetz, F.W. & F. Cetta, 1983. Ovarian and plasma PGE and PGF levels in naturally ovulating brook trout, (Salvelinus fontinalis) and the effects of indomethacin on prostaglandin levels. Prostaglandins 26: 387-395.

Goetz, F.W., Berndtson, A.K. & M. Ranjan, 1991. Ovulation: mediators at the ovarian level. In: Vertebrate Endocrinology, Fundamentals and Biochemical Implications. P.M. Pang and M. Schreibman (Eds.), Vol. VI (A). Academic Press, New York, N.Y. p.127-203.

Ogata, H., Kitamura, S. & F. Takashima, 1993. F Prostaglandins in the holding water of female loach. Nippon Suisan Gakkaishi 59:1259.

Parsons, W.G. & L.J. Roberts, 1988. Transformation of prostaglandin D2 to isomeric PGF2α compounds by human eosinophils. J. Immunol. 141: 2413-2419.

Sorensen, P.W. & F. W. Goetz, 1993. Pheromonal and reproductive function of F prostaglandins and their metabolites in teleost fish. J. Lipid Mediators 6: 385-393.

Sorensen, P.W., Chamberlain, K.J., Stacey, N.E. & T.J. Hara, 1987. Differing roles of prostaglandin F2α and its metabolites in goldfish reproductive behavior. In: Proceedings of the Third International Symposium on the Reproductive Physiology of Fish. D.R. Idler, L.W. Crim & J.M. Walsh (Eds.): Memorial University Press, St. John's, Canada p.164.

Sorensen, P.W., T.J. Hara, Stacey, N.E. & F. W. Goetz, 1988. F prostaglandins function as potent olfactory stimulants comprising the post-ovulatory female sex pheromone in goldfish. Biol. Reprod. 39: 1039-1050.

Stacey, N.E., 1981. Hormonal regulation of female reproductive behavior in fish. Amer. Zool. 21:305-316.

Stacey, N.E., 1987. Roles of hormones and pheromones in fish reproductive behavior. In: Psychobiology of Reproductive Behavior. D. Crews, (Ed.), Prentice Hall, N.Y. p.28-69.

Stacey, N.E. & F.W. Goetz, 1982. Role of prostaglandins in fish reproduction. Can. J. Fish. Aquat. Sci. 39:92-98.

Stacey, N.E. & N.R. Liley, 1974. Regulation of spawning behavior in the female goldfish. Nature 247: 71.

Stacey, N.E. & S. Pandey, 1975. Effects of indomethacin and prostaglandins on ovulation in goldfish. Prostaglandins 9:597-607.

Stacey, N.E., A.L. Kyle & N.R. Liley, 1986. Fish reproductive pheromones. In: Chemical Signals in Vertebrates IV. D. Duvall, D. Muller-Schwarze & R.M. Silverstein (Eds.): Plenum Press, New Jersey, pp.117-133.

(Supported by the National Science Foundation (BNS9109027), the Minnesota Agricultural Experiment Station, and the National Institutes of Health (Prostaglandin Core Laboratory; Grant # HD05797)

F-SERIES PROSTAGLANDINS HAVE A PHEROMONAL PRIMING EFFECT ON MATURE MALE ATLANTIC SALMON (*Salmo salar*) PARR

C.P. Waring [1] and A. Moore[2]

[1] School of Biological Sciences, University of East Anglia, Norwich, Norfolk, NR4 7TJ, U.K.

[2] MAFF, Fisheries Laboratory, Pakefield Rd, Lowestoft, Suffolk, NR33 0HT, U.K.

Summary

Ovulated female salmon urine contains at least one compound which elevates plasma 17,20βP concentrations in conspecific male parr. In an attempt to try and identify the compound(s) , we tested a variety of hormones or hormone metabolites which are known to be potent olfactory stimulants to mature males. Waterborne testosterone and 17,20βP-S had no effect on plasma 17,20βP concentration in males, whereas prostaglandins $F_{1\alpha}$ and $F_{2\alpha}$ when added to the holding water caused an elevation in male parr plasma 17,20βP levels. This effect increased as the season progressed.

Introduction

Recently it has been shown by Scott *et al* (1994) that the urine from mature female rainbow trout, *Oncorhynchus mykiss*, contains a priming pheromone that significantly increases the plasma concentrations of gonadotropin and 17α,20β-dihydroxy-4-pregnen-3-one (17,20βP) in conspecific males, although the chemical identity of this pheromone has not yet been established. In Atlantic salmon, however, it has been shown previously that testosterone is a potent odorant to mature male salmon parr (Moore & Scott, 1991). Moreover, 17,20βP-sulphate (17,20βP-S) also is a potent odorant to mature male parr, but only after the olfactory epithelium has been perfused with the urine from ovulated females (Moore & Scott, 1992).

To date, however, the significance of these findings to the reproductive physiology of Atlantic salmon has not been established. The present study examined the possibility that female salmon urine may also contain a priming pheromone. We also sought to determine if any of the hormonal compounds known to be odorants to mature male parr, i.e. testosterone, 17,20βP-S and F-series prostaglandins (PGFs) had any priming-like activity.

Materials and Methods

Spermiating mature male parr (length 127 ± 2 mm; GSI 6.6 ± 0.1 %) were obtained in August, September, and November 1994. Groups of 5 males were placed into one of a series of 63 l glass tanks supplied with flow-through dechlorinated tap-water (1 l min^{-1}) and left to recover for at least 72 h. Three experiments were carried out. In experiment 1 (October 1994) males were either exposed to ovulated female salmon urine (630 μl: to give an initial dilution of one part in 10^5), testosterone (at an initial dilution of 10^{-9} M), 17,20βP-S (at 10^{-8} M), or ethanol control. Blood and milt were sampled 3 h after exposure and plasma 17,20βP concentrations were measured by radioimmunoassay (Scott *et al* 1982).

In experiment 2 (November 1994) groups of males were exposed to water, immature male urine, mature male urine and ovulated female urine (all at an initial dilution of one part in 10^5), and female urine with ethanol or female urine with 17,20βP-S (at 10^{-8}M). The males were sampled 3 h after exposure as described above. In experiment 3 (October-December 1994) groups of males were exposed to $PGF_{1\alpha}$, $PGF_{2\alpha}$ (both at an initial dilution of 10^{-8} M) or ethanol. Males were sampled 3 h after exposure as described above.

Results

Plasma 17,20βP concentrations were significantly elevated in the parr exposed to

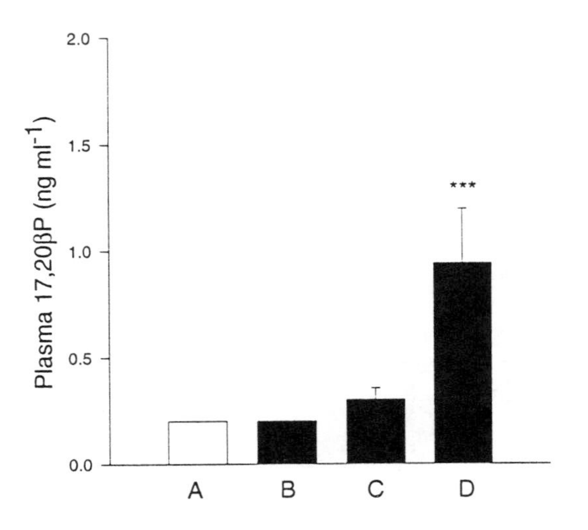

Fig. 1. The effect of exposure to various test substances on plasma 17,20βP concentrations in male salmon. A= ethanol controls, B= testosterone, C= 17,20βP-S, D= female urine. Data represents mean ± sem of 7 observations. *** $p<0.001$ compared to controls (A).

female urine, but were not altered in the parr exposed to testosterone or 17,20βP-S (Figure 1). When 17,20βP-S was added to the water along with female urine, no significantly greater elevation in male 17,20βP levels was apparent compared to the males exposed to female urine only (Figure 2). Exposure of mature male parr to the urines from both immature and mature males had no significant priming effect on plasma levels of 17,20βP (Figure 2).

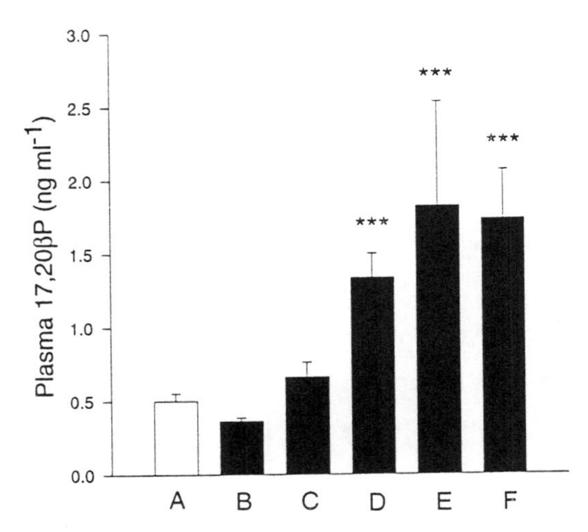

Fig. 2. The effect of various waterborne urines on plasma 17,20βP concentrations of male salmon parr. A= water control, B= immature male urine, C= mature male urine, D= female urine, E= female urine with ethanol, F= female urine with 17,20βP-S. Data represents means ± sem of 8-10 observations. *** $p<0.001$ compared to controls (A).

The effects of exposure to waterborne PGFs are shown in figure 3. Initially, at the first sampling point, both PGFs tested had no effect on male plasma 17,20βP levels. However, as the reproductive season progressed exposure to PGFs significantly elevated the plasma 17,20βP concentrations of males, and there was some indication that $PGF_{2\alpha}$ was a more potent stimulator of male physiology than $PGF_{1\alpha}$.

Discussion

The results from these experiments clearly shows that the urine from ovulated female salmon contains at least one priming pheromone which elevates plasma 17,20βP concentrations in conspecific males. The priming pheromone does not occur in the urine of immature or mature male salmon but specifically only in the female urine.

Neither of the two steroids tested (testosterone and 17,20βP-S), at concentrations where the olfactory receptors approach saturation (Moore & Scott, 1991, 1992),

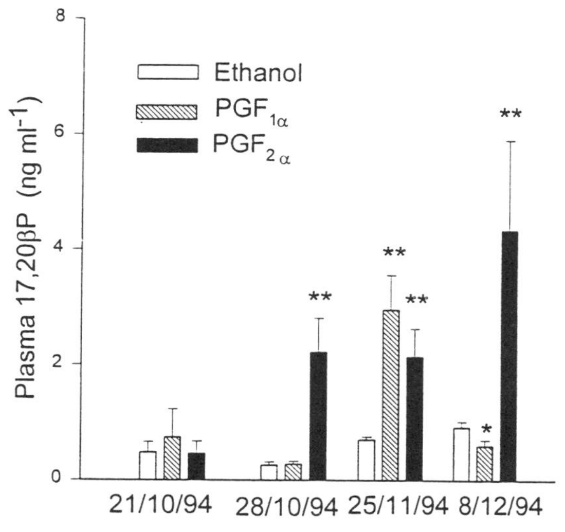

Fig. 3. The effect of PGFs on plasma 17,20βP concentrations of male salmon parr. Data represents mean ± sem of 5 observations. * $p<0.05$, ** $p<0.01$ compared to ethanol controls.

elevated plasma 17,20βP concentrations in males (nor expressible milt or plasma testosterone or 11-ketotestosterone; Waring & Moore, Unpubl. Obs.), indicating that the priming pheromone is unlikely to be either of these steroids. The fact that testosterone is only detected by male parr early on in the reproductive season (Moore & Scott, 1991) and is a potent stimulator of upstream swimming by males (Moore, 1991), but has no effect on male reproductive physiology, may indicate that it functions as an attractant for males to bring them to the spawning grounds before females have ovulated.

However, as the season progressed the PGFs became progressively more stimulatory. A similar pattern of increasing sensitivity to these compounds was apparent in the electro-olfactogram data (Moore & Waring, this volume), and in levels of expressible milt and plasma androgen concentrations (Waring & Moore, Unpubl. Obs.). However, we have no information as yet as to whether the urine from ovulated female salmon contains PGFs and, if so, at what concentrations. Our data as it stands is correlative but it does suggest that a PGF-like compound has a priming effect on male salmon and may be the priming pheromone in female urine.

References

Moore, A. (1991). Behavioural and physiological responses of precocious male Atlantic salmon (*Salmo salar* L.) parr to testosterone. In 'Reproductive Physiology of Fish 'Fishsymp 91', (ed: A.P. Scott, J.P. Sumpter, D.E. Kime, & M.S. Rolfe), University of Sheffield, pp. 194-196.

Moore, A. & Scott, A.P. (1991). Testosterone is a potent odorant in precocious male Atlantic salmon (*Salmo salar* L.) parr. Phil. Trans. R. Soc. Lond. B 332: 241-244

Moore A. & Scott, A.P. (1992). 17α,20β-dihydroxy-4-pregnen-3-one 20-sulphate is a potrent odorant in precocious male Atlantic salmon (*Salmo salar* L.) parr which have been pre-exposed to the urine of ovulated females. Proc. R. Soc. Lond. B 249: 205-209.

Scott, A.P., Liley, N.R., & Vermeirssen, E.L.M. (1994). Urine of reproductively mature female rainbow trout, *Oncorhynchus mykiss* (Walbaum), contains a priming pheromone which enhances plasma levels of sex steroids and gonadotrophin II in males. J. Fish Biol. 44: 131-147.

Scott, A.P., Sheldrick, E.L., & Flint, A.P.F. (1982). Measurement of 17α,20β-dihydroxy-4-pregnen-3-one in plasma of trout (*Salmo gairdneri* Richardson): seasonal changes and response to salmon pituitary extract Gen.Comp.Endocrinol. 46: 444-451.

ALTERNATIVE LIFE HISTORY STRATEGIES AND DIMORPHIC MALES IN AN ACOUSTIC COMMUNICATION SYSTEM

Andrew H. Bass
Section of Neurobiology and Behavior, Cornell University, Ithaca, New York 14853.

Summary

The plainfin midshipman (*Porichthys notatus*) has two male reproductive morphs: Nest-building "Type I" males generate long duration, quasi-sinusoidal-like, advertisement calls ("hums") to attract females to nests and trains of short duration agonistic calls ("grunts") in defense of their egg clutch and nest against potential intruder males. Sneak/satellite-spawning "Type II" males do not build nests, guard eggs or acoustically court females; like females, they infrequently generate isolated grunts in non-spawning contexts. Studies of vocal motor traits indicate distinct, non-overlapping developmental trajectories for Type I and II males, while otolith analyses show that Type II males are sexually precocious compared to Type I males. Thus, compared to Type II males, Type Is have an extended juvenile stage during which they are investing in body growth and a vocal motor system. In contrast, Type II males are investing in earlier reproduction.

Alternative Male Reproductive Morphs - Overview

Since the turn of the century, midshipman (*Porichthys notatus*) have been known to generate an unusual sinusoidal-like vocalization ("hum") during the breeding season (review: Bass, 1990). Studies of captive populations of midshipman show that hums apparently function as acoustic courtship displays used by parental males to attract gravid females to their nests (Brantley and Bass, 1994). This hypothesis was confirmed in playbacks of computer-synthesized hums through underwater speakers (McKibben et al., 1995; also see Ibara et al., 1983). Parental males also generate long trains of 50-150 msec duration "grunts" when challenged by intruder males (Type I or II; see Brantley and Bass, 1994). Playbacks of grunts are not attractive to gravid females (McKibben et al., 1995).

Early in our studies, we discovered alternative male reproductive phenotypes in midshipman on the basis of a suite of somatic traits - body size, gonadosomatic index (GSI, ratio of gonad weight to body weight), and the size of vocal muscles and vocal motoneurons (Bass and Marchaterre, 1989a, b; also see Brantley et al., 1993a). Only reproductively active adults - Type I and II males and females - are found in nest sites (juveniles are only collected by seining in offshore feeding sites). Studies of nesting males and females revealed that parental mate-calling males, which we designated as "Type I", had an eight-fold larger body mass than a second group of "Type II" males. By contrast, Type II males had on average a nine-fold larger GSI. Lastly, there were dramatic dimorphisms in vocal musculature. The vocal muscles of midshipman are a pair of skeletal muscles attached to the lateral walls of their swimbladder; their contraction rate establishes a vocalization's fundamental frequency, which is under the control of a brainstem motor pathway (Bass and Baker 1990, 1991). The mass of the Type I male vocal musculature was several fold larger than that of Type II males (or females), consistent with the robust ability of Type I males to produce hums continuously for long periods of time, often on the order of minutes. The relatively small vocal muscle in Type II males and females is paralleled by their apparent ability to infrequently generate isolated, low amplitude grunts (Bass and Baker, 1990; Brantley and Bass, 1994). Sexual polymorphisms in the vocal muscle were paralleled by dimorphisms in the sizes of motoneurons that innervate the muscles (Bass and Marchaterre, 1989b).

The early studies also showed dimorphisms in the most fundamental characters of skeletal muscle (Z-lines, sarcoplasmic reticulum, mitochondria density), motor axons (cross sectional area, the size of neuromuscular junctions, and synaptic vesicle density) and the morphology of physiologically-identified vocal motor neurons and their presynaptic, pacemaker neurons (review: Bass, 1992). Intracellular staining and recording studies demonstrated dimorphisms in the soma-dendritic dimensions and the oscillatory-like firing frequency of individual motoneurons and pacemaker neurons. All vocal traits are greater in magnitude among the mate-calling Type I males; this includes the firing frequency of the central motor pathway which is 20% higher and matches the higher fundamental frequency of their vocalizations (Bass and Baker, 1990; Brantley and Bass, 1994).

Neuroendocrine studies of midshipman have focused on two traits: gonadal steroids and GnRH-like immunoreactive (ir) neurons. There are morph-typical circulating levels of gonadal steroids (Brantley et al., 1993b). Type I and II males and females exhibited contrasting steroid hormone profiles. 11-ketotestosterone was the predominant androgen in Type I males ; testosterone alone was detectable in Type II males and females, while 17ß-estradiol was detectable only among females. A separate immunocytochemical study identified age-, sex- and male morph-specific patterns in the size and number of GnRH-ir neurons in the preoptic area (POA) (Grober et al., 1994). The data suggested that the POA-pituitary/ GnRH-gonadotropin axis initiates sexual maturation events at different ages in Types I and II males (also see Bass, 1993). Otolith studies have confirmed this and show that Type II males mature about 1 year earlier than Type I males (A. Bass and E. Brothers, unpub. observ.).

Ontogeny of Androgen-Sensitive Vocal Traits

Among juvenile midshipman, there is a divergence in muscle phenotype at about 1 year of age (Brantley et al., 1993a): Only juvenile Type I males experience a pre-maturational, 4-fold increase in fiber number followed at sexual maturity by a sudden increase in fiber size. Relative to body mass, this results in a 6-fold greater vocal muscle mass in adult Type I males. Androgen, and not estrogen or cholesterol, treatments of juvenile males and females markedly increased sonic muscle size which included changes in muscle fiber number and structure (Brantley et al., 1993c).

Biocytin, a low molecular weight complex of biotin and lysine, is transported transneuronally in the vocal system of midshipman. Thus, a single application of biocytin to the cut end of a sonic nerve at the level of the swimbladder results in a labelling, we propose, of all neurons in the brain forming a vocal motor network (Bass et al., 1994). This includes a pacemaker-motoneuron circuit and a ventral medullary nucleus that links the pacemaker circuitry bilaterally. Importantly, surgical isolation of the brainstem region inclusive of motor, pacemaker and ventral medullary neurons appears both necessary and sufficient for generating a rhythmic vocal motor discharge. We have now exploited the biocytin method for mapping to study the ontogeny of the proposed vocal motor circuit. The results show that increases in motoneuron soma size among (nascent) juvenile Type I males parallel the development of dimorphisms in vocal muscle fiber number. There were also parallel increases in the size of pacemaker and ventral medullary neurons, although more modest in magnitude. Thus, sexual maturation of the Type I male's mate calling circuit parallels the ontogeny of its target muscle (Horvath et al., 1994).

Androgen pellet implants have also been found to induce alterations in morpho-physiological traits (Marchaterre et al., in press). The biocytin method identified significant (Student t-test; $p<0.05$) increases in the cross-sectional area of motor (30%), pacemaker (30%), and ventral medullary (15%) neurons among androgen-treated juvenile males ($n=9$) compared to controls matched for body size ($n=10$; $p>0.7$). Androgen treatments included the non-aromatizable androgens 11-ketotestosterone and dihydrotestosterone, and testosterone propionate; their effects did not differ ($p>0.2$).

Neurophysiological studies next identified increases in two traits: fundamental discharge frequency (FDF) and "fatigue resistance" (FR) among androgen-treated juveniles. FDF refers to the highly stable (standard deviation of 1.1-1.2 Hz) firing frequency of the central motor circuit (Bass and Baker, 1990). As noted earlier, a Type I male's vocal phenotype is characterized by vocalizations and a motor volley with a FDF 20% higher than females, Type II males, or juveniles. Androgens induced a significant ($p<0.05$) 13% increase in FDF in androgen-treated juveniles ($n=4$)) compared to untreated controls ($n=7$; there was no significant difference in body size between the test groups, $p>0.3$). The ability of Type I males to generate long duration mate calls is paralleled by the ability to evoke a rhythmic motor volley for long time periods using central brain stimulation - the system seemingly does not fatigue. The following experiment was carried out to quantify the vocal system's FR: A train of low amplitude, midbrain stimuli were delivered 10 times at 1 ms intervals during a trial (see Bass and Baker, 1990 for methods). Ten trials, each separated by a 10 minute rest period, constituted a single experiment. FR was defined as the percentage of stimulus trains over the entire experiment that evoked a rhythmic volley. As predicted by the vocal behavior of each reproductive morph, FR is 1-3 fold higher in Type I males compared to females and Type II males. There is also a significant ($p<0.05$) 47% increase in FR among androgen-treated juveniles compared to controls. FR and FDF were unchanged in two 17ß-estradiol-treated juveniles ($p>0.3$).

Together, the data suggests that androgen-sensitive events can account for the development of the Type I male vocal motor phenotype.

Concluding Comments

Growth patterns, age and size at sexual maturity are examples of life history traits (Stearns, 1992). The data presented here suggests that neurobiological traits, namely those of the vocal motor system, can characterize the development of alternative life history strategies in midshipman. Thus, nascent Type I males adopt a growth trajectory characterized by delayed maturation coupled to increased body size and androgen-dependent, sexual differentiation of a vocal motor system that functions in the production of mate calls. By contrast, Type II males undergo precocious maturation coupled to gonadal hypertrophy and sneak or satellite spawning without acoustic courtship. The results imply that the ontogenetic event governing the onset of Type I vs. Type II growth patterns involves a tradeoff between primary (gonadal) and secondary (vocal) sexual characteristics.

References

Bass A.H., 1990. Sounds from the intertidal zone: "Vocalizing" fish. Bioscience 40:247-258.

Bass, A.H., 1992. Dimorphic male brains and alternative reproductive tactics in a vocalizing fish. Trends Neurosci. 15:139-145.

Bass, A.H., 1993. From brains to behavior: Hormonal cascades and alternative mating tactics in teleost fishes. Rev. Fish Biol. Fisheries 3:181-186.

Bass A.H. and R. Baker , 1990. Sexual dimorphisms in the vocal control system of a teleost fish: Morphology of physiologically identified cells. J. Neurobiol. 21:1155-1168.

Bass A.H. and R. Baker , 1991. Adaptive modification of homologous vocal control traits in teleost fishes. Brain Behav. Evol. 38:240-254.

Bass A.H. and M. A. Marchaterre, 1989a. Sound-generating (sonic) motor system in a teleost fish (*Porichthys notatus*): Sexual polymorphism in the ultrastructure of myofibrils. J. Comp. Neurol. 286:141-153.

Bass A.H. and M. A. Marchaterre, 1989b. Sound-generating (sonic) motor system in a teleost fish (*Porichthys notatus*): Sexual polymorphisms and general synaptology of a sonic motor nucleus. J. Comp. Neurol. 286:154-169.

Bass A.H., M. A. Marchaterre and R. Baker, 1994. Vocal-acoustic pathways in a teleost fish. J. Neurosci. 14:4025-2039 .

Brantley R.K. and A. H. Bass, 1994. Alternative male spawning tactics and acoustic signalling in the plainfin midshipman, *Porichthys notatus*. Ethology 96: 213-232.

Brantley R.K., J. Tseng and A. H. Bass, 1993a. The ontogeny of inter- and intrasexual vocal muscle dimorphisms in a sound-producing fish. Brain Behav. Evol. 42:336-349.

Brantley R.K., M. A. Marchaterre and A. H. Bass, 1993b. Androgen effects on vocal muscle structure in a teleost fish with inter and intrasexual dimorphisms. J. Morph. 216:305-318.

Brantley R.K., J. Wingfield J. and A. H. Bass, 1993c. Hormonal bases for male teleost dimorphisms: Sex steroid levels in *Porichthys notatus*, a fish with alternative reproductive tactics. Horm. Behav. 27:332-347.

Grober M.S., S. Fox, C. Laughlin and A. H. Bass, 1994. GnRH cell size and number in a teleost fish with two male reproductive morphs: Sexual maturation, final sexual status and body size allometry. Brain Behav. Evol. 43:61-78.

Horvath B.J., M. A. Marchaterre and A. H. Bass, 1994. Transneuronal biocytin delineates ontogeny of a sexually dimorphic, androgen-sensitive vocal pacemaker circuit. Soc. Neurosci. Abstr. 20:372.

Ibara, R. M., L. T. Penny, A. W. Ebeling, G. van Dykhuizen and G. Cailliet, 1983. The mating call of the plainfin midshipman fish, *Porichthys notatus*. In: Predators and Prey in Fishes. D. l. G. Noakes, D. G. Lindquist, G. S. Helfman and J. A. Ward (Eds.): Dr W Junk Publishers, The Hague, p. 205-212.

Marchaterre M. A., B. Horvath, D. Bodnar and A. H. Bass, in press. Androgen-sensitive, brainstem vocal pacemaker. Soc. Neurosci. Abstr. 21:

McKibben J., D. Bodnar and A. H. Bass, in press. Everybody's humming but is anybody listening: Acoustic communication in a marine teleost fish. Fourth Intl. Cong. Neuroethol.

Stearns, S. C. (1992) The Evolution of Life Histories. Oxford Univ. Press, Oxford, New York, Tokyo.

BEHAVIOURAL AND ENDOCRINE STUDY OF *OREOCHROMIS AUREUS*, WITH SPECIAL REFERENCE TO SEX-REVERSED MALES.

P. Poncin,[1] M. Ovidio,[1] G. Skoufas,[1] V. Gesquière[1], C. Mélard,[2] D. Desprez,[2] K. Mol,[3] N. Byamungu[3], B. Cuisset,[4] E.R. Kühn,[3] J.C. Philippart[2] and J.C. Ruwet.[1]

[1]Dept. Ethology, University of Liège, 22 quai Van Beneden B-4020 Liège, Belgium; [2]Lab. Fish Demography & Aquaculture, 10 Chemin de la Justice B-4500 Tihange, Belgium; [3]Lab. Comp. Endocrinol., Catholic University of Leuven, Naamsesstraat 61, B-3000 Leuven, Belgium. [4]Gr. de Biol. de la Rep. des Poissons, UA-INRA, Univ. Bordeau I, Avenue des Facultés, F-33405 Talence.

Summary

The relationship between social status and hormonal profiles (T, E2, 11-KT, T3 and T4) was analysed in normal tilapias *Oreochromis aureus*. Dominant females and dominant males exhibited respectively T and 11-KT plasma level higher than subordinate fish of the same sex. Moreover, the behaviours of females and 17α-ethynylestradiol sex-reversed males (pseudofemales) were studied in mixed pairs or assemblages (4, 6 or 8 fish in 500 l and 10 fish in 100 l). The results suggested that pseudofemales were more aggressive and dominant that females. Activity rhythms were similar in females and pseudofemales.

Introduction

Relatively little is known about the hormonal control of aggressive behaviour and social dominance in tilapia species. Similarly, the social status and sexual behaviour of sex-reversed fish have received scant attention despite growing interest of such fish in aquaculture and the possible discrepancies of their physiology with respect to normal fish. For instance, Desprez *et al.* (1995) described significantly lower spawning frequency in sex-reversed *Oreochromis aureus*. This paper presents an overview of a three-year research programme aiming to characterise behavioural and endocrine profiles of normal and sex-reversed tilapia *O. aureus*. Fish were studied under controlled conditions in an attempt to improve knowledge on cultured cichlid fish.

Material and methods

Mature tilapia *O. aureus* (male, female, pseudofemale: i.e. 17α-ethynylestradiol sex-reversed males; see Mélard, 1995) were obtained from the Tihange Fish Breeding Centre. The fish were maintained in 100- to 1000-litre aquaria at a constant 26°C temperature and under a 12L/12D photoperiod. They were daily fed ad libitum with trout grower pellets. Experiments were conducted in 1992, 1993 and 1994, from March to June.

Blood samples were taken from the caudal artery. T (Testosterone) and E2 (Estradiol-17ß) antisero were purchased respectively from Byck-Sangtes, Diagnostica (D) and Bio-Tecx Laboratorium Inc. (USA) and these steroids assayed by the method described by Lamba *et al.* (1982) and Rinchard *et al.* (1993). Radioactive T4 (thyroxine) and T3 (triiodothyronine) were purchased from Amersham International (UK). T3 and T4 were assayed as described by Byamungu *et al.* (1990). 11-KT (11-Ketotestosterone) was measured by an enzyme immunoassay using acetylcholine esterase as a tracer (Cuisset *et al.* 1992).

In *experiment 1*, the relationship between social status (dominant - subordinate) and hormonal profiles (T, E2, 11-KT, T3 and T4) were analysed in 87 fishes (720 ± 136 g and 640 ± 191g, in males and in females respectively). The following groups were used : i) isolated fish, ii) pairs of males or females, randomly selected, introduced to each other by removal of an opaque separator dividing the aquarium and iii) males under «crowded» conditions (9 fish in 1000 l) (Skoufas *et al.*, 1993).

In *experiment 2*, the aggressive behaviours of 6 females (weight 533 ± 178 g), 15 pseudofemale (454 ± 71 g) and 5 mâles (636 ± 126 g) *O. aureus* were studied in mixed pairs. Moreover 18 females (359 ± 31 g) and 18 pseudofemales (334 ± 50 g) were studied in assemblages of 4, 6 or 8 fish in 500 l and 10 fish in 100 l (Ovidio *et al.*, 1994).

The data on the behaviour were recorded with a video camera system. The social status was determined on the basis of colour and behavioural patterns described by Falter (1987).

Results

Experiment 1

The reproductive behaviours of *O. aureus* were similar to those described by Falter (1986) in *Oreochromis niloticus*. However, *O. aureus* displayed body quiver sequences which are uncommon in other *Oreochromis* species.

Dominant and isolated females exhibited a T plasma level higher than subordinates or in mouthbreeding females: there was a positive linear correlation between the frequency of aggressive patterns and individual plasma levels of T (Fig. 1). E2 plasma level was significantly lower in the 'mouthbreeding' females compared to the other groups. T3, T4, and 11-KT levels in females were independant of social status.

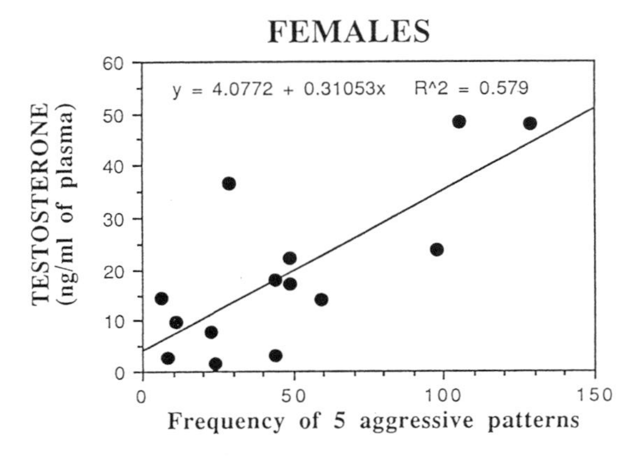

Fig. 1. Relationship between testosterone levels and frequency of 5 aggressive patterns (lateral display, tail beating, biting, jagen, mouth fighting) in the females.

In contrast, 11-KT levels were significantly higher in dominant males than in subordinate or isolated individuals whilst T, T3, T4 and E2 plasma levels were not affected by social status. Similarly, fish stocked at higher densities (9 fish in 1000l), known to inhibit or lower the frequency of aggressive behaviours, showed significantly lower 11-KT levels than dominant males.

Experiment 2

In pairs, the frequency of aggressive behaviours was higher in pseudofemales (PS) which turned out to be dominant towards females (F) in 69 % of the tests (Fig. 2). The average number of mouth fighting in pairs during a 30 min. period is significantly higher between females and pseudofemales than in the other pairs (PS x PS; Males x PS; M x F).

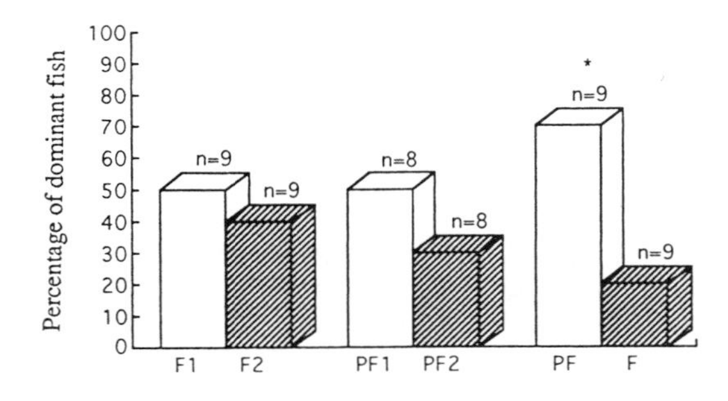

Fig. 2. Percentage of dominant fish in pairs of females (F) and/or pseudofemales (PF), randomly selected, introduced to each other by removal of an opaque separator dividing the aquarium. * Significant results (P<0.05).

Similar results were obtained at higher densities. Pseudofemales were significantly more aggressive than females in groups of 6 fish/500l and 10 fish/100l. When both female and pseudofemales were placed in an aquarium together with a territorial male, only females were observed spawning. Still no significant differences in hormonal profiles (T, E2, T3, T4) were observed between female and pseudofemale tilapia. Similarly, no difference was observed between the activity rhythm patterns of females and pseudofemales: both consistently behaved as typical diurnal individuals throughout the different experiments (Fig.3).

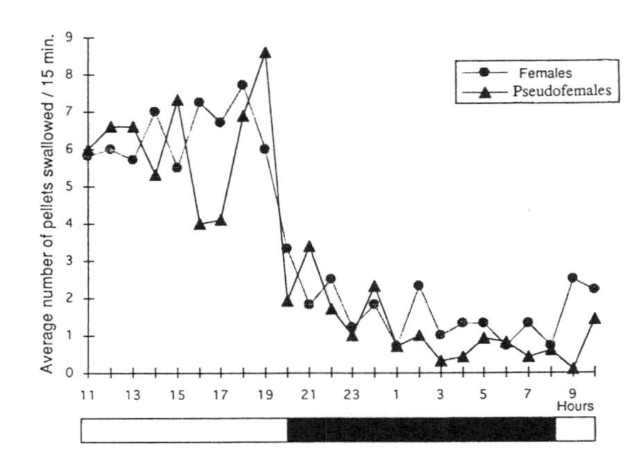

Fig. 3. Daily rhythm of feeding activity (average number of pellets swallowed / 15 min.) in females and pseudofemales.

Discussion and conclusions

This study demonstrated that dominant *O. aureus* females have higher plasma T concentration than subordinate females. When studying reproductive behaviour of the same species, Rothbard *et al.* (1991) suggested that increase of circulating T in females during the acquisition of nuptial coloration and pairing in the nest might be associated with the development of aggressive behaviour towards other females. The lower level of E2 observed in mouthbreeding females is in agreement with the results of Smith and Haley (1988) in *O. mossambicus*.

With respect to male *O. aureus*, the higher concentration of 11-KT in dominant individuals is consistent with the results of Rouger (1991) showing increase in 11-KT and T plasma levels in dominant *O. niloticus*.

The results on sex-reversed fish suggest that pseudofemales were more aggressive and dominant than females (in fighting pairs or under different stocking densities). Moreover, similar to the results of Meriwether and Shelton (1986), pseudofemales never developp sexual behaviours. The similarity between the body and gonad morphology - or GSI - in female and pseudofemale tilapia (Desprez *et al.*, 1995), as well as between their respective thyroidal and steroidal hormonal profiles (Desprez, pers. comm.) suggests that the behavioural differences seen in this study may have another origin. Possibly such variations in agonistic behaviour could originate from neuroanatomical or neurophysiological differences between the brains of normal and sex-reversed fish though little evidence is presently at hand to support such a functional hypothesis.

References

Byamungu, N., Corneillie, S., Mol, K., Darras, V. and Kühn, E.R., 1990. Stimulation of thyroid function by several pituitary hormones results in an increase in plasma thyroxine and reverse triiodothyronine in tilapia (*Tilapia nilotica*). Gen. Comp. Endocrinol. 80: 33-40.

Cuisset, B., Paradelles, P., Kime, D.E., Kühn, E.R. and Le Menn, F., 1992. Enzyme immunoassay for 11-Ketotestosterone using acethylcholine esterase tracer. In: Abstracts of the 2nd Inter. Symp. of Fish Endocrinol., p. 76, Saint Malo, France.

Desprez, D., Mélard C. and Philippart J.C., 1995. Production of a high percentage of male offspring with 17a-ethynylestradiol sex-reversed *Oreochromis aureus*. II. Comparative reproductive biology of females and F2 pseudofemales and large-scale production of male progeny. Aquaculture 130: 35-41.

Falter, U., 1986. Fluctuations journalières dans le comportement territorial chez *Oreochromis niloticus* (Teleostei: Cichlidae). Annls Soc. Roy. Belg. 2: 175-190.

Lamba, V.J., Goswami, S.V. and Sundarajai, B.I. (1982). Radioimmunoassay for plasma cortisol, testosterone, estradiol-17ß and estrone in catfish, *Heteropneustes fossilis* (Bloch): development and validation. Gen. Comp. Endocrinol. 47: 170-181.

Mélard, C., 1995. Production of a high percentage of male offspring with 17a-ethynylestradiol sex-reversed *Oreochromis aureus*. I. Estrogen sex-reversed and production of F2 pseudofemales. Aquaculture 130: 25-34.

Meriwether, F.H. and Shelton, W.L., 1986. Observation on aquarium spawning of estrogen-treated and untreated tilapia. Proc. Ann. Conf. S.E. Assoc. Fish Wildl. Agencies 34: 81-87.

Ovidio, M., Desprez, D., Poncin, P., Mélard, C., Mol, K. and Kühn, E., 1994. Comparison of the behaviour of females and sex-reversed females tilapias *Oreochromis aureus* in aquaria. In: Lecture and Posters Abstracts of the 1st Benelux Congress of Zoology, p. 80, Leuven, Belgium.

Rinchard, J., Kestemont, P., Kühn, E.R., and Fostier, A. (1993). Seasonal changes in plasma levels of steroid hormones in an asynchronous fish, the gudgeon *Gobio gobio* L. (Teleostei, Cyprinidae). Gen. Comp. Endocrinol. 92: 168-178.

Rothbard, S., Ofir, M., Levavi-Sivan, B. and Yaron, Z., 1991. Hormonal profile associated with breeding behaviour in *Oreochromis aureus*. In: Reproductive Physiology of Fish (A.P. Scott, J.P. Sumpter, D.E. Kime & Rolfe M.S. eds); 185-187. Published by FishSymp 91, Sheffield, U.K.

Rouger, Y., 1991. Déterminisme du comportement territorial et sexuel ches les mâles de *Oreochromis mossambicus* et *O. niloticus*. *In* «Abstracts of the Third International Symposium on Tilapia in Aquaculture», Ed. ICLARM, p. 74.

Skoufas, G., Poncin, P., Byamungu, N., Cuisset, B., Mol, K., Kühn, E.R., Mélard, C. and Ruwet, J.C., 1993. Sexual and social behaviour in *Oreochromis aureus* (Pisces: Cichlidae): Endocrine profiles. Bel. J. Zool. 122: p. 251.

Smith, C. J. and Haley, S.M., 1988. Steroid profiles of the female tilapia, *Oreochromis mossambicus*, and correlation with oocyte growth and mouthbrooding behavior. Gen. Comp. Endocrinol. 69: 88-89.

BEHAVIOR, BRAINS, AND BIOPHYSICS: STEROIDAL MODULATION OF COMMUNICATION SIGNALS IN ELECTRIC FISH

H. Zakon, M.B. Ferrari, and J. Schaefer

Department of Zoology, University of Texas, Austin, TX 78712

Summary

The electric organ discharges (EODs) of electric fish are sexually-dimorphic signals that convey information on sex, reproductive status, and motivational state. The neurons and effectors which generate these signals are simple, accessible for biophysical examination, and modified by sex steroids. In the genus *Sternopygus* males make longer EOD pulses than females and the EOD pulse is broadened by androgens. We have identified a voltage-dependent Na^+ current in the electric organ which shuts off rapidly in females, slowly in males, and whose voltage-dependent kinetics are altered by androgen treatment. In the genus *Apteronotus*, in which the EOD frequency of females is higher than males and in which EOD frequency is lowered by estradiol 17-ß (E_2) and raised by 11 ketotestosterone, we found that E_2 lowers and 11 KT raises the endogenous firing frequency of pacemaker and electromotorneurons.

Introduction

A primary role of gonadal steroids in reproduction in vertebrates is to coordinate reproductive behaviors. A complete understanding of how hormones accomplish this necessitates identifying all of the neurons involved in the generation of the behaviors, and the synaptic interactions and intrinsic ion conductances of each type of neuron. This is a daunting task since many behaviors are controlled by distributed circuits with numerous cell types and complex synaptic interactions. However, the steroid-sensitive circuits controlling the communication signals of weakly electric fish are ideal for this analysis.

We have been studying hormonal modulation of the EOD of two species that generate sinusoidal EODs that are sexually dimorphic in their frequency and waveshape. In *Sternopygus macrurus*, the gold-lined black knifefish, sexually mature males discharge at low frequencies (50-90 Hz) and sexually mature females discharge at higher frequencies (110-200 Hz). In *Apteronotus leptorhynchus* mature males discharge at high frequencies (800-1,000 Hz) while mature females discharge at lower frequencies (600-800 Hz).

Treatment of adult or juvenile *Sternopygus* of both sexes with testosterone (T) or dihydrotestosterone (DHT) lowers EOD frequency and broadens the duration of each EOD pulse. In keeping with the opposite direction of the sexual dimorphism in EOD frequency in *Apteronotus*, the androgen 11 ketotestosterone (11 KT) raises EOD frequency and E_2 lowers it.

The EOD is generated by a medullary nucleus, called the pacemaker nucleus (PMN), which comprises two neuron types: pacemaker and relay neurons. The pacemaker neurons are endogenously active at exceedingly stable rates. These drive relay neurons which send their axons down the spinal cord to activate electromotorneurons (EMNs). In *Sternopygus*, the axons of the EMNs innervate a muscle-derived electric organ. The waveform of its EOD is jointly determined by the pacemaker nucleus, which determines EOD frequency, and the cells of the electric organ, the electrocytes, which determine the duration of each pulse. In *Apteronotus* the axons themselves form the electric organ.

Sternopygus Electric Organ

Electrocytes were depolarized by current injection to generate an action potential (AP). We found that electrocyte AP duration was negatively correlated with EOD frequency: fish that produced a high-frequency EOD had APs that lasted 4-5 msec, while those that produced a low-frequency EOD had APs that lasted up to 12 or 14 msec. Chronic (2 weeks) treatment of fish with DHT caused an increase in AP duration when compared with baseline AP durations before hormone treatment (Mills and Zakon, 1991). No difference in AP duration was observed in the control group (Fig. 1).

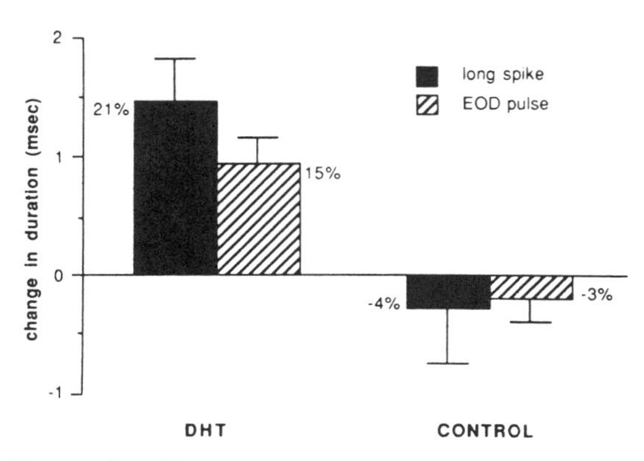

Figure 1. The mean change in duration of EOD pulse and long spike in all fish, with SE bars, in the DHT-treated and control fish. *DHT*: mean pulse duration increase, 0.95 ± 0.21 msec; mean spike duration increase, 1.48 ± 0.35 msec. *Control*: mean pulse change, –0.19 ± 0.21 msec; mean spike change, 0.28 ± 0.46 msec.

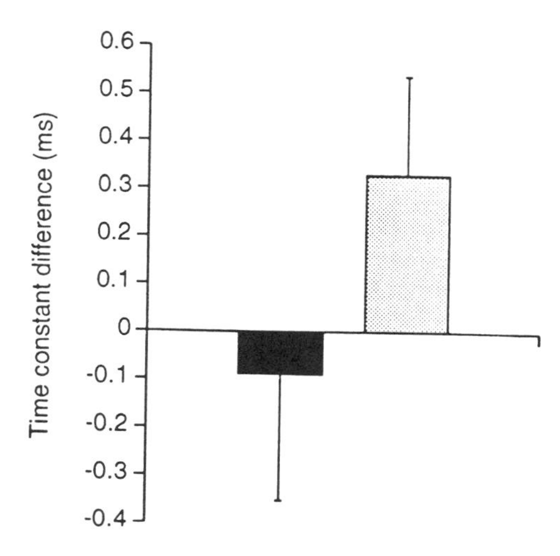

Figure 2. The effects of DHT on inactivation time constants of the peak Na^+ current. Baseline measures of inactivation time constants were calculated from one to five electrocytes per fish and averaged. These parameters were measured again 3 - 4 weeks after androgen or control implant. Pre- and postimplant mean values were subtracted for each fish, and the group difference scores were plotted.

In order to identify ion currents which DHT might modulate, we isolated each ion current and studied its voltage- and time-dependent kinetic properties with the voltage clamp. We determined that the electrocytes have two K^+ currents and a Na^+ current (Ferrari and Zakon, 1993). Na^+ currents are self-limiting: they are activated by membrane depolarization and then inactivated (shut off) by these same depolarizations. The Na^+ current of *Sternopygus* electrocytes showed the unusual property of inactivating at different rates depending on EOD frequency. That is, a fish with a high EOD frequency and short AP duration had a Na^+ current that shut off rapidly (time constant of < 1 msec) while those with a low EOD frequency and long duration AP had Na^+ currents that inactivated slowly (time constants > 2.5 msec) (Ferrari and Zakon, 1995).

We presumed, then, that this current was a target of DHT and studied Na^+ current kinetics in electrocytes of fish that had been implanted with DHT or control capsules. Na^+ current inactivation of fish implanted with DHT was significantly slower than their baseline inactivation values ($p < 0.008$, Mann-Whitney) whereas Na^+ current kinetics of fish implanted with control capsules were no different (Fig. 2).

We are currently attempting to understand how DHT exerts its effects on the Na^+ current. Given the long time course of its actions and the presence of androgen receptors in the electrocytes (Gustavson et al., 1994), we presume that it acts *via* a genomic route. One hypothesis is that DHT up- or down-regulates the transcripion or splicing of different molecular isoforms of the Na^+ channel gene each of which have different inactivation kinetics. To this end, using the polymerase chain reaction, we have cloned and sequenced Na^+ channel mRNA from low EOD frequency males (G. Lopreato, N. Atkinson, H. Zakon, in prep.). We will do the same for a high EOD frequency female and see if there is any evidence of molecular variants of the channel.

Apteronotus Electromotorneurons

In most electric fish so far studied, the EMNs are electrically silent and are driven by the PMN. The EMNs of *Apteronotus*, on the contrary, oscillate at each individual's EOD frequency when acutely severed from the PMN *in vivo* or in an *in vitro* spinal cord slice. To test whether this is dependent on long-term stimulation of the EMNs by the PMN we recorded EOD frequency, transected the spinal cord, and recorded the oscillation frequency of EMNs in the slice two weeks later. We found that the EMNs still oscillate at the pre-transection EOD frequency.

We then wished to determine whether steroid hormones shift the endogenous firing rate of neurons in the PMN and the EMNs. After recording EOD frequency we implanted fish with either empty capsules or capsules containing E_2 or 11 KT and recorded from neurons in the PMN two weeks later. The firing rate of PMN cells in the E_2-implanted fish was lower than their initial EOD frequencies, it was higher in the 11 KT-implanted fish, and no different in the controls ($p < .001$, Mann-Whitney U test). We then determined that the same was true for the EMNs using the same design except that the spinal cord was transected before the capsules were implanted. Thus, PMN and EMN oscillation rates are raised by androgen and lowered by estrogen treatments but are independently controlled since the EMNs showed similar changes to the PMN despite being synaptically disconnected from it (Schaefer and Zakon, 1995).

Our next goal is to develop a whole-cell voltage clamp to identify the ion currents that underlie the oscillatory behavior of these cells, and to determine which ion currents are modulated by steroids. Once these are known, we will clone and sequence the genes for them and study how steroids act on a molecular level.

References

Ferrari, M.B. and H.H. Zakon, 1993. Conductances contributing to the action potential of Sternopygus electrocytes. J. Comp. Physiol. A 173: 281-292.

Ferrari, M.B., and H.H. Zakon, 1995. Individual in and androgen-modulation of the sodium current in electric organ. J. Neurosci. 15: 4023-4032.

Gustavson, S., H. Zakon and G.Prins, 1994. Androgen receptors in the brain, electroreceptors and electric organ of a wave-type electric fish. Abstr. Soc. Neurosci. 20: 371.

Mills A, Zakon HH (1991) Chronic androgen treatment increases action potential duration in the electric organ of Sternopygus. J Neurosci **11**: 2349-2361.

Schaefer, J. and H. H. Zakon, 1995. Antagonistic actions of androgen and estrogen on the intrinsic firing rate of a specialized motoneuron. Abstr. Soc. Neurosci. (in press).

ANDROGEN-INDUCED CHANGES IN CHIRPING BEHAVIOR ARE CORRELATED WITH CHANGES IN SUBSTANCE P-LIKE IMMUNOREACTIVITY (SPl-ir) IN THE BRAIN OF THE WEAKLY ELECTRIC FISH, *Apteronotus leptorhynchus*

Joseph G. Dulka

Department of Biology & The Nebraska Behavioral Biology Group, Creighton University, Omaha, NE 68178 U.S.A.

Summary

The steroidal regulation of sex differences in electrocommunicatory behavior ("chirping") and brain SPl-ir were examined in the weakly electric fish, *Apteronotus leptorhynchus*. The results suggest that androgens regulate the onset of chirping in females, and cause both specific and wide spread increases in brain SPl-ir. Androgen-induced changes in SPl-ir may underly behavioral changes in chirping, since treated females showed a male-like pattern of SPl-ir in regions normally devoid of the peptide, including the prepacemaker nucleus (PPn), the command center for chirping behavior.

Introduction

Sexually dimorphic reproductive behaviors and their corresponding neuroregulatory systems have been described in a number of vertebrates (Breedlove, 1992). Many of the brain regions that control sexually dimorphic behaviors are known to be influenced by the actions of gonadal steroids. Two mechanisms of steroid action on brain and behavior have been proposed: 1.) steroids may act early during critical periods to *organize* neural systems for the expression of specific behaviors later in life, or 2.) they may *activate* neural systems which are already present in adulthood (Arnold and Breedlove, 1985). In both cases, the specific behaviors in question normally do not occur without steroidal modulation.

This report summarizes recent findings on a teleost social communication system that shows prominent sex differences in brain and behavior, and a high degree of steroidal sensitivity in adulthood. The neuroethological model to be described involves the production of electric social signals, or "chirps", by the brown ghost knifefish, *Apteronotus leptorhynchus*.

Electric Organ Discharge, Chirping Behavior and Social Communication

The brown ghost knife fish emits and detects a quasi-sinusoidal electric organ discharge (EOD) which is used for both electrolocation and intraspecific communication. During the reproductive season, the frequency of the EOD is both species-specific and sex-specific, with females emitting at lower EOD frequencies than males (Meyer et al., 1987). Moreover, the gradual lowering of female discharge rate during the reproductive season appears to be mediated by seasonal changes in circulating levels of gonadal steroids (Zakon et al. 1991).

The brown ghost can also modulate its EOD waveform to produce discrete communicatory signals, or chirps, which are characterized by a brief increase in EOD frequency and a simultaneous decrease in EOD amplitude. Although chirps are produced only during the performance of agonistic or reproductive behaviors, the types of chirps emitted under these two conditions differ with respect to their structure or quality. Reproductive chirps have longer durations and more pronounced frequency and amplitude modulations than aggressive chirps (Hagedorn and Heiligenberg, 1985).

Sex Differences in Stimulus-Evoked Chirping

The brown ghost will also produce chirps in response to a artificial sinusoidal signal that mimics the presence of a conspecific; such signals are normally delivered to the water at frequencies 1-10 Hz above or below an animal's own discharge frequency. However, when tested under these conditions, males and females show clear differences in their propensity to chirp; males readily chirp in response to artificial electrosensory stimuli, whereas females generally do not (Dulka and Maler, 1994). The general lack of chirping in females does not appear to be due an inability to detect the stimulus wave form, since they readily perform the jamming avoidance response (JAR) in response to a stimulus frequency 3 Hz below their own EOD frequency. The JAR is a behavior in which an animal gradually shifts the frequency of its EOD away from an interfering signal of similar frequency. An animal will perform the JAR to minimize the detrimental effects of a neighbor's EOD on its own ability to electrolocate.

The finding that females chirp less than males under laboratory testing conditions was perplexing, since both sexes have been reported to chirp during courtship and spawning (Hagedorn and Heiligenberg, 1985). However, since we could not be certain that our females were in peak reproductive condition, (i.e. they complete gonadal recrudescence, but rarely spawn in the laboratory), it remained to be determined if acute changes in gonadal maturation and/or endocrine state play a role in regulating female chirp propensity. If so, gonadal steroids might be considered as likely mediators of behavioral shifts in female chirping activity. To examine this possibility, we tested the effects of androgens on stimulus-evoked chirping in females, since testosterone (T) is known to increase in the blood of a variety of teleosts during the periovulatory period.

Androgen Effects on Chirping Behavior in Females

The following information summarizes a series of experiments that examined whether androgens play a role in regulating female chirp responsiveness (Dulka and Maler, 1994; Dulka et al. 1995). In all cases, females were first screened to establish pretreatment values for basal EOD frequency (Hz), magnitude of the JAR (Hz), and the incidence of chirping behavior (chirps/ 30 s). The fish were then randomly assigned to different groups and implanted (ip.) with either silastic capsules alone (control) or capsules containing sufficient testosterone (T) or dihydrotestosterone (DHT) to create a final dose of 100 mg/g body weight. Treated females were then repeatedly tested over a 4-5 week period for their propensity to chirp in response to standardized electrosensory stimuli; the JAR and basal EOD were also monitored. In some experiments, the behavioral responses were digitized and stored on video tape for off-line computer analysis of chirp structure.

The results were similar in all experiments. Females treated with T and DHT produced significantly more chirps than controls (Dulka and Maler, 1994; Dulka et al., 1995). Moreover, the chirps recorded from the androgen-implanted females had longer durations and more dramatic frequency and amplitude modulations compared to controls (Dulka et al. 1995), and appeared similar to those reported to be produced by spawning individuals (Hagedorn and Heiligenberg, 1985). In addition, T, but not DHT, caused a gradual lowering of basal EOD frequency similar to that observed in reproductive females. The effects of T on chirping behavior and basal EOD frequency appear specific, since the magnitude of the JAR was not affected by hormone treatment. Taken together, the results suggest that androgens cause both an enhancement of female chirping and an alteration in chirp structure that may be related to reproduction. However, it remains to be determined if androgen treatment is mimicking changes in chirp rate and chirp structure that normally occur during the reproductive season.

Brain-Steroid Interactions and Chirping Behavior

In a second series of experiments, we began to ask questions about the central mechanisms that underly androgen-induced changes in chirping behavior. Although many possibilities exist, we focused on steroid-neuropeptide interactions, and more specifically, on androgen-Substance P interactions, for two reasons. First, substance P-like immunoreactivity (SPl-ir) is sexually dimorphic in the brain of brown ghosts: this peptide is present in a number of diencephalic nuclear groups of males but not females (Weld and Maler, 1992). One of these regions is the prepacemaker nucleus (PPn), the command center for chirping behavior. The PPn of males, which normally chirp, receives a substantial SPl-ir projection whereas the same region in females, which normally do not chirp, is completely devoid of SPl-ir. Second, micro-injections of SP into the vicinity of the PPn are effective at evoking chirping in acute preparations (L. Maler, personal communication). Taken together, these findings implicate SP as a potential neuromodulator of chirping behavior. To test this possibility, we repeated the above experiments and examined whether androgen treatment causes changes in SPl-ir in the brains of females (Dulka et al., 1995).

Androgen Effects on Brain SPl-ir in Females

As in our previous experiments, androgen treatment (T and DHT) caused both an induction of chirping behavior and an alteration of chirp structure in females. Moreover, androgen-induced changes in chirping were correlated with increased expression of SPl-ir within specific brain nuclei of females (Table 1). These changes may underly behavioral changes in chirping, since treated females showed a male-like pattern of SPl-ir in the PPn. However, alterations in SPl-ir were not restricted to the PPn, but also occurred in diencephalic regions related to pituitary function

(PPa, Ha, Hv, Hl) and reproductive behavior (PPa, CP; Table 1). These changes appear specific since only diencephalic brain regions known to be sexually dimorphic for the peptide were affected by the treatments.

Table 1. Qualitative ranking of SPl-ir in the diencephalon of normal males and in female *A. leptorhynchus* implanted (45 days) with either silastic capsules alone (CON) or capsules containing testosterone (T) or dihydrotestosterone (DHT).

BRAIN REGION	FEMALE CON	FEMALE T	FEMALE DHT	MALE
PPa	_	+	+	++
CP	_	+	+	++
PPn	_	+	+	++
Ha	_	+	+	++
Hv	_	++	++	++
Hl	++	++	++	++
Hc	++	++	++	++

Each symbol represents the average pattern of staining observed in 4 animals; -, complete lack of staining; +, weak to moderate staining; ++, moderate to strong staining. Abbreviations: CP, central-posterior thalamic nucleus; Ha, hypothalamus anterioris; Hc, hypothalamus caudalis; Hl, hypothalamus lateralis; Hv, hypothalamus ventralis; PPa, nucleus preopticus periventricularis anterioris; PPn, prepacemaker nucleus (modified from Dulka et al., 1995)

Discussion

The results suggest that androgens modulate chirping activity and cause both specific and wide spread changes in SPl-ir that may relate to a functional system which interrelates pituitary function, reproductive behavior and chirping.

We assume that androgen-induced changes in SPl-ir to the PPn may underly behavioral changes in female chirping activity. However, although this species exhibits clear sexual dimorphisms in both chirping behavior and SPl-ir staining, the androgen-induced increases in SPl-ir alone cannot account for all the behavioral changes seen in treated females. For example, although SPl-ir is more pronounced in males than in females, males rarely, if ever, give chirps with high frequencies, long durations and large amplitude modulations (Zupanc and Maler, 1994). However, control females with little or no SPl-ir in the PPn rarely, if ever, chirp when tested under the same conditions. Therefore, androgen-induced increases in SPl-ir input to the PPn may be mainly responsible for the induction of female chirping behavior. In contrast, concomitant changes in chirp structure may be mediated by a separate physiological mechanism, possibly involving the actions of different neurotransmitters or direct hormonal effects on the neural circuitry that controls chirping.

References

Arnold AP, Breedlove SM (1985) Organizational and activational effects of sex steroids on brain and behavior: a reanalysis. Horm. Behav. 19: 469-498

Breedlove SM (1992) Sexual dimorphism in the vertebrate nervous system. J Neurosci 12: 4133-4142

Dulka JG, Maler L (1994) Testosterone modulates female chirping behavior in the weakly electric fish, *Apteronotus leptorhynchus*. J Comp Physiol A 174:331-343.

Dulka JG, Maler L, Ellis W (1995) Androgen-induced changes in electrocommunicatory behavior are correlated with changes in substance P-like immunoreactivity in the brain of the electric fish *Apteronotus leptorhynchus*. J. Neurosci. 15: 1879-1890

Hagedorn M, Heiligenberg W (1985) Court and spark: electric signals in the courtship and mating of gymnotoid fish. Anim Behav 33: 254-265

Meyer JH, Leong M, Keller CH (1987) Hormone-induced and maturational changes in electric organ discharges and electroreceptor tuning in the weakly electric fish *Apteronotus*. J Comp Physiol A 160: 385- 394

Weld MM, Maler L (1992) Substance P-like immunoreactivity in the brain of the gymnotiform fish *Apteronotus leptorhynchus*: presence of sex differences. J Chem Neuroanat 5: 107-129

Zakon HH, Thomas P, Yan H-Y (1991) Electric organ discharge frequency and plasma sex steroid levels during gonadal recrudescence in a natural population of the weakly electric fish *Sternopygus macrurus*. J Comp Physiol A 169: 493-499

Zupanc GKH, Maler L (1994) Evoked chirping in the weakly electric fish, *Apetronotus leptorhynchus*. Can J Zool 71: 2301-2310

FEMALE GOLDFISH APPEAR TO RELEASE PHEROMONALLY-ACTIVE F-PROSTAGLANDINS IN URINARY PULSES

C.W. Appelt, P.W. Sorensen, and R. G. Kellner

Dept. Fisheries & Wildlife, University of Minnesota, St. Paul, MN 55108 U.S.A.

Introduction

Prostaglandin $F_{2\alpha}$ ($PGF_{2\alpha}$) and its metabolite 15-keto prostaglandin $F_{2\alpha}$ (15K-$PGF_{2\alpha}$) are potent olfactory stimulants with pheromonal functions in the goldfish (*Carassius auratus*; Sorensen, et al. 1988). Sorensen (this symposium) has shown that both compounds are released into the water by ovulated and $PGF_{2\alpha}$-injected females along with 5 other PGF metabolites which have little olfactory activity. However, the routes and temporal pattern of F-prostaglandin (PGF) release are unknown. In moths, pulsed release of pheromones is essential for males to locate females effectively (Mafra-Neto and Cardé, 1994). Thus, the manner by which PGFs are released by fish may directly determine their function as behavioral stimuli. Using $PGF_{2\alpha}$-injected goldfish as a model, this study sought to determine: 1) routes of PGF release and, 2) whether females release PGFs in pulses.

1) Determining routes of PGF release

We hypothesized that PGFs are released via the urine and gills. Nine female goldfish (26-39 g) were anesthetized, implanted with urinary catheters (PE tubing), injected with 1 µg/g body wt. tritiated-$PGF_{2\alpha}$, and placed into 1 L of 18°C water for 2 h. PGFs from urine and water (i.e. gills only) were extracted with C-18 columns (Sep-Pak). To check catheter results, silicone-sealed urinary plugs were sutured into the urogenital opening of 5 other females to prevent urine release. HPLC was used to identify released PGFs (Sorensen, this symposium). Approximately 73 ± 5% (mean ± S.E.M.) of all PGFs released were in urine and 27 ± 5% in gill water (Fig. 1). 15K-$PGF_{2\alpha}$ made up 25 ± 4% of the PGFs in the urine but was identified in only one water sample (1 ± 1%). $PGF_{2\alpha}$ comprised 14 ± 2% of all PGFs in the urine and 54 ± 10% and 59 ± 11% in the catheter and urinary plug (not shown) waters, respectively. The similarity of catheter and urinary plug water composition indicates that the catheter results are valid. Gill dam results showed the same trend (data not shown).

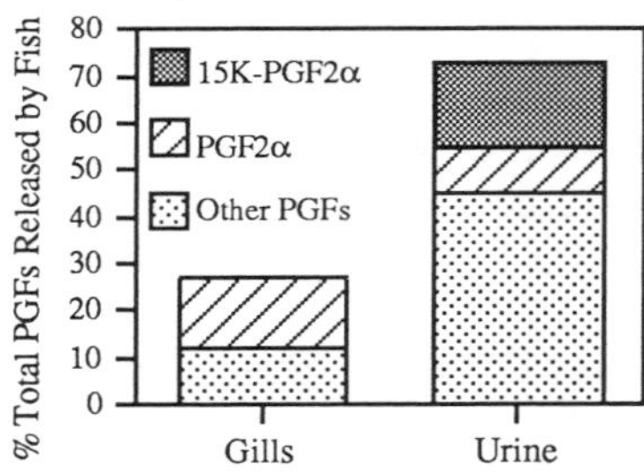

Fig. 1. Proportion of PGFs released and their sources (catheter experiment)

2) Determining whether urination is pulsed

Four males and females were held in 2 L of water for at least 3 days. Each female (18-25 g) was injected with 20 µg tritiated $PGF_{2\alpha}$, and placed with a male, spawning substrate, and 3 airstones (for maximum mixing) in 1.5 L of 18°C water. Water samples (5 ml) were taken at 1 min. intervals from 10 to 35 min. after injection. Urination was assumed to occur when radiation levels increased by more than what could be accounted for by gill release for that interval (3,500 cpm) and sampling error (±5,000 cpm). The latter was determined by manually adding radiation to the water, sampling the water, and then doubling the largest range between consecutive samples. We defined a urinary pulse as a urination event (>8,500 cpm) followed by an interval (1+ min.) of no urination (see Fig. 2). Three pairs of fish exhibited spawning behavior and released 10 ± 1 pulses (average urination = 27,645 ± 2,339 cpm.) per 35 min. Our design could not identify pulse rates greater than 1 pulse/ 2 min. and likely resulted in underestimating the number of pulses.

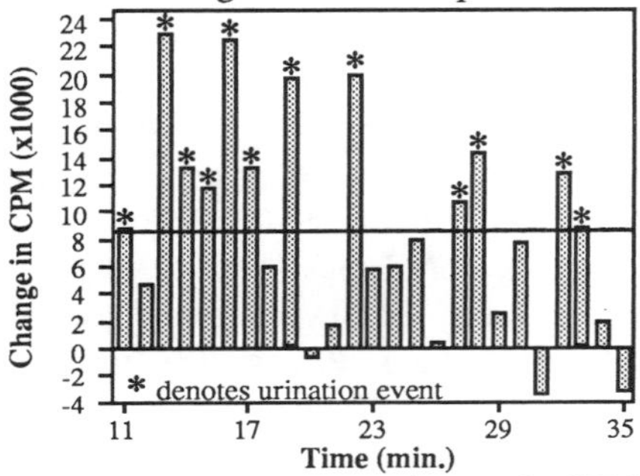

Fig. 2. Minute-to-minute changes in CPM released by injected fish. Urination events (*) are changes in CPM exceeding that expected from gill release plus sampling error (shown by horizontal line).

Discussion and References

Three times more PGFs was released via urine than gills. 15K-$PGF_{2\alpha}$ was released exclusively via urine and seemed to be pulsed, while $PGF_{2\alpha}$ was released via both urine and gills. Ongoing studies using a visible dye also support the hypothesis that urine release is pulsed. Since naturally ovulated females release far more 15K-$PGF_{2\alpha}$ (Sorensen, this symposium), which is more potent than $PGF_{2\alpha}$ (Sorensen, et al. 1988), release of urine in pulses is likely an important factor in modifying male behavior.

Mafra-Neto, A. & R.T. Cardé. 1994 Fine-scale structure of pheromone plumes modulates upwind orientation of flying moth *Nature* 369:142-144.

Sorensen, P.W., T.J. Hara, N.E. Stacey, & F.W. Goetz. 1988. F prostaglandins function as potent olfactory stimulants that comprise the post ovulatory female sex pheromone in goldfish. *Biol. of Reprod.* 39:1039-1052.

Supported by the National Science Foundation and Minnesota Agricultural Experiment Station

SPERMIATED SEA LAMPREY RELEASE A POTENT SEX PHEROMONE

R. Bjerselius[1], W. Li[2], P.W. Sorensen[1] and A.P. Scott[3]

[1]Dept. Fisheries and Wildlife, Univ. Minnesota, 200 Hodson Hall, 1980 Folwell Ave., St. Paul, MN 55108, USA; [2]Monell Chemical Senses Center, 350 Market Street, Philadelphia, PA 19104, USA; [3]Ministry of Agriculture Fisheries and Food, Fisheries Laboratory, Lowestoft, Suffolk, UK

INTRODUCTION

The sea lamprey, *Petromyzon marinus*, is a primitive, jawless fish which has a complex and dramatic life history. After hatching in streams, they spend 5+ years as filter-feeding larvae before metamorphosing into a parasitic stage which then moves into oceans or lakes to feed. Parasitic lamprey grow rapidly for a year and then enter a migratory phase. Inland migration is dramatic and brief (4-8 weeks), associated with sexual maturation, and culminates in spawning and death. Male sea lamprey build nests in rivers and are later joined by females. French fishermen have been using mature males to attract and trap female sea lamprey for decades, suggesting the existence of a pheromonal communication system (Teeter, 1980). A plethora of evidence now suggests that teleost fish use both free and conjugated steroids to function as sex pheromones. This study pursues this possibility by characterizing the olfactory sensitivity of sea lamprey to a variety of synthetic, crude, and partially purified odorants.

MATERIALS AND METHODS

Adult sea lamprey were trapped on their spawning grounds and shipped to Minnesota, USA. Electro-olfactogram recording (EOG) was used to test the olfactory responsiveness of adult sea lamprey (see Li *et al.*, 1995). Odorants were diluted in well water for testing and all responses were compared to that elicited by our standard, 10^{-5} M L-arginine.

1) *Characterizing olfactory sensitivity to washings of adult male and female conspecificss.* Washings from recently ovulated females, non-ovulated females, spermiated males, and non-spermiated males were collected and tested by EOG (1 fish/10 L/4 h). **2)** *To elucidate the possibility of conjugation.* An aliquot of male water was fractionated by injecting onto a DEAE column as described by Scott (1994), fractions collected at times corresponding to where free, glucuronidated, and sulfated sex steroids elute, and tested by EOG.

RESULTS

1) Potency of lamprey washings

The odor of spermiated sea lamprey was considerably more potent than washings of immature males and females which had equivalent potencies (Fig. 1). Olfactory sensitivity was not sexually dimorphic. The detection threshold of mature male water was 1:1000000.

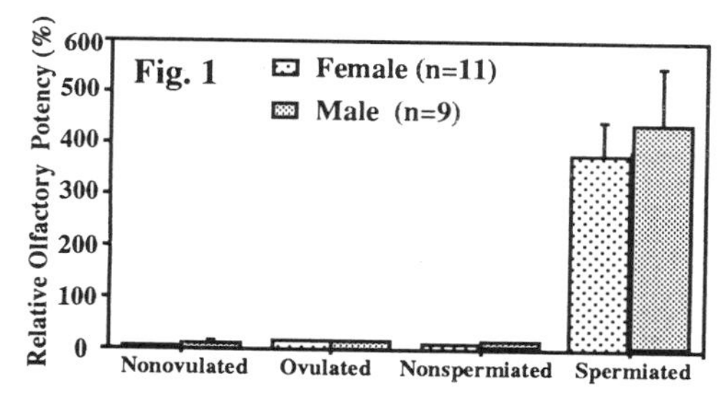

2) Potency of spermiated male washings after DEAE fractionation. Although most of the fractions had some olfactory activity, those fractions eluting where sulfated steroids were expected, were the most potent (Fig. 2). Cross-adaptation between the fractions and the crude pheromone suggested that only one compound was present (data not shown).

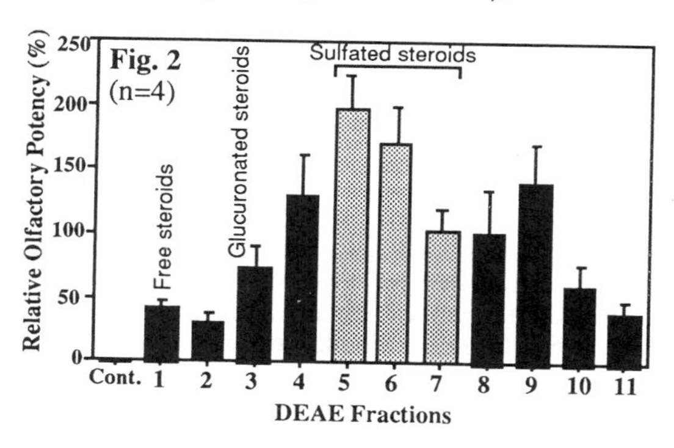

DISCUSSION

It is clear that male sea lamprey release potent odorant(s) when sexually mature. Our finding supports several behavioral studies which have shown females to be attracted to this odor (Li, 1994). Although the identity of the male sex pheromone is unclear, a sulfated gonadal steroid, which may have more than one conjugating group is suspected. This possibility is also supported by the fact that urine, which is known also to contain sulfates (Scott, 1994), is a strong female attractant for this species (Teeter, 1980). Sex steroid pathways are poorly understood in male lamprey but appear extremely unusual (Kime & Rafter, 1981; Sower, 1989). This, and the fact that EOG tests with 60 synthetic commercially available steroids were inconclusive (Li, 1994), leads us to believe the pheromone is a novel compound.

Kime, D & Rafter, J. 1981. Biosynthesis of 15-hydroxylated steroids by gonads of the river lamprey, *Lampetra fluviatilis, in vitro*. Gen. Comp. Endocrinol. 44: 69-76.

Li, W. 1994. The olfactory bilogy of adult sea lamprey (*Petromyzon marinus*). Unpublished Ph.D. thesis, University of Minnesota, USA. 184 pp.

Li, W. et al. 1995. The olfactory system of the migratory sea lamprey (*Petromyzon marinus*) is specifically and acutely sensitive to unique bile acids released by conspecific larvae. J. Gen. Physiol. 105: 569-587.

Scott, A.P. 1994. The production of conjugated steroids by teleost gonads and their roles as pheromones. In: Perspectives in Comparative Endocrinology (eds. Davey, K.G. *et al.*). pp. 438-448. National Research Council of Canada, Ottawa.

Sower, S. A. 1989. Effects of lamprey gonadotropin-releasing hormone and analogs on steroidogenesis and spermiation in male sea lampreys. Fish Physiol. Biochem. 7:101-106.

Teeter, J. 1980. Pheromone communication in sea lampreys (*Petromyzon marinus*): Implications for population management. Can. J. Fish. Aquat. Sci. 37: 2123-2132.

Supported by the Great Lakes Fisheries Comission and Minnesota Sea Grant

OLFACTORY HYPERSENSITIVITY TO SEX PHEROMONES IN BLIND CAVE FISH

J.R. Cardwell and N.E. Stacey

Department of Zoology, University of Alberta, Edmonton, AB, T6C 3K9, Canada

Introduction

Despite demonstrations that the use of hormones or their metabolites as sex pheromones is widespread in teleosts (Stacey and Cardwell, this volume), there is little evidence to suggest how this has come about. Most fishes are likely preadapted to use released hormones as sex pheromones since fish routinely void hormones to the water as part of their metabolic activity. Subsequently, some species may evolve mechanisms to detect released sex hormones that are rich in information about the sender's reproductive condition. This form of *chemical spying* (Stacey and Sorensen, 1991) allows the receiver to predict reproductive events and to synchronize gamete and behavioral readiness with those of the sender.

Unfortunately, it has not been shown that selection promotes changes in the ability to detect released hormones. In this paper, we use electro-olfactogram recording (EOG, methods in Cardwell *et al.*, 1995) to examine peripheral olfactory responses to prostaglandin-$F_{2\alpha}$ (PGF) in the mexican blind cave tetra, an eyeless, unpigmented form of *Astyanax mexicanus* (Characiformes). This form is believed to be derived from sighted, riverine conspecifics since the two forms interbreed (Mitchell *et al.*, 1977). Since compensatory selection for enhanced olfactory sensitivity to sex pheromones might be expected in the absence of visual cues, comparison of the EOG responses to sex pheromones in the blind and sighted forms of *A. mexicanus* may provide evidence for the evolution of chemical spying.

Results and Discussion

The blind cave tetra exhibited extraordinary large responses to PGF (Fig. 1), a pheromone that initiates male sexual behavior in this species (Cardwell, unpublished). The detection threshold (sensitivity) was lower in males than in females (Table 1). In some individuals, the response magnitude was greater than 50 mV, more than 50-fold higher than responses recorded from goldfish, (*Carassius auratus;* eg. Fig. 1) in which PGF pheromones are well characterized. However, the sighted form of this species was also extremely sensitive to PGF (Table 1).

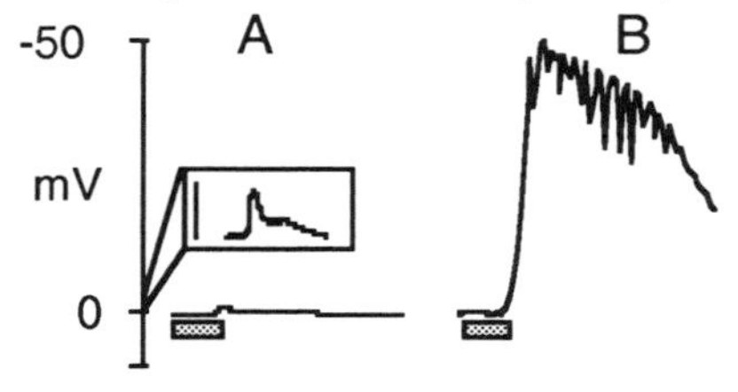

Figure 1. Representative 10-second EOG recordings of responses to 10 nM PGF in male goldfish (A) and (B) blind cave tetra. Horizontal bars are stimulus periods. Inset: magnification of A, scalebar = 1 mV.

Table 1. Magnitude and sensitivity of responses to $PGF_{2\alpha}$ in females of the blind form and males of both blind and sighted forms of *Astyanax mexicanus.*

	MAGNITUDE at 10 nM			SENSITIVITY
	Mean	SEM	(n)	
Female (blind)	15.27	3.97	mV (5)	1 pM
Male (blind)	38.44	1.31	mV (5)	0.1 pM
Male (sighted)	41.5	4.53	mV (5)	0.1 pM

Two hypotheses may account for the observed similarity of response in blind and sighted male *A. mexicanus*. First, there may have been selection for enhanced olfactory sensitivity in the cave-dwelling population, but genes coding for this feature may have then introgressed into the sighted population through interbreeding. Second, a high degree of olfactory sensitivity may have existed prior to cave colonization, preadapting *A. mexicanus* for the light-free subterranean environment.

If extreme peripheral olfactory sensitivity was a preadaptation for cave life, closely-related species may also be highly sensitive to sex pheromones. To examine this, we compared EOG responses to PGFs in 14 species in the same sub-family as *A. mexicanus* (Tetragonopterinae) and 2 within the related Cheirodontinae. All species detected PGFs, and at least one exhibited marked sensitivity; responses to PGF in the Buenos Aires tetra (*Hemigrammus caudovittatus*) were as high as 33mV, with sensitivity of 1 pM. Thus, peripheral olfactory hypersensitivity to sex pheromones in *A. mexicanus* is not unique to that species, and may have preceeded colonization of caves. It will be interesting to determine whether *A. mexicanus* shares attributes of its spawning ecology with that of *H. caudovittatus* and whether these conditions have led to selection for enhanced peripheral olfactory sensitivity to PGF in both species.

A. mexicanus remains conspicuous in having the largest, most sensitive olfactory responses to hormonal sex pheromones, but the present data do not provide evidence that selection enhanced the ability to detect (spy on) released hormones to compensate for the absence of visual cues.

References

Cardwell, J.R., N.E. Stacey, E.S.P. Tan, D.S.O. McAdam and S.L.C. Lang. 1995. Androgen increases olfactory receptor response to a vertebrate sex pheromone. J. Comp. Physiol. A. **176**:55-61.

Mitchell, R.W., W.H. Russell, and W.R. Elliot. 1977. Mexican eyeless fishes, genus *Astyanax*: environment, distribution and evolution. Spec. Publ. Mus. Texas Tech. Univ. **12**:1-89.

Stacey, N.E. and P.W. Sorensen, 1991. Function and evolution of fish hormonal pheromones. In: The Biochemistry and Molecular Biology of Fishes, Vol.1, P.W. Hochachka & T.P. Mommsen, Eds. Elsevier, Amsterdam, p. 109-135.

SEASONAL CHANGES IN OLFACTORY SENSITIVITY OF MATURE MALE ATLANTIC SALMON (*Salmo salar* L.) PARR TO PROSTAGLANDINS.

A. Moore[1] and C.P. Waring[2]

[1] MAFF, Fisheries Laboratory, Pakefield Rd. Lowestoft, Suffolk NR33 0HT, UK.
[2] School of Biological Sciences, University of East Anglia, Norwich, Norfolk NR4 7TJ.

Introduction

Recent studies have indicated that a number of teleosts release hormones (steroids and prostaglandins) which may function as pheromones and synchronise reproductive behaviour and physiology in conspecifics. In Atlantic salmon certain F-series prostaglandins (PGF's) have a priming effect on levels of plasma steroids and gonadotrophin II in mature male parr (Waring & Moore this volume). The present study examined the olfactory sensitivity of male Atlantic salmon to four PGF's either considered, or known, to have a pheromonal role in salmonid reproduction.

Materials and methods

Immature male parr, length 124 ± 11 mm, GSI 0.04 ± 0.002 % (mean ± sem) and mature male salmon parr, length 135 ± 15 mm, GSI 8.4 ± 1.4 % (mean ± sem) were studied using the electro-olfactogram (EOG) technique. Serial dilutions of $PGF_{1\alpha}$ and $PGF_{2\alpha}$, 15-keto$PGF_{2\alpha}$ and 13,14-dihydro-15-keto$PGF_{2\alpha}$ were studied. Recording procedure, testing procedure and data analysis were as described by Moore & Scott (1992).

Results

The olfactory epithelium of mature male parr was acutely sensitive to $PGF_{1\alpha}$ (Fig. 1) and $PGF_{2\alpha}$, less sensitive to 15-keto$PGF_{2\alpha}$ and did not respond to 13,14-dihydro-15-keto$PGF_{2\alpha}$. The response thresholds for $PGF_{1\alpha}$ and $PGF_{2\alpha}$ were 10^{-11} M. and for 15-keto$PGF_{2\alpha}$ was 10^{-8} M. There was also an increase in sensitivity of male salmon parr to both $PGF_{1\alpha}$ (Fig. 1) and $PGF_{2\alpha}$ corresponding to sexual maturity. The response threshold for both $PGF_{1\alpha}$ and $PGF_{2\alpha}$ increased 100 fold between immature (10^{-9} M.) and mature (10^{-11} M.) stages. During November there was also a significant increase in the amplitude of the response to $PGF_{1\alpha}$ recorded from spermiating male parr (Fig. 1).

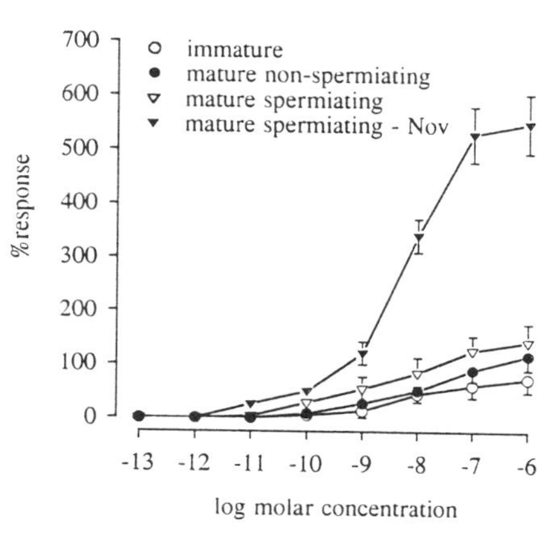

Fig. 1. EOG responses of male Atlantic salmon parr to $PGF_{1\alpha}$.

Discussion

The olfactory epithelium of mature male parr is extremely sensitive to both $PGF_{1\alpha}$ and $PGF_{2\alpha}$. Sensitivity to these PGF's appears to be related to the sexual maturity of the fish as previously described for the steroid testosterone (Moore & Scott 1991). The rapid increase in the size of the recorded response to $PGF_{1\alpha}$ during November may suggest a window of sensitivity which is related to a specific pheromonal function during reproduction. The increase in olfactory sensitivity to $PGF_{1\alpha}$ and $PGF_{2\alpha}$ corresponded with a priming effect by these compounds on levels of certain plasma steroids, GtH II and expressible milt in mature male salmon parr (Waring & Moore this volume). It is suggested that these prostaglandins may have a role in synchronising reproductive physiology in male Atlantic salmon.

References

Moore, A. & Scott, A.P. 1991. Phil.Trans. R. Soc. Lond B 332: 241-244.

Moore, A. & Scott, A.P. 1992. Proc. R. Soc. Lond. B 249: 205-209.

Waring, C.P & Moore (This volume)

CHARACTERIZATION OF STEROIDAL SEX PHEROMONES IN THE ROUND GOBY, (*Neogobius melanostomus*)

C.A. Murphy, J.R. Cardwell, and N.E. Stacey

Department of Biological Sciences, University of Alberta, Edmonton, AB, T6G 2E9, Canada

Introduction

The round goby (*Neogobius melanostomus*) recently brought to the Great Lakes from the Black or Caspian Seas in ship ballast, is a possible cause of declining mottled sculpin (*Cottus bairdi*) populations (Crossman *et al.*, 1992). Since gobies and cottids use similar mating systems in the same habitats, and use steroidal sex pheromones (Colombo *et al.*, 1982; Dmitrieva and Ostroumov, 1986), the previously allopatric *Neogobius* and *Cottus* may interfere reproductively. To address this possibility, we first used electro-olfactogram (EOG) recordings to characterize olfactory response of the round goby to steroids and prostaglandins (PGs), many known to be pheromones in other fish (Stacey and Cardwell, this volume).

Methods

EOG responses of mature *Neogobius* (collected at Windsor, Ontario) first were recorded (Cardwell *et al.*, 1995) during 2-sec exposures to 8 PGs and approximately 150 steroids (10^{-8}M). Dose response relationships were then determined for all compounds that induced a response, and finally the most potent of the detected compounds were used in cross-adaptation studies (Cardwell *et al.*, 1995).

Results and Discussion

Our EOG studies showed that *Neogobius* detects a variety of free and conjugated steroids. Surprisingly, none of the PGs tested elicited a response, although these compounds are the most commonly detected water-borne hormones in other fish groups (Stacey and Cardwell, this volume).

Cross-adaptation studies indicate that *Neogobius* has four types of olfactory receptor mechanisms detecting steroid compounds. Based on the most potent compounds, the four types can be termed: A- estrone (which also detects estradiol-17β); B- estradiol-glucuronide; C- etiocholanolone (etio; which also detects etio-glucuronide; D - dehyro-epiandrosterone-glucuronide (DHEA-gluc; which also detects DHEA-sulphate). Dose response relationships for these four most potent compounds are shown in Figure 1.

Interestingly, etio-glucuronide was reported to be a behavioural pheromone attracting females to males in another European goby (*Gobius jozo*) (Colombo *et al.*, 1982). Although the behavioural or physiological function of water-borne steroids remains to be determined in *N. melanostomus*, our findings suggest this species has evolved a complex pheromone system using released sex steroids.

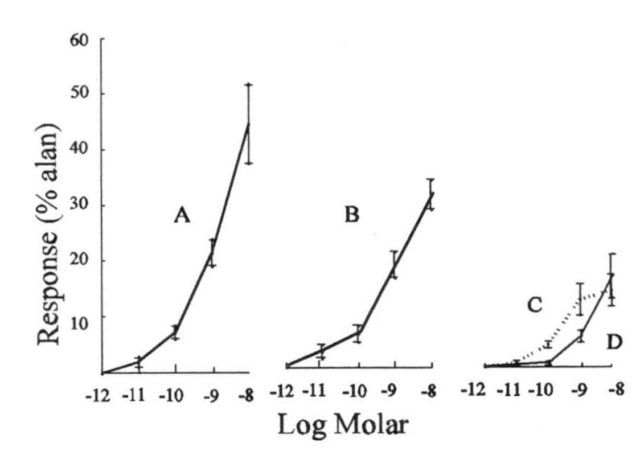

Figure 1 Dose responses for the most potent compound in each receptor class. A) estrone (n=6), B)estradiol-gluc (n=5), C) etiocholanolone (n=4), D) Dehydroepiandrosterone-gluc (n=5)

References

Cardwell, J.R., N.E. Stacey, E.S.P.Tan, D.S.O. McAdam and S.L.C. Lang, 1995. Androgen increases olfactory receptor response to a vertebrate sex pheromone. J.Comp. Physiol A 176: 55-61.

Colombo, L., P.C. Belvedere, A. Marconata and F. Bentivegna, 1982. Pheromones in teleost fish. In Proc. 2nd Intl. Symp. Reprod. Physiol. Fish. C.J.J. Richter and H.J.TH Goos (Eds.):. Pudoc, Wageningen.pp 84- 89.

Crossman, E.J., E. Holm, R. Cholmondeley and K. Tuininga, 1992. First records for Canada of the rudd, *Scardinius erythrophthalmus*, and the round goby, *Neogobius melanostomus*. Can. Field Nat. 106: 206-209.

Dmitrieva, T.M. and V.A. Ostroumov, 1986. The role of chemical communication in the organization of spawning behavior of the yellowfin Baikal sculpin (*Cottocomephorus grewingki*). Nauki 10: 38-42.

Stacey, N.E., and Cardwell, J.R.1995. Hormones as sex pheromones in fish:widespread distribution among freshwater species. This volume.

Supported by NSERC grant to NES

GAMETOGENESIS OR BEHAVIOR? THE ROLE OF SEX STEROIDS IN HERMAPHRODITE REPRODUCTION

A.S. Oliver[1]*, P. Thomas[2], and C.V. Sullivan[3]

[1]Duke Marine Laboratory, Beaufort, NC 28516; [2]The University of Texas at Austin, Marine Science Institute, P.O. Box 1267, Port Aransas, TX; [3]Department of Zoology, North Carolina State University, Campus Box 7617, Raleigh, NC 27695

Summary

Levels of the sex steroid 17α,20ß-dihydroxy-4-pregnen-3-one (DHP) are correlated with mating behavior and and 11-ketotestosterone (11KT) is correlated with size in the belted sandfish (*Serranus subligarius*), a simultaneous hermaphrodite. Estradiol-17ß (E2), testosterone (T), and 17α,20ß,21-trihydroxy-4-pregnen-3-one (20ßS) do not vary with behavior and are probably more important in regulating gametogenesis.

Introduction

Individual sandfish trade eggs, switching between male and female spawning roles with each of multiple partners. All individuals spawn in three roles: female pair spawn, male pair spawn, and streak spawn. Streakers dash in to release sperm just as a pair is releasing gametes. Frequency of streaking and of female pair spawning decrease with individual body size, while male pair spawning increases with size ($p < 0.001$). These behavioral trends intersect at approximately 75 mm SL, therefore fish ≥ 73 mm SL are classified as "large" and fish ≤ 72 mm as "small."

We tested the hypothesis that aggressive behavior, relative testis size, and circulating sex steroids differ between sandfish practicing alternative spawning behaviors at different frequencies. Sex steroid levels vary between fish practicing streaker *vs.* territorial male mating strategies in sex-changing parrotfish and wrasses (Cardwell & Liley, 1991; Hourigan, et al., 1991), but no studies have examined simultaneous hermaphrodites. Underwater observations of naturally spawning sandfish were conducted in the Gulf of Mexico at St. Andrew Bay, Florida during June - August, 1991 and 1992. Plasma levels of E2, T, 11KT, DHP, and 20ßS were quantified by radio-immunoassays validated for sandfish plasma. Gonads were excised for determination of relative testis size (testis to ovotestis weight ratio).

Results

Aggressive behavior (biting and chasing conspecifics that approach a courting pair) increased with size ($p = 0.006$) and with frequency of male pair spawning ($p = 0.02$). Relative testis size was not correlated with body size ($p = 0.12$) or with frequency of male pair spawning ($p = 0.65$). Neither aggressive behavior nor relative testis size was correlated with plasma T or 11KT.

Plasma levels of E2, T, and 20ßS did not differ with spawning role or body size. In fish sampled immediately after a single spawn, circulating DHP levels were significantly lower in small streakers than in small pair spawners (Fig. 1). Small fish that streaked at least once during a 30 min observation period had the lowest frequency of male pair spawns ($p = 0.059$). Plasma 11KT levels increased significantly with size (Fig. 2).

Figure 1. DHP in fish that streaked as males *vs.* fish that pair spawned as males or females. 2-way ANOVA, size * spawning role interaction, **$p = 0.005$.

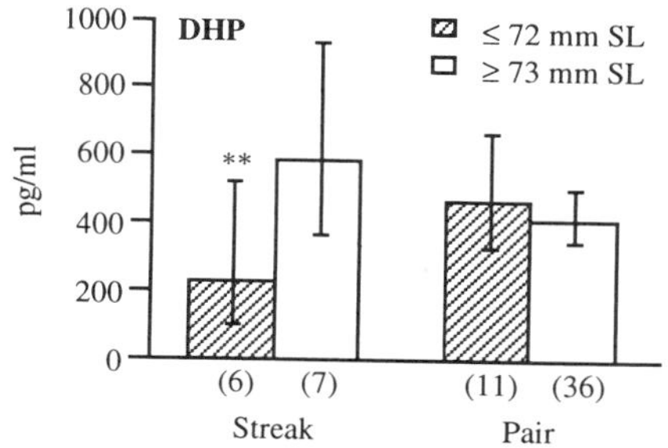

Figure 2. 11KT in relation to body size. Larger points indicate multiple data points. $r^2 = 0.30$, $p = 0.003$.

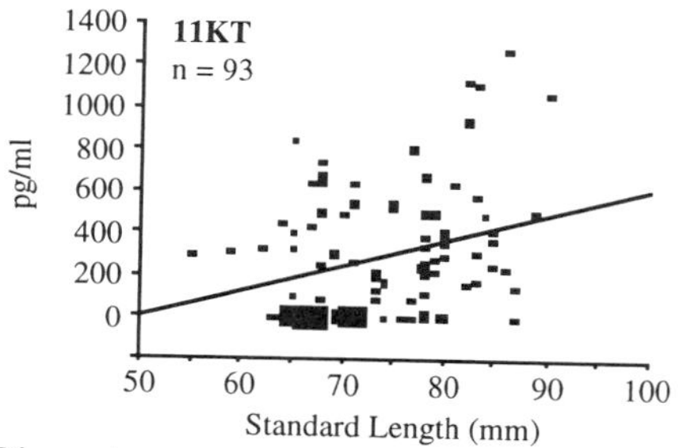

Discussion

Mating behaviors vary with body size in sandfish and body size increases with age ($r = 0.96$, Hastings & Bortone, 1980). Aggressive behavior varies in relation to body size and male pair spawning behavior, but relative testis size is unrelated to size or behavior. The increase in DHP levels with frequency of male pair spawning and of 11KT levels with body size suggests a relationship between DHP, 11KT, and size-influenced changes in male mating behavior over the lifetime of a hermaphrodite. E2, T, and 20ßS profiles are similar among all individuals, indicating that these steroids are probably more important for regulating gametogenesis.

References

Cardwell, J.R. and N.R. Liley. 1991. *Gen. Comp. Endocrinol.* 81: 7 - 20.

Hastings, P.A. and S.A. Bortone. 1980. Observations on the life history of the belted sandfish, *Serranus subligarius* (Serranidae). *Env. Biol. Fish.* 5(4): 365 - 374.

Hourigan, T.F., M. Nakamura, Y. Nagahama, K. Yamauchi and E.G. Grau. 1991. *Gen. Comp. Endocrinol.* 83: 193 - 217.

*Dept. of Science and Mathematics, Univ. Texas of the Permian Basin, Odessa, TX 79762

RELATIONS BETWEEN MATERNAL BEHAVIOUR, OVARIAN DEVELOPMENT, AND ENDOCRINE STATUS, IN THE MOUTHBROODING FEMALE OF *OREOCHROMIS NILOTICUS*

P. Tacon, A. Fostier, P.Y. Le Bail, P. Prunet and B. Jalabert

INRA, Laboratoire de physiologie des Poissons, Campus de Beaulieu, 35042 Rennes cedex, France

Summary

Endocrine status and ovarian development were studied in two groups of *Oreochromis niloticus* females allowed or prohibited performing mouthbrooding behaviour. Measurements of Gonadosomatic Index (GSI), and Estradiol and Testosterone levels show that vitellogenesis is accelerated when parental care is prevented. Furthermore, plasma GH levels are significantly higher in mouthbrooding females than in non-mouthbrooding females and show a dramatic increase at the end of the fry incubation. Our results do not show any role of PRL in parental cares.

Introduction

Parental care is one of the most important features of the tilapia's reproductive strategy. Our purpose was to analyse links between the expression of maternal behaviour, ovarian development and possible endocrinological cues . We were interested in 17ß-estradiol (E2), Testosterone (T) prolactin, (tiPRL) and growth hormone (tiGH).

Material and Methods

Fish (50-150 g) were maintained in aquaria at ambiant temperature (24-29°C) for spontaneous reproduction with a ratio of 3 or 4 females for each male. After spawning (day 0), each female was separated from the other fish. Two groups of females were set up : females with a normal cycle (INC), and females in which eggs have been artificially removed after spawning (NI). In each group, ten females were sampled and killed each 3 days, from day 1 to day 27 for NI, and from day 1 to day 15 for INC. Blood and pituitaries were sampled for hormonal quantification, and ovaries were removed for GSI determination. tiPRLs isoforms (tiPRL I and II), tiGH, E2 and T were all measured by homologous RIAs (Fostier and Jalabert 1986, Aupérin *et al* 1994, Ricordel *et al* 1995).

Results and discussion

The mean interspawning interval (ISI) for INC females was 27 days but only 15 days for NI females. GSI increase in both groups during the ISI (Fig 1a). However, GSI of INC females reach a plateau from day 12 to day 21. Steroids plasma levels follow roughly the same pattern than GSIs, but are higher in the NI group than in the INC group, especially at the beginning of the cycle (up to 40 ng/ml for E2 and 53 ng/ml for T in NI).

The great variability between fish and the large range of plasma PRLs levels in both groups (from 0,3 to 10 ng/ml for PRl I and from 1 to 25 ng/ml for PRL II) precluded us from noting significant variations during the cycles. In contrast, the concentrations of both tiPRLs decreased in the pituitary,at day 3 in the INC group.

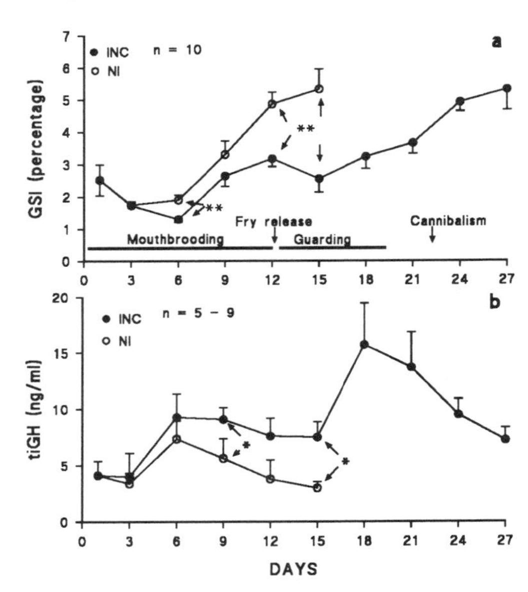

Fig 1. : GSI (a) and GH (b) plasma levels.

Plasma GH levels are higher in the INC group than in the NI group (Fig. 1b), but pituitary concentrations do not show any changes during the ISI in either group. It has been shown in *O. mossambicus* that fasting increases GH release by pituitary *in vitro* (Rodgers *et al* 1992), therefore we suggest that the GH increase during mouthbrooding might be related to the fasting that occurs during this period.

In conclusion, data on GSI and steroid hormones suggest that the vitellogenesis process is accelerated when parental care is suppressed, as shown in *O. mossambicus* (Smith and Haley 1988). However, our results are not sufficient to suggest any role for either tiPRL forms in maternal behaviour.

References

Aupérin B., F. Rentier Delrue, J.A. Martial and P. Prunet, 1994, *J. Mol. Endoc.*, **12**, 13-21

Fostier A .and B. Jalabert, 1986. *Fish Physiol Biochem*, **2**, 87-99

Ricordel M.J., J. Smal, and P.Y. LeBail, *Aquat. Living Resour.*, **8**, 153-160

Rodgers B.D., L.M.H. Helms and E.G. Grau, 1992. *Gen. Comp Endoc.*, **86**, 344-351.

Smith C.J., C.R. Haley, 1988. *Gen.Comp.Endoc.*, **69**, 88-98.

A SEXUAL PARADOX: ANDROGEN AND ESTROGEN SYNTHESIS IN THE TILAPIA KIDNEY

Stephen A. Watts, Kristina M. Wasson and Gene A. Hines

Department of Biology, University of Alabama at Birmingham, Birmingham, AL 35294-1170, USA

Summary

In this study we report an apparent paradox in the steroid synthesizing capability in kidneys of reproductively mature adult *Oreochromis niloticus*. Male kidneys synthesized high quantities of estrogens, female kidneys synthesized relatively high quantities of androgens, including those normally associated with the testis in males. Suggestively, these compounds may be involved in chemical communication strategies between sexes.

Introduction

Sex steroids, their derivatives, prostaglandins, and potentially other unidentified compounds are involved in the regulation of reproductive behavior and physiology in teleosts (reviewed by Stacey et al., 1993). Several free and conjugated C21 or C19 steroids have been implicated as having endocrine and/or pheromonal function. Although many of those steroids are presumed to be released in the urine, the site of synthesis of those steroids is not well understood.

Results

In this study, adult tilapia were exposed to various social treatments in individual recirculating aquaria including single males or females (SM, SF, n=3, held for 2wk), grouped males or grouped females (GM, GF, n=5/group, held for 2 wk), or paired males with females (PM, PF, n=3 pairs held together for 2 d). Total kidney tissues were removed and incubated with ^{3}H-androstenedione for up to 3 hr. Synthesized metabolites were recovered by organic or aqueous extractions and identified by TLC, derivatizations and recrystalizations. Within 3 hr, ca. 30% of the precursor was incorporated into unidentified water-soluble metabolites in SM and GM; less than 13% was incorporated in all other male or female treatments (Table 1). Male kidneys, regardless of social treatments, produced high quantities of estrone and estradiol (Table 2). Female kidneys, regardless of social treatments, produced high quantities of 5ß-androstane-3α,17ß-diol.

Table 1. Percentage (Mean and SE) of ^{3}H-androstenedione converted to aqueous-soluble metabolites after 3 hr in kidneys of *Oreochromis niloticus*.

Treatment	% Aqueous Metabolites
Single Males	29.3 ± 2.3
Grouped Males	26.7 ± 4.6
Paired Males	13.0 ± 2.2
Single Females	7.0 ± 1.6
Grouped Females	11.0 ± 2.3
Paired Females	11.8 ± 3.0

Table 2. Percent (Mean and SE) estrogens relative to total metabolites recovered from the organic extracts of kidney tissues of *Oreochromis niloticus* incubated with androstenedione for 3 hr.

Treatment	% Estrogens
Single Males	26.5 ± 9.7
Grouped Males	64.2 ± 8.9
Paired Males	64.0 ± 7.3
Single Females	0
Grouped Females	5.4 ± 1.1
Paired Females	1.5 ± 1.5

Discussion

The presumed paradoxical production of androgens vs estrogens in kidney tissues has not been reported previously in teleosts. The biological significance of these differential biosynthetic patterns is unknown. However, we do not believe that this difference is related to classical kidney function (filtration). Speculatively, it is possible that males may be producing estrogens, and females may be producing androgens, as a chemical communication strategy. This assumes that the steroids are released in the urine. This further assumes that males and females have the respective receptors in order to physiologially "interpret" the steroidal signals from the opposite sex. The identities of the steroids found in the aqueous fractions are not known, although preliminary results indicate they are conjugated homologs of those steroids observed in the organic extracts. We hypothesize that the paradoxical production of sex-related steroids in kidneys is related to either (i) an undefined intra-organismal endocrine function and/or (ii) pheromonal processes associated with, but not confined to, sexual identification and other reproductive parameters.

References

Stacey, N., J. Cardwell, N. Liley, A. Scott and P. Sorenson. 1994. Hormones as sex pheromones in fish. In: Perspectives in Comparative Endocrinology. K. Davey, R. Peter and S. Tobe (Eds.): National Research Council of Canada, Ottawa, p. 438-448.

Gonadal Physiology

INSULIN-LIKE GROWTH FACTOR I (IGF-I) IN THE FISH OVARY.

Maestro, M. A.[1], Planas, J. V.[2], Swanson, P.[3], and Gutiérrez, J.[1]

[1]Department of Physiology, University of Barcelona, Barcelona, Spain, [2]Department of Pharmacology, University of Washington, Seattle, WA, and [3]Northwest Fisheries Science Center, 2725 Montlake Blvd. East, Seattle, WA, USA.

Summary

Insulin and insulin like growth factor I (IGF-I) receptor binding and intrinsic tyrosine kinase activity were characterized in receptor-enriched fractions of fish ovarian membranes (carp, brown trout and coho salmon) that were obtained by WGA-agarose affinity chromatography. Specific ovarian receptors for insulin and IGF-I were found with tyrosine kinase activity similar to that of other fish tissues. In membranes of both granulosa cells and theca-interstitial layers, IGF-I receptors had higher binding, number, affinity and specificity than insulin receptors. Insulin and IGF-I binding changed throughout the reproductive cycle in carp and brown trout, with maximum binding during primary oocyte growth. The lowest binding was found at the end of the vitellogenic period, however a second peak in of binding appeared prior to ovulation. In preovulatory ovaries, prior to germinal vesicle break down (GVBD), IGF-I had an inhibitory effect on basal and gonadotropin II (GTH II)-stimulated testosterone (T) and 17α-hydroxyprogesterone (17OH-P) production by theca-interstitial layers *in vitro*. However, in granulosa cells at this stage, IGF-I stimulated the conversion of 17OH-P to 17α,20β-dyhydroxy-4-pregnen-3-one (17,20β-P) and enhanced the effects of GTH II on this process *in vitro*. After GVBD, IGF-I showed the same inhibitory effect in theca-interstitial layers, and stimulated 17,20β-P production by granulosa layers. Immunoreactive IGF-I was detected by radioimmunoassay in incubation media from both theca-interstitial layers and granulosa cells. These findings, together with previous studies by other laboratories, suggest an important role of IGF-I in the fish ovary.

Introduction

The insulin like growth factors are important mediators of growth, development and differentiation. These peptides are related to insulin and their structural homology suggests a common origin. Receptors for IGF-I and insulin are also closely related in both structure and function. In fact, these receptors are heterotetramers with 2 α extracellular subunits and 2 β transmembrane subunits. The α subunits bind the ligand while β subunits exhibit tyrosine kinase activity. The phosphorylation of different proteins by tyrosine kinase initiates the signal transduction cascade. This mechanism is similar for both insulin and IGF-I receptors.

Insulin and IGF-I are increasingly recognized as potentially important regulators of ovarian function. Steroidogenic, gonadotrophic and maturational functions have been attributed to insulin and IGF-I (Poretsky and Kalin 1987; Hammond et al., 1991). Furthermore, a large body of information supports the existence of an ovarian IGF system, including ligand, binding proteins and receptors. This system appears to facilitate follicular growth and development.

Information regarding IGF-I function in fish is scarce (reviewed by Siharath and Bern, 1994; Plisetskaya et al., 1994). The first report of IGF-I receptors in fish was published by Maestro et al. (1991) for IGF-I receptors in carp ovaries and later in other tissues (Gutiérrez et al., 1993; 1995). However, the function of both insulin and IGF-I in ovarian physiology in fish has not been fully elucidated. Lessman (1985) demonstrated that oocyte maturation was stimulated by insulin, alone or in combination with progestagens. Later Tyler et al. (1987) reported that insulin stimulated vitellogenin incorporation into rainbow trout oocytes. More recently, Kagawa et al. (1994) have shown that IGF-I stimulated GVBD in red sea bream and that the ovarian follicles were more sensitive to the effects of IGF-I in the presence of maturation induced steroid, 17,20β-P. In these studies, insulin and IGF-II weakly stimulated GVBD, and this effect was also enhanced by 17,20β-P. Immunocytochemical studies by Kagawa et al. (1995) have shown that granulosa cells are the main site of intraovarian IGF-I production. Kagawa et al. (1995) suggested that IGF-I may be involved in regulating the proliferation and/or differentiation of granulosa cells and steroid biosynthesis in early vitellogenic stages. In the goldfish, Srivastava and Van Der Kraak (1994) suggested that insulin, but not IGF-I, is involved in regulating ovarian steroid biosynthesis. Clearly, the roles of insulin and IGF-I in ovarian physiology requires further investigation.

Results

Insulin and IGF-I receptors.

Insulin and IGF-I receptors have been found in the ovary of all fish species studied so far, and the insulin binding is similar to that found in other fish tissues. Interestingly, the percentage of specific binding of IGF-I was 9-10 times higher than insulin. In Table 1 the binding characteristics of IGF-I and insulin in the fish ovary are shown. Receptors for IGF-I had higher affinity than insulin receptors. Specificity of this binding was tested by displacement of radiolabeled insulin and IGF-I by heterologous nonradioactive hormone. IGF-I receptors were highly specific since insulin did not displace binding of radiolabeled IGF-I to ovarian receptors. In contrast, ovarian insulin receptors showed weak binding to IGF-I. In conclusion, insulin and IGF-I receptors are present in the fish ovary, but IGF-I receptors are greater in number and affinity, and are more specific than insulin receptors.

Insulin

	R_0	K_D	% Bsp
Carp	282 ± 88	0.35 ± 0.1	2.7 ± 0.48
Trout	65 ± 9	0.38 ± 0.03	0.83 ± 0.14

IGF-I

	R_0	K_D	% Bsp
Carp	845 ±135	0.21 ± 0.03	22.8 ± 3.6
Trout	176 ± 23	0.17 ± 0.02	9.36 ± 0.88

Table 1: Binding characteristics of insulin and IGF-I to semipurified receptors from carp and brown trout ovaries. R_0, receptor number (binding capacity, fm/mg glycoprotein eluted), K_D, dissociation constant (affinity of the receptors, nM) and %Bsp, percentage of specific binding per 20 µg glycoprotein at near saturating hormone concentrations.

Tyrosine kinase activity

IGF-I and insulin receptors showed tyrosine kinase activity (TKA) and exhibited autophosphorylation in a dose-dependent manner. Autoradiography of this reaction suggested that the β subunit of the receptor molecule is similar in size (95 KDa), to that described in other species. Both IGF-I and insulin receptors can also phosphorylate exogenous substrates under ligand stimulation. The maximum percentage of phosphorylation over basal levels (100 %) ranged between 200-250 % .

Binding in theca-interstitial layer and granulosa cells

The next step in this study was to localize ovarian IGF-I and insulin receptors. Insulin and IGF-I receptors were studied by ligand binding to semipurified receptors from isolated theca-interstitial layers and granulosa cells of coho salmon and brown trout ovaries at the preovulatory stage. In trout, the number and affinities of insulin and IGF-I receptors in theca-interstitial layers and granulosa cells were similar. However, in coho salmon, more insulin and IGF-I receptors per mg of protein were found in granulosa cells than in theca-interstitial layers. It is noteworthy that the affinity of the IGF-I receptors in both follicular cell-layers in salmon (Kd= 0.03-0.04 nM) was the highest found among the fish tissues studied so far.

Insulin and IGF-I binding throughout the reproductive cycle.

In carp, the binding of insulin and IGF-I varied throughout the reproductive cycle (Fig. 1). Although the magnitude of binding differed, the patterns of changes in insulin and IGF-I binding were similar, with a maximum level of binding in early stages of oocyte growth and a decline in binding as vitellogenesis proceeded. Lowest values of binding were observed at the end of vitellogenesis and were followed by a small peak just before ovulation. The same pattern was observed for the both carp and brown trout, although in trout the highest binding occurred in March compared to September for carp. The lowest binding occurred in August and May for trout and carp, respectively. However, the pattern of receptor binding relative to gonadal stage was similar between the two species. In the two studied species, the changes in percentage of binding were due to variations in receptor number while changes in affinity were not observed.

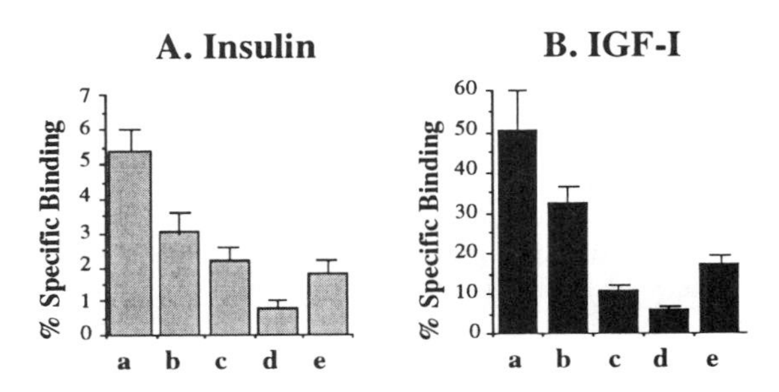

Figure 1: Changes in specific binding of insulin (A) and IGF-I (B) to carp ovarian receptors (percent specific binding per 20 µg protein) at different stages of ovarian development: a. primary oocyte growth; b. cortical granule formation and early vitellogenesis; c. vitellogenesis; d. late vitellogenesis; e. preovulatory stage. Data are mean ± standard errors of triplicates.

Production of IGF-I

Using a homologous radioimmunoassay for recombinant salmon IGF-I (Moriyama et al., 1994) production of IGF-I by theca layers and granulosa cells was measured *in vitro*. After a 24-hour incubation period, levels of IGF-I in incubation medium ranged from 0.3 to 0.8 ng/ml for granulosa cells and from 0.8 and 1.2 ng/ml for theca-interstitial layers.

Effects of IGF-I effects on steroidogenesis of the preovulatory ovarian follicle

The effects of IGF-I on basal and gonadotropin (GTH II)-stimulated steroid production was studied at two stages of final oocyte maturation: prior to germinal vesicle breakdown (GVBD) and after GVBD. These experiments were conducted using isolated theca-interstitial layers and granulosa cells. The production of 17OH-P and T by theca-interstitial layers was measured. Granulosa layers were incubated in the presence of both T and 17OH-P, and the production of E and 17,20β-P was measured.

Prior to GVBD, IGF-I inhibited both basal and GTH II-stimulated production of T and 17OH-P by theca-interstitial layers. In granulosa cells, IGF-I stimulated the conversion of T to E and 17OH-P to 17,20β-P. Although GTH II inhibited conversion of T to E by granulosa cells at this stage, this effect was not observed in the presence of IGF-I. At this stage, the stimulatory effects of GTH II on the conversion of 17OH-P to 17,20β-P were higher in the presence of

IGF-I.

After GVBD, IGF-I inhibited 17OH-P and T production by theca-interstitial layers, and suppressed the stimulatory effects of GTH II on the production of these steroids. In granulosa cells, aromatase activity was low and unaltered by IGF-I or GTH II. At this stage, the conversion of 17OH-P to 17,20β-P was higher than prior to GVBD, and IGF-I enhanced the stimulatory effects of GTH II on 17,20β-P production in a concentration dependent manner.

	Theca-Interstitial Layer		Granulosa Cells	
	T	17αOH-P	E	17,20β-P
Pre-GVBD	-	-	+	+
Post-GVBD	-	-	NE	+

Table 2: Effects of IGF-I on steroid production by isolated theca-interstitial layers and granulosa cells of the preovulatory coho salmon ovary before and after GVBD. - , inhibitory effects ; +, stimulatory effects; NE, no effects

Discussion

IGF-I and insulin receptors have been found in mammalian ovaries (Poretsky et al., 1985) and in *X. laevis* oocytes (Hainaut et al., 1991). Recently, insulin and IGF-I receptors in carp ovaries were characterized (Maestro et al., 1991; Gutiérrez et al., 1993). The data reported in the present study on ovarian receptors of trout and coho salmon are consistent with the data obtained in carp. Characteristics of insulin and IGF-I receptors are similar to those observed in other tissues of the same species (Párrizas et al., 1995a; Gutiérrez et al, 1995). In addition, the ratio of insulin/IGF-I receptors in the ovary is similar to that observed in muscle or heart (Párrizas et al., 1995 b).

Like the mammalian ovary, the fish ovary has a higher number of IGF-I receptors than insulin receptors. The presence of IGF-I receptors in different cells of the human ovarian follicles has been reported (Poretsky et al.,1985), which suggests that IGF-I acts on both theca and granulosa cells. Adashi (1988; 1991) suggested that granulosa cells are the predominant target and site of production of IGF-I in mammals. Data from our studies in salmon indicate that IGF-I receptors are present in both cells and IGF-I has effects on steroid production by both theca-interstitial layers and granulosa cells.

The β subunits of insulin and IGF-I receptors in the fish ovary possess TKA and undergo insulin and IGF-I-stimulated autophosphorylation, followed by phosphorylation of other proteins to initiate the signal transduction cascade. This type of activity has already been observed in carp ovarian receptors (Gutiérrez et al., 1993) and is similar to that found in X. laevis oocytes (Hainaut et al., 1991). In fact, insulin and IGF-I increased the incorporation of ^{32}P-ATP 2-2.5-fold over basal values. This is lower than the 5-fold increase found in the human ovary (Poretsky et al., 1985), but agrees with the differences in TKA observed between mammals and fish in other tissues (Párrizas et al., 1995b).

Maximum levels of insulin and IGF-I binding were found in the fish ovary just before oocyte growth and at the beginning of vitellogenesis. Tyler et al. (1987) demonstrated that insulin, like gonadotropin, stimulated vitellogenin uptake by the trout ovary. In the mammalian ovary, FSH exerts its differentiation-inducing effect at least partly through the action of IGF-I. IGF-I induces differentiation of granulosa cells (Kanzaki et al., 1994). When follicles reach the primary stage of development, granulosa cells begin to secrete IGF-I, which promotes the proliferation of theca cells (Magoffin and Weitsman, 1993). Therefore, IGF-I produced by granulosa cells may play an autocrine role as well as a paracrine signal to theca cells (Hernández et al., 1991). Interestingly, Kagawa et al. (1995) found no immunoreactive(ir) IGF-I in red seabream oocytes at the perinucleolar stage, but staining intensity for ir-IGF-I in granulosa cells of primary yolk globule stage oocytes was high. During final oocyte maturation the amount of ir-IGF-I decreased relative to that of early vitellogenic stages. It is possible that IGF-I plays several roles during vitellogenesis including: regulation of steroid biosynthesis, vitellogenin uptake, and the proliferation and differentiation of follicular cells.

As vitellogenesis ends, the number of insulin and IGF-I receptors decreased, in parallel with the completion of the period of vitellogenin uptake. However, a second peak in binding appeared just before ovulation. Lessman (1985) suggested that insulin may have a role in reinitiation of meiosis in the goldfish oocyte. IGF-I also has been shown to stimulate GVBD in seabream (Kagawa et al., 1994 and these proceedings).

There is convincing evidence that mammalian granulosa cells express the IGF-I gene (Oliver et al., 1989) and can secrete IGF-I in response to FSH and estradiol (Hsu et al., 1987). Of all adult rat organs tested, the ovary has the third highest level of IGF-I gene expression, the uterus and liver being the most active in this regard (Adashi et al., 1991). In agreement with this information, Duan et al., (1994) found IGF-I mRNA in the ovary of coho salmon. Detection of IGF-I in culture media in the present study provides evidence for the secretion of this peptide by both theca-interstitial layers and granulosa cells. The levels of IGF-I released into the incubation medium by salmon ovarian follicular cells are in a range similar to those reported by Kanzaki et al., (1994) for rat granulosa cells. However, the results in salmon are in contrast to that of immunocytochemical studies in red seabream (Kagawa et al., 1995) where ir-IGF-I was found in granulosa cells, but not the theca cells.

The effects of IGF-I on ovarian steroidogenesis are well known in mammals (Adashi et al., 1991; 1993), while in the fish relatively little information is available. Srivastava and Van Der Kraak (1994) found that the responsiveness to insulin changes during gonadal development and consequently, the period during the reproductive cycle when the experiments are performed must be taken into account. Several studies suggest that IGF-I plays a role in theca cell function in mammals, potentiating the LH-induced synthesis of androgenic precursors. In our study, during the preovulatory stage in coho salmon, pronounced inhibitory effects of IGF-I on steroid production in theca-interstitial layers were observed either before or after GVBD.

Information on mammalian granulosa cell responsiveness to IGF-I is more abundant (Spicer et al., 1994; Kanzaki et al., 1994). IGF-I is involved in FSH-induced differentiation of granulosa cells (Adashi et al., 1988). IGF-I is capable of augmenting progesterone and estrogen biosynthesis, FSH-mediated acquisition of LH receptors and mitosis in granulosa cells (Adashi et al., 1991). IGF-I synergizes with FSH to induce aromatase activity. Similarly, in preovulatory salmon granulosa layers, IGF-I appears to be involved in the regulation of steroidogenesis. In particular, the stimulatory effects of IGF-I on the production of 17,20β-P, the maturation-inducing hormone, suggests that this growth factor may have an important role during final oocyte maturation. Clearly IGF-I has been shown to stimulate GVBD in red seabream (Kagawa et al. 1994). However, in this species, it was suggested by Kagawa et al. (1994) that IGF-I acts directly on oocytes to induce maturational competence and final oocyte maturation, not through production of maturation-inducing steroid.

In conclusion, the present work indicates the existence of an IGF-I system in fish ovaries , including reception, production and action. Ovarian IGF-I receptors, with high affinity and specificity were found. Production of endogenous IGF-I by ovarian follicular cells was demonstrated. Finally, effects of IGF-I on basal and gonadotropin-regulated ovarian steroidogenesis were observed.

Acknowledgments

This work was supported by the grants from Direcci'on General de Ciencia y Tecnologia (PB 91-0471 and PTR93-0026), CIRIT (GRQ94-1051) and North Atlantic Treaty Organization (5-2-0.5/RG 921175) to JG and grants from FPI and CIRIT to M.A.M. Support from the U.S. Dept. of Agriculture (93-37203-9409) is also acknowledge. We wish to thank the Piscifactoria de Baga (Dept. Medi Natural, Generalitat de Catalunya) as well as Barcelona Zoological Park, for providing the trout and carp. We acknowledge the gift of human recombinant IGF-I by CIBA-GEIGY (Basel) and of salmon recombinant IGF-I by S. Moriyama and E.M. Plisetskaya (University of Washington).

References

Adashi, E. Y., C. E. Resnick, A.J. D'Ercole, M.E. Svoboda, and J.J. Van Wyk. 1985. Insulin-like growth factors as intraovarian regulators of granulosa cell growth and function. Endocr. Rev. 6, 400-420.

Adashi, E.Y., C.E. Resnick, E.R. Hernandez, J.V. May, M. Knecht, E. Svoboda and J.J Van Wyk.1988. Insulin-like growth factor-I as an amplifier of follicle-stimulating hormone action: studies on mechanism(s) and site (s) of action in cultured rat granulosa cells. Endocrinology 122, 1583-1591.

Adashi, E.Y., C.E. Resnick, A. Hurwitz, E. Ricciarelli, E.R. Hernandez, C. T. Roberts, D. Leroith and R. Rosenfeld. 1991. Insulin-like growth factors: the ovarian connection. Human Reproduction. vol. 6, 1213-1219.

Adashi, E.Y. 1993. The intraovaian insulin-like growth factor system. In: The ovary (E.Y. Adashi and P.C. Leung, Eds.) pp. 319-335, Raven Press, New York.

Duan, C., S. Duguay and E.M. Plisetskaya. 1993. Insulin-like growth factor-I (IGF-I) mRNA expression in coho salmon, *Oncorhynchus kisutch*: tissue distribution and effects of growth hormone/prolactin family proteins. Fish. Physiol. Biochem. 11, 371-399.

Gutiérrez , J., M. Parrizas, N. Carneiro., J.L. Maestro, M.A. Maestro and J. Planas. 1993. Insulin and IGF-I receptors and tyrosine kinase activity in carp ovaries: changes with reproductive cycle. Fish. Physiol. Biochem. 11, 247-254.

Gutiérrez, J., M. Parrizas, M.A. Maestro, I. Navarro, E.M. Plisetskaya. 1995. Insulin and IGF-I binding and tyrosine kinase activity in fish heart. J. Endocrinol. 145 (in press).

Hainaut, P., A. Kowalski, S. Giorgetti, V. Baron and E. Van Obberghen. 1991. Insulin and insulin-like growth factor-I (IGF-I) receptors in Xenopus laevis oocytes. Biochem. J. 237, 637-678.

Hammond, J.M., J.S., Mondschein, S.E., Samaras, and S.F. Canning. 1991. The ovarian insulin-like growth factors, a local amplification mechanism for steroidogenesis and hormone action. J. Steroid. Biochem. Mol.Biol. 40: 411-416.

Hernández , E.R., A. Hurwitz, L. Botero, E. Ricciarelli, H. Werner, C. T. Roberts, D. Leroith, E.Y. Adashi. 1991. Insulin-like growth factor receptor gene expression in the rat ovary: divergent regulation of distinct receptor species. Mol. Endocrinol. 5, 1799-1805.

Hsu, C.J. and J.M. Hammond. 1987. Concomitant effects of growth hormone on secretion of insulin-like growth factor I and progesterone by cultured porcine granulosa cells. Endocrinology, 1121, 1343-1348.

Kagawa, H., M. Kobayashi, Y. Hasegawa, and K. Aida. 1994. Insulin and insulin-like growth factors I and II induce final oocyte maturation of oocytes of red seabream, *Pagrus major, in vitro.* Gen. Com. Endocrinol. 95, 293-300.

Kagawa, H., S. Moriyama, and H. Kawauchi. 1995. Immunocytochemial localization of IGF-I in the ovary of red seabream, *Pagrus major*. Gen. Comp. Endocrinol. 99, 307-315.

Kanzaki, M., M-A. Hattori, R. Horiuchi and I. Kojima. 1994. Co-ordinate actions of FSH and insulin-like growth factor-1 on LH receptor expression in rat granulosa cells. J. Endocrinol. 141, 301-308.

Lessman, Ch. A. 1985. Effect of insulin on meiosis reinitiation induced in vitro by progestogens in oocytes of the goldfish (Carassius auratus). Dev. Biol. 107: 259-263.

Maestro, J.L.,M. Parrizas, I. Navarro, and J. Gutierrez. 1991. Insulin and IGF-I receptors in carp oocytes. International Congres: Research for Aquaculture: Fubdamental and Applied Aspects, p. 190. Antibes-Juan les Pins (Abstract).

Magoffin, D.A. and S. R. Weitsman. 1993. Insulin-like growth factor-I stimulates the expression of 3β-hydroxysteroid dehydrogenase messenger ribonucleic acid in ovarian theca-intersticial cells. Biol. Reprod. 48, 1166-1173.

Moriyama, S., P. Swanson, M. Nishii, A. Takahashi, H. Kawauchi, W. W. Dickhoff, E. M. Plisetskaya. 1994. Development of a homologous radioimmunoassay for coho salmon insulin-like growth factor I. Gen. Comp. Endocrinol. 96, 149-161.

Oliver, J.E., T.J. Aitman, J.F. POwell, C.A. Wilson, and R.N. Clyton. 1989. Insulin-like growth factor I gene expression in the rat ovary is confined to the granulosa cells of developing follicles. Endocrinology 124, 22671-2679.

Párrizas, M., E. M. Plisetskaya, J. Planas and J. Gutierrez. 1995a. Abundant insulin-like groth factor 1 (IGF-I) receptor binding in fish skeletal muscle. Gen. Comp. Endocrinol. 98, 16-25.

Párrizas, M., M.A. Maestro, N. Banos, I. Navarro, J. Planas and J. Gutierrez. 1995 b. Insulin/IGF-I binding ratio in skeletal and cardiac muscles of vertebrates: a phylogenetic approach. Amer. J. Physiol. (in press).

Plisetskaya, E.M., S.J. Duguay and C.Duan. 1994. Insulin and insulin-like growth factor 1 in salmonids: comparison of structure, function and expression. In: Perspectives in Comparative Endocrinology, Natl. Res. Council of Canada, 226-233.

Poretsky, L., F. Grigorescu, F., Seibel, M., Moses, A.C. and J.S. Flier. 1985. Distribution and characterization of insulin and insulin-like growth factor-1 receptor in normal human ovary. J. Clin. Endocrinol. Metab. 61, 728-734.

Poretsky, L. and M.F Kalin. 1987. The gonadotropic function of insulin. Endocrine Rev. 3, 132-141.

Siharath, K. and H.A.,Bern. 1994. The physiology of insulin-like growth factor (IGF) and its binding proteins in teleost fishes. Proc. Zool. Soc. of Calcuta, Haldane Comm. Vol., 113-124.

Spicer, L.J., E. Alpizar and R.K. Vernon. 1994. Insulin-like growth factor-I receptors in ovarian granulosa cells: effect of follicle size and hormones. Mol. Cel. Endocrinol. 102, 69-76.

Srivastava, R. K. and G. Van Der Kraak. 1994. Insulin as an amplifier of gonadotropin action on steroid production: mechanism and sites of action in goldfish prematurational full-grown ovarian follicles. Gen. Comp. Endocrinol. 95, 60-70.

Tyler, C.,J. Sumpter,and N. Bromage.1987. The hormonal control of vitellogenin uptake into cultured oocytes of the rainbow trout. Proc. Third Int. Symp. Reproductive Physiology of Fish. p. 142. Memorial University Press. St. John's.

ACTIVIN B IS A MAJOR MEDIATOR OF HORMONE-INDUCED SPERMATOGONIAL PROLIFERATION IN THE JAPANESE EEL

T. Miura,[1] C. Miura,[1] K. Yamauchi,[1] and Y. Nagahama[2]

[1]Dept. of Biol., Fac. of Fisheries, Hokkaido Univ., Hakodate, Japan and [2]Lab. of Reprod. Biol., Natl. Inst. for Basic Biology, Okazaki, Japan.

Introduction

Spermatogenesis is an extended developmental process. It begins with the proliferation of spermatogonia. Then it proceeds through meiosis and spermatogenesis during which the haploid spermatid is converted into sperm. Although it is generally accepted that the principal stimuli for vertebrate spermatogenesis are pituitary gonadotropins and androgens, the specific role played by individual hormones has not been clarified (Steinberger, 1971; Callard et al., 1978; Hansson et al., 1976; Billard et al., 1982). Progress in this field has been hampered by the complex organization of the testis of higher vertebrates in which seminiferous tubules contain several successive generations of germ cells.

Under cultivated conditions, the male Japanese eel, *Anguilla japonica*, has immature testes containing only premitotic spermatogonia (type A and early type B spermatogonia) (Miura et al., 1991a). Spermatogenesis is induced by human chorionic gonadotropin (HCG) injection *in vivo* and *in vitro* organ culture supplement with HCG or 11-ketotestosterone which is a major androgen in male eel (Miura et al., 1991b and c). Using these *in vivo* and *in vitro* systems of eel, the following control mechanisms of spermatogenesis have been clarified. Hormonal induction of spermatogenesis in eel testes is via gonadotropin stimulation on Leydig cells to produce 11-ketotestosterone. In turn, 11-ketotestosterone induces complete spermatogenesis from premitotic spermatogonia to spermatozoa through the activation of Sertoli cells. In this predicted control mechanism, many interesting questions remain unsettled. Especially, there seem to be complex control mechanisms in the initiation of spermatogenesis (Miura et al. 1991a). This review describes our recent experimental observations, which focus on the initiation of spermatogenesis controlled by hormone (HCG and 11-ketotestosterone) stimulation.

Cloning of the spermatogenesis-related gene

Our pervious findings strongly suggest that Sertoli cells play a key role in the hormone-induced spermatogenesis in eels. We anticipated some Sertoli cell factors involved in the regulation of spermatogenesis were expressed or suppressed by gonadotropin and/or androgen during HCG-induced spermatogenesis. To test this possibility, mRNA was extracted from testes of eels with or without a single injection of HCG for 1 day. Subtractive cDNA libraries were constructed according to Wang and Brown (1991) with minor modifications. With this approach we hoped to clone specific complementary DNAs (cDNAs) expressed in each stage. From these libraries, two up-regulated cDNAs and six down-regulated cDNAs have been obtained (manuscript in preparation). To date, one of the up-regulated cDNAs was analyzed. The nucleotide sequence of this cDNA clone was 3.3 kb long. A methionine codon at nucleotide 1572 initiates a long open reading frame specifying a protein of 395 amino acids. The similarity search of the predicted amino acid sequence of this cDNA clone was practiced using the Swiss Protein Sequence Data Bank. A computer search showed that this cDNA clone had an amino acid sequence similar to activin ßB subunit. Especially, the C-terminal (115 residues) of the mature protein has 90 % homology with those of mammals, chicken and *Xenopus* (Mason et al. 1985; Feng et al. 1989; Mitrani et al. 1990; Thomsen et al. 1990). To confirm that this cDNA clone codes activin ßB subunit, its recombinant protein was constructed using Chinese hamster ovarian cells (CHO cells). With introduction of the expression vector containing this up-regulated clone into CHO cells, these cells secreted 22kd protein into the medium. Purified recombinant protein had EDF (erythroid differentiation factor) activity, which is an activin-specific activity (manuscript in preparation). These findings indicate that this up-regulated cDNA clone codes eel activin ßB subunit.

Expression of activin ßB subunit in eel testis

Northern blot analysis and *in situ* hybridization techniques were employed to examine the sequential changes in the transcripts of testicular activin ßB mRNA during HCG-induced spermatogenesis. Activin ßB subunit mRNA transcripts were hardly detected in the testes prior to HCG injection. By contrast, 3.3 kb mRNA transcripts were prominent in testes 1 day after the injection; however, the transcripts began to decrease 3 days after the injection, followed by a sharp decrease by 9 days. HCG-dependent activin ßB mRNA expression was confirmed by *in situ* hybridization. The site of activin ßB mRNA hybridization signals was restricted to Sertoli cells in testes treated with HCG for 1-3 days. These findings suggest that the transcription of activin ßB mRNA was accompanied by the initiation of spermatogenesis.

As mentioned above, HCG-induced spermatogenesis in eel testes is mediated by testicular production of 11-ketotestosterone. To determine whether HCG action on transcription of activin ßB subunit mRNA is direct or mediated through 11-ketotestosterone production, we examined the effects of HCG and 11-ketotestosterone on activinB mRNA expression using an organ culture system for eel testis. Before cultivation, activin ßB subunit mRNA transcripts were hardly detected in testes. On the 6th day after cultivation with HCG and 11-ketotestosterone, 3.3 kb mRNA transcripts were prominent in testicular fragments, but these transcripts were hardly detected in the cultivation without hormones. Thus, the activin ßB subunit is regulated not only by gonadotropin (HCG) but also 11-ketotestosterone; a spermatogenesis-inducing androgen controlled by gonadotropin.

Testicular production of activin

The activin ßB subunit is one of the components of dimeric growth factors which belong to the transforming growth factor-ßs (TGF-ßs) family (Mason et al., 1985). The activin ßB subunit from eel testis was estimated to be a component of activin B, activin AB and/or inhibin B protein. Therefore, we examined which molecule was composed of activin ßB subunit using a testicular organ culture system. Testicular cultured medium with or without 11-ketotestosterone was fractionated by reversed-phase HPLC, and the activin activity in each fraction was assayed by an EDF assay. In the control group without hormones, the activin activities of medium were detected in two peaks after fractionation. The retention time of these peaks corresponds to those of recombinant human activin A and B, respectively. However, the concentration of activin B was much lower than that of activin A. Supplement with 11-ketotestosterone into the culture medium induced marked production of activin B in cultured testes. On the other hand, the concentration of activin A was not changed by the treatment with 11-ketotestosterone. These findings indicate that eel activin ßB subunit constructs activin B protein, and the production of this protein was induced by 11-ketotestosterone stimulation in the testis.

Biological action of activin B

Activins are developmentally and physiologically important factors. In spermatogenesis, activin A also induces DNA synthesis of rat spermatogonia and preleptotene spermatocytes (Mather et al., 1990; Hakovirta et al., 1993), whereas the detailed actions of activin B have not been clarified yet. To understand the biological action of activin B in eel spermatogenesis, we examined the effect of human recombinant activin B, which has a high homology (90 %) with eel activin B, using a testicular organ culture system (Miura et al. 1995). Before cultivation, most of the germ-cells in the

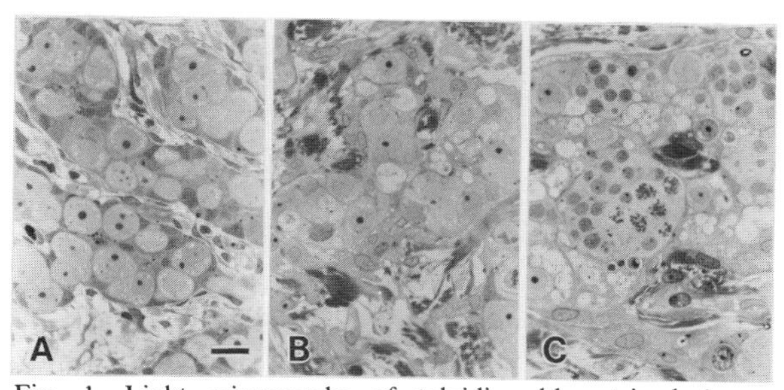

Fig. 1. Light micrographs of toluidine blue-stained 1-μm sections of testes before culture (A), cultured in medium without hormones (B) and human recombinant activin B at 10 ng/ml (C) for 15 days. Arrowheads indicate late type B spermatogonia. (Bar=10 μm)

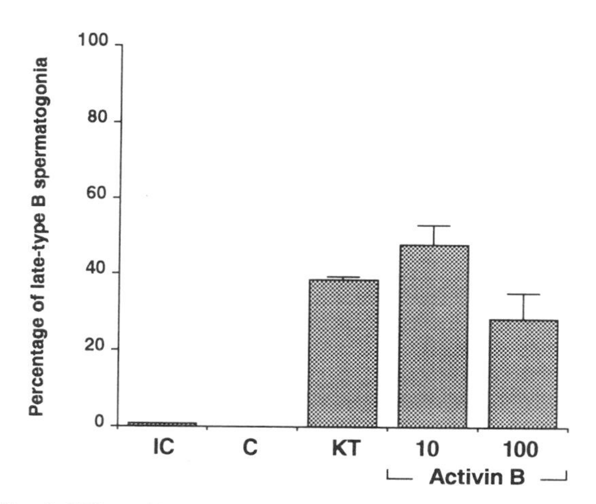

Fig. 2. Effect of human recombinant activin B on induction of proliferation of spermatogonia *in vitro*.

Testicular fragments were cultured in the medium with human recombinant activin B (10 ng and 100 ng/ml) and 11-ketotestosterone (10 ng/ml; for positive control) for 15 days. IC, initial control; C, control; KT, 11-ketotestosterone. The vertical bars represent the mean ± SEM.

eel testis were type A and early type B spermatogonia, which were premitotic spermatogonia (Fig.1-A). Addition of human recombinant activin B (10 and 100 ng/ml) to the culture medium induced proliferation of spermatogonia, producing late type B spermatogonia (Fig.1-C and Fig. 2), within 15 days in the same manner as addition of 11-ketotestosterone (10 ng/ml) for the positive control (Fig. 2). However, cultured testes without activin B, the proliferation of spermatogonia was not observed (Fig. 1-B and Fig.2). Eel activin B, partially purified by HPLC from testicular cultured medium and eel recombinant-activin B from cDNA clone, also induced proliferation of spermatogonia *in vitro* in the same manner as human recombinant-activin B (manuscript in preparation). These findings suggest that activin B induces initiation of spermatogenesis in the Japanese eel.

Summary

Activin ßB subunit cDNA was isolated from a subtractive cDNA library between HCG injected and uninjected eel testicular mRNA. Activin ßB subunit mRNA transcripts were found in eel testis at initiation of spermatogenesis after HCG stimulation, with their expression site restricted in Sertoli cells. Both transcription and translation of eel activin B was induced by 11-ketotestosterone stimulation *in vitro*. Furthermore, human-recombinant activin B and partially purified eel activin B induced proliferation of spermatogonia. Taken together, it is concluded that activin B is one of the initiators of spermatogenesis after 11-ketotestosterone stimulation in eel (Fig 3).

Although activin B induced proliferation of spermatogonia, whether the actions of activin B on spermatogenesis were direct or indirect for germ cells in these experiments remains unknown. Further studies are needed to solve this problem.

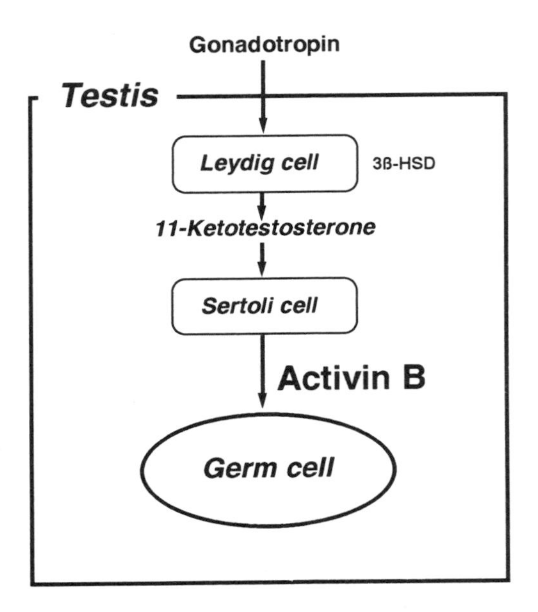

Fig. 3. A schematic representation summarizing the present data on the possible control mechanisms of spermatogenesis in Japanese eel.

Acknowledgments

We are grateful to Dr. Yuzuru Eto (Ajinomoto Co., Ltd. Japan) for providing human recombinant activin B, and for his helpful advice at various stages of this work.

References

Billard, R., A. Fostier, C. Weil, and B. Breton, 1982. Endocrine control of spermatogenesis in teleost fish. *Can. J. Fish. Aquat. Sci.*, **39**: 65-79.

Callard, I. P., G. V. Callard, V. Lance, J. L. Bolaffi, and J. S. Rosset, 1978. Testicular regulation in nonmammalian vertebrates. *Biol. Reprod.*, **18**: 16-43.

Feng, Z.-N., Bardin, C. W., and Chen, C.-L. C. 1989. Characterization and regulation of testicular inhibin b-subunit mRNA. *Mol. Endocrinol.* **3**: 939-948.

Hakovirta, H., A. Kaipia, O. Soder, and M. Parvinen: Effects of activin-A, inhibin-A, and transforming growth factor-ß1 on stage-specific deoxyribonucleic acid synthesis during rat seminiferous epithelial cycle. *Endocrinology,* **133**, 1664-1668 (1993).

Hansson, D., R. Calandra, K. Purvis, M. Ritzen, and F. S. French, 1976. Hormonal regulation of spermatogenesis. *Vitam. Horm.*, **34**: 187-214.

Mason, A. J., Hayflic, J. S., Ling, N., Esch, F., Ueno, N., Ying, S.-Y., Guillemin, R., Niall, H., and Seeburg, P. H. 1985. Complementary DNA sequences of ovarian follicular fluid inhibin show precursor structure and homology with transforming growth factor-ß. *Nature* **318**: 659-663.

Mather, J. P., K. M. Attie, T. K. Woodruff, G. C. Rice, and D. M. Phillips, 1990. Activin stimulates spermatogonial proliferation in germ-cell co-cultures from immature rat testis. *Endocrinology,* **127**: 3206-3214.

Miura, T., C. Miura, K. Yamauchi and Y. Nagahama, 1995. Human recombinant activin induces proliferation of spermatogonia *in vitro* in the Japanese eel *Anguilla japonica. Fish. Sci.,* **63:** 434-437.

Miura, T., K. Yamauchi, Y. Nagahama, and H. Takahashi, 1991a. Induction of spermatogenesis in male Japanese eel, *Anguilla japonica*, by a single injection of human chorionic gonadotropin. *Zool. Sci.*, **8**: 63-73.

Miura, T., K. Yamauchi, H. Takahashi and Y. Nagahama, 1991b. Human chorionic gonadotropin induces all stages of spermatogenesis *in vitro* in the male Japanese eel (*Anguilla japonica*). *Dev. Biol.*, **146**: 258-262.

Miura, T., K. Yamauchi, H. Takahashi and Y. Nagahama, 1991c. Hormonal induction of all stages of spermatogenesis *in vitro* in the male Japanese eel (*Anguilla japonica*). *Proc. Natl. Acad. Sci. U.S.A.*, **88**: 5774-5778.

Mitrani, E., Ziv, T., Thomsen, G., Shimoni, Y., Melton, D. A., and Bril, A. 1990. Activin can induce the formation of axial structures and is expressed in the hypoblast of the chick. *Cell* **63**: 495-501.

Steinberger, E., 1971. Hormonal control of mammalian spermatogenesis. *Physiol. Rev.*, **51**: 1-22.

Thomsen, G., Woolf, T., Whitman, M., Sokol, S., Vaughan, J., Vale, W., and Melton, D. A. 1990. Acivins are expressed early in *Xenopus* embryogenesis and can induce axial mesoderm and anterior structures. *Cell* **63**: 485-493.

Wang, Z., and Brown, D. D. 1991. A gene expression screen. *Proc. Natl. Acad. Sci. U.S.A.* **88**: 11505-11509.

OVULATION SPECIFIC TRANSCRIPTION OF AN ANTILEUKOPROTEINASE-LIKE mRNA IN THE FISH OVARY

S-Y. Hsu[1] and F.W. Goetz

Department of Biological Sciences, University of Notre Dame, Notre Dame, IN, 46556

Summary

Ovulatory specific cDNAs were isolated from the brook trout ovary using subtractive cDNA screening. One of these cDNAs, designated TOP-1, was sequenced and the deduced protein sequence is similar to a group of mammalian protease inhibitors called antileukoproteinases. RNA transcripts hybridizing with TOP-1 were observed in ovarian theca/stroma and granulosa tissue but not in other tissues. These transcripts were highly upregulated at the time of ovulation.

Introduction

A number of peptide factors have been isolated from the vertebrate ovary. However, the function of many of these factors and their relationship with specific maturational processes in the ovary has not always been clear. We used subtractive complimentary DNA (cDNA) screening in an attempt to isolate and characterize messenger RNAs (mRNAs) for peptide factors that might specifically be involved in the process of ovulation. With this technique we were able to isolate a unique group of mRNAs that are highly upregulated in the trout ovary just prior to ovulation.

Results

Preovulatory-specific brook trout cDNA was prepared by hybridizing ovarian poly(A)$^+$ RNA, obtained from ovaries taken prior to meiotic maturation, with ovarian cDNA prepared from ovarian mRNA taken just prior to the time of ovulation. This subtracted cDNA probe was used to screen a brook trout preovulatory ovarian ZAP II cDNA library. The positive phage plaques obtained from this screening were purified to homogeneity by subscreening with the same probe. This initial screening resulted in 8 positive clones. Partial sequence analysis showed that four of the eight clones encoded a similar transcript, however none contained a complete open reading frame. The largest of these 4 repetitive clones was then used to rescreen the library. From this screening, the largest clone obtained, designated TOP-1 (trout ovulatory protein-1), was completely sequenced. The TOP-1 clone contains 714 bp and a poly(A)$^+$ tail of 18 residues (Fig. 1). An open reading frame extends from a Met codon at nucleotide 6 to a TGA stop codon at nucleotide 369 and presumably encodes a protein of 121 amino acid residues. This deduced protein contains two similar peptide domains with high cysteine content (13.93 %) (Fig. 1).

```
1        MetAsnLeuSerAlaArgCysAlaLeuValLeuSerLeuLeuAlaPheValAlaLeuLysIleValSerA
    tggacatgaatttgtcagcccgttgtgctttggttctttctctgttggcatttgtagctttgaaaatagtctctg    75
                          ├Domain 1→
26  laAlaGluThrGlyGlyIleSerThrAlaLysProGlyValCysProArgArgArgTrpGlyIleGlyIleCysA
    ctgcagaaacgggaggcatatctacagcaaagcctggagtgtgccctcgtagacgatggggcatagggatatgtg    150

51  laGluLeuCysSerSerAspSerAspCysProAsnAspGluLysCysCysHisAsnGlyCysGlyHisValCysI
    cagagttatgttccagtgacagtgactgccccaatgatgaaaaatgctgtcacaacggatgtgggcatgtctgca    225
               ├Domain 2→
76  leAlaProTyrThrAlaLysProGlyValCysProArgArgArgTrpGlySerGlyIleCysAlaGluLeuCysS
    ttgcaccttacacagcaaagcctggagtgtgccctcgtagacgatggggctcagggatatgtgcagagttatgtt    300

101 erAsnAspSerAspCysProAsnAspGluLysCysCysHisAsnGlyCysCysIleThrProThrGln***
    ccaatgacagtgactgccccaatgatgaaaaatgctgccacaatggatgttgcattacacctacacagtgaagcc    375
    cggtcgctgtgtcctgcccaagggacctacatgtgtgccgagtactgttacaagatggccagtgccgaggaacag    450
    aagtgtttccggatatgttgattctacgcctgcactgagccattgaagctttattggaaattcttaataattata    525
    agaaataagtagaaaatacaaattattacaagtacagttatgtctttgaaaactgatgattttatgcaataataa    600
    aaatgaaaaatgatcatagcttttatggaatctggaccaacttcatggacggtatgtcaattgacaaactcatgt    675
    ggactgctgaatttgaataaaacaagattttgaacacctaaaaaaaaaaaaaaaaaa                      732
```

Fig. 1. Nucleotide and deduced amino acid sequences of the brook trout TOP-1 cDNA, indicating presumed signal peptide (underline), beginning of repeating peptide domains (├Domain) and stop codon (***). Genbank accession # U03890. Nucleotides numbered on right, amino acids on left (from Hsu and Goetz, 1995).

[1] Present Address: Department of Gynecology and Obstetrics, Stanford University, Stanford, CA, 94305

Using the TOP-1 cDNA as a probe, four major species of RNA, 0.8, 1.1, 1.4 and 1.7 kb, can be detected on Northern blots of total RNA from ovarian theca and stroma (Fig. 2). Expression of these transcripts is very low in ovaries prior to germinal vesicle breakdown (GVBD-resumption of meiosis; #s 1-4, Fig. 2). However, the transcripts are highly upregulated in the ovaries of certain fish (#s 5 & 7, Fig. 2) following GVBD but prior to ovulation. The ovaries of two fish taken just at the beginning of ovulation (#s 9 & 10, Fig. 2) also show a significant increase in all four RNA bands. Besides the results shown here, we have hybridized TOP-1 against the ovarian RNA from a number of other individuals at stages prior to and following GVBD. In no case have we ever observed upregulation of these transcripts prior to GVBD. Further, given that upregulation of TOP-1 transcripts was observed only in the ovarian tissue of a few individuals (#s 5 & 7-Fig. 2) following GVBD (but in advance of ovulation), it is likely that the transcripts are very specifically upregulated very close to the time of ovulation. Presumably, these positive individuals were much closer to ovulation than were the others taken at that stage.

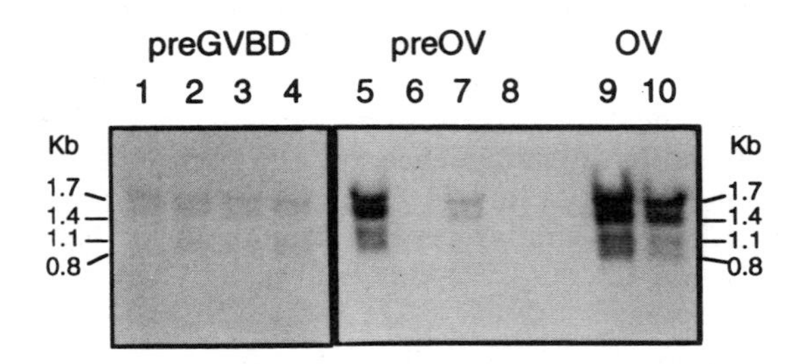

Fig. 2. Northern blot of total RNA from brook trout ovarian theca/stromal tissue taken from individual brook trout females after vitellogenesis but prior to the breakdown of the germinal vesicle (preGVBD; #s 1-4); following GVBD but prior to ovulation (preOV; #s 5-8) and just at the beginning of ovulation (OV; #s 9 & 10). While all samples were electrophoresed on the same gel and probed (with full-length TOP-1) on the same Northern blot, the figure for preGVBD samples is taken from a 4 day X-ray exposure while the figure for the preOV and OV samples is taken from an 18 hour exposure (from Hsu and Goetz, 1995).

The TOP-1 cDNA also hybridizes with RNA extracted from isolated granulosa (results not shown). In contrast, hybridization is not observed on Northern blots of kidney, intestine, heart, skin, gills, liver, muscle (Fig. 3), and brain (not shown).

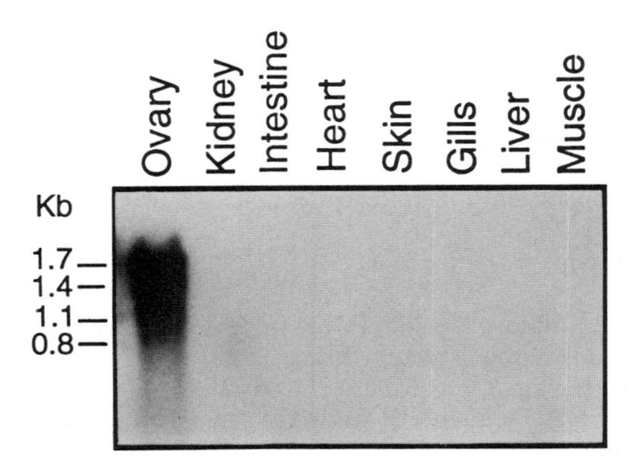

Fig. 3. Northern blot of total RNA extracted from tissues taken from a preovulatory brook trout female. Blot was probed with the full-length TOP-1 cDNA (from Hsu and Goetz, 1995).

Comparison of the deduced amino acid sequence of TOP-1 with all sequences in GenBank indicates that the TOP-1 polypeptide is similar to mammalian antileukoproteinases (ALPs) (Heinzel et al., 1986; Farmer et al., 1990) (Fig. 4). TOP-1 is 34% identical and 51% similar (with conserved substitutions) to pig ALP and 32% identical and 55% similar to human ALP. In comparison, the pig and human APLs are 67% identical and 79% similar. A distinct feature shared by the deduced TOP-1 brook trout polypeptide and the ALPs is the presence of two consecutive, repeating domains and the number and placement of cysteine residues in each domain. In the mammalian ALPs this results in a structure called a "four-disulfide bond core" arrangement of cysteines in each domain. In addition to the conserved cysteines, there are also a number of conserved prolines. Several other proteins such as caltrin II protein (Coronel et al., 1990) also show amino acid similarity with TOP-1; however, these proteins contain only one "four disulfide bond core" domain.

ALPs are inhibitors of leukocytic proteases such as elastase, cathepsin and chymotrypsin-like proteases. They are found in human mucosal fluids including seminal plasma, cervical mucus, and parotid and bronchial secretions. Their function may be to protect mucosal epidermis from nonspecific degradation by proteases released from granulocytic lysosomes.

Discussion

The very strong upregulation of the TOP transcripts just prior to ovulation would suggest that these are important ovulatory factors.

```
            1                                                                      70
Human       ....MKSSGL FPFLVLLALG TLAPWAVEGS GKSFKAGVCP PKKSAQCLRY KKPECQSDWQ CPGKKRCCPD
Pig         .......... .......... .MAPWAVEGA ENALKGGACP PRKIVQCLRY EKPKCTSDWQ CPDKKKCCRD
TOP-1       MNLSARCALV LSLLAFVALK IVSA.AETGG ISTAKPGVCP RRRWGIGICA EL..CSSDSD CPNDEKCCHN
Consensus   ..........................A..G. ....K.G.CP .......... ....C.SD.. CP....CC..

            71                                                          136
Human       TCGIKCLDPV DTPNPTRRKP GKCPVTYGQC LMLNPPNFCE MDGQCKRDLK CCMGMCGKSC VSPVKA
Pig         TCAIKCLNPV AITNPVKVKP GKCPVVYGQC MMLNPPNHCK TDSQCLGDLK CCKSMCGKVC LTPVKA
TOP-1       GCGHVCIAP. .....YTAKP GVCPRRRWGS GICAE..LCS NDSDCPNDEK CCHNGC...C ITPTQ.
Consensus   .C...C..P. ........KP G.CP...... ........C. .D..C..D.K CC...C...C ..P...
```

Fig. 4. Comparison of the presumed TOP-1 peptide with human (Heinzel et al., 1986; Thompson and Ohlsson, 1986) and pig (Farmer et al., 1990) antileukoproteinases.

If TOP is produced at ovulation and acts as a protease inhibitor similar to ALPs, then one of the functions may be to control and contain proteolysis that is commonly observed in the follicle at the time of ovulation. Vertebrate ovulation has been hypothesized to be a type of inflammatory reaction (Espey, 1980). Changes observed during inflammation, including increased vascular permeability and tissue infiltration by leukocytes, have also been observed in the ovary at ovulation (Cavender and Murdoch, 1988). Leukocytes that enter preovulatory follicles could presumably release a variety of substances including proteolytic enzymes, cytokines, or eicosanoids, and through these agents could contribute to the process of ovulation. If TOP functions as an ALP then it could control nonspecific proteolysis arising from invading leukocytes.

In addition, given the types of proteases that ALPs inhibit, it is also possible that TOP could be produced to regulate other serine proteases produced by the ovary. We have recently cloned a kallikrein-like cDNA in the trout ovary and transcripts are abundant both in theca and granulosa at the time of ovulation (Goetz and Hsu, 1994). Kallikrein is a serine protease and it is possible that TOP may regulate the activity of ovarian kallikrein. Alternatively, it is possible that TOPs have some other function in the ovary that is unrelated to the inhibition of proteolysis. For example, another protein that shares structural similarity to ALPs is caltrin II protein (Coronel et al., 1990). This protein inhibits calcium transport into spermatozoa.

Northern analysis indicates that there are several RNA bands hybridizing with TOP-1 and, therefore, the possibility of multiple transcripts. We have progressively rescreened the ovarian library and now have cDNAs that correspond in size with each RNA band from 0.8-1.7 kb. Partial sequence analysis of these other cDNAs indicate that they are also arranged in repeating domains.

References

Cavender, J. L., and Murdoch, W. J. 1988. Morphological studies of the microcirculatory system of periovulatory ovine follicles. Biol. Reprod. 39:989-997.

Coronel CE, San Agustin J, Lardy HA. Purification and structure of caltrin-like proteins from seminal vesicle of the guinea pig. J. Biol. Chem. 265:6854-6859.

Espey, L. L. 1980. Ovulation as an inflammatory reaction - a hypothesis. Biol. Reprod. 22:73-106.

Farmer, S.J., Fliss, A.E. and Simmen, R.C.M. 1990. cDNA cloning and regulation of expression of the messenger RNA encoding a pregnancy-associated procine uterine protein related to human antileuko-proteinase. Mol. Endoc. 4:1095-1104.

Goetz, F.W. and Hsu, S-Y. 1994. Expression of kallikrein gene family mRNAs in the brook trout ovary. Biol. Reprod. 50 (suppl. 1):114 (abstract)

Heinzel, R., Appelhans, H., Gassen, G., Seemuller, U., Machleidt, W., Fritz, H., and Steffens, G. 1986. Molecular cloning and expression of cDNA for human antileukoprotease from cervix uterus. Eur. J. Biochem. 160:61-67.

Hsu, T. and F.W. Goetz 1995. Ovulation specific transcription of antileukoproteinase-like mRNAs in the brook trout ovary. Proceedings of the IIIrd Sapporo International Symposium on Ovarian Function. Sept. 2-3, 1995, Sapporo, Japan. Cold Spring Harbor Laboratory Press. (in press).

Thompson, R. C., and Ohlsson, K. 1986. Isolation, properties, and complete amino acid sequence of human secretory leukocyte protease inhibitor, a potent inhibitor of leukocyte elastase. Proc. Nat. Acad. Sci. U.S.A. 83:6692-6696.

SEX STEROIDS DURING THE OVULATORY CYCLE OF GILTHEAD SEABREAM (*SPARUS AURATA*).

A.V.M. Canario[1], E. Couto[1], P. Vília[1], D.E. Kime[2], S. Hassin[3], Y. Zohar[3]

[1]Center of Marine Sciences, UCTRA, University of Algarve, 8000 Faro, Portugal,[2] Department of Animal and Plant Sciences, University of Sheffield, Sheffield S10 2UQ, UK, [3]Center of Marine Biotechnology, 600 East Lombard Street, Baltimore, Maryland 21202, USA.

Summary

In vitro and *in vivo* evidence is provided to suggest that 17α,20β,21-trihydroxy-4-pregnen-3-one is the most likey candidate as oocyte maturation-inducing steroid in gilthead seabream

Introduction

The female gilthead seabream is a marine sparid which has an asynchronous ovary and is a serial spawner with individual cycles of 24 hours (Kadmon *et al.*, 1984). Previous studies on the steroidogenic potential of the female gonad have identified as main products progesterone and 17α-hydroxyprogesterone from a pregnenolone precursor (Colombo *et al.*, 1972), although 17α,20β-dihydroxy-4-pregnen-3-one (17α,20β-P) has been sugested to be the potential maturation-inducing steroid (MIS) (Kadmon *et al.*, 1984). The objective of this study was to identify ovarian steroids with potential MIS function during the ovulatory cycle of gilthead seabream and to follow their plasmatic changes in vivo.

Results

Ovarian metabolism-Incubations of ovarian fragments of gilthead seabream with tritiated progesterone or 17α-hydroxyprogesterone precursors during the ovulatory cycle produced a complex of up to 20 different metabolites of which the major ones are shown in Table 1. The main feature of these results is the absence of 17α,20β-P from the incubation media. In contrast, significant conversion to 17α,20β,21-trihydroxy-4-pregnen-3-one (17α,20β,21-P) and its 3α,5β reduced metabolite was observed. There was only a negligible production of sulphated or glucuronidated metabolites.

In Vivo Stimulation of Steroidogenesis-In order to confirm the *in vivo* production of the steroids found in the *in vitro* incubations with labelled precursors, blood plasma samples were obtained from 6 females at late vitellogenesis which had received EVAC implants with [D-Ala^6-Pro^9-Net]-luteinizing hormone releasing hormone (LHRHa) and 3 females which received implants without analogue.

Levels of free and conjugated forms of 17α,20β,21-P and 3α,17α,21-trihydroxy-5β-pregnan-20-one (3α,17α,21-P-5β) and 3α,17α,20β,21-tetrahydroxy-5β-pregnane (3α,17α,20β,21-P-5β) were relatively high (Figs 1-3).

Individual fish showed levels in some samples of several hundred ng/ml. However, the changes observed were not statistically significant owing to the high variability among individuals as illustrated for 3α,17α,20β,21-P-5β in Fig. 4.

The steroid 3α,17α,20β,21-P-5β was measured with a radioimmunoassay developped for 3α,5β–reduced progestins and optimised for this steroid (Canario and Scott, 1990). In the gilthead seabream the immunoactivity detected by the antiserum was ca 75% 3α,17α,20β,21-P-5β as demonstrated after separation of plasma samples on tlc (Fig. 5).

Levels of 17α,20β-P, including conjugates, were always low (ca 1ng/ml or below), as were levels of 11-deoxycortisol and conjugates (ca 5ng/ml or below).

Daily Cycle of Steroid Hormones- Ovulating females sampled at 9:00 am and 3:00 pm during 24 hr showed only low levels of steroids in blood plasma and no significant alterations were detected during sampling (Table 2).

Table 1. Relative quantities of metabolites extracted from seabream ovarian incubations according to precursor.

Metabolites	P	17P
Progesterone (P)	54	-
17α-hydroxyprogesterone (17P)	12	11
Estradiol-17β	0.04	31
11-Deoxycortisol	0.6	6
17α,20β,21-Trihydroxy-4-pregnen-3-one	1	5.5
3α,17α,21-Trihydroxy-5β-pregnan-20-one	trace	12
3α,17α,20β,21-tetrahydroxy-5β-pregnane	0.05	9
Unidentified	33	25.5

Note: each precursor corresponds to a different fish. Biopsies of ovarian tissue were taken from spawning females and incubated with precursor for 6 hours at 18°C. Fractions of free, sulphated and glucuronidated steroids were obtained using the methods described by Scott and Canario (1992). Identifications of metabolites were based on isopolarity with cold standards on silica gel thin layer chromatography (tlc) before and after microchemical reactions.

Sponsored partly by NATO's Scientific Affairs Division in the framework of the Science for Stability Programme and by STRIDE Programme, FEDER, JNICT, Portugal

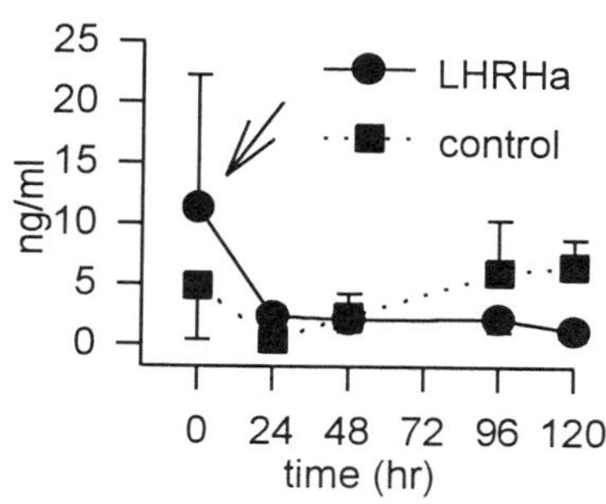

Fig 1. Changes in free 17α,20β,21-trihydroxy-4-pregnen-3-one blood plasma levels in female gilthead seabream which received EVAC implant at 0 hr (arrow) either with 75-100 μg [D-Ala[6]-Pro[9]-Net]-luteinizing hormone releasing hormone (LHRHa) or implant only (control). Error bars are standard error of means.

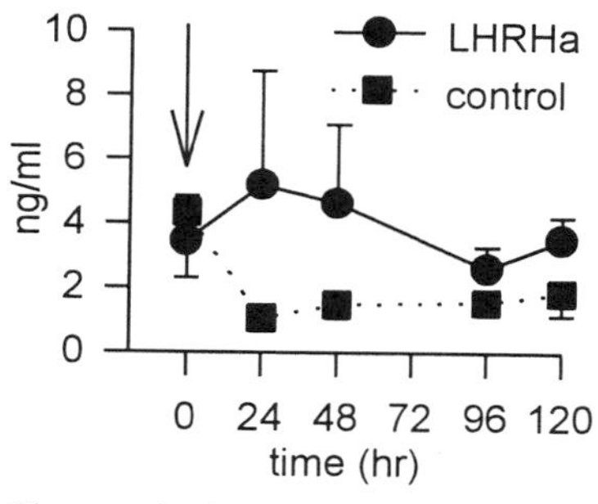

Fig . 2. Changes in free 3α,17α,21-trihydroxy-5β-pregnan-20-one plasma levels in female gilthead seabream stimulated with a LHRH analogue (see Fig 1).

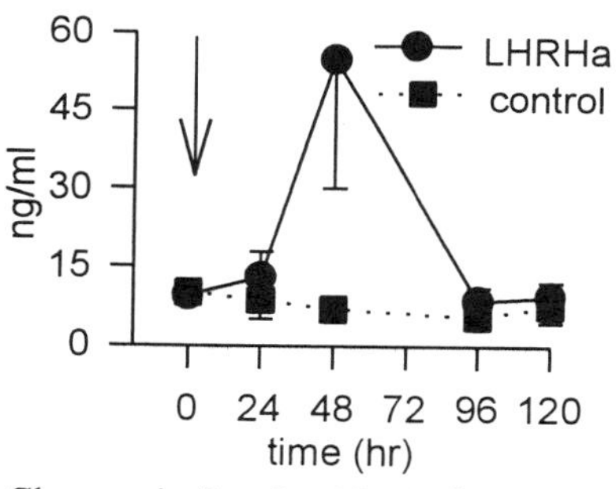

Fig . 3. Changes in free 3α,17α,20β,21-tetrahydroxy-5β-pregnane plasma levels in female gilthead seabream stimulated with a LHRH analogue (see Fig 1).

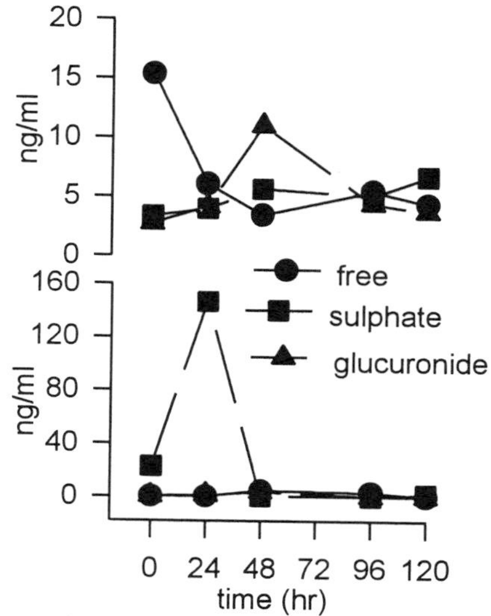

Fig . 4. Changes in free and conjugated 3α,17α,20β,21-tetrahydroxy-5β-pregnane plasma levels in two separate gilthead seabream females stimulated with a LHRH analogue.

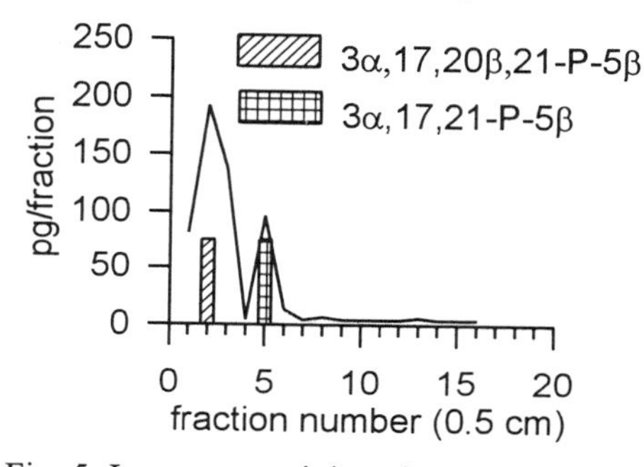

Fig. 5. Immunoreactivity of 0.5 cm thin layer chromatography fractions of gilthead seabream blood plasma with a radioimmunoassay optimized for 3α,17α,20β,21-tetrahydroxy-5β-pregnane (3α,17α,20β,21-P-5β). 3α,17α,21-P-5β (3α,17α,21-trihydroxy-5β-pregnan-20-one).

Table 2. Plasma steroid levels (ng/ml, mean±standard deviation, n=15) of female seabream sampled in the morning and noon of two consecutive days.

Steroid	Sampling time		
	03.00 pm	09.00 am	03.00 pm
17,20β-P Free	1.3±0.21	1.4±0.24	1.3±0.17
11-Deoxycortisol			
Free	3.7±0.96	3.8±1.19	4.0±1.35
Glucuronide	2.7±1.42	3.6±0.90	3.0±1.25
3α,17,21-P-5β			
Free	1.5±0.46	1.5±0.38	1.4±0.25
Sulphate	4.8±2.71	5.9±3.37	7.4±4.4
17,20β,21-P Free	2.0±0.81	2.4±1.05	1.8±0.47

Discussion and Conclusions

It is now apparent that teleosts are able to synthesize at least two oocyte maturation-inducing steroids: 17,20β-P identified in most of the fish studied (Scott and Canario, 1987) and 17,20β,21-P which has been identified in Atlantic croaker (*Micropogonias undulatus*, Trant *et al.*, 1986), spotted seatrout (*Cynoscion nebulosus*, Thomas and Trant, 1989), toadfish (*Halobatrachus didactylus*, Modesto and Canario, unpublished observations) and gilthead seabream (results from this study). The gilthead seabream ovary has the enzymes required for producing 17,20β,21-P and for its further reduction to the reduced 3α,5β tetrol. The absence of a 17,20β-P intermediate suggests that 11-deoxycortisol and 3α,17,21-P-5β are precursors of the 20β-hydroxy progestins and not a parallel route of metabolism.

The results obtained in the *in vitro* incubations with radiolabelled precursors were to a large extent confirmed by measurements *in vivo* under LHRHa stimulation during the spawning period. High levels of 17,20β,21-P, 3α,17,21-P-5β and 3α,17,20β,21-P-5β and their sulphated and glucuronidated conjugates were found although the stimulatory effect of the analogue on steroid levels was not statistically significant. This was due largely to a large variation of steroid levels with time and among individuals, possibly caused by short lived hormonal peaks (with a duration of hours or minutes). This may also explain the lack of success in finding alterations in steroid hormones in the fish sampled only twice a day during their natural ovulatory cycle.

In a previous study using radioimmunoassay, levels of 18-60 ng/ml of conjugated 3α,17,20β,21-P-5β were measured in seabass, *Dicentrarchus labrax*, in comparison to 6-14 ng/ml of conjugated 17,20β,21-P (Scott *et al.*, 1990). In the present study levels of the reduced metabolite (free or conjugated) were also generally 2-3 times higher than the levels of 17,20β,21-P. Similar results have been obtained for other progestin metabolites that have been assayed to date, suggesting that, at least in cases where the active steroid is difficult to detect, measurement of sex steroid metabolites may be a useful tool in endocrinological studies. It is still not clear, however, the physiological role of reduction and conjugation in the gonads.

In conclusion, on the basis of *in vitro* and *in vivo* studies it is suggested that 17,20β,21-P (and not 17,20β-P) is the likely maturation-inducing steroid in gilthead seabream. However, it is apparent that details of the daily endocrine ovulation cycle of this species can only be elucidated with more frequent sampling.

References

Canario, A. V. M., Scott, A. P., and Flint, A. P. F., 1989. Radioimmunoassay investigations of 20β-hydroxylated steroids in maturing/ovulating female rainbow trout (*Salmo gairdneri*). Gen. Comp. Endocrinol., 74, 77-84.

Canario, A. V. M., and Scott, A. P., 1990. Identification of, and development of radioimmunoassays for 17α,21-Dihydroxy-4-pregnene-3,20-dione and 3α,17α ,21-trihydroxy-5β -pregnan-20-one in the ovaries of mature plaice (*Pleuronectes platessa*). Gen. Comp. Endocrinol. 78, 273-285.

Colombo, L., Del Conte, E. and Clemenze, P., 1972. Steroid biosynthesis in vitro by the gonads of *Sparus auratus* L. (Teleostei) at different stages during natural sex reversal. Gen. Comp. Endocrinol., 19: 26-36.

Kadmon, G., Yaron, Z. and Gordin, H., 1984. Patterns of estradiol and 17α,20β-dihydro-progesterone in females *Sparus aurata*. Gen. Comp. Endocrinol., 53: 453. Abstract.

Scott, A. P., and Canario, A. V. M.,1987. Status of oocyte maturation-inducing steroids in teleosts. In" Proceedings of the 3rd International Symposyum on Reproductive Physiology of Fish.",(D. R. Idler, L. W. Crim, and J. M. Walsh, Ed.), (pp. 224-234). St John's, Newfoundland, Canada: Memorial University of Newfoundland.

Scott, A. P., and Canario, A. V. M., 1992. 17α,20β-Dihydroxy-4-pregnen-3-one 20-sulphate; a major new metabolite of the teleost oocyte maturation-inducing steroid. Gen. Comp. Endocrinol., 85, 91-100.

Scott, A.P., Canario, A.V.M. and Prat, F., 1990. Radioimmunoassay of ovarian steroids in plasmas of ovulating female sea bass (*Dicentrarchus labrax*). Gen. Comp. Endocrinol., 78: 299-302.

Thomas, P. and Trant, J.M., 1989. Evidence that 17α,20β,21-trihydroxy-4-pregnen-3-one is a maturation-inducing steroid in spotted seatrout. Fish Physiology Biochemistry 7, 185-191.

Trant, J. M., Thomas, P., and Shackleton, C. H. L., 1986. Identification of 17α,20β,-trihydroxy-4-pregnen-3-one as the major ovarian steroid produced by the teleost *Micropogonias undulatus* during final oocyte maturation. Steroids 47, 89-99.

REGULATION OF AROMATASE ACTIVITY IN THE RAINBOW TROUT, ONCORHYNCHUS MYKISS, OVARY

A. Fostier

Laboratoire de Physiologie des Poissons, Institut National de la Recherche Agronomique, Campus de Beaulieu, Rennes, France.

Summary

An assay based on the specific release of tritiated water after the conversion of 1β-tritiated androstenedione to estrone has been used to measure aromatase activity in pieces of rainbow trout, Oncorhynchus mykiss, ovary. In vitro incubations of ovarian follicles have been performed to study the regulation of aromatase activity by various hormones. (GtH2, GtH1, rtGH, rhIGF1, 17,20βP). Only the maturational gonadotropin, GtH2, had a significant and strong inhibiting effect on aromatase activity both on vitellogenic and pre-mature ovaries. This effect was suppressed by cycloheximide. The other hormones had neither a clear stimulatory nor inhibitory action.

Introduction

The decrease in plasma estradiol (E2) levels before female spawning in salmonids is now well known (Jalabert et al, 1991). This endocrine event facilitates oocyte maturation by releasing E2 inhibition on maturation inducing steroid synthesis (Jalabert and Fostier, 1984; Fostier and Baek, 1994). It occurs when the maturational gonadotropin (GtH2) increases, and, at this time of the reproductive cycle, GtH2 does not stimulate but even reduces E2 production by the ovary. (Young et al., 1983; Fostier and Jalabert, 1986).

This decrease in E2 levels may be related both to the acceleration of E2 metabolic clearance (Baroiller et al., 1987) and to the decrease of E2 synthesis by the ovary (Van der Kraak and Donaldson, 1986). A major fact for the arrest of E2 synthesis is a decrease in aromatase activity in granulosa cells (Young et al., 1983). Aromatase activity has been measured by quantifying the yield of estrogens from labelled androstenedione (Δ4) or testosterone (T). In this way, it has been suggested that GtH2 could inhibit aromatase activity (Sire and Depeche, 1981, De Mones and Fostier, 1987). This quantification raised some methodological problems: a low sensitivity, possible further metabolism of estrogens. In the present study, we have used a specific aromatase assay based on the release of tritiated water from Δ4 tritiated at position 1β (Gore Langton and Dorrington, 1981). Hormonal regulation of the aromatase activity in the ovary of rainbow trout, Oncorhynchus mykiss, has been studied in vitro using this method.

Materials and methods

In vitro incubations : Isolated follicles (just before oocyte maturation) or pieces of ovary (vitellogenic oocytes) were incubated at 12°C for 2 to 3 days according to Jalabert and Fostier (1984). When necessary, the following hormones were added at the beginning of incubation: GtH2, GtH1 (purified by B. Breton and M. Govorum), recombinant trout growth hormone (rtGH), recombinant human insulin-like growth factor (rhIGF1), and 17,20β-dihydroxy-4-pregnen-3-one (17,20βP).

Tissue homogenization: At the end of the incubation, tissues were homogenized in a 20 mM phosphate buffer pH 7.55 (NaCl 0.2 M, KCl 0.15 M, saccharose 0.25 M, dithiothreitol 5 mM, phenyl methylsulfonyl fluoride 1mM) with an Ultraturax then in a glass-teflon Potter (1 to 2 ml of buffer per g of ovary). Subcellular fractions were prepared by successive centrifugations: 800 g for 15 minutes (supernatant collected), 10,000 g for 20 minutes (supernatant collected) and 150,000 g for 1 hour (microsomes in pellet). 150,000 g pellets were suspended in buffer. Homogenization and centrifugation were performed at 0-4°C.

Aromatase assay: Less than 500 µl of the sample was put into a 5 ml glass tube. The final volume was adjusted to 500 µl when adding cofactor (NADPH, 1mM). The assay was started (at 12°C) by adding 18 kBq of $1\beta^3$H-androstenedione (NEN, sa = 795 Gbq/mmol) mixed with unlabelled Δ4 (to give 100 nM as a final concentration for the standardized assay). It was stopped by adding 50 µl of 3M trichloroacetic acid. After adding 500 µl water, samples were centrifuged (3,500 g , 15 min). 800 µl of the supernatant was transfered to a 5 ml glass tube containing a charcoal pellet (50 mg/tube). Tubes were shaken overnight at 4°C, then centrifuged (3,500 g, 15 min) and 600 µl of the supernatant was collected for radioactivity counting in 4 ml Picofluor (Packard). Estradiol production was estimated according to Gore Langton and Dorrington (1981).

Results

Aromatase assay: Release of tritiated water could be followed both in 800 g and 10,000 g supernatants and in microsomal fractions. In all cases, the measured activity was linearly correlated to the volume of samples and to the amount of protein in the incubate (Fig. 1). Besides, 1,4,6-androstatrien-3,17-dione (1 µM), an aromatase inhibitor, suppressed the release of tritiated water, showing the specificity of the assay. Further studies were performed on 10,000 g supernatants.

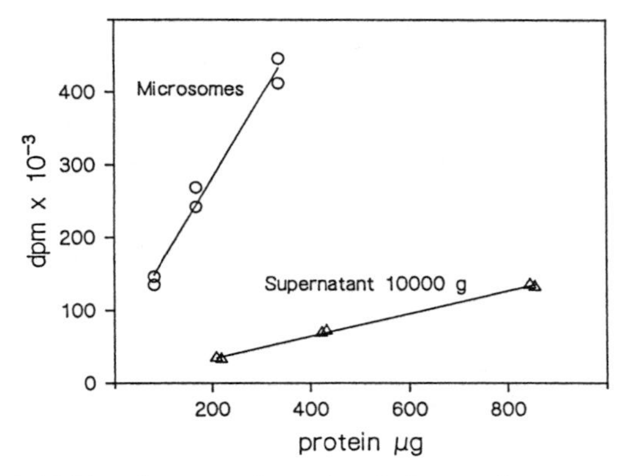

Fig.1: Tritiated water release in relation to the amount of protein in incubates.

The released of tritiated water was linearly correlated with the duration of incubation until 2 hours, showing that NADPH was not limiting. Thus, we chose an 1 hour incubation for the studies on aromatase regulation.

When adding increasing quantity of unlabelled Δ4, we observed a saturating curve and 100 nM was usually enough to saturate the enzyme. This concentration was chosen for the next studies.

Regulation of aromatase activity by GtH2: In prespawned females, 30 ng/ml GtH2 gave a 55 % inhibition of aromatase activity (Fig. 2). This GtH2 dose was sufficient to induce oocyte maturation in vitro. Such an inhibition was also observed at the beginning of exogenous vitellogenesis (Fig. 3).

Further studies were done on vitellogenic ovaries. When cycloheximide (10^{-3} M) was added to the in vitro incubation of the ovary, no more inhibition of aromatase activity by GtH2 could be detected (Fig. 4).

Supernatants from control incubates, to which no hormone had been added, were also mixed with

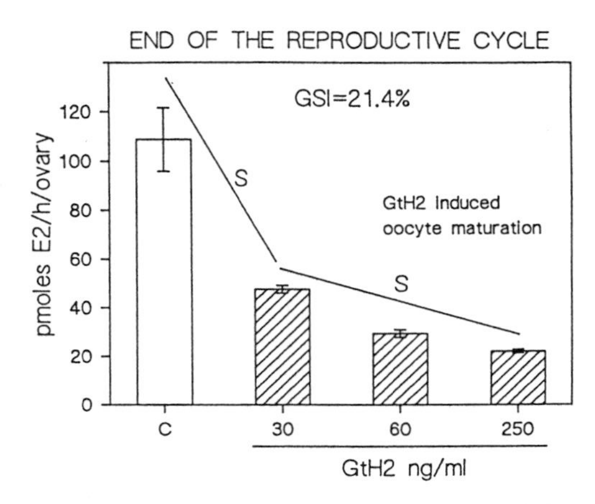

Fig.2: In vitro effect of GtH2 on aromatase activity of ovarian follicles sampled just before oocyte maturation.

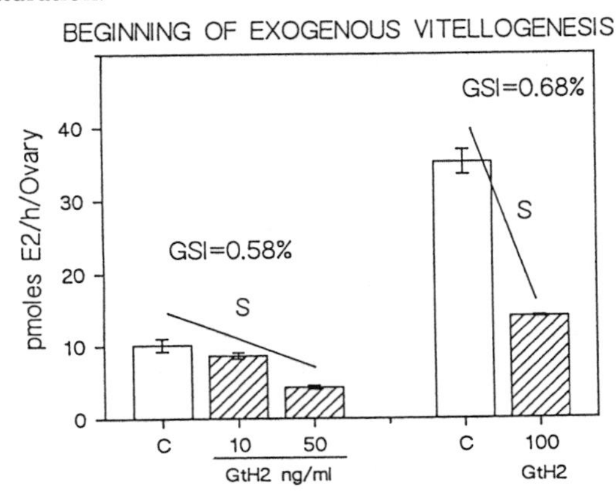

Fig. 3: In vitro effect of GtH2 on aromatase activity of vitellogenic follicles.

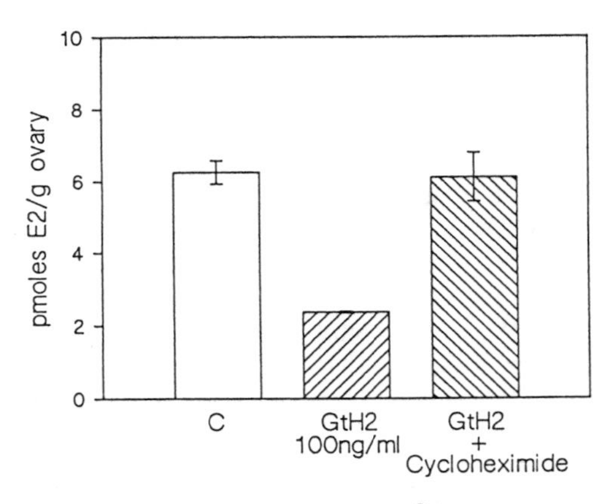

fig. 4: Effect of cycloheximide (10^{-3} M) on GtH2 inhibition.

increasing volumes of supernatants from GtH2 treated ovarian pieces. Aromatase activity decreased in relation to the volume of inhibiting supernatants. This effect remained even after the removal of endogenous steroids from the supernatants with charcoal (25 mg/ml, 15 minutes).

Regulation of aromatase activity by other hormones: GtH1 (50 ng/ml) had a weak inhibitory effect on vitellogenic ovaries, in comparison to GtH2. 17, 20βP (10, 50 and 100 ng/ml), rtGH (10 and 100 ng/ml) and IGF1 (50 ng/ml) had no significant effects on aromatase activity in vitellogenic ovaries.

Discussion

The assay based on the release of tritiated water after the conversion of 1β-^{3}HΔ to estrone can be used on 10,000 g supernatants of ovarian homogenates. This method gives an easier and more accurate quantification of aromatase than by using microsomal fractions, because of the difficulty in estimating the efficiency of their recovery. Besides, small quantities of tissues can be processed and several assays can be run at the same time within 2 days which is helpful to study the regulation of aromatase in in vitro treatments.

Using this method, we confirm that GtH2 can inhibit aromatase activity. This effect, combined with the increase of E2 metabolic clearance at the end of the reproductive cycle (Baroiller et al., 1987), could contribute to the drastic decrease of E2 plasma levels before oocyte maturation. In addition, these data suggest that other factors than GtH2 are necessary to stimulate estradiol synthesis during vitellogenesis.

References

Baroiller, J.F., Fostier, A., Zohar, Y. and O.Marcuzzi, 1987. The metabolic clearance rate of estradiol-17β in rainbow trout, Salmo gairdneri R., estimated by both single injection and constant infusion methods: increase during oocyte maturation. Gen. Comp. Endocrinol. 66: 85-94.

De Mones, A. and A. Fostier, 1987. Characterization and GtH regulation of microsomal activity in rainbow trout. In Proceedings of the Third International Symposium on Reproductive Physiology of Fish. St John's, Newfoundland. 71.

Fostier, A. and B. Jalabert., 1986. Steroidogenesis in rainbow trout (Salmo gairdneri) at various preovulatory stages: changes in plasma hormone levels and in vivo and in vitro responses of the ovary to salmon gonadotropin. Fish Physiol. Biochem. 2: 87-99.

Fostier, A. and H. Baek, 1994. Induction of maturation inducing steroid production in rainbow trout granulosa cells: effect of estradiol on gonadotropin stimulated 20β-hydroxysteroid dehydrogenase activity. Reprod. Nutr. Develop. 33: 81-82.

Gore Langton, R.E. and J.H. Dorrington, 1981. FSH induction of aromatase activity in cultured rat granulosa cells measured by a radiometric assay. Mol. Cell. Endocrinol. 22: 135-151.

Jalabert, B. and A. Fostier, 1984. The modulatory effect in vitro of estradiol-17β, testosterone or cortisol on the output of 17α-hydroxy-20β-dihydroprogesterone by rainbow trout (Salmo gairdneri) ovarian follicles stimulated by the maturational gonadotropin s-GtH. Reprod. Nutr. Develop. 24: 127-136.

Jalabert, B., Fostier, A., Breton, B. and C.Weil, 1991. Oocyte maturation in vertebrates. In: Vertebrate Endocrinology: Fundamentals and Biomedical Implications. P.K.T. Pang and M.P. Schreibman (Eds), Academic Press, New York. p. 21-90.

Sire, O. and J. Depêche, 1981. In vitro effect of a fish gonadotropin on aromatase and 17α-hydroxysteroid dehydrogenase activities in the ovary of the rainbow trout (Salmo gairdneri R.) Reprod. Nutr. Develop. 21: 715-726.

Van Der Kraak , G. and E.M. Donaldson, 1986. Steroidogenic capacity of coho salmon ovarian follicles throughout the periovulatory period. Fish Physiol. Biochem. 1: 179-186.

Young, G., Kagawa, H. and Y. Nagahama, 1983. Evidence of a decrease in aromatase activity in the ovarian granulosa cells of amago salmon (Oncorhynchus rhodurus) associated with final oocyte maturation. Biol. Reprod. 29: 310-315.

Regulation of Ovarian Steroidogenesis *In Vitro* by Gonadotropins during Sexual Maturation in Coho Salmon (*Oncorhynchus kisutch*)

J. V. Planas[1,2], J. Athos[3], and P. Swanson[3]

[1]School of Fisheries, University of Washington, Seattle, WA 98195; [3]Northwest Fisheries Science Center, National Marine Fisheries Service, Seattle, WA 98112

Introduction

Since the initial discovery of two distinct gonadotropins (GTH I and GTH II) in salmonids (Suzuki et al., 1988a; Swanson et al., 1991), a growing number of non-salmonid species (e.g. carp, seabream, killifish, bonito, tuna, croaker) have also been found to have two distinct GTHs (Lin et al., 1992; Van Der Kraak et al., 1992; Copeland and Thomas, 1993; Koide et al., 1993; Tanaka et al., 1993; Okada et al., 1994; see also these proceedings). Therefore, the duality of pituitary gonadotropins appears to have been well conserved during evolution, from fish to mammals. In mammals, it is well known that FSH and LH have quite distinct functions and that they bind to specific receptors. Two important questions that arise are whether teleost GTH I and GTH II have different functions, and whether these functions are homologous to those of FSH and LH.

Initial studies of the steroidogenic activities of teleost GTHs have shown that GTH I and GTH II share similar activities *in vitro*, despite some differences in potency (Suzuki et al., 1988b; Swanson et al., 1989; Van Der Kraak et al., 1992). In both male and female teleosts, GTH I and GTH II appear to be equipotent in stimulating steroid production during early phases of gametogenesis. However, during late stages of spermatogenesis, at spermiation and during final oocyte maturation, GTH II appears to have substantially higher steroidogenic potency than GTH I (Suzuki et al., 1988b; Planas and Swanson, 1995). These initial studies suggested that there are developmental changes in gonadal responsiveness to the steroidogenic actions of GTH I and GTH II, and possible functional differences during late stages of gametogenesis .

Developmental Changes in Gonadal Steroidogenesis

In salmonids, the marked temporal differences in the circulating levels of GTH I and GTH II during sexual maturation suggested the possibility that GTH I and GTH II may have different functions. In several salmonid species it is known that plasma GTH I levels are elevated during secondary oocyte growth and spermatogenesis, when the circulating levels of GTH II are extremely low or non-detectable. On the other hand, plasma GTH II levels increase during spermiation and final oocyte maturation when GTH I levels decline (Suzuki et al., 1988c; Swanson et al., 1989; Swanson, 1991).

During the transition from vitellogenesis to final oocyte maturation in female salmonids, changes occur in plasma GTH I and GTH II levels as well as in the production of two major ovarian steroids: estradiol-17β (E_2) and 17α,20β-dihydroxy-4-pregnen-3-one (17,20β-P). These changes in plasma steroid levels occur as a result of a steroidogenic shift in the salmonid ovary, from the production of E_2 to the production of 17,20β-P (reviewed by Nagahama, 1987). Due to the significant positive correlation between the plasma levels of GTH I and E_2, as well as between the plasma levels of GTH II and 17,20β-P (Swanson, 1991; Oppen-Berntsen et al., 1994; Slater et al., 1994), the question arises whether GTH I and GTH II play important roles in regulating the steroidogenic shift in the ovary. Previous studies have investigated the developmental changes in gonadotropic regulation of ovarian steroidogenesis in salmonids (Van Der Kraak and Donaldson, 1986; Kanamori et al., 1988). Although these studies did not use purified GTH I and GTH II, several interesting findings were reported. First, the gradual decrease in the production of E_2 during follicular development was attributed to a decrease in aromatase activity. Second, the increase in the basal as well as GTH-stimulated production of 17,20β-P *in vitro* prior to final maturation appeared to be dependent on the ability of the theca layers to produce 17α-hydroxyprogesterone (17OH-P) in response to GTH stimulation. However, the possible roles of GTH I and GTH II in regulating the change in steroidogenesis which occurs in follicular cells during final oocyte maturation remain to be determined.

GTH I and GTH II Actions during Final Oocyte Maturation

Previous studies in male salmon by our laboratory have shown that the effects of GTH I and GTH II on *in vitro* steroid production by testicular tissue change during the course of spermatogenesis (Planas and Swanson, 1995). The testicular sensitivity to the steroidogenic actions of GTH II, but not GTH I, increased significantly from stage IV of spermatogenesis through spermiation, suggesting a specific role for GTH II during male final maturation.

In view of these results it was important to determine whether the steroidogenic actions of GTH I and GTH II could also change during the course of ovarian development, particularly during final maturation when shifts in the steroidogenic pathways occur. Therefore, we examined the steroidogenic response of isolated coho salmon follicular layers to GTH I and GTH II at three different stages of final oocyte maturation: migrating germinal vesicle (GV) stage; peripheral GV stage; and after breakdown of the GV (GVBD). At each stage, the effects of GTHs (12.5-200 ng/ml) on *in vitro* steroid production of testosterone (T) and 17OH-P by theca layers as well as the conversion of T to E_2,

[2] Present Address: Department of Pharmacology, Box 357280, University of Washington, Seattle, WA 98195

or 17OH-P to 17,20β-P, by granulosa layers were determined.

In isolated theca layers, both GTH I and GTH II stimulated the production of 17OH-P in a concentration-dependent manner at all three stages in ovarian development. GTH II was more potent than GTH I in stimulating 17OH-P production at all stages, and the sensitivity to the steroidogenic effects of GTH II slightly increased at the post-GVBD stage (Fig. 1A). Similarly, GTH I and GTH II stimulated the production of T by theca layers and the sensitivity to GTH II, but not to GTH I, increased markedly during final maturation (Fig. 1B). The difference between the stimulatory effects of GTH II on 17OH-P and T production suggests that GTH II may be specifically stimulating the activity of the enzymes involved in the conversion of 17OH-P to T.

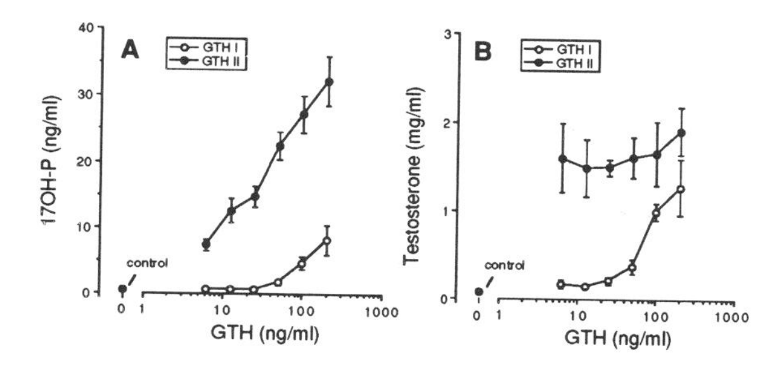

Figure 1: Effects of GTH I and GTH II on (A) 17α-hydroxyprogesterone (17OH-P) and (B) testosterone production by theca layers after germinal vesicle breakdown.

In isolated granulosa layers, no significant stimulation of aromatase activity by GTH I or GTH II was detected at any developmental stage, although the basal production of E_2 gradually declined during final maturation. Interestingly, high concentrations of GTH II significantly inhibited aromatase activity at the peripheral GV stage (Fig. 2A). A clear developmental decrease in the stimulation of 17,20β-P production by GTH I in granulosa layers was detected: from minimal stimulatory effects of GTH I on 17,20β-P production at the migrating GV stage to a complete lack of effects at later stages. On the other hand, GTH II stimulated the production of 17,20β-P at the migrating and peripheral GV stages, but not at the post-GVBD stage. The stimulation of 17,20β-P production by GTH II, but not by GTH I, just prior to GVBD (Fig. 2B) has been also seen in other studies (Maestro et al., 1995).

It is possible that the loss of steroidogenic activity by GTH II after GVBD is related to a possible pre-exposure of the granulosa cells used in the bioassay to high levels of GTH II *in vivo*, desensitizing the receptors, or to a change in the post-receptor binding mechanisms involved in GTH regulation of 20β-HSD activity .

The underlying mechanisms involved in the developmental changes in the effects of GTH I and GTH II on *in vitro* steroid production by isolated follicular layers are unknown. It is possible that they may partially be due to changes in receptor number, type and distribution. However, other interpretations cannot be ruled out, such as changes in post-receptor mechanisms (generation of different second messenger molecules) or changes in the activity of steroidogenic enzymes.

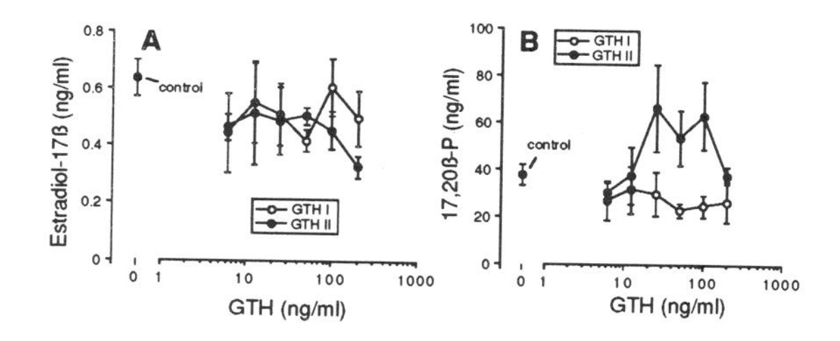

Figure 2: Effects of GTH I and GTH II on (A) estradiol-17β and (B) 17α,20β-dihydroxy-4-pregnen-3-one (17,20β-P) production by theca layers at the migrating germinal vesicle stage. Granulosa layers were incubated in the presence of 17α-hydroxyprogesterone and testosterone.

In granulosa layers, the unique steroidogenic effects of GTH II on 17,20β-P production, as well as its inhibitory effects on E_2 production during final maturation appear to be temporally associated with the exclusive appearance in granulosa cells of a specific receptor for GTH II, designated type II (Yan et al., 1992; Miwa et al., 1994). A second type of GTH receptor (type I), which binds both GTH I and GTH II, is found in both theca and granulosa layers at all stages of oogenesis, except in the late preovulatory follicle, when type I receptors are not detected in granulosa cells (Yan et al., 1992; Miwa et al., 1994). The loss of the type I receptor from the granulosa layers during final oocyte maturation could explain the decline in effectiveness of GTH I to stimulate 17,20β-P production. The higher steroidogenic potency of GTH II over GTH I in theca layers (primarily on T production) during final maturation could perhaps be explained, among other things, by the presence of type II receptors. However, previous binding studies were unable to detect type II receptors in theca layers.

Our observation that GTH II directly inhibits the production of E_2 by granulosa layers, albeit not completely, supports earlier reports of GTH-inhibition of aromatase activity (Sire and Depeche, 1981; Suzuki et al., 1988a) and could provide a mechanism to partially explain the decreasing ability of the ovary to produce E_2 during final maturation. Interestingly, LH specifically inhibits aromatase in luteinizing granulosa cells of the rat (Richards, 1994).

Conclusions

In this study, changes in the response of isolated follicular layers to the steroidogenic effects of GTH I and GTH II *in vitro* have been detected during final maturation. These results suggest that the change in ovarian steroidogenesis which occurs during final oocyte maturation may be regulated in part by changes in plasma levels of GTH I and GTH II and in GTH receptor type and distribution. The increase in circulating GTH II levels during final maturation appears to be essential in that it serves a double purpose. On one hand, GTH II uniquely stimulates the production of 17,20β-P, the maturation-inducing steroid. On the other hand, GTH II appears to inhibit the production of E_2 by granulosa layers. These important and specific actions of GTH II during final maturation support the hypothesis of differential functions for GTH I and GTH II. In addition, these actions of GTH II are similar to those of LH in the mammalian ovary, supporting the hypothesis that GTH II is structurally and functionally homologous to LH.

References

Copeland, P. A. and P. Thomas. 1993. Isolation of gonadotropin subunits and evidence for two distinct gonadotropins in atlantic croaker (*Micropogonias undulatus*). Gen. Comp. Endocrinol. 91:115-125.

Kanamori, A., S. Adachi and Y. Nagahama. 1988. Developmental changes in steroidogenic responses of ovarian follicles of amago salmon (*Oncorhynchus rhodurus*) to chum salmon gonadotropin during oogenesis. Gen. Comp. Endocrinol. 72:13-24.

Koide, Y., H. Itoh and H. Kawauchi. 1993. Isolation and characterization of two distinct gonadotropins, GTH I and GTH II, from bonito (*Katsuwonus plelamis*) pituitary glands. Int. J. Pept. Protein Res. 41:52-65.

Lin, Y.- W. P., B. A. Rupnow, D. A. Price, R. M. Greenberg and R. A. Wallace. 1992. *Fundulus heteroclitus* gonadotropins. 3. Cloning and sequencing of gonadotropic hormone (GTH) I and II β-subunits using the polymerase chain reaction. Mol. Cell. Endocrinol. 85:127-139.

Maestro, M. A., J. V. Planas , J. Gutierrez, S. Moriyama and P. Swanson. 1995. Effects of insulin-like growth factor I (IGF-I) on steroid production by isolated ovarian theca and granulosa layers of preovulatory coho salmon. Netherlands J. Zool. in press.

Miwa, S., L. Yan and P. Swanson. 1994. Localization of two gonadotropin receptors in the salmon gonad by *in vitro* ligand autoradiography. Biol. Reprod. 50:629-642.

Nagahama, Y. 1987. Gonadotropin action on gametogenesis and steroidogenesis in teleost gonads. Zool. Sci. 4: 209-222.

Okada, T, I. Kawazoe, S. Kimura , Y. Sasamoto, K. Aida and H. Kawauchi . 1994. Purification and characterization of gonadotropin I and II from pituitary glands of tuna (*Thunnus obesus*). Int. J. Pept. Protein Res. 43:69-80.

Oppen-Berntsen, D. O., S. O. Olsen, C. J. Rong, G. L. Taranger, P. Swanson and B. T. Walther. 1994. Plasma levels of eggshell zr-proteins, estradiol-17β, and gonadotropins during an annual reproductive cycle of atlantic salmon (*Salmo salar*). J. Exp. Zool. 268:59-70.

Planas, J. V. and P. Swanson. 1995. Maturation-associated changes in the response of the salmon testis to the steroidogenic actions of gonadotropins (GTH I and GTH II) *in vitro*. Biol. Reprod. 52:697-704.

Richards, J. S. 1994. Hormonal control of gene expression in the ovary. Endocr. Rev. 15:725-751.

Sire, O. and J. Depeche. 1981. *In vitro* effect of a fish gonadotropin on aromatase and 17β-hydroxysteroid dehydrogenase activities in the ovary of rainbow trout. Reprod. Nutr. Dev. 21:715-726.

Slater, C. H., C. B. Schreck and P. Swanson. 1994. Plasma profiles of the sex steroids and gonadotropins in maturing female spring chinook salmon (*Oncorhynchus tshawytscha*). Comp. Biochem. Physiol. 109A:167-175.

Suzuki, K., H. Kawauchi and Y. Nagahama. 1988a. Isolation and characterization of two distinct gonadotropins from chum salmon pituitary glands. Gen. Comp. Endocrinol. 71:292-301.

Suzuki, K., Y. Nagahama and H. Kawauchi. 1988b. Steroidogenic activities of two distinct gonadotropins. Gen. Comp. Endocrinol. 71:452-458.

Suzuki, K., A. Kanamori, Y. Nagahama and H. Kawauchi. 1988c. Development of salmon GTH I and GTH II radioimmunoassays. Gen. Comp. Endocrinol. 71:459-467.

Swanson, P., M. Bernard, M. Nozaki, K. Suzuki, H. Kawauchi and W. W. Dickhoff. 1989. Gonadotropins I and II in juvenile coho salmon. Fish Physiol. Biochem. 7:169-176.

Swanson, P., K. Suzuki, H. Kawauchi, W. W. Dickhoff. 1991. Isolation and characterization of two coho salmon gonadotropins, GTH I and GTH II. Biol. Reprod. 44:29-38.

Swanson, P. 1991. Salmon gonadotropins: reconciling old and new ideas. In: A. P. Scott, J. P. Sumpter, D. E. Kime and M. S. Rolfe (Eds.), Reproductive Physiology of Fish. Sheffield. FishSymp. 2-7.

Tanaka, H., H. Kagawa, K. Okuzawa and K. Hirose. 1993. Purification of gonadotropins (PmGTH I and II) from red seabream (*Pagrus major*) and development of a homologous radioimmunoassay for PmGTH II. Fish Physiol. Biochem. 11:409-418.

Tyler, C., J. P. Sumpter, H. Kawauchi and P. Swanson. 1991. Involvement of gonadotropin in the uptake of vitellogenin into vitellogenic oocytes of the rainbow trout, *Oncorhynchus mykiss*. Gen. Comp. Endocrinol. 83:337-344.

Van Der Kraak, G. and E. M. Donaldson. 1986. Steroidogenic capacity of coho salmon ovarian follicles throughout the periovulatory period. Fish Physiol. Biochem. 1:179-186.

Van Der Kraak, G., K. Suzuki, R. E. Peter, H. Itoh and H. Kawauchi. 1992. Properties of common carp gonadotropin I and II. Gen. Comp. Endocrinol. 85:217-229.

Yan, L., P. Swanson and W. W. Dickhoff. 1992. A two-receptor model for salmon gonadotropins (GTH I and GTH II). Biol. Reprod. 47:418-427.

STEROIDOGENESIS BY MILT & TESTIS OF ROACH IS AFFECTED BY SUBSTRATE CONCENTRATION

M. Ebrahimi, P. B. Singh and D. E.Kime

Department of Animal and Plant Sciences, The University of Sheffield, Sheffield S10 2TN, United Kingdom.

Summary

Substrate affects *in vitro* steroidogenesis in roach testis and sperm. Increased substrate induces a shift from 11-ketotestosterone, and 17,20β-dihydroxy-4-pregnen-3-one (17,20βP) to 17,20α-dihydroxy-4-pregnen-3-one (17,20αP). Glucuronidation is important only at low substrate concentration and long incubation time. Three hours was confirmed as the optimal time for incubation. Evidence is presented that 20αHSD activity may not be specific to the gonads of cyprinid fish.

Introduction

The roach (*Rutilus rutilus*, Cyprinidae) is an important sport fish in the United Kingdom but, because of its presence in many of the slow flowing rivers, may also be of importance as a biomonitor of water purity. As a basis for future studies on the effects of pollution on fecundity of these fish we have examined gonadal steroidogenesis in testis and milt of this species. Recent studies (Abdullah and Kime, 1994) have shown that substrate concentration may affect the nature of the *in vitro* products, and in particular may induce a switch from 11-oxygenated androgens to progestogens. We have therefore also examined the effect of substrate concentration on steroidogenesis in both sperm and testis of the roach. Since *in vitro* incubations are artificial systems, allowing the build up of metabolites, we have also examined the effect of incubation time on steroidogeneic profile.

Methods

For each incubation 100 mg of testis or 50μl of milt from 3 individual fish was shaken with tritiated 17-hydroxyprogesterone plus 0, 0.1, 1 or 10 μg/ml unlabeled 17-hydroxyprogesterone (17P) at 20° in 2 ml medium for 1, 3, 6 or 18 hr for testis, or 3 hr for sperm. Steroids were extracted with dichloromethane, conjugates hydrolysed with β-glucuronidase or acid solvolysis and metabolites separated by TLC and HPLC. Details of identification by microchemical reactions and crystallization to constant specific activity will be reported elsewhere.

Results

Three 11-oxygenated androgens: 11-ketotestosterone (KT), 11β-hydroxyandrostenedione (11βAD), and androstenetrione (AT) were identified, but in the most mature spermiating fish these were either not detectable or present in only low concentration. Both 17,20αP and 17,20βP were identified, although the ratio of the 20α:20β-epimers showed major variations. 11-Deoxycortisol was found in only very low yield in a few incubations, but was not fully characterised since its yield indicated that it did not play a significant role in roach testes.

Utilisation of 17P substrate increased with time, and decreased with increased substrate concentration (Fig. 1) with over 60% of substrate metabolised after 1 hour even at initial concentrations of 10μg/ml indicating highly active metabolising enzymes.

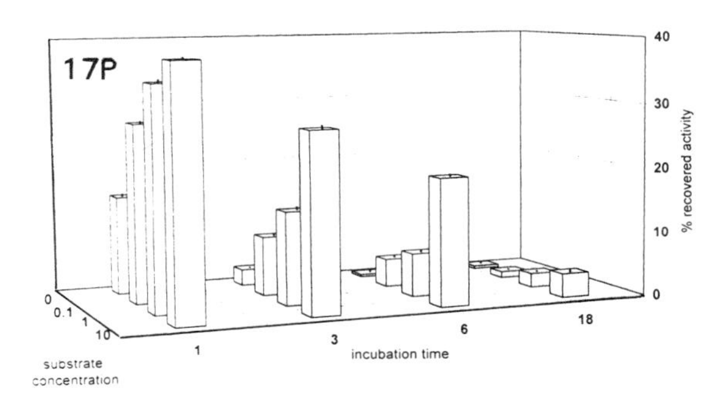

Fig. 1. Effect of substrate concentration on recovery of 17-hydroxyprogesterone

Glucuronides (Fig. 2) increased with incubation time and decreased with increased substrate concentration. Sulphates showed very similar patterns but yields never exceeded 10%. 11-Oxygenated androgens (mainly 11-ketotestosterone) were major metabolites at low substrate concentration only, with undetectable yields at the highest substrate concentration (Fig. 3). Androstenetrione yields followed a similar pattern to 11-ketotestosterone but were generally lower.

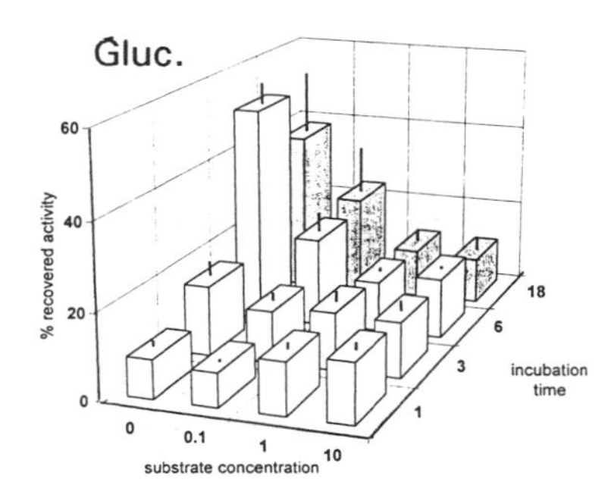

Fig. 2. Effect of substrate concentration on yields of glucuronides

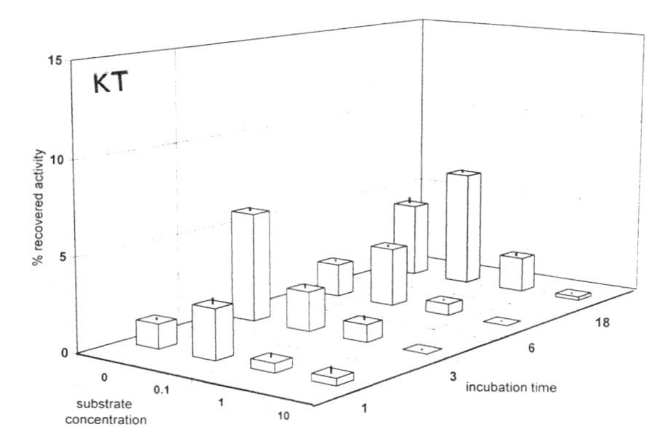

Fig. 3. Effect of substrate concentration on yields of 11-ketotestosterone

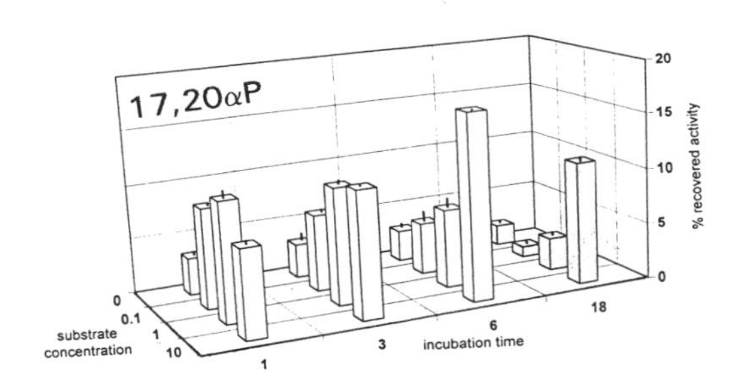

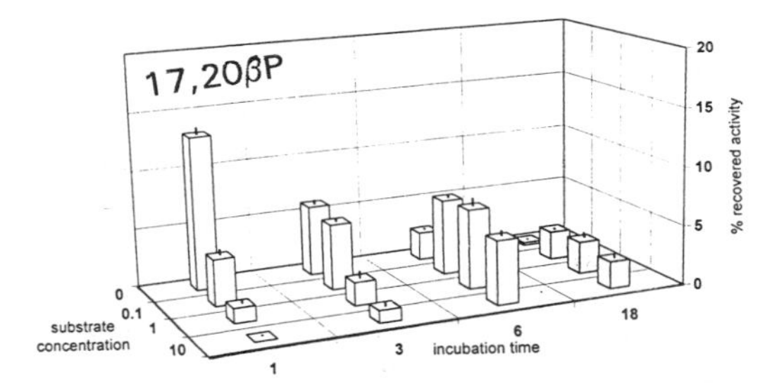

Fig. 4. Effect of substrate concentration on yields of 17,20αP and 17,20βP.

While 17,20βP followed a similar pattern to 11-ketotestosterone with generally highest concentrations at low substrate, 17,20αP increased with substrate (Fig. 4). In all incubations the trends were most marked at 3 hr incubation time which we consider optimal for such studies.

In sperm 11-ketotestosterone predominated at low substrate and 17,20αP at high substrate in incubations of 17-hydroxyprogesterone. We also found evidence for 17,20βP production at low substrate concentrations. Sperm did not convert progesterone into 17-hydroxyprogesterone, and only unchanged substrate was recovered.

Discussion

In its general pattern of metabolites of 17-hydroxyprogesterone, especially in the production of 17,20αP, the roach testis showed a close similarity to other cyprinids (Kime and Scott, 1993), and to some sparids and serranids (Kime *et al.*, 1995). There was a very clear effect of substrate concentration on the profile of steroid hormone production - with a shift from 11-ketotestosterone and 17,20βP at low substrate to 17,20αP at high substrate. In this respect the roach differs from the goldfish in which the 17,20αP:17,20βP ratio showed little change with substrate and suggests a lower level for saturation of the 20βHSD enzyme. Conjugation, especially as glucuronides, was particularly important at low substrate concentration and long incubation times. In this respect the roach resembles the goldfish (Abdullah and Kime, 1994).

Steroidogenesis is generally similar in both testis and sperm - although sperm is unable to convert progesterone into 17-hydroxyprogesterone. Incubations of sperm with 17-hydroxyprogesterone gave predominantly 11-ketotestosterone at low substrate concentrations and 17,20αP at high substrate. To our knowledge this is the first demonstration of synthesis of 11-ketotestosterone by teleost sperm. Our results indicate that previous suggestions (Asahina *et al.*, 1990, 1993; Sakai *et al.*, 1989) that sperm was only able to convert 17P into 17,20Ps and could not form 11-KT may be incorrect and that sperm, like the testis, may change its steroidogenic pattern with maturation. As shown by Abdullah and Kime (1994), and supported by the present results, this change in pathway may be due to changes in endogenous substrate concentration. Our work confirms previous suggestions from the rainbow trout (Sakai *et al.*, 1989) that sperm is unable to convert progesterone to 17-hydroxyprogesterone and requires synthesis of this precursor by the testicular somatic cells.

The results suggest that the 3 hour incubation commonly employed is optimal and gives a balance between insufficient utilisation of substrate and excessive further catabolism (especially to glucuronides).

Our results show that 20α-Hydroxysteroid dehydrogenase (20αHSD) is very active in both testis and sperm of roach. But this raises the question as to the role of 17,20αP in reproduction and whether there is a similarity in function between cyprinid 20αHSD and salmonid 20βHSD? There is, however, so far no evidence of maturational or pheromonal activity for 17,20αP in cyprinids, and in both functions 17,20βP is much more potent. It is therefore not clear whether 17,20αP could be an artifact of the artificial *in vitro* system or whether it has an as yet undefined role in reproduction?

In mammals porcine testicular 20αHSD has been shown by gene sequencing to be identical to aldol reductase, and 20βHSD to carbonyl reductase (Tanaka *et al.*, 1992; Warren *et al.*, 1993). It is not clear, however, whether these mammalian enzymes which act on progesterone to give 20α- or 20β-hydroxy-4-pregnen-3-ones are identical to the teleost enzymes which convert 17-hydroxyprogesterone to 17,20αP and 17,20βP or how similar the salmonid 17,20βHSD and cyprinid 17,20αHSD are. The published chromatograms of Asahina *et al.* (1990, 1993) in cyprinids and Sakai *et al.* (1989) in salmonids in fact show trace peaks of the epimeric progestogen. Aldol and carbonyl-reductases would not be expected to have great tissue specificity, but non-gonadal tissue has not been tested for 20HSD activity in teleosts. To test this specificity, we incubated eyeball, heart, muscle, testis and sperm from three male trout and three male goldfish with the same range of 17-hydroxyprogesterone concentrations used for roach. In trout, 20βHSD was present only in testis and sperm and there was no significant activity in the other organs, but in goldfish conversion of 17-hydroxyprogesterone to 17,20αP was high (20-30%) in eyeball, heart, testis and sperm , but low in muscle (6%). This suggests that cyprinid 20αHSD, unlike salmonid 20βHSD is not a specific gonadal steroidogenic enzyme. The lack of extragonadal 20HSD activity in trout, however, suggests that the cyprinid 20αHSD is not simply an aldol reductase. Since 17-hydroxyprogesterone is likely to be present only in reproductive tissue, a role for 17,20αP in cyprinid reproduction cannot be precluded. The nature and function of the cyprinid extragonadal 20αHSD remains to be clarified.

References

Abdullah, M. A. S. and D. E. Kime, 1994. Increased substrate concentration causes a shift from production of 11-oxygenated androgens to 17,20-dihydroxyprogestogens during the in vitro metabolism of 17-hydroxy- progesterone by goldfish testes. Gen. Comp. Endocrinol. 96: 129-139.

Asahina, K., T. P. Barry, K. Aida, N. Fusetani, and I. Hanyu, 1990. Biosynthesis of 17,20α-dihydroxy-4-pregnen-3-one from 17α-hydroxyprogesterone by spermatozoa of the common carp, *Cyprinus carpio.* J. Exp. Zool. 255: 244-249.

Asahina, K., K. Aida and T. Higashi, 1993. Biosynthesis of 17α,20α-dihydroxy-4-pregnen-3-one from 17α-hydroxyprogesterone by goldfish (*Carassius auratus*) spermatozoa. Zool. Sci. 10: 381-383.

Kime, D. E. and A. P. Scott, 1993. In vitro synthesis of 20α-reduced and of 11- and 21-oxygenated steroids and their sulfates by testes of the goldfish (*Carassius auratus*): Testicular synthesis of corticosteroids. Fish Physiol. Biochem. 11: 287-292.

Kime, D. E., M. A. S Abdullah and K. P. Lone, 1995. Effects of substrate concentrations on steroidogenesis in ovaries and testes of three species of ambisexual fishes. This volume.

Sakai, N., H. Ueda, N. Suzuki and Y. Nagahama, 1989. Involvement of sperm in the production of 17,20α-dihydroxy-4-pregnen-3-one in the testis of spermiating rainbow trout, *Salmo gairdneri.* Biomed. Res. 10: 131-138.

Tanaka, M., S. Ohno, S. Adachi, S. Nakajin, M. Shinoda and Y. Nagahama, 1992. Pig testicular 20β-hydroxysteroid dehydrogenase exhibits carbonyl reductase-like structure and activity. J. Biol. Chem. 267: 13451-13455.

Warren, J. C., G. L. Murdock, Y. Ma, S. R. Goodman and W. E. Zimmer, 1993. Molecular cloning of testicular 20α-hydroxysteroid dehydrogenase: identity with aldose reductase. Biochem. 32: 1401-1406.

STUDIES OF THE NUCLEAR PROGESTOGEN RECEPTOR IN THE OVARY OF THE SPOTTED SEATROUT: REGULATION OF THE FINAL STAGES OF OVARIAN DEVELOPMENT

J. Pinter and P. Thomas

University of Texas Marine Science Institute, Port Aransas, TX 78373

Summary

A novel, nuclear progestogen receptor with high affinity for the two teleostean maturation-inducing steroids (MIS's) has previously been characterized in the ovary of the spotted seatrout. The current results indicate that this receptor demonstrates steroid specificity intermediate between progesterone and glucocorticoid receptors, with characteristics of each.

During gonadal recrudescence, receptor levels increased two-fold on a whole gonad basis, although receptor concentrations per tissue weight decreased significantly. This indicates that the ovary is growing at a faster rate than the increase in receptor concentrations. *In vitro* studies demonstrate that addition of either teleostean MIS to post-vitellogenic oocytes results in disappearance of 90% of the receptor from the cytosol. This disappearance demonstrates kinetics similar to steroid binding and is presumed to represent translocation to the nucleus.

Investigations into the physiological role of this receptor demonstrate that the MIS induces both oocyte hydration and ovulation via genomic mechanisms. We propose a model of final oocyte maturation, hydration and ovulation wherein the MIS acts through a membrane receptor to induce final oocyte maturation, and through the nuclear receptor to induce hydration and ovulation.

Introduction

Previous studies have characterized a membrane receptor for the MIS 17α,20β,21-trihydroxy-4-pregnen-3-one (20β-S) in the ovary of the spotted seatrout, *Cynoscion nebulosus* (Patiño and Thomas, 1990). This receptor is upregulated by gonadotropin and is necessary for MIS-induced final oocyte maturation (FOM) (Thomas and Patiño, 1991).

During characterization of the ovarian membrane receptor, a second, 'soluble', binding component was found that had high affinity for both teleost MIS's, 20β-S and 17α,20β-dihydroxy-4-pregnen-3-one (17α,20β-P) (Patiño and Thomas, 1990). Subsequent investigations have characterized a nuclear progestogen receptor in this tissue (Pinter and Thomas, 1995)

Previous research on goldfish and yellow perch has demonstrated that the MIS acts via a genomic mechanism to induce ovulation (Goetz and Theofan, 1979, Theofan and Goetz, 1981). Therefore, this study focused on the physiological characterization, regulation and role(s) of this receptor in the final stages of ovarian development in the spotted seatrout.

Results

Steroid Specificity

Previous investigations have demonstrated that the nuclear progestogen receptor has steroid specificity similar to previously characterized progesterone receptors (PR's) in a wide variety of vertebrates (Pinter and Thomas, 1995), with the exception that the primary ligands are different (17α,20β-P and 20β-S vs progesterone). Receptor binding studies were subsequently carried out with synthetic progestins and corticosteroids to further elucidate any differences in steroid specificity. Interestingly, the majority of synthetic agonists (e.g.: megestrol actetate, norethindrone, prednisolone, dexamethasone) and antagonists (ORG 31710, ZK98299, ZK112993) demonstrated very low affinity for this receptor. Addition of a hydroxyl or ketone group at the 11 position caused an order of magnitude decrease in binding affinity. In addition, the 17, 20 and 21 hydroxyl groups also demonstrated a similar sensitivity to modification, with the 20 hydroxyl group being particularly important. For example, 20β-S had a relative binding affinity of 65%, while 11-deoxycortisol had an RBA of 1.6% and cortisol had an RBA of 0.10%. Likewise 17α-hydroxyprogesterone had an RBA of 6.8% while medroxyprogesterone (17-acetyl-progesterone) had an RBA of 0.89%. The compounds demonstrating the highest affinities were the 11- deoxy, 20-hydroxy (11-dehydro) derivatives of various corticosteroids.

Physiological Regulation of Receptor Levels

Receptor levels during gonadal recrudescence increased two-fold on a whole gonad basis, from 65 pmol/gonad in regressed individuals to 130 pmol/gonad in fully developed individuals. However, receptor levels per tissue weight demonstrated an asymptotic decrease from 1 pmol/g ovary to 0.1 pmol/g ovary during recrudescence.

Analysis of the effects of gonadotropin (20 IU/ml) and various steroids (1 - 300 nM) on receptor levels *in vitro* indicates that both teleost MIS's (17α,20β-P and 20β-S) caused a 90% decrease in cytoplasmic levels with a $T_{1/2}$ = 2. 24 ± 0.09 h (n = 3) and an EC_{50} of 2.96 ± 0.23 nM (n = 4). Gonadotropin and all other steroids had no effect. While the appearance of the receptor in the nucleus could not be demonstrated, the kinetics of disappearance from the cytoplasm were similar to the kinetics of steroid binding (Pinter and Thomas, 1995).

Physiological Role

Addition of hCG to post-vitellogenic oocytes results in the development of maturational competence. Subsequent addition of 20β-S causes both oocyte maturation and hydration (Thomas and Trant, 1989). We were able to confirm the previously reported results of Lafleur and Thomas (1991) that oocyte hydration is dependent on genomic action (RNA transcription).

Further incubation with MIS induces ovulation of mature oocytes within 9 - 12 h, as determined by histological examination. The RNA transcription inhibitor Actinomycin D significantly inhibited MIS-induced ovulation, and the protein synthesis inhibitor cycloheximide consistently decreased the rate of ovulation. This indicates that induction of ovulation by the MIS is dependent on a genomic mechanism. The requirements for maturation, hydration and ovulation are summarized in Table 1.

Discussion

We have shown that the previously characterized nuclear progestogen receptor demonstrates steroid specificity intermediate between progesterone and glucocorticoid receptors. The decrease in affinity caused by addition of hydroxyl or ketone groups at the 11 position is characteristics of progesterone receptors (PR's), while the sensitivity to modification of the 17 position is characteristic of glucocorticoid receptors (GR's). This result is not surprising given the structure of the seatrout MIS 20β-S, which is also intermediate.

The results of the recrudescence study indicate that this receptor is slightly upregulated during ovarian growth, suggestting that the receptor is most likely active during the later stages of oocyte development. The *in vitro* studies provide indirect evidence that the receptor is translocated, and therefore activated, by both MIS's. Analysis of steroid levels in the blood show that 20β-S is elevated only during FOM (unpublished results), further confirming the inference from the field study that the receptor is active at this time.

Investigations into the physiological role of this receptor indicate that it may have a dual function. First, we confirmed previous results (Lafleur and Thomas, 1991) demonstrating that oocyte hydration is dependent on MIS acting through a genomic mechanism. Regulation of oocyte hydration is via ion channels and osmotic pressure, indicating that the MIS may be acting through the nuclear progestogen receptor to induce glucocorticoid-like activity. Second, we have demonstrated that ovulation can be induced by addition of MIS to mature, hydrated oocytes and is also dependent on RNA and protein synthesis. This indicates that the receptor also demonstrates progestogen activity. Studies with other fish species indicate that prostaglandin production is a likely target of progestin activity (Goetz, 1989).

Finally, in conjunction with previous studies regarding initiation of final oocyte maturation and the role of the 20β-S membrane

Table 1. Summary of physiological requirements for maturation, hydration and ovulation of fully-grown oocytes *in vitro*.

	Maturation[a]	Hydration[b]	Ovulation
MIS Synthesis	Yes	Yes	Yes
RNA Synthesis	No	Yes	Yes
Protein Synthesis	Yes	Yes	Yes
cAMP regulation	Decrease	?	Decrease
Na^+/K^+-ATPase	?	Increase	?
20β-S membrane Receptor	Yes	No	No
Nuclear Receptor	No	Yes	Yes

[a]From Thomas and Trant, 1989; Thomas and Patiño, 1991.
[b]From Lafleur and Thomas, 1991.

receptor (Thomas and Patino, 1991), we propose a model of the final stages of ovarian development in the spotted seatrout (See Fig. 1). When vitellogenesis is complete and spawning conditions are appropriate, gonadotropin II (GtH II) is released from the pituitary. The maturational gonadotropin first acts to upregulate the concentration of 20β-S membrane receptors and induce maturational competence. Subsequently, GtH II induces synthesis of the MIS, which acts via three pathways. First, 20β-S demonstrates progestational activity by binding to the 20β-S membrane receptor and initiating a cascade of second messengers resulting in FOM. Secondly, 20β-S demonstrates corticosteroid-like activity by binding to the nuclear receptor and initiating oocyte hydration, possibly through regulation of Na^+,K^+-ATPase. Finally, the MIS demonstrates progestational activity and induces ovulation via a genomic mechanism.

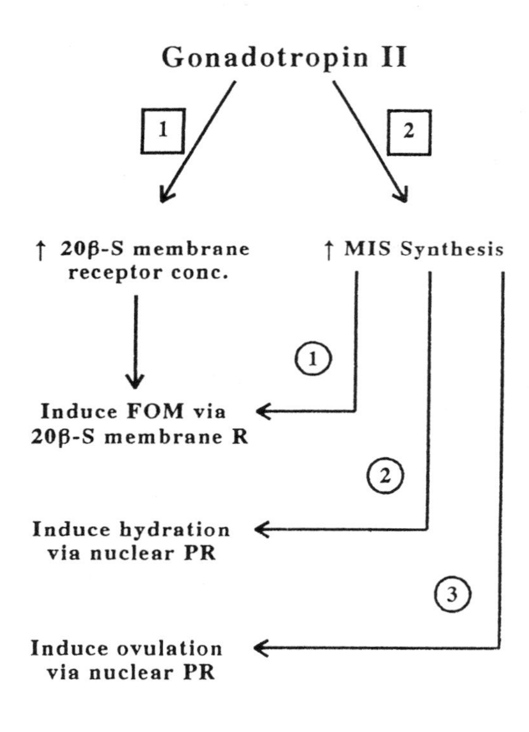

Fig. 1. Model of the final stages of ovarian development in the spotted seatrout.

References

Goetz W., P. Duman, A. Berndtson, and E.G. Janowsky. The role of prostaglandins in the control of ovulation in yellow perch, *Perca flavescens*. Fish Physiol. Biochem. 1989; 7:163-168.

Goetz, F.W. and G. Theofan, 1981. *In vitro* stimulation of germinal vesicle breakdown and ovulation of yellow perch (*Perca flavescens*) oocytes: Effects of 17α-hydroxy-20β-dihydroprogesterone and prostaglandins. Gen. Comp. Endocrinol. 37:273-285.

Lafleur, G.J. and P. Thomas, 1991. Evidence for a role of Na^+,K^+-ATPase in the hydration of Atlantic croaker and spotted seatrout oocytes during final maturation. J. Exp. Zool. 258:126-136.

Patiño, R. and P. Thomas, 1990. Characterization of membrane receptor activity for 17α,20β,21-trihydroxy-4-pregnen-3-one in ovaries of spotted seatrout (*Cynoscion nebulosus*). Gen. Comp. Endocrinol. 78:204-217.

Pinter, J. and P. Thomas, 1995. Characterization of a progestogen receptor in the ovary of the spotted seatrout, *Cynsocion nebulosus*. Biol. Reprod. 52:667-675.

Theofan, G. and F.W. Goetz, 1979. The *in vitro* effects of actinomycin D and cycloheximide on germinal vesicle breakdown and ovulation of yellow perch (*Perca flavescens*). Comp. Biochem. Physiol. A 69:557-561.

Thomas, P. and R. Patiño, 1991. Changes in 17α,20β,21-trihydroxy-4-pregnen-3-one membrane receptor concentrations in ovaries of spotted seatrout during final oocyte maturation. In: Scott AP, Sumpter JP, Kime DE, Rolfe MS (eds), Proceedings of the Fourth International Symposium on Reproductive Physiology of Fish; 1991; University of East Anglia, Norwich, United Kingdom, 122-124.

Thomas, P. and J. Trant, 1989. Evidence that 17α,20β,21-trihydroxy-4-pregnen-3-one is a maturation-inducing steroid in the spotted seatrout. Fish Physiol. Biochem. 7:185-191.

EFFECTS OF 17α,20β-DIHYDROXY-4-PREGNEN -3-ONE ON TESTICULAR ANDROGENS IN MATURE ATLANTIC SALMON (SALMO SALAR) MALE PARR IN VIVO AND IN VITRO STUDIES.

E. Antonopoulou, S. Jakobsson, I. Mayer and B. Borg

Department of Zoology, University of Stockholm, S-106 91 Stockholm, Sweden.

Introduction

During natural sexual maturation in male salmonids a peak in plasma androgen levels at the period of spermatogenesis is followed by a rise in 17α, 20β-dihydroxy-4-pregnene-3-one (17,20P) at the time of final maturation (spermiation) and a concomitant decline in androgens. The aim of the present investigation was to find out whether the decline in androgen levels is due to a suppressive effect of 17,20P. To that end, the effects of 17,20P administration *in vivo* and *in vitro* were studied.

Experimental procedure

In vivo experiments:

Mature Atlantic salmon male parr were implanted with Silastic capsules filled with 17,20P or empty capsules in early July (experiment 1) or in late August (experiment 2). Both experiments were terminated in September, when natural androgen levels are at their highest.

In vitro experiments:

Testicular fragments from mature salmon parr were incubated with 17,20P in increasing doses of 3, 30, 300 and 3000 ng/ml, in combination with various substrates, i.e. testosterone (T, 100 ng/ml), 17α-hydroxy-4-pregnene-3-one (17P, 100 ng/ml) or medium alone.

Androgen levels in plasma and incubation media were measured using radioimmunoassay (RIA).

Results

In vivo experiments:

Both experiments yielded similar results: 17,20P treatment raised plasma levels of 17,20P, whereas levels of T and 11-ketotestosterone (11KT) were not influenced. No difference in gonadosomatic index, testes histology (exp. 1), or time of spermiation was observed between the controls and the treated fish.

In vitro experiments:

11KT was formed in testes incubates containing T or 17P as substrate, whereas medium levels of 11KT were usually non-detectable in those containing medium alone. A very high dose of 17,20P (3000 ng/ml) suppressed 11KT production, whereas 3, 30 or 300 ng/ml were without effect.

Discussion

Increased levels of plasma 17,20P after *in vivo* treatment did not influence androgen levels, indicating an absence of a suppressive action of 17,20P on androgen production exerted via extratesticular sites (eg. hypothalamus-pituitary).

At the testicular level, a suppressive action of 17,20P on 11KT production was only present at a very high dose. This finding, which is in contrast to the situation in carp (Barry *et al.*, 1989), suggests that physiological levels of 17,20P do not actively suppress 11KT production intratesticularly. The decline in androgens when 17,20P levels rise may instead be a consequence of the substrate 17P being used for 17,20P production rather than for androgens.

Reference

Barry, T. P., Aida, K. and Hanyu, I. (1989). Effects of 17α,20β-dihydroxy-4-pregnen-3-one on the *in vitro* production of 11-ketotestosterone by testicular fragments of the common carp, *Cyprinus carpio.* J. Exp. Zool. 251: 117-120.

STEROIDOGENESIS DURING ESTROGEN-INDUCED SEX INVERSION IN THE SEABREAM, *SPARUS AURATA*.

J.A.B. Condeça and A.V.M. Canario

Center of Marine Sciences, UCTRA, University of Algarve, 8000 Faro, Portugal.

Summary

Sex inversion was artificially induced in the sea bream by feeding estrogen to juveniles. While 17α-ethynilestradiol ($ethE_2$) was ineffective to induce sex inversion, 15 mg/kg of estradiol-17β (E_2) during 14 weeks fully accomplished this process. Feminized fish showed an increase in 17β-hydroxysteroid dehydrogenase (17β-HSD) activity and a reduction in 11β-hydroxylase (11β-OH) in comparison with transitional fish. We suggest that 11β-OH and aromatase are probably directly involved in natural sex change, although extragonadal factors might regulate aromatase activity.

Introduction

In order to gain some insight on the steroidogenic alterations occurring during sex inversion we have artificially induced sex reversal in the sea bream *Sparus aurata* with estrogen and followed the changes in the potential of the gonads to produce E_2, T and 11-ketotestosterone (11-KT).

Methods

In exp.1 fish were fed a commercial diet containing 15mg/kg of $ethE_2$ for 0, 37 or 112 days while on exp. 2 the diet contained 0 (CTL2), 2 (T3) or 15mg/Kg of E_2 (T4) and all fish were fed for 98 days.

Results

Sex inversion was not achieved in exp.1 (data not shown). However, it was obtained in exp.2 when fish were fed with the higher dose of E_2 (fig.1). In contrast, gonads from control fish were largely dominated by male tissue. Comparison of gametogenesis in homologous tissue showed that oogenesis was more advanced in the estrogen treated animals as well as spermatogenesis in the controls. Steroid blood plasma levels were always very low or undetectable.

E_2, together with T, were highly correlated to the cross-sectional area of female tissue in the gonads. 11KT conversion was higher in the groups at the transitional stage of sex reversal (T3) than in control (CTL2) or sex reversed groups (T4).

T glucuronide and E_2 glucuronide and sulfate were the major products of feminized gonads.

Discussion

Accumulation of T in the feminized fish was associated with an increase in 17β-HSD (increased total steroid conversion) and reduction in 11β-OH in comparison to transitional fish. This observation is consistent with the hypothesis of 11β-OH regulation

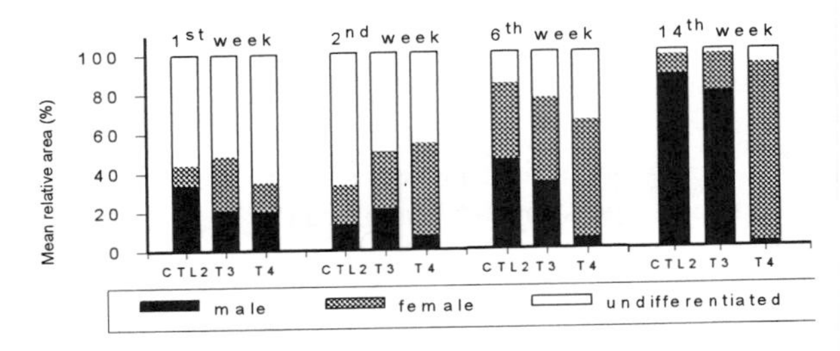

Fig.1- Evolution of gonad cross-sectional mean relative areas during experiment 2.

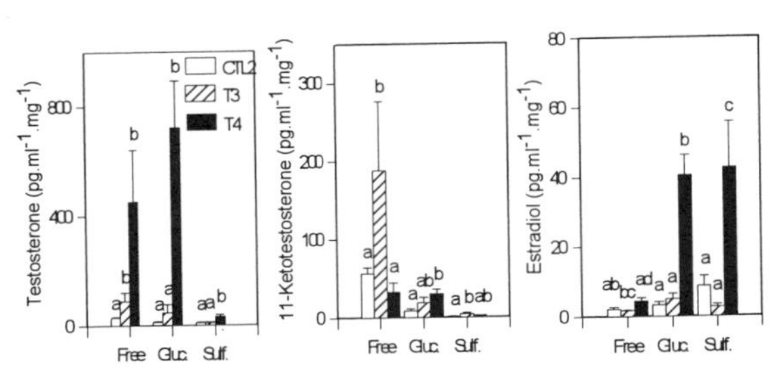

Fig.2- Concentrations of T, 11KT and E2 (mean+SEM) in incubations of gonads from exp.2 with androstenedione. Different letters above bars represent statistically significant differences among groups within a form of the steroid .

during sex change (Reinboth *et al.*, 1986).

The increase in E_2 production *in vitro* was correlated to an increase in proportion of female tissue in the gonad. This suggests that aromatase was not stimulated by the exogenous addition of E_2. As estrogen is the probable endogenous direct mediator of sex change, it is likely that other factors, possibly of extragonadal origin (eg. GtH) are involved in the regulation of this key enzyme.

Although there seems to exist a correlation between steroid conjugation and sex inversion, the role of these metabolites is not clear.

References

Reinboth, R., Becker, B. & Latz, M., 1986. *In Vitro* Studies on Steroid Metabolism by Gonadal Tissues from Ambisexual Teleosts. II. Conversion of [^{14}C] androstenedione by the heterologous tissues of the protandric sea bream *Pagellus acarne* (Risso). Gen. *Comp. Endocrinol.* **62**, 335-340.

This work was supported by JNICT- Program CIENCIA and NATO- Project POSEABREAM

CLONING AND FUNCTIONAL EXPRESSION OF THE COHO SALMON (*ONCORHYNCHUS KISUTCH*) TYPE II GONADOTROPIN RECEPTOR

A.H. Dittman[1,2], L. Yan[2], and P. Swanson[2]

[1]School of Fisheries, Box 357980, University of Washington, Seattle WA 98195; [2]Northwest Fisheries Science Center, 2725 Montlake Blvd. E., Seattle, WA 98112.

Summary

A 2.8 kb cDNA corresponding to a putative gonadotropin receptor was cloned from a coho salmon gonadal cDNA library. Sequence analysis indicated that the cDNA was highly homologous to the mammalian glycoprotein hormone receptors (TSH-R, LH/CG-R, FSH-R). The responses of HEK-293 cells transiently transfected with this cDNA to purified coho salmon gonadotropins, GTH I and GTH II, indicated that the cDNA encodes for a functional receptor with pharmacological properties most consistent with that of the coho salmon GTH II-specific receptor (GTH-RII).

Introduction

The functional effects of the salmon gonadotropins, GTH I and GTH II are thought to be mediated via G protein-coupled glycoprotein hormone which stimulate intracellular cAMP production in the gonads. Based on previous binding studies in our laboratory, we have proposed a two-receptor model for gonadotropin action wherein there are two GTH receptors with distinct spatial and temporal distribution: 1) GTH-RI, which binds both GTH's but with higher affinity for GTH I and 2) GTH-RII which is highly specific for GTH II (Yan et al., 1992; Miwa et al., 1994). GTH-RI is present in both thecal and granulosa cells whereas the GTH-RII is present only in granulosa cells (Yan et al., 1992; Miwa et al., 1994). Furthermore, GTH-RI is detectable during all stages of gonadal development while GTH-RII was only detected in the preovulatory stage (Miwa et al., 1994). The objectives of this study were to further characterize the distinct properties of these receptors by cDNA cloning, sequencing and functional expression of receptor proteins.

Results and Discussion

Probes for the salmon GTH receptors were prepared by PCR amplification of reverse-transcribed mRNA from immature coho salmon testes using degenerate primers derived from conserved regions of the mammalian glycoprotein hormone receptors (LH/CG-R, FSH-R, TSH-R). The resulting 250 probe was used to screen a coho salmon gonadal cDNA library enriched in granulosa cell mRNA. Three overlapping clones were identified and used to generate a full-length 2.8 Kb cDNA encoding a putative GTH receptor which is 40-60% identical to mammalian LH, FSH, and TSH receptors (Table I).

To determine whether this cDNA encoded a functional GTH receptor, HEK-293 cells were transiently transfected with an expression vector containing the full-length cDNA and assayed for responsiveness to purified coho salmon GTH I and GTH II. Functional expression was determined by measuring changes in intracellular cAMP production as

	Extracellular	Transmembrane	Intra-cellular
	5'		3'
rTSHR	54.6%	74.0%	25.3%
rLHR	45.3%	71.0%	25.7%
rFSHR	35.3%	67.9%	22.6%

Table 1. Comparison of the amino acid sequence identity between the salmon GTH-R and the rat TSH-R, LH-R and FSH-R.

indicated by a β-galactosidase reporter. In HEK-293 cells expressing the putative receptor, cAMP production increased in a dose-dependent manner in response to partially purified coho salmon gonadotropins (SG-G100) while control cells showed no response. Furthermore, in cells expressing the putative receptor, intracellular cAMP increased in response to purified GTH II but was unaffected by GTH I (Fig. 1.).

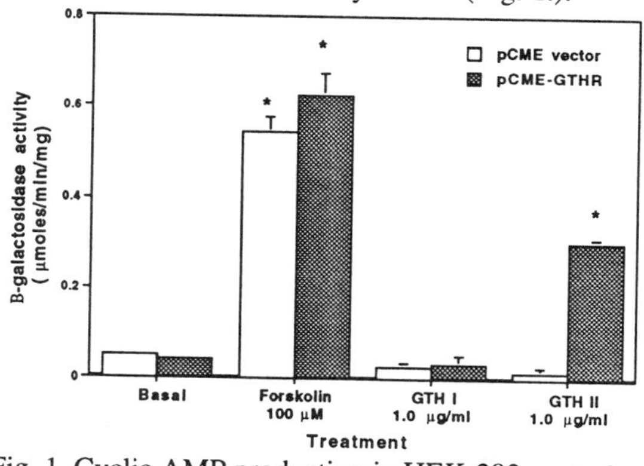

Fig. 1. Cyclic AMP production in HEK-293 control cells and cells expressing putative GTH-RII.

Cyclic AMP production in control cells transfected with vector alone was unaffected by either GTH I or II. These results suggest that the cloned cDNA encodes for the coho salmon GTH II receptor and are consistent with the two receptor model for gonadotropin action.

References

Yan, L., Swanson, P. and W. W. Dickhoff. 1992. A two receptor model for salmon gonadotropins (GTH I and II). Biol. Reprod. 47:418-427.

Miwa, S., Yan, L., and Swanson, P. 1994. Localization of two gonadotropin receptors in the salmon gonad by in vitro ligand autoradiography. Biol. Reprod. 50: 629-642.

Supported by U.S. Dept. of Agriculture Grant # 93-37203-9409.

GONADAL ANDROGEN RECEPTORS IN FISHES

M.S. Fitzpatrick, W.L. Gale, C.H. Slater, and C.B. Schreck

Department of Fisheries & Wildlife, Oregon State University, Corvallis, OR 97331 USA

Summary

We have identified androgen binding sites in the gonads of Nile tilapia and Pacific lamprey. The characteristics of these sites in tilapia suggest a receptor-mediated mechanism of masculinization by 17α-methylated androgens. The lamprey binding site indicates that gonadal steroid receptors may have evolved very early.

Introduction

Gonadal sex differentiation in many fishes can be manipulated by treatment with certain steroids. If steroids act directly on the gonads to induce differentiation, then receptors should be present in the gonads, as found in the ovaries of coho salmon (Fitzpatrick et al. 1994). This receptor had particular affinity for 17α-methylated androgens which are also potent masculinizing agents in fishes (Torrans et al. 1988; Piferrer et al. 1993). We investigated whether similar receptors exist in phylogenetically disparate species such as tilapia (Oreochromis niloticus) and lamprey (Lampetra tridentata).

Methods

Cytosol was prepared from adult Nile tilapia and Pacific lamprey gonads as described for coho salmon (Fitzpatrick et al. 1994). Binding kinetics were determined for tilapia using ^{3}H-mibolerone (^{3}H-Mb) and for lamprey using ^{3}H-Mb and ^{3}H-testosterone (^{3}H-T). Specificity of binding was determined by incubating ^{3}H-steroid with increasing concentrations of radioinert steroids.

Results

Tilapia cytosol bound ^{3}H-Mb with high affinity (testes: K_d=1.0 nM and B_{max}=5.6 fmol/mg protein; ovaries K_d=0.1 nM and B_{max} =9.6 fmol/mg protein). In lampreys, ^{3}H-T binding was much higher than ^{3}H-Mb binding in both ovaries and testes. ^{3}H-T binding showed high affinity in lamprey ovarian cytosol (K_d=0.8 nM and B_{max}=16.8 fmol/mg protein) and testicular cytosol (K_d=2.5 nM and B_{max}=478.6 fmol/mg protein).

In tilapia testicular cytosol, only 17α-methyltestosterone (MT), 11-keto-testosterone (KT), 5α-dihydrotestosterone (DHT), testosterone (T) and 17α-ethynyl-estradiol displaced ^{3}H-Mb binding below 50% specific binding. Estradiol (E2), 17α,20β-dihydroxy-4-pregnen-3-one (DHP), and progesterone (P4) were ineffective competitors. In lamprey testicular cytosol, MT was potent at displacing bound ^{3}H-T, but other 17α-methylated steroids were ineffective. DHT was the most potent naturally-occurring competitor; KT, E2, P4, and 15α-hydroxytestosterone were poor competitors.

Discussion

Our results confirm that androgen binding sites occur in the gonads of non-salmonid fishes, namely Nile tilapia and Pacific lamprey. The tilapia binding site shares many characteristics with the gonadal androgen receptor from coho salmon (Fitzpatrick et al. 1994); however, the lamprey gonadal binding site differs in that only ^{3}H-T binding showed saturation. Both tilapia and lamprey binding sites demonstrate a high degree of ligand specificity. This specificity is consistent with the potency of 17α-methylated androgens in causing masculinization in tilapia, and indicates potential avenues of research on sex inversion in lampreys.

References

Fitzpatrick, M.S., W.L. Gale, and C.B. Schreck, 1994. Binding characteristics of an androgen receptor in the ovaries of coho salmon, Oncorhynchus kisutch. Gen. Comp. Endocrinol. 95:399-408.

Piferrer, F., I.J. Baker, and E.M. Donaldson, 1993. Effects of natural, synthetic, aromatizable, and non-aromatizable androgens in inducing male sex differentiation in genotypic female chinook salmon (Oncorhynchus tshawytscha). Gen. Comp. Endocrinol. 91:59-65.

Torrans, L., F. Meriwether, and F. Lowell, 1988. Sex-reversal of Oreochromis aureus by immersion in mibolerone, a synthetic steroid. J. World Aquaculture Soc. 19:97-102.

Funded by the Pond Dynamics/Aquaculture CRSP and the Great Lakes Fishery Commission.

BINDING CHARACTERISTICS OF 17α,20β,21-TRIHYDROXY-4-PREGNEN-3-ONE (20β-S) TO ATLANTIC CROAKER SPERM MEMBRANE PREPARATIONS.

S. Ghosh and P. Thomas.

The University of Texas at Austin, Marine Science Institute, Port Aransas, Texas 78373.

Summary

There is mounting evidence for the presence of progestogen membrane receptors on sperm in several vertebrate species. However, to date there are no reports on the binding characteristics of these putative receptors. We report here a preliminary characterization of a binding site for the progestin, 20β-S on the plasma membrane of sperm in a teleost, Atlantic croaker.

Introduction

The presence of a progesterone receptor on sperm plasma membrane has been proposed based on considerable evidence of direct and rapid actions of progesterone on human sperm leading to activation (Ravelli *et al.*, 1994). To our knowledge, however, to date the binding characteristics of this putative receptor on sperm have not been reported in any vertebrate species. In the present study, the presence of binding sites for the progestin 20β-S on Atlantic croaker (*Micropogonias undulatus*) sperm was examined. A modification of the radio-receptor assay protocol developed for the ovarian 20β-S membrane receptor in spotted seatrout (Patiño & Thomas, 1990) was used to quantify 20β-S binding to sperm. Specific binding of 20β-S to croaker testicular membrane preparations was detected in a preliminary study using this technique.

Results

Both the association and dissociation rates were very rapid with $T_{1/2}$'s of 2 min and 2.5 min, respectively.

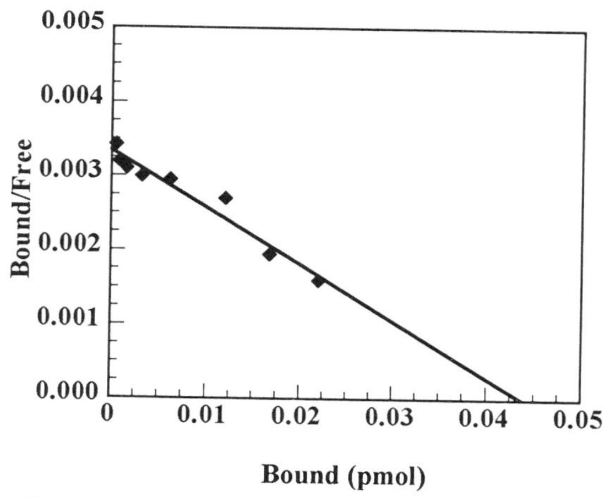

Fig. 1. Scatchard plot of specific binding of 20 β-S to the croaker sperm membrane preparation.

The specific binding was saturated between 5-10 nmol 20β-S. A Scatchard plot of specific binding (fig. 1) reveals the presence of a single class of binding sites for 20β-S with a K_D in the 10^{-8} M range. 20β-S was the most effective steroid in displacing [^{3}H]20β-S from its binding sites. None of the other steroids tested had relative binding affinities greater than 11%. Progesterone, 17,20β-P, and 11-DOC displaced 20β-S from the binding sites but estradiol and cortisol were ineffective.

Discussion

The present study demonstrates the presence of a single class of saturable and displaceable binding sites with high affinity (K_D in 10^{-8} M range) for 20β-S in the plasma membrane preparation of Atlantic croaker sperm. In addition, the binding is highly specific for the 20β-S, a major progestin in this species.

The rank order of steroid affinities of this binding moiety is similar to that of the ovarian plasma membrane 20β-S receptor in seatrout. However, unlike the ovarian receptor, the affinity of progesterone for this binding site was one order of magnitude lower (K_D in 10^{-7} range) than that of 20β-S. The extremely rapid association and dissociation kinetics of this binding site also resemble those of the ovarian plasma membrane 20β-S receptor in seatrout.

Progesterone has been shown to exert direct and rapid membrane effects on human sperm (e.g., capacitation and Ca++ influx; Ravelli *et al.*, 1994) resulting in activation. We have detected high circulating levels of 20β-S in spermiating male croaker. Therefore, the presence of 20β-S binding sites on the membrane of croaker sperm raises the possibility that 20β-S may be involved in some membrane-mediated functions related to sperm activation.

References

Ravelli, A., Modotti, M., Piffaretti-Yanez, A., Massobrio, M., and Balerna, M. 1994. Steroid receptors in human spermatozoa. Human Reproduction. 9,760-766.

Patiño, R., and Thomas, P. 1990. Characterization of membrane receptor activity for 17α,20β,21-trihydroxy-4-pregnen-3-one in ovaries of spotted seatrout (*Cynoscion nebulosus*). General Comparative Endocrinology 78,204-217.

EXPRESSION OF KALLIKREIN GENE FAMILY mRNAS IN THE TROUT OVARY

F.W. Goetz[1], M.A. Garczynski[1] and S-Y. Hsu[2]

[1] Department of Biological Sciences, University of Notre Dame, Notre Dame, IN, 46556; [2] Department Gynecology and Obstetrics, Stanford University, Stanford, CA 94305.

Summary

A 2.4 kb clone (designated KT-14) was obtained from a brook trout preovulatory ovarian cDNA library. KT-14 contains an open reading frame encoding a protein with significant homology to several members of the mammalian kallikrein gene family.

Introduction

Tissue kallikrein is a serine protease that stimulates the conversion of kininogen to kinins. During the characterization of several complementary DNA (cDNA) clones, obtained from screening a brook trout preovulatory cDNA library, we obtained a sequence for a 1.8 kb cDNA that was homologous with several members of the mammalian kallikrein gene family. Northern analysis of ovarian RNA, using this cDNA as a probe, revealed transcripts larger than 1.8 kb. Therefore, we used this cDNA to rescreen the library to obtain larger clones.

Results

Rescreening of the ovarian cDNA library resulted in the isolation of several 2.4 kb clones. One of these, designated KT-14, was sequenced and used for Northern analysis. KT-14 has a presumed open reading frame of 801 nucleotides, encoding a protein of 267 amino acids. KT-14 also contains an extended 5' region of 673 nucleotides and approximately 1.0 kb of untranslated 3' sequence. The 3' region has not been fully sequenced since it is composed of a 66 base pair sequence that is repeated a number of times. The protein presumably encoded by KT-14 is homologous with several member of the kallikrein gene family. It is approximately 55% similar and 35% identical to mammalian pancreatic kallikrein and adipsin/complement factor D. On Northern blots of total RNA from ovarian theca and stroma, KT-14 hybridizes with two RNA bands of 2.4 and 3.2 kb (Fig. 1). In some individuals a distinct 1.8 kb band can also be observed (Fig. 1). These RNAs are present in ovaries prior to and following germinal vesicle breakdown (GVBD) and at the time of ovulation. However, there does not appear to be a change in the level of these bands throughout GVBD and ovulation in this particular tissue. Both RNA bands are also observed in granulosa and on blots of RNA from kidney and heart.

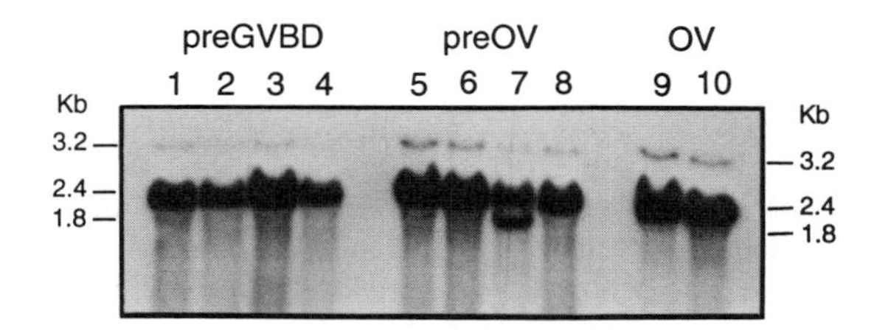

Fig. 1. Northern blot of total RNA (25 ug/lane) from ovarian theca/stroma obtained from females (1-8) at stages prior to germinal vesicle breakdown (preGVBD), following GVBD but prior to ovulation (preOV) and just at the time of ovulation (OV). Blot was probed with full-length KT-14.

Discussion

Kinins can stimulate ovulation in the perfused rat and rabbit ovary and a number of investigations have observed changes in ovarian kallikrein, kininogen and kinins at the time of ovulation (Goetz et al., 1991). Using an in vitro whole follicle assay system, we have shown that mammalian kallikrein is a potent stimulator of brook trout follicle wall contraction (Goetz and Hsu, 1994). However, in the same assay, mammalian kinins are curiously ineffective in stimulating contraction while angiotensin is as potent as kallikrein. Thus, one possibility is that in fish, an ovarian kallikrein-like protease stimulates the conversion of angiotensinogen to angiotensin, not kininogens to kinins.

References

Goetz, F.W. and S-Y. Hsu, 1994. Expression of kallikrein gene family mRNAs in the brook trout ovary. Biol. Reprod. 50 (Suppl. 1):114.

Goetz, F.W., A. Berndtson and M. Ranjan, 1991. Ovulation: Mediators at the ovarian level. In "Vertebrate Endocrinology, Fundamentals and Biomedical Implications" Vol.IV (A)-Reproduction. eds. P. Pang, M. Schreibman. Academic Press, N.Y., 127-203.

MONITORING FINAL OOCYTE MATURATION IN FEMALE PLAICE (*PLEURONECTES PLATESSA*) USING RIAS WHICH DETECT METABOLITES OF 17α,20β-DIHYDROXY-4-PREGNEN-3-ONE

R. Moses Inbaraj, A. P. Scott and E. L. M. Vermeirssen

MAFF Fisheries Laboratory, Pakefield Rd., Lowestoft, Suffolk, NR33 0HT, UK

Summary

Radioimmunoassays have been developed which detect the main metabolites of the maturation-inducing steroid, 17α,20β-dihydroxy-4-pregnen-3-one in female plaice. Plasma levels of these metabolites increase dramatically during final oocyte maturation.

Introduction

In many teleost species, it is possible to monitor the effects of gonadotropins, gonadotropin-releasing hormones and pheromones on oocyte maturation, by measuring plasma levels of 17α,20β-dihydroxy-4-pregnen-3-one (17,20β-P). In some other species, however, despite *in vitro* bioassay evidence that this steroid is a potent inducer of oocyte maturation, its levels do not change. In one such species, the North Sea plaice (*Pleuronectes platessa*), we have long-suspected that 17,20β-P is the maturation-inducing steroid but that, because this species is a 'batch ovulator' (and thus needs to exert fine control over the number of oocytes which are matured at any one time), the 17,20β-P is metabolised before it is released into the bloodstream. Based on earlier studies which showed that plaice ovaries contained highly active 5β(3α-hydroxy)-reducing and sulphating enzymes, we hypothesised that the main end-product of 17,20β-P metabolism was most likely to be 20-sulphated 5β-pregnane-3α,17,20β-triol (3α,17,20β-P-5β).

In this study, we have set out to measure the levels of this steroid by radioimmunoassay (RIA) with the ultimate aim of developing a 'quick and easy' method for monitoring maturation in this species. In order to carry out this study, two powerful new 'broad-spectrum' RIAs have been developed - one of which cross-reacts with C21 steroids with a 5β(3α-OH)-reduced configuration and the other which only cross-reacts with C21 steroids with a 17,20β-diOH configuration. Acid solvolysis was used to convert sulphates to free steroids. Steroids were identified by their elution positions on HPLC and TLC and comparative cross-reactivity in the RIAs.

Results

17,20β-P RIA. This assay, which only cross-reacted with a single peak (17,20β-P) on HPLC, detected 10-15 ng/ml of sulphated material in maturing female plaice. However, non-maturing females had relatively high 'background' levels (4-5 ng/ml).

RIA for '5β(3α-OH)-reduced' steroids. When this assay was applied to solvolysed plasmas (using 3α,17,20β-P-5β as a standard), it detected levels of 40 ng/ml in non-maturing females and 200-500 ng/ml in maturing females. Only *c.* 40% of the activity was found to be due to 3α,17,20β-P-5β, however. The other major compound was 3α,17,21-trihydroxy-5β-pregnan-20-one. Both steroids were > 95% sulphated, via the 20- and 21- positions, respectively. These sulphates were also found to cross-react in the RIA (i.e. without prior solvolysis), which meant that they could be measured directly (and thus quickly) in diluted plasma samples.

RIA for '17,20β–diOH' steroids. When this assay was applied to solvolysed plasmas (using 3α,17,20β-P-5β as a standard), it detected levels of 60-270 ng/ml in maturing females and, unlike the other assays, very low background levels (< 2 ng/ml) in non-maturing females. The bulk of the activity (> 95%) was due to 3α,17,20β-P-5β. The remainder was due to 17,20β-P (and possibly also 3β,17,20β-P-5β, but this needs to be confirmed).

Discussion

The results confirm that female plaice release substantial quantities of metabolised 17,20β-P into the bloodstream at the time of final oocyte maturation. The predominant metabolite is 3α,17,20β-P-5β 20-sulphate. Two RIAs (one of which has a high degree of specificity) have been developed to measure this compound. Both RIAs form potentially powerful tools for investigating the mechanism of control of oocyte maturation in other species.

ANDROGEN BINDING IN THE STICKLEBACK KIDNEY

Staffan Jakobsson1, Ian Mayer1, Rüdiger W. Schulz2, Marinus A. Blankenstein3, and Bertil Borg1.

1. Department of Zoology, Stockholm University, S-106 91 Stockholm, Sweden

2. Department of Experimental Zoology, University of Utrecht, 3584 CH Utrecht, The Netherlands.

3. Department of Endocrinology, Academic Hospital Utrecht, 3584 CX Utrecht, The Netherlands.

Introduction

In teleosts the most effective hormone in stimulating male secondary sexual characters is 11-ketotestosterone (11KT). However, no receptors for 11KT have so far been found. In the male three-spined stickleback (Gasterosteus aculeatus) the kidney hypertrophies during the breeding season and produces a "glue" which is used in building of the nest. This hypertrophy is androgen dependent with 11KT being particulary effective (Borg et al. 1993). The aim of the present study was to investigate the possible presence of specific binding for 11KT in the kidney of sticklebacks. Tissue fragments, membrane fractions and cytosol extracts were investigated.

Experimental procedure

In the tissue fragment study, kidneys were excised from males in breeding condition that had been castrated two days earlier. The purpose of this short-term castration was to reduce the levels of endogenous androgens. Tissue fragments were incubated in L-15 medium containing tritiated 11KT with and without unlabelled steroids at different concentrations. The incubations were terminated after two hours and the activity in the tissue determined.

Displacement of tritiated 11KT was also studied in membrane fractions. The kidneys were homogenized and thereafter centrifuged at low speed (600 g). After centrifugation the supernatant was decanted and centrifuged again at high speed (120,600 g), after which the supernatant was drawn off and the pellet (containing the membrane fraction) resuspended and incubated for 1 hour with tritiated 11 KT with and without unlabelled 11KT. The incubation was terminated by adding charcoal and centrifuged at low speed in order to separate the unbound steroids from the membrane bound steroids. The activity in the supernatant was measured.

The cytosol receptor assay was performed on extracts from the kidneys and brains of intact sticklebacks and the skin from brown trout. Cytosolic fractions were incubated with either tritiated 11KT or tritiated testosterone (T) at increasing concentrations. Following incubation the bound and unbound steroids were separated by adding charcoal and centrifuged. The activity in the supernatant was determined. The data was plotted, and the equlibrium dissociation constant (KD) calculated according to Scatchard.

Results and Discussion

A specific binding of 11KT was demonstrated in the stickleback kidney fragments but not in control tissues; muscle and liver. Unlabelled 11KT, 17β-hydroxy-5α-androstane-3,11-dione and 5α–dihydrotestosterone displaced specific binding at similar ED50 values, whereas T and progesterone were less effective. Estradiol, 17α-20β-dihydroxyprogesterone and the antiandrogens, flutamide and cyproteroneacetate did not displace 11KT. All the substances that displaced specific binding have a stimulatory effect on kidney hypertrophy. Tritiated 11KT was displaced by unlabelled 11KT at concentrations within the natural plasma level range.

A displacement of tritiated 11KT with unlabelled 11KT was also detected in the membrane fractions. Cytosolic receptor-like binding was found for T in the skin of brown trout and in the brain of sticklebacks but not in the kidney. Further, no specific binding of 11KT was detected in any of the cytosol extracts investigated.

The present study demonstrates the presence of specific 11KT binding in the stickleback kidney. It cannot be said with certainity if this binding represents a receptor binding. If it is indeed a receptor binding the absence of specific binding in cytosolic extracts and the presence of binding in membrane fractions suggests that 11KT binds to non-genomic receptors.

Reference

Borg, B., Antonopoulou, E., Andersson, E. Carlberg, T., and Mayer, I. 1993. Effectivness of several androgens in stimulating kidney hypertrophy, a secondary sexual character, in castrated male three-spined stickleback. Can. J. Zool. 71:2327-2329

EFFECTS OF SUBSTRATE CONCENTRATIONS ON STEROIDOGENESIS IN OVARIES AND TESTES OF THREE SPECIES OF AMBISEXUAL FISHES

D. E. Kime[1], M. A. S. Abdullah[2] and K. P. Lone[2]

[1]Department of Animal and Plant Sciences, The University of Sheffield, Sheffield S10 2TN, United Kingdom, and [2]Department of Mariculture and Fisheries, Kuwait Institute for Scientific Research, PO Box 1638 Salmiya, Kuwait.

Summary

Increased substrate concentration induces a switch from androgen (aetiocholanolone and androst enedione in females; 11β-hydroxyandrostenedione in males) to 17,20αP in serranid and sparid gonads.

Introduction

Ambisexual sparid and serranid species are of major commercial importance in the Arabian Gulf but the identity of their reproductive steroids and their role in sex inversion is not known. We have therefore incubated gonads of 2-3 sexually mature sheim (*Acanthopagrus latus*, Sparidae), sobaity (*Sparidentex hasta*, Sparidae) and hamoor (*Epinephalus coioidus*, Serranidae) with ^{3}H-17-hydroxyprogesterone (17P). Since previous studies (Kime and Abdullah, 1994) have shown that substrate concentration affects the pattern of metabolites 0, 0.1, 1 or 10μg/ml unlabelled substrate were added to the incubations.

Results

In ovarian tissue of both sparids, androstenedione and aetiocholanone decreased with substrate concentration while 17,20α-dihydroxy-4-pregnen-3-one (17,20αP) was found only at high substrate levels (Fig. 1). Hamoor ovaries gave similar results but additional unidentified products constituted 50% of the products at low substrate concentration.

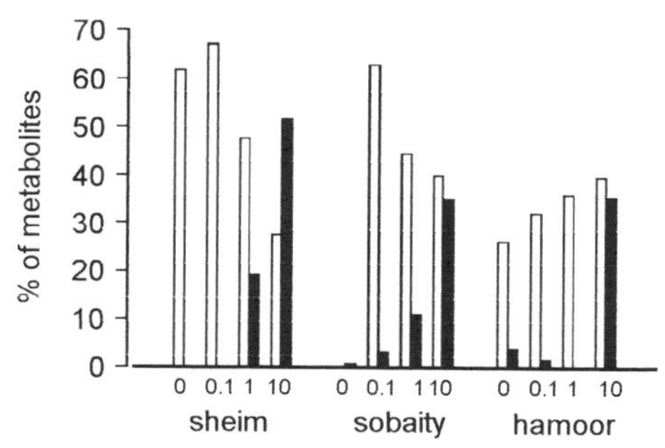

Fig. 1. Yields of androgens (□) and 17,20αP (■) from ovaries of sobaity, sheim and hamoor.

In testes, 11β-hydroxyandrostenedione (β AD), the major identified androgen, decreased with substrate, while 17,20αP predominated at high substrate levels (Fig. 2). A polar metabolite was present (32-37% at 0 μg/ml 17P; 15-17% at 20 μg/ml 17P) in all species and in hamoor 28-14% of a less polar product was also found.

11-Deoxycortisol was also present in both sexes, but yields of 17,20βP were very low.

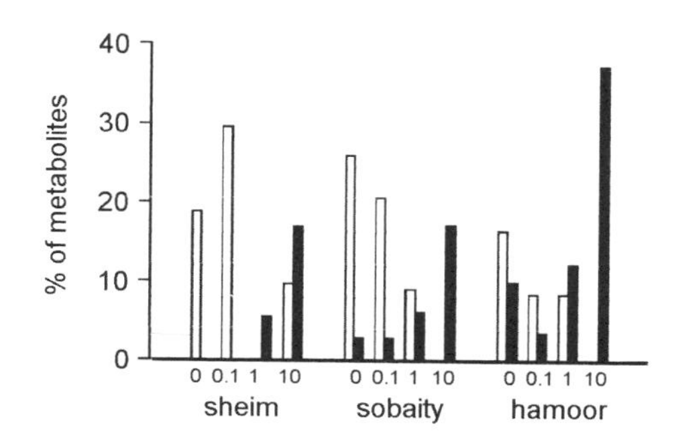

Fig. 2. Yields of βAD (□) and 17,20αP (■) from testes of sobaity, sheim and hamoor.

Discussion

Gonadal steroidogenesis in the three ambisexual fish showed a similar change with substrate concentration to that found in cyprinid species with a switch from androgen to 17,20αP with increased substrate concentration. In sparid ovaries, androstenedione was the predominant androgen with lower yields of aetiocholanolone, but in hamoor there were high yields of two reduced metabolites. The major androgen in testes was βAD with only trace amounts of 11-ketotestosterone, and in all species there were high yields of a polar reduced metabolite.

References

Kime, D.E. and M. A. S. Abdullah, 1994. The *in vitro* metabolism of 17-hydroxyprogesterone by ovaries of the goldfish, *Carassius auratus*, is affected by substrate concentration. Gen. Comp. Endocrinol. 95: 109-116.

OVARIAN RECEPTORS FOR 17α, 20β, 21-TRIHYDROXY-4-PREGNEN-3-ONE (20β-S) IN STRIPED BASS

W. King V,[1] S. Ghosh,[2] P. Thomas,[2] and C.V. Sullivan.[1]

[1]Department of Zoology, North Carolina State University, Raleigh, North Carolina 27695-7617, USA.

[2] The Marine Science Institute, University of Texas at Austin, P.O. Box 1267, Port Aransas, Texas 78373, USA.

Summary

A receptor for 20β-S was found on ovarian membranes of striped bass (Morone saxatilis) undergoing final oocyte maturation (FOM). It exhibited characteristics expected of a steroid hormone receptor including: high affinity, limited capacity, ligand and tissue specificity, and relation to biological response. 17α,20β-dihydroxy-4-pregnen-3-one (DHP) showed no ability to saturate binding sites on ovarian membranes and was a poor competitor in the ligand specificity assay. The results indicate that 20β-S, and not DHP, is the maturation-inducing steroid hormone (MIH) in striped bass.

Introduction

In teleosts, FOM is regulated by a C_{21} steroid hormone known as the MIH. Binding of the MIH to a receptor on the oocyte plasma membrane initiates FOM (Nagahama et al., 1993). DHP has been identified as the MIH in salmonids while 20β-S is the MIH in sciaenids. Previous studies demonstrated that both immunoreactive DHP and 20β-S are produced during FOM in striped bass (King et al., 1994a) and both are potent inducers of FOM in vitro (King et al., 1994b). The specific objective of this study was to identify DHP or 20β-S as the MIH in striped bass based on their ability to bind to ovarian receptors.

Materials and Methods

Semi-purified membranes were prepared by homogenization and centrifugation. Receptor binding assays were performed by incubating membranes with tritiated DHP or 20β-S. Vacuum filtration was used to separate bound and free hormone (Patino and Thomas, 1990). All membrane preparation and assay procedures were carried out at 0-4 °C.

Results and Discussion

20β-S receptors on ovarian membranes were ~90% saturated after 10 minutes of incubation and the binding was pH dependent. Dissociation of 20β-S from its receptor was complete after 30 minutes. Scatchard analysis revealed a single class of high affinity, limited capacity binding sites for 20β-S (Fig. 1). Saturable binding sites for DHP were not found on ovarian membranes. In a competitive binding assay, 20β,21-dihydroxy-4-pregnen-3-one was very effective for displacing 20β-S from its binding sites (Kd ~3 nM); however, it is a relatively weak inducer of GVBD in vitro. DHP was a poor competitor (Kd ~800 nM) for 20β-S binding sites. Of the 12 remaining steroids tested, only progesterone competed for 20β-S binding sites within a physiologically relevant concentration (Kd ~80 nM). Only low levels of specific 20β-S binding were detected on membranes prepared from testes, liver, brain and muscle when compared to membranes prepared from ovary. Ovarian membranes prepared from vitellogenic fish also had low levels of specific 20β-S binding. Taken together with prior studies, these data indicate that 20β-S, and not DHP, is the MIH in striped bass.

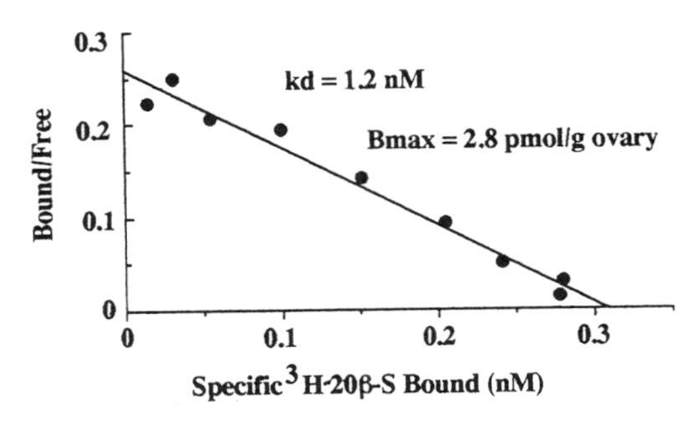

Figure 1. Representative Scatchard analysis of 3H-20β-S binding to ovarian membranes. Membranes were incubated in triplicate with 3H-20β-S in concentrations from 0.156 nM-20 nM with or without 100 fold molar excess cold 20β-S. Kd = dissociation constant. Bmax = maximum binding capacity of membranes for 20β-S.

Literature Cited

King V, W., P. Thomas, R. Harrell, R.G. Hodson and C.V. Sullivan, 1994a. Plasma levels of gonadal steroids during final oocyte maturation of striped bass, Morone saxatilis L. Gen. Comp. Endocrinol. 95:178-191.

King V, W., P. Thomas and C.V. Sullivan, 1994b. Hormonal regulation of final maturation of striped bass oocytes in vitro. Gen. Comp. Endocrinol. 96: 223-233.

Nagahama, Y., M. Yoshikuni, M. Yamashita and M. Tanaka, 1993. Regulation of oocyte maturation in fish. In: "Fish Physiology" Vol. XIII. (N.M. Sherwood and C.L. Hew, Eds.), pp. 393-439. Academic Press, New York.

Patino, R. and P. Thomas, 1990. Characterization of membrane receptor activity for 17α,20β,21-trihydroxy-4-pregnen-3-one in ovaries of spotted seatrout (Cynoscion nebulosus). Gen. Comp. Endocrinol. 78:204-217.

This work was supported by a grant to CVS from the University of North Carolina Sea Grant College Program (NA90AA-D-SG-062) and a PHS grant (ESO 4216) to PT.

PURIFICATION AND CHARACTERIZATION OF A PLASMA SEX-STEROID BINDING PROTEIN IN THE SPOTTED SEATROUT (*Cynoscion nebulosus*)

Charles W. Laidley[1] and Peter Thomas[2]

[1] Environmental Chemistry & Toxicology Laboratory, Dept.of Environmental Science, Policy & Management, University of California, Berkeley, CA 94720-3112.
[2] The University of Texas at Austin, Marine Science Institute, Port Aransas, TX 78373.

Summary

A plasma sex-steroid binding protein (SBP) was purified and characterized in the spotted seatrout, *Cynoscion nebulosus*. It was shown to be a high affinity estrogen/androgen binding protein. The purified protein had a native mass of 135 *kDa* on native PAGE and dissociated into two identical subunits with masses of 49 to 52 *kDa* on SDS-PAGE. The *N*-terminal sequence and amino acid composition were not similar to the mammalian SBP. The plasma levels of this protein increased in association with ovarian recrudescence suggesting an integral role in the reproductive physiology of this species. Competition studies demonstrate that the seatrout SBP binds several important "environmental estrogens" including o,p'-DDE, methoxychlor and PCBs.

Introduction

Plasma proteins with high affinities for androgens, and with somewhat variable affinities for estrogens, have been identified and partially characterized in many but not all vertebrate groups. Several mammalian SBPs have been purified and sequenced but the physiological role of this protein remains controversial. The discovery of receptors for the SBP dimer on the plasma membranes of several steroid target tissues has resulted in renewed interest in the role of this protein in the mechanism of steroid action. Given the extensive study of steroids in fish and the recent interest in environmental estrogen and androgen mimics we decided to examine the plasma SBP in a perciform teleost, the spotted seatrout.

Results & Discussion

A single protein was shown to bind both testosterone and 17β-estradiol with affinities in the low nanomolar range. The plasma concentration of this protein ranged from about 200 to 800 nM.

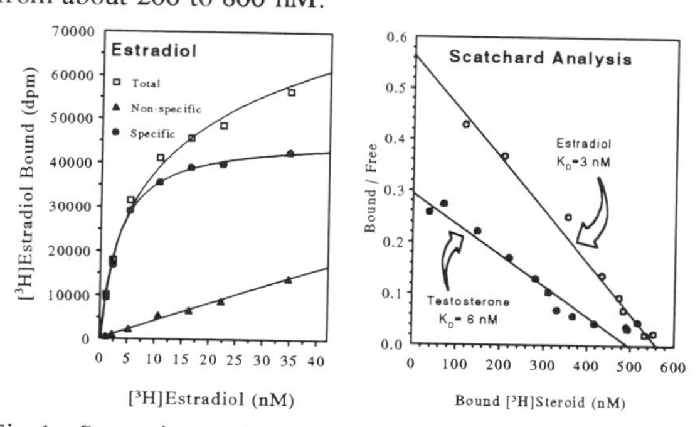

Fig.1. Saturation and Scatchard plots for the binding of estrogen and testosterone to seatrout plasma.

Competition studies demonstrated that the seatrout SBP had high affinities for a number of estrogens and androgens, with the exception of 11-ketotestosterone, and little affinity for C_{21}-steroids including cortisol and the maturation-inducing steroid, 20β-S. A number of "environmental estrogens" including methoxychlor, o,p'-DDE and PCBs were also shown to bind to the seatrout SBP. The rates of steroid association and dissociation were very rapid with a $t_{1/2}$ of less than 30s for ligand association and 90 s for ligand dissociation.

Plasma SBP concentration and binding affinity did not differ significantly ($P \leq 0.05$) between male (n=14) and female (n=81) spotted seatrout. SBP levels increased with the stage of ovarian recrudescence in the females, with the lowest levels (300 nM) in regressed females and the highest levels (470 nM) in females with fully developed ovaries. In addition, the steroid dissociation constant (K_D) increased from less than 5 nM in fish with regressed ovaries to greater than 6.5 nM in vitellogenic fish.

The seatrout SBP was purified over 2400-fold by acetone and $(NH_4)_2SO_4$ precipitation, anion exchange and gel filtration chromatography, preparative native PAGE and reverse phase HPLC. The purified SBP migrated as a dimeric protein with a molecular mass of 135 *kDa* on native PAGE and dissociated into subunits with molecular masses of 49 to 52 *kDa* on SDS-PAGE.

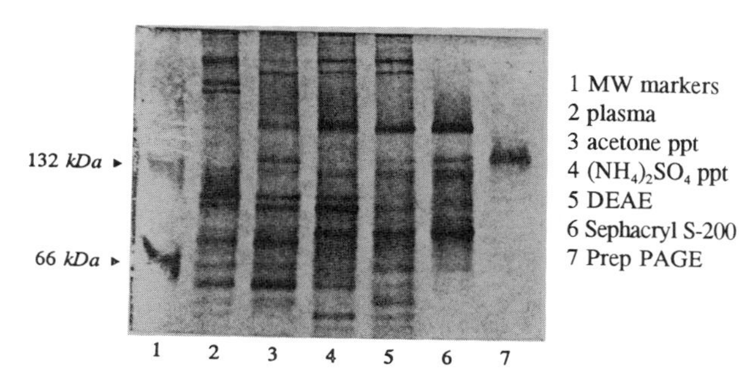

Fig.2. Native PAGE (7.5%) of fractions obtained during the purification of the seatrout SBP.

The SBP subunits consisted of a single protein with the N-terminal 37 amino acids having little "homology" to any presently known protein sequences. The amino acid composition also appears substantially different from SBPs characterized thus far in mammals.

IN VITRO STEROIDOGENESIS BY THE GONADS AND SPERMATOZOA OF THE GROUPER (*EPINEPHELUS TAUVINA*) IMPLANTED WITH 17α-METHYLTESTOSTERONE.

S.T.L. Lee[1], A.M.C. Tan[1], D.E. Kime[2], T.M. Chao[3], H.S. Lim[3], R. Chou[3], T.J. Lam[1], and C.H. Tan[1].

[1]Laboratory for Hormone Research, Department of Zoology, National University of Singapore, Singapore 0511; [2]Department of Animal & Plant Sciences, University of Sheffield, Sheffield S10 2TN, United Kingdom; [3]Marine Aquaculture Section, Primary Production Department, Ministry of National Development, Singapore 1749.

Summary

The major metabolites produced *in vitro* by gonadal tissues of the female and secondary male grouper (*Epinephelus tauvina*) are the 5β-reduced androgens. 11-oxoandrogens are biosynthesized only by the testicular tissue. 17,20α-dihydroxy-4-pregnen-3-one (17,20α-P) and 5β-pregnane-triols were produced by the sperm, indicating that 17,20α-P may play a role in regulating spermiation in the grouper.

Introduction

The greasy grouper (*Epinephelus tauvina*) is protogynous and natural sex inversion rarely occurs in captivity. To obtain functional males for aquaculture, mature females were implanted with 17α-methyltestosterone (MT) to induce sex change. However, the milt volume produced by these secondary males is small. Milt production is inconsistent and spontaneous spawning behavior is also absent. This study examines the possible reasons for these observations.

Methods

Ten mature female fish were implanted with MT in silastic capsules.Ten control fish were implanted with similar capsules but without MT. Gonadal tissue was surgically removed from each fish after six months when the treated fish had completely inverted sex. Tissue from each fish was individually incubated with chromatographically pure [^{3}H]testosterone for 30 hr. Sperm collected from the secondary males were incubated with pure [^{3}H]17α-hydroxy-progesterone for 6 hr using the same incubation conditions. The steroids in the media were extracted and identified using HPLC and TLC methods.

Results

One month after MT-implantation, the ovaries of the treated fish regressed and the fish were observed to become morphologically more "male-like". Milt could be expressed from half of the treated fish during the six months of treatment. The milt volume varied from 0.1 to 1.5 ml, and the spermatocrit ranged from 30 to 51%. Four of the fish were still spermiating at the time of biopsy for tissue incubation. All control fish remained female during the course of the experiment. Figure 1 shows the HPLC radiochromatograms of steroid metabolites resulting from incubations of gonadal tissues of the fish after treatment with or without MT (Control). Peak 1 is the precursor peak. 5β-dihydrotestosterone (Peak 2) and 5β-androstane-3β,17β-diol (Peak 3) were produced as free and conjugated steroids in both cases. However, there was a shift towards the production of 11-oxoandrogens (Peak 4) as the major free metabolites after sex inversion. The 11-oxoandrogens were subsequently resolved and identified as 11β-hydroxy-testosterone and 11-ketotestosterone. The HPLC radiochromatograms of steroids biosynthesized by the sperm show that 17,20α-P and 5β-pregnane-triols were produced.

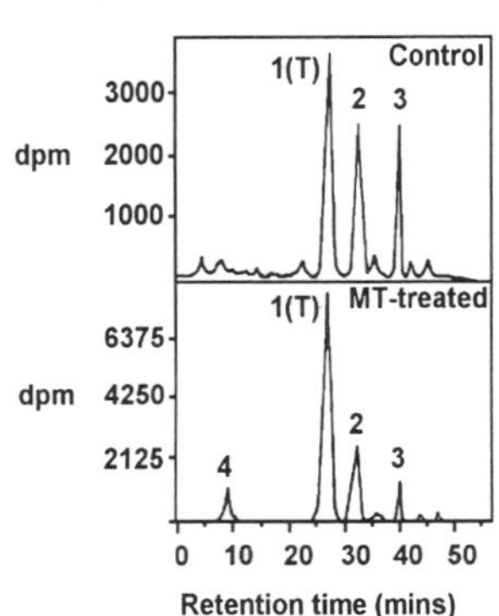

Fig. 1. HPLC radiochromatogram of steroid metabolites produced by gonadal tissue of *E. tauvina*.

Discussion

The results show significant synthesis of 11-oxoandrogens by the testes of the MT-induced males. However, the milt volume could not be positively correlated to the production of these steroids as the highest yield of 11-oxoandrogens were derived from fish that were not spermiating at the time of biopsy. Instead, the 11-oxoandrogens could play a role in inducing spermatogenesis and/or secondary sexual characteristics. The biosynthesis of 17,20α-P by the sperm suggests that it may play a role in regulating spermiation in the grouper rather than the 20β-isomer.

IDENTIFICATION OF OVARIAN STEROIDS PRODUCED DURING THE REPRODUCTIVE SEASON IN CHANNEL CATFISH (*Ictalurus punctatus*)

Julie M. Lehrter and John M. Trant

Department of Zoology and Physiology, Louisiana State University, Baton Rouge, LA, USA

Summary

The ovary of the channel catfish, *Ictalurus punctatus*, produces a large number of steroids, *in vitro*. Twelve ovarian steroids, produced during the vitellogenic phase of the reproductive season in the female channel catfish, have been identified by HPLC and TLC. A novel steroid, 7α-hydroxypregnenolone (7α-P5), was the major product yet its biological action is unknown. At least five additional steroids exist for which complete structural information is unavailable.

Introduction

Appropriate changes in gonadal steroidogenesis are necessary for successful reproduction in all vertebrates. In spite of the economic importance of the channel catfish, the steroidogenic nature of its gonads has not been thoroughly investigated. Seasonal changes in plasma titers of estradiol and testosterone have been quantified by radioimmunoassay (MacKenzie et al., 1989). The present study demonstrates that a large number of ovarian steroids are produced during the reproductive season in channel catfish. Once identified, the biological function of these steroids can be elucidated.

Results

Freshly harvested, deyolked ovarian tissue (~100 mg) was incubated with 8 μCi [7-^{3}H]-pregnenolone (sp. act. of 21.1 Ci/mmol) without the addition of cofactors. Changes in the pattern of ovarian steroidogenesis were evident as the reproductive season progressed (data not shown). The following steroids were identified by HPLC and TLC: estriol; 11β-hydroxytestosterone; 11β-hydroxyandrostenedione; 7α-hydroxypregnenolone; estradiol; testosterone; androstenedione; 17α-hydroxy-pregnenolone; 17α-hydroxyprogesterone; 5α(?)-dihydrotestosterone; progesterone; and 20β-dihydro-progesterone. Five additional steroids were isolated but unidentified.

The most abundant metabolite found throughout the vitellogenic growth phase was 7α-P5. Large quantities (10 μg) of this steroid were purified from incubations of three grams of fresh ovarian follicular tissue with 250 mg of pregnenolone and 25 μCi of tritiated pregnenolone in 250 ml of 75% L-15 for 10 hours. This steroid was positively identified as 7α-P5 by GC-MS. Known inhibitors of cytochromes P450 (CO, miconazole, clotrimazole) reduced the rate of 7α-P5 synthesis to 0 - 20% of control.

Discussion

A previously unidentified steroid, 7α-P5, was the most abundant steroid produced during the vitellogenic stage in the channel catfish ovary. Other 7α-hydroxylated steroids have been identified in cyprinids, guppies, and hagfish (e.g. Kime et al., 1992). The biological function of 7α-P5 remains unknown. Enzyme inhibitors indicate that this 7α-hydroxylase is a member of the cytochrome P450 superfamily. With further studies, we hope to characterize 7α-P5 secretion, and illustrate its biological function in the channel catfish.

The channel catfish ovary produced many steroids during the growth phase of the oocytes. Activity of the mitochondrial cytochrome P450, 11β-hydroxylase, was demonstrated by the presence of 11-oxygenated androgens. 11β-HSD, which is evident in the testis of channel catfish, is not evident in the ovary. In addition, 20β-HSD activity is evident with the identification of 20β-dihydroprogesterone, and 5α or 5β reductase activity is evident with the production of dihydrotestosterone. The cytochrome P450, 16α-hydroxylase, is necessary for estriol synthesis. The steroidogenic nature of the channel catfish ovary is highly complex, as indicated by the multitude of steroids produced and steroidogenic enzymes expressed. With additional studies, we hope to fully characterize the enzymes active in the channel catfish ovary and the steroids secreted.

References

Kime, D.E., Scott, A.P., and Canario, A.V.M., 1992. *In vitro* biosynthesis of steroids, including 11-deoxycortisol and 5α-pregnane-3β,7α,17,20β-tetrol, by ovaries of the goldfish *Carassius auratus* during the stage of oocyte final maturation. Gen. Comp. Endocrinol. 87:375-384.

MacKenzie, D.S, Thomas, P., and Farrar, S.M., 1989. Seasonal changes in thyroid and reproductive steroid hormones in female channel catfish (*Ictalurus punctatus*) in pond culture. Aquaculture. 78:63-80.

Dr. Cedric Shackleton conducted the GC-MS analysis. This research was funded by a grant from the USDA (9401429).

EPIDERMAL GROWTH FACTOR ENHANCES OVARIAN PROSTAGLANDIN SYNTHESIS IN THE GOLDFISH *Carassius auratus*

T. MacDougall and G. Van Der Kraak

Department of Zoology, University of Guelph, Guelph, Ontario, Canada, N1G 2W1

Summary

Epidermal growth factor was shown to enhance the stimulatory action of arachidonic acid on prostaglandin-E_2 biosynthesis by goldfish ovarian follicles.

Introduction

Previous studies have shown that goldfish vitellogenic ovarian follicles are responsive to epidermal growth factor (EGF) and transforming growth factor alpha (TGFα). These peptides inhibit gonadotropin stimulated steroid production while enhancing DNA synthesis (Srivastava 1994). EGF and TGFα may play a role in other aspects of ovarian function as these growth factors affect prostaglandin (PG) biosynthesis in a variety of mammalian tissues (Chudaska and Schlegel 1993, Zhang et al 1992). A series of experiments were conducted to test the effects of murine EGF on PGE_2 synthesis by vitellogenic ovarian follicles. Vitellogenic follicles (approx. 600μm diam.) were isolated in a modified Cortland's saline solution and icubated with test compounds in tissue culture plates (20 follicles per well) at 18°C. Follicles were exposed to EGF alone and in combination with known stimulators of PG synthesis in fish. PGE_2 produced and released to the medium was quantified by radioimmunoassay.

Results and Discussion

While EGF alone had no effect on PGE_2 production, it caused a dose related increase in the conversion of exogenous arachidonic acid (AA) to PGE_2. This effect was seen when incubation times exceeded 12 hours. Indeed, EGF enhanced the conversion of graded dosages of AA to PGE_2 (Fig 1.). By comparison EGF had no effect on calcium ionophore-A23187 and/or phorbol myristate acetate-stimulated PGE_2 production. As the ability of EGF to enhance AA stimulated PGE_2 synthesis was blocked with chloroquine, a phospholipase A_2 (PLA_2)-inhibitor(not shown), it seems that one site of EGF action is at the level ofPLA_2 which regulates the cellular levels of free AA (the substrate for PG synthesis). In light of past studies, these findings implicate EGF as an integral regulator of ovarian development in goldfish.

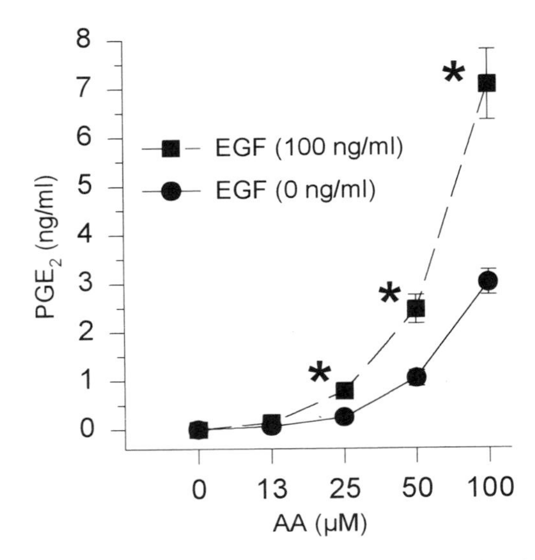

Fig 1. *In vitro* effect of EGF on AA stimulated PGE_2 production by goldfish vitellogenic ovarian follicles incubated at 18°C for 18hrs. Asterix indicates significant enhancement ($p<0.05$) compared to AA alone.

References

Chudaska, C., and Schlegel, W. 1993. The influence of growth factors on the prostaglandin synthesis in cultured rabbit luteal cells. Horm Metab Res 25:301-304.

Srivastava, R.K. 1994. Multifactorial regulation of steroidogenesis and DNA synthesis by ovarian follicles of goldfish (Carassius auratus). PhD Thesis, University of Guelph, Guelph, ON, CAN.

Zhang, Z., Kause, M., and Davis, D.L. 1992. Epidermal growth factor receptors in porcine endometrium: binding characteristics and the regulation of prostaglandin E and F2α production. Biol Reprod 46:932-936.

IN VITRO STEROID BIOSYNTHESIS BY GONADS OF A HYBRID STURGEON, BESTER, AT DIFFERENT DEVELOPMENTAL STAGES

M. Maebayashi[1], B. Mojazi Amiri[2], N. Omoto[1], T. Kawakita[1], T. Yamaguchi[1], S, Adachi[2] and K. Yamauchi[2]

1:Department of Research and Development, The Hokkaido Electric Power Co., Inc., Sapporo 004, Japan

2:Department of Biology, Faculty of Fisheries, Hokkaido University, Hakodate 041, Japan

Summary

Relationship between steroid production and gonadal development in a hybrid sturgeon, bester, was examined to obtain basic data necessary for further studies on the mechanisms controlling gonadal maturation.

Introduction

Bester, a hybrid sturgeon (*Huso huso* female ×*Acipenser ruthenus* male), does not spermiate or ovulate in the absence of exogenous hormone treatment, similar to other sturgeons (Steffens *et al.*, 1990). In this study, *in vitro* steroidogenic ability of gonads to produce 11-ketotestosterone (11KT) in males, estradiol-17β (E_2) in females, and 17,20β -dihydroxy-4-pregnen-3-one (DHP) in both sexes at the presence of (1, 10, 100, 1000ng/ml) of pregnenolone, 17α-OHprogesterone (17α-OHP), Testosterone (T), and also forskolin and HCG was examined. In addition, LHRHa-induced spermiation in males and ovulation in females as well as *in vitro* induction of oocyte maturation were studied to find clues regarding control of gonadal maturation.

Results

Higher concentrations of 11KT during late spermatogenesis in male and of E_2 during vitellogenesis in female, and DHP at the final stage of gonadal development in both sexes were detected in media incubated with all doses of pregnenolone, 17α-OHP and T (Fig.1). LHRHa treatment caused spermiation and ovulation in majority of fish with gonads at the late stage of its development. Serum concentrations of DHP were elevated in these spermiated and ovulated fish. GVBD occurred at the presence of several doses of DHP and other preparations. Sensitivities in oocyte maturation *in vitro* decreased gradually during long period of nucleus migration from early stage of postvitellogenic oocyte (July) to the end of nucleus migration (April, Table 1).

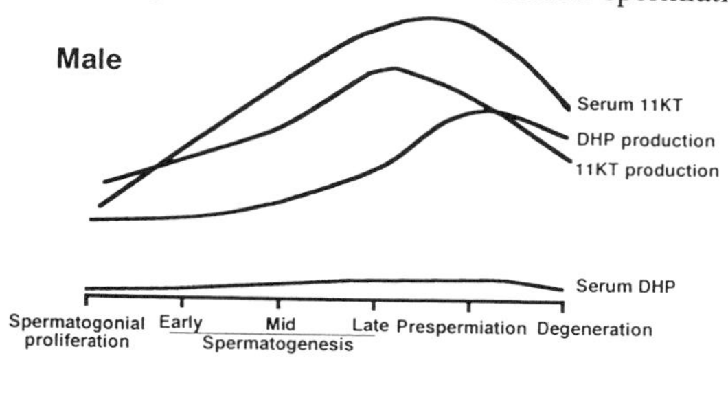

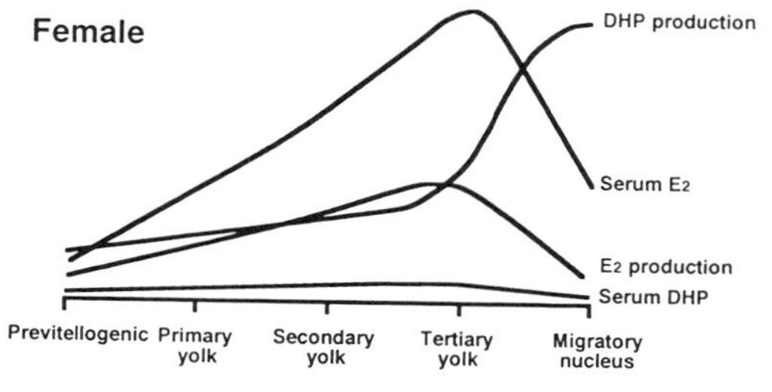

Fig.1 Relationship between serum and *in vitro* steroid production in the male and female bester

Table 1. Monthly changes in the rates of *in vitro* oocyte maturation (GVBD) in the presence of different precursors in the same individual

	GVBD (%)											
	Concentration (ng/ml)											
Treatment	1000				100				10			
	Jul 93	Oct 93	Jan 94	Apr 94	Jul 93	Oct 93	Jan 94	Apr 94	Jul 93	Oct 93	Jan 94	Apr 94
A	0	100	100	100	0	73	7*	13	0	13	0	0
B	0	100	100	100	0	100	80+	7+	0	0	0	0
C	0	100	100	100	0	100	0*	0	0	0	0	0
D	0	100	100	100	0	53	0*	7	0	0	0	0
E	0	100	100	100	0	100	20*	7*	0	10	0	0
F	0	100	100	100	0	20	7*	0	0	7	0	0
G	0	100	100	100	0	26	0*	0	0	0	0	0
H	0	100	100	100	0	13	7	0	0	0	0	0
I	0	100	0*	-	0	0	0	0	0	0	0	0
J	0	0	0	-	0	0	0	0	0	0	0	0

*:Caused more than 50% pre-maturation

+:Caused less than 50% pre-maturation

A:17,20β-diOH-4-pregnen -3-one B:17α-OH progesterone C:Deoxycorticosterone D:Progesterone E:17α,20β,21-triOH -4 -pregnen -3-one F:17α,20α-diOH-4-pregnen -3-one G:Pregnenolone H:17α-OH pregnenolone I:Testosterone J:Estradiol-17β

Discussion

The present results indicate a direct correlation between serum levels and *in vitro* production of 11KT in males and E_2 in females. Significant amounts of DHP in incubation media at the final stage of gonadal development indicate the high ability of gonads to synthesis DHP and also suggest a possible role of this steroid in the final steps of gonadal maturation. The presence of maximal 20β-HSD activities at the final stage of oocyte development indicated that the failure of bester to reach final maturation may be due to lack of enough secretion of precursors. *In vitro* positive oocyte maturation response provides a practical tool for selecting the appropriate fish for artificial spawning.

References

Steffens, W., Jahnichen, H. Fredrich, F. 1990. Possibilities of sturgeon culture in central europe. Aquaculture, 89:101-122

GONADOTROPIC AND SECOND MESSENGER CONTROL OF STEROIDOGENIC ENZYMES IN THE CHANNEL CATFISH TESTES (*Ictalurus punctatus*)

Kathryn F. Medler and John M. Trant

Department of Zoology & Physiology, Louisiana State University, Baton Rouge, LA, USA

Summary

This study focussed on the changes exerted by gonadotropin and various second messenger agents on key steroidogenic enzymes: cholesterol side chain cleavage (SCC), 17α-hydroxylase (c17), and 3ß-hydroxysteroid dehydrogenase (3ß-HSD). Results show that 3ß-HSD mRNA abundance, but not activity, was enhanced via activation of protein kinase C. Activity of 3ß-HSD was doubled by dibutyryl cAMP but mRNA abundance remained stable. The mRNA abundance for SCC was elevated by the calcium ionophore, A23187. No treatments affected mRNA abundance of c17.

Introduction

SCC, c17, and 3ß-HSD are important steroidogenic enzymes for the production of reproductive steroids. Gonadotropins have been shown to stimulate steroidogenesis in mammalian Leydig cells through the second messenger, cAMP (Payne and Youngblood, 1995). In goldfish testes, Wade and Van der Kraak (1991) have shown that phorbol ester (an activator of protein kinase C) and calcium ionophores modulate steroidogenesis. The present study examines the gonadotropic control of chronic changes in the steroidogenesis of channel catfish testes. Changes in mRNA abundance of SCC, c17 and 3ß-HSD in response to dibutyryl cAMP, phorbol ester, and calcium ionophore along with pituitary homogenate and hCG were determined.

Materials and Methods

Anterior testes, harvested late in the reproductive season, were incubated in modified L-15 media as the control or supplemented with hCG (5 units/mL), dibutyryl cAMP (1 mM), phorbol 12-myristate 13-acetate (PMA) (1 μM), the Ca^{++} ionophore A23187 (4 μM) or pituitary homogenate (0.01 pituitary/mL). After 0 and 18 hours of incubation, total RNA was immobilized on nitrocellulose using a slot blot. The blot was sequentially probed with ^{32}P-labelled cDNAs for catfish SCC and c17, rainbow trout 3ß-HSD, and stingray actin and quantitated by densitometry of the autoradiograph. The 3ß-HSD activity was determined using tritiated DHEA (0.1 μM) and 1 mM NAD^+ Samples were analyzed by HPLC.

Results

No treatments affected the mRNA abundance of c17. The Ca^{++} ionophore, A23187, caused a two fold increase in SCC mRNA abundance as compared to control (Table 1). The enzyme activity of 3ß-HSD in the presence of cAMP was increased two fold but did not affect mRNA abundance. PMA treatment resulted in a 4.4 fold increase in 3ß-HSD mRNA abundance but caused no change in activity of 3ß-HSD. Pituitary homogenate caused a two fold increase in mRNA abundance for this enzyme without enhancing 3ß-HSD activity.

Table 1. Relative mRNA abundance and 3ß-HSD enzyme activity (in parentheses). Control activity was 23 pmol/min/g.

Treatment	3ß-HSD	c17	SCC
0 Hour	0.9 (1.0)	0.4	0.5
Control	1.0 (1.0)	1.0	1.0
hCG	0.8 (1.0)	1.0	1.4
cAMP	1.1 (2.1)	0.6	0.8
PMA	4.4 (1.1)	0.7	1.2
Ionophore	1.6 (---)	0.9	2.0
Pituitary	2.0 (1.2)	0.9	1.0

Discussion

This initial study indicates a complicated regulatory mechanism that utilizes both protein kinase C and protein kinase A pathways. 3ß-HSD is regulated at both transcriptional and translational levels. Levels of SCC mRNA were doubled with Ca ionophore, A23187. Abundance of c17 mRNA was not affected by any treatments, indicating a lack of regulation by a protein kinase A or protein kinase C pathway. This differs from mammalian systems and requires further study.

References

Payne, A. and G. Youngblood, 1995. Regulation of expression of steroidogenic enzymes in Leydig cells. Biol. Reprod. 52: 217-225.

Wade, M. and G. Van der Kraak, 1991. The control of testicular androgen production in the goldfish: Effects of activators of different intracellular signalling pathways. Gen. Comp. Endocrinol. 83: 337-344.

The 3β-HSD cDNA was a gift from Dr. Y. Nagahama. Research was supported by a grant from USDA (9401429).

EVIDENCE FOR PEROXISOMAL INVOLVEMENT IN OVARIAN STEROIDOGENESIS IN TELEOSTS

F. Mercure and G. Van Der Kraak

Department of Zoology, University of Guelph, Guelph, Ontario, Canada, N1G 2W1.

Summary

The presence of peroxisomes in ovarian follicles of teleosts was confirmed. Drugs known to induce peroxisome proliferation inhibited testosterone production suggesting a role for this organelle in ovarian steroidogenesis.

Introduction

Peroxisomes are organelles involved in lipid, eicosanoid, and cholesterol metabolism. They have been identified in rat Leydig cells, and luteinizing hormone injections changed their sterol carrier protein content suggesting a role in cholesterol availability for steroidogenesis (Mendis-Handagama et al., 1990). Free fatty acids, known to regulate peroxisomal enzymes in mammals (Keller and Wahli, 1993), attenuate ovarian steroid production in teleosts by affecting cholesterol availability (Mercure and Van Der Kraak, 1995). The purpose of this study was to determine whether peroxisomes are present in ovarian follicles of the goldfish and the rainbow trout and to test the effects of drugs known as peroxisome proliferators on steroid production.

Results and Discussion

Fig. 1 shows the activity of the peroxisomal marker catalase measured in an ovarian tissue homogenate using the methods described by Cohen et al. (1970).

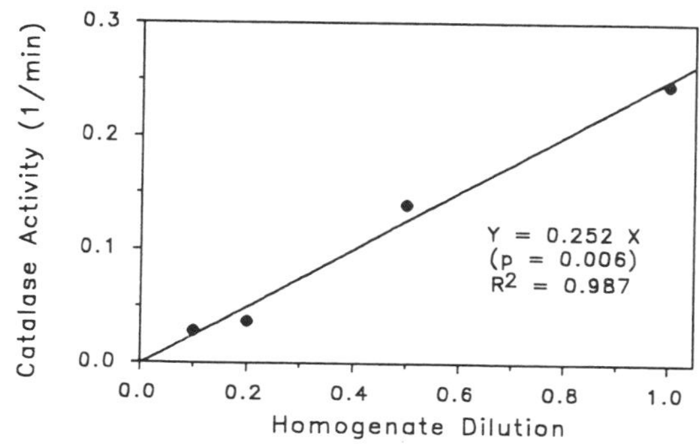

Fig. 1. The relationship between catalase activity and homogenate concentration in rainbow trout ovarian follicles.

For both goldfish and rainbow trout, catalase activity was concentrated by centrifugation in a fraction corresponding to the expected density of peroxisomes (data not shown).

The treatment of ovarian follicles with the peroxisome proliferator clofibric acid inhibited testosterone production in a dose related manner (Table 1).

Table 1. The effects of clofibric acid (CA) on hCG-stimulated (10 IU/ml) testosterone production (pg/ml) by goldfish ovarian follicles. Values represent mean ± s.e., and the symbol "*" indicates significant difference ($p < .05$) from hCG alone. (ND; below detection level).

CA (μM)	Basal	hCG
0	1.2 ± 1.0	227.8 ± 10.8
16	ND	221.4 ± 8.8
63	ND	*142.9 ± 18.0
250	ND	*135.5 ± 17.3
1000	ND	*122.5 ± 8.9

The presence of peroxisomes in ovarian follicles of two species of teleosts was confirmed by measuring the activity of the peroxisomal marker catalase. Together with the findings of Mendis-Handagama et al. (1990) and Mercure and Van Der Kraak (1995), the demonstration that peroxisome proliferators affect ovarian steroid production suggests that peroxisomes are involve in steroidogenic processes. Peroxisomes may play a role in the supply of cholesterol to steroidogenic enzymes.

References

Mendis-Handagama, S.M.L.C., P.A. Watkins, S.J. Gelber, T.J. Scallen, B.R. Zirkin, and L.L. Ewing, 1990. Endocrinology 127: 2947-2954.

Mercure, F. and G. Van Der Kraak, 1995. Lipids 72: (in press).

Cohen, G., D. Dembiec, and J. Marcus, 1970. Analyt. Biochem. 34: 30-38.

Keller, H. and W. Wahli, 1993. Trends Endocrinol. Metab. 4: 291-296.

STEROID LEVELS DURING OVULATION AND SPERMIATION IN TOADFISH *HALOBATRACHUS DIDACTYLUS*

T. Modesto and A. V. M. Canario

Center of Marine Sciences, UCTRA, University of Algarve, 8000 Faro, Portugal.

Summary

Plasma levels of 11-ketotestosterone (11KT) and 3α,17α,21-trihydroxy-5β-pregnan-20-one (3α,17α,21-P-5β) in males of toadfish and 17α,20β,21-trihydroxy-4-pregnen-3-one (17,20β,21-P) and 3α,17α,21-P-5β in females were highest during the spawning season. LHRH had no significant effect on levels of these steroids.

Introduction

The objective of this work was to study the profile of sex steroids during the reproductive cycle and during LHRHa-induced ovulation and spermiation in toadfish.

Results

Seasonal cycle- Levels of sex steroids peaked in June coinciding with the decrease in GSI (Fig. 1).

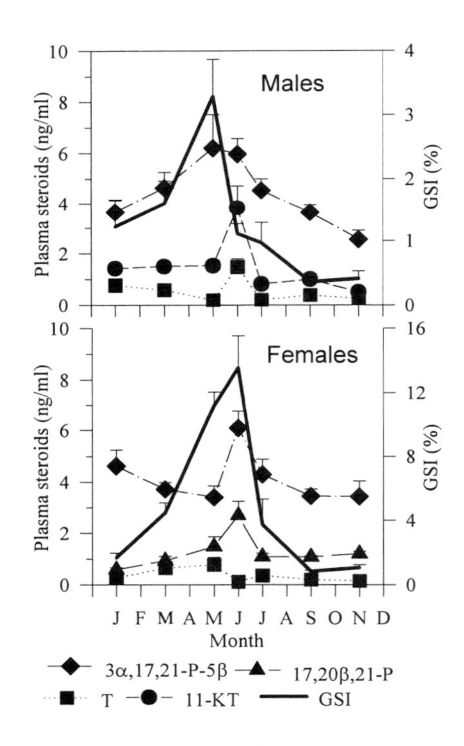

Fig. 1 - Seasonal variation of plasma levels (mean+SE) steroids in male and female toadfish.

Effect of LHRH analogue- Most females ovulated and most males produced sperm in both LHRHa-injected and saline fish. 11KT in the males saline group and 17,20β,21-P in the LHRHa-injected females had a slight but non-significant increase in levels (Fig. 2).

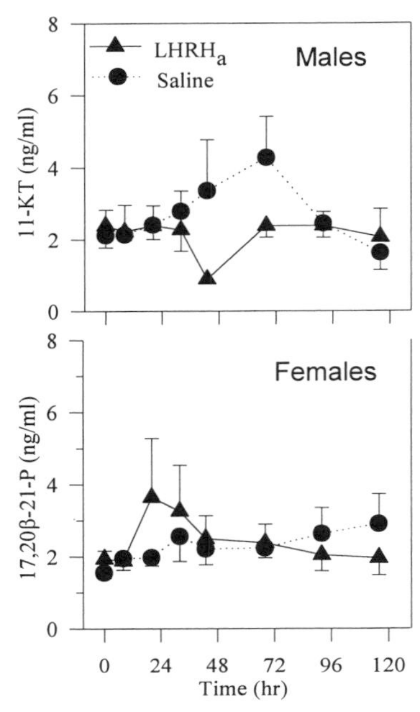

Fig. 2 - Changes in plasma concentration (mean±SE) of 11-KT (males) and 17,20β,21-P (females) in toadfish injected with LHRHa in July.

Discussion

11KT showed a similar seasonal profile to that found in other teleosts (Yoshikuni and Nagahama, 1991). The maturation-inducing steroid 17,20β,21-P (Trant *et al.*, 1986) is likely to be involved in reproduction also in toadfish. The steroid with highest levels both in males and females was 3α,17α,21-P-5β, a gonadal steroid found also in other marine teleosts (Canario and Scott, 1990).

References

Canario, A.V.M. and Scott, A.P., 1990. *General and Comparative Endocrinology*, 78: 273-285.

Trant, J.M., Thomas, P. and Shackleton, C.H.L., 1986. *Steroids*, 47: 89-99.

Yoshikuni, M. and Nagahama, Y., 1991. *Bull. Inst. Zool., Academia Sinica, Monograph*, 16: 139-172.

Supported by STRIDE programme, FEDER and JNICT, Portugal

MATURATION INDUCING STEROID IN TURBOT, SCOPHTHALMUS MAXIMUS L.

C. Mugnier[1,3], J.L. Gaignon[2], E. Lebegue[3], C. Fauvel[4], and A. Fostier[1]

[1]Laboratoire de Physiologie des Poissons, INRA, Campus de Beaulieu, 35042 Rennes cedex, France; [2]Département des Ressources Vivantes, IFREMER, BP70, 29280 Plouzané; [3]Noirmoutier Aquaculture Techniques Avancées, BP 305, 85330 Noirmoutier. [4]Station Expérimentale IFREMER, chemin de Maguelone, 34250 Palavas-les Flots.

Summary

Ovarian steroidogenic capacity has been analysed in turbot during spawning season. Besides, 15 steroids have been tested in vitro for their potency to induce oocyte maturation. 17,21-dihydroxy-4-pregnene-3,20-dione (S) and 17,20β,21-trihydroxy-4-pregnene-3-one (20β-S), but not 17,20β-dihydroxy-4-pregnene-3-one (17,20β-P), were found as metabolites of 17-hydroxy-4-pregnene-3-one (17-P). 20β-S was the most active inducer of oocyte maturation and ovulation. We suggest that 20β-S is a maturation inducing steroid (MIS) in turbot.

Introduction

Turbot is a multiple spawning flatfish. Plasma levels of 17,20β-P, which has been shown to be a MIS in several species (Jalabert et al., 1991), are low in mature females, and not correlated with oocyte maturation and ovulation (Howell and Scott, 1989). The aim of this work was to identify hydroxylated pregnenes synthesized in vitro by the ovary when incubated with tritiated or radioinert 17-P. On the other hand, potency of metabolites, and some other steroids, in inducing oocyte maturation was tested in vitro.

Material and methods

Three mature females were used for this experiment.

* *Metabolism studies* : Medium of incubates were analysed by thin layer chromatography (TLC), high pressure liquid chromatography (HPLC) and microchemical reactions (acetylation and oxydation). TLC zone from incubate with radioinert 17-P, suspected to contain 20β-S, was analyzed by HPLC coupled with atmospheric pressure chemical ionization-mass spectrometry (LC-APCI-MS).

* *In vitro oocyte maturation* : Pieces of ovaries from the three females were incubated at 12°C in an incubation medium adapted to turbot with several steroids, including C_{19} steroids (testosterone, androstenedione), estradiol-17β and C_{21} 4-ene,3-one steroids (17-P, 17,20α-P, 17,20β-P, S, 20β-S), or 5-ene,3β-one steroids.

Results and conclusion

* *Metabolism* : C_{21} steroids identified in incubates with tritiated 17-P are listed in table 1. 17,20β-P could not be detected.

Table 1. Percentages of C_{21} steroids identified in the medium after two hours of incubation, in relation to the total radioactivity of the medium after extraction.

	Female 1	Female 2	Female 3
S	21.7	4.0	9.7
20β-S	< 0.1	< 0.1	1.7
17,20α-P	0.9	0.4	1.0

Besides, 20β-S was identified by LC-MS in medium of incubates with radioinert 17-P.

* *In vitro maturation and ovulation* : Female 3, which was sampled between two spawnings, was the only one giving a positive response under steroid stimulation. 6% of the total number of oocytes were undergoing maturation and ovulation. Testosterone and estradiol had no effect. 20β-S and 17,20β-P were the most active, however, after 48 hours of incubation, 20β-S showed the highest activity (figure 1).

Figure 1 : Number of ovules obtained in ovarian fragments of female 3 after 48h incubation with increasing concentrations of steroids (means of duplicates).

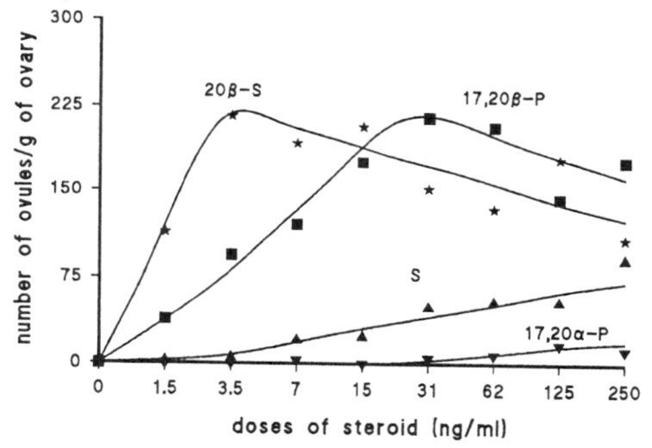

Conclusion : Although 17,20β-P is able to induce oocyte maturation and ovulation in vitro, it was not detected as a metabolite of 17-P. 20β-S was the most efficient pregnene tested in vitro. It has been detected and identified using both tritiated and radioinert 17-P as a substrate in incubation with pieces of ovary. Thus, 20β-S is a potential MIS for turbot. The low yield found for this steroid could be explained by its high efficiency to induce ovulation, by the small percentage of oocytes undergoing maturation within one spawn, and by its transient synthesis in a multiple spawner, like turbot.

References

Howell R. and Scott A.P., 1989. Ovulation cycles and post-ovulatory deterioration of eggs of the turbot (Scophthalmus maximus L.). Rapp. Réun. cons. int. Explor. Mer 191:21-26.

Jalabert B., Fostier A., Breton B. and Weil C., 1991. Oocyte maturation in vertebrate in « vertebrate endocrinology ». P.K.J. Pang and M.P. Schreibman eds. Academic Press. 23-90.

REGULATION OF STEROIDOGENESIS IN THE ELASMOBRANCH INTERRENAL

Scott Nunez and John Trant

Department of Zoology, Louisiana State University, Baton Rouge, LA, 70803

Summary

In order to better understand how interrenal steroidogenesis is regulated in elasmobranchs, cDNA clones of SCC and 3β-HSD were isolated from the interrenal of a stingray, *Dasayatis americana*. Interrenal RNA was analyzed by Northern blot and RT-PCR.

Introduction

The elasmobranch interrenal is a discrete encapsulated gland composed entirely of steroidogenic cells which produces a unique steroid, 1α-hydroxycorticosterone. The elasmobranch interrenal is therefore an excellent and inherently interesting model for the study of the regulation of interrenal steroidogenesis. ACTH and angiotensin II (AII) stimulate steroidogenesis in elasmobranch interrenal tissue (Armour *et al*, 1993). The interrenal may also be influenced by other hormones (Schreck *et al*, 1989). In order to investigate the manner in which hormones such as ACTH and AII increase interrenal steroidogenesis in elasmobranchs, cDNA clones for two key steroidogenic enzymes, $P450_{scc}$ and 3β-HSD, were isolated from a stingray (*Dasayatis americana*) cDNA library. The effect of ACTH and AII on the level of mRNA transcription of these two enzymes in the interrenal of *Dasayatis sabina* was subsequently investigated.

Results

Preliminary studies indicated ACTH (0.3 μM) and homologous pituitary extract (0.1 pit / ml) increased the synthesis of 1α-hydroxycorticosterone over uninduced controls (3 and 10 fold, respectively). Cycloheximide (70 μM) inhibited this induction in the presence of 25-hydroxycholesterol (50 μM), indicating the induction of steroidogenic enzymes. To investigate this possibility further, partial cDNA clones of $P450_{scc}$ and 3β-HSD were isolated from a *Dasayatis americana* interrenal library. Sequence was obtained for greater than 80% of the encoding region for both of these enzymes (1557 bp for $P450_{scc}$ and 950 bp for 3β-HSD). The deduced amino acid sequence for these enzymes was only 30-40% homologous to other known forms. However, in conserved regions of these proteins, homology was increased to 60-80%. Approximately 3440 bp of 3' untranslated sequence (including three putative polyadenylation sites and a poly(A)-tail) was obtained for $P450_{scc}$. No poly(A)-tail was obtained for 3β-HSD, however, 1030 bp of the 3' untranslated (including two putative polyadenylation sites) was obtained. Northern blot analysis of interrenal RNA indicated two transcripts for 3β-HSD (2600 bp and 1600 bp).

Preliminary RT-PCR analysis of RNA isolated from *Dasayatis sabina* interrenal tissue exposed to ACTH and AII for 14 hours indicated that transcription of $P450_{scc}$ and 3β-HSD was induced by these hormones. After 30 PCR cycles, $P450_{scc}$ and 3β-HSD amplification products could be detected in ACTH and AII reactions, but not in control reactions. Only after 50 cycles could amplification be seen in control reactions.

Discussion

ACTH appears to stimulate interrenal steroidogenesis in elasmobranchs by a process other than cholesterol mobilization. Preliminary data suggested that stimulation by ACTH and AII induced the transcription of two key steroidogenic enzymes, $P450_{scc}$ and 3β-HSD. The deduced amino acid sequence of partial cDNA clones of $P450_{scc}$ and 3β-HSD displayed low homology to other known forms, but in regions known to be highly conserved, homology was greatly increased, confirming the identity of these clones. The presence of multiple polyadenylation sites in the cDNA clones of each of these enzymes and two distinct 3β-HSD transcripts in the elasmobranch interrenal indicate that the regulation of these enzymes may be intricate. The cDNA clones described in this study will therefore be useful in the further study of the regulation of interrenal steroidogenesis.

Literature cited

Armour, K.J.; L.B. O'Toole and N. Hazon. 1993. Mechanisms of ACTH- and angiotensin II-stimulated 1α-hydroxycorticosterone secretion in the dogfish, Scyliorhinus canicula. J. Mol. Endocrinol. 10:235-244.

Schreck, C.B., C.S. Bradford, M.S. Fitzpatrick and R. Patiño. 1989. Regulation of the interrenal of fishes: non-classical control mechanisms. Fish Physiol. Biochem. 4:359-265.

This work was supported by a grant from the National Science Foundation (IBN-9219926).

17α,20β-DIHYDROXY-4-PREGNEN-3-ONE STIMULATES CORTISOL PRODUCTION BY RAINBOW TROUT INTERRENAL TISSUE *IN VITRO*: MECHANISM OF ACTION

T. Barry, J. Riebe, J. Parrish[1], and J. Malison

University of Wisconsin Aquaculture Program and [1]Department of Meat and Animal Science, University of Wisconsin-Madison, Madison, WI 53706

Summary

17α,20β-dihydroxy-4-pregnen-3-one (17,20-P) rapidly stimulated cortisol production from rainbow trout (*Oncorhynchus mykiss*) head kidney preparations (interrenal tissue) *in vitro*. This effect of 17,20-P was blocked by H-89, a potent inhibitor of cAMP-dependent protein kinase (PKA), suggesting that 17,20-P stimulates cortisol production via a rapid, cAMP-dependent mechanism.

Introduction

Plasma cortisol levels are elevated during the spawning period in salmonid fishes. The physiological significance of this rise is unclear, although in semelparous species cortisol hypersecretion is known to mediate programmed death (Dickhoff, 1989; Stein-Behrens and Sapolsky, 1992). Evidence that gonadectomy blocks the normal increase in cortisol in prespawning sockeye salmon (*Oncorhynchus nerka*) suggests that a gonadal factor plays an important role in regulating hypercortisolism (Donaldson and Fagerlund, 1970). Androgens and estrogens, however, probably do not play important roles in this regard (reviewed by Dickhoff, 1989). We recently showed that the gonadal steroid 17α,20β-dihydroxy-4-pregnen-3-one (17,20-P) directly stimulates cortisol production by rainbow trout interrenal tissue *in vitro* with a potency equivalent to that of ACTH (unpublished data, Fig. 1). The present study was conducted to determine the cellular mechanism by which 17,20-P stimulates cortisol production by the rainbow trout interrenal.

Results

17,20-P stimulated a dose-dependent increase in cortisol production by interrenal tissue of sexually immature rainbow trout *in vitro*. Cortisol production was significantly enhanced above control levels within one hour by 17,20-P. Estradiol-17β, testosterone, and 11-ketotestosterone had no effect on cortisol production, whereas 17α,20α-dihydroxy-4-pregnen-3-one was 50% as effective as 17,20-P. Radiotracer studies using [3H]17,20-P indicated that no more than 20% of the total cortisol in the medium could be accounted for by enzymatic conversion of 17,20-P. Neither actinomycin-D nor cycloheximide (10, 100, or 1000 ng/ml) inhibited the stimulatory effects of 17,20-P. Neither the calcium channel blocker $CoCl_2$ (0.1, 1 or 5 mM) nor incubation in low calcium medium altered 17,20-P-stimulated cortisol production. H-89, a potent inhibitor of PKA, inhibited 17,20-P-stimulated cortisol production (Fig. 1). The less potent PKA inhibitor H-8 was ineffective at the doses tested (1, 10, 33 and 100 μM).

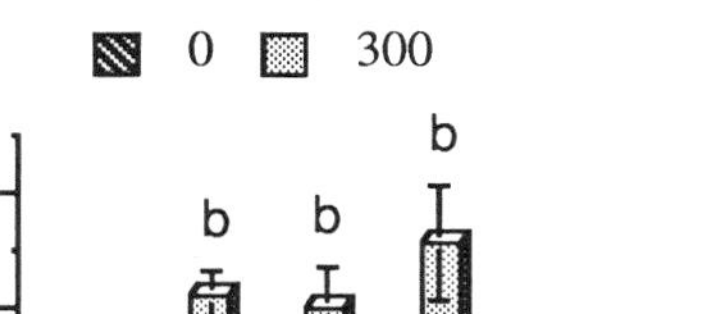

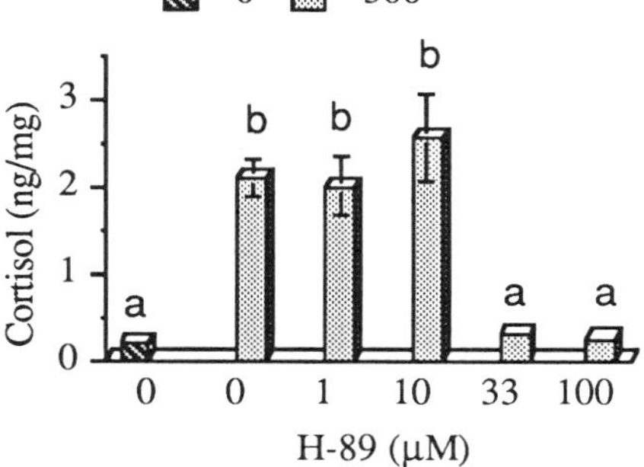

Fig. 1. Effects of H-89 on 17,20-P-stimulated cortisol production by rainbow trout interrenal tissue *in vitro*. Tissues were preincubated for 3 hours with medium changes each hour and then incubated for 18 hours under static conditions. Values represent the mean ± SEM (n = 4). Letters show statistical groupings as determined by a protected LSD test following ANOVA ($p < 0.05$).

Discussion

Our results suggest that 17,20-P stimulates interrenal corticosteroidogenesis via a rapid, cAMP-dependent mechanism, and not through a classical nuclear receptor-mediated mechanism involving new mRNA and protein synthesis. The data suggest that 17,20-P could be a physiological regulator of cortisol secretion in peri-ovulatory salmonids. We hypothesize that at the time of spawning regulatory control of corticosteroidogenesis may shift from the hypothalamic-pituitary-interrenal axis to the gonad. Such a regulatory change could explain why highly elevated cortisol levels are not reduced by classical negative feedback mechanisms in spawning salmonids.

References

Dickhoff, W.W., 1989. Salmonids and annual fishes: Death after sex. In: Development, Maturation, and Senescence of Neuroendocrine Systems: A Comparative Approach. M.P. Schreibman and C.G. Scanes (Eds.): Academic Press, New York, p. 253-266.

Donaldson, E.M., and U.H. Fagerlund, 1970. Effect of sexual maturation and gonadectomy at sexual maturity on cortisol secretion rate in sockeye salmon. Fish. Res. Bd. Can. 26:2287-2296.

Stein-Behrens, B.A. and R.M. Sapolsky, 1992. Stress, glucocorticoids, and aging. Aging Clin. Exp. Res. 4:197-210.

Supported by the Wisconsin Sea Grant Institute (NA90AA-D-SF469, R/AQ-21) and a Wisconsin/Hilldale Undergraduate Fellowship to J.R.

MONITORING TESTICULAR ACTIVITY IN SPAWNING MALE PLAICE (*PLEURONECTES PLATESSA*)

A. P. Scott[1], E. L. M. Vermeirssen[1], C. C. Mylonas[2] and Y. Zohar[2]

[1]MAFF Fisheries Laboratory, Pakefield Rd., Lowestoft, Suffolk, NR33 0HT, UK
[1]Center of Marine Biotechnology, University of Maryland, Baltimore, MD, USA

Summary

Testicular activity and sperm production of male plaice can be monitored by assaying levels of the 5β-reduced, sulphated metabolites of 17α,20β-dihydroxy-4-pregnen-3-one in plasma.

Introduction

Generally, in male teleosts, at the time of spawning, the testes change from synthesising mainly C18 and C19 steroids to mainly C21 steroids. The predominant C21 steroid produced by male rainbow trout is 17α,20β-dihydroxy-4-pregnen-3-one (17,20β-P). However, when this steroid is assayed in blood plasma of spermiating male plaice (*Pleuronectes platessa*; a marine flatfish), only very low levels (< 2 ng/ml), and only small differences between spermiating and non-spermiating fish, are found. Assays for 17α,20β,21-trihydroxy-4-pregnen-3-one and 17α,20α-dihydroxy-4-pregnen-3-one are equally unproductive. In the course of studies on female plaice, however, we have developed two novel radioimmunoassays (RIAs) - one of which cross-reacts with a variety of 5β(3α-OH)-reduced C21 steroids and the other which cross-reacts with a variety of 17,20β-dihydroxylated steroids. In both RIAs, 5β-pregnane-3α,17,20β-triol (3α,17,20β-P-5β) is used as a standard. Both RIAs have been used to assay plasmas collected from male plaice which have been caught on or off their spawning grounds or which have been brought back to the laboratory and then injected with Gonadotrophin-releasing hormone analogue (GnRHa). The nature of the cross-reacting steroids has also been examined using HPLC and TLC.

Results

RIA for 5β(3α-OH)-reduced steroids This assay was carried out directly on diluted plasma samples which had been heated to 90 °C for 10 min to denature any enzymes and binding proteins. Very significant differences in levels were found between males caught on (123 ± 10 ng/ml) or off (23 ± 4 ng/ml) their spawning grounds. Very significant increases were also induced by sustained administration of GnRHa by polymeric microspheres (Fig. 1). Following HPLC, the activity resolved into at least 6 peaks, the largest of which was identified as 3α,17,20β-P-5β 20-sulphate.

RIA for '17,20β–di-OH' steroids This assay was carried out on pools of plasma which had been treated by acid solvolysis to free all the sulphated steroids. Low levels of activity (<0.8 to 5 ng/ml) were found in non-spermiating and non-spawning males, rising to as high as 220 ng/ml in spermiating males caught in the middle of their spawning grounds. Following HPLC, the activity resolved into at least three peaks, two of which were identified as 3α,17,20β-P-5β and 17,20β-P, but the largest of which was identified as 5β-pregnane-3β,17,20β-triol.

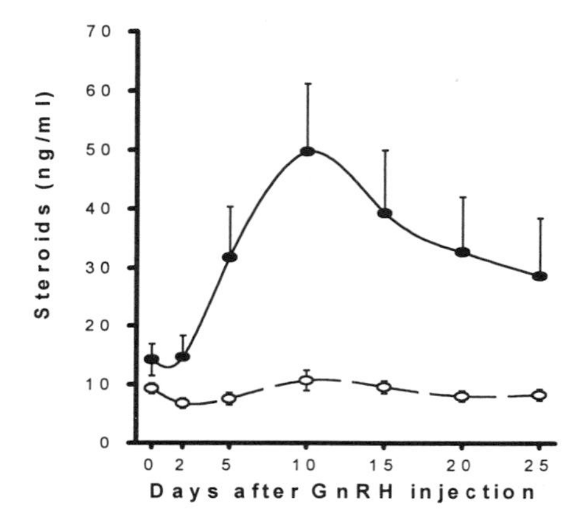

Fig. 1 Changes which take place in plasma levels of 5β(3α-OH)-reduced steroids in male plaice after i.m. injection of a suspension of microspheres containing *c.* 45 μg of D-Ala[6], Pro[9]-NEt-LHRH (●; n=10) or with vehicle only (O; n=10).

Discussion

The results show that, in common with mature females, spermiating male plaice have substantial amounts of 3α,17,20β-P-5β 20-sulphate in their plasma. They also contain 3β,17,20β-P-5β 20-sulphate. Both these compounds are metabolites of 17,20β-P.

CHARACTERIZATION OF AN ANDROGEN RECEPTOR IN SALMONID LEUKOCYTES

Caleb H. Slater, Martin S. Fitzpatrick, and Carl B. Schreck

Oregon Cooperative Fishery Research Unit, Department of Fisheries and Wildlife, Oregon State University, Corvallis, Oregon 97331-3803

Summary

The objective of the present study was to describe the androgen binding characteristics of salmonid leukocytes and to determine if a true androgen receptor was responsible.

Introduction

All Pacific salmon (genus *Oncorhynchus*) are semelparous (die after spawning) and their bodies degenerate greatly during sexual maturation. Immunosuppressive effects of elevated testosterone during sexual maturation may be partly responsible for determining programmed post-spawning death, and for the increase in infection and mortality which often occurs before spawning. This speculation is based on anecdotal evidence which suggests that gonadectomy prevents post-spawning mortality, and our evidence that testosterone reduces the immune response in chinook salmon (Slater and Schreck, 1993). The immunosuppressive effects of cortisol on juvenile salmonids have been linked to direct action on the leukocytes through glucocorticoid receptors (Maule and Schreck, 1990). Based on this evidence we have hypothesized that testosterone's immuno-suppressive action is mediated by an androgen receptor in the salmonid leukocyte.

Results

Cytosol of rainbow trout leukocytes demonstrated specific and saturable binding of ^{3}H-testosterone (K_d = 0.99 nM $\pm$ 0.17 nM and B_{max} = 30.4 $\pm$ 4.9 fmol/mg protein; n=6). Specific binding of ^{3}H-testosterone was high in leukocytes and other tissues with known androgen binding affinity (plasma, skin and liver), and low in other tissues (heart, muscle and red blood cells). Specific binding of ^{3}H-testosterone was displaced by testosterone and dihydrotestosterone. Androstenedione displaced 50% of specifically bound ^{3}H-testosterone between 10 and 100-fold excess, while 17α-methyltestosterone, 11-ketotestosterone, and progesterone displaced 50% of specifically bound ^{3}H-testosterone between 100 and 500-fold excess. Cortisol, 17β-estradiol, 17α,20β-dihydroxyprogesterone, the synthetic androgen mibolerone, and the synthetic estrogen ethenyl-estradiol did not displace ^{3}H-testosterone binding, even at 500-fold excess. Treatment of cytosol with proteolytic enzyme significantly reduced the specific binding of ^{3}H-testosterone. HPLC analysis determined that ^{3}H-testosterone was not metabolized during assay incubation with cytosol.

Discussion

Our results support the hypothesis that androgen receptors exist in the leukocytes of salmonid fish. The cytosolic binding sites characterized in rainbow trout leukocytes fulfill the basic description of steroid receptors, i.e., high affinity, low capacity, steroid and tissue specificity, and correlation with biological response. The fact that specific binding was reduced by treatment with trypsin also indicates that binding is due to a proteinaceous receptor.

Androgen receptors with similar binding affinities have been described in tissues of a number of species; the skin of brown trout, *Salmo trutta*, (Pottinger, 1987) and the ovaries of coho salmon, *O. kisutch*, (Fitzpatrick *et al.*, 1994).

References

Fitzpatrick, M. S., Gale, W. L., and Schreck, C. B. (1994). Binding characteristics of an androgen receptor in the ovaries of coho salmon, *Oncorhynchus kisutch*. *Gen. Comp. Endocrinol.* 95, 399-408.

Maule, A. G. and Schreck, C. B. (1990). Glucocorticoid receptors in leukocytes and gills of juvenile coho salmon *(Oncorhynchus kisutch)*. *Gen. Comp. Endocrinol.* 77, 448-455.

Pottinger, T. G. (1987). Androgen bindind in the skin of the mature brown trout, *Salmo trutta* L. *Gen .Comp. Endocrinol.* 66, 224-232.

Slater, C. H. and Schreck, C. B. (1993). Testosterone alters the immune response of chinook salmon *(Oncorhynchus tshawytscha)*. *Gen. Comp. Endocrinol.* 89, 291-298.

ISOLATION AND CHARACTERIZATION OF THE cDNA ENCODING THE SPINY DOGFISH SHARK (*Squalus acanthias*) FORM OF CYTOCHROME P450c17

John M. Trant

Department of Zoology & Physiology, Louisiana State University, Baton Rouge, LA, USA

Summary

A full-length cDNA (1964 bp) encoding the dogfish shark cytochrome P450c17 (17α-hydroxylase) was isolated. The predicted protein is 509 residues with a molecular mass of 57.2 kDa. The shark P450c17 is 43-46% identical to mammalian forms of this enzyme, whereas the rainbow trout and chicken forms are 59% and 57% identical, respectively. The cDNA was used to direct heterologous expression of recombinant shark P450c17 in *E. coli* (for antisera production) and yeast (for enzymatic characterization).

Introduction

The timely and appropriate secretion of gonadal steroids is a vital component of successful reproduction in all vertebrates and is controlled via expression of steroidogenic enzymes. Cytochrome P450c17 is a microsomal steroidogenic enzyme found primarily in gonads and adrenals. The one enzyme is responsible for both hydroxylation (17α-hydroxylase) of the C21-steroid nucleus and cleavage of the 2-carbon side-chain (C17,20-lyase) to a C19 androgen. In elasmobranchs, changes in cytochrome P450c17 activity have been correlated with gamete development in both testicular (Callard *et al.*, 1985) and ovarian (Fasano, *et al.*, 1989) tissue. In teleosts, Sakai, *et al.* (1992) have recently demonstrated increased mRNA abundance of P450c17 in follicles of developing oocytes.

Results and Discussion

A full-length cDNA encoding the dogfish shark testicular cytochrome P450c17 (17α-hydroxylase) was isolated from a lambda library. The cDNA contained a poly$(A)_{26}$ tail 18 bp downstream from a consensus polyadenylation signal (AATAAA). The 1964 bp cDNA encoding the shark P450c17 has an open reading frame of 1527 bp flanked by 23 bp of the 5'-untranslated and 411 bp of the 3'-untranslated regions. Northern blot analysis of shark testicular RNA indicated the presence of a single transcript. The cDNA predicts a 509 residue protein with a molecular mass of 57.2 kDa. The shark P450c17 is 43-46% identical to the mammalian forms (rat, mouse, human, bovine, and pig), whereas the rainbow trout and chicken forms are 59% and 57% identical, respectively.

Heterologous expression of the shark P450c17 in *E. coli* required modification of amino terminus as described by Barnes *et al.* (1991). Fusion of a hexa-histidinyl peptide to the carboxyl terminus facilitated purification of the recombinant protein with a single Ni^{++}-chelated column. This system yielded an average of 1.8 mg of purified, bioengineered protein per 100 ml culture. This protein was used to raise a specific antiserum for a western blot procedure that can readily detect P450c17 in 10 µg of shark testis microsomes.

The native cDNA was used to direct heterologous expression of a functional enzyme in yeast (*Pichia pastoris*). P450c17 activity was supported in living cells by the endogenous yeast P450 reductase. Progesterone was metabolized twice as fast as pregnenolone. The shark P450c17 was capable of converting both of the Δ4 and Δ5 pregnenes to their respective androgens (C17,20 lyase), unlike many of the mammalian forms.

This is the first report of a cDNA encoding a steroidogenic enzyme isolated from an elasmobranch species, in spite of this taxa's evolutionary importance. Both the cDNA and antibody probes will be used to demonstrate the control of P450c17 at the transcriptional, translational, and post-translational level in future studies.

References

Barnes, H.J., M.P. Arlotto, and M.R. Waterman (1991) Expression and enzymatic activity of recombinant cytochrome P450 17α-hydroxylase in *Escherichia coli*. *Proc. Nat. Acad. Sci. USA* **88:** 5597-5601.

Callard, G.V., J.A. Pudney, P. Mak, and J.A. Canick (1985) Stage-dependent changes in enzymes and estrogen receptors during spermatogenesis in the testis of the dogfish, *Squalus acanthias*. *Endocrinology* **117:** 1328-1335.

Fasano, S., M. D'Antonio, R. Pierantoni, and G. Chieffi (1992) Plasma and follicular tissue steroid levels in the elasmobranch fish, *Torpedo marmorata*. *Gen. Comp. Endocrinol.* **85:** 32-333.

Sakai, N., M. Tanaka, S. Adachi, W.L. Miller, and Y. Nagahama (1992) Rainbow trout cytochrome P450c17 (17α-hydroxylase/17,20-lyase) cDNA cloning, enzymatic properties and temporal pattern of ovarian P-450c17 mRNA expression during oogenesis. FEBS 301: 60-64.

Shark testis was a gift of Dr. Gloria Callard, Boston University. This work was supported by NSF (IBN-9219926).

EFFECT OF LHRH ON SEX REVERSAL AND STEROID LEVELS ON GILTHEAD SEABREAM (*SPARUS AURATA*)

P. Vília and A.V.M. Canario

Center of Marine Sciences, UCTRA, University of Algarve, 8000 Faro, Portugal.

Summary

The long term effect of LHRH on spermiation and steroid levels in male gilthead seabream (*Sparus aurata*) was investigated. LHRH injections had no significant effect in the number of spermiating fish or in steroid levels. However the average percentage volume of male gonadal tissue was 70% in the control group as compared to 30% in the LHRH group. The results suggest that LHRH may be involved in the process of sex change in gilthead seabream.

Introduction

In most teleost species studied so far LHRH plays a major role in reproduction (Peter *et al*, 1991). The endocrinological mechanisms controlling male gilthead seabream reproduction are little known. In the female it has been shown that LHRH and LHRH analogues are very effective in inducing ovulation (Zohar *et al.*, 1995). In this work we studied the long term effect of an analogue of LHRH in spermiation and steroid levels.

Methods

Seven one-year-old fish obtained from the same stock received over a period of 50 days, starting in October, at the beginning of the spawning season, 3 injections of 4 μg/kg of des-Gly^{10},[D-Ala^{6}]-Neth-LHRH ($LHRH_A$) in 250μl of 0.8% NaCl (days 0, 22 and 38). Another group received saline only on the same dates. At the end of the experiment the gonadosomatic index (GSI) and the male/female gonadal tissue volume were determined. Changes in plasma levels of testosterone (T), 11-ketotestosterone (11-KT), estradiol-17β (E_2), 17α,20α-dihydroxy-4-pregnen-3-one (17,20α-P) and 17α,20β-dihydroxy-4-pregnen-3-one (17,20β-P) in response to treatment were measured by radioimmunoassay.

Results

At the end of the experiment the GSIs of the $LHRH_A$-injected group (0.89±0.27%) and of the control group (0.95±0.14%) were not statistically different. However, as shown in Figure 1, the male/female tissue percentage in the gonads was significantly different between groups, 70/30% in the control group and the reverse in $LHRH_A$ group.

Levels of 11-KT, E_2 and 17,20β-P were generally very low (bellow 1ng/ml) and did not change significantly during the experiment. Generally higher but variable levels were found for 17,20α-P and T but neither showed statistically significant diferences between groups, except for T on day 50, which was higher in the control group. Differences in steroid levels were not related to the effect of the treatment (Fig. 2).

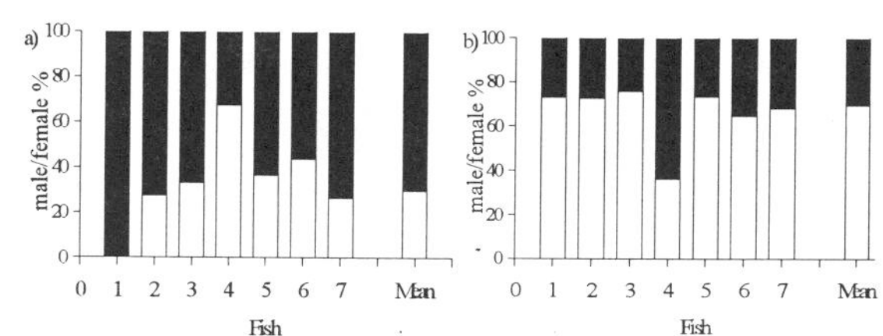

Figure 1 - Percentage of male/female gonadal tissue of $LHRH_A$ injected fish (a) and control fish (b). ■ % of female tissue; □ % de male tissue.

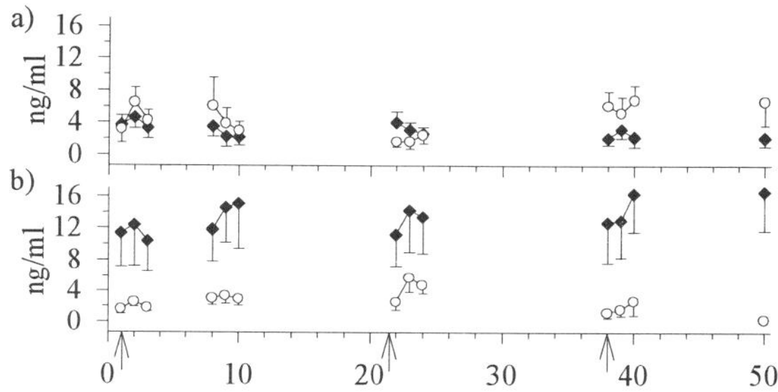

Figure 2 - Changes in plasma concentration of 17,20α-P (a) and T (b) (mean± s.e.). ○ - $LHRH_A$ group ◆ control group; ↑ - injection dates.

Discussion

The effects of the $LHRH_A$ injections in male gilthead seabream were unexpected. First, although there were important morphological changes in the gonads, there were no significant differences in steroid levels and second, $LHRH_A$ did not stimulate spermiation at the begining of the spawning season but instead it seems to have accelerated sex reversal. These results suggest that LHRH is involved in gilthead seabream sex reversal acting directly on the gonad or indirectly through the brain-pituitary-gonad axis

References

Peter, R., Trudeau, V. & Sloley, B. 1991 Brain regulation of reproduction in Teleosts. *Bull. Inst. Zool.,Academia Sinica, Monograph,* pp.89-118.

Zohar, Y., Harel, M., Hassin, S. & Tandler, A. 1995. Gilthead seabream (*Sparus aurata*). *In* "Broodstock management and egg and larval quality"(N.Bromage & R.Roberts, Ed), pp 94-117. Blackwell Science, Cambridge

This work was supported by JNICT- Program CIENCIA and NATO- Project POSEABREAM

Gametogenesis

VITELLINE ENVELOPE PROTEINS IN TELEOST FISH

S.J. Hyllner[1], and C. Haux[2].

[1]Roche Institute of Molecular Biology, Roche Research Center, Nutley, NJ 07110, U.S.A.; [2]Dept. of Zoophysiology, University of Göteborg, Medicinaregatan 18, S-413 90 Göteborg, Sweden.

Introduction

During ovarian growth, the teleost oocyte becomes surrounded by an acellular envelope, often referred to as the vitelline envelope (VE; Dumont & Brummet, 1985). The envelope is deposited between the plasma membrane of the oocyte and the surrounding follicle cells. Electronmicroscopical studies show that the VE is composed of different layers, clearly distinguishable in terms of electron density and structural appearance (Laale, 1980). Many teleosts have a VE that is subdivided into an outer and an inner layer. The inner layer is often relatively thick and tends to be fibrillar or filamentous in its structure, whilst the outer layer is thin and granular (Dumont & Brummet, 1985; Guraya, 1986). In some species, the VE is adorned with certain specializations, like adhesive fibrils or jelly coats (Guraya, 1986).

The VE performs important functions during fertilization and embryonic development. Teleost eggs have the location of sperm entry restricted to the micropyle, a channel through the VE, which is formed together with the VE during oocyte development (Guraya, 1986; Hart, 1990). The narrow inner diameter of the micropyle normally prevents polyspermy (Hart, 1990). Fertilization triggers the cortical reaction and leads to the hardening process of the VE (Yamagami et al., 1992). The hardened VE protects the developing embryo against mechanical damage, desiccation, rapid chemical changes in the environment and the envelope also exerts bactericidal and fungicidal activity (Dumont & Brummet, 1985; Kudo, 1992).

Biochemical characterization

Detailed study of the composition of the VE was precluded until recently, because solubilization procedures were not effective. The first method to be presented utilized a solution of NaOH, 0.05 or 1 M, which solubilized the inner layer of the VE of medaka (*Oryzias latipes*) and the whole envelope of rainbow trout (*Oncorhynchus mykiss*), respectively (Hamazaki et al., 1987a; Iuchi et al., 1991). A method that solubilized more than 80 % of the total dry weight of the VE in cod (*Gadus morhua)* and rainbow trout (Oppen-Berntsen et al., 1990; Hyllner & Haux, unpublished results) involved incubation of the envelope in 100 mM Tris-HCl, pH 8.8, 8 M urea, 2% SDS, 300 mM EGTA for 30 min at 70 °C. Some of the remaining, not solubilized VE, was shown to be peptides of low molecular mass (Oppen-Berntsen et al., 1990). Other similar solutions have been used to solubilize the VE from different teleosts with comparable results (Cotelli et al., 1988; Begovac & Wallace, 1989; Scapigliati et al., 1994). One important aspect to consider for the best results is that only preparations from eggs that have not been fertilized can be used (Oppen-Berntsen et al., 1990). Thus, hardening effectively prevents solubilization.

Initial work on the medaka, goldfish (*Carrassius auratus*), pipefish (*Syngnathus scovelli*) and cod (Hamazaki et al., 1987a; Cotelli et al., 1988; Begovac & Wallace, 1989; Oppen-Berntsen et al., 1990) demonstrated that the VE was composed of two to four major proteins. In 1991, three groups independently reported the protein composition of the VE in rainbow trout (Hyllner et al., 1991; Brivio et al., 1991; Iuchi et al., 1991). The VE of rainbow trout consists of three major proteins with molecular masses of 60 kDa, 55 kDa and 50 kDa. (Hyllner et al., 1991). Interestingly, Brivio et al. (1991) and Iuchi et al. (1991) include a fourth protein, about 110 kDa, as a major constituent, but the relative amount of this fourth protein varies from sample to sample (Iuchi et al., 1991). Similarly, the VE of medaka is composed of three (Hamazaki et al., 1987a; Masuda et al., 1992) or four major proteins (Masuda et al., 1991). Beside these major VE proteins, an interesting minor component of the rainbow trout VE has been isolated and characterized (Tezuka et al., 1994). The component is a glycoprotein containing about 80 % carbohydrate, with a molecular weight of around 2900 kDa (Tezuka et al., 1994).

The VE has by now been isolated and solubilized from eleven species. The VE is composed of 2 to 4 major proteins (Hamazaki et al., 1987a; Cotelli et al., 1988; Begovac and Wallace 1989; Oppen-Berntsen et al., 1990, 1994; Hyllner et al., 1991, 1994a, 1995; Scapigliati et al., 1994). Thus, the recently proposed "three major protein" model of the VE of teleosts does not seem to be general among teleost fish (Oppen-Berntsen et al. 1992b, 1994).

In rainbow trout, the 50 kDa protein is glycosylated and belongs to the "asparagine-linked" glycoprotein family (Brivio et al., 1991). Similar results were also obtained by Iuchi et al. (1991) on the same species and these findings indicate that the 60 and 55 kDa proteins are not glycoproteins. However, all major VE proteins in medaka and pipefish and two VE proteins in European sea bass (*Dicentrarchus labrax*) contain carbohydrates (Hamazaki et al., 1987a; Begovac & Wallace, 1989; Scapigliati et al., 1994).

The amino acid composition of the whole VE has been reported for 8 species (Young & Smith, 1956; Kaighn, 1964; Iuchi & Yamagami, 1976; Hagenmaier et al., 1976 and Ohzu & Kusa, 1981; Kobayashi, 1982; Begovac & Wallace, 1989; Hyllner et al., 1995) and the composition is characterized by a high content of proline and glutamine/glutamic acid and by a low content of cysteine, except in pipefish where the cysteine content is slightly higher. The amino acid composition of the teleost VE is different from collagen, elastin and keratin, and is suggested to be composed of a new type of structural proteins (Hagenmaier, 1985). Analysis of the amino acid composition of each of the VE proteins of rainbow trout, European sea bass and gilthead sea bream

(*Sparus aurata*), reveals that each protein shares characteristics with the total VE (Hyllner et al., 1991, 1995). In all three species, the amino acid composition indicates that each protein is related but certain differences can be distinguished, suggesting that each protein is distinct. Similar observations were reported for the major protein component of the VE of medaka eggs (Hamazaki et al., 1987a).

Recently, blocks of identity were found between a major region of the gene of a possible VE protein of the winter flounder (*Pseudopleuronectes americanus*) and two mammalian zona pellucida protein genes, rc55 of rabbit and ZP2 of mouse (Lyons et al., 1993). The identity between this region of the deduced teleost amino acid sequence and the deduced amino acid sequences of the mammalian zona pellucida genes was 37 % and 28 %, respectively, and a common ancestry of all three proteins was suggested. Further, a recent study demonstrated that the deduced amino acid sequence of the low-molecular-weight VE protein of medaka contains a region similar to a region of a zona pellucida protein, ZP3, of three mammalian species (Murata et al., 1995). The identity between this region of the deduced medaka amino acid sequence and the deduced amino acid sequences of ZP3 of mouse, hamster and human was 38, 39 and 42 %, respectively.

Lyons et al. (1993) found that another part of the deduced amino acid sequence of the possible teleost VE protein consisted of a repeat region that was composed primarily of proline, glutamine and/or glutamic acid. A similar region is also reported in this volume for one of the VE protein clones isolated from *Fundulus heteroclitus* expression library (LaFleur et al.). This repeat region was not found in the deduced amino acid sequence of the low-molecular-weight VE protein of medaka (Murata et al., 1995). However, the hardened VE of medaka releases proline-and glutamine/glutamic acid-rich polypeptides when incubated with a homologous hatching enzyme (Lee et al., 1994) indicating that the repeat region may exist in the high-molecular-weight VE proteins of medaka. Moreover, a very high content of these amino acids characterizes the higher molecular weight proteins of the VE of rainbow trout (Hyllner et al., 1991), European sea bass and gilthead sea bream (Hyllner et al., 1995), indicating that some or several of the proteins that constitute the VE in teleosts have a region that consists primarily of proline and glutamine/glutamic acid.

In conclusion, the general pattern is that the teleost VE is composed of a few major proteins that are related but distinct. The suggestion that these proteins represent a separate class of structural proteins is sustained. It is further suggested that teleost VE proteins share some homology with mammalian zona pellucida proteins.

Estradiol-17ß induction

About ten years ago, a "spawning female specific substance", which resembled one major protein of the inner layer of the VE of medaka eggs, was suggested to be formed in the liver under the action of estradiol-17ß (E_2; Hamazaki et al. 1984, 1985, 1987a, 1987b, 1989). The two remaining major constituents of the VE were believed to be synthesized in the oocytes and no endocrine regulation for these was proposed (Hamazaki et al., 1989). However, direct observation of the VE proteins was not possible due to the complete loss of immunoreactivity on SDS-PAGE (Hamazaki et al. 1985). These studies were the first to propose that one of the VE proteins is induced by E_2.

More recently, E_2 treatment of rainbow trout was found to induce the synthesis of all three major proteins that constitute the VE (Hyllner et al., 1991). The VE proteins were immunologically detected, using Western blot technique and homologous antiserum, in the plasma of hormone treated rainbow trout of both sexes but not in the plasma of control fish (Hyllner et al., 1991). These results were the first to demonstrate that E_2 induces all the major VE proteins in teleost fish. Furthermore, the induction of the major VE proteins by E_2 is dose-dependent (Hyllner, 1994). Further findings demonstrated that all three major VE protein-like substances are induced by E_2 also in medaka (Murata et al., 1991). E_2 has been shown to induce VE proteins in brown trout, turbot (*Scophthalmus maximus*), cod, Atlantic halibut (*Hippoglossus hippoglossus*), Atlantic salmon (*Salmo salar*), European sea bass and gilthead sea bream (Hyllner et al., 1991, 1994a, 1995; Oppen-Berntsen et al., 1992a, 1994).

In fourteen teleost species treated with E_2, three different heterologous antisera were used to detect VE proteins in plasma (Larsson et al., 1994). An E_2 dependency of the synthesis of VE proteins was demonstrated in ten of the species. The ten species belong to three taxa that represent the majority of teleost species (62%; Nelson, 1984).

The demonstrated induction of the VE proteins in 17 species supports the hypothesis that the physiological regulation of the synthesis of the VE proteins in teleost fish is controlled by E_2.

Site of synthesis of vitelline envelope proteins

Based mainly on morphological and histochemical data, the prevailing notion in the literature has been that the teleost VE originates from within the ovary (Chaudry, 1956; Anderson, 1967; Wourms, 1976; Wourms & Sheldon, 1976; Tesoriero, 1978; Stehr & Hawkes, 1979; Begovac & Wallace, 1989). However, as males of a dozen species all have the potential to synthesize VE proteins, it is evident that the site of synthesis is not confined to the ovary (Hyllner et al., 1991, 1994a, 1995; Murata et al., 1991; Oppen-Berntsen et al., 1992a, 1994; Larsson et al., 1994).

Several studies on medaka suggest that a VE protein-like substance is formed in the liver (Hamazaki et al. 1984, 1985, 1987a, 1987b, 1989). More recently, it was shown that the liver from males treated with E_2 and spawning females was immunoreactive with specific antisera that cross-react with all three major VE proteins, while the liver from untreated males was not (Murata et al., 1991). Similarly, liver tissue from sexually maturing female rainbow trout cross-reacted strongly with a homologous antiserum directed against VE proteins whereas the liver from sexually immature females showed no immunoreactivity (Hyllner & Haux, 1991). A time-course study on juvenile rainbow trout showed that the major VE proteins were first detected in the liver after treatment with E_2 (Hyllner & Haux, 1991). Comparable results have been reported for the medaka (Murata et al., 1994). Noticeably, in the three species investigated, an additional immunoreactive protein has been detected in liver tissue but not in other tissues (Hyllner & Haux, 1991; Murata et al., 1991, 1994; Oppen-Berntsen et al., 1992a). It has been suggested that this protein could be a precursor to a VE protein (Murata et al., 1991, 1994; Oppen-Berntsen et al., 1992a). Further evidence for a hepatic origin was obtained in rainbow trout that were treated with E_2 *in vivo* (Oppen-Berntsen et al., 1992b). 18 days after the initial treatment, hepatocytes were isolated and shown to synthesize the major VE proteins.

Apparently, these recent findings contradict earlier studies where the observations suggested that the constituents of the VE are synthesized within the ovary. However, it must be kept in mind that before the advent of specific antisera early studies were predominantly based on morphological and histochemical data and direct observations of the VE constituents were not possible. Finally, there still may exist alternatives for the site of synthesis of the VE proteins in certain teleost species. Thus, Begovac & Wallace (1989) reported that the two major VE proteins in pipefish are synthesized within the intact follicle. These results appear more difficult to assimilate with the proposed hepatic origin of VE proteins, but there could exist species-related differences due to the pipefish highly specialized strategy for taking care of the fertilized eggs.

Together, these recent findings provide convincing evidence that the synthesis of the teleost VE proteins is not confined to the ovary in the majority of teleost species and the results strongly indicate that the liver is the tissue that synthesizes the major VE proteins.

Regulation of vitelline envelope proteins during oocyte development

Ultrastructural observations in several teleost species suggest that the formation of the VE begins with the outer layer and proceeds with the inner layer, as the oocyte grows from the primary oocyte stage of development (Busson-Mabillot, 1973; Wourms, 1976; Flegler, 1977; Matsuyama et al., 1991). The VE then continues growth during oocyte development until the final maturation of the oocyte (Guraya, 1986).

Specific antisera have made it possible to use an immunohistochemical approach to study the timing of events during the formation and growth of the VE and the appearance and incorporation of vitellogenin during oocyte development. These events can be related to the relative concentration of VE proteins, E_2 and vitellogenin in plasma and the diameter of oocytes. Combined data from two studies make it possible to follow the oocyte development in rainbow trout (Hyllner & Haux, 1992; Hyllner et al., 1994b). Immunohistochemistry revealed that the formation of the immuno-reactive part of the VE started when the oocytes reached a diameter of about 450 µm. Oocytes of this size were first observed in females sampled about a year before ovulation, which coincided with an increase in plasma E_2 from 0.16 ± 0.04 (SE) to 0.62 ±0.07 ng/ml. A second increase in plasma E_2, from 0.56 ± 0.19 to 0.99 ± 0.40 ng/ml, occurred 9 months before ovulation and appeared to be connected with the startof the active uptake of vitellogenin, i.e. the receptor mediated endocytosis of vitellogenin. This uptake was first observed in oocytes with a size of 600 µm, which were found in females 8-9 months before ovulation. These oocytes had an immunoreactive VE with a thickness of about 3 µm. It was apparent that the initial formation of the VE started before the active uptake of vitellogenin and that the early sequential increases of E_2 in plasma seem to be of physiological importance.

When the oocytes had a diameter of about 1.0 mm, the immunoreactive VE had a thickness of about 5 µm and an increase in the active uptake of vitellogenin was observed. During the rapid growth phase of the oocytes, when the gonadal somatic index (GSI) increased from 1, half a year before ovulation, to more than 13 just prior to ovulation, the amount of VE proteins and vitellogenin further increased in plasma. Just prior to ovulation, the oocytes had reached a diameter of about 5.0 mm and the VE had a thickness of 30 µm. These changes of VE proteins in female plasma coincide with, and can be explained by, the changes of plasma E_2, because E_2 is an inducer of the major VE proteins, as discussed previously. This would partly parallel the regulation of vitellogenin, which has previously been investigated in rainbow trout (Scott & Sumpter, 1983; Copeland et al., 1986). It thus seems that the amount of VE proteins and vitellogenin in plasma vary in a manner that depends on the changes of plasma E_2 during female sexual maturation.

These findings are further strengthened by recent studies on Atlantic halibut and Atlantic salmon (Hyllner et al., 1994b; Oppen-Berntsen et al., 1994). Female fish were followed during sexual maturation and the relative amount of VE proteins in plasma was measured. The increase of VE proteins in plasma coincided with elevated levels of plasma E_2 in both species. In Atlantic salmon, the plasma VE proteins were positively correlated with GTH I but not with GTH II. This relation seems reasonable because GTH I is reported to be a more potent inducer of E_2 than GTH II in other salmonids (Sumpter et al., 1991).

These results together provide strong evidence that the physiological regulation of the synthesis of the VE proteins in teleosts is controlled by E_2. Altogether, it can be concluded that our knowledge about the endocrine control of oocyte development in teleosts is extended and that the low plasma levels of E_2 observed in females before the active uptake of vitellogenin are of physiological significance.

Concluding summary

All vertebrate eggs are surrounded by an extracellular envelope and the knowledge about the biochemistry, development and functions of the envelope has become wider in recent years. There are several major similarities between the inner part of the egg envelopes of different vertebrate groups, including ultrastructure, biochemical composition and developmental characteristics (Hyllner, 1994). These similarities suggest that the inner egg envelope of vertebrate eggs are homologous structures. As of today, an endocrine regulation of the envelope proteins has only been described in teleost fish, where the synthesis of VE proteins is regulated by E_2 during oocyte development, and the proteins have a proposed hepatic origin.

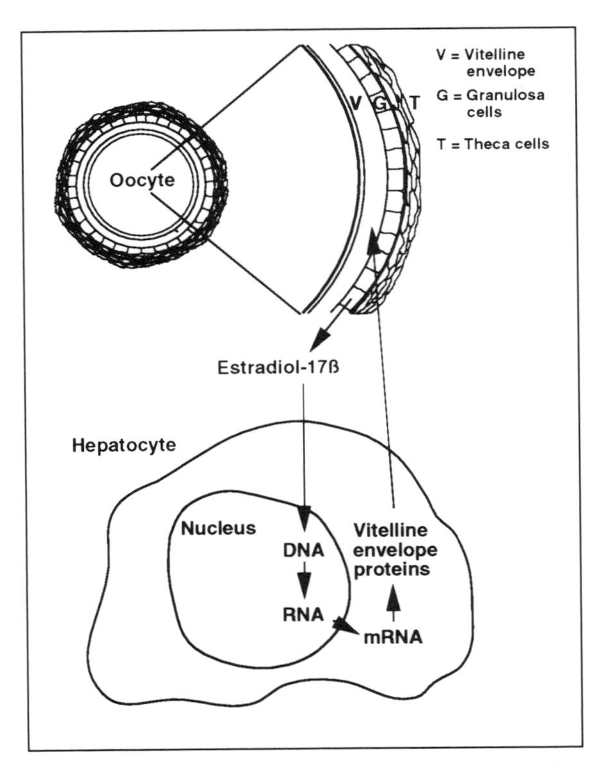

Figure 1. Overview of the synthesis of vitelline envelope proteins in teleost fish

The function of the teleost VE remains more uncertain, particularly the role during fertilization. In most vertebrates, the envelope provides sperm binding components that are generally species-specific, and the sperm have to penetrate the envelope prior to fertilization. In most teleost fish however, sperm reach the plasma membrane of the egg through the micropyle and a sperm receptor has not been identified. The minor glycoprotein of the rainbow trout VE is a possible candidate because it can specifically bind to the surface of sperm (Tezuka et al., 1994). If, however, teleosts do not have a sperm receptor, it seems reasonable to assume that some protection for species-specificity is needed. Indeed, a sperm guidance factor is associated to the VE at the micropylar area, and facilitates the entry of homologous sperm into the micropyle of salmonid and herring eggs (Yanagimachi et al., 1992). After fertilization, the chemistry and morphology of the VE change during a hardening process called the cortical reaction of the egg (Oppen-Berntsen et al., 1990; Yamagami et al., 1992). The cortical reaction leads to a change of the structural design of the micropyle which prevents sperm from further access to the plasma membrane of the egg (Hart, 1990). These fundamental events are all, in one way or another, associated with the VE and despite the recent progress, we are only beginning to understand the functions of the VE.

References

Anderson, E., 1967. The formation of the primary envelope during oocyte differentiation in teleosts. J. Cell Biol. 35:193-212.

Begovac, P.C. & R.A. Wallace, 1989. Major vitelline envelope proteins in pipefish oocytes originate within the follicle and are associated with the Z3 layer. J. Exp. Zool. 251:56-73.

Brivio M.F., R. Bassi & F. Cotelli, 1991. Identification and characterization of the major components of the *Oncorhynchus mykiss* egg chorion. Mol. Reprod. Dev. 28:85-93.

Busson-Mabillot, S., 1973. Evolution des envelopes de l'ovocyte et de loeuf chez un poisson teleosteen. J. Microscopie 18:23-44.

Chaudry, H.S., 1956. The origin and structure of the zona pellucida in the ovarian eggs of teleost. Z. Zellforschung 43:478-485.

Copeland P.A., J.P. Sumpter, T.K. Walker & M. Croft, 1986. Vitellogenin levels in male and female rainbow trout (*Salmogairdneri* Richardson) at various stages of the reproductive cycle. Comp. Biochem. Physiol. 83:487-493.

Cotelli, F., F. Andronico, M. Brivio & C.L. Lamia, 1988. Structure and composition of the fish egg chorion (*Carassius auratus*). J. Ultrastruct. Res. 99:70-78.

Dumont, J.N. & A.R. Brummet, 1985. Egg envelopes in vertebrates. In: Developmental Biology: A Comprehensive synthesis Vol 1. L.W. Browder (Ed.): Plenum Press, New York, p. 235-288.

Flegler, C., 1977. Electron microscopic studies on the development of the chorion of the viviparous teleost *Dermogenys pusillus* (Hemirhamphidae). Cell Tissue Res. 179:255-270.

Flügel, H., 1967. Licht- und Elektronmikroskopische untersuchungen an Oozyten und Eiern einiger Knochenfische. Z. Zellforsch. 83:82-116.

Guraya, S.S., 1986. The cell and molecular biology of fish oogenesis. Monographs in developmental biology, Vol 18. Karger, Basel.

Hagenmaier, H.E., I. Schmitz & J. Föhles, 1976. Zum Vorkommen von Isopeptidbindungen in der Eihülle der Regenbogenforelle (*Salmo gairdneri* Rich.). Z. Physiol. Chem. 357:1435-1438.

Hagenmaier, H.E., 1985. The hatching process in fish embryos VIII. The chemical composition of the trout chorion (zona radiata) and its modification by the action of the hatching enzyme. Zool. Jb. Physiol. 89:509-520.

Hamazaki, T., I. Iuchi & K. Yamagami, 1984. Chorion glycoprotein-like immunoreactivity in some tissues of adult female medaka. Zool. Sci. 1:148-150.

Hamazaki, T., I. Iuchi & K. Yamagami, 1985. A spawning female-specific substance reactive to anti-chorion (egg envelope) glycoprotein antibody in the teleost, *Oryzias latipes*. J. Exp. Zool. 235:269-279.

Hamazaki, T.S., I. Iuchi & K. Yamagami, 1987a. Isolation and partial characterization of a "spawning female-specific substance" in the teleost, *Oryzias latipes*. J. Exp. Zool. 242:343-349.

Hamazaki, T.S., I. Iuchi & K. Yamagami, 1987b. Production of a "spawning female-specific substance" in hepatic cells and its accumulation in the ascites of the estrogen-treated adult fish, *Oryzias latipes*. J. Exp. Zool. 242:325-332.

Hamazaki, T.S., Y. Nagahama, I. Iuchi & K. Yamagami, 1989. A glycoprotein from the liver constitutes the inner layer of the egg envelope (zona pellucida interna) of the fish, *Oryzias latipes*. Dev. Biol. 133:101-110.

Hart, N.F., 1990. Fertilization in teleost fishes: mechanism of sperm-egg interactions. Int. Rev. Cytol. 121:1-66.

Hosokawa, K., 1985. Electron microscopic observation of chorion formation in the teleost, *Navodon modestus*. Zool. Sci. 2:513-522.

Hyllner, S.J. & C. Haux, 1991. Vitelline envelope proteins in rainbow trout (*Oncorhynchus mykiss*). In: Proc. 4th Int. Symp. Reproductive Physiology of Fish. A.P. Scott, J.P. Sumpter, D.E. Kime & M.S. Rolfe (Eds.): Fish Symp 91, Sheffield, p. 323.

Hyllner, S.J., D.O. Oppen-Berntsen, J.V. Helvik, B.T. Walther & C. Haux, 1991. Oestradiol-17ß induces the major vitelline envelope proteins in both sexes in teleosts. J. Endocrinol. 131:229-336.

Hyllner, S.J. & C. Haux, 1992. Immunochemical detection of the major vitelline envelope proteins in the plasma and oocytes of the maturing female rainbow trout, *Oncorhynchus mykiss*. J. Endocrinol. 135:303-309.

Hyllner, S.J., 1994. Vitelline envelope proteins in teleost fish. Ph.D. Thesis, Department of Zoophysiology, University of Göteborg, Medicinaregatan 18, S-413 90 Göteborg, Sweden, ISBN 91-628-1285-8.

Hyllner, S.J., B. Norberg & C. Haux, 1994a. Isolation, partialcharacterization, induction, and the occurrence in plasma of the major vitelline envelope proteins in the Atlantic halibut (*Hippoglossus hippoglossus*), during sexual maturation. Can. J. Fish. Aquat. Sci. 51:1700-1707.

Hyllner, S.J., C. Silversand & C. Haux, 1994b. Formation of the vitelline envelope precedes the active uptake of vitellogenin during oocyte development in the rainbow trout, *Oncorhynchus mykiss*. Mol. Reprod. Dev. 39:166-175.

Hyllner, S.J., H. Fernández-Palacios Barber, D.G.J. Larsson & C. Haux, 1995. Amino acid composition and endocrine control of vitelline envelope proteins in European sea bass (*Dicentrarchus labrax*) and gilthead sea bream (*Sparus aurata*). Mol. Reprod. Dev. 41:339-347.

Iuchi, I. & K. Yamagami, 1976. Major glycoproteins solubilized from the teleostean egg membrane by the action of the hatching enzyme. Biochim. Biophys. Acta 453:240-249.

Iuchi, I., K. Masuda & K. Yamagami, 1991. Change in component proteins of the egg envelope (chorion) of rainbow trout during hardening. Develop. Growth Differ. 33:85-92.

Kaighn, M.E., 1964. A biochemical study of the hatching process in *Fundulus heteroclitus*. Dev. Biol. 9:56-80.

Kobayashi, W., 1982. The fine structure and amino acid composition of the envelope of the chum salmon egg. *J. Fac. Sci. Hokkaido Univ. Ser. VI Zool.* **23**:1-14.

Kudo, S., 1992. Enzymatic basis for protection of fish embryos by the fertilization envelope. Experientia 48:277-281.

Laale, H.W., 1980. The perivitelline space and egg envelopes of bony fishes: A review. Copeia 2:210-226.

Larsson, D.G.J., S.J. Hyllner & C. Haux, 1994. Induction of vitelline envelope proteins by estradiol-17ß in 10 teleost species. Gen. Comp. Endocrinol. 96:445-450.

Lee, K.-S., S. Yasumasu, K. Nomura & I. Iuchi, 1994. HCE, a constituent of the hatching enzymes of *Oryzias latipes* embryos, releases unique proline-rich polypeptides from its natural substrate, the hardened chorion. FEBS 339:281-284.

Lyons, C.E., K.L. Payette, J.L. Price & R.C.C. Huang, 1993. Expression and structural analysis of a teleost homolog of a mammalian zona pellucida gene. J. Biol. Chem. 268:21351-21358.

Masuda, K., I. Iuchi & K. Yamagami, 1991. Analysis of hardening of the egg envelope (chorion) of the fish, *Oryzias latipes*. Develop. Growth Differ. 33:75-83.

Masuda, K., K. Murata, I. Iuchi & K. Yamagami, 1992. Some properties of the hardening process in chorions isolated from unfertilized eggs of medaka, *Oryzias latipes*. Develop. Growth Differ. 34:545-551.

Matsuyama, M., Y. Nagahama & S. Matsuura, 1991. Observations on ovarian ultrastructure in the marine teleost, *Pagrus major*, during vitellogenesis and oocyte maturation. Aquaculture 92:67-82.

Murata, K., T.S. Hamazaki, I. Iuchi & K. Yamagami, 1991. Spawning female-specific egg envelope glycoprotein-like substances in *Oryzias latipes*. Develop. Growth Differ. 33:553-562.

Murata, K., I. Iuchi & K. Yamagami, 1994. Synchronous production of the low- and high-molecular-weight precursors of the egg envelope subunits, in response to estrogen administration in the teleost fish *Oryzias latipes*. Gen. Comp. Endocrinol. 95:232-239.

Murata, K., T. Sasaki, Y. Yasumasu, I. Iuchi, J. Enami, I. Yasumasu & K. Yamagami, 1995. Cloning of cDNA for the precursor protein of a low-molecular-weight subunit of the inner layer of the egg envelope (chorion) of the fish *Oryzias latipes*. Dev. Biol. 167:9-17.

Nelson, J.S., 1984. Fishes of the world, 2nd ed.: John Wiley & Sons, Inc., New York.

Ohzu, E. & M. Kusa, 1981. Amino acid composition of the eggchorion of rainbow trout. Annot. Zool. Japon. 54:241-244.

Oppen-Berntsen, D.O., J.V. Helvik & B.T. Walther, 1990. The major structural proteins of cod (*Gadus morhua*) eggshells and protein crosslinking during teleost egg hardening. Dev. Biol. 137:258-265.

Oppen-Berntsen, D.O., S.J. Hyllner, C. Haux, J.V. Helvik & B.T. Walther, 1992a. Eggshell zona radiata-proteins from cod (*Gadus morhua*): extra-ovarian origin and induction by estradiol-17ß. Int. J. Dev. Biol. 36:247-254.

Oppen-Berntsen, D.O., E. Gram-Jensen & B.T. Walther, 1992b. Zona radiata proteins are synthesized by rainbow trout (*Oncorhynchus mykiss*) hepatocytes in response to oestradiol-17ß. J. Endocrinol. 135:293-302.

Oppen-Berntsen, D.O., S.O. Olsen, C.J. Rong, G.L. Taranger, P. Swanson & B.T. Walther, 1994. Plasma levels of eggshell zr-proteins, estradiol-17ß, and gonadotropins during an annual reproductive cycle of Atlantic salmon (*Salmo salar*). J. Exp. Zool. 268:59-70.

Scapigliati, G., M. Carcupino, A.R. Taddel & M. Mazzini, 1994. Characterization of the main egg envelope proteins of the sea bass *Dicentrarchus labrax* L. (Teleostea, Serranidae). Mol. Reprod. Dev. 38:48-53.

Scott, A.P. & J.P. Sumpter, 1983. A comparison of the female reproductive cycles of autumn-spawning and winter-spawning strains of rainbow trout (*Salmo gairdneri* Richardson). Gen. Comp. Endocrinol. 52:79-85.

Stehr, C.M. & J.W. Hawkes, 1979. The comparative ultrastructure of the egg membrane and associated pore structures in the starry flounder, *Platichtys stellatus* (Pallas), and pink salmon, *Oncorhynchus gorbuscha* (Walbaum). Cell Tissue Res. 202:347-356.

Sumpter, J.P., C.R. Tyler & H. Kawauchi, 1991. Actions of Gth I and Gth II on ovarian steroidogenesis in the rainbow trout, *Oncorhynchus mykiss*. In: Proc. 4th Int. Symp. Reproductive Physiology of Fish. A.P. Scott, J.P. Sumpter, D.E. Kime & M.S. Rolfe (Eds.): Fish Symp 91, Sheffield, p. 27.

Tesoriero, J.V., 1978. Formation of the chorion (zona pellucida) in the teleost, *Oryzias latipes*. III. Autoradiography of [3H]proline incorporation. J. Ultrastruct. Res. 64:315-326.

Tezuka, T., T. Taguchi, A. Kanamuri, Y. Muto, K. Kitajima, Y. Inoue & S. Inoue, 1994. Identification and structural determination of the KDN-containing N-linked glycan chains consisting of bi- and triantennary complex-type units of KDN-glycoprotein previously isolated from rainbow trout vitelline envelopes. Biochemistry 33:6495-6502.

Wourms, J.P., 1976. Annual fish oogenesis. I. Differentiation of the mature oocyte and formation of the primary envelope. Dev. Biol. 50:338-354.

Wourms, J.P. & H. Sheldon, 1976. Annual fish oogenesis. II. Formation of the secondary envelope. Dev. Biol. 50:355-366.

Yamagami, K., T.S. Hamazaki, S. Yasumasu, K. Masuda & I. Iuchi, 1992. Molecular and cellular basis of formation, hardening, and breakdown of the egg envelope in fish. Int. Rev. Cytol. 136:51-92.

Yanagimachi, R., G.N. Cherr, M.C. Pillai & J.D. Baldwin, 1992. Factors controlling sperm entry into the micropyles of salmonid and herring eggs. Develop. Growth Differ. 34:447-461.

Young, E.G. & D.G. Smith, 1956. The amino acid in the ichthulokeratin of salmon eggs. J. Biol. Chem. 219:161-164.

LIVER-DERIVED cDNAs: VITELLOGENINS AND VITELLINE ENVELOPE PROTEIN PRECURSORS (CHORIOGENINS)

Gary J. LaFleur,[1,2] Jr., B. Marion Byrne,[1] Carl Haux,[1,3] Robert M. Greenberg,[1] and Robin A. Wallace[1,2]

[1]Whitney Laboratory, University of Florida, St. Augustine FL 32086, USA, [2]Department of Anatomy and Cell Biology, College of Medicine, University of Florida, Gainesville, FL 32610, USA, and [3]Department of Zoophysiology, University of Göteborg, Göteborg, Sweden

Summary

We have substantially increased our chances of isolating cDNAs that encode reproductively significant proteins by constructing a λgt10 library from the hepatic poly(A$^+$)-RNA of *Fundulus heteroclitus* treated with estradiol-17β. Here we describe cDNAs that encode two vitellogenins (Vtgs) and two vitelline envelope protein precursors (choriogenins). Besides providing valuable tools to study the evolution, regulation, and synthesis of ovarian constituents, these findings provide evidence for the potential investment by the liver in the production of ovarian follicles.

Introduction

Previous studies in our lab had documented the synthesis and transport of Vtg from the liver to the oocyte (Selman & Wallace, 1983; Kanungo et al., 1990). It was reported that a suite of yolk proteins appeared during vitellogenesis, and these were further processed as the oocyte increased in size (Wallace & Selman, 1985). Though it was evident that the yolk proteins were derived from Vtg, the large yolk protein precursor, a description of the definitive relationship between precursor and product was impossible until the entire primary structure of Vtg had been resolved. To achieve this goal, we set out to construct a cDNA library and screen it for Vtg. To increase the likelihood of Vtg transcripts being represented in the library, we adopted the traditional strategy of artificially inducing Vtg synthesis by the treatment of males with estrogen. A cDNA coding for Vtg I was isolated quickly thereafter (GB:U07055), and we in turn began using it as a heterologous probe in search of other Vtgs. This led to the isolation of a cDNA encoding *F. heteroclitus* Vtg II, which is similar but not identical to Vtg I. While screening for regions flanking the initial Vtg II clone by PCR, we serendipitously isolated two other cDNAs that did not appear to be Vtg-related. A BLAST search revealed that our first cDNA shared highest identity with a *Pseudopleuronectes americanus* "zona pellucida gene product" (Lyons et al., 1993), while the second cDNA was most similar to an *Oryzias latipes* "low-molecular-weight spawning female-specific substance" (L-SF) (Murata et al., 1995), both of which were described as precursors to either egg envelope or chorion proteins. Accordingly, we designate these two new cDNAs as "choriogenin" cDNAs, encoding hepatically synthesized precursors to vitelline envelope proteins. Thus, the *F. heteroclitus* choriogenin I cDNA is similar to a cDNA reported from *P. americanus*, while the *F. heteroclitus* choriogenin II cDNA is similar to a cDNA reported from *O. latipes*. In this paper, we present sequence alignments with selected portions of the *F. heteroclitus* Vtg and choriogenin cDNAs.

Materials and Methods

Five *F. heteroclitus* males were injected IP twice, 4 days apart, with estradiol-17β and the liver poly(A$^+$)-RNA was subsequently used for construction of a λgt10 (Promega, Madison, WI) library as previously described (LaFleur et al., 1995). The initial Vtg I cDNA was isolated by probing plaque lifts with a P^{32}-labelled degenerate 17-mer, MB-6, designed according to the N-terminal a.a. sequence obtained from a 45-kDa yolk protein (Fig. 1). The Vtg II cDNA was amplified in anchored PCR using a λgt10 vector primer and the degenerate 21-mer, ROW-19. The choriogenin I cDNA was isolated in anchored PCR with the 21-mer ROW-45, and the choriogenin II cDNA was amplified using ROW-55, both of which were designed as specific Vtg II primers (Fig. 1).

N-terminal a.a. sequencing and oligonucleotide synthesis were performed by the UF ICBR facility. Sequencing data were compiled using the program PC/GENE (Intelligenetics, Mountain View, CA). The program ALIGN Plus (S&E Software, State Line, PA) was used for pairwise alignments. Initial protein homologies were revealed by the BLAST network service available from the National Center for Biotechnology Information.

Research supported by NSF Grant No. IBN-9306123 and the Division of Sponsored Research, Univ. Florida

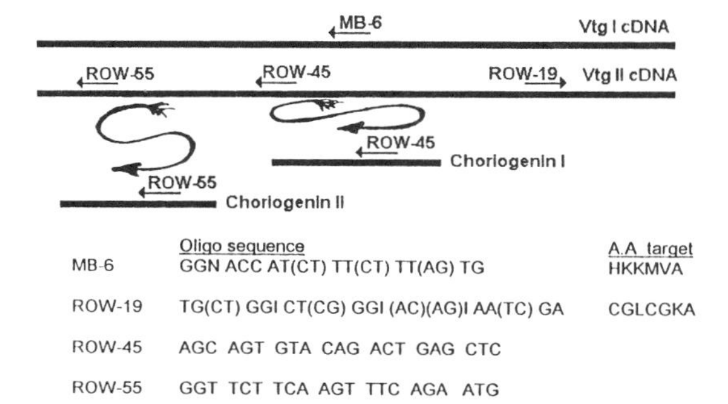

Fig. 1. Positions of oligonucleotide primers along cDNAs isolated from a *F. hetereroclitus* liver library.

Results

I. Vitellogenins:

The cDNA sequence of Vtg I appears in LaFleur et al. (1995). Overlapping cDNAs encoding Vtg II have been cloned and are near completion. The conceptual translations of the N- and C-termini of two *F. heteroclitus* Vtg cDNAs are presented in Figure 2. The C-termini translations of Vtg cDNAs previously reported from *Oncorhyncus mykiss* (GB:M27651) and *Oreochromis aureus* (Ding et al., 1990) are also provided. The two *F. heteroclitus* Vtgs share 60% sequence identity with each other and are 30-40% identical to other vertebrate Vtgs. The polyserine domains of Vtgs I and II are displayed in Figure 3. The AGY and TCX codons in these domains appear to be clustered in Vtg I, but more dispersed in Vtg II.

PREDICTED N-TERMINI FROM *F. heteroclitus* VTG cDNAs

```
F.h. Vtg I  MKAVVLALTLAFVAGQNF--APEFAAGKTYVYKYEA 34
F.h. Vtg II MRVLVLALTVALVAGNQVRYAPEFAPGKTYEYKYEG

F.h. Vtg I  LILGGLPEEGLARAGLKISTKLLLSAADQNTYMLKL 70
F.h. Vtg II YILGGLPEEGLAKAGVKIQSKVLIGAAGPDSYILKL
```

PREDICTED C-TERIMINI FROM TELEOST VTG cDNAs

```
                                                 1668
F.h. Vtg I  LESVQLEKQLTIHGEDSTCFSVEPVPRCLPGCLPVKTTPVTVG
F.h  Vtg II LESVKLEKQISLEGEESKCYSVEPVVRCLPGCAPVRTTSVTVG
O.m. Vtg    LEPAK-QVIVDD--RESKCYSVEPVLRCLPGCSPVRTTPITIG
O.a. Vtg    S---ELMVPILKVSEPNATLLSPCCSAC-PACIPVRTTTVNVG

                                                 1704
F.h. Vtg I  FSCLA--------SDPQTSVYDRSVDLRYTTQAHLACSCNTKCS
F.h. Vtg II --TRVSLDSNLNRSDSLSSIYQKSVDVSETAESHLACRCTPQCA
O.m. Vtg    --HCLPFDSNLNRSEGLSSIYIKSVDLMEKAEAHVACRCSEQCM
O.a. Vtg    FYGCLPSDTT-VDRSGLSSFFEKSIDLRDTAEAHLACRCTPQCA
```

Fig. 2. Alignment of putative Vtg N- and C-termini as translated from cDNA sequences. F.h. = *F. heteroclitus*; O.m. = *O. mykyss*; O.a. = *O. aureus*. Shading indicates identity; double underlining indicates conserved cysteine residues.

II. Choriogenins:

The choriogenin I cDNA encodes a 449 a.a. putative protein product. BLAST analysis indicates that choriogenin I is most closely related to the *P. americanus* cDNA, described as a "zona pellucida gene product" (Fig. 4). Both of these cDNAs contain a striking region of PQQ repeats, a motif not apparent in mammalian zona pellucida proteins. The next highest BLAST score is obtained by alignment with the pig (*Sus scrofa*) ZP3-α sequence (GB:L11000), suggesting a common ancestry of envelope components.

The choriogenin II cDNA encodes a 427 a.a. putative protein product. BLAST analysis reveals that the choriogenin II cDNA is closely related to the *O. latipes* L-SF cDNA, which was also isolated from an estrogen induced liver library. Neither of these two cDNAs code for a substantial PQQ region, though slight traces of this motif are apparent. The next highest identity with choriogenin II is the cDNA for *S. scrofa* ZP3-β (GB:L22169), a protein that has been implicated in porcine sperm recognition (Bagavant et al.,1993).

Preliminary results from SDS-PAGE of solubilized *F. heteroclitus* vitelline envelopes suggest the presence of three major vitelline envelope proteins, with apparent sizes of 65, 52, and 48 kDa (not shown). The calculated size of choriogenin I is 61 kDa, providing the closest match for the 65-kDa band (not shown). The calculated size of choriogenin II is 47 kDa, a candidate match for the 48-kDa band. From these results we can predict the existence of a third cDNA encoding a protein of approximately 52 kDa, keeping in mind that post-translational modifications of the 48-kDa protein, as well as proteolytic processing of the 65-kDa protein may be the source of the intermediate species.

Polyserine Domain

	Number of Ser codons	
Vtg I	AGY	62
	TCX	35
Vtg II	AGY	29
	TCX	40

I = one AGY serine codon
I = one TCX serine codon
20 codons

Fig. 3. A comparison of *F. heteroclitus* Vtg I and Vtg II polyserine domains [as defined by LaFleur et al., 1995)].

Conclusions

We report isolations of two separate Vtg and two separate choriogenin cDNAs. All four of these cDNAs were obtained from an estrogen-induced liver library, providing indirect evidence that their respective genes are transcribed under estrogen induction. These results also indicate that at least two protein components of the vitelline envelope are in fact synthesized by the liver, and transported

```
Choriogenin I        MTMKLIYCCLLAVAIHGYLVGA-------- 22
P.americanus zp      MakrwsansLvAqvvliYLVwtnvevlgsr 30

Choriogenin I        ----------------QPGKPQYPSK---- 32
P. americanus zp     rrsrssergg riivqQtGhyhpagKgqryv 60

Choriogenin I        -------------------PQQPQQPQYPS 43
P. americanus zp     qqrrrlhhdfspqnpgaepPQtPQQPtYPq 90

Choriogenin I        KPQQPQQPQYPSKPQQPQQPQYPQQP---Q  70
P. americanus zp     qPQQPQQPQqPkyPQQPQQPQqPQQPkypQ 120

Choriogenin I        QPQQPQYPSKPQYPSKPQQPQQPQYPSKPQ 100
P. americanus zp     QPQQPQqPqqPkYPqqPQQPQQPQqPkyPQ 150

Choriogenin I        QPQQPQYPQKPQQPQQPQYPQKPQTPTETF 130
P. americanus zp     QPQQPknPQ-PknPQpPQ-PQKnpqPTkqq 178

Choriogenin I        HTCD-----VPAPFRIQCGAPTISNTECEA 155
P. americanus zp     vsdDrifcgVdpylRIQCGvddItaaECEA 208

Choriogenin I        INCCFDGRMCYYGKSVTLQCTKDGQFIIVV 185
P. americanus zp     lkCCFeGyqCffGKaVTvQCTKDaQFvvVV 238

Choriogenin I        ARDATLPHIDLESISLLGGGPNCGPVGTTS 215
P. americanus zp     AkDATLPnliintISLqGeGqqCtaVdsnS 268

Choriogenin I        AFAIYQFPADCCGTIMTEEPGVIIYENRMA 245
P. americanus zp     eFAIfQFPvlaCGsvvTEEPGtIIYsNRMt 298

Choriogenin I        SSYEVAVGPYGAITRDSQYELFVQCRYIGT 275
P. americanus zp     SSYEVdVGPnGvITRDSffELqfQCRYtGl 328

Choriogenin I        SIEALVIEVGLLP---PPPGVAAPGPLRVE 302
P. americanus zp     SIEtvVIE--iLPsntPPrpVAAlGPiRVq 356

Choriogenin I        LRLGNGECSVKGCTEEQVAYTSYYTDADYP 332
P. americanus zp     LRLGNGECetKGCnEveaAYTSYYTegDYP 386

Choriogenin I        VTKILRDPVYVEVRILERTDPNIVLTLGRC 362
P. americanus zp     VTKvLRDPVYVEVRlLEkrDPNlVLTLGRC 416

Choriogenin I        WATASPFPQSLPQWDLLINGCPYQDDRYRT 392
P. americanus zp     WvTnSPnPhhqPQWDLLIdGCPYaDDRYis 546

Choriogenin I        NLIPVDSSSGLLFPTHYRRFVFKMFTFVSG 422
P. americanus zp     sLvPVgpSSGvnFPTHYkRFiFKMFTFVds 476

Choriogenin I        GGGASDATKKTPSDPSWNPLHEKGY------TL 449
P. americanus zp     stlepqrrrcTftvvqlsaLvtqaapvsrhaTg 509
```

Fig. 4. *F. heteroclitus* choriogenin I shares 60% identity with the *P. americanus* zp gene product.

```
Choriogenin II       MMMKWTVFCVVALALLGSFCDAQG-YAKPG 29
O. latipes L-SF      -MMKfTavClVvLALLdgFCDAQhnYgKPs 29

Choriogenin II       KPSKPQSPPTQNQQQLQTFEKELTWKYPDD 59
O. latipes L-SF      yPptgsktPqdptQQkQlhEKELTWKYPaD 59

Choriogenin II       PQPDPKPNVPFELRYPVPAATVAVECRESI 89
O. latipes L-SF      PQPeaKPvVPFEqRYPVPAATVAVECREdl 89

Choriogenin II       AHVEVKKDMFGTGQPINPNDLTLGNCAPVG 119
O. latipes L-SF      AHVEaKKDlFGiGQfIdPaDLTLGtCpPsa 119

Choriogenin II       EDSAAQVLIYEAELHQCGSQLMMTNDALVY 149
O. latipes L-SF      EDpAAQVLIfEspLqnCGSvLtMTeDsLVY 149

Choriogenin II       TFVLNYNPTPLGSVPVVRTSQAAVIVECHY 179
O. latipes L-SF      TFtLNYNPkPLGSaPVVRTSQAvVIVECHY 179

Choriogenin II       PRKHNVSSLPPGSPLVPFSAVKMAEEFLYF 209
O. latipes L-SF      PRKHNVSSLaldplwVPFSAaKMAEEFLYF 209

Choriogenin II       TMKLMTDDWMYERPSYQYFLGDLIRIEVTV 239
O. latipes L-SF      TlKLtTDDfqfERPSYQYFLGDLIhIEaTV 239

Choriogenin II       KQYFHVPLRVYVDRCVATLSPDVTSSPNYA 269
O. latipes L-SF      KQYFHVPLRVYVDRCVATLSPDanSSPsYA 269

Choriogenin II       FIDNFGCLIDARTTGSDSKFMARTQENHLQ 299
O. latipes L-SF      FIDNyGCLlDgRiTGSDSKFvsRpaENkLd 299

Choriogenin II       FHVEAFRFQNSDSGVIYITSSLKATSTSQA 529
O. latipes L-SF      FqlEAFRFQgaDSGmIYITchLKATSaayp 529

Choriogenin II       IDSQHRACSYTGGWREASGVDGACGSCETN 559
O. latipes L-SF      lDaeHRACSYiqGWkEvSGaDpiCaSCEsg 559

Choriogenin II       VTPYTAPAVTFASPPVVVHDGGGVTLPAPG 589
O. latipes L-SF      gfevhAnA--------VVsHgtstlsggghG 582

Choriogenin II       SPKVPYNPRKVRDVTQAEILEWEGVVSLGP 419
O. latipes L-SF      tgKpsdpsRKtReaaktEvLEWEGdVtLGP 412

Choriogenin II       IPIMEKKL 427
O. latipes L-SF      IPIeErrv 420
```

Fig. 5. *F. heteroclitus* choriogenin II shares 65% identity with the L-SF protein from *O. latipes*. Shading indicates sequence identity.

to the ovary, where they are subsequently organized into the vitelline envelope.

These data verify recent reports regarding the similarities between Vtgs and choriogenins (VEPs), namely that both Vtgs and choriogenins are induced by estrogen, synthesized by the liver, and targeted to the ovary, where they are eventually used to protect and maintain the developing embryo, respectively (e.g., Hamazaki and Murata, 1992; Hyllner et al., 1994; Oppen-Berntsen et al., 1994). On the other hand, these data also reveal an important difference between Vtgs and choriogenins in evolutionary history. The Vtgs possessed by a given species appear to be closely related isoforms, suggesting somewhat recent divergence, whereas the choriogenins appear to be conserved across species lines more strictly than in relation to each other, suggesting either a more ancient divergence of the various forms, or a specialization of the choriogenins for different, as yet unknown functions.

References

Bagavant, H., E.C. Yurewicz, A.G. Sacco, G.P. Talwar, and S.K. Gupta, 1993. J. Reprod. Immunol. 25:277-283.

Ding, J.L., B. Ho, Y. Volataire, K. LeGuellec, E.H. Lim, S.P. Tay, and T.J. Lam, 1990. Biochem. Int. 20:843-852.

Hamazaki, T.S., and K. Murata, 1992. Fish Biol. J. MEDAKA 4:19-25.

Hyllner, S.J., B. Norberg, and C. Haux, 1994. Can. J. Fish. Aquat. Sci. 51:1700-1707.

Kanungo, J., T.R. Petrino, and R.A. Wallace, 1990. J. Exp. Zool. 254:313-321.

LaFleur, G.L. Jr., B.M. Byrne, J. Kanungo, L.D. Nelson, R.M. Greenberg, and R.A. Wallace, 1995. J. Mol. Evol., in press.

Lyons, C.E., K.L. Payette, J.L. Price, and R.C.C. Huang, 1993. J. Biol. Chem. 268:21351-21358.

Murata, K., T. Sasaki, S. Yasumasu, I. Iuchi, J. Enami, I. Yasumasu, and K. Yamagami. 1995. Dev. Biol. 167:9-17.

Oppen-Berntsen, D.O., S.O. Olsen, C.J. Rong, G.L. Taranger, P. Swanson, and B.T. Walthers, 1994. J. Exp. Zool. 268:59-70.

Selman, K., and R.A. Wallace, 1983. J. Exp. Zool. 226:441-457.

Wallace, R.A., and K. Selman, 1985. Dev. Biol. 110:492-498.

TOWARDS THE DEVELOPMENT OF GENETIC PROBES TO THE VITELLOGENIN RECEPTOR IN THE RAINBOW TROUT, *ONCORHYNCHUS MYKISS* : CHARACTERISATION OF OVARIAN RECEPTOR PROTEINS SPECIFIC FOR VITELLOGENIN.

C.R.Tyler, K. Lubberink, S. Brooks and K. Coward.

Fish Physiology Research Group, Department of Biology and Biochemistry, Brunel University, Uxbridge, Middlesex. UB8 3PH. U.K. Tel. 01895 274000 Fax. 01895 274348.

Abstract

Membrane proteins from ovarian follicles of rainbow trout were isolated on PVDF following SDS gel-electrophoresis. Vitellogenin [VTG] receptor proteins were visualised using protein staining and hybridisation with ^{125}I-VTG. Four follicle proteins with molecular masses of 220kDa, 210kDa, 110kDa and 100kDa showed a strong affinity for VTG and were specific to the ovary. Other homologous lipoproteins [very low density lipoprotein (VLDL), low density lipoprotein (LDL) and high density lipoprotein (HDL)] did not hybridise to the receptor proteins binding VTG. VTG receptor proteins were digested with modified trypsin and the peptides separated by reverse phase HPLC and sequenced. A number of the peptides indicated homology with receptors in the LDL family.

Introduction

Vitellogenin sequestration is central to the oocyte growth process in teleosts and in the rainbow trout may account for as much as 80% of the final egg size [Tyler, 1991]. In fish, as in other oviparous vertebrates, VTG is sequestered by receptor mediated endocytosis involving specific cell surface receptors.The expression and modulation of the VTG receptor therefore are key determinants in the oocyte growth process. Establishing what controls VTG receptor function, hormonal or otherwise, is likely to come from studies on the expression of the receptor gene and this in turn requires specific genetic probes.

In birds, where developing oocytes sequester significant amounts of lipoproteins other than VTG, it has been established that a single receptor effects the ovarian uptake of both VLDL and VTG [Stifani *et al.*,1988]. Recently this receptor has been cloned and sequenced [Bujo *et al.*, 1994]. The chicken VTG/VLDL receptor displays sequence and structural features common to receptors in the LDL family. An 8-ligand, cysteine -rich, binding repeat at the N-terminus hallmarks the chicken VTG receptor as akin to the mammalian VLDL receptor [see figure 1] .

There is little evidence of yolk proteins derived from lipoprotein precursors other than VTG in fish, and none in salmonids. Our previous work on the rainbow trout has indicated that there is more than one oocyte membrane protein capable of interacting with VTG [Tyler and Lancaster, 1993]. This study set out to establish clearly the number of oocyte membrane receptor proteins capable of binding VTG and further whether these proteins were also able to bind other homologous serum lipoproteins. Isolated receptor proteins binding vitellogenin alone were subjected to modified trypsin digests and selected peptides were sequenced with a view to their use in the genesis of specific genetic probes.

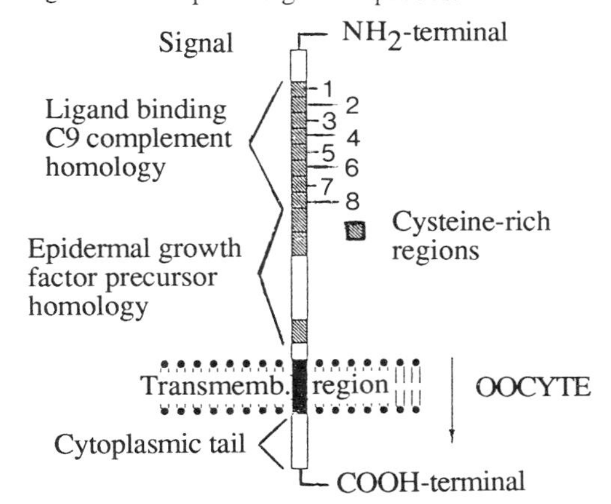

Figure 1. Structure of the chicken VTG/VLDL receptor.

Materials and Methods

Isolation of membrane receptor proteins.

Membrane receptor proteins from rainbow trout follicles were extracted using a non-ionic detergent and ultracentrifugation according to Stifani *et al.* [1988] and run on electrophoretic gels. Separated proteins were either stained with Coomassie Brilliant Blue R-250 and subjected to densitometry or trans-blotted to PVDF membranes. Isolates on PVDF membranes were subjected to either ligand blotting, or protein staining for receptor protein collection. Membrane proteins extracted from the liver, muscle, spleen and testis were run as controls.

Isolation and radiolabelling of lipoproteins

VTG was purified from the plasma of vitellogenic females according to Tyler and Sumpter [1990]. Other homologous lipoproteins [HDL, LDL and VLDL] were purified from VTG-free male plasma using density gradient ultracentrifugation [modified from Babin, 1987]. All lipoproteins were radiolabelled using the Iodogen method [Salackinski *et al.*, 1981]. Labelled proteins were separated from free ^{125}Iodine by gel filtration on

Sephadex-G25. Specific activities of the labelled protein were between 200-600 dpm/ng. Protein concentrations were determined using bicinchoninic acid. The integrity of radiolabelled lipoproteins were assessed on electrophoretic gels followed by autoradiography.

Ligand blotting

Proteins isolated on PVDF membranes were incubated overnight in 5% non-fat milk buffer to saturate non-specific binding sites, then radiolabelled lipoprotein for 8h [500 000dpm/ml], and finally, again overnight with the milk buffer, before exposure of the membranes to Kodak XAR-5 film at -70°C.

Microsequencing of VTG receptor proteins

Ovarian receptor proteins isolated on PVDF membranes binding VTG alone could not be sequenced directly [all were N-terminally blocked]. Receptor proteins were therefore subjected to enzyme digests, using modified trypsin, and the resulting peptides separated on a microbore HPLC [containing an Aquapore RP-300 C8 cartridge] using acetonitrile. Selected peptides were subjected to automated sequencing on an Applied Biosystems 477A instrument with a 120A on-line phenylthiohydantoin analyser, according to Hayes *et al.* [1989]. Resulting peptide sequences were screened in sequence data banks.

Results and Discussion

Ovarian follicle, liver, spleen, muscle and testis membrane proteins separated on SDS gels under non reducing conditions and stained with Coomassie Brilliant Blue R-250 are shown in figure 2. Densitometry of the ovarian follicle separations detected 10 membrane proteins. Ligand blotting studies with ^{125}I-VTG showed 5 of these follicle membrane proteins were capable of interacting with VTG, eluting with molecular masses of 400 kDa [the weakest of the hybridisation signals], 220kDa [VRP-1] 210kDa [VRP-2], 110 kDa [VRP-3] and 100kDa [VRP-4; fig. 2 - confirmation that both of the proteins in the doublets eluting at around 100kDa and 200kDa bound VTG, was obtained by excising the individual proteins from a stained gel and subjecting them to ligand blotting with ^{125}I-VTG]. None of the proteins in non-ovarian membrane extracts hybridised with ^{125}I-VTG, confirming that the VTG receptor binding sites were specific to the ovary (fig.2).

Size estimates of receptor proteins for VTG in other oviparous species vary between 97-200kDa (see Tyler and Lancaster, 1993).Attempts to isolate the VTG receptor in its pure form in all species studied so far has resulted in several protein bands on electrophoretic gels. VRP-1 and 2 may be dimeric forms of VRP-3 and 4 - dimer formation has been shown to occur with LDL receptors in mammals and the VTG/VLDL receptor in birds.The 400kDa receptor protein binding VTG may be an oligomeric form of the VTG receptor or alternatively an analogue of the LDL receptor -related protein isolated in both oviparous and non-oviparous animals [(Barber *et al.*, 1991); in this study however, this protein did not appear to bind homologous lipoproteins other than VTG (see below)].

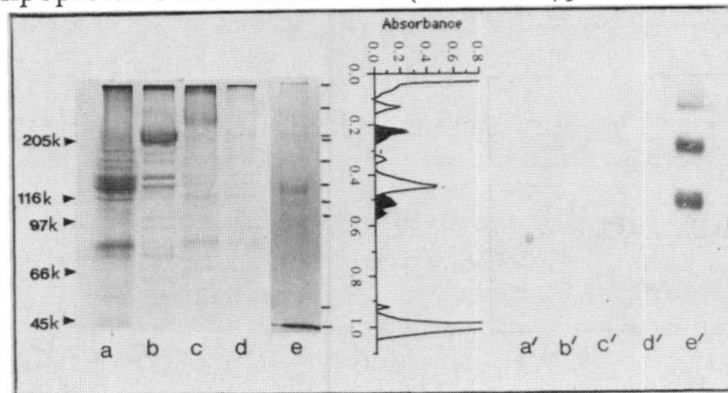

Figure 2. SDS-polyacrylamide gel [5%] electrophoresis of membrane extracts from ovarian and non-ovarian tissues. Proteins were either stained with Coomassie Brilliant Blue R-250 or transferred to PVDF for ligand blotting with ^{125}I-VTG- a/a'- liver; b/b'- muscle; c/c'- spleen; d/d'- testis [all ~50ug/lane] ; e/e'- ovary [30ug/lane]. Autoradiography was for 24h. The densitometry trace is for the ovary.

Purified VTG eluted on electrophoretic gels [under reducing conditions] as a single band with a molecular mass of 170kDa. VLDL was comprised of a series of proteins eluting with molecular masses of 260kDa, 240kDa, 76kDa and <20kDa. LDL was comprised of only 3 major proteins with molecular masses of 260kDa, 240kDa and 76kDa. HDL contained an array of proteins, all with molecular masses of less than 80kDa [fig.3]. The purified lipoproteins showed similar gel-elution profiles to that established previously in the rainbow trout by Babin [1987]. Radiolabelling with ^{125}Iodine did not affect the integrity of any of the lipoproteins [fig.3].

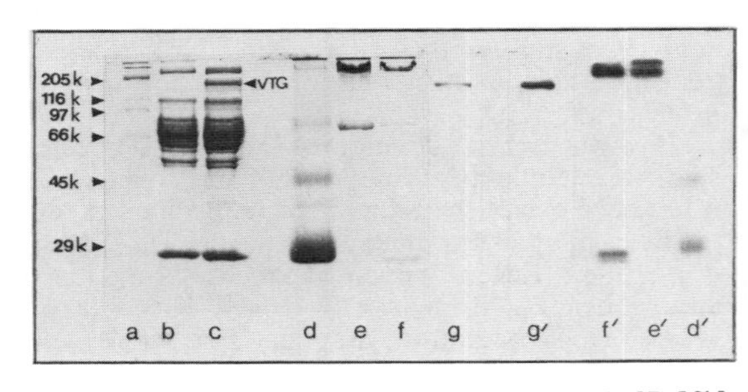

Figure 3. SDS-polyacrylamide gel [7.5%] electrophoresis of plasma and purified lipoproteins under reducing conditions. Proteins were stained with Coomassie Brilliant Blue R-250. Radiolabelled lipoproteins were transferred to PVDF and autoradiographed for 24h. a- molecular weight markers; b, male [non-vitellogenic] plasma; c, maturing female [vitellogenic] plasma; d [protein stained]/d'[radiolabelled]-HDL; e/e'-LDL; f/f'-VLDL; g/g'- VTG.

Ovarian VRPs binding ^{125}I-VTG did not bind any of the other radiolabelled lipoproteins [fig. 4]. This is at variance with the studies on the domestic fowl,where a single oocyte membrane receptor has been shown to bind both VTG and VLDL. The specificity of the VTG receptor proteins for VTG alone, and not other homologous lipoproteins, is consistent with the absence of significance amounts of VLDL in rainbow trout oocytes. It is likely that oocyte membranes in trout contain receptors for lipoproteins other than VTG, but they are at far lower concentrations than the VTG receptor. Binding of LDL to its receptor occurred in the liver when levels of protein loadents on gels were 200µg or above [data not shown].

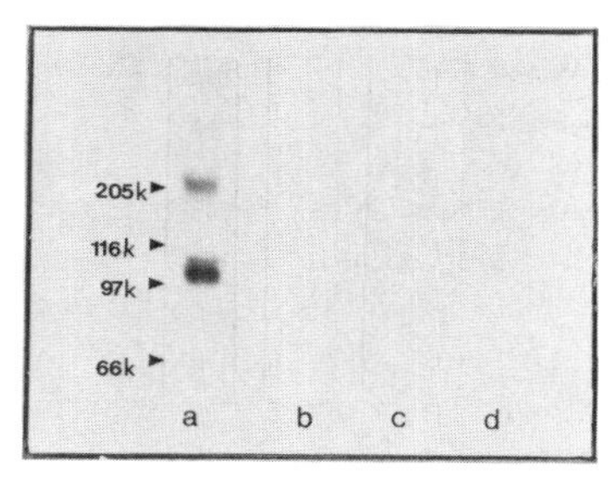

Figure 4. SDS-polyacrylamide gel [5%] electrophoresis of ovarian membrane extracts [30ug/lane] transferred to PVDF and ligand blotted with ^{125}I-lipoproteins: a- ^{125}I-VTG; b - ^{125}I-VLDL; c- ^{125}I-LDL; ^{125}I-HDL.]. Autoradiography was for 48h [longer exposure times did not yield additional hybridisation signals]

HPLC traces of peptides generated from trypsin digests of VRPs-1 and 2 were almost identical suggesting VRP-1 and 2 were similar, if not the same protein. Digests of VRP-3 and 4 also produced a similar array of peptides. Yields from digests were low and the maximum sequence length obtained for any peptide was 13 amino acids. VRP-1 and 2 had sequences similar to a variety of binding proteins and receptors [including, epidermal growth factor, mineralcorticoid and LDL receptors].Two VRP-1/2 peptides had strong similarities with human LDL scavenger receptors [one short sequence was identical: Val-Asn-Ala-Asp-Ile]. Another VRP-1 peptide had a 5/ 7 amino acid identity with a region in the cytoplasmic tail of the human LDL receptor [Asn-Gln-Phe-Gly-Tyr-Phen-Tyr]. Peptides sequenced from digests of VRP-3 and 4, showed similarities with a variety of lipoprotein receptors. A VRP-3 peptide contained a 6/7 amino acid identity with regions in human LDL scavenger receptors [- Gly-Pro-Leu-Gly-Ala-Tyr-Gly]. A VRP-4 peptide contained a sequence identical to a region in the cytoplasmic domain of mammalian and amphibian LDL receptors [and the VTG/VLDL receptor in the chicken] believed to act as a internalisation signal [-Phen-Asp-Asn-Phen-Tyr-]. A further peptide in VRP-4 contained a 6/7 sequence identity with the chicken VTG/VLDL receptor [Ser-Glu-Leu-Tyr-Glu-Pro-Ala].Together the ligand blotting and protein sequence data support the contention that the 4 ovarian membrane proteins isolated are receptor proteins for VTG. Primers developed to these peptides are presently being employed in rtPCR in an attempt to generate a specific genetic probe to the rainbow trout VTG receptor.

Acknowledgement

This work was supported by a bbsrc grant.

References

Babin, P.J. 1987. Apolipoproteins and the association of egg yolk proteins with plasma high density lipoproteins after ovulation and follicular atresia in the rainbow trout [*Salmo gairdneri*]. J.Biol. Chem. 262:4290-4296.

Barber, D.L., E.J.Sanders, R. Aebersold, W.J. Schneider. 1991. The receptor for yolk lipoprotein deposition in the chicken oocyte. J. Biol. Chem. 266:18761-18770.

Bujo, H., M. Hermann, M.O. Kaderli, L. Jacobsen, S. Sugawara, J. Nimpf, T. Yamamoto, W.J. Schneider, 1994. Chicken oocyte growth is mediated by an eight ligand binding repeat member of the LDL receptor family. The EMBO Journal. 13:5165-5175.

Hayes, J.D., L.A.Kerr and A.D. Cronshaw, 1989. Evidence that glutathione S-transferase B_1B_1 and B_2B_2 are the products of separate genes and that their expression in human liver is subject to inter-individual variation. Biochem. J. 264:437-445.

Salacinski, P.R., C. Mclean, J.E. Sykes, V.V. Clement-Jones and P.J. Lowry, 1981. Iodination of proteins, and peptides using a solid phase oxidising agent, 1,3,4,6-tetrachloro-3a,6a,-diphenyl glycoluril (Iodogen). Anal. Biochem. 117:136-146.

Stifani, S.T., R. George and W.J. Schneider, 1988. Solubilisation and characterisation of the chicken oocyte vitellogenin receptor. Biochem. J. 250:467-475.

Tyler, C.R.1991. Vitellogenesis in Salmonids. In: Reproductive Physiology of Fish. A.P. Scott, J.P.Sumpter, D. Kime, J. Rolfe (Eds.): University of East Anglia Press, UK.p.295-299.

Tyler, C.R. and P.M. Lancaster, 1993.Isolation and characterisation of the receptor for vitellogenin from follicles of the rainbow trout, *Oncorhynchus mykiss*. J. Comp. Physiol. 163:219-224.

Tyler, C.R. and J.P.Sumpter, 1990. The purification and partial characterisation of carp, *Cyprinus carpio*, vitellogenin. Fish Physiol. Biochem. 8:111-120.

CONNEXIN GENES, GAP JUNCTIONS, AND OVARIAN MATURATIONAL COMPETENCE

G. Yoshizaki,[1] W. Jin,[1] R. Patiño,[1] and P. Thomas[2]

[1]Texas Cooperative Fish and Wildlife Research Unit, Texas Tech University, Lubbock, TX 79409-2125;
[2]The University of Texas Marine Science Institute, Port Aransas, TX 78373-1267.

Introduction

Pituitary maturational gonadotropin (GtH-II in fishes, or LH in tetrapods) is the primary hormone controlling the onset of oocyte maturation in vertebrates. In fishes and amphibians, GtH-II (LH) acts on the follicle cells to stimulate the production of maturation-inducing steroid (MIS). The MIS then binds to a receptor on the surface of the oocyte to induce maturation (Redding and Patiño, 1993). Late vitellogenic and full-grown oocytes before the GtH-II surge are often unresponsive or relatively insensitive to MIS stimulation. However, oocyte maturational competence (OMC) can be induced by maturational GtH (Patiño and Thomas, 1990a; Patiño and Purkiss, 1993; Kagawa *et al.*, 1994; Zhu *et al.*, 1994). In Atlantic croaker (*Micropogonias undulatus*), we have shown that the GtH-dependent acquisition of OMC requires *de novo* ovarian RNA and protein synthesis (Patiño and Thomas, 1990a). This requirement was later confirmed by Kagawa *et al.* (1994) for red seabream (*Pagrus major*).

Our studies with spotted seatrout (*Cynoscion nebulosus*) and *Xenopus laevis* suggested that the development of OMC is at least partly due to increased MIS receptor activity (Patiño and Thomas, 1990b; Thomas and Patiño, 1991; Liu and Patiño, 1993). However, whether this increased receptor activity is the result of increased receptor synthesis is unknown.

Although an earlier study with *X. laevis* indicated that GtH can stimulate granulosa cell-oocyte gap junction (GJ) coupling (Browne *et al.*, 1979), the significance of this observation in regards to oocyte maturation was not explored. Interestingly, since GtH also induces OMC in *Xenopus* ovarian follicles (Patiño and Purkiss, 1993), it seems possible that a cause-effect relationship exists between heterocellular GJ coupling and OMC. In mammals, the function of granulosa cell-oocyte GJ during oocyte maturation is controversial (Schultz, 1991). Our current research with GJ in the fish and amphibian ovarian follicle aims to clarify their role in oocyte maturation. In this review, we discuss our findings with Atlantic croaker.

Gap junctions and connexins

Gap junctions are aggregates of intercellular channels between adjacent cells. An intercellular channel is formed by two hemichannels or connexons (one provided by each of two cells), and each connexon is a hexamer of connexin protein (Cx) subunits. Intercellular channels allow direct cell-to-cell passage of small molecules up to about 1000 daltons in mass (Beyer *et al.*, 1990).

Heterocellular GJ dynamics in the ovarian follicle during OMC acquisition

Full-grown ovarian follicles obtained from laboratory-maintained Atlantic croaker remain maturationally incompetent until stimulated with GtH, thus making this species an ideal model for studies of OMC. As part of a study on GJ dynamics during oogenesis, we prepared and analyzed ultra-thin sections of ovarian follicles from incompetent and hCG-induced, competent ovaries by transmission electron microscopy (York *et al.*, 1993). Incompetent follicles had relatively low levels of heterocellular associations and GJ contacts, whereas competent follicles showed high levels of association and GJ contacts between oocyte microvilli and the granulosa cell layer. These changes seemed to be mainly due to alterations in oocyte microvilli length and surface contact with granulosa cell rather than to alterations in the numbers of microvilli. Therefore, the induction of OMC coincided with the establishment of granulosa cell-oocyte GJ in full-grown ovarian follicles.

Molecular characterization of Cx mRNA in the teleost ovary

Prior to our study with Atlantic croaker (Yoshizaki *et al.*, 1994), there were no reports on the molecular structure of fish Cx. Thus, we chose an RT-PCR approach to obtain croaker Cx cDNA. Using highly degenerate Cx primers designed according to known tetrapod Cx sequences, we were able to amplify two major Cx cDNA fragments from the hCG-induced competent ovary of Atlantic croaker. We used these two cDNA fragments as probes to screen a competent ovary cDNA library. Respective positive clones contained the full cDNA sequence for two distinct Cx. One clone encoded 282 amino acids (32,693 daltons; Cx32.7), and the other encoded 285 amino acids (32,169 daltons; Cx32.2). Both Cx showed good homology to tetrapod Cx (about 53%

Study supported by U.S. Department of Agriculture Grant No. 91-37203-6680

amino acid identity with chicken Cx56). Also, hydropathicity analysis of their predicted amino acid sequences indicated that croaker Cx have four major hydrophobic regions, representing transmembrane domains, and four hydrophilic regions, representing two extracellular loops and two intracellular domains (N and C terminus). This predicted topology of Cx32.2 and Cx32.7 is similar to that of all known Cx (Beyer *et al.*, 1990). Furthermore, all major Cx consensus sequences were identified in both Cx32.2 and Cx32.7. Therefore, the amino acid sequences, predicted topologies, and consensus sequences of croaker Cx are in excellent agreement with the structural traits of tetrapod Cx.

Croaker Cx contained consensus sequences for casein kinase II and protein kinase C (pKC) phosphorylation sites. In general, activation of pKC pathways disrupts GJ coupling between cells (Stagg and Fletcher, 1990). Thus, phosphorylation of pKC target sites on Cx32.2 and Cx32.7 protein may serve to modulate GJ function in Atlantic croaker ovary. In *Fundulus* ovarian follicles, pKC activation downregulates heterocellular GJ coupling (Cerdá *et al.*, 1993).

Cx mRNA levels during OMC acquisition

Northern blot analysis for poly$(A)^+$ RNA extracted from maturationally incompetent and competent ovaries was carried out using Cx32.2 and Cx32.7 cDNAs as probes (Yoshizaki *et al.*, 1994). The mRNA level of Cx32.7 was similar between maturationally competent and incompetent ovaries. However, Cx32.2 mRNA levels were negligible in incompetent ovaries and increased substantially upon induction of competence by hCG. Therefore, the induction of OMC coincided with a selective increase in ovarian Cx32.2 mRNA levels.

Functional characterization of teleost Cx

We have functionally analyzed croaker Cx32.2 and Cx32.7 with conductance assays based on paired *Xenopus* oocytes (Bruzzone *et al.*, 1995). Croaker Cx32.2 and Cx32.7 were transcribed *in vitro* and their cRNA injected into *Xenopus* oocytes. Both Cx were expressed at comparable levels in the oocytes, but only Cx32.2 was competent to induce the formation of functional, intercellular channels. Cx32.7 connexons could not form homotypic channels with other Cx32.7 connexons nor heterotypic channels with Cx32.2 connexons. Also, Cx32.7 did not act as dominant negative inhibitor of Cx32.2 channel formation when the two Cx were coexpressed in both members of the oocyte pair. On the basis of its voltage dependent character, it appeared that Cx32.2 was the functional homolog of rat Cx43, which is also present in the rat ovary (Stagg and Fletcher, 1990). These findings indicated that the GtH-induced increase of Cx32.2 level in the croaker ovary can lead to the *de novo* formation of functional, intercellular channels.

Cx32.2 gene structure

Although we have not determined Cx mRNA turnover rates in croaker ovaries, it seems likely that the large, selective increase in Cx32.2 mRNA levels that accompany the GtH induction of OMC is due to Cx32.2 gene activation. Also, since most well-known actions of GtH in the ovary are mediated by changes in cAMP production in target cells, it seems reasonable to expect that the effect of GtH on Cx32.2 mRNA levels is mediated by cAMP-dependent pathways. To study this question further, we determined *cis*-acting regulatory elements associated with the Cx32.2 gene (unpublished data). We made and screened a croaker genomic DNA library using Cx32.2 cDNA probe. We cloned and sequenced the Cx32.2 gene including about 1.8 kb upstream of the transcription start site. We found that the Cx32.2 gene is composed of two exons and one intron, a structure is similar to that reported for mammalian Cx genes (Hennemann *et al.*, 1992). We identified one cAMP responsive element consensus sequence in the intron and two more in the 5' upstream region. Also, an activation protein-1 (AP-1) binding site consensus sequence was found in the 5' upstream region. Therefore, these findings are consistent with the proposal that transcription of the Cx32.2 gene may be positively regulated by cAMP-dependent pathways. However, the presence of predicted AP-1 binding sites in the 5' upstream region indicates that pKC-dependent mechanisms may also be operational.

Conclusions

Our results suggest that maturational GtH selectively activates the Cx32.2 gene in croaker ovary via cAMP-dependent regulation of transcriptional activity. We have also shown that Cx32.2 connexons can form functional, intercellular channels. Furthermore, we have shown that increased ovarian Cx32.2 mRNA levels coincide with increased heterocellular (granulosa cell-oocyte) GJ contacts in full-grown follicles and with the development of OMC. Therefore, in addition to increased oocyte MIS receptor levels, another essential element of the mechanism of GtH induction of OMC in Atlantic croker may be the *de novo* synthesis of Cx32.2 for use in the formation of heterocellular GJ. This increased granulosa cell-oocyte GJ coupling is likely required for the establishment of transduction mechanisms for maturational signals (*i.e.*, MIS transfer or action; see also Patiño and Purkiss, 1993).

References

Beyer, E.C., D.L. Paul, and D.A. Goodenough. 1990. Connexin family of gap junction proteins. J. Membr. Biol. 116:187-194.

Browne, C.L., H.S. Wiley, and J.N. Dumont. 1979. Oocyte follicle cell gap junctions in *Xenopus leavis* and the effects of gonadotropins on their permeability. Science 203:182-183.

Bruzzone, R., T.W. White, G. Yoshizaki, R. Patiño, and D.L. Paul. 1995. Intercellular channels in teleosts: functional characterization of two connexins from Atlantic croaker. FEBS Lett. 358:301-304.

Cerdá, J.L., T.R. Petrino, and R.A. Wallace. 1993. Funtional heterologous gap junctions in *Fundulus* ovarian follicles maintain meiotic arrest and permit hydration during oocyte maturation. Dev. Biol. 160:228-235.

Hennemann, H., G. Kozjek, E. Dahl, B. Nicholson, and K. Willecke. 1992. Molecular cloning of mouse connexins26 and -32: similar genomic organization but distinct promoter sequences of two gap junction genes. Eur. J. Cell Biol. 58:81-89.

Kagawa, H., H. Tanaka, K. Okuzawa, and K. Hirose. 1994. Development of maturational competence of oocytes of red seabream, *Pagrus major*, after human chorionic gonadotropin treatment *in vitro* requires RNA and protein synthesis. Gen. Comp. Endocrinol. 94:199-206.

Liu, Z. and R. Patiño, 1993. High-affinity binding of progesterone to the plasma membrane of *Xenopus* oocytes: characteristics of binding and hormonal and developmental control. Biol. Reprod. 49:980-988.

Patiño, R., and R.T. Purkiss. 1993. Inhibitory effects of *n*-alkanols on the hormonal induction of maturation in follicle-enclosed Xenopus oocytes: implications for gap junctional transport of maturation-inducing steroid. Gen. Comp. Endocrinol. 91:189-198.

Patiño, R. and P. Thomas, 1990a. Effects of gonadotropin on ovarian intrafollicular processes during the development of oocyte maturational competence in a teleost, the Atlantic croaker: evidence for two distinct stages of gonadotropic control of final oocyte maturation. Biol. Reprod. 43:818-827.

Patiño, R., and P. Thomas. 1990b. Characterization of membrane receptor activity for 17α,20β,21-trihydroxy-4-pregnen-3-one in ovaries of spotted seatrout *(Cynoscion nebulosus*). Gen. Comp. Endocrinol. 78:204-217.

Redding, J.M., and R. Patiño. 1993. Reproductive Physiology. In: The Physiology of Fishes, D. H. Evans (Ed). CRC Press, Boca Raton, pp. 503-534.

Schultz, R.M. 1991. Meiotic maturation of mammalian oocytes. In: Elements of Mammalian Fertilization, vol 1, P.M.Wassarman (Ed). CRC Press, Boca Raton, pp. 77-104.

Stagg, R.B. and W.H. Fletcher. 1990. The hormone-induced regulation of contact-dependent cell-cell communication by phosphorylation. Endocr. Rev. 11:302-325.

Thomas, P. and R. Patiño. 1991. Changes in 17α,20β,21-trihydroxy-4pregnen-3-one membrane receptor concentrations in ovaries of spotted seatrout during final oocyte maturation. In: Proceedings of the Fourth International Symposium on the Reproductive Physiology of Fish. A.P. Scott, J.P. Sumpter, D.E. Kime, and M.S. Rolfe (Eds.). University of East Anglia Press, United Kingdom, pp. 122-124.

York, W.S., R. Patiño, and P. Thomas. 1993. Ultrastructural changes in follicle cell-oocyte associations during development and maturation of the ovarian follicle in Atlantic croaker. Gen. Comp. Endocrinol. 92:402-418.

Yoshizaki, G., R. Patiño, and P. Thomas. 1994. Connexin messenger ribonucleic acids in the ovary of Atlantic croaker: molecular cloning and characterization, hormonal control, and correlation with appearance of oocyte maturational competence. Biol. Reprod. 51:493-503.

Zhu, Y., M. Kobayashi, K. Furukawa, and K. Aida. 1994. Gonadotropin develops sensitivity to maturation-inducing steroid in the oocytes of the daily spawning teleosts, Tobinumeri-dragonet *Repomucenus beniteguri* and kisu *Sillago japonica*. Fisheries Sci. 60:541-545.

EFFECTS OF INSULIN-LIKE GROWTH FACTOR-I ON FINAL MATURATION OF OOCYTES OF RED SEABREAM, *PAGRUS MAJOR, IN VITRO*

H. Kagawa and S. Moriyama

National Research Institute of Aquaculture, Nansei, Mie 516-01, Japan

Summary

IGF-I (10 nM) alone induced GVBD in both 17α, 20β-dihydroxy-4-pregnen-3-one (DHP)-insensitive (maturational incompetent) and DHP- sensitive (maturational competent) oocytes. Cyanoketone (0.1 μg/ml), an inhibitor of 3β-hydroxysteroid dehydrogenase, slightly reduced % of GVBD induced by IGF-I in DHP-insensitive oocytes but did not block GVBD in DHP-sensitive oocytes. These data indicate that IGF-I acts both directly on oocytes to induce GVBD and MIS production. DHP-insensitive oocytes underwent GVBD in response to DHP when oocytes were incubated with low doses of IGF-I, which alone were ineffective. GVBD also occurred in response to DHP after oocytes were preincubated with IGF-I. Actinomycin D blocked HCG-induced maturational competence, but not IGF-I-induced maturational competence. These results show the possibility that IGF-I can induce maturational competence of oocytes of the red seabream, a mechanism different from HCG-induced maturational competence.

Introduction

Recent studies in mammals and *Xenopus* report the involvement of growth factors, acting through autocrine and paracrine mechanisms in follicular development and oocyte maturation. In the goldfish, insulin, alone or in combination with progestogens, induces maturation of oocytes (Lessman, 1985). In teleosts, IGF-I is present in the plasma of rainbow trout (Niu et al., 1993) as are IGF-I receptors in carp ovary (Gutierrez et al., 1993), suggesting that growth factors may be involved in the control of oocyte maturation. However, physiological significance of growth factors on the oocyte maturation in teleosts has not been examined. The present study, thus, aimed to assess the effects of IGF-I on maturation of red seabream oocytes, especially on production of maturation-inducing steroid and the development of maturational competence.

Results

Effects of of various growth factors on GVBD

Sexually mature female red seabream spawn daily in the evening during the spawning season (April and May). Oocytes with follicular layers at the yolk globule stage were collected from the fish obtained at 0800 hr. Approximately 60 oocytes were incubated in 24-well culture plates containing 1 ml of medium with various growth factors. After incubation for 24 hr at 20 °C, the number of oocytes which had undergone germinal vesicle breakdown (GVBD) was counted. Oocytes underwent GVBD in response to HCG (10 IU/ml) but not to DHP (maturational incompetent). Among growth factors used (Table 1), insulin (1 μM) and IGF-I (10 nM) and -II (13 nM) induced GVBD in oocytes of red seabream. IGF-I was the most potent inducer of GVBD. Salmon recombinant IGF-I (rsIGF-I) was almost as effective as human recombinant IGF-I (rsIGF-I) in oocytes of red seabream (Fig. 1).

Table 1. Effects of various growth factors on GVBD in oocytes of red seabream.

Growth factors	% GVBD
IGF-I, human recombinant (10 nM)	70.9 ± 2.3
IGF-II, human recombinant (13 nM)	12.5 ± 2.3
Insulin, bovine (1 μM)	51.1 ± 4.2
EGF, mouse (100 nM)	0
Inhibin A, bovine (200 ng/ml)	0
Activin A, bovine (200 ng/ml)	0

Each value represents the mean ± SEM of three replicates

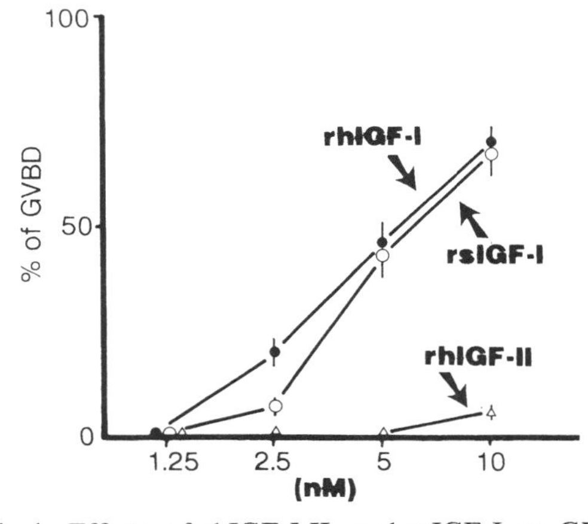

Fig 1. Effects of rhIGF-I,II, and rsIGF-I on GVBD

Effects of cyanoketone (CK, an inhibitor of 3β -hydroxysteroid dehydrogenase) on IGF-I-induced GVBD

Maturational incompetent (DHP-insensitive) and competent (DHP-sensitive) oocytes were incubated in the medium containing IGF-I, HCG, or DHP in the presence or absence of CK (Fig. 2). CK (10, 100 ng/ml) inhibited both HCG- and IGF-I-induced GVBD in maturational incompetent oocytes. However, CK (10 and 100 ng/ml) inhibited HCG-induced GVBD but did not inhibit IGF-I-induced GVBD in maturational competent oocytes.

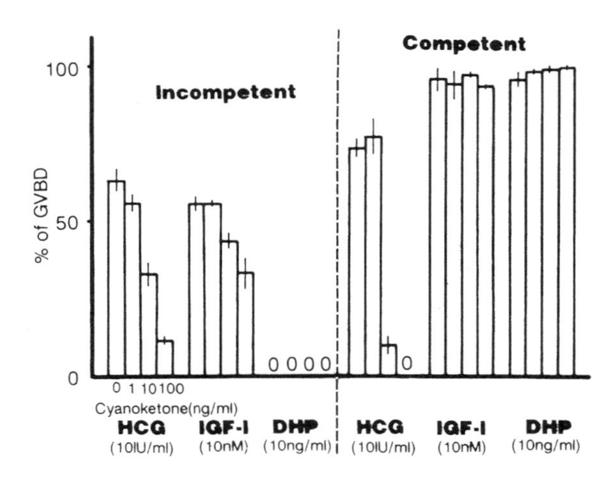

Fig. 2. Effects of cyanoketone on IGF-I-induced GVBD

Effects of IGF-I on GVBD in the presence or absence of DHP

IGF-I induced GVBD in maturational incompetent oocytes at concentrations of 3.3 and 10 nM. The presence of DHP (10 ng/ml) significantly increased the percentage GVBD induced by IGF-I at concentrations of 0.37, 1.1, and 3.3 nM but DHP did not increase GVBD in incubations with 10 nM IGF-I (Fig. 3).

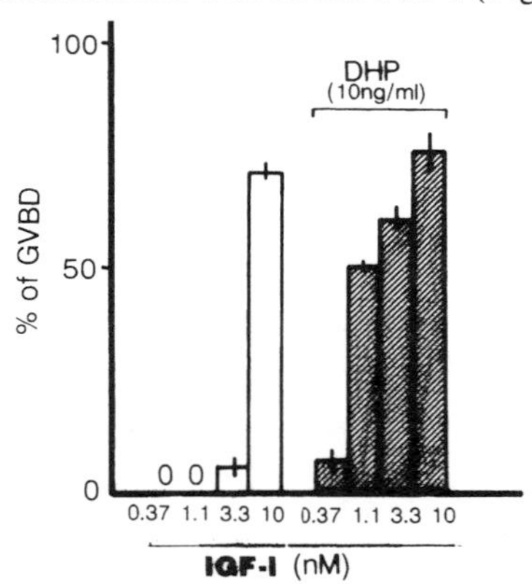

Fig. 3. Effects of IGF-I on GVBD in the presence or absence of DHP

Maturational incompetent oocytes were preincubated with HCG (10 IU/ml) or IGF-I (10 nM) for 30, 60, and 120 min, washed three times with medium, and then incubated with DHP (10 ng/ml) or hormone free medium for 24 hr. Oocytes, preincubated with HCG or IGF-I, underwent GVBD in response to DHP but no GVBD occurred in hormone free medium (Fig. 4).

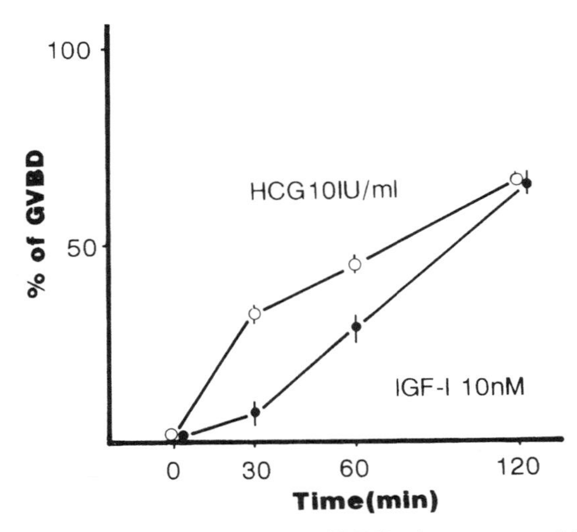

Fig. 4. Effects of DHP on GVBD in oocytes which were preincubated with HCG and IGF-I for 30,60, and 120 min.

Effects of actinomycin D and cycloheximide on IGF-I- and DHP-induced GVBD (Fig. 5)

Maturational incompetent oocytes were incubated in the medium containing IGF-I alone (1 and 10 nM) or in combination with DHP (10 ng/ml), and with or without actinomycin D (1 μg/ml) or cycloheximide (1 μg/ml). Actinomycin D did not block GVBD induced by IGF-I alone or in combination with DHP. Cycloheximide totally inhibited induction of GVBD by IGF-I alone or in combination with DHP (data not shown).

Discussion

Of the growth factors used, IGF-I was the most effective growth factor in inducing GVBD of the red seabream. These data agree with those on induction of GVBD in *Xenopus* oocytes by IGF-I and insulin (Hainaut et al., 1991). In mammals, stimulatory effects of EGF on GVBD have also been reported (Reed et al., 1993) but EGF did not induce GVBD in oocytes of red seabream at a concentration of 100 nM which is higher than the effective dose for *in vitro* induction of GVBD in porcine oocytes (Reed et al., 1993).

Cyanoketone inhibited both HCG- and IGF-I-induced GVBD in maturational incompetent oocytes.
Cyanoketone completely abolished gonadotropin-stimulated MIH production in ovarian follicles of amago salmon and inhibited GVBD (Young et al.,

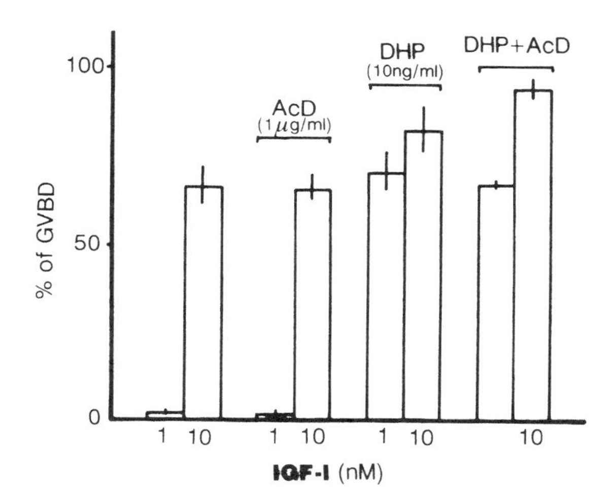

Fig. 5. Effects of actinomycin D (AcD) on IGF-I- and DHP-induced GVBD

1982).Thus IGF-I may be involved in MIH production in follicles of the red seabream. However, in maturational competent oocytes, CK did not block IGF-I-induced GVBD, although CK inhibited HCG-induced GVBD. Moreover, AcD blocked HCG-induced GVBD but did not block IGF-I-induced GVBD in maturational incompetent oocytes of the red seabream. Since AcD blocks GTH-induced DHP production in ovarian follicle of amago salmon (Nagahama et al., 1985), IGF-I may also directly act on oocytes through IGF-I receptor, as reported in *Xenopus* (Hainaut et al., 1991).

Oocytes underwent GVBD in response to DHP when oocytes were incubated with doses of IGF-I that alone were ineffective. Moreover, preincubation of DHP-insensitive oocytes with IGF-I resulted in oocytes undergoing GVBD in response to DHP. IGF-I can thus induce maturational competence of oocytes of the red seabream. Since the action of IGF-I was not blocked by AcD, RNA synthesis is not needed for this induction of GVBD. However, AcD blocked HCG-induced protein synthesis mediated by RNA synthesis, related to development of maturational competence (Kagawa et al., 1994; Patiño and Thomas, 1990). Thus, these results show the possibility that IGF-I can induce maturational competence of oocytes of the red seabream, through a mechanism different from HCG-induced maturational competence.

Acknowledgments

This study was supported in part by a grant-in aid (Bio Media Program, BMP-95-II-2-5) from the Ministry of Agriculture, Forestry, and Fisheries.

References

Gutierrez, J., M. Parrizas, N. Carneiro, J.L. Maestro, M.A. Maestro, and J. Planas, 1993. Insulin and IGF-I receptors and tyrosine kinase activity in carp ovaries: Changes with reproductive cycle. Fish Physiol. Biochem. 11: 247-254.

Hainaut, P., S. Giorgetti, A. Kowalski, R. Ballotti, and E. Van Obberghen, 1991. Antibodies to phosphotyrosine injected in *Xenopus laevis* oocytes modulate maturation induced by insulin/IGF-I. Exp. Cell Res. 195: 129-136.

Kagawa, H. H. Tanaka, K. Okuzawa, and K. Hirose, 1994. Development of maturational competence of oocytes of red seabream, *Pagrus major*, after gonadotropin treatment *in vitro* requires RNA and protein synthesis. Gen. Comp. Endocrinol. 94: 199-206.

Lessman, C.A., 1985. Effect of insulin on meiosis reinitiation induced in vitro by three progestogens in oocytes of the goldfish (*Carassius auratus*). Dev. Biol. 107: 259-263.

Nagahama, Y., G. Young, and S. Adachi, 1985. Effect of actinomycin D and cycloheximide on gonadotropin-induced 17α,20β-dihydroxy-4-pregnen-3-one production by intact ovarian follicles and granulosa cells of the amago salmon, *Oncorhynchus rhodurus*. Dev. Growth Differ. 27: 213-221.

Niu, P.D., J. Perez-Sanchez, and P.Y. Le Bail, 1993. Development of a protein binding assay for teleost insulin-like growth factor (IGF)-like: Relationships between growth hormone (GH) and IGF-like in the blood of rainbow trout (*Oncorhynchus mykiss*). Fish Physiol. Biochem. 11: 381-391.

Patiño, R., and P. Thomas, 1990. Effects of gonadotropin on ovarian intrafollicular processes during the development of oocyte maturational competence in a teleost, the Atlantic croaker: Evidence for two distinct stages of gonadotropic control of final oocyte maturation. Biol. Reprod. 43: 818-827.

Reed, M. L., J.L. Estrada, M.J. Illera, and R.M. Petters, 1993. Effects of epidermal growth factor, insulin-like growth factor-I, and dialyzed porcine follicular fluid on porcine oocyte maturation *in vitro*. J. Exp. Zool. 226: 74-78.

Young G., H. Kagawa, and Y. Nagahama, 1982. Oocyte maturation in the amago salmon (*Oncorhynchus rhodurus*): *In vitro* effects of salmon gonadotropin, steroids, and cyanoketone (an inhibitor of 3β-hydroxy-Δ^5-steroid dehydrogenase). J. Exp. Zool. 224: 265-275.

A PERTUSSIS TOXIN SENSITIVE GTP-BINDING PROTEIN IS INVOLVED IN THE SIGNAL TRANSDUCTION PATHWAY OF THE MATURATION-INDUCING HORMONE (17α,20β-DIHYDROXY-4-PREGNEN-3-ONE) OF RAINBOW TROUT (*ONCORHYNCHUS MYKISS*) OOCYTES

M. Yoshikuni, Y. Oba and Y. Nagahama
Laboratory of Reproductive Biology, National Institute for Basic Biology, Okazaki 444, Japan

Summary

This paper briefly reviews information on the signal transduction mechanism of maturation-inducing hormones (MIH) and presents evidence that a pertussis toxin-sensitive inhibitory G-protein (Gi) is involved in the signal transduction pathway of 17α,20β-dihydroxy-4-pregnen-3-one (MIH in salmonid fishes) in rainbow trout oocytes.

Introduction

Meiotic maturation in amphibian and fish oocytes is induced by maturation-inducing hormone (MIH) which is secreted from ovarian follicle cells under the influence of pituitary gonadotropin (GTH)(Nagahama, 1987). MIH has been identified in amphibians (progesterone) and fishes (17α,20β-dihydroxy-4-pregnen-3-one, 17α,20β-DP for salmonid fishes; 17α,20β,21-trihydroxy-4-pregnen-3-one, 20β-S for sciaenid fishes)(Nagahama, 1987; Smith, 1989; Thomas, 1994). It was established that unlike classical steroid hormone action, these maturational steroids act at the level of the oocyte membrane (Nagahama *et al.*, 1995). Specific plasma membrane receptors for these progestogens have been characterized in *Xenopus* (Sadler and Maller, 1982), spotted seatrout (Patino & Thomas, 1990) and rainbow trout (Yoshikuni *et al.*, 1993). The MIH-initiated signal is transduced across the oocyte plasma membrane to the cytoplasm, where a maturation-promoting factor (MPF) is formed; MPF is composed of cdc2 kinase and cyclin B (Maller, 1991; Nagahama *et al.*, 1995).

Because of the intense interest in MPF as a universal regulator of mitosis, less attention has been paid in the past few years to molecular details of the mechanisms coupling MIH action to the appearance of MPF. A decrease in cAMP appears to be necessary for the MIH-induced oocyte maturation in *Xenopus* (Sadler and Maller, 1985) and fish (Jalabert and Finet, 1986; DeManno and Goetz, 1987). The possible involvement of cAMP in hormone-induced oocyte maturation suggested that a guanine nucleotide binding protein (G-protein) might be involved in the pathway coupling MIH action at the cell surface to meiotic maturation. The purpose of this article is to review information on the signal transduction mechanism of fish MIHs and discuss this information with that of *Xenopus* and starfish.

17α,20β-DP receptors

Recently, we developed an ultracentrifugation method to prepare plasma membrane fractions from defolliculated oocytes of rainbow trout and used them to characterize the 17α,20β-DP receptor (Yoshikuni *et al.*, 1993). A specific receptor for 17α,20β-DP exists in the plasma membrane of rainbow trout oocytes. Scatchard analysis yielded two classes of high-affinity (Kd = 18 nM; Bmax = 0.2 pmol/mg protein) and low-affinity (Kd = 0.3 μM; Bmax = 1 pmol/mg protein) receptors. More recently, a 17α,20β-DP binding protein (DBP) was also purified and characterized from plasma of prespawning female rainbow trout. The rainbow trout DBP had an apparent molecular weight of 110 kDa on native-PAGE and was composed of two subunits with molecular weights of 50 and 55 kDa on SDS-PAGE (Yoshikuni *et al.*, 1994). Scatchard analysis exhibited a single binding component with a Kd of 10 nM and a Bmax of 1.49 mmol/liter plasma. Thus, the physicochemical characteristics of the DBP and 17α,20β-DP membrane receptors differ significantly.

Maneckjee *et al.* (1991) described a 17α,20β-DP receptor-like protein in the zona radiata membranes by photoaffinity labeling in brook trout oocytes during final stages of maturation. Furthermore, a specific, high-affinity (Kd = 1 nM) receptor for 20β-S was demonstrated in oocyte membranes from spotted seatrout (Patino and Thomas, 1990).

Involvement of cAMP in 17α,20β-DP action

Inhibitors of phosphodiesterase such as IBMX and theophylline and activators of adenylate cyclase such as forskolin and cholera toxin (CT) which increase the intracellular cAMP have been reported to inhibit steroid-induced oocyte maturation *in vitro* in rainbow trout (Jalabert and Finet, 1986) and brook trout (DeManno and Goetz, 1987). Furthermore, 17α,20β-DP-induced oocyte maturation was followed by a transient decrease in oocyte cAMP in rainbow trout, although the decrease varied over time and between fish (Jalabert and Finet, 1986). These results are consistent with an important role of a transient decrease in cAMP in steroid-induced oocyte maturation. Supportive evidence for the role of cAMP comes from our recent finding that 17α,20β-DP inhibited adenylate cyclase activity in oocyte membrane preparations of rainbow trout (Yoshikuni *et al.*, 1994).

Involvement of G-proteins in 17α,20β-DP action

Plasma membrane preparations of rainbow trout oocytes were subjected to ADP-ribosylation by CT or pertussis toxin (PT) and to immunoprecipitation with anti-G-protein antibodies (Yoshikuni *et al.*, 1994). Plasma membrane preparations from rainbow trout postvitellogenic oocytes contain two G-proteins, Gs (43 kDa) and Gi (40 kDa), normally associated with regulation of adenylate cyclase. Preincubation of the membrane with 17α,20β-DP decreased the PT-catalyzed ADP ribosylation of the 40 kDa protein, without affecting CT-catalyzed ADP-ribosylation of the 43 kDa protein. The specific binding of 17α,20β-DP to membrane fractions was decreased by PT, but not by CT. 17α,20β-DP binding was also inhibited in the presence of nonhydrolyzable GTP analogs such as GTPγS and GppNHp, but not by either ATP or GDPβS. Scatchard analysis revealed that GppNHp-induced decrease in 17α,20β-DP binding is due to the decrease in binding affinity between 17α,20β-DP and its receptors.

Discussion

Despite the recent progress in the studies of MPF, the mechanisms by which MIHs initiate oocyte maturation have remained elusive. The results from each of the experiments presented here are consistent with the important role of PT-sensitive inhibitory Gi-proteins in the signal transduction pathway of 17α,20β-DP in rainbow trout oocytes. We also attempted to inhibit 17α,20β-DP-induced maturation in rainbow trout oocytes by external application or microinjection of PT; however, either treatment was unsuccessful probably due in part to the presence of a large amount of yolk, which may cause PT to be restrictedly distributed within the oocyte cytoplasm (Yoshikuni *et al.*, unpublished). More recently, we have shown that treatment with an antibody against the α-subunit of mammalian Gi partially prevented 17α,20β-DP-induced decrease in adenylate cyclase activity in oocyte membranes of rainbow trout (Yoshikuni, unpublished). A similar signal transduction pathway has previously been reported in the MIH of starfish, 1-methyladenine (Tadenuma *et al.*, 1992). In this case, exposure to 1-methyladenine activates Gi, and the dissociated βγ subunit of Gi activates the subsequent events of oocyte maturation, although the effecter protein for βγ is unknown. Microinjection of PT inhibits 1-methyladenine-induced oocyte maturation (Shilling *et al.*, 1989). The possible involvement of G-proteins in progesterone-induced oocyte maturation has also been suggested in *Xenopus* (Sadler and Maller, 1985). However, progesterone does not appear to regulate cAMP levels through these proteins in a manner consistent with that shown in other cell systems. For example, PT inhibits or slows oocyte maturation in response to progesterone but does not affect adenylate cyclase activity inhibition by progesterone in isolated oocyte membranes (Sadler *et al.*, 1984). Based on these unusual properties, Sadler and Maller (1985) suggested that progesterone might act by inhibiting Gs. This hypothesis was supported by the observation by Gall *et al.* (1995), who reported that oocytes of *Xenopus* underwent maturation when injected with an affinity-purified antibody against the COOH-terminal decapeptide of the a subunit of the G-protein Gs, an antibody that inhibits Gs activity. Since oocyte membranes of rainbow trout also contain Gs, the role of this G-protein in transducing the 17α,20β-DP signal needs to be determined. In addition to the protein kinase A pathway, other G-protein linked pathways such as the protein kinase C pathway have been suggested to be involved in progesterone-induced meiotic maturation in amphibian oocytes (Smith, 1989). For example, in *Rana pipiens* oocytes, progesterone appears to activate a G-protein-linked, phosphatidylcholine-specific phospholipase C in the oocyte plasma membrane (Kostellow *et al.*, 1993). Therefore, although G-proteins may be involved in the signaling pathway for progesterone action on amphibian oocyte maturation, the specific second messenger mechanisms involved are still unclear.

It should be emphasized that evidence for plasma membrane steroid hormone receptors is not limited to the amphibian and fish oocyte system. Others have reported the presence of non-genomic progesterone receptors on human sperm plasma membranes (Revelli *et al.*, 1994). It is also becoming increasing apparent that some steroid hormones may effect rapid alterations in brain function independently of classical intracellular receptors. For example, certain neuroactive steroids have been shown to alter central nervous system excitability through an interaction with receptors for the major inhibitory neurotransmitter in brain, γ-aminobutyric acid (GABA) (Paul and Purdy, 1992). A novel membrane-bound corticosteroid receptor has been demonstrated in the newt brain that appears to be mediated by G-proteins (Orchinik *et al.*, 1992). It is of great interest to determine whether these membrane steroid receptors share sequence and structural homology with classical cytosolic or nuclear steroid receptors. Further isolation and characterization of cDNAs encoding these membrane steroid receptors should facilitate studies of the signal transduction pathways of MIHs leading to oocyte maturation. This information will undoubtedly be essential to understanding not only the hormonal regulation of oocyte maturation, but also the mechanisms of steroid hormone action in general.

Reference

Nagahama Y., 1987. In Hormone and Reproduction in Fishes, Amphibians, and Reptiles. D.O. Norris and R. E. Jones (Eds.): Plenum Publishing, New York, p. 171-202.

Smith L. D., 1989. The induction of oocyte maturation: Transmembrane signaling events and regulation of the cell cycle. Development 107: 685-699.

Thomas P., 1994. Hormonal control of final oocyte maturation in sciaenid fishes. XII. ICCE (Toronto, Canada) p. 619-625.

Nagahama Y., Yoshikuni M., Yamashita M., Tokumoto T. and Katsu Y., 1995. In Current Topics in Developmental Biology. R. A. Pedersen and G. P. Scatten (Eds.), Vol. 30: Academic Press, San Diego, p. 103-146.

Sadler S. and Maller J. L., 1982. Identification of a steroid receptor on the surface of *Xenopus* oocytes by photoaffinity labeling. J. Biol. Chem. 257: 355-361.

Patino R. and Thomas P., 1990. Characterization of membrane receptor activity for 17α,20β,21-trihydroxy-4-pregnen-3-one in ovaries of spotted seatrout (*Cynoscion nebulosus*). Gen. Comp. Endocrinol. 257: 35-361.

Yoshikuni M., Shibata N. and Nagahama Y., 1993. Specific binding of [^{3}H]17α,20β-dihydroxy-4-pregnen-3-one to oocyte cortices of rainbow trout (*Oncorhynchus mykiss*). Fish Physiol. Biochem. 11: 15-24.

Maller J. L., 1991. Mitotic control. Curr. Opin. Cell Biol. 3: 269-275.

Sadler S. and Maller J. L., 1985. Inhibition of *Xenopus laevis* oocyte adenylate cyclase by progestin: a novel mechanism of action. Adv. Cyclic Nucleotide Protein Phosphorylation Res. 19: 179-194.

Jalabert B. and Finet B., 1986. Regulation of oocyte maturation in rainbow trout, *Salmo gairdneri*: Role of cyclic AMP in the mechanism of action of the maturation inducing steroid (MIS), 17α-hydroxy-20β-dihydroprogesterone. Fish Physiol. Biochem. 2: 65-74.

DeManno D. A. and Goetz F. W., 1987. Steroid-induced final maturation in brook trout (*Salvelinus fontinalis*) oocytes *in vitro*: The effects of forskolin and phosphodiesterase inhibitors. Biol. Reprod. 36: 1321-1332.

Yoshikuni M., Matsushita H., Shibata N. and Nagahama Y., 1994. Purification and characterization of 17α,20β-dihydroxy-4-pregnen-3-one binding protein from plasma of rainbow trout, *Oncorhynchus mykiss*. Gen. Comp. Endocrinol. 96: 189-196.

Maneckjee A., Idler D. R. and Weisbart M., 1991. Demonstration of putative membrane and cytosol steroid receptors for 17α,20β-dihydroxy-4-pregnen-3-one in brook trout *Salvelinus fontinalis* oocytes by photoaffinity labeling using synthetic progestin 17,20-dimethyl-19-nor-pregn-4,9-diene-3,20-dione (R5020). Fish Physiol Biochem. 9: 123-135.

Yoshikuni M. and Nagahama Y., 1994. Involvement of an inhibitory G-protein in the signal transduction pathway of maturation-inducing hormone (17α,20β-dihydroxy-4-pregnen-3-one) action in rainbow trout (*Oncorhynchus mykiss*) oocytes. Dev. Biol. 16: 615-622.

Tadenuma H., Takahashi K., Chiba K., Hoshi M. and Katada T., 1992. Properties of 1-methyladenine receptors in starfish membrane: Involvement of pertussis toxin-sensitive GTP-binding protein in the receptor-mediated signal transduction. Biochem. Biophys. Res. Commun. 186: 114-121.

Shilling F., Chiba K., Hoshi M., Kishimoto T. and Jaffe L. A., 1989. Pertussis toxin inhibits 1-methyladenine-induced maturation in starfish oocytes. Dev. Biol. 13:605-608.

Sadler S., Maller J. L. and Cooper D. M. F., 1984. Progesterone inhibition of *Xenopus* oocytes adenylate cyclase is not mediated *via* the *Bordetella pertussis* toxin substrate. Mol. Pharmacol. 26: 526-531.

Gallo C. J., Hand A. R., Jones T. L. and Jaffe L. A., 1995. Stimulation of *Xenopus* oocyte maturation by inhibition of the G-protein αs subunit, a component of the plasma membrane and yolk platelet membranes. J. Cell Biol. 130: 275-284.

Kostellow A. B., Ma G.-Y. and Morill G. A., 1993. Steroid action at the plasma membrane: progesterone stimulation of phosphatidylcholine-specific phospholipase C following release of the prophase block in amphibian oocytes. Mol. Cell. Endocrinol. 92: 33-44.

Revelli A., Modotti M., Piffareti-Y. A., Massobrio M. and Balerna M., 1994. Steroid receptors in human spermatozoa. Human Reprod. 9: 760-766.

Paul S. and Purdy R. H., 1992. Neuroactive steroids. FASEB J. 6: 2311-2322.

Orchinik M., Murray T., Franklin P. H. and Moore F. L., 1992. Guanyl nucleotides modulate binding to steroid receptors in neuronal membranes. Proc. Natl. Acad. Sci. USA 89: 3830-3834

INHIBITORY REGULATION OF SPERMATOGENESIS IN SHARK TESTIS

F. Piferrer[1] and G. Callard

Department of Biology, Boston University, Boston, MA 02215

Summary

In the dogfish shark (*Squalus acanthias*), the epigonal organ (EO), a testis-associated lymphomyeloid tissue, contains a bioactivity capable of inhibiting premeiotic DNA synthesis, which we termed EO growth inhibitory factor (EGIF). Here, we show that the effects of EGIF could not be mimicked by sex steroids, prostaglandins and various growth factors. We also show that EGIF is water- and acid-soluble, heat-stable and lyophilizable. Trypsin and pronase did not completely eliminate EGIF. After precipitation of the large proteins, solvent separation and chromatography revealed that the activity was mainly detected in the eluate of the aequous fraction, suggesting that EGIF may comprise a peptide of low molecular weight. Since the properties of EGIF do not correspond to those of factors of lymphomyeloid origin implicated in spermatogenesis so far, these results raise the possibility that EGIF may include a novel substance.

Introduction

Despite intense research interest, factors and mechanisms involved in regulating spermatogenesis are not well understood (Mullaney & Skinner, 1994). Compared to those of conventional laboratory animals, the shark testis is advantageous for studying spermatogenesis due to (a) a simple linear progression and clear spatial segregation of different spermatogenic stages, and (b) organization of germ cells and Sertoli cells into follicle-like units termed spermatocysts. This allows stage-by-stage analysis of germ cell development *in vivo* and *in vitro* (Callard *et al.*, 1994). In sharks and other cartilaginous fishes, hemopoiesis occurs in several organs distributed throughout the body. The epigonal organ (EO) is one of the largest and encapsulates the testis, with which it shares a common vascular pathway. Thus substances produced in the EO can reach the testis.

Recently we showed that the EO of the dogfish shark (*Squalus acanthias*) contained a bioactivity, termed EO growth inhibitory factor (EGIF), which was able to inhibit premeiotic (PrM; stem cells, spermatogonia, preleptotene spermatocytes) DNA synthesis, as measured by a decrease in [^{3}H]thymidine incorporation. The inhibitory response was specific, dose- and time-dependent, had a short response latency, was completely reversible and counteracted but did not block the stimulatory effects of insulin on thymidine incorporation (Piferrer & Callard, 1995). Although EGIF was tissue-specific in origin and present year-round, during the period of spermatogenic arrest we detected a gradient of EGIF-like bioactivity within the testis, with maximal levels adjacent to the EO. These data provide evidence for a novel, functional interaction between the reproductive and immune systems early in vertebrate evolution. Here we attempted to duplicate EGIF effects with several substances implicated in male reproduction and report initial efforts to determine the nature of EGIF.

Methods

Inhibition of DNA synthesis by PrM spermatocysts was used as a bioassay, as described earlier (Piferrer *et al.*, 1993; Piferrer & Callard, 1995). Test substances were added directly to culture wells for 32-48 h and [^{3}H]thymidine was added (5 μCi/ml) for the last 16-24 h.

Cytosolic extracts of EO pooled from 10 sharks were prepared in 50 mM potassium phosphate buffer (PB; 1:6.6 w/v) at 4°C as described previously (Piferrer & Callard, 1995), and aliquots were tested directly or after various treatments: (a) four freeze-thaw (F/T) cycles; (b) heating to 100°C for 5 min; (c) lyophilization; (d) trypsin digestion; (e) pronase digestion. Trypsin (Sigma type III, from bovine pancreas; 11,700 BAEE U/mg protein) was added to a final concentration of 500 μg/ml in the presence of 0.5 mM $CaCl_2$ for 1.5 h at 37°C. Proteolysis was terminated by adding trypsin inhibitor (Sigma type II, from soybean, STI; 10,000 U/mg protein) at 1 mg/ml for 30 min at r.t. Pronase (Sigma, 4 U/mg) was added at 250 μg/ml o.n. at 37°C. The reaction was stopped by adding 2 U/ml aprotinin, 1 mM phenyl-methyl-sulfoxylfluoride and 500 μg/ml STI. Samples were then clarified by centrifugation and tested for their effect on PrM DNA synthesis. Controls were extracts of muscle or PB alone. Alternatively, pooled EO were homogenized in 1 M HCl/acetone (1:33 v/v) at 4°C at (1:4 w/v). The homogenate was adjusted to pH 3.0 and centrifuged at 10,000 rpm for 10 min at 4°C. The precipitate was resuspended with 2 vol of HCl/acetone and 400 μl dH_2O per g of original weight and

[1]Current address: Consejo Superior de Investigaciones Científicas, Inst. de Acuicultura, 12595 Torre de la Sal, Castellón, Spain.

homogenized. Pooled supernatants were combined and fractionated twice with 1 vol petroleum ether (b.p. 40-60°C). The solvent and aqueous fractions were bubbled with nitrogen, vacuum dried (solvent) or lyophilized o.n. (aqueous) and reconstituted with media. media. Some fractions were applied to a C18 reverse phase SepPak column, washed with 10 ml methanol and 10 ml of 0.1% trifluoroacetic acid (TFA). The samples were applied in TFA, washed with 0.1% TFA and then eluted with 75% acetonitrile in 0.1% TFA. The pH of fractions was adjusted to 7.2 before addition to PrM cultures. In dose-response experiments, inhibitory bioactivity of extracts was expressed as percent of [^{3}H]thymidine incorporation in control cultures. In additional experiments, steroids and prostaglandins were dissolved in ethanol ($<0.5\%$ in media) and added to cultures. Activin-A and inhibin-A, a gift of Dr. J.P. Mather (Genentech, S. San Francisco, CA), were dissolved directly in media. Experiments were repeated at least 3 times, with 3-5 replicate wells per treatment. Effects were analyzed by ANOVA followed by Tukey´s multiple range test.

Results

As shown in Fig. 1, at the maximum in-assay concentration tested, crude extracts of EO in PB inhibited DNA synthesis to $<25\%$ of PrM spermatocysts receiving buffer, but crude muscle extracts were ineffective (not shown; see Piferrer & Callard, 1995 and Fig. 2). None of the physical or chemical pretreatments used here significantly altered the inhibitory bioactivity of this dosage of EO extract, nor did these treatments of tissue-free buffer affect control rates of DNA synthesis.

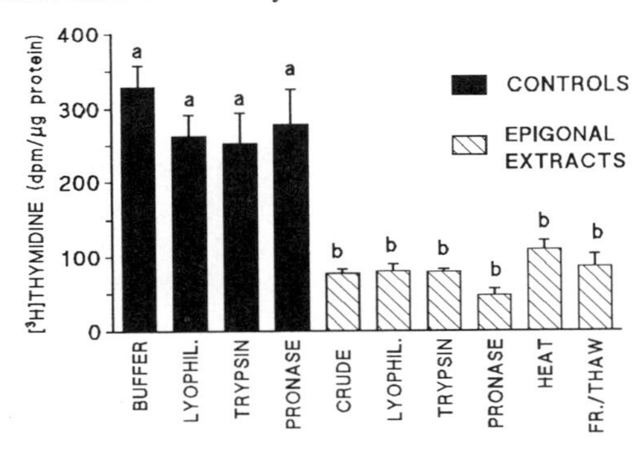

Fig. 1. Effects of physical and chemical treatments of PB extracts of EO (20% final vol) on the resulting capacity to inhibit PrM DNA synthesis. Data shown as mean+SEM of quadruplicate determinations and representative of 2 separate experiments. Letters indicate significant differences ($p<0.05$).

Table 1 shows results when the same extracts were used in dose-response experiments to obtain the specific bioactivity (SB) and the ID_{50} dose necessary for half-maximal inhibition. Pronase, trypsin and lyophilization incompletely destroyed protein but had quantitatively the same effect on bioactivity. By contrast, heat removed protein but not bioactivity, increasing SB two fold, while F/T cycles had greater effect on bioactivity than on protein. The SB in epigonal tissue was calculated from several tests (n=4) with both PB and acid extracts and found to be 439 ± 20 (mean $\pm$SE) U/g tissue equivalent.

Table 1. Bioactivity of EGIF after various treatments

	Protein		Bioact.*		Sp.Bioact.
Treatment	mg/ ml	yield (%)	U/ ml	yield (%)	U/ mg prot.
Crude Extr.	3.51	100	62.5	100	17.8
Fr./Thaw	3.14	89	25.6	41	8.2
Pronase	1.87	53	31.2	50	16.7
Trypsin	1.74	49	31.2	50	17.9
Lyophilizat.	2.37	67	47.6	76	20.1
Heat	1.79	51	58.8	94	32.9

*1 Unit (U) defined as the amount of epigonal extract that resulted in 50% maximal inhibition.

Results of acid extraction compared to PB extracts are shown in Fig. 2. Increasing doses of PB extracts progressively inhibited PrM DNA synthesis, with a maximal response of about 10% of control values. A similar dose-response effect was obtained with acid extracts, with twice the inhibitory bioactivity detected in the aequous compared to the solvent fraction, although both fractions showed a clear dose-response curve (Fig. 2). Control cultures for the acid extraction procedure (dH_2O) and for a generalized protein effect (muscle extracts in PB) did not influence PrM DNA synthesis. Column fractionation showed that the aequous eluate had more inhibitory bioactivity than the solvent eluate with no activity in the flow-through (data not shown). Addition of key steroids or their precursors, including 25-OH-cholesterol (0.06-60μM) progesterone, testosterone (1μM) and estradiol-17ß (0.01-10 μM) did not significantly influence PrM DNA synthesis, nor was a dose-response relationship obtained. Similarly, a variety of prostaglandins (E_1, E_2, $F_{1\alpha}$ $F_{2\alpha}$; 0.002-20 μg/ml) and activin and inhibin (100-400 ng/ml) did not influence PrM DNA synthesis (data not shown).

Research carried out at the Mount Desert Island Biological Laboratory, Salsbury Cove, ME 04672.

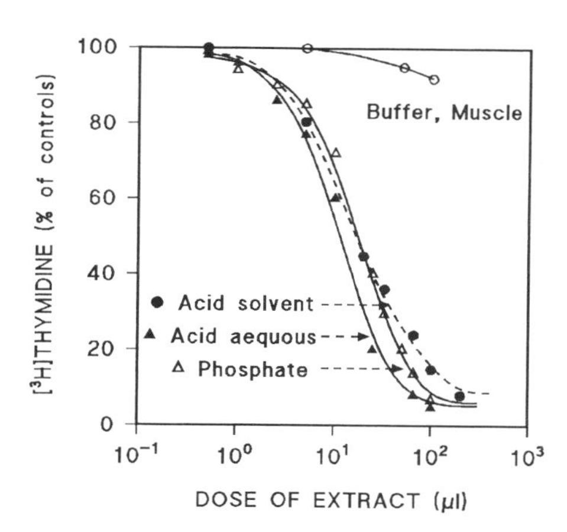

Fig. 2. Dose-response and tissue specificity of inhibitory bioactivity of EO homogenate. Activity of acid extracts is shown for both the aqueous and the solvent fractions after partition. Controls are plotted together. Curves are representative of 3 (PB extract) or 4 (acid extract) experiments.

Discussion

The data presented here confirms our previous results that the *Squalus* EO has a growth inhibitory factor (EGIF) capable of inhibiting PrM DNA synthesis (Piferrer & Callard, 1995). Together with earlier observations (Piferrer *et al.*, 1993), data show that the effects of EGIF could not be mimicked by substances with a defined role in male reproduction, including steroids , prostaglandins and peptide growth factors of the IGF family, the TGF-ß family, the EGF/TGF-α family and the cytokine family. The possibility that the effect of EGIF might be a generalized protein effect was ruled out previously and confirmed here by including other tissues and the appropriate controls in each case. PDE inhibitors also decrease DNA synthesis in this system (Redding, this meeting) and may be part of the effector pathway in which EGIF operates, which may include the pathway leading to apoptosis (Callard *et al.*, 1995).

The potency of EGIF was further exemplified by its ability to inhibit PrM DNA synthesis after exposure to heat, lyophilization, F/T cycles and pronase or trypsin digestion. The partial stability to trypsin indicates that EGIF does not have a trypsin-sensitive site, or that it maintains activity despite trypsin cleavage. Our data suggests that EGIF may be a small molecular weight (mw) peptide or have a peptide moiety because (a) when high mw proteins are selectively precipitated by heat, SB increases two fold, (b) the effectiveness of acid/acetone extraction, designed to separate high and low mw (< 10 kDa) peptides, and (c) the adherence to SepPak columns, designed to separate peptides from amines and other low mw substances and buffer components. The apparent low mw of EGIF and the fact that it is not fully inactivated by pronase or trypsin, distinguishes it from other factors of lymphomyeloid origin such as TNF and IL-1α that can influence spermatogonial mitoses (Mullaney & Skinner, 1994), and from the <5 kDa rat spermatogonial chalone, which is sensitive to heat and trypsin (Thumann *et al.*, 1981). However, we cannot rule out more than one factor in the crude extract. In summary, these studies exemplify the use of the shark testis model to identify growth factors potentially relevant in male reproduction. EGIF may be a paracrine factor regulating germ cell mitosis. To address this and other questions it will be necessary to purify EGIF to homogeneity.

References

Callard, G.V., J.C. Jorgensen and J.M. Redding, 1995. Biochemical analysis of programmed cell death during premeiotic stages of spermatogenesis *in vivo* and *in vitro*. Develop. Genet., 16: 140-147.

Callard, G.V., M. Betka and J.C. Jorgensen, 1994. Stage-related functions of Sertoli cells: lessons from lower vertebrates. In: Function of Somatic Cells in the Testis. A. Bartke (Ed.): Springer-Verlag, New York, p. 27-54.

Mullaney B.P. and M.K. Skinner, 1994. Growth factor regulation of testicular function. In: Function of Somatic Cells in the Testis. A. Bartke (Ed.): Springer-Verlag, New York, p. 55-61.

Piferrer F., M. Redding, W. DuBois and G. Callard, 1993. Stimulatory and inhibitory regulation of DNA synthesis during spermatogenesis: studies in *Squalus acanthias*. Fish Physiol. Biochem. 11:293-298.

Piferrer F.C. and G.V. Callard, 1995. Inhibition of DNA synthesis during premeiotic stages of spermatogenesis by a factor from testis-associated lymphomyeloid tissue in the dogfish shark (*Squalus acanthias*). Biol. Reprod. (in press).

Thumann A., R. Carboni and E. Bustos-Obregón, 1981. *In vitro* characterization of rat G_1 spermatogonial chalone. Andrología 13:583-589.

Supported by NIH grant HD 16715 to G.C. and a postdoctoral fellowship from the Spanish government to F.P.

INSULIN LIKE GROWTH FACTOR EXPRESSION, BINDING AND ACTION IN TROUT TESTIS (Oncorhynchus mykiss).

F. Le Gac and M. Loir

Laboratoire de Physiologie des Poissons, I.N.R.A., Campus de Beaulieu, 35042 RENNES, France.

Summary

This paper summarizes data obtained in trout testis in our laboratory, concerning IGF-I and II mRNA expression and regulation, IGF receptors on testicular cells and IGF effects on germ cell proliferation *in vitro*.

Introduction

Evidence for growth hormone (GH) binding to specific receptors in trout testes (Le Gac et al., 1992) and GH involvement in the regulation of fish gonadal functions have been obtained *in vivo* and *in vitro* (review : Le Gac et al., 1993). Whether these GH effects are mediated through modification of insulin-like growth factor I (IGF-I) production and/or IGF binding in fish gonadal tissues remained to be studied. In other respects, spermatogonial proliferation during the first phase of gametogenesis is regulated by pituitary gonadotropins and by locally produced factors with a paracrine role. Among these factors activin could be involved in eel spermatogenesis through a yet unknown mechanism (Nagahama, 1994), while stimulatory effects of IGF-I on premeïotic germ cell proliferation have been recently reported in trout (Le Gac and Loir, 1993; Loir, 1994) and in dogfish (Dubois and Callard, 1993). However, few data exist concerning IGF system in fish testes, with the exception of IGF-I and -II mRNA expression in salmonid testes (Duguay et al., 1992 ; Shamblott and Chen, 1994).

For these reasons, the occurrence of a potentially functional IGF system and aspects of the possible role of IGFs have been investigated in the trout testis.

Results

IGF mRNA expression:

Total and poly A+ RNA extracted from rainbow-trout liver and testicular tissue or cells in various physiological or experimental situations have been tested for relative content in IGF mRNA.

Using northern blot hybridization with a specific ^{32}P labelled coho salmon IGF-I-cDNA (sIGF I P388 29, Duguay et al., 1992), we show that IGF-I transcription occurs at low or undetectable levels in unstimulated testes. However, a major 3.9 - 4 kb mRNA was obtained in testicular poly A+ RNA from GH treated immature fish, while smaller mRNA species were also found at other stage of gonadal development (2.8 - 1.8 kb) (Le Gac and Loir, 1993).

Slot blots of total RNA from control, GH or gonadotropin (GtH2) treated fish (beginning of gametogenesis) were also hybridized with the sIGF-I probe. *In vivo,* rtGH treatment tended to increase the relative amount of IGF-I mRNA in the testes. However this increase was small (1.3 to 3 fold control level) compared to the liver response to GH (10 to 12 fold control level). Treatment with GtH2 (3 successive injections at 12 hour interval and tissue sampling 6 and 24 hours after the last injection) had no detectable effect on the relative abundance of IGF-I mRNA in testes at the beginning of spermatogenesis (Le Gac et al. 1996); this is in aggreement with the absence of detectable GtH effect on juvenile salmon ovary IGF-I mRNA (Duguay et al., 1994). Possible effects of GH, GtH1, GtH2 or steroids are under investigation *in vitro,* in cultured testicular cells or explants, but no clear effect has been obtained yet.

Reverse transcription of testicular RNA, followed by amplification (RT- PCR) of specific cDNAs using trout IGF-I or IGF-II specific primers allowed demonstration of IGF-I and also IGF-II mRNA expression in testes at different stages of maturation : immature, beginning and end of spermatogenesis. However, the size of the specific IGF-II transcripts have not been described yet by northern blot.

IGF receptors:

Recombinant human IGF-I and -II peptides labelled with ^{125}Iodine were used to study IGF receptors in testicular preparations and cultured cells (fig. 1) (Le Gac et al. 1996).

Binding sites with high affinity (Ka= 0.2 to 0.7 10^{10} M^{-1}) and low capacity for ^{125}I-IGF-I have been found in whole testes preparations and in isolated testicular cells (10 to 20 fmoles / 10^7 mixed testicular cells). These sites were also found in enriched membrane preparations (≈70 fmoles / mg prot.). Binding specificity shows an order of affinity - IGF-I ≈(QAYL)IGF-I > IGF-II >> INSULIN, characteristic of type 1 IGF receptor described in vertebrates. It is different from the binding specificity of a fraction of trout blood plasma containing IGF binding proteins - IGF-I ≈ IGF-II >>(QAYL)IGF-I or INSULIN-. ((QAYL)IGF-I is a potent analog of IGF-I that recognizes IGF receptor but not IGF binding proteins of higher vertebrates).

In cross-linking studies, we found that the size of the testicular protein α-subunit, affinity labelled with ^{125}I-IGF-I, is ≥ 130 kDa in reducing conditions, which is compatible with the molecular size of α-subunits of mammalian type 1 IGF receptor (fig. 1A).

These data demonstrate for the first time, the existence of IGF-I receptors in fish testes, with the binding and size characteristics of type 1 IGF receptors. Type 1 receptors have been recently identified in the brain and liver of fish (Drakenberg et al., 1993); IGF-I and IGF-II specific binding sites have also been found in trout brain and pituitary (Blaise et al., 1995), and IGF-I binding sites with associated tyrosine kinase activity have been described in the carp ovary (Gutiérrez et al., 1993).

Our studies of radiolabelled rhIGF-II binding to testicular membrane also revealed specific binding of this factor to type 1 IGF receptor (as shown by ligand specificity and molecular size after affinity labelling) but not to a "mannose-6-phosphate" receptor type. To our knowledge, this type of receptor has never been described in fish. It may not bind IGF-II in these species (as previously observed in chicken and xenopus), or may not recognise the human peptide used as ligand in this study.

Localisation of IGF mRNA and IGF receptors:

To start identifying IGF producing cells and IGF target cells in the testis, we studied the distribution of IGF mRNA and IGF receptors in trout testicular cells. Gonads in mid-spermatogenesis (stages 3 to 5) were dissociated by perfusion with collagenase. "Total cells" were further separated to obtain populations enriched in Sertoli cells (percoll centrifugation) or enriched in spermatogonia (Go+CI), in primary spermatocytes (CI), in spermatids (STD) or spermatozoa (SPZ), obtained by centrifugal elutriation. We applied the RT-PCR approach and ^{125}I-IGF binding studies to these populations of cells from the tubular compartment. Both, IGF-I mRNA and IGF receptors (fig.1 C) were preferentially observed in Sertoli cells and Go+CI. Their lower levels in post-meïotic cell populations could be, at least in part, attributed to the presence of contaminating premeïotic germ-cells in these preparations.

IGF-II mRNA was detected by RT-PCR in every cell population tested, except a STD+SPZ population.

These data suggest that IGFs are potential paracrine/autocrine regulators inside the spermatogenic compartment.

IGF mRNA expression and regulation, IGF binding and IGF action in separated trout interstitial cells remain to be investigated to understand GH action on fish testicular steroidogenesis described previously.

IGF effects on premeïotic germ cell proliferation:

When Go+CI prepared from trout testis were cultured for 3 days in the presence of rh IGF-I, rh IGF-II, (QAYL)IGF-I, bovine or salmon insulin, all these peptides stimulated the incorporation of ^{3}H-thymidine in the germ cell DNA. This growth factor effect was observed with early germ cells obtained from gonads at all stages of the reproductive cycle (stage 3 to 9).

IGF-I and (QAYL)IGF-I, were the most potent stimulators of DNA synthesis (maximum stimulation = +100 to +300% ; ED50 ≈ 6ng/ml). IGF-II was 3 to 5 fold less potent, and salmon and bovine insulins 100 to 300 fold less potent than IGF-I. IGF receptors were found on Go+CI, with hormonal binding specificity similar to this biological effect specificity. We conclude that, *in vitro*, IGFs stimulate DNA synthesis of early germ cells by interacting directly with these cells, through type 1 IGF receptors (Loir and Le Gac, 1994).

Coculture of Go+CI with a cell fraction enriched with Sertoli cells, or in the presence of Sertoli cell conditioned culture medium, could be inhibitory or stimulatory on *basal* germ cell DNA synthesis. However, it always reduced the *stimulatory effect* of IGF-I (Loir, 1994). Recombinant human IGF-BP3 inhibited the IGF-I stimulatory effect but not that of (QAYL)IGF-I. Whether interaction between Sertoli and germ cells occurs through IGF and/or IGF-BP production by the Sertoli cells remains to be investigated.

Conclusion

IGF-I and IGF-II are expressed and bind to a type 1 IGF receptor in the trout testis.

The relative abundance of testicular IGF-I mRNA may be influenced by GH treatment: therefore we cannot exclude tht IGF-I is a potential mediator of GH action at the testicular level in fish. Further investigation is needed to understand the influence of gonadotropins, especially of GtH1.

Furthermore, the possible implication of IGFs in local Sertoli cell/germ cell, paracrine or autocrine regulations inside the spermatogenic compartment can be postulated; in particular IGF-I role in early germ cell proliferation/differentiation is probable.

References

Blaise, O., Weil, C. and Le Bail, P.Y. 1995. Role of IGF-I in the controle of GH secretion in Rainbow trout . *Growth Regulation* (in press)

Drakenberg, K., Sara VR., Falkmer, S., Gammeltoft, S., Maake, C. and Reinecke, M. 1993 Identification of IGF-1 receptors in primitive vertebrates. *Regulatory Peptides* 43:73-81.

Dubois, W. and Callard, G. 1993. Culture of intact Sertoli/germ cell units and isolated Sertoli cells from *Squalus* testis. II. Stimulatory effects of insulin and IGF-I on DNA synthesis in premeiotic stages. *J. Exp. Zool.*, 267, 233-244.

Duguay, S. J., Park, L. K., Samadpour, M. and Dickoff, W. W. 1992. Nucleotide sequence and tissue distribution of three insulin-like growth factor I prohormones in salmon. *Mol. Endocrinol.* 6: 1202-1210.

Duguay, S. J., Swanson, P .and Dickoff, W. W. 1994. Differential expression and hormonal regulations of alternatively spliced IGF-I mRNA transcripts in salmon. *J. Mol. Endocrinol.*, 12: 25-37.

Gutierrez, J., Parrizas, M., Carneiro, N., Maestro, M. and Planas, J. 1993. Insulin and IGF-1 receptors and tyrosine kinase activity in carp ovaries : changes with reproductive stage. Fish. Physiol. Biochem.11: 247-254

Le Gac, F., Blaise, O., Fostier, A., Le Bail P.Y., Loir, M., Mourot B., Weil, C. 1993. Growth hormone (GH) and reproduction: a review. Fish. Physiol. Biochem., 11: 219-232.

Le Gac, F. and Loir, M. 1993 Expression of insulin-like growth factor (IGF-I) and action of IGF-I and II in the trout testis. Reprod. Nutr. Develop., 33: 80-81.

Le Gac, F., Loir, M., Le Bail, P. Y. and Ollitrault, M. 1996. Insulin -like growth factor (IGF-I) mRNA and and IGF-I receptors in isolated spermatogenic and Sertoli cells. Mol. Biol. Reprod. Dev. (In press).

Le Gac, F., Ollitrault, M., Loir, M. and Le Bail, P.Y. 1991. Binding and action of salmon growth hormone (sGH) in the mature trout testis. In*:* A.P. Scott, J.P. Sumpter, D.E. Kime and M.S. Rolf (eds) « Reproductive Physiology of Fish » Sheffield : Fish Symp 91, pp. 117-119.

Le Gac, F., Ollitrault, M., Loir, M. and Le Bail, P.Y. 1992. Evidence for binding and action of growth hormone in trout testis. Biol. Reprod. 46 : 949-957.

Loir, M. 1994. *In vitro* approach to the control of spermatogonia proliferation in the trout. Mol. Cell. Endocrinol., 102: 141-150.

Loir, M. and Le Gac, F. ,1994. Insulin-like growth factors-I and II binding and action on DNA synthesis in rainbow trout spermatogonia and spermatocytes. *Biol. Reprod.*51, 1154-1163

Nagahama Y. 1994. Endocrine regulation of gametogenesis in fish. Int. J. Dev. Biol.38: 217-229.

Shamblott, M. J. and Chen, T. T. 1993. Age related and tissue-specific levels of five forms of Insulin-like growth factor mRNA in a teleost. *Mol. Mar. Biol. Biotechnol. 2:* 351-361

Figure 1: IGF receptors in cultured testicular cells:
A: Receptor subunit size determination: total cell membranes were submitted to affinity hybridization with 125I-IGF-I, in the presence or absence of competitor, followed by cross-linking, protein solubilisation, electrophoresis in reducing conditions and revelation by autoradiography. B:. 125-I IGF-I binding to populations enriched in germ cells (see text) or Sertoli cells, and specific inhibition with increasing concentrations of competitors: rh.IGF-I, rh.IGF-II (1 to 1000 ng/ml) and b. Insulin (10 µg/ml).

A)

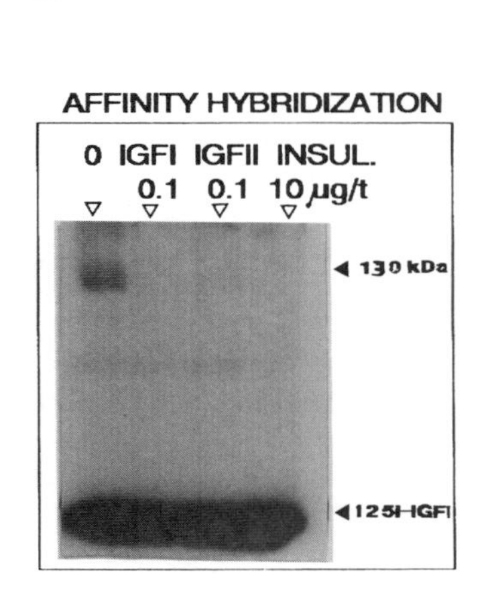

B)

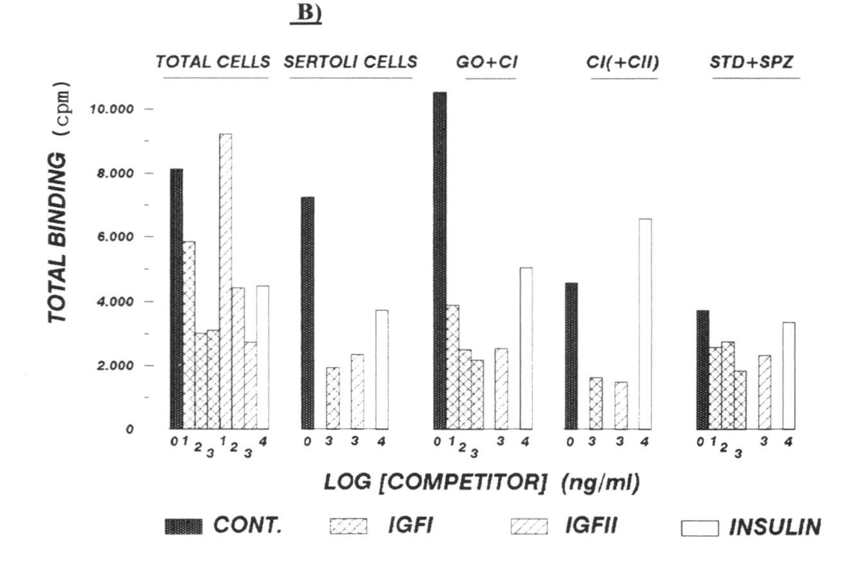

OVARIAN DEVELOPMENT IN TRIPLOID BROOK TROUT (Salvelinus fontinalis)

Tillmann J. Benfey

Department of Biology, University of New Brunswick, Fredericton, New Brunswick, Canada, E3B 6E1

Introduction

Numerous studies have demonstrated that triploid males undergo substantial testicular development, but that ovarian growth is greatly retarded in triploid females (Benfey, 1991). These studies have generally been terminated at the time of first sexual maturation in diploid controls. The purpose of this research was to test the hypothesis that ovarian growth is delayed in triploids due to an abnormally long period of vitellogenesis, and that older triploids would therefore eventually develop ovaries.

Methods

Triploid brook trout were produced by applying 9500 psi (65,500 kPa) pressure to eggs for 5 min, beginning 20 min after fertilization at 10°C. Diploid controls came from the same egg lots, but were not subjected to pressure treatment. Ploidy level of each fish was confirmed by erythrocyte flow cytometry. Fish of two year classes were killed at the time of ovulation in diploids (first-time as 2-year-olds and second-time as 3-year-olds), and their gonads removed for weight measurement and histology.

Results

With one exception, there was no difference in ovarian weight, when expressed as a percentage of total body weight, between 2- and 3-year-old triploids (0.06 ± 0.03% [SD] for both, n = 13 and 10, respectively). These values were significantly lower ($P < 0.0001$) than for diploids (22.1 ± 4.2%, n = 6). Maximum oocyte diameter was no different between the two year classes of triploids, with no oocytes developed beyond the cortical alveolus stage in these fish. The exception was a single 3-year-old triploid which ovulated numerous eggs that were greatly variable in size, as were unovulated oocytes remaining within the ovaries at the time this fish was killed (Fig. 1). The ovulated eggs were fertilized with normal (haploid) sperm, but none survived through the pre-hatch period. Many of the larger eggs appeared to be over-ripe. Although ovarian and ovulated egg weights were not measured for this "atypical" fish, their sum would have been intermediate between that of "typical" triploids and of normal diploids.

Fig. 1. Ovaries and ovulated eggs from an "atypical" triploid brook trout (bar = 1 cm).

Discussion

Triploid brook trout generally showed no further ovarian development between the ages of 2 and 3 years old, suggesting that slowed vitellogenesis does not account for the lack of ovarian growth in triploids up to the time of first sexual maturation in diploids. However, a single triploid showed substantial, but apparently asynchronous, ovarian development by the age of 3 years. This female may have had a greater amount of steroidogenic tissue development than is typical for triploids, thus allowing vitellogenesis to proceed at a slow rate and leading to asynchronous oocyte development.

Reference

Benfey, T.J., 1991. The physiology of triploid salmonids in relation to aquaculture. Can. Tech. Rep. Fish. Aquat. Sci. 1789: 73-80.

ISOLATION AND CHARACTERIZATION OF A VITELLOGENIN RECEPTOR IN WHITE PERCH, *MORONE AMERICANA*

D.L. Berlinsky, Y. Tao, and C.V. Sullivan.

Department of Zoology, North Carolina State University, Raleigh, North Carolina 27695-7617, USA.

Summary

Receptors for white perch vitellogenin (wVTG) were characterized by binding wVTG, labeled in vivo with ^{3}H-leucine or in vitro with ^{125}I, to semipurified ovarian membranes. The binding was saturable, ligand specific, and showed appropriate affinity (Kd) for a VTG receptor. When compared to plasma levels of VTG, the Kd suggests VTG receptors are normally saturated during vitellogenesis. The Kd did not change during vitellogenesis and the maximum binding capacity (MBC) increased only slightly in the preovulatory stage. These data suggest that changes in plasma VTG levels and wVTG receptor affinity or capacity are unlikely to regulate rates of oocyte growth in white perch.

Introduction

Most oocyte growth in teleosts is due to the incorporation of VTG, a hepatically synthesized lipoglycophosphoprotein, produced in response to circulating estrogens, primarily estradiol-17β (E_2). The primary objective of this study was to characterize the ovarian VTG receptor of a highly evolved perciform teleost, the white perch.

Methods and Results

Synthesis of ^{3}H-wVTG was induced in males injected with E_2 and ^{3}H-leucine. The wVTG was purified on DEAE-agarose (Tao et al., 1993). wVTG was also radiolabeled with ^{125}I using Iodogen. The integrity of radiolabeled wVTG's was verified by SDS-PAGE and autoradiography. Nearly 80% of the ^{125}I-wVTG was able to bind to a polyclonal VTG antiserum. Specific activity of the ^{125}I-VTG was determined by self-displacement using the antiserum. Specific binding of the ^{3}H-wVTG to ovarian membranes was proportional to the quantity of membrane present and saturable. The Kd of the wVTG binding sites (≃676 nM) for ^{3}H-wVTG was consistent with circulating wVTG levels (540-2700 nM) in maturing females (Jackson and Sullivan 1995). In a ligand specificity assay, wVTG reduced total binding of ^{3}H-wVTG to the membranes by more than 70% whereas white perch yolk proteins, striped bass VTG and chicken egg yolk very low density lipoprotein reduced binding by 40%, 63% and 69% respectively. Specific binding of ^{125}I-wVTG reached equilibrium after 4 hours at 24° C. As for ^{3}H-wVTG, Scatchard analysis indicated a single class of high affinity binding sites (Kd≃390 nM) for ^{125}I-wVTG and the MBC was 51 pmol wVTG/mg membrane protein. Specific binding of ^{125}I-wVTG was detected for membranes prepared from ovary, liver and muscle. The Kd for liver and ovary were similar, (345 vs. 390 nM) but the Kd for muscle was ≃ 4 fold greater (1440 nM). Scatchard analysis indicated that the Kd for ovarian membranes did not change during vitellogenesis and the MBC increased only slightly during the preovulatory period. Ligand blotting experiments revealed a wVTG receptor protein (Mr ≃ 157 KD) and a smaller protein likely to be its degradation product.

Discussion

The results of this study have demonstrated the presence of a specific receptor for wVTG on ovarian membranes of white perch supporting the concept that VTG is taken up into perch oocytes by receptor-mediated endocytosis. Comparison of the Kd of the receptor with circulating levels of wVTG during vitellogenesis suggests that wVTG receptors are normally saturated in situ and changes in circulating wVTG levels are not involved in regulating rates of oocyte growth. The finding that the Kd and MBC did not change during vitellogenesis suggests that the modulation of VTG receptor affinity and density is also unlikely to regulate oocyte growth rates. Membranes prepared from liver and muscle also contained specific binding sites for wVTG. We postulate that the hepatic receptor is involved in recycling yolk proteins from atretic oocytes and that wVTG may not be the preferred lipoprotein ligand for the muscle "VTG receptor."

Literature Cited

Jackson, L.F. and C.V. Sullivan, 1995. Reproduction of white perch: The annual gametogenic cycle. Trans. Am. Fish. Soc. in press.

Tao, Y., A. Hara, R.G. Hodson and C.V. Sullivan, 1993. Purification, characterization and immunoassay of striped bass (Morone saxatilis) vitellogenin. Fish Physiol. Biochem. 12:31-46.

This work was supported by grants from the UNC Sea Grant College Program (NA86AA-D-SG062 and NA90AA-D-SG062) and the National Coastal Resources Research and Development Institute (NA87AA-D-SG065, #2-5606-22-2).

IN VITRO HORMONAL CONTROL OF VITELLOGENIN SYNTHESIS IN TWO MARINE SPECIES, *DICENTRARCHUS LABRAX* AND *SPARUS AURATA*

Carnevali O., Mosconi G., Zanuy Doste S*. and Polzonetti-Magni A M

Department of Biology MCA, University of Camerino, Italy ; *Inst.Aquaculture of Torre de la Sal, Castellon, Spain

SUMMARY

The results indicate that, in male seabream, estradiol-17ß (E_2) induces vitellogenin (VTG) synthesis. By a time-course experiment, it was found that soon after three days' E_2 treatment, VTG appears in the plasma reaching the maximum within 30 days' treatment.

Pituitary hormones like GH and PRL as well homologous pituitary homogenates (HPH), induced *in vitro* vitellogenin (VTG) synthesis in both male and female sea bass, while they did not have any effects on the liver taken from spawning seabream.

INTRODUCTION

The accumulation of yolk proteins, i.e. vitellogenesis, is a characteristic shared by all oviparous vertebrates including birds, amphibians and teleosts (Wallace, 1985). Vitellogenin (VTG) is synthesized in the liver and transported in the blood to the oocytes, where it is selectively taken up by receptor-mediated endocytosis.

The aim of the present work was to investigate the hormonal regulation of the biosynthesis of VTG in marine teleosts. Therefore, seabream (*Sparus aurata*) VTG was characterized and its antibody validated using ELISA. To investigate the role of estradiol-17ß on VTG synthesis, a time-course experiment was performed. Lastly, the involvement of pituitary hormones in inducing hepatic VTG synthesis was considered, using *in vitro* experiments in which European sea bass (*Dicentrarchus labrax*) liver was cultured in presence of bream growth hormone (f-GH, GroPep).

RESULTS

Sparus aurata VTG purified from estrogenized male plasma, in SDS-PAGE showed one component of 180 kDa; in native PAGE, two bands were found both stainable with Coomassie Blue, Stains-all and concanavalin A-binding proteins. The antiserum raised against seabream VTG did not react with male plasma; serial dilutions of female plasma and that of estradiol-treated male ran parallel with the standard curve obtained with purified VTG.

The appearance of VTG in the seabream male plasma started three days after stimulation; VTG levels increased reaching the maximum at 30 days' treatment.

Hepatic VTG synthesis in *in vitro* incubation was significantly stimulated by E_2, HPH, f-GH and o-PRL in both male and female reproductive but not yet spawning sea bass; in females, the basal levels of VTG in the media (CM) were about 1 µg/ml (Fig. 1). On the contrary, no stimulation was observed in male or in female seabream liver during spawning, when the basal levels of female VTG in the media were found to be very high (10 µg/ml).

Fig. 1- February incubation of female and male sea bass liver.

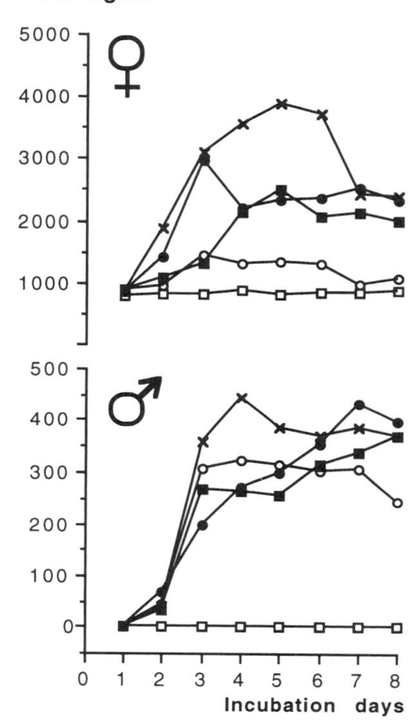

DISCUSSION

This study provides evidence that *in vivo* stimulation of seabream male by estradiol-17ß, as in other teleostean species, induces VTG synthesis, and that VTG plasma content reaches a maximum 30 days after treatment. In addition, a multihormonal control of vitellogenin synthesis was observed in sea bass in *in vitro* liver culture, since stimulatory effects of both f-GH and PRL were found.

REFERENCES

Wallace R.A. 1985. Vitellogenesis and oocyte growth in non-mammalian vertebrates. In: "Developmental Biology" 127-177 Ch. 3. Ed. by L.W. Bowder, Plenum Publ. Corp., New York.

Acknowledgments- This study was supported by National Research Council of Italy, Special Project RAISA

SEXUAL STEROIDS AND REGULATION OF PUBERTY IN MALE AFRICAN CATFISH (*Clarias gariepinus*)

J.B.E. Cavaco, R.W. Schulz, V.L. Trudeau* J.G.D. Lambert, H.J.Th. Goos

Utrecht University, Res. Group Comp. Endocrinol., Padualaan 8, 3584 CH Utrecht, The Netherlands; * Dept. of Zoology, Univ. of Aberdeen, Aberdeen AB9 2NT, UK

Introduction

In male African catfish, the first wave of spermatogenesis is initiated between 10-12 weeks of age. The brain-pituitary-gonad axis (BPG axis) is the endocrine system of prevailing importance for the regulation of reproductive processes, and hence for puberty. In fish, sex steroids initiate and/or accelerate the maturation of the BPG axis. Therefore the hypothesis was adopted that prepubertally produced sex steroids have stimulatory effects on the development of the BPG axis.

Material and Methods

Animals at the onset of puberty (10 weeks old) were implanted with solid silastic pellets containing 11β-hydroxyandrostenedione (OHA), 11-ketoandrostenedione (OA), testosterone (T), 17β-estradiol (E2) or no steroid (control). OHA was applied as it is the main product of testicular steroidogenesis at all stages of puberty. 11-Ketotestosterone (11-KT) and T are prominent plasma steroids. Their levels increase during puberty. OA was used instead of 11-KT, since OA has been shown to be converted effectively to 11-KT *in vivo*. E2 has been included in this study as it is known to stimulate pituitary gonadotropin (GTH) expression. Two weeks after implantation, we have examined the gonadosomatic index (GSI, Fig. 1) and the GTH II levels in the plasma and pituitary (Fig. 2).

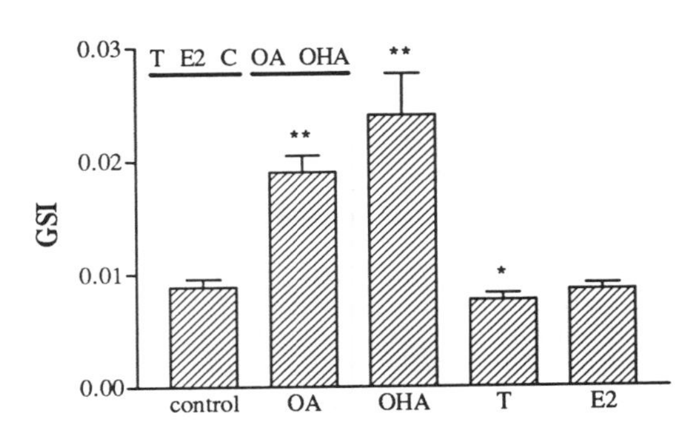

Fig. 1 - GSI (mean ± S.E.M., n = 18-35) of male African catfish 2 weeks after implantation of different steroids. ** - seminal vesicles were present in more than 50% of the animals; * seminal vesicles present in 5 animals. Groups sharing the same underscore are not significantly different (ANOVA followed by Fisher's PLSD, $p<0.05$).

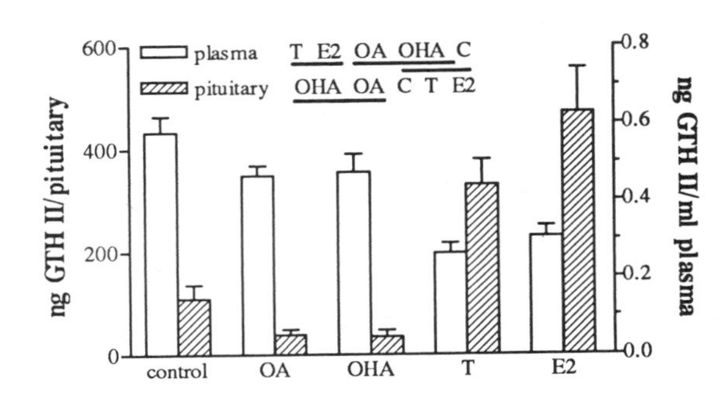

Fig. 2 - GTH II plasma levels (n = 18-35) and pituitary content (n = 10) (mean ± S.E.M.) 2 weeks after implantation of different steroids.

Results and Discussion

We have observed a clear difference between T and E2 on the one hand, and 11-oxygenated androgens on the other hand, regarding all parameters studied. The 11-oxygenated androgens significantly stimulated testicular growth and induced precocious differentiation of the seminal vesicles. However, both the pituitary GTH II content and circulating GTH II levels were reduced. This indicates that 11-oxygenated androgens may have a direct stimulatory effect on the testicular level, as has been shown for 11-KT in the male Japanese eel [1;2]. T and E2 did not affect testicular development, but led to an increase in the pituitary GTH II content. Interestingly, this was associated with reduced circulating GTH II levels, indicating that under the present experimental conditions, GTH II synthesis was stimulated, while its secretion was inhibited.

The present data suggest that different steroids have differential effects on the development of the BPG axis. As sex steroids stimulate the expression of GnRH in the brain [3;4], the GnRH system may be involved in the stimulation of pubertal development as well. Thus, our future experiments will include a combined treatment of steroids plus GnRH. In addition, the effects of a treatment with GTH alone will be studied.

Supported by J.N.I.C.T. - Portugal, grant BD/2603/93

References

[1] Miura *et al.* (1991) - *Proc.Natl.Acad.Sci.USA* 88: 5774-5778
[2] Nagahama,Y. (1994) - *Int.J.Dev.Biol.* 38: 217-229
[3] Amano *et al.* (1994) - *Gen.Comp.Endo.* 95: 374-380
[4] Montero *et al.* (1995) - *Neuroendocrinology* 61: 525-535

INHIBITION OF *FUNDULUS HETEROCLITUS* OOCYTE MATURATION *IN VITRO* BY SEROTONIN

J. Cerdá[1], T. R. Petrino[2], Y.-W. P. Lin[2] and R. A. Wallace[1]

[1]Whitney Laboratory, St. Augustine, FL 32036, USA; [2]Barry University, Miami Shores, FL 33161, USA.

Introduction

Increasing attention has recently been paid to brain peptides and growth factors that may regulate ovarian functions in a paracrine and/or autocrine fashion. Evidence also indicates that some neurotransmitters, such as serotonin (5-hydroxytryptamine, 5-HT), inhibit progesterone-induced oocyte maturation in amphibian follicles (Buznikov et al., 1993).

In fish, however, Iwamatsu et al. (1993) have reported that a low molecular weight serum factor isolated from various animals, identified as 5-HT, is able to induce oocyte maturation in medaka (*Oryzias latipes*) preovulatory follicles *in vitro*, by stimulating granulosa cell steroid production.

In view of these contradictory results, we investigated the influence of 5-HT on steroidogenesis and oocyte maturation *in vitro* in full-grown *F. heteroclitus* ovarian follicles.

Results

Preincubation of follicles with 5-HT for 5 min inhibited both *F. heteroclitus* pituitary extract (FPE, 0.5 pituitary equivalents/ml)- and 17α,20β-dihydro-4-pregnen-3-one (DHP, 0.1 μg/ml)-induced oocyte maturation in a dose-dependent manner. Maximum inhibition occurred at $5x10^{-7}$ and $5x10^{-6}$ M 5-HT for the FPE- and DHP-induced maturation, respectively; the continuous presence of 5-HT alone did not induce oocyte maturation.

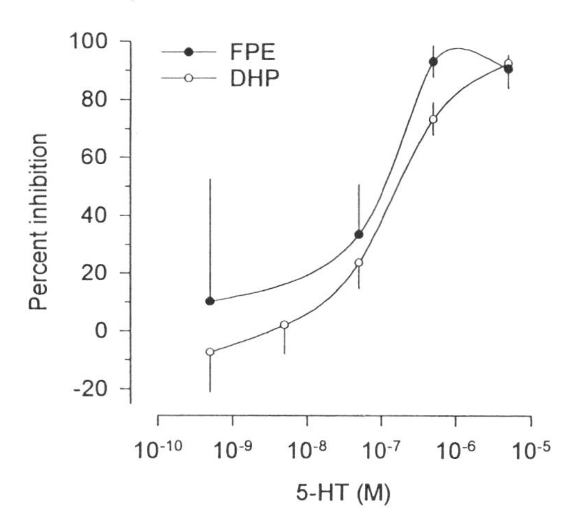

Fig. 1. Inhibition of oocyte maturation by 5-HT.

When the effect of 5-HT was plotted as % of inhibition, both FPE- and DHP-induced GVBD showed an IC_{50} (5-HT concentration to induce 50% inhibition) of approximately 10^{-7} and $2x10^{-7}$ M, respectively (Fig. 1). This inhibition proved to be reversible and independent of FPE-induced steroid production by the follicle, since the levels of 17β-estradiol and DHP released into the culture medium were not modified by 5-HT.

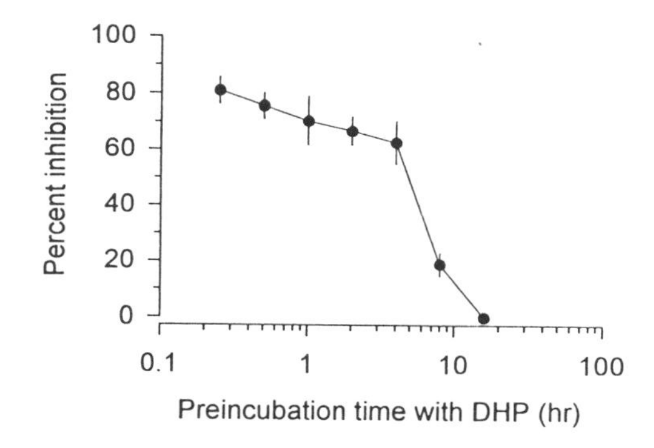

Fig. 2. 5-HT action on DHP-preincubated follicles.

Finally, when follicles were preincubated with DHP for increasing times, subsequently washed, and transferred to steroid-free, 5-HT-containing medium, inhibition was still observed in follicles preincubated for up to 4 h; however, inhibition by 5-HT was reduced or eliminated after 8-h and 16-h preincubation with DHP, respectively (Fig. 2).

Conclusions

1. 5-HT reversibly inhibits both FPE- and DHP-induced oocyte maturation *in vitro* in *F. heteroclitus* ovarian follicles, similar to what has been found for amphibian follicles.
2. Inhibition of FPE-induced oocyte maturation by 5-HT is not due to a negative effect on steroid production by granulosa cells.
3. It appears that 5-HT receptors are present in the ovarian follicle and that they are able to inhibit steroid-induced maturation at some point downstream in the steroid-transduction pathway.

References

Buznikov et al. (1993). Int. J. Dev. Biol., 37:363-364.
Iwamatsu et al. (1993). Develop. Growth Differ., 35:625-630.

Participation of JC was financed by a postdoctoral fellowship from the Ministry of Education and Science (Spain).

THE EFFECTS OF EXOGENOUS ESTRADIOL-17β ON GONADAL DEVELOPMENT IN JUVENILE MALES OF PROTANDROUS BLACK PORGY, *ACANTHOPAGRUS SCHLEGELI*

C.F. Chang, E.L. Lau and B.Y. Lin

Department of Aquaculture, National Taiwan Ocean University, Keelung 10224, Taiwan, Republic of China

Summary

Juvenile black porgy were divided into 4 groups, fed with control diet or diet mixed with estradiol-17β (E_2, 0.25, 1.0, or 4.0 mg/kg feed) for 7 mo. Higher GSI was observed in 0.25 and 1.0 mg E_2 groups compared to the control and 4.0 mg E_2 groups during spawning season. E_2 (4.0 mg) completely suppressed spermiation, while low doses of E_2 (0.25 and 1.0 mg) delayed spermatogenesis but increased the number of spermiating fish and spermiation volume. After 5 mo of 4 mg E_2 treatment, testicular tissue regressed and ovarian tissue with primary oocytes developed. Elevated levels of plasma E_2 were observed only in the 4.0 mg E_2 group while low E_2 and testosterone (T) were observed in the control and other E_2-treated groups. Higher 11-ketotestosterone (11-KT) levels were observed in the 0.25 and 0.1 mg E_2-treated groups. The data suggest that exogenous E_2 could stimulate the development of either testicular or ovarian tissue depending on the dosage of E_2. Plasma 11-KT but not T closely correlates with testicular development in black porgy.

Introduction

Black porgy is a marine protandrous hermaphrodite. Fish are males for the first two years of life but begin to reverse to females after the third year. High levels of plasma E_2 in the prespawning and spawning season are likely correlated with the natural sex reversal of 3-year-old black porgy (Chang *et al.*, 1994). Therefore, the objectives are to investigate the responses of gonadal development, spermiation, the concentration of plasma sex steroids and vitellogenin by oral administration (from Sep to Apr) of E_2 in yearling black porgy.

Results and Discussion

Higher GSI ($p < 0.05$) were observed in Feb in the groups treated with low doses of E_2 (0.25 and 1.0 mg). In these two groups, the maximal value of GSI (3.2% in Feb) was 10 times higher than the maximal value observed in control (0.32% in Jan). As indicated by the occurrence of spermiation, the beginning of the spawning season (Dec in control) was delayed by two wk in 0.25 mg E_2 group and by 1 1/2 mo in 1.0 mg E_2 group. The duration of the spawning season was increased from 2 1/2 mo in control to 3 1/2 mo in the 0.25 mg E_2 group. Low doses of E_2 also increased the percentage of spermiating fish from 60% in control to 100% in low doses of E_2 group as well as amoung of milt per spermiating fish from a maximum volume of 0.22 ml in control to 1.6 ml in the 0.25 mg E_2 dose (4.0 mg/kg) completely suppressed spermiation.

Low levels of plasma E_2 (< 100 pg/ml) were observed in all groups except the 4.0 mg E_2 group in Nov (400 pg/ml). T slightly increased in Mar (270 pg/ml) only in control. 11-KT increased in Nov and became plateau in control group. In the 0.25 mg E_2 group, 11-KT increased in Nov but kept increasing up to Feb (maximal value 2.5 ng/ml 6 times higher than maximal value 0.42 ng/ml in control). In 1.0 mg E_2 group, the start of increase in plasma 11-KT levels was delayed until Dec, such larger increase of 11-KT (up to 5.0 ng/ml) was observed in Feb.

In conclusion, low doses of E_2 (0.25 and 1.0 mg/kg) stimulated testicular development and spermiation. On the contrary, high dose (4.0 mg/kg) of E_2 suppressed spermiation and induced the sex reversal in one-year-old black porgy. Plasma levels of 11-KT but not T were closely associated with testicular development and spermiation.

The mechanism of stimulatory effects of low doses of E_2 on testicular development and activity deserves further study in black porgy. The success of the induction of sex reversal by high dose of E_2 further supports the important role of E_2 in the sex change of black porgy (Chang *et al.*, 1994).

References

Chang, C.F., Lee, M.F., and Chen, G.R. 1994. Estradiol-17β associated with the sex reversal in protandrous black porgy, *Acanthopagrus schlegeli*. J. Exp. Zool. 268:53-58.

UNIVERSAL VERTEBRATE VITELLOGENIN ANTIBODIES.

S. A. Heppell[1], N.D. Denslow[2], L.C. Folmar[3], and C.V. Sullivan[1]

[1]Department of Zoology, North Carolina State University, Raleigh; [2]Department of Biochemistry and Molecular Biology, University of Florida, Gainesville; [3]United States Environmental Protection Agency, Gulf Breeze, FL

Summary

Anti-vitellogenin (anti-VTG) antibodies were generated against conserved epitopes of the VTG molecule using both monoclonal and polyclonal techniques. Antibody screening indicates that these antibodies recognize VTG from a wide variety of vertebrates. An enzyme-linked immunosorbant assay (ELISA) is being developed for use as an indicator of maturation in fishes and as a test for the estrogenicity of compounds being released into the environment.

Introduction

Vitellogenesis, the hepatic synthesis and secretion of egg-yolk precursor (VTG), is estrogen inducible and can be used as a marker for the onset of maturation in fish and as an indicator of the estrogenicity of xenogenic compounds. In order to reduce the effort necessary to generate homologous VTG assays for every species of interest, we attempted to create a "universal" immunoassay capable of detecting VTG in all oviparous vertebrates. Monoclonal antibodies (mAbs) were generated against purified rainbow trout VTG. A polyclonal antiserum was generated against a peptide representing the consensus N-terminal amino acid sequence of VTG (Folmar et al. 1994). Resulting antibodies were screened by ELISA and Western blot.

Methods

MAbs were generated in Balb/C mice using VTG purified from rainbow trout (Oncorhynchus mykiss; Heppell 1994). Hybridoma colonies were selected that showed immunological cross-reactivity in an ELISA between VTG purified from rainbow trout (Hara et al. 1993) and striped bass (Morone saxatilis; Tao et al. 1993), but not to male plasma. Positive mAbs were subsequently screened by ELISA and Western blot against VTG's from a wide variety of fish and other vertebrates. The polyclonal antiserum was raised in New Zealand white rabbits against the 15-amino acid synthetic consensus peptide, conjugated to keyhole limpet hemocyanin (Heppell et al. 1995). Computer aided analysis of protein structure indicated that this segment was strongly antigenic. The antiserum was screened in the same manner as the mAbs, with the same VTG peptide conjugated to ovalbumin to ensure reactivity only to the peptide part of the conjugate.

Results

Three fusions resulted in the generation of seven positive colonies. All hybridomas demonstrating diverse cross-reactivity were of the IgM class. The most promising hybridoma is hybridoma 2D8, which has been selected for use in current and future research. Western blot analyses show that the 2D8 mAb is specific to a large molecular weight, estrogen inducible plasma protein which we identify as VTG in the plasma of rainbow trout and striped bass. ELISA and Western blot analyses of plasma from other species demonstrate that our mAb recognizes high molecular weight, estrogen inducible proteins in all species of teleost fish screened to date, including the rainbow trout, striped bass, brown bullhead, hardhead catfish, and pinfish. The mAb also recognizes a high-molecular weight plasma protein in female and/or estrogen-treated chicken, black rat snake, and bullfrog. The anti-VTGpeptide antiserum directed against the consensus N-terminal region of VTG reacts well with VTG in the plasma of striped bass, brown bullhead, hardhead catfish, pinfish, and black rat snake. There is only slight reactivity to bullfrog VTG, and the antiserum fails to recognize VTG in the plasma of rainbow trout, carp, or chicken.

Discussion

Our results indicate that we have generated a mAb capable of recognizing VTG from all vertebrate classes and species tested to date. The mAb demonstrates a more diverse array of cross-reactivity than the polyclonal antiserum. The conserved site(s) recognized by the mAb is likely only slightly antigenic, as the immune response of the mouse failed to induce a class shift from IgM to IgG in antibodies directed at the relevant epitope(s). Further research is in progress to develop a "universal" VTG assay for use in fisheries, aquaculture, and environmental toxicology.

References

Folmar, L.C., N.D. Denslow, R.A. Wallace, G. LaFleur, S. Bonomelli, and C.V. Sullivan. 1995. J. Fish Biology. 46:255-263.

Hara, A., C.V. Sullivan, and W.W. Dickhoff. 1993. Zool. Sci. 10:245-256.

Heppell, S.A. 1994 M.S. Thesis. North Carolina State University, Raleigh. 56 pp.

Heppell, S.A., N.D. Denslow, L.C. Folmar, and C.V. Sullivan. 1995. Env. Health Perspect, in press.

Tao, Y., A. Hara, R.G. Hodson, L.C. Woods III, and C.V. Sullivan. 1993. Fish Physiol. Biochem. 12:31-46.

This work was supported by the North Carolina Biotechnology Center, the Electric Power Research Institute, and a cooperative agreement between the USEPA and the University of Florida.

THREE EGG YOLK PROTEINS ARE DERIVED FROM SALMONID VITELLOGENIN

N. Hiramatsu and A. Hara

Nanae Fish Culture Experimental Station, Faculty of Fisheries, Hokkaido University,
Kameda, Hokkaido 041-11 Japan.

Summary

Vitellogenin (Vg) and its 3 egg yolk protein products lipovitellin (Lv), phosvitin (Pv) and β'-component (β') were isolated from mature female Sakhalin taimen (*Hucho perryi*). The antiserum against Vg reacted with Lv and β'. Furthermore, an antiserum against Pv was prepared and it reacted with dephosphorylated Pv and Vg. These results show 3 egg yolk proteins (Lv, Pv and β') are derived from Vg in salmonid fishes.

Introduction

Vg has been well-characterized as a precursor for egg yolk. In amphibians and birds, it has been proved conclusively that Vg is converted into two egg yolk proteins, Lv and Pv. In salmonid fishes, Lv, Pv and β' have also been purified and identified as egg yolk proteins. Hara and Hirai (1978) showed that rainbow trout Vg is converted into two egg yolk proteins, egg protein 1 (Lv) and egg protein 2 (Ep2), and that these two proteins cross react with an antiserum to Vg. Ep2 consists of a complex of β' and Pv. It has previously been unidentified as such because the Ep2 complex had only one antigenicity. There have been few studies on the direct relationship between serum Vg and egg yolk proteins in fishes. The present study describes the purification and characterization of taimen Vg and its three egg yolk protein products. The relation between Vg and the yolk proteins was also examined using antisera raised against Vg and Pv.

Materials and Methods

Vg and egg yolk proteins were purified according to Hara *et al.* (1993) and Markert and Vanstone (1971). Antisera were raised in rabbits against taimen Vg, Lv and β' by intradermal injection of each protein. An antiserum against taimen Pv (a-Pv) was also prepared as follows. Alkaline phosphatase (Ap) and 0.2M Na_2HPO_4 were added to purified Pv, the mixture was incubated for 4 hours at 37 ℃ to dephosphorylate Pv, and the solution was injected into rabbits to generate a-Pv. The antiserum was pre-absorbed by Ap before use. Purified proteins were analyzed by biochemical, immunological, and electrophoretical techniques.

Results and Discussion

Vg and egg yolk proteins of taimen were highly purified immunologically and electrophoreetically. Mol. mass (Table 1) and amino acid composition of these proteins were similar to those reported for other salmonids. The similarlity supports the hypothesis that the Vg gene and Vg structure are highly conserved in salmonids (Hara *et al.*, 1993). Lv and β' reacted with antiserum against Vg, but Pv did not. In addition, Pv showed peculiar characteristics; strong negative charge, non-stainability against Coomassie Brilliant Blue (CBB) in electrophoresis, non-antigenicity and low absorbancy at 280nm. We suspect these properties are caused by extensive phosphorus binding to serine residues on Pv. Pv dephosphorylated by Ap stained with CBB and an antiserum (a-Pv) could be prepared against it (Fig.1). The a-Pv reacted with Vg and dephosphorylated Pv, but not with Lv, β' and Ap in Western blots. These results confirm that Vg includes Pv in its structure. It can be suggested that Vg is incorporated into salmonid oocytes, and then converted to three egg yolk proteins, Lv, Pv and β'.

Table1. Molecular mass of Vg and its related egg yolk proteins in Sakhalin taimen.

Mol. mass (kDa)		Vg	Lv	β'	Pv
Gel filtration		540	330	30	23
SDS-PAGE	2ME-	240	150	34	23
	2ME+	165	92,29	17	23

2ME+: reduced samples, 2ME-: non-reduced samples.

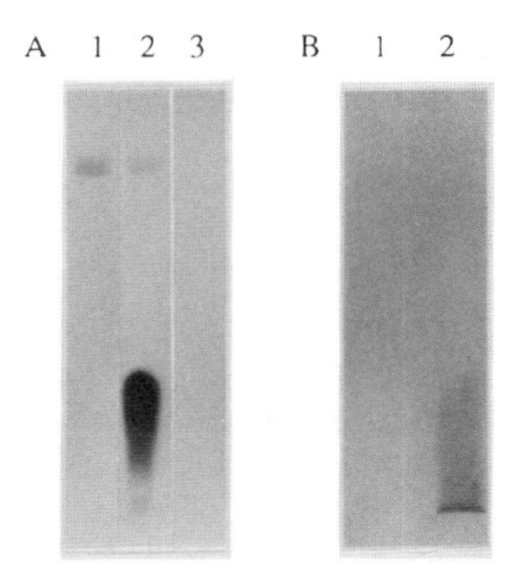

Fig.1. SDS-PAGE (A) and Western blotting (B) of phosvitin. 1 : alkaline phosphatase, 2 : dephosphorylated phosvitin, 3 : phosvitin. Western blot analysis was carried out using antiserum against phosvitin.

References

Markert, J. R. and W. E. Vanstone, 1971. Egg proteins of coho salmon (*Oncorhynchus kisutch*): chromatographic separation and molecular weights of the major proteins in the high density fraction and their presence in salmon plasma. J. Fish. Res. Bd. Can. 28:1853-1856.

Hara, A. and H. Hirai, 1978. Comparative studies on immunochemical properties of female-specific serum protein and egg yolk proteins in rainbow trout (*Salmo gairdneri*). Comp. Bichem. Physiol. 48B: 389-399.

Hara, A., C. V. Sullivan and W. W. Dickhoff, 1993. Isolation and some characterization of vitellogenin and its related egg yolk proteins from coho salmon (*Oncorhynchus kisutch*). Zool. Sci. 10: 245-256.

INFLUENCE OF NATURAL VITELLOGENESIS AND ESTRADIOL-17β TREATMENT ON PROTEIN SYNTHESIS AND GLUCONEOGENESIS IN ISOLATED HEPATOCYTES OF *ZOARCES VIVIPARUS*.

Bodil Korsgaard, Institute of Biology, Odense University, Denmark

Summary

Estradiol-17β treatment of male *Z. viviparus* induces *de novo* synthesis of the yolk-precursor protein vitellogenin and a simultaneous shift in overall metabolic performance of isolated hepatocytes A decrease was observed in the rate of gluconeogenesis, the activity of associated enzymes and the oxidative metabolic flux correlated with an increase in the protein synthesis capacity of the cells.

Introduction

In *Zoarces viviparus* males overall amino acid metabolism was observed to be affected by estradiol-17β as indicated by marked changes in biochemical cellular components associated with protein synthesis in the liver and alterations in the amount of free amino acids in the blood and liver (Korsgaard, 1990). The present study attemps to elucidate the effect of estradiol-induced protein synthesis activity on alanine gluconeogenesis in isolated hepatocytes.

Results

During natural vitellogenesis an increase was observed in the hepatic protein synthesis activity. Estradiol-treatment of male or female fish resulted in a marked increase in the capacity of isolated hepatocytes for incorporating labeled amino acid into protein correlated with a decrease in the gluconeogenic capacity of the cells (Table 1). Activities of enzymes associated with amino acid metabolism were also decreased by estradiol-treatment (Figure 1).

Table 1. Hepatic protein synthesis activity and gluconeogenesis from ^{14}C labelled amino acids. Mean (±SE). Estradiol (500 μg) was implanted intraperitoneally in a coconut oil pellet 3 weeks before sampling.

	^{14}C incorpor. pmol/min/g	Glucose μmol/g/h	CO_2
Females			
Control	23.6 (5.1)	0.108 (0.04)	0.42 (0.10)
Estradiol	69.9 (8.6)	0.03 (0.004)	0.17 (0.04)
Males			
Control	8.23 (1.58)	0.11 (0.06)	0.33 (0.13)
Estradiol	22.9 (2.93)	0.005 (0.00)	0.13 (0.01)

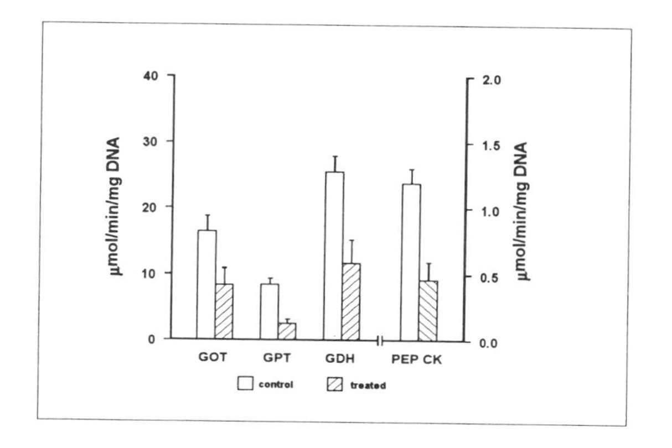

Figure 1. Activities of enzymes associated with amino acid metabolism in estradiol-treated males.

Estradiol-treatment of male fish resulted in a dramatic decrease in testicular weight and a concomitant disappearance of the lobule boundary (sertoli) cells in the testicular tubules.

Discussion

Vitellogenin (partly purified) was identified in the plasma by native gel electrophoresis. Estradiol-treatment was observed to enhance the incorporation of amino acids into protein and specifically altered the concentration of amino acids in liver and plasma of male *Zoarces viviparus (*Korsgaard, 1990). Amino acids are important precursors for protein synthesis but also a source of carbon for synthesis of glucose by gluconeogenesis. The present results indicate that under influence by estradiol hepatic incorporation of amino acid into protein dominate over *de novo* glucose synthesis by amino acid derived carbon as also observed in hepatocytes from estradiol-treated salmonid fish (Korsgaard and Mommsen, 1993). Estradiol-treatment had a profound effect on testicular weight and cytology.

References

Korsgaard, B (1990). Estrogen treatment and its influence on protein synthesis and amino acid metabolism in *Zoarces viviparus* males. Fish Physiol. Biochem. 8, 121-127.

Korsgaard, B and Mommsen T.P. (1993). Gluconeogenesis in hepatocytes of immature rainbow trout *(Oncorhynchus mykiss):* control by estradiol. Gen. Comp. Endocrinol. 89:17-27

CORRELATION BETWEEN PLASMA AND EGG STEROID HORMONE CONTENT OF ARCTIC CHARR

M. N. Khan, R. Renaud, and J. F. Leatherland

Department of Biomedical Sciences, Ontario Veterinary College, University of Guelph, Guelph, ON N1G 2W1, Canada

Summary

HPLC methods were used to identify the major steroid classes in eggs of Arctic charr, Salvelinus alpinus and compare the steroid profile with that of maternal plasma; although progestogens, predominantly MIH, were the principal steroids in eggs, they were not found in plasma. By the time of hatching, embryos had non-detectable total body steroid levels and exhibited a marked ability to metabolize [^{3}H]P4 in vitro.

Introduction

During gonadal maturation, the thecal and granulosa cells secrete large amounts of steroid hormones; the transfer of these steroid hormones to the maturing oocyte is likely. This study uses HPLC methods to determine which steroids are transferred to the eggs, and the ability of the developing embryos to metabolize those steroids.

Results

The plasma HPLC steroid profiles of 5 year old charr (3-4 kg) sampled prior to, during and 3 weeks after ovulation comprised small C, CS, E_2, 11-keto-A and T peaks in pre-ovulatory fish, additional 11-β-OH-A and AD peaks in the ovulatory fish, and a reduction in the size of peaks and loss of 11β-OH-A in post-ovulatory fish. No progestins were present in the chromatographs. By comparison, unfertilized eggs exhibited a major MIH peak and smaller peaks of 17α-OH-P, 11β-OH-P, 20β-DHP, P4 and possibly T; these steroids were not measurable in embryos at the time of hatch.

Minces of newly-hatched embryos incubated in Medium 199 metabolized [^{3}H]P4; after 20 min radioactive (ra) 20α-DHP and 20β-DHP were seen. By 24h, there was no ra P4; the major ra was found eluting between 9.6 and 12 min, concomitant with the unlabelled CSS, 11β-OH-A, 11-keto-A and 11-DOC.

Discussion

The results support work carried out in salmon in which steroid hormone levels were measured by RIA in eggs (Feist et al., 1990). In both salmon and charr, the predominant steroid hormones in the yolk are progestogens, particularly MIH, but these steroids were not evident in the plasma of adult females sampled during the pre-, ovulatory and post-ovulatory period, probably reflecting the brief appearance of these steroids in the circulation.

The absence of steroid hormones in extracts of embryonic Arctic charr at the time of hatch suggests that the embryos can metabolize yolk steroid hormones, agreeing with similar work in carp, Cyprinus carpio (Kime, 1990); this hypothesis was supported by evidence of the metabolism of P4 by newly-hatched Arctic charr embryos in vitro, beginning within 20 min and completed by 24 h.

The results of the study to date suggest that unlike thyroid hormones, which appear to be eliminated from the embryo at approximately the same rate as yolk is absorbed, the steroid hormones may be more actively eliminated, by metabolism and conjugation followed by excretion.

Abbreviations

AD, androstenedione; C, cortisol; CS, cortisone; CSS, corticosterone; 11-DOC, 11-deoxycortisol; 20β-DHP, 20β-dihydroxyprogesterone; E_2, 17β-estradol; 11β-OH-P, 11β-hydroxy-progesterone; 17α-hydroxy-progesterone; 11-keto-A, 11-ketoandrostenedione; MIH, 17α,20β-dihyroxy-4-pregnen-3-one; P4, progesterone; T, testosterone;

References

Feist, G., C. B. Schreck, M. S. Fitzpatrick and J. M. Redding, 1990. Sex steroid profiles of coho salmon (Oncorhynchus kisutch) during early development and sexual differentiation. Gen. Comp. Endocrinol. 80:299-313.

Kime, D. E. 1990. In vitro metabolism of progesterone, 17-hydroxyprogesterone, and 17,20β-dihydroxy-4-pregnen-3-one by ovaries of the common carp Cyprinus carpio: production rates of polar metabolites. Gen. Comp. Endocrinol. 79:406-414.

Acknowledgements : The work was supported by the OMAFRA Program 42.

ANALYSIS OF DIFFERENTIALLY EXPRESSED mRNAS DURING SPERMATOGENIC DEVELOPMENT IN THE SHARK TESTIS

H. L. Krasnow and G.V. Callard
Department of Biology, Boston University, Boston, MA 02215

Summary

In this report, we describe the isolation and characterization of stage-specific cDNA markers with utility for the elucidation of gene expression and regulation during spermatogenic development.

Introduction

Spermatogenesis is a unique developmental sequence, involving the functional interdependence of germ cells and Sertoli cells. Due to a cystic mode of spermatogenesis and the simple linear arrangement of succeeding germ cell stages across the diameter of the testis, the dogfish shark (Squalus acanthias) is an ideal model for studying the stepwise regulation of spermatogenesis (1). Dissected tissues in defined spermatogenic stages (PrM, premeiotic; M, meiotic; PoM, postmeiotic) can be used for structural or biochemical analysis. Also, staged tissues can be used to isolate intact spermatocysts (germ cell/Sertoli cell units), which continue to express many stage-related morphological and functional traits for at least 7 days in culture. This provides an in vitro system for studying direct effects of steroids and other factors on spermatogenic development. To obtain stage-specific and cell-type specific maturational markers for use in regulatory studies, we have taken two approaches: (1) use available probes representing genes known or suspected to vary during spermatogenesis to isolate homologous cDNAs from stage-specific testicular libraries; and (2) identify unknown cDNAs which are differentially expressed during development.

Methods

RNA was isolated from staged tissues (PrM, M, PoM) and the Superscript Kit (BRL, Bethesda MD) was used to construct cDNA libraries which yielded 10^6 pfu's(2). An aliquot (300,000 clones) of the PrM library was screened using a rockcod ß-tubulin cDNA (W. Detrich, Northeastern) and positive clones isolated by standard methodology and sequence analysis. Northern analysis was used to study number and size of ß-tubulin transcripts in staged tissues (30 µg each). Polymerase chain reaction-differential display (PCR-DD) is a method that employs a series of arbitrary 10 mer primer pairs to identify differentially expressed mRNAs . RNA (1 µg) from staged tissues were reverse transcribed and amplified using different sets of 5' and 3' primers and ^{35}S-labeled nucleotides (Genhunter, Brookline MA). Reactions were carried out in triplicate and products size-separated on sequencing gels.

Results

A 500 bp clone with high sequence homology to B-tubulins of other species was isolated. Northern blot analysis using either the rockcod (Fig. 1) or shark testis-specific (not shown) ß-tubulin cDNA as hybridization probe revealed that steady state transcript levels varied quantitatively and qualitatively by stage. After PCR-DD, 180 different bands were visualized, of which 30% were reproducibly stage-specific or stage-dependent (data not shown). Cloning and sequencing of selected bands is now in progress.

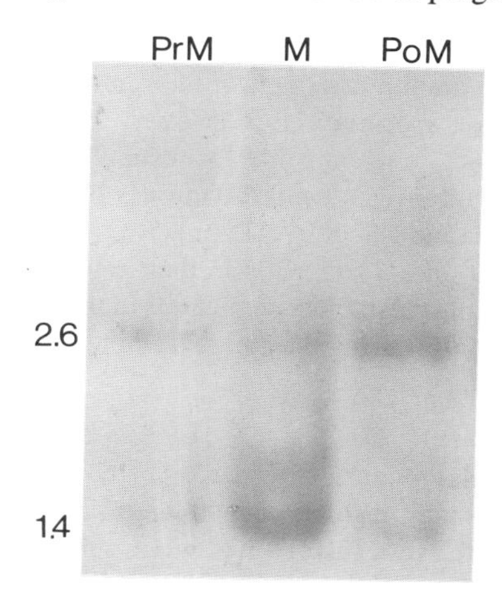

Figure 1.Stage-dependence of ß-tubulin transcripts in shark testis. RNA from premeiotic (PrM), meiotic (M), and postmeiotic (PoM) stages was hybridized with a rockcod ß-tubulin cDNA. Two stage-dependent transcripts (1.4, 2.6 kb) were seen, and an additional PrM-specific band (2.9 kb) was seen using our shark specific cDNA (not shown).

Discussion

Results show that a dual approach for obtaining known and unknown cDNAs as markers of spermatogenesis is feasible using the shark testis model. Additional studies are required to identify the genes encoding the products obtained by PCR-DD and to determine the germ cell vs Sertoli cell origin of these cDNAs and the different ß-tubulin transcripts. This panel of cDNAs will have utility for in vivo and in vitro regulatory studies.

References

1. Callard GV, Betka M and J. Jorgensen In: Bartke, A. (ed): Function of Somatic Cells in the Testis, pp.27-54, 1994
2. Krasnow HL and GV Callard (1995) Cloning of a B-Tubulin cDNA From Shark(Squalus acanthias) Testis. The Bulletin MDIBL 34: 111
3. Liang P and A Pardee, 1992 Differential Display of Eukaryotic Messenger RNA by Means of the Polymerase Chain Reaction. Science 257:967

Supported by NICHD16715 (GVC) and the Endocrine Society (HLK)

REGULATION OF HYDROMINERAL BALANCE AND SPERMIATION BY mGtH AND PROLACTIN IN FRESHWATER CATFISH, HETEROPNEUSTES FOSSILIS

Bechan Lal, T.P. Singh, H.N. Singh and S. Harikrishnan

Fish Endocrinology Laboratory, Department of Zoology, Banaras Hindu University, Varanasi - 221 005, India

Summary

Present study indicates that spermiation in *H. fossilis* requires massive testicular hydration which occurs due to high osmotic tension created by active Na^+ and K^+ influx under the influence of prolactin and mGtH.

Introduction

In teleosts, spermiation is hormone dependent thinning or hydration of semen. However, no regulatory mechanism of spermiation is known. Therefore, to explore the factors responsible for testicular hydration, seasonal variations in water content, Na^+ and K^+ concentrations and Na^+ K^+ ATPase activity in testes and role of o-PRL and *Mystus* GtH in their regulation were monitored through established methods in the freshwater catfish, *H. fossilis* during the early spawning phase of its annual reproductive cycle.

Results

Testicular water, Na^+ and K^+ concentrations exhibited seasonal fluctuations with their peak values in spawning phase coinciding with spermiation. Hypophysectomy caused significant reduction in testicular levels of water, Na^+ and K^+ as well as Na^+ K^+ ATPase activity and inhibited spermiation. PRL treatment recovered water and K^+ partially, but Na^+ and Na^+ K^+ ATPase activity fully. Administration of mGtH also recovered water and K^+ partially, but had no effect on Na^+ and Na^+ K^+ ATPase. Combined treatment of PRL and mGtH restored normal levels of all parameters studied.

Discussion

Present study demonstrated massive testicular hydration during reproductively active phases which took place under the control of PRL and GtH as they not only compensated water loss folloing hypophysectomy but also restored spermiation which was inhibited due to hypophysectomy. The massive water drive in testes appeared to occur due to high osmotic tension created by active influx of Na^+ & K^+ perhaps under PRL & GtH. K^+ influx in testes was regulated by PRL as well as GtH while Na^+ by PRL only. PRL influenced Na^+ & K^+ influx through Na^+ K^+ ATPase activity. However, GtH seemed to mediate its effect on K^+ influx through any other mechanism which remains to be determined as it did not alter the activity of testicular Na^+ K^+ ATPase.

Table 1 : Testicular water content, Na^+ & K^+ conc. (mEq/kg wet testes, ± SEM) during different reproductive phases

Phases	% Water content	Na^+	K^+
Preparatory	61.92 ± 0.64*	46.26 ± 0.87*	19.40 ± 0.72*
Prespawning	69.90 ± 0.68*	60.98 ± 0.93*	24.51 ± 0.51*
Spawning	81.72 ± 0.96*	70.84 ± 1.36*	36.68 ± 0.43*
Postspawning	58.54 ± 0.54*	45.29 ± 1.41*	15.29 ± 0.26*

* Denotes Significant to Preceeding Values

Table 2 : Effect of Hypo and oPRL & mGtH therapy on testicular hydromineral balance and Na^+ K^+ ATPase acticvity (uM Pi/h/mg protein). Values are Mean ± SEM.

Treatments	% Water content	Na^+	K^+	Na^+ K^+ ATPase
Unoperated	82.80 ± 0.25	71.01 ± 0.72	29.44 ± 2.07	51.39 ± 1.79
Sham-operated	82.81 ± 0.40	69.56 ± 1.33	28.67 ± 2.91	49.12 ± 2.29
Hypo	68.01 ± 0.42[a]	58.27 ± 4.04[a]	17.50 ± 0.88[a]	40.86 ± 1.52[a]
Hypo + Saline	68.31 ± 0.71	58.31 ± 2.64	16.50 ± 0.83	38.11 ± 1.15
Hypo + oPRL	76.48 ± 0.42[b]	69.40 ± 2.89[b]	22.91 ± 1.04[b]	49.12 ± 2.22[b]
Hypo + mGtH	77.62 ± 0.38[b]	59.83 ± 1.30	23.79 ± 0.66[b]	38.90 ± 1.42
Hypo + oPRL+mGtH	83.01 ± 0.93[b]	70.21 ± 1.01[b]	31.41 ± 1.79[b]	50.93 ± 1.32[b]

Significant at $P < 0.05$, a - Sham-operated vs. Hypo; b - Hypo+Saline vs. Hypo+Hormones

ESTRADIOL-17ß INDUCES VITELLINE ENVELOPE PROTEINS IN 15 TELEOST SPECIES

D. G. J. Larsson[1], S.J. Hyllner[1], H. Fernández-Palacios Barber[2], B. Norberg[3] and C. Haux[1]

[1]Department of Zoophysiology, Göteborg University, Göteborg, Sweden
[2]Seccíon de Cultivos Marinos, Instituto Canario de Ciencias Marinas, Gran Canaria, Spain
[3]Institute for Marine Research, Austevoll Aquaculture Research Station, Storebø, Norway

Introduction

Estradiol-17ß (E_2) has been demonstrated to induce the vitelline envelope proteins in teleosts. The vitelline envelope proteins, which are suggested to be synthesized in the liver, are found in plasma of maturing females as well as E_2-treated fish of both sexes. This summary presents species for which we have investigated the induction (Hyllner et al., 1991, 1994, 1995; Larsson et al., 1994) and puts the results in an evolutionary perspective.

Materials and methods

Fish of 19 species were injected with a total dose of 1-20 mg E_2/kg body weight and plasma was sampled 6-21 days later. Induced vitelline envelope proteins were detected by Western blot technique, using homologous antisera for rainbow trout, brown trout, turbot, Atlantic halibut, European sea bass and gilthead sea bream. The remaining 13 species were analysed using three heterologous antisera directed against vitelline envelope proteins from Atlantic halibut, turbot and rainbow trout respectively.

Results and Discussion

Vitelline envelope proteins were induced by E_2 in all 6 species where homologous antisera were used. In 9 of the remaining 13 species, envelope proteins were detected by one or more of the heterologous antisera (Fig 1). Thus, including medaka and Atlantic salmon, induction of vitelline envelope proteins has today been demonstrated in a total of 17 species. Induction was demonstrated in males of 10 species, suggesting an extra ovarian origin of the proteins. The molecular weight of detected proteins in plasma ranged from 45 to 110 kDa.

Using heterologous antisera, vitelline envelope proteins were not detected in species belonging to Elopomorpha or Ostariophysi (Fig 1). This may be due to lack of immunoreactivity of the proteins, or possibly the induction mechanism evolved after these groups separated from other teleosts.

Vitelline envelope proteins have been induced in all investigated species belonging to Acanthopterygii, Paracanthopterygii and Salmoniformes. These groups constitute 62% of all teleost species. Altogether, this indicates that E_2 induces vitelline envelope proteins in the majority of teleost species.

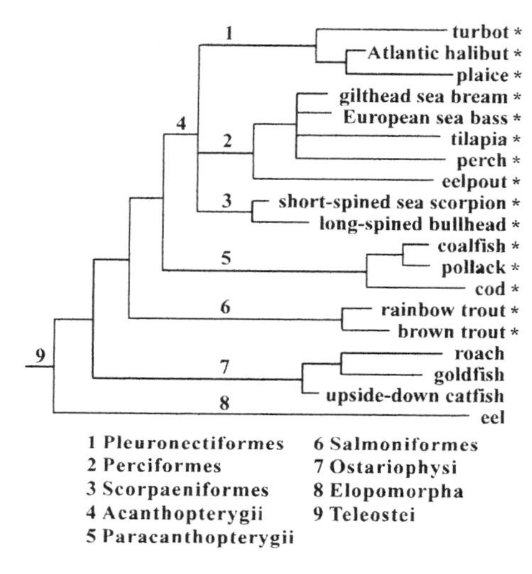

Fig 1. Evolutionary tree of the investigated species

References

Hyllner S.J., Fernàndez-Palacios Barber H., Larsson D.G.J. and Haux C. 1995. Amino acid composition and endocrine control of vitelline envelope proteins in European sea bass (*Dicentrarchus labrax*) and gilthead sea bream (*Sparus aurata*). Mol. Reprod. Dev. In press.

Hyllner S.J., Norberg B. and Haux C. 1994. Isolation, partial characterization, induction and the occurrence in plasma of the major vitelline envelope proteins in Atlantic halibut (*Hippoglossus hippoglossus*) during sexual maturation. Can. J. Fish. Aquat. Sci. **51**. 1700-1707.

Hyllner S.J., Oppen-Berntsen D.O., Helvik J.V., Walther B.T. and Haux C. 1991. Oestradiol-17ß induces the major vitelline envelope proteins in both sexes in teleosts. J. Endocrinol. **131**: 229-336

D.G.J. Larsson, Hyllner S.J. and Haux C. 1994. Induction of vitelline envelope proteins by estradiol-17ß in 10 teleost species. Gen. Comp. Endocrinol **96**: 445-450.

EARLY GONADAL DEVELOPMENT AND SEX DIFFERENTIATION IN MUSKELLUNGE (*ESOX MASQUINONGY*)

F. Lin[1], K. Dabrowski[1], and L. P. M. Timmermans[2]

1. School of Natural Resources, The Ohio State University, Columbus, OH 43210, USA.
2. Department of Experimental Animal Morphology and Cell Biology, Agricultural University, Wageningen, The Netherlands.

Summary

The primordial germ cells (PGCs) were first identified in fish of 20 mm in length. PGCs started proliferation at a fish size of 52-82 mm. Two types of gonads were histologically identified at a fish size of 138 mm. Gonads of females contained clusters of germ cells at this stage, whereas gonads of males still resembled the morphology of earlier stages, undifferentiated. This stage marked the onset of the morphological sex differentiation. The gonads developed from an undifferentiated stage directly into an ovary or a testis in muskellunge. Germ cells had proliferated in both sexes of fish at a size of 211 mm.

Introduction

Understanding the early gonadal development and sex differentiation is essential to determine the hormonal treatment period for sex reversal in fish species. A general review of early development and differentiation in teleost fish was reported by Timmermans (1987). Unfortunately, little information on gonadal development and sex differentiation was available for esocids. Our goal was to describe the early gonadal development and sex differentiation in muskellunge by histological methods.

Materials and Methods

Samples were collected periodically from fish after hatching until they reached 52 mm and from fish larger than 82 mm in total body length in 1992 and 1993, respectively. Fish were fixed in Bouin's solution and then embedded in paraffin. Cross sections of 5 µm were cut and stained with Mayer's hematoxylin and eosin or with Crossmon.

Results

The PGCs were first identified in fish of 20 mm in length (4 weeks post fertilization). The PGCs were accompanied with a few somatic cells. Gonads did not form yet a complete string at this stage. At a fish size of 32 mm, gonads formed a typical spherical shape in cross sections and the PGCs were completely enveloped by somatic cells. Gonads hung on the dorsal peritoneum at the lateral side of the swimming bladder and the gonadal strings were complete. Blood vessels were first found inside the gonads with Crossmon staining at a fish size of 52 mm. Some of the PGCs underwent mitotic division at this stage. Gonads of the fish up to 82 mm were still considered as sex undifferentiated. At a fish size of 138 mm, one type of the gonad contained dispersed germ cells among stromal cells, similar to that of the early stages, whereas the other contained clusters of the germ cells (Fig. 1). Gonads with clusters of germ cells developed into ovaries. This stage marked the onset of the morphological sex differentiation. As in carp and rosy barb, the gonads developed from an undifferentiated stage directly into an ovary or a testis in muskellunge, whereas in a number of gonochoristic teleost species gonad differentiation is of the protandrous or protogynous. Germ cells had proliferated in both sexes of fish examined at a size of 211 mm. Female gonads contained lobes of germ cells, some of them enlarged. Germ cells of testes still resembled the PGCs in morphology.

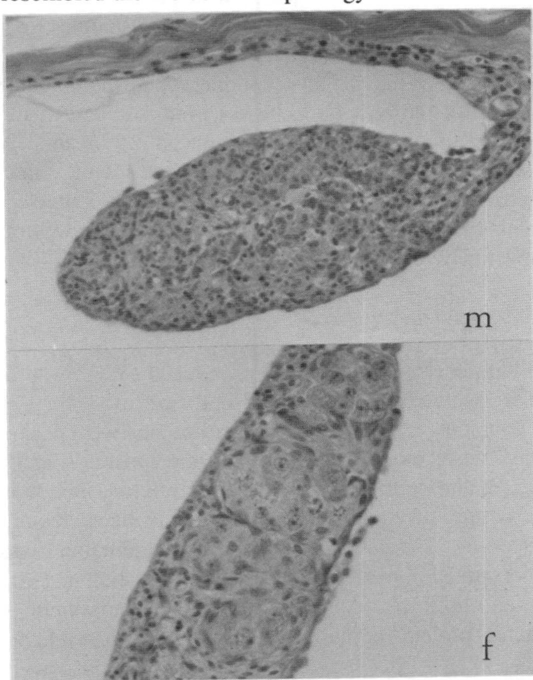

Fig. 1. Gonads of fish at a size of 138 mm in length. m-male; f-female.

References

Timmermans, L. P. M., 1987. Early development and differentiation in fish. Sarsia, 72: 331-339.

LIPID CONTENTS OF FEMALE STRIPED BASS PLASMA AND OVARIES EXHIBIT SEASONAL CHANGES ASSOCIATED WITH OOCYTE MATURATION.

E. D. Lund[1], A. R. Place[1] and C. V. Sullivan[2]. [1]Center of Marine Biotechnology, Baltimore, MD. 21202. [2]North Carolina State University, Raleigh, NC. 27695-7601.

Summary

Captive striped bass were studied to determine the dynamics of lipid mobilization and deposition in the ovaries during oocyte maturation. Plasma lipid levels in females dropped relative to those of males just prior to and during spawning however the fatty acyl composition of plasma lipids did not vary between sexes. Lipid accumulation in striped bass ovaries was primarily due to an increase in wax ester content.

Introduction

Total protein and lipid concentrations in plasma of eight male and eight female captive striped bass (Morone saxatilis) were monitored monthly over the course of two reproductive cycles as part of an effort to investigate the time course of lipid mobilization and subsequent deposition in the oocytes of mature females. Vitellogenin levels in the plasma of these fish had been previously determined for each sampling date used in the study. The lipid class and fatty acyl composition of vitellogenic lipids were determined from lipids extracted from purified striped bass vitellogenin. Changes in lipid content of ovaries of captive four-year-old female striped bass were followed from the approximate onset of vitellogenesis in September through spawning in April.

Results

Total lipid concentrations in the plasma of both males and females showed seasonal fluctuations with the highest levels in the late Spring after spawning and during the Fall. Total lipid concentrations in the plasma of females were significantly lower than those of the males during the four months preceding spawning when vitellogenin levels were at their highest. Analysis of the lipid class composition of the plasma lipids by Iatroscan TLC/FID revealed that the decrease in plasma lipid concentrations in females prior to spawning is primarily due to a decrease of up to 50% in the phospholipid content of the plasma relative to males. No wax esters were detected in the plasma of females. The fatty acyl composition of plasma lipids in these fish did not vary seasonally or with sex. Total protein levels showed seasonal fluctuation, but did not vary with sex.

Changes in lipid content of oocytes of captive female striped bass consisted of increases in total lipid from October until February when total lipid content stabilized at 850-900 mg lipid/g dry weight (Fig. 1.). The increase in total lipid content of the ovaries during oocyte maturation was due primarily to increases in triglyceride and wax ester content. Both triglycerides and wax esters started accumulating with the onset of vitellogenesis in October. Triglyceride content continued to increase until February, but then decreased from February until April. Wax ester content increased continuously into April at which time it comprised 70 percent of total lipids.

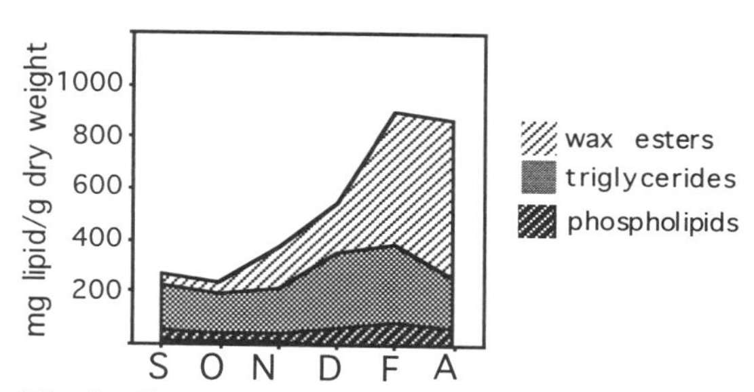

Fig. 1. Changes in ovarian lipid content during oocyte maturation.

Discussion

Mature striped bass eggs contain a large lipid droplet comprised primarily of wax esters, yet wax esters could not be detected in the plasma lipids of females. This strongly suggests that wax esters are synthesized within the maturing oocytes as has been shown in other fish species containing wax esters in their eggs (Sand et al., 1969).

The reduction in female plasma lipid content prior to spawning has been reported in other species of fish (Sargent et al., 1989). In the case of striped bass this reduction appears to occur once ovarian lipid content has already peaked.

Analysis of lipids isolated from striped bass vitellogenin revealed that over 75% of lipids bound to vitellogenin are phospholipids, yet the phospholipid content of ovarian lipids rose only slightly during oocyte maturation. The rise and fall of triglyceride content of ovaries during vitellogenesis while wax esters continued to accumulate is consistent with triglycerides acting as a temporary intermediate for the storage of deposited lipids from vitellogenin and VLDLs. Conversion of triglycerides to wax esters may explain the concurrent drop in triglyceride content and rise in wax ester content.

References

Sand, D. M., Hehl, J. L. and Schenk, H., 1969. Biosynthesis of wax esters in fish. Reduction of fatty acids and oxidation of alcohols. Biochemistry. 8: 4851.

Sargent, J. S., Henderson, R. J. and Tocher, D.R. 1989. The lipids. In: Fish Nutrition. H. E. Halver (Ed.): Academic Press, London, p. 153-218.

PROTEOLYTIC CLEAVAGE OF YOLK PROTEINS DURING OOCYTE MATURATION IN BARFIN FLOUNDER

T. Matsubara and Y. Koya[1]

Hokkaido National Fisheries Research Institute, 116, Katsurakoi, Kushiro, Hokkaido 085, Japan

Introduction

Proteolytic cleavage of yolk proteins during oocyte maturation was first documented in *Fundulus* oocytes (Wallace & Begovac, 1985; Wallace & Selman, 1985). Occurrence of the yolk proteolysis was further demonstrated in other species, particularly in marine fish which lay pelagic eggs. The present study examined structural changes in three classes of vitellogenin-derived yolk proteins during oocyte maturation in barfin flounder, *Velasper moseri*, a pelagic egg spawner.

Results

Native dimeric lipovitellin (Lv), estimated to be 410 kDa in vitellogenic oocytes, cleaved into 170 kDa monomeric Lv during oocyte maturation (Fig. 1A,B). Both ß'-component at 19 kDa in native form (Fig. 1A) and a highly phosphorylated band of phosvitin (Pv) in SDS-PAGE (data not shown) identified in vitellogenic oocytes became undetectable after oocyte maturation. SDS-PAGE analysis indicated that the Lv in vitellogenic oocyte yields two major bands corresponding to 105 kDa and 90 kDa (Fig. 1C). The 105 kDa band completely disappeared and several smaller bands newly appeared in ovulated eggs (Fig. 1D).

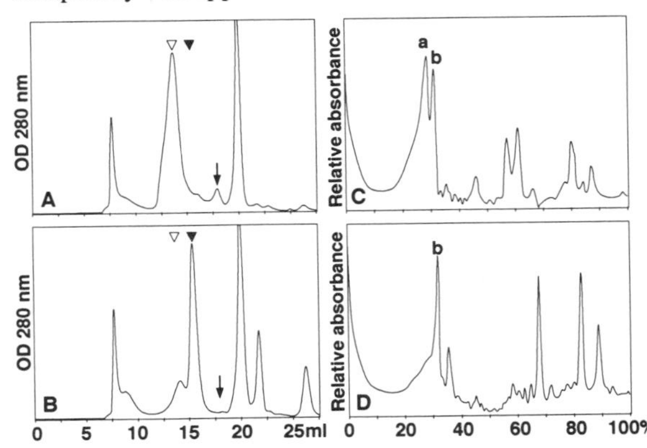

Fig. 1. Elution profiles after Superose 6 chromatography (A, B) and densitograms after 5-20% gradient SDS-PAGE (C, D) of the homogenates from vitellogenic oocytes (A, C) and ovulated eggs (B, D). Closed and open triangles indicate the peak positions of 410 kDa Lv (dimer) and 170 kDa Lv (monomer), respectively. Arrows indicate the peak position of 19 kDa ß'-component. a, 105 kDa band; b, 90 kDa band.

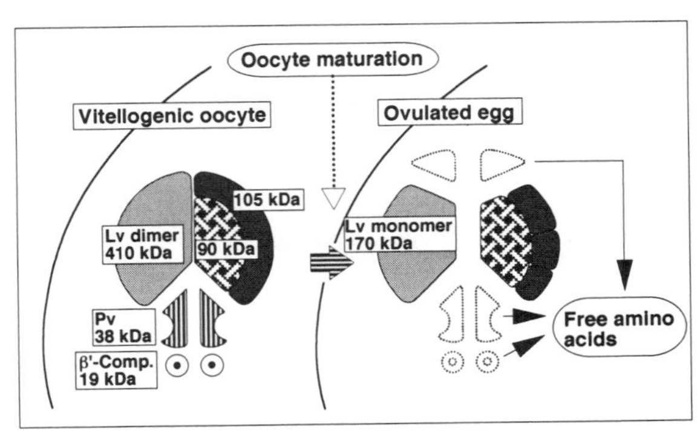

Fig. 2. Schematic drawing of the structural models of three classes of yolk proteins in vitellogenic oocyte and ovulated eggs of barfin flounder.

Discussion

Barfin flounder vitellogenin (520 kDa) appears to decompose into 410 kDa Lv, and two molecules each of ß'-component and Pv during yolk accumulation into oocytes (Matsubara & Sawano, 1995). All the three classes of yolk proteins thus undergo proteolysis again during oocyte maturation, as shown in the model of Figure 2. Together with the results of quantitative analysis of free amino acids in vitellogenic oocytes (3.1 μg/egg) and ovulated eggs (46.0 μg/egg), the yolk proteolysis suggests that free amino acids are utilized as the osmotic effecter for oocyte hydration and a stock of nutrient during early development.

References

Matsubara, T and K. Sawano, 1995. Proteolytic cleavage of vitellogenin and yolk proteins during vitellogenin uptake and oocyte maturation in barfin flounder (*Verasper moseri*). J. Exp. Zool., 272: 34-45.

Wallace, R.A. and P.C. Begovac, 1985. Phosvitins in *Fundulus* oocytes and eggs: Preliminary chromatographic and electrophoretic analyses together with biological considerations. J. Biol. Chem., 260: 11268-11274.

Wallace, R.A. and K. Selman, 1985. Major protein changes during vitellogenesis and maturation of *Fundulus* oocytes. Dev. Biol., 110: 492-498.

[1] Y. Koya is now at Dept. Biol. Fac. Educ. Gifu Univ, Gifu, Gifu, 502,Japan.

EFFECT OF GnRH PEPTIDES ON HISTONE H-1 KINASE ACTIVITY IN THE FOLLICLE-ENCLOSED GOLDFISH OOCYTES, *IN VITRO*

D. Pati [1] and H. R. Habibi

Department of Biological Sciences, University of Calgary, Calgary, Alberta, Canada T2N 1N4

Summary

GnRH peptides directly affect germinal vesicle breakdown (GVBD) and reinitiation of oocyte meiosis in the follicle-enclosed goldfish oocytes, in vitro. The present findings indicate that the effect of GnRH peptides on reinitiation of oocyte meiosis involves activation of maturation promoting factor (MPF) in goldfish.

Introduction

Gonadotropin hormone (GTH) stimulates synthesis of maturation inducing steroid (17α,20β-dihydroxy-4-pregnen-3-one; DHP) which in turn stimulates production of maturation promoting factor (MPF). MPF, a complex of cdc2 kinase and cyclin B mediates germinal vesicle breakdown (GVBD) and reinitiation of oocyte meiosis in goldfish and other vertebrates. Recently, we have shown that salmon GnRH (sGnRH) and chicken GnRH-II (cGnRH-II) which are native in goldfish brain affect oocyte meiosis and follicular steroidogenesis (Habibi and Pati, 1993). Treatment with sGnRH or cGnRH-II alone stimulates the reinitiation of oocyte meiosis as indicated by GVBD. In the presence of GTH, however, sGnRH was found to inhibit GTH-induced oocyte meiosis while cGnRH-II had no effect on GTH-induced response. In the present study, we investigated the effect of both sGnRH and cGnRH-II on MPF activity in the follicle-enclosed goldfish oocytes by determining histone H-1 kinase activity as previously described (Lohka et al., 1988).

Results

Treatment with GTH, sGnRH or cGnRH-II alone significantly increased histone H-1 kinase activity. Concomitant treatment with sGnRH significantly reduced GTH-induced H-1 kinase activity (Fig. 1 A). In accordance with the observed GVBD response (Fig. 1 B), treatment with cGnRH-II had no effect on GTH-induced H-1 kinase activity. Subsequent time course studies revealed that the effect of GTH on H-1 kinase activity is significantly slower than GnRH by approximately 4 hours (results not shown). Even in the presence of GTH, sGnRH initially stimulated H-1 kinase activity, followed by a reduction corresponding to its inhibitory effect on GVBD response. In this regard, the time course of the inhibitory action of sGnRH on GTH-induced GVBD response is consistent with the reduced histone H-1 kinase activity.

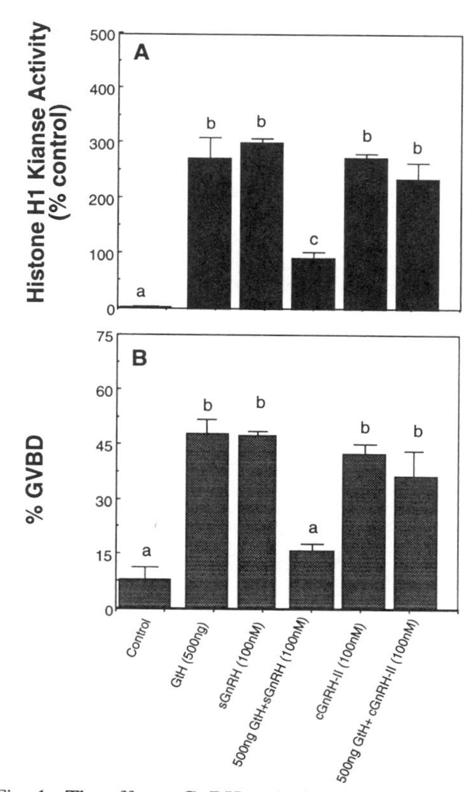

Fig. 1. The effect sGnRH and cGnRH-II on germinal vesicle breakdown (GVBD) and histone H-1 kinase activity in the follicle-enclosed goldfish oocytes, *in vitro*. Columns displaying different symbols are significantly different.

Discussion

From the present study it is evident that the stimulatory effect of GnRH peptides alone on meiosis correlates with enhanced kinase activity while the inhibitory effect of sGnRH on GtH-induced meiosis involves inhibition of MPF production.

References

Habibi, H.R. and Pati, D., 1993. Endocrine and paracrine control of ovarian function: Role of compounds with GnRH-like activity. In: Cellular Communication in Reproduction. F. Facchinetti, I.W. Henderson, R. Pierantoni, and A.M. Polzonetti-Magni (Eds.): J. of Endocrinology Ltd, Bristol, U.K., p. 59-70.

Lohka, M.J., Hayes, M.K., and Maller, J.L., 1988. Purification of maturation-promoting factor, an intracellular regulator of early mitotic events. Proc. Natl. Acad. Sci. USA 85, 3009-13.

[1] Present address: Baylor College of Medicine, Texas Children's Cancer Center, 6621 Fannin St., Houston, Texas 77030, USA.

NEGATIVE REGULATION OF DNA SYNTHESIS BY PHOSPHODIESTERASE INHIBITORS IN TESTIS OF THE SHARK *SQUALUS ACANTHIAS*

J.M. Redding[1,3] and G.V. Callard[2,3]

[1]Dept. Biology, Tennessee Tech University, Cookeville, TN 38505
[2]Dept. Biology, Boston University, Boston MA 02215
[3]Mount Desert Island Biological Laboratory, Salsbury Cove, ME 04672

Introduction

Preliminary studies showed that 3-isobutyl-1-methylxanthine (IBMX), a commonly used inhibitor of intracellular phosphodiesterases (PDE), rapidly and reversibly inhibited DNA synthesis in spermatocysts of the dogfish shark *Squalus acanthias*. Inhibition of PDE would likely lead to the elevation of intracellular cyclic nucleotides, cAMP or cGMP. Numerous studies on the meiotic maturation of vertebrate oocytes have shown inhibitory regulation by cyclic nucleotides. The purpose of this study was to assess the role of PDE and cyclic nucleotides as possible regulators of spermatogenesis in *Squalus* testis.

Methods

Spermatocysts, containing both premeiotic germ cells and Sertoli cells, were isolated from *Squalus* testis and cultured in vitro as in DuBois and Callard (1993). DNA synthesis was estimated by the incorporation rate of (3H)-thymidine in treated and control spermatocysts.

Results and Discussion

Three commonly used non-specific PDE inhibitors, IBMX, theophyllin (THEO), and papaverine (PAPA) all inhibited DNA synthesis in a dose-responsive manner. PAPA was about 100 times more potent than IBMX and 400 times more potent than THEO. Several other purine derivatives also inhibited DNA synthesis but with lower potency.

There are at least five different types of PDE present in the mammalian testis (Thompson 1991). Several selective PDE inhibitors were tested for their effects on DNA synthesis to identify specific mechanisms of action. Dipyridamole, a selective inhibitor of Type V PDE (cGMP-specific), was the most effective inhibitor of DNA synthesis, showing a potency over 100 times greater than IBMX. Rolipram and RO-20-1724, selective inhibitors of Type IV PDE (cAMP-specific), both inhibited DNA synthesis with similar potency to IBMX. Millirone and Piroximone, selective inhibitors of Type III PDE (cGMP-inhibited) were ineffective. These results suggest that increased intracellular cGMP or cAMP may be mediating the inhibitory effects of PDE inhibitors on DNA synthesis.

Forskolin (25 mM), an adenyl cyclase activator, failed to inhibit DNA synthesis despite elevating the intracellular cAMP concentration at least an order of magnitude greater than the four-fold increase observed with 1 mM IBMX. Dibutyryl cAMP (db-cAMP) alone appeared to stimulate DNA synthesis at lower concentrations (0.1 mM) and inhibit it at higher levels (10 mM). However, a normally ineffective dose of db-cAMP (1.0 mM) synergistically enhanced the inhibitory effects of IBMX on DNA synthesis. Cyclic GMP (0.1-2.3 mM) was ineffective. Thus, our results are equivocal regarding the role of cAMP as a negative regulator of spermatogenesis. The literature on this subject is likewise ambiguous. Both stimulation and inhibition of DNA synthesis have been noted in response to cyclic nucleotide treatment in various tissues.

Less well known effects of PDE inhibitors, i.e., blockage of adenosine receptors or adenosine transport, may suggest plausible mechanisms leading to decreased DNA synthesis. IBMX and the two most potent inhibitors used in this study, PAPA and dipyridamole, can block adenosine uptake. A commonly used adenosine transport inhibitor, nitrobenzyl-thio-inosine, also inhibited DNA synthesis in our system, a maximal effect apparent at the lowest dose tested (0.1 µM). Inasmuch as adenosine is required for DNA synthesis, blockage of its transport into the cell by PDE inhibitors could indirectly decrease the formation of DNA.

Conclusions

Negative regulation of DNA synthesis by non-specific and cAMP-specific (type IV) PDE inhibitors suggests a regulatory role for cAMP in shark spermatocysts; however, the general ineffectiveness for forskolin and cAMP analogs argues against this interpretation. Since spermatocysts are aggregates of both germ and Sertoli cells, differential sensitivity of these cells to PDE inhibitors and cAMP may complicate our analysis. An alternative mechanism involving blockage of the adenosine transport seems applicable to the most effective inhibitors used in this study.

References

DuBois, W. and G. V. Callard, 1993. Culture of intact Sertoli/germ cell units and isolated Sertoli cells from *Squalus* testis. (II). Stimulatory effects of IGF I and other factors on DNA synthesis. J. Exp. Zool. 267:233-244.

Thompson, W. J., 1991. Cyclic nucleotide phosphodiesterases: Pharmacology, biochemistry and function. Pharmac Ther. 51:13-33.

Supported by grants from NIH (HD16715, GVC) and the Burroughs Wellcome Fund (JMR)

DIETARY INFLUENCE ON THE FATTY ACID COMPOSITION OF VITELLOGENIN AND THE SUBSEQUENT EFFECT ON THE EGG COMPOSITION IN COD (GADUS MORHUA)

C. Silversand,[1] B. Norberg,[2] J.C. Holm,[2] Ø. Lie,[3] and C. Haux.[1]

[1]Dept. Zoophysiology, Göteborg University, Medicinaregatan 18, S-413 90 Göteborg, Sweden; [2]Institute of Marine Research, Austevoll Aquaculture Research Station, N-5392 Storebø, Norway; [3]Institute of Nutrition, Directorate of Fisheries, P.O. Box 1900, N-5024 Bergen, Norway.

Summary

The fatty acid composition of dietary lipids clearly influences the fatty acid composition of cod liver lipids. In contrast, lipids of cod vitellogenin and eggs are highly consistent, in spite of differences in dietary fatty acid composition. Only when a diet with an extreme composition is used, is the fatty acid composition of vitellogenin and eggs markedly influenced.

The fatty acid composition of eggs is strikingly similar to the fatty acid composition of vitellogenin, suggesting that vitellogenin plays a fundamental role in the process of lipid accumulation in the growing cod oocytes.

Taken together, the selectivity during the accumulation of fatty acids into the growing oocytes is suggested to be mainly due to a specific selection of fatty acids in the liver during the synthesis of vitellogenin.

Introduction

A large gap exists in our knowledge about the maternal source of the egg lipids and the influence of maternal nutrition on reproduction in teleost fish. The accumulation of lipids in growing oocytes are believed to occur mainly during vitellogenesis via the uptake of the lipoprotein, vitellogenin. In the present study, we examined the influence of dietary fatty acids on the composition of total lipids from liver, vitellogenin and eggs of cod.

Materials and methods

Cod (Gadus morhua), hatched and reared at the Austevoll Aquaculture Research Station in Norway (60 °N), were distributed into three sea-cages in October and fed three different diets for five months, until sexual maturation in February. The cod were fed pellets of dry feed, coated with either soya bean oil (diet A), capelin oil (diet B) or sardine oil (diet C). Diet A lipids were characterised by high levels of 18:2(n-6) (30%). Diet B lipids contained high levels of 20:1(n-9) and 22:1(n-11), and diet C lipids had the highest levels of 20:5(n-3) and 22:6(n-3). (n-3) fatty acids accounted for 15, 19 and 28 % of the total fatty acids in the lipids of diet A, B and C, respectively.

The cod were sampled on two occasions: **1.** After six weeks, males from each dietary group were collected and injected with estradiol for 20 days in order to induce vitellogenin synthesis. Blood was taken and the livers were dissected. **2.** After five months, at the time of ovulation, females from each dietary group were collected. Blood, freshly ovulated eggs and livers were sampled.

Vitellogenin was isolated from plasma by precipitation with EDTA-Mg^{2+} and distilled water. Total lipid was extracted from liver, vitellogenin and eggs and the fatty acid methyl esters were analysed by gas-liquid chromatography (Silversand and Haux, 1995).

Results and Discussion

There were no apparent signs of nutritional imbalance during the feeding period, indicating that the dietary levels of fatty acids were adequate to allow normal growth in all groups of cod.

A clear dietary effect on liver fatty acid composition was observed. The fatty acid profile of liver lipids readily took on the characteristics of the corresponding dietary lipids. In contrast, the fatty acid composition of the lipid moiety of vitellogenin was remarkably conserved in spite of the different dietary fatty acid composition. A homogenous fatty acid profile was also observed in the eggs from the three dietary groups. In all dietary groups, lipids from vitellogenin and eggs were highly unsaturated with constant levels of (n-3) fatty acids, particularly 22:6(n-3). The only major dietary effect on vitellogenin and egg fatty acid composition was the higher level of 18:2(n-6) and lower level of 20:5(n-3) in cod fed diet A.

The fatty acid composition of vitellogenin and livers from the first sampling occasion was almost identical to the composition after five months, showing that cod respond rapidly to the intake of dietary lipids. Furthermore, it also revealed that the fatty acid profile of vitellogenin induced by hormone injections were highly similar to that of naturally induced vitellogenin.

Finally, the fatty acid composition of total egg lipids was strikingly similar to the fatty acid composition of vitellogenin from the same female.

REGULATION OF THE MATURATION-INDUCING RECEPTOR IN SPOTTED SEATROUT OVARIES

P. Thomas and S. Ghosh.

The University of Texas at Austin, Marine Science Institute, Port Aransas, Texas 78373.

Introduction

Previous *in vitro* experiments have shown that up-regulation of the ovarian 20β-S membrane receptor occurs during the initial gonadotropin-controlled priming phase of final oocyte maturation (FOM) in the spotted seatrout, *Cynoscion nebulosus* (Thomas and Patiño, 1990). The primed oocytes are then capable of undergoing the later stages of FOM including germinal vesicle breakdown (GVBD) in response to the maturation-inducing steroid, 20β-S (Patiño and Thomas, 1990a). In the present study, the effects of insulin, IGF-1 and other agents which influence FOM in amphibians and fish (e.g., Kagawa *et al.* 1994), on 20β-S receptor concentrations and the development of oocyte maturational competence of seatrout oocytes were investigated in an *in vitro* ovarian incubation system.

Methods

Ovarian fragments (5 g) were incubated with the hormones and drugs for 12 hours; most of the tissue was removed for 20β-S membrane receptor measurement (Patiño and Thomas, 1990b) and the remaining oocytes were assessed for maturational competence (i.e., their ability to complete FOM in response to 20β-S) in an *in vitro* bioassay.

Results and Discussion

The induction of maturational competence of seatrout oocytes *in vitro* by a variety of hormones and drugs was invariably associated with an increase in ovarian 20β-S membrane receptor concentrations. Moreover, there was a close correlation between the relative increase in receptor concentrations and the percentage of oocytes that became maturationally competent after gonadotropin treatment *in vitro*. These studies suggest that up-regulation of the MIS receptor is a critical regulatory step in the hormonal control of FOM.

Possible intermediaries in gonadotropin up-regulation of receptor concentrations were investigated by incubating ovarian tissues with various drugs during the priming phase. Both forskolin and dbc-AMP at concentrations of 1-100 μM caused dramatic increases in receptor concentrations (up to 12 fold) which suggests that induction of the MIS receptor by gonadotropin is mediated by increases in cyclic AMP levels. Ouabain, a Na^+/K^+ channel blocker, (10-100 μM) attenuated the gonadotropin induction of the MIS receptor, whereas the Ca^{2+} channel blocker verapamil was ineffective. Other Na+/K+ channel blockers need to be tested to confirm the ouabain results.

Interestingly, insulin and IGF-1 caused dose-dependent increases in receptor concentrations over the range of 0.1-100 μM, although the increase was modest (2-4 fold) compared to that induced by gonadotropins. These peptides have previously been shown to induce maturational competence of *Pagrus major* oocytes (Kagawa *et al.*, 1994). The present results demonstrate that insulin and IGF-1 induce maturational competence of seatrout oocytes by up-regulating the ovarian membrane receptor for the MIS.

References

Kagawa, H., M. Kobayashi, Y. Hasegawa and K. Aida. 1994. Insulin and insulin-like growth factors I and II induce final maturation of oocytes of red seabream, *Pagrus major*, *in vitro*. Gen. Comp. Endocrinol. 95, 293-300.

Patiño, R. and P. Thomas. 1990a. Effects of gonadotropin on ovarian intrafollicular processes during the development of oocyte maturational competence in a teleost, the Atlantic croaker: Evidence for two distinct stages of gonadotropic control of final oocyte maturation. Biol. Reprod. 43, 818-827.

Patiño, R. and P. Thomas. 1990b. Characterization of membrane receptor activity for 17α,20β,21-trihydroxy-4-pregnen-3-one in ovaries of spotted seatrout (*Cynoscion nebulosus*). Gen. Com. Endocrinol. 78, 204-217.

Thomas P., and R. Patiño. 1991. Changes in 17α, 20β,21-trihydroxy-4-pregnen-3-one membrane receptor concentrations in ovaries of spotted seatrout during final maturation. *In* Proc. 4th Internat. Symp. Rep. Physiol. Fish. pp. 122-124. Fish symp. 91, Sheffield.

SPERMATOGENESIS IN THE YELLOW PERCH (*PERCA FLAVESCENS*) - COMPARISON OF RELATIVE GERM CELL TYPES IN 2 SUBPOPULATIONS OF YOUNG-OF-YEAR FISH

G.P. Toth[1], S.A. Christ[1], R.E. Ciereszko[2], and K. Dabrowski[2]

[1]U.S. Environmental Protection Agency (MD642), Cincinnati, OH, 45268

[2]School of Natural Resources, The Ohio State University, Columbus, OH 43210

Summary

An "accelerated growth rate" (AGR) subpopulation of young-of-year yellow perch (*Perca flavescens*) finished production of mature spermatozoa at least two weeks earlier than a "normal growth rate" (NGR) subpopulation as assessed by semi-quantitative testis histology methods.

Introduction

The wide range in rates of sexual maturation in the yellow perch (*Perca flavescens*)(Tanasichuk and Mackey, 1989) can best be understood in the context of size-dependent sexual maturation (Malison et al., 1986) and density-dependent prey availability (Post and McQueen, 1994). In the present study young-of-year stock from artificial insemination were stocked in ponds in May 1994 in Southern Ohio (Piketon). Evidence of two size subpopulations of perch in August led to an examination of possible differences in germ cell maturation rates using testicular histology.

Results

Two subpopulations of young-of-year perch developed in the same pond (body lengths: AGR-94.4 $\pm$15.1 mm; NGR- 40.4$\pm$ 4.9mm[Aug 24]). AGR fish were subsequently tank cultured with natural diet. Results from the histological evaluation of the hematoxylin/eosin stained testis samples showed that the percentage of spermatozoa in AGR fish was greater than NGR fish in early October (36% [Oct4] vs. 2% [Oct7]; $p \leq 0.01$)(Fig 1). In AGR males, spz comprised 90-95% of the germ cells beginning at the Nov 15 sampling, while a comparable percentage of spz was not observed in the NGR males until Dec 7.

Conclusions

Sexual maturation in male young-of-year yellow perch is shown for the first time under field conditions. NGR fish began germ cell maturation at approximately the same body length (80 mm) as reported by Malison et al. (1986) for tank cultured young-of-year perch. Appearance of spermatocytes was first observed in the AGR fish on Aug 24 when the mean body length was 94.4 $\pm$15.1 mm. These findings support the evidence presented by Malison et al. (1986) for size dependent sexual maturation.

Figure 1. Relative germ cell distributions

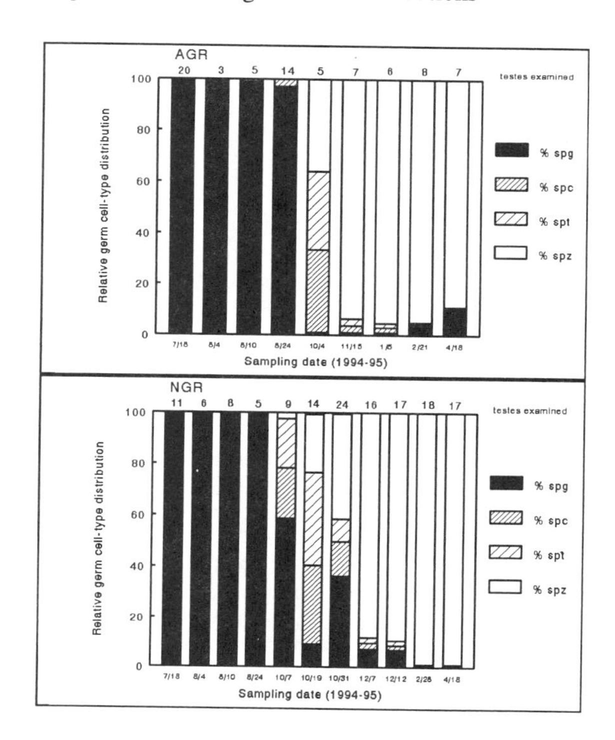

References

Malison, J.A., Kayes, T.B., Best, C.D., and Amundsen, C.H., 1986. Sexual differentiation and use of hormones to control sex in yellow perch (*Perca flavescens*). Can.J.Fish.Aquat.Sci. 43:26-35.

Post, J.R. and McQueen, D.J., 1994. Variability in first-year growth of yellow perch: Predictions from a simple model, observations, and an experiment. Can.J.Fish.Aquat.Sci. 51:2501-2512.

Tanasichuk, R.W. and Mackey, W.C., 1989. Quantitative and qualitative characteristics of somatic and gonadal growth of yellow perch (*Perca flavescens*) from Lac Ste. Anne, Alberta. Can.J.Fish.Aquat.Sci. 46:989-994.

Acknowledgements: Susan Dawes, Dyncorp. and DOE-SERDP program.

EFFECT OF SOME STEROIDS AND PROSTAGLANDINS ON GVBD AND OVULATION IN CATFISH, HETEROPNEUSTES FOSSILIS

V. Tripathi and T.P. Singh

Fish Endocrinology Laboratory, Department of Zoology, Banaras Hindu University, Varanasi- 221 005, India.

Summary

PGF2a increased ovulation but did not affect maturation while PGE1 & PGE2 not only blocked ovulation but also the final maturation of oocytes.

Introduction

Earlier evidence in teleosts, indicates that PGF's stimulate ovulation while PGEs had varied effect on it. But such evidences come from *in vitro* studies where PGs were incubated with oocytes having germinal vesicle broken. Infact, hormonal regulation of GVBD and its relation with control of ovulation are not yet clear. Therefore, in order to establish this relation, an attempt was made to observe the effect of PGs on GVBD & ovulation, when administered with MIS in *H. fossilis*. As this fish is group synchronous breeder, the understanding of such interplay between hormonal control of maturation and ovulation require immediate attention.

Results

17a,20B-DHP was most effective in inducing GVBD. Other progestins and corticoids were also effective. In presence of 17a,20B-DHP, PGF2a increased ovulation, while the PGE1 & PGE2 stopped completion of GVBD and ovulation in such incubations. PGF2a also increased ovulation of the oocytes with GVBD obtained from carp GtH primed fish. Indomethacin partially blocked 17a,20B-DHP induced ovulation but not spontaneous ovulation of oocytes obtained from carp GtH primed fish.

Table 1: *In vitro* effect of PGs on GVBD and ovulation of 17a,20B-DHP (P) induced oocytes. (± SEM)

Treatments	% GVBD	% Ovulation
P	90± 2.1	28± 2.64
P + PGF1α	93± 1.8	36± 1.76
P + PGF2α	94± 1.2	51± 3.03*
P + PGE1	nil*	nil*
P + PGE2	nil*	nil*

*Significant at $P \leq 0.05$

This work was supported by CSIR award to V.T.

Table 2 : *In vitro* effect of PGs on ovulation of oocytes obtained from GtH primed fish. (± SEM)

Treatments	% Ovulation
Control	59± 2.61
PGF1α	62± 3.05
PGF2α	92± 2.89*
PGE1	60± 3.05
PGE2	54± 4.16

* Significant to respective control at $P \leq 0.05$

Table 3: *In vitro* effect of indomethacin (IM) on ovulation. (± SEM)

Treatments	GtH Primed	P- induced
Control	59± 2.61	36± 2.18
IM	58± 2.00	22± 2.16*
IM + PGF2α	87± 3.12*	54± 3.10*

*Significant at $P \leq 0.05$

Discussion

In vitro incubation of oocytes with 17a,20B-DHP induced not only the GVBD but also caused some spontaneous ovulation unlike salmonids. Failure of PGs in inducing *in vitro* ovulation of oocytes obtained from gravid female *H. fossilis* could be due to the fact that oocytes were not fully mature. However, PGFs induced ovulation when such oocytes were incubated with 17a,20B-DHP. Interestingly, PGEs not only checked ovulation but GVBD also which was induced by 17a,20B-DHP in normal incubation. This fish being group synchronus breeder, needs a mechanism through which it could sustain the first few oocytes that are fully grown, and also to advance the maximum immature oocytes to fully grown state before final maturation resumes. Thus, in *H. fossilis*, relationship between mechanism controlling final maturation and ovulation appears to be moderately different than that of salmonids.

References

Goetz, F.W., A.K. Berndtson & M. Ranjan, 1991. Ovulation: Mediators at the ovarian level. In: Pang, P. and Schreibman, M. (eds.), Vertebrate Endocrinology, Fundamentals and Biomedical Implications, Academic Press, n.Y. pp.127-203.

EFFECTS OF ESTROGEN TREATMENT ON RED DRUM (*SCIAENOPS OCELLATUS*) THYROID FUNCTION

Cinnamon L. Moore VanPutte and Duncan S. MacKenzie

Department of Biology, Texas A&M University, College Station, TX 77843, U.S.A.

Introduction

Although evidence exists for reciprocal influences between thyroid hormones and reproductive activity in teleost fish (Leatherland, 1994), the mechanisms through which reproductive steroid hormones affect thyroid function are not well established.

To identify potential locations at which steroids influence the thyroid axis, red drum, maintained at 25°C received silastic implants containing 0, 5 or 50mg Estradiol-17β (E_2)/g silastic for up to 16 days. Blood samples were collected at Day 0 and blood and liver samples were collected 3 and 16 days after implantation. Plasma was analyzed for total T_4 and T_3 levels by RIA, total protein levels by Bradford protein assay, T_4 and T_3 binding to blood proteins by radioiodinated hormone elution from Sephadex G-25 minicolumns, and changes in protein composition by PAGE. Liver samples were analyzed for hepatic 5'-outer ring deiodinase (5'-ORD) activity (VanPutte *et al.*, 1994).

Results and Discussion

Electrophoretic analysis of plasma revealed substantial amounts of circulating vitellogenin from Day 3 to Day 16 in 50mg/g E_2-treated animals along with a significant increase in total plasma protein levels in the same groups (Table 1). At Day 16 both T_4 and T_3 for 50mg/g E_2-treated animals were significantly higher than controls, whereas at Day 3 (data not shown) T_3 only for 50mg/g E_2-treated fish was significantly elevated over controls. Associated with increased vitellogenin, there was a significant increase in T_4 binding to blood proteins at Day 16 (and Day 3 for T_4, data not shown). These results indicate that elevated thyroid hormone levels may be due, in part, to increased capacity for thyroid hormone binding in the circulation. A similar situation has been reported for channel catfish, *Ictalurus punctatus* (MacKenzie *et al.*, 1987). Paradoxically, although there was a significant increase in total T_3 levels, 5'-ORD activity was unchanged at Day 3 and was significantly reduced at Day 16. This indicates that increased total thyroid hormone levels or estrogen may act to inhibit 5'-ORD and that thyroid hormone binding to plasma proteins may be a more important determinant of circulating thyroid hormone levels than activity of hepatic 5'-ORD.

References

Leatherland, J.F., 1994. Reflections on the thyroidology of fishes: from molecules to humankind. Guelph Ichthyol. Rev. 2: 1-67.

MacKenzie, D.S., J. Warner, and P. Thomas, 1987. Thyroid-reproductive relationships in the channel catfish, *Ictalurus punctatus*: evidence for estradiol-induced changes in plasma thyroid hormone binding. Proc. 3rd Int. Repro. Physiol. Fishes. 205.

VanPutte, C.L.M., D.S. MacKenzie and J.G. Eales. 1994. 5'-iodothyronine outer ring deiodinase in red drum (*Sciaenops ocellatus*):Characterization and environmental influences. Am. Zool. 34:26A.

Table 1. Effects of Estrogen Implants on Red Drum Thyroid Function

Treatment	N	Total T_4	Total T_3	T_4Binding*	T_3Binding*	5'-ORD#	Total Protein
Day 0	37	7.80±0.57	12.94±0.41	58.70±1.94	84.20±1.55		64.60±4.14
Day 16 (0)	6	6.10±1.54	14.18±0.82	49.43±3.44	84.66±1.64	0.14±0.01	45.01±1.33
Day 16 (5)	7	8.94±2.22	15.16±1.11	62.13±4.44+	86.82±0.94	0.11±0.01+	66.89±6.11+
Day 16 (50)	6	12.77±1.69+	19.50±1.84+	84.87±1.79+	89.02±0.95	0.10±0.01+	119.5±11.1+

(Dose of E_2 in mg/g silastic); +$p<0.05$ vs. (0) at same time; values are mean±s.e., units are ng/ml for total hormones, *% radiolabeled hormone eluted from column by plasma diluted 1:25, #pmol T_4 converted to T_3/h/mg protein for hepatic 5'-ORD and mg/ml for total protein.

Supported by Texas Advanced Technology Program. We thank Angie Lott and David Campbell for excellent technical assistance.

ASPECTS OF SPERMATOGENESIS AND SPERMIOGENESIS IN THE OCEAN POUT *Macrozoarces americanus*

Z. WANG, and L.W. CRIM

Ocean Sciences Centre, Memorial University of Newfoundland, St. John's, NF., Canada, A1C 5S7

Summary

The testis of ocean pout is paired and identified as the lobular or unrestricted type. Spermatogonia are located along the testicular tubules throughout the whole testis. The "semi-cystic" type of spermatogenesis in ocean pout results in the presence of a mixture of spermatids and spermatozoa in the vas deferens, a characteristic distinct from some other internally fertilizing teleosts. Spermatozoa, with a well developed mid-piece and biflagellated tail, are free and motile in the vas deferens, rather than being grouped.

Introduction

The ocean pout forms an important evolutionary link between oviparity and ovoviviparity by virtue of the fact that it is an internal fertilizing oviparous marine teleost. Studies of reproductive physiology in this species will contribute to a better understanding of the mechanism of the evolution from oviparous to ovoviviparous forms. Preliminary studies in our laboratory demonstrated that the sperm of this fish become motile in seminal plasma while remaining in the male reproductive tract; however, the motility disappeared instantly upon addition of seawater to the milt. The duration of ocean pout sperm motility lasts extremely long *in vitro* (several days or longer). Moreover, it was found that, early in the spermiation, the milt contained two types of germ cells, big non-motile spermatids and small mature sperm with motility. The present study examined both spermatogenesis and spermiogenesis in the ocean pout by histological (Ehrlich Hematoxylin stain) and ultrastructural (Fixed in 5% glutaraldehyde and 2% osmium tetroxide) approaches.

Results and discussion

The ocean pout testes are elongate paired organs attached to the dorsal body wall. The vas deferens which is enlarged during spermiation, functions as a spermatozoa reservoir. At the centre of the testis is a cavity-testicular canal. Testicular tubules irradiate from the canal to the perimeter. The organization of the testis is similar throughout the anterior, mid, and posterior regions. Although spermatogonia are situated along the testicular tubules throughout the testis, most of them are distributed at the blind end of tubules (Fig. 1A). Based on the distribution of spermatogonia, the testicular structure of the ocean pout corresponds to the lobular type (Billard, 1990), or the unrestricted type (Grier et al., 1980).

When spermatocytes develop into spermatids, germinal cysts break down to discharge spermatids into the tubular cavity (Fig. 1A). Spermiogenesis takes place in the cavity of tubules and the canal of the testis, resulting in spermatids at different maturational stages and mature spermatozoa in the vas deferens and milt (Fig. 1B). Such characteristics of spermatogenesis and spermiogenesis in ocean pout were compatible with the description of the "semi-cystic" type of spermatogenesis proposed by Mattei et al. (1993).

Spermatozoa, which are composed of an elongated head, well develop mid-piece and biflagellated tail, are free and motile in suspension in seminal plasma, rather than being grouped into spermatophores or spermatozeugmata. Such a characteristic of free spermatozoa in ocean pout is probably correlated with its special intromittent organ, the tubular gonopodium or the pseudopenis.

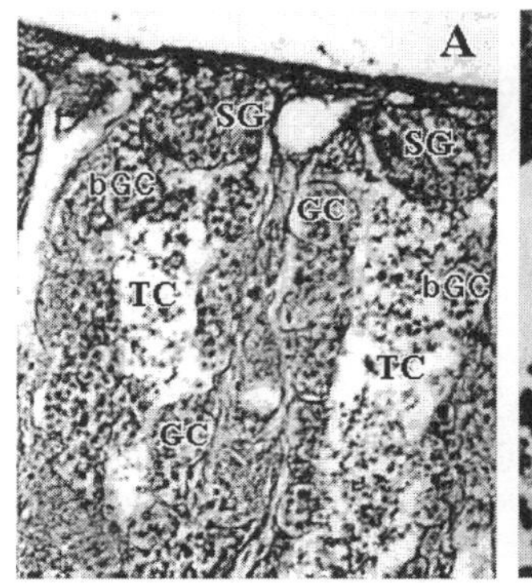

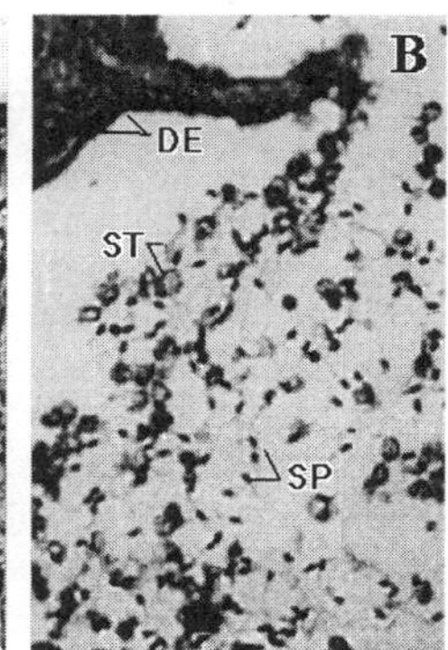

Fig. 1A. A cross-section of testis showing the testicular structure and the characteristics of spermatogenesis and spermiogenesis in the ocean pout (*Macrozoarces americanus*). x 170. Fig. 1B. A cross-section of the vas deferens showing a mixture of free spermatids and spermatozoa. x 415. Arrow heads - the distal end, SG - spermatogonia, ST - spermatids, SP - spermatozoa, GC - germinal cyst, bGC - broken germinal cyst, TC - the tubular cavity, DE - deferent epithelium.

References

Billard, R.,1990. Reproduction in males (Lamming, G. E. ed.), Vol.2: 183-212.

Mattei, X., Y. Siau, O.T. Thiaw and D. Thiam, 1993. Journal of Fish Biology 43:931-937.

Grier, H.J., J.R. Linton, J.F. Leatherland and V.L. De Vlaming, 1980. Amercian Journal of Anatomy 159:331-345.

CHARACTERISTICS OF SEMEN AND OVARY IN RAINBOW TROUT (*ONCORHYNCHUS MYKISS*) FED FISH MEAL AND/OR ANIMAL BY-PRODUCTS BASED DIETS.

J-Z. Pan[1], K. Dabrowski[1]*, L. Liu[1], and A. Ciereszko[1,2]

[1]School of Natural Resources, The Ohio State University, Columbus, OH 43210; [2] Centre of Agrotechnology and Veterinary Sciences, Polish Academy of Sciences, 10-718, Olsztyn-Kortowo, Poland.

Summary

In the present study, rainbow trout were fed five different diets in which fish meal protein was replaced by an animal by-product mixture. No significant differences in the brookstock growth were found. Based on the results of free amino acids, protein concentration, activity of aspartate aminotransferase in semen and ovary, as well as sperm density, we suggest that there was not a significant effect of diet composition on the quality of gametes. However, sperm motility from fish fed a fish meal based diet was significantly lower than in other groups.

Introduction

The success in using fish meal alternatives in diet formulation for salmonids is still limited and use of animal by-products will be vital for the continued development of aquaculture in terms of the increasing costs and scarcity of fish meal. The effect of brookstock nutrition on reproductive performance in fish have received little attention. For example, Cerdà et al. (1994) analyzed whether two diets with 35 and 51% of protein influence the sea bass growth and reproductive performance.

Results

The fish meal-based diet (36% protein, 18% lipid) and four other diets in which 25, 50, 75 and 100% fish meal protein was replaced with animal by-product mixture were fed to fish (triplicate groups per dietary treatment of 35 fish per group) of an initial weight 207±7g. There was no significant difference in weight gain when fish reached the average weight of 631± 31g, and most of the males matured. Sperm (n=8-17 per dietary treatment) density did not differ among groups ($10.1\text{-}12.1 \times 10^9$ ml^{-1}) whereas sperm motility in group fed fish meal-based diet was significantly lower (20.6±5.3%) than in other groups (39.4-47.5%). Protein concentration varied in seminal plasma (0.74±0.26 mg/ml) and in ovarian tissue (128±46 mg/g); so did the activities of aspartate aminotransferase in seminal plasma (97.8±63.2 U/g. protein) and in ovarian tissue (81.8±73.8 U/g. protein). The differences were not significant among dietary treatments. In ovarian tissue, total free amino acids in the group fed diet with 75% of by-product was significantly higher (27.9±2.2 μmole/g) than in other groups (19.6-22.5 μmole/g) (Fig. 1).

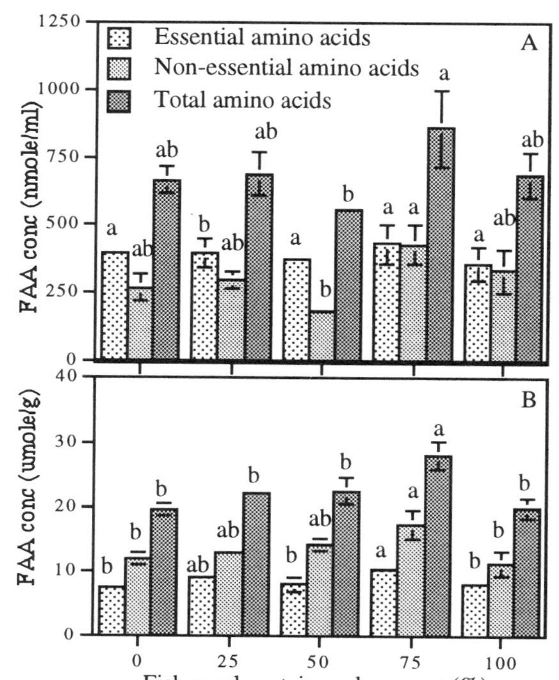

Fig 1. Free amino acids in the seminal plasma (A) and ovarian tissue (B). Bars in each group marked with the same superscripts are not significantly different ($p<0.05$).

Discussion

The change in nutrient composition and their availability in brookstock diets may result in alteration of the biochemical composition of gonads. Biochemical indicators may correlate to the fertilization success (Kjørsvik et al., 1990). The results obtained in the present study demonstrated that an optimal formulation of low-cost, animal by-product-based diets for rainbow trout did not have significant effect on the brookstock growth although total free amino acids in ovary and sperm motility were significantly different among dietary treatments.

References

E, Kjørsvik., A, Mangor-Jensen. and I, Holmefjord., 1990. Egg quality in fishes. In: Advances in Marine Biology, 26: Academic Press, p. 71-105.

Cerdà, J., Carrillo, M., Zanuy, S., Ramos, J. and Higuera, M. de. la., 1994. Influence of nutritional composition of diet on sea bass, *Dicentrachus labrax* L., reproductive performance and egg and larval quality. Aquaculture 128: 345-361.

Partly supported by the North-Central Regional Aquaculture Research Center, USDA.

MEASUREMENT OF LIPID PEROXIDATION IN RAINBOW TROUT (*ONCORHYNCHUS MYKISS*) SPERMATOZOA BY THIOBARBITURIC ACID (TBA) ASSAY

L. Liu, A. Ciereszko, and K. Dabrowski

School of Natural Resources, The Ohio State University, Columbus, Ohio 43210 (L.L., K.D., A.C.), Center for Agrotechnology and Veterinary Sciences, Polish Academic of Sciences, 10-718 Olsztyn, Poland (A.C.)

Summary

This study was designed 1) to optimize TBA assay for rainbow trout (*Oncorhynchus mykiss*) spermatozoa lipid peroxidation, 2) to analyze rainbow trout spermatozoa peroxidation with and without ferrous ion stimulation by the optimized TBA method, and 3) to evaluate the effect of graded level of dietary antioxidant (ascorbic acid) on TBA values of spermatozoa. We found significant differences ($p<0.001$) in total ascorbic acid concentrations in seminal plasma of the fish fed diet with different ascorbic acid status. Significant differences ($p<0.05$) in peroxidation values (measured by production of malondialdehyde,MDA) of sperm both with and without ferrous ion stimulation using semen from the same experimental fish were found in spermiation on day 100. Thus, ascorbic acid supplemented in fish diet could transfer to seminal plasma and protect spermatozoa from peroxidative damages during reproductive season and consequently might affect sperm fertilizing ability.

Introduction

Spermatozoa of fish and several mammalian species including humans contain high concentrations of unsaturated fatty acid. The abundance of unsaturated lipid makes spermatozoa sensitive to a peroxidative damage and this constitutes a potential hazard to the functional integrity of spermatozoa. Ascorbic acid is an essential nutrient in some species of mammal and teleost fish. Its antioxidant effect on peroxidation has been documented in many different tissues and organs. The primary goal of this work was to pursue the role of ascorbic acid in the fish male reproductive system, especially its antioxidant effect on sperm.

Materials and Methods

The experimental rainbow trout were fed diets supplemented with 0, 110, 870 ppm of ascorbyl phosphate magnesium salt for 5-8 months prior to a reproductive season. Modified TBA reaction for human sperm peroxidation (Aitken et al., 1993) was used for rainbow trout. Data were analyzed by ANOVA using Proc GLM procedure in SAS (TM) version 6.07.

Results

The modified TBA reaction for human sperm peroxidation could be applied to rainbow trout sperm. The optimal condition for pH and ascorbic acid concentration were in the range of 0.59-1.44 and 0.1-0.2 mM, respectively. Within 0.1-0.2 mM ascorbic acid concentration, ferrous sulfate concentration did not affect peroxidation to a great extent within the range of 0.02-0.16 mM. No significant differences in peroxidation of sperm using semen from fish fed diets with different ascorbic acid status were found on spermiation day 1 and day 64 (Fig 1). Significant differences were found in peroxidation of sperm on spermiation day 100. Sperm from fish fed ascorbic acid had significantly lower peroxidation level than those fed no ascorbic acid. Significant differences ($p<0.001$) of total ascorbic acid in seminal plasma of the different fish groups were found and the ascorbic acid concentrations decreased toward the end of the season.

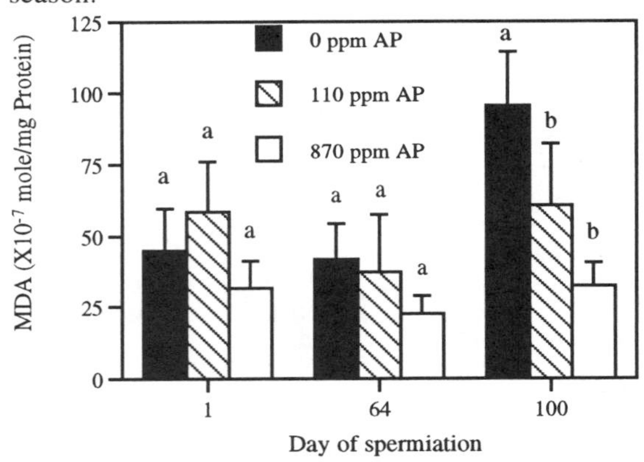

Fig. 1 Peroxidation of sperm from experimental fish with ferrous ion stimulation.

Discussion

Our results demonstrated that the dietary ascorbic acid could be transferred to the reproductive system and the seminal plasma and protected spermatozoa from peroxidative damage during reproductive season. In rainbow trout spermatogenesis is completed about two months before the spawning season and spermatozoa are stored in sperm ducts. During this long period of spermatozoa storage, ascorbic acid in seminal plasma is an excellent candidate to protect spermatozoa from the attack of reactive oxygen species generated by disintegrated sperm and somatic tissues.

References

Aitken, RJ, Harkiss, D, and Buckingham, DW (1993): Analysis of lipid peroxidation mechanisms in human spermatozoa. Molecular Reprod. and Develop. 35:302-315.

Endocrine control of sex differentiation in XX female, and in XY and XX male common carp (Cyprinus carpio, L.)

J.Komen[1], J.G.D. Lambert[2], C.J.J. Richter[1], and H.J.Th. Goos[2].
[1] Department of Fish Culture and Fisheries, Wageningen Agricultural University, and
[2] Research group on Reproductive Endocrinology, Department of Experimental Zoology, University Utrecht, The Netherlands.

Introduction

The common carp is a gonochorist species with XX/XY sex determination. However, in XX animals which are homozygous for a recessive mutation in a putative sex determining gene *mas* (-culinization), female to male sex reversal occurs (1,2). We investigated the role of sex steroids in sex reversal of XX (*mas/mas*) common carp by comparing clones of XY males, XX(*mas/mas*) males and XX(*mas/+*) females. The period of sex differentiation was characterized by histology. The production of endogenous steroids during the period of sex differentiation was studied by incubating gonads with ^{3}H-Androstenedione. The effects of exogenous steroids were investigated by administration of 17ß-Estradiol to XY and XX (*mas/mas*) males and 11-keto-Androstenedione to XX (*mas/+*) females during the period of sex differentiation.

Materials and Methods

XX (*mas/mas*) animals were androgenetic progeny of a homozygous XX(*mas/mas*) male (3). XY offspring was produced by crossing a YY male with a homozygous XX(+/+) female. This female was also crossed with a XX(*mas/mas*) male to produce XX(*mas/+*) female progeny. Fish in each of these clones are genetically uniform and of the same sex. Samples of 20 gonads (from 10 fish), collected at 10 day intervals between day 60 and day 120, were incubated with labelled ^{3}H-Androstenedione for 2 hrs. Steroid conver-sions were analysed by TLC. Exogenous steroids were administered after ethanol incor-poration in a commercial pellet (final concentration 50 ppm).

Results and Discussion

Normal female sex differentiation is characterised by the formation of an oviduct between day 50 and 60, and the occurrence of oogenesis from day 60 onward. Male XY gonads are quiescent until day 90 when spermatogenesis commences. In XX(*mas/-mas*) gonads, oviduct formation is inhibited. Spermatogenesis starts at day 60 concomittant with oogenesis. At day 120 these gonads have developed into testes.

From day 60 onward female gonads are able to synthesize estrogens (table 1). In contrast male XY gonads

Table 1. Conversion of ^{3}H-androstenedione by common carp gonads

gonads	60 days			80 days			100 days		
	T	E	A	T	E	A	T	E	A
XY ♂	10.2	1.5	48.8	11.9	0.0	49.5	32.8	0.0	51.5
XX *mas/mas*	21.2	24.1	19.9	21.2	29.9	7.8	15.8	34.2	18.1
XX *mas/+*	14.8	51.4	0.0	20.7	29.5	0.0	55.8	75.4	0.0

T = converted androstenedione (%); E = estrogens formed (% of T);
A = 11-oxygenated androgens formed (% of T).

only synthesize 11-oxygenated androgens. XX (*mas/mas*) sex reversing gonads initially synthesize *both* estrogens and 11 oxygenated androgens, but the capacity to produce estrogens subsequently decreases.
Exogenous 17ß estradiol induced the forma-tion of an oviduct and oogenesis in both XY and XX (*mas/mas*) gonads. However the ef-fects were only temporary since at day 160 all treated XY animals contained testes, while most XX (*mas/mas*) animals contained inter-sex gonads. In contrast, exogenous 11-keto-Androstenedione induced permanent male sex reversal in XX (*mas/+*) females.
Apparently, 17ß-Estradiol can support oogenesis in germ cells but can *not* induce sex reversal of male steroid producing cells. 11-keto-Androstenedione on the other hand induces both germ cell sex reversal and endocrine sex reversal. Together these results indicate that in XX (*mas/mas*) animals, sex reversal might be caused by precoccious production of 11-oxy-genated androgens, which gradually overrule the effects of estrogens by male sex reversal of female steroid producing cells.

References

1) Komen, J., de Boer, P. and Richter, C.J.J. 1992. J. Heredity 83: 431-434.
2) Komen, J., Yamashita, M. and Nagahama, Y. 1992. Develop. Growth & Differ. 34: 535-544.
3) Bongers, A.B.J., in 't Veld, E.P.C., Abo-Hashema, K., Bremmer, I.M., Eding, E.H., Komen, J. and Richter, C.J.J. 1994. Aquaculture 122: 119-132.

THE EFFECT OF PHOTOPERIOD ON VITELLOGENIN SYNTHESIS AND OOCYTE ENDOCYTOSIS IN RAINBOW TROUT (*ONCORHYNCHUS MYKISS*)

E. BON [1,2], J. NUÑEZ RODRIGUEZ [1,3] and F. LE MENN [1]

[1] UA INRA de Biologie de la Reproduction des Poissons, Université Bordeaux I, avenue des Facultés, 33405 Talence CEDEX, France

[2] Les Salmonidés d'Aquitaine, route de Taller, 40260 Castets, France

[3] ORSTOM, CRO, BP V18, Abidjan, Ivory Coast

Introduction

Oocyte vitellogenesis is one of the best examples of cell specialization for specific endocytosis of proteins. Vitellogenin (VTG) a hepatically synthetized glycophospholipoprotein, the main plasma yolk precursor, is taken up selectively from blood and internalized into oocytes by a receptor-mediated mechanism.

Fish VTG receptors undergo seasonal fluctuations in both concentration and binding affinity during oocyte growth. This study deals with the influence of photoperiod on the binding characteristics of VTG to its specific receptors by comparing two groups of precocious winter rainbow trout (*Oncorhynchus mykiss)* reared at constant temperature in spring water on a spawner diet :

- group 1 (**N**) with a simulated natural photoperiod cycle of 12 months, spawning in early winter
- group 2 (**S**) with a cycle shortened by a daily reduction of the light from an initial 16h to 8h over a period of 7 months, resulting in summer spawning.

Material and methods

Ovaries were taken from freshly sacrificed rainbow trout of each of the two groups at the same stage of the reproductive cycle. This was during the rapid development phase when mean oocyte diameters were around 3.5 mm, in July for the **S** group and in September for the **N** group. Follicles from 6 fish per group were taken from the ovaries, separated and prepared following Le Menn and Nuñez Rodriguez (1990). Rainbow trout VTG was purified using one step DEAE Biogel ion-exchange chromatography, and iodinated using iodogen. Characterization of binding was performed by filter assay (Stifani *et al.*, 1990) with VTG specific activity not exceeding 100,000 cpm/pM.

Results

Changes in the gonadosomatic index (GSI) of **S** and **N** fish had a similar profile. After a postovulatory phase, the vitellogenesis of rainbow trout showed three successive gonad development phases: a very slow development (VSD), a slow development (SD) and a rapid development (RD). Compared with group **N** fish the only phase modified in the **S** group was the SD which was halved. The oocyte diameter in both **N** and **S** groups increased in direct proportion with the gonadosomatic index during the vitellogenic stages. However, at spawning, there was a significant decrease (p=0.005) in ovule size in the **S** group.

For both groups an increase in E2 levels appeared along with the VSD phase and peaked one-and-a-half months before spawning, with maximum values around 30% greater for **S** than for **N**. Similarly, in both groups, increasing VTG levels appeared along with the VSD phase and peaked two weeks before spawning with maximum values around 30% greater for **S** than for **N**.

Transformation of VTG binding data to Scatchard plots indicated for both groups a single class of binding site for VTG. In spite of photoperiod modifications, VTG receptors remained saturable with **N** receptor preparations saturable faster than those of **S**. Binding data per 100 mm^2 of oocyte surface indicated in the **S** group a significant decrease (p=0.002) of over 2.4 fold in the affinity (Ka) and a significant increase (p=0.005) of over 2.1 fold in the number of binding sites (Bmax) (Fig.1) The consequence on the average ovule diameter was a significant decrease (p=0,005) in ovule size in the S group.

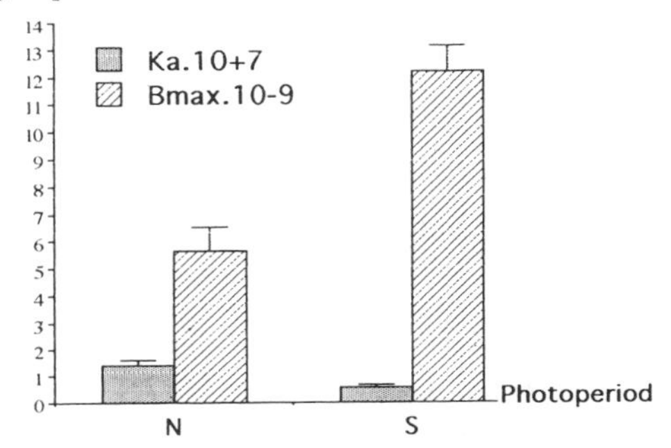

Fig.1 : VTG-receptors binding chatacteristics

Conclusion

The characteristics of VTG binding to its specific membrane receptors are clearly photoperiod dependent during the oocyte rapid growth phase. Shortening the reproductive cycle in rainbow trout over a period of 7 months induces a modification of the capacity of oocyte receptors for their specific binding with VTG, leading to a decrease in the amount of yolk sequestrated and thus in the size of spawned ovules.

References

Le Menn and Nuñez Rodriguez J. (1990) Liaisons de vitellogenines de poissons et d'oiseau avec des récepteurs specifiques homologues et hétérologues. Coll. GIS-BBA, Guidel, France.

Stifani S, Le Menn F., Nuñez Rodriguez J. and Schneider W. (1990) Regulation of oogenesis. The piscine receptor for vitellogenin. Biophys. Biochem. Acta, 1045, 271-279.

Author Index